Jost Functions in Quantum Mechanics

Sergei A. Rakityansky

Jost Functions in Quantum Mechanics

A Unified Approach to Scattering, Bound, and Resonant State Problems

Springer

Sergei A. Rakityansky
Department of Physics
University of South Africa
Pretoria, South Africa

ISBN 978-3-031-07760-9 ISBN 978-3-031-07761-6 (eBook)
https://doi.org/10.1007/978-3-031-07761-6

This Springer imprint is published by the registered company Springer Nature Switzerland AG
The registered company address is: Gewerbestrasse 11, 6330 Cham, Switzerland

Preface

It is long since quantum mechanics became the central part of modern science that deals with all forms of inorganic matter. Basic principles of quantum mechanics are well established, and nowadays, nobody can question its ability to accurately describe various microscopic phenomena in physics and chemistry. The properties of nuclei, atomic and molecular systems, as well as of solids, liquids, gases, and plasmas can be calculated (or at least reasonably estimated) by solving an appropriate Schrödinger equation.

Under reasonable physical assumptions, most of the practical problems can be formulated in terms of this equation in a rigorous mathematical way. However, its exact solution can be found not in all the cases. Even modern computers fail when a Schrödinger equation becomes multidimensional and has complicated boundary conditions. Such cases require some physical insight into the problem, which enables the researcher to either simplify the underlying physical model or to search for a solution in an approximate way with a simplified mathematical structure. Usually, as a result of such a simplification, initial problem is reduced to the Schrödinger equation describing a single particle moving in an effective potential. After the partial-wave expansion, one obtains a system of coupled one-dimensional radial Schrödinger equations.

There are three types of physical problems associated with these equations, namely *bound*, *scattering*, and *resonant* state problems. They differ in the boundary conditions imposed on the wave function at large distances. These three different types of quantum-mechanical problems can be unified if one follows the ideas of Res Jost who suggested to consider two unphysical solutions of the Schrödinger equation with simple asymptotic behaviour. At large distances, one of these *Jost solutions* becomes the in-going and the other out-going unperturbed wave.

Since formally the Jost solutions constitute a complete basis in the space of all possible solutions, any physical solution can be represented as a linear combination of them with the combination coefficients determined by the appropriate physical boundary conditions. These coefficients depend on the value of the energy and are called *Jost functions*. They have a remarkable property that their zeros (or the zeros of their determinants in the multichannel problems) are at those complex energies at which the system has bound or resonant states. In principle, this offers an elegant way of solving physical problems by locating those zeros. But how to calculate the Jost function?

On the bookshelves of any university library one can find plenty of textbooks on quantum mechanics and scattering theory. Most of them cover the standard list of themes more or less in the same fashion. The Jost function as well as the Jost solutions of the Schrödinger equation are indispensable parts of most of these textbooks. However, after reading even a most comprehensive text on this subject, one usually gets a feeling that Jost function is a pure mathematical entity, very elegant, very useful in formal theory, but impractical for those who are intended to apply quantum mechanical methods for solving real physical problems. This is because none of the texts offer the reader any practical method for calculating the Jost function.

This book is an attempt to collect under one cover all that is known about Jost functions as well as to fill the gap between the mathematical theory of these functions and their practical applications. The theory is revisited here and systematized. The central part of the revised theory is the set of linear differential equations whose solutions give the Jost function and the physical wave function directly and concurrently.

These differential equations are also valid for complex values of the angular momentum, which enables one to locate the Regge poles and trajectories. The theory is extended to the multichannel problems as well as to the class of few-body problems which can be treated within the hyperspherical approach. The Jost function method is also generalized for use in one- and two-dimensional problems associated, for example, with semiconductor nano-structures.

This book is not a historical review of the Jost function theory. This is why only a few references are given. They appear at some places where the discussion touches the subjects that are outside the scope of this book. In a sense, the book can be considered as an alternative textbook on the quantum scattering theory. In contrast to the traditional presentation of such a theory, it is given here within the stationary picture and exclusively in the coordinate representation. Some of the facts known in the scattering theory are re-derived here using the Jost functions and Jost matrices (in the multi-channel case).

The book is mostly intended for postgraduate students. All the derivations are therefore given in detail and are accompanied by detailed explanations. The presentation goes up the steps of increasing complexity. The basic ideas are introduced using the most simple single-channel problem with a short-range potential. Then, the Coulomb forces are added, coupling of the partial waves appears, and different channels are introduced. At the end of the book, there are several Appendices explaining some mathematical facts that are needed for understanding of the theory.

Pretoria, South Africa Sergei A. Rakityansky

Contents

1 The Basic Concepts

The basic concepts of quantum mechanics that are necessary to understand the rest of the book are given here in a very concise way. This is certainly neither a full account of the quantum mechanical concepts nor an introductory course. Rather, this is a kind of "admission test": If you understand everything written in this chapter, you may safely proceed; otherwise it would be better to go back to your favourite quantum mechanical textbook and refresh your knowledge. As a possible option, you can take one of the famous courses [1–3]. The formulae given in this chapter can also be used together with the Appendices as reference material when reading the main part of the book.

1.1 Quantum Vectors

Physical systems of atomic and subatomic particles behave in accordance with the stochastic quantum laws. Throughout this book, we restrict our consideration to such quantum systems that can be treated non-relativistically. Most problems concerning molecules, atoms, nuclei, and solids belong to this category.

At a given instant of time, a quantum system is said to be in the state $|\Psi_a\rangle$ determined by a full set of the so-called quantum numbers $\{a\} = \{\alpha_1, \alpha_2, \dots, \alpha_N\}$. These numbers are the eigenvalues of the set $\{A\} = \{A_1, A_2, \dots, A_N\}$ of the operators, i.e.

$$A_i|\Psi_a\rangle = \alpha_i|\Psi_a\rangle\ , \qquad i = 1, 2, \dots, N\ , \tag{1.1}$$

that are linear, Hermitian, commute with each other,

$$[A_i, A_j] \equiv A_iA_j - A_jA_i = 0\ , \qquad \forall \quad i, j\ , \tag{1.2}$$

S. A. Rakityansky, *Jost Functions in Quantum Mechanics*,
https://doi.org/10.1007/978-3-031-07761-6_1

and represent a full set of physical variables that can, in principle, be measured simultaneously and therefore have definite values α_i, $i = 1, 2, \ldots, N$, where N is equal to the number of the degrees of freedom of the system under consideration.

Each of the quantum numbers has its own range of possible values, within which it can be either discrete or continuous (or both). Every combination of these numbers (permitted by the conservation and symmetry laws) determines certain state $|\Psi_a\rangle$. When the quantum numbers run over their respective ranges, covering all permitted combinations $a_1, a_2, \ldots$, the corresponding set of the states $\left\{|\Psi_{a_1}\rangle, |\Psi_{a_2}\rangle, \ldots\right\}$ constitutes a basis in the quantum space describing the physical system.

Any possible state of this system is a linear combination of the basic states,

$$|\Phi\rangle = \sum\!\!\!\!\!\!\int_a C_a |\Psi_a\rangle \,, \tag{1.3}$$

where the sum transforms into integration when a runs over the states representing the continuous part of the basis. A quantum state can therefore be considered as a vector in the abstract (Hilbert) space, with C_a being its projections onto the basic vectors. The symbol $|\ \rangle$ is called the ket-vector. Together with its Hermitian conjugate partner, the bra-vector $\langle\ |$, they form full "bracket" $\langle\ |\ \rangle$ which is the scalar product in the vector space of quantum states.

An important property of the basic vectors belonging to the same basis is their orthogonality, i.e.

$$\langle\Psi_a|\Psi_{a'}\rangle = 0 \,, \qquad \text{if} \quad a \neq a' \,. \tag{1.4}$$

The basic vectors can always be normalized in such a way that

$$\langle\Psi_a|\Psi_{a'}\rangle = \delta_{aa'} \,, \tag{1.5}$$

where the Kronecker symbol $\delta_{aa'}$ transforms into the corresponding δ-function for the continuum states.

The combination coefficients C_a in Eq. (1.3) are the amplitudes of the probability $P_\Phi(\Psi_a) = |C_a|^2$ that the system can be found in the corresponding basic states $|\Psi_a\rangle$ while it is known with certainty that it is in the state $|\Phi\rangle$. The ortho-normality relation (1.5) enables us to find C_a as the scalar product

$$C_a = \langle\Psi_a|\Phi\rangle \,. \tag{1.6}$$

While the "bra-ket" $\langle\ |\ \rangle$ is a scalar product, the inverse combination $|\ \rangle\langle\ |$ of the same vectors is an operator in the quantum space. As is seen from Eqs. (1.3) and (1.5), $|\Psi_a\rangle\langle\Psi_a|$ is the operator that makes projection onto the state $|\Psi_a\rangle$. Since an arbitrary vector $|\Phi\rangle$ is a sum of its components along all basic vectors, the sum of all projection operators is the identity operator, i.e.

$$\sum_a |\Psi_a\rangle\langle\Psi_a| = I \, , \tag{1.7}$$

which is known as the closure relation.

Quantum vector $|\Psi_a\rangle \equiv |\alpha_1, \alpha_2, \ldots, \alpha_N\rangle$ defined in the "full space" can be also viewed as the direct product

$$|\Psi_a\rangle \equiv |\alpha_1\rangle|\alpha_2\rangle \ldots |\alpha_N\rangle \tag{1.8}$$

of the "incomplete" vectors $|\alpha_i\rangle$ defined in the subspaces associated with the corresponding variables. Each vector $|\alpha_i\rangle$ in this product should be understood as a state in which the variable α_i has certain value while all the other variables are unknown. The representation (1.8) implies that the operator A_i in Eq. (1.1) acts only in the corresponding subspace, i.e.

$$A_i|\alpha_i\rangle = \alpha_i|\alpha_i\rangle \, . \tag{1.9}$$

The complete scalar products of the type (1.6) are (complex) numbers while an incomplete product

$$\langle\alpha_i|\Psi_a\rangle = |\alpha_1\rangle|\alpha_2\rangle \ldots |\alpha_{i-1}\rangle|\alpha_{i+1}\rangle \ldots |\alpha_N\rangle \tag{1.10}$$

remains a vector in the complementary subspace.

A variable β, represented by the operator $B \notin \{A\}$, cannot have a definite value when the system is in the state $|\Psi_a\rangle$. If measured, β assumes all its possible values with certain probabilities $P_{\Psi_a}(\beta)$. For discrete states, these probabilities can be found by making the incomplete projections of $|\Psi_a\rangle$ onto all possible $|\beta\rangle$ and folding all the other variables, namely

$$P_{\Psi_a}(\beta) = |\langle\beta|\Psi_a\rangle|^2 = \langle\Psi_a|\beta\rangle\langle\beta|\Psi_a\rangle \, , \tag{1.11}$$

where $\langle\Psi_a|\beta\rangle$ and $\langle\beta|\Psi_a\rangle$ are the bra- and ket-vectors in the same complementary subspace. Since $\langle\Psi_a|\beta\rangle^* = \langle\beta|\Psi_a\rangle$, their scalar product is a real non-negative number. For continuum states, the quantities of the type (1.11) are densities of the corresponding probabilities. Apparently,

$$\sum_i P_{\Psi_a}(\beta_i) = \sum_i \langle\Psi_a|\beta_i\rangle\langle\beta_i|\Psi_a\rangle = 1 \, , \tag{1.12}$$

which implies that the closure relation

$$\sum_i |\beta_i\rangle\langle\beta_i| = I \tag{1.13}$$

is valid in the subspaces associated with each individual variable.

The expectation (mean or average) value $\langle\beta\rangle$ of the variable β can be found as

$$\langle\beta\rangle = \langle\Psi_a|B|\Psi_a\rangle\,. \tag{1.14}$$

Indeed, with the help of the closure relation and the fact that $B|\beta_i\rangle = \beta_i|\beta_i\rangle$, this equation can be re-written as

$$\langle\beta\rangle = \sum_i \langle\Psi_a|B|\beta_i\rangle\langle\beta_i|\Psi_a\rangle = \sum_i \beta_i\,|\langle\beta_i|\Psi_a\rangle|^2 = \sum_i \beta_i\,P_{\Psi_a}(\beta_i)\,, \tag{1.15}$$

which coincides with the definition of the expectation value in the probability theory.

The quantum vectors are abstract quantities. When doing calculations, we deal with numbers representing them. For a given vector, these numbers are its projections on the basic vectors of a chosen basis. The set of such projections is called the wave function. Thus, the set of numbers C_a given by Eq. (1.6) is the wave function Φ in the "a"-representation, which has usual meaning as the mathematical function $\Phi(a)$ of N variables. For discrete variables, the number $|\Phi(a)|^2$ is the probability that in the quantum state $|\Phi\rangle$ the physical variables associated with the set of operators $\{A\}$ have values $a = \{\alpha_1, \alpha_2, \ldots, \alpha_N\}$. For continuous variables (like space coordinates, for example), $|\Phi(a)|^2 da$ is the probability that these variables are within the N-dimensional interval $[a, a + da]$.

The same state-vector $|\Phi\rangle$ can be represented in many different ways by using different bases. The transformation of the wave function from one to another representation is done with the help of the closure relation. If, for example, $\{|\Psi_a\rangle\}$ and $\{|\Theta_n\rangle\}$ are two different bases in the same quantum space, then the wave function $\langle\Theta_n|\Phi\rangle$ can be expressed in terms of $\langle\Psi_a|\Phi\rangle$ by inserting the identity operator between the bra- and ket-vectors as follows

$$\langle\Theta_n|\Phi\rangle = \sum\!\!\!\!\!\!\int_a \langle\Theta_n|\Psi_a\rangle\langle\Psi_a|\Phi\rangle\,.$$

An operator B acting on a quantum vector $|\Phi\rangle$ transforms it into another vector

$$|\Phi'\rangle = B|\Phi\rangle\,. \tag{1.16}$$

In a chosen basis $\{|\Psi_a\rangle\}$, both $|\Phi\rangle$ and $|\Phi'\rangle$ are represented by the sets of their projections, i.e. by the sets of complex numbers C_a and C'_a given by Eq. (1.6). Therefore the transformation (1.16) can be viewed as the transformation of $\{C_a\}$

into $\{C'_a\}$ by a square matrix. This can be written explicitly if we multiply Eq. (1.16) from the left by $\langle\Psi_a|$ and insert the identity operator (1.7) between B and $|\Phi\rangle$, i.e.

$$\langle\Psi_a|\Phi'\rangle = \sum_{a'}\langle\Psi_a|B|\Psi_{a'}\rangle\langle\Psi_{a'}|\Phi\rangle\ , \tag{1.17}$$

where $B_{aa'} = \langle\Psi_a|B|\Psi_{a'}\rangle$ is the matrix representation of the operator B.

1.2 Schrödinger Equation

In general, a given state-vector $|\Psi_a\rangle_t$ changes with time (hence the subscript t). Its time evolution is governed by the fundamental quantum mechanical operator, the Hamiltonian

$$H = H_0 + U\ , \tag{1.18}$$

consisting of the kinetic energy operator H_0 (the free motion Hamiltonian) and the potential energy operator U (sum of inter-particle potentials plus a potential generated by an external field, if any). The principal equation of quantum mechanics, the Schrödinger equation,

$$i\hbar\frac{\partial}{\partial t}|\Psi_a\rangle_t = H|\Psi_a\rangle_t\ , \tag{1.19}$$

determines $|\Psi_a\rangle_t$ at any time t in the past or future if it is known at an instant $t = t_0$. Here $\hbar \approx 6.5822 \times 10^{-22}$ MeV $\cdot$ sec is the Planck's constant, which is the minimal quantity (quantum) of the orbital angular momentum (it has also the units of mechanical action).

Natural and most frequently used is the coordinate representation of quantum state-vectors. The basis of this representation consists of the continuous states $|\vec{r}\,\rangle$ in which the system of particles has certain space configuration determined by the full set $\{\vec{r}\,\}$ of the particle coordinates (in general, the symbol $\vec{r}$ should be understood as a multi-dimensional vector in the N-body configuration space, i.e. $\vec{r} \equiv \{\vec{r}_1, \vec{r}_2, \ldots, \vec{r}_N\}$).

In the coordinate representation, the dynamics is totally described by the complex-valued wave function

$$\langle\vec{r}\,|\Psi_a\rangle_t = \Psi_a(\vec{r}, t)$$

depending on the coordinates and the time. At any fixed instant of time the squared absolute value of the wave function, $P_{\Psi_a}(\vec{r}\,) \equiv |\Psi(\vec{r}, t)|^2$, is the density of the probability to find the particles in the configuration $\{\vec{r}\,\}$. There is another interpretation of this probability, based on the notion of quantum ensemble, i.e. a statistically large set of identical quantum systems being in the same state. In the ensemble picture, the quantity $|\Psi(\vec{r},t)|^2\,\mathrm{d}^{3N}r$ is understood as the fraction of this set, constituted by the systems whose configurations are within the infinitesimal volume, $\mathrm{d}^{3N}r$, around the tip of vector $\vec{r}$.

For many physical systems the Hamiltonian is independent of time. Since H is the total energy operator, the total energy of such systems is conserving. The opposite can take place, for example, when an external field is present and it is varying with time. We shall always be considering only conservative systems governed by time-independent Hamiltonians. For such systems, the total energy E can always be chosen as one of the quantum numbers, i.e. $E \in \{a\}$, and the solution of Eq. (1.19) can be formally written as the following relation:

$$\Psi_a(\vec{r},t) = \exp\left(-\frac{it}{\hbar}H\right)\Psi_a(\vec{r},t_0)\,, \tag{1.20}$$

which determines time evolution of the system from t_0 to any subsequent ($t > t_0$) or previous ($t < t_0$) instant of time.

Since Ψ_a is an eigenfunction of H, the total wave function can be factorized into the time-independent spatial part, $\psi_a(\vec{r}\,)$, and the time-evolution exponential function,

$$\Psi_a(\vec{r},t) = \exp\left(-\frac{i}{\hbar}Et\right)\psi_a(\vec{r}\,)\,. \tag{1.21}$$

Quantum states described by wave functions of the type (1.21) are called the stationary states, and $\psi_a(\vec{r}\,)$ is the stationary wave function that obeys the stationary Schrödinger equation,

$$H\psi_a = E\psi_a\,. \tag{1.22}$$

If we write the spatial part of the wave function (which is a complex number) in the polar representation,

$$\psi_a(\vec{r}\,) = |\psi_a(\vec{r}\,)|\,e^{i\theta}\,, \tag{1.23}$$

where $\theta = \arg(\psi_a)$, then at each point $\{\vec{r}\,\}$ of the configuration space the point representing the value of

$$\Psi_a = |\psi_a|\exp\left[i\left(\theta - \frac{i}{\hbar}Et\right)\right] \tag{1.24}$$

on the complex plane moves with time in the clockwise direction along the circle of the radius $|\psi_a|$. In other words, wave function (1.21) oscillates between pure real and pure imaginary values with the frequency $\omega = E/\hbar$. Its absolute value, however, remains unchanged. This is why the probability density for a stationary state does not depend on time.

For a single particle of mass μ, the kinetic energy operator H_0 is represented by the following continuous matrix

$$\langle \vec{r}\,|H_0|\vec{r}\,'\rangle = -\frac{\hbar^2}{2\mu}\delta(\vec{r}-\vec{r}\,')\Delta_{\vec{r}}\,. \tag{1.25}$$

If this particle moves in the potential U, Eq. (1.22) in the coordinate representation is generally integro-differential,

$$-\frac{\hbar^2}{2\mu}\Delta_{\vec{r}}\,\psi_a(\vec{r}\,)+\int\langle\vec{r}\,|U|\vec{r}\,'\rangle\psi_a(\vec{r}\,')\mathrm{d}^3r\,' = E\psi_a(\vec{r}\,)\,. \tag{1.26}$$

For local potentials that, by definition, are represented by diagonal matrices

$$\langle\vec{r}\,|U|\vec{r}\,'\rangle = \delta(\vec{r}-\vec{r}\,')U(\vec{r}\,)\,, \tag{1.27}$$

the stationary Schrödinger equation is more simple (differential),

$$-\frac{\hbar^2}{2\mu}\Delta_{\vec{r}}\,\psi_a(\vec{r}\,)+U(\vec{r}\,)\psi_a(\vec{r}\,) = E\psi_a(\vec{r}\,)\,. \tag{1.28}$$

In many cases the potential $U(\vec{r}\,)$ does not depend on the direction of vector $\vec{r}$ (central potential) which implies that the wave function (for spinless particles) can be factorized in the radial and angular parts,

$$\psi_a(\vec{r}\,) = \frac{u_a(r)}{r}Y_{\ell m}(\theta,\varphi)\,, \tag{1.29}$$

where $Y_{\ell m}(\theta,\varphi)$ are the so-called spherical harmonics depending on the spherical angles θ and φ as well as on the angular momentum ℓ and its third component m, which are conserving and together with the energy E are part of the set $\{a\} = \{E,\ell,m,\dots\}$ of the quantum numbers characterizing the quantum state. The radial wave function $u_a(r)$ obeys the radial Schrödinger equation

$$\left[\frac{d^2}{dr^2}+k^2-\frac{\ell(\ell+1)}{r^2}-V(r)\right]u_a(r) = 0\,, \tag{1.30}$$

where $k^2 = 2\mu E/\hbar^2$ and $V(r) = 2\mu U(r)/\hbar^2$.

1.3 Boundary Conditions

The boundary conditions determining various physical solutions of Eq. (1.30) are imposed at both ends of the interval $r \in [0, \infty)$, i.e. at $r = 0$ and at large r. Since $\psi_a(\vec{r}\,)$ is the amplitude of the probability density, it must be finite everywhere. Then Eq. (1.29) implies that

$$u_a(0) = 0 \,. \tag{1.31}$$

This condition is the same for all types of solutions, while the conditions at $r \to \infty$ are different for the bound, scattering, and resonant states. If the potential vanishes at large distances fast enough (faster than $1/r^2$) then Eq. (1.30) becomes the so-called free radial Schrödinger equation

$$\left[\frac{d^2}{dr^2} + k^2 - \frac{\ell(\ell+1)}{r^2}\right] u_a(r) = 0 \,, \qquad r \to \infty \,. \tag{1.32}$$

This is a second-order differential equation and therefore it has two linearly independent solutions. There are many ways of choosing such solutions, but the most convenient choice is the pair of Riccati–Hankel functions $h_\ell^{(\pm)}(kr)$ asymptotically behaving as

$$h_\ell^{(\pm)}(kr) \underset{r\to\infty}{\longrightarrow} \mp i \exp[\pm i(kr - \ell\pi/2)] \,. \tag{1.33}$$

The general solution of Eq. (1.32) can be constructed as their linear combination

$$u_a(r) \underset{r\to\infty}{\longrightarrow} C_1 h_\ell^{(-)}(kr) + C_2 h_\ell^{(+)}(kr) \,, \tag{1.34}$$

where an appropriate choice of the combination coefficients C_1 and C_2 determines the asymptotic behaviour of the solution, i.e. the solution type (bound, scattering, or resonant).

1.3.1 Bound States

In a bound state, the particle cannot leave the source of the attractive field. This means that the probability density wanes when $r \to \infty$, i.e.

$$u_a(r) \underset{r\to\infty}{\longrightarrow} 0 \,. \tag{1.35}$$

Since the total probability of finding the particle somewhere around the source of the potential is 1,

$$\int |\psi_a(\vec{r}\,)|^2 \, \mathrm{d}^3 r = 1 \,, \tag{1.36}$$

the attenuation of the wave function in (1.35) is such that $\psi_a(\vec{r}\,)$ is square-integrable. This can only be the case if C_1 in Eq. (1.34) is zero. Indeed, the choice of the energy scale in such a way that the potential at infinitely large r is zero, means that the energy is counted from this value and the bound states (if any) have negative energies $E < 0$, which implies that the corresponding momentum $k = \sqrt{2\mu E/\hbar^2} = i|k|$ is pure imaginary. According to Eq. (1.33), this means that the first term in the general solution (1.34) is exponentially growing while the second term is exponentially vanishing,

$$h_\ell^{(-)}(kr) \underset{r\to\infty}{\sim} e^{+\gamma r} \,, \qquad h_\ell^{(+)}(kr) \underset{r\to\infty}{\sim} e^{-\gamma r} \,, \tag{1.37}$$

where the parameter $\gamma \sim \sqrt{|E|}$ increases with the growth of the binding energy. Therefore the function $u_a(r)$ vanishes at large distances

$$u_a(r) \underset{r\to\infty}{\sim} h_\ell^{(+)}(kr) \sim e^{-\gamma r} \,, \tag{1.38}$$

if and only if $C_1 = 0$. For each bound state, an appropriate value of C_2 can be found from the normalization condition (1.36).

1.3.2 Scattering States

Scattering is a physical process during which the particle approaches the interaction region from afar and after passing through the potential, moves far away again but perhaps in a different direction and (for inelastic scattering) with a different energy and some other quantum numbers changed. Since in the asymptotically far regions this particle moves with a real physical momentum, the energy of a scattering state is real and positive $E > 0$.

Long before and long after the collision ($t \to \mp\infty$) the particle is far away from the source of force. At such distances the potential vanishes and the particle moves free, i.e. the time evolution of its quantum state is determined by the free Hamiltonian H_0,

$$\exp\left(-\frac{it}{\hbar}H_0\right)|\phi_a^{(\mathrm{in})}\rangle \underset{-\infty\leftarrow t}{\longleftarrow} \exp\left(\frac{it}{\hbar}H\right)|\psi_a\rangle \underset{t\to+\infty}{\longrightarrow} \exp\left(-\frac{it}{\hbar}H_0\right)|\phi_a^{(\mathrm{out})}\rangle \,, \tag{1.39}$$

where $|\phi_a^{(\text{in/out})}\rangle$ are the incoming and outgoing stationary asymptotic states. As is seen from Eq. (1.39), for each possible "in" and "out" asymptotic states, the corresponding stationary state at the time of actual scattering ($t = 0$) can be found using the so-called Møller operators $\Omega_\pm$, defined as,[1]

$$|\psi_a\rangle = \lim_{t\to-\infty} \exp\left(\frac{it}{\hbar}H\right)\exp\left(-\frac{it}{\hbar}H_0\right)|\phi_a^{(\text{in})}\rangle \equiv \Omega_+|\phi_a^{(\text{in})}\rangle\,, \tag{1.40}$$

$$|\psi_{a'}\rangle = \lim_{t\to+\infty} \exp\left(\frac{it}{\hbar}H\right)\exp\left(-\frac{it}{\hbar}H_0\right)|\phi_{a'}^{(\text{out})}\rangle \equiv \Omega_-|\phi_{a'}^{(\text{out})}\rangle\,. \tag{1.41}$$

If the initial state was $|\phi_a^{(\text{in})}\rangle$, then the actual state is $|\psi_a\rangle$. If we want to obtain the final state $|\phi_{a'}^{(\text{out})}\rangle$, the actual state must be $|\psi_{a'}\rangle$. The more these two actual states overlap, the greater is the probability that the given initial state will evolve into the chosen final state. In other words, the quantity

$$\langle\psi_{a'}|\psi_a\rangle = \langle\phi_{a'}^{(\text{out})}|\Omega_-^\dagger\Omega_+|\phi_a^{(\text{in})}\rangle = S_{a'a} \tag{1.42}$$

is the amplitude of probability for the transition $a \to a'$. This is the so called S matrix, i.e. the matrix element of the S operator,

$$S = \Omega_-^\dagger\Omega_+\,. \tag{1.43}$$

For physical scattering energies ($E > 0$), it can be shown that because of the conservation of the flux of the incoming particles the S matrix is unitary,

$$S^\dagger S = SS^\dagger = 1\,, \qquad E > 0\,. \tag{1.44}$$

1.3.2.1 Plane and Spherical Waves

When considering quantum scattering, we should (strictly speaking) solve full time-dependent problem for the motion of a wave packet (that describes the motion of the projectile) with a randomly chosen impact parameter and then do the average over this parameter. However, this is rather complicated approach. The difficulty stems from the fact that a wave packet is a mixture of the states with different momenta, i.e. with different energies,

$$|\psi\rangle = \int \chi(\vec{k})|\vec{k}\rangle\,\mathrm{d}^3k\,, \tag{1.45}$$

[1] A mathematically rigorous analysis of these limits and of the properties of $\Omega_\pm$ can be found, for example, in the book by J.R. Taylor: [4].

where the amplitude function $\chi(\vec{k})$ describes the contribution of each plane wave to the total wave packet. This means that the time evolution of such a state cannot be described by a single common factor $\exp\left(-\frac{i}{\hbar}Et\right)$, because there is no single energy E for it.

It would be more convenient to reduce the problem to a stationary, i.e. time-independent, Schrödinger equation, like we do it for bound states. This is possible if the function $\chi(\vec{k})$ is strongly peaked around a central momentum $\hbar\vec{k}_0$. Actually, this is always the case in the scattering experiments, where one always tries to generate a flux of particles that move with the same speed.

If the distribution of the momenta in the wave packet is strongly peaked, it can be viewed as the δ-function and thus the quantum state of each particle in the beam is approximated by the state with the momentum $\hbar\vec{k}_0$,

$$|\psi\rangle \approx \int \delta(\vec{k} - \vec{k}_0)|\vec{k}\,\rangle\, \mathrm{d}^3k = |\vec{k}_0\rangle \;. \tag{1.46}$$

The only principal restriction on using this approximation may be imposed by the uncertainty principle,

$$\Delta k \cdot \Delta x \geqslant 1/2 \;, \tag{1.47}$$

according to which the spread of the momenta, $\Delta p = \hbar\Delta k$, cannot be zero unless the length of the beam, Δx, is infinite. However, for the macroscopic beams generated in particle accelerators, this restriction does not play any practical role. This can be seen if we estimate the theoretical restriction on the sharpness (accuracy) of the energy distribution for a pulse (bunch) of electrons of the length, say, $\Delta x \sim 1$ cm. Such a bunch is schematically shown in Fig. 1.1. The corresponding wave packet describing a single electron is practically a harmonic wave whose wavelength is many orders of magnitude shorter than Δx.

From the uncertainty principle, we have

$$\Delta p \geqslant \frac{\hbar}{\Delta x} \;,$$

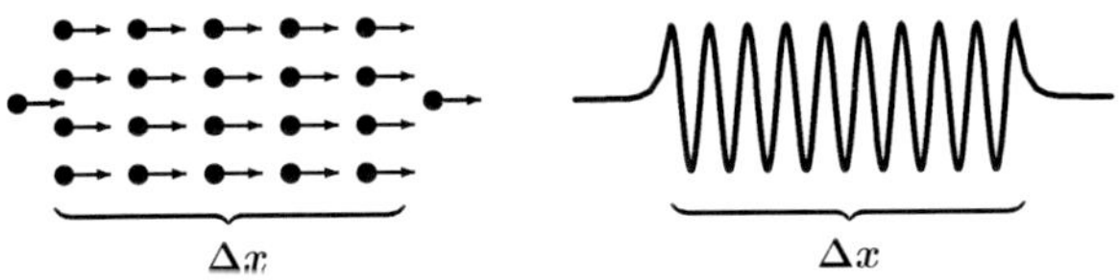

Fig. 1.1 Schematic (out of scale) picture of a bunch of accelerated particles (left) and the corresponding wave packet (right), which is practically a plane wave

and thus the minimal energy spread can be estimated as follows:

$$\Delta E \geqslant \frac{(\Delta p)^2}{2m} \geqslant \frac{\hbar^2}{2m(\Delta x)^2} = \frac{(\hbar c)^2}{2mc^2(\Delta x)^2}$$

$$\approx \frac{(197\,\text{MeV}\cdot\text{fm})^2}{1\,\text{MeV}\cdot(10^{13}\,\text{fm})^2} \geqslant \frac{10^4}{10^{26}}\,\text{MeV} = 10^{-22}\,\text{MeV}\,.$$

Therefore the energy spread for such a beam of particles cannot be made smaller than 10^{-16} eV. To all practical purposes, this surely corresponds to the δ-function in Eq. (1.45). Actually, the equipment generates so many much higher errors that we may safely ignore the quantum limitation on the sharpness of the energy distribution.

The above reasoning leads us to the conclusion that in a theoretical analysis the projectile particle can be treated as having the "δ-sharp" momentum distribution, i.e. its motion is described by the wave function of the state with a definite value of the momentum,

$$\Psi_{\vec{k}}(\vec{r}, t) = \exp\left(-\frac{i}{\hbar}Et\right)\psi_{\vec{k}}(\vec{r}\,)\,, \tag{1.48}$$

where the stationary wave function,

$$\psi_{\vec{k}}(\vec{r}\,) = \frac{e^{i\vec{k}\vec{r}}}{(2\pi)^{3/2}}\,, \tag{1.49}$$

is the plane wave. This wave is called plane because its phase, namely the scalar product $\vec{k}\vec{r}$ is constant on each infinite plane orthogonal to the momentum, as is shown in Fig. 1.2.

It should be understood that, when speaking about the motion of a wave, we actually have in mind the total wave function (1.48) involving the time dependence. The total phase, ϕ, of this wave,

$$\phi = \vec{k}\vec{r} - \omega t\,,$$

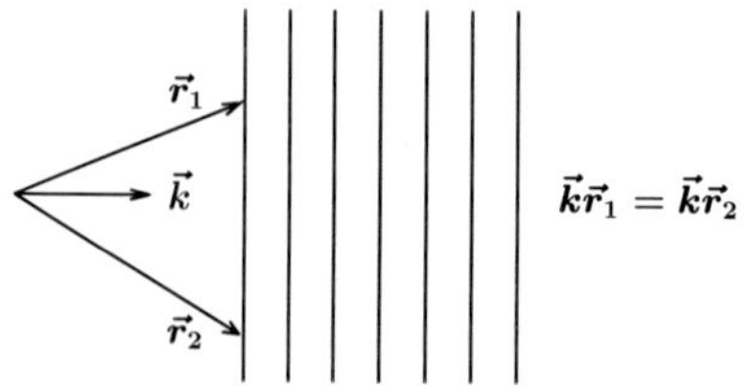

Fig. 1.2 The plane wave has a constant phase, $\vec{k}\vec{r}$, on any plane that is orthogonal to the momentum (because the projections of $\vec{r}_1$ and $\vec{r}_2$ onto the horizontal direction, i.e. onto $\vec{k}$, are the same for any $\vec{r}_1$ and $\vec{r}_2$)

where $\omega = E/\hbar$, remains constant with increasing time, if the product $\vec{k}\vec{r}$ is also increasing. This is possible if we shift the observation point in space in the direction of $\vec{k}$. In other words, a wave crest moves in the direction of $\vec{k}$. For the wave $\sim \exp(-i\vec{k}\vec{r} - i\omega t)$ the situation is opposite, i.e. the wave crest moves in the opposite direction.

Since we consider the scattering process as a stationary state problem (after separating the time-dependent exponential), we usually forget about the time dependence. For example, in the one-dimensional problem the exponentials $\exp(+ikx)$ and $\exp(-ikx)$ are called the waves moving to the right and to the left, respectively. This, however, is a "sloppy" terminology. These exponentials become moving waves only if we take into account the time-dependent factor.

Similarly to the plane waves, we can introduce the spherical waves that are moving from and toward the centre. For them the total phase,

$$\phi = \pm kr - \omega t \ ,$$

is constant on a sphere. Such waves,[2] i.e. $\sim \exp(\pm ikr)$, appear when a wave function is decomposed in the so-called partial waves, i.e. expanded in a series over the spherical harmonics. The spherical waves are also needed when the conservation of the particle-flux is considered. The most convenient way to formulate such a conservation consists in equalizing the fluxes carried by the waves entering and exiting a sphere.

The plane wave (1.49) goes in and out of any given sphere. This implies that it can be split in two terms describing the incoming and outgoing spherical waves. And indeed, it is known the following expansion (see Appendix D or the book [5]):

$$\begin{aligned}\frac{e^{i\vec{k}\cdot\vec{r}}}{(2\pi)^{3/2}} &= \psi_{\vec{k}}^{(\mathrm{in})}(\vec{r}\,) + \psi_{\vec{k}}^{(\mathrm{out})}(\vec{r}\,) \\ &= \frac{1}{\sqrt{2\pi}kr}\sum_{\ell m} i^{\ell} h_{\ell}^{(-)}(kr) Y_{\ell m}^{*}(\Omega_{\vec{k}}) Y_{\ell m}(\Omega_{\vec{r}}) \\ &+ \frac{1}{\sqrt{2\pi}kr}\sum_{\ell m} i^{\ell} h_{\ell}^{(+)}(kr) Y_{\ell m}^{*}(\Omega_{\vec{k}}) Y_{\ell m}(\Omega_{\vec{r}}) \ ,\end{aligned} \tag{1.50}$$

[2] The intensity (amplitude) of a spherical wave is diminishing with increasing radius, which is taken into account by the factor $1/r$ in Eq. (1.29). Such a factor is always present in the final formulae, but within the "sloppy" terminology the exponentials $\exp(\pm ikr)$ without the factor $1/r$ are called spherical waves as well.

where the Riccati–Hankel functions $h_\ell^{(\mp)}(kr)$ describe the incoming and outgoing spherical waves,

$$h_\ell^{(\mp)}(kr) \underset{|kr|\to\infty}{\longrightarrow} (\pm i)^{\ell+1} \exp(\mp ikr) \; . \tag{1.51}$$

1.3.2.2 Scattering Amplitude and Cross Section

In the spirit of the quantum ensemble picture, the stationary scattering states are understood as describing not an individual particle but a flux of such particles. Mathematically, this flux, $\vec{j}$, is defined via the corresponding wave function (1.29) as

$$\vec{j} = \frac{\hbar}{2i\mu} \left[\psi_a^*(\vec{r}\,) \,\vec{\nabla}\, \psi_a(\vec{r}\,) - \psi_a(\vec{r}\,) \,\vec{\nabla}\, \psi_a^*(\vec{r}\,) \right] . \tag{1.52}$$

Beams of particles usually used in the scattering experiments, are practically monochromatic (i.e. all the particles in the beam have almost the same energy). In other words, the particles of the beam constitute a quantum ensemble, which is described by the plane-wave wave function

$$\psi_{\vec{k}}^{(\mathrm{pw})}(\vec{r}\,) = \frac{e^{i\vec{k}\vec{r}}}{(2\pi)^{3/2}} \; . \tag{1.53}$$

This is an eigenfunction of H_0 and describes free particle motion with the velocity $\vec{v} = \hbar\vec{k}/\mu$. Substituting this wave function in Eq. (1.52), we find that the corresponding incoming flux is

$$\vec{j}^{\,(\mathrm{pw})} = \frac{\hbar\vec{k}}{(2\pi)^3\mu} = \frac{\vec{v}}{(2\pi)^3} \; . \tag{1.54}$$

The potential scatters this incoming wave in all directions and thus forms the scattered wave

$$\psi_{\vec{k}}^{(\mathrm{sc})}(\vec{r}\,) = \frac{\mathsf{f}(\hat{\vec{r}} \leftarrow \hat{\vec{k}}, E)}{(2\pi)^{3/2}} \frac{e^{ikr}}{r} \; , \tag{1.55}$$

i.e. the outgoing spherical wave e^{ikr}/r with different intensities in different directions. The intensities are determined by the amplitude $\mathsf{f}(\hat{\vec{r}} \leftarrow \hat{\vec{k}}, E)$, which depends on the spherical angles $\hat{\vec{r}} = \{\theta, \varphi\}$ of vector $\vec{r}$, i.e. on the direction of scattering relative to the incident momentum $\vec{k}$.

The scattered particles move along all the radial directions relative to the centre of the potential, i.e. perpendicular to the spherical surface surrounding the interaction region. It is therefore natural to use the spherical coordinates to find the corresponding quantum flux (1.52). Among the three components of the gradient operator

$$\vec{\nabla} = \{\partial_r,\ (1/r)\partial_\theta,\ (1/r\sin\theta)\partial_\varphi\}\ ,$$

we should use only the radial one. This gives the radial component

$$j_r^{(\mathrm{sc})} = \frac{|\vec{v}||\mathsf{f}(\hat{\vec{r}} \leftarrow \hat{\vec{k}}, E)|^2}{(2\pi)^3 r^2} \tag{1.56}$$

of the scattered flux $\vec{j}^{\,(\mathrm{sc})} = \{j_r^{(\mathrm{sc})}, j_\theta^{(\mathrm{sc})}, j_\varphi^{(\mathrm{sc})}\}$, which is perpendicular to the spherical surface and directed outwards.

Therefore the fraction of the incoming flux, scattered into the solid angle $d\Omega$ that is directed to $\hat{\vec{r}}$ and spanned by the element $r^2 d\Omega$ of the spherical surface, is given by

$$\frac{j_r^{(\mathrm{sc})} r^2 d\Omega}{|\vec{j}^{\,(\mathrm{pw})}|} = \left|\mathsf{f}(\hat{\vec{r}} \leftarrow \hat{\vec{k}}, E)\right|^2 d\Omega = d\sigma\ . \tag{1.57}$$

As is easily seen, this ratio is measured in the units of area and can be interpreted as the fraction $d\sigma$ of the incoming beam cross section, such that all the particles passing through it, end up in the said solid angle $d\Omega$. The function $\mathsf{f}(\hat{\vec{r}} \leftarrow \hat{\vec{k}}, E)$ is called the scattering amplitude and the quantity

$$\frac{d\sigma}{d\Omega} = \left|\mathsf{f}(\hat{\vec{r}} \leftarrow \hat{\vec{k}}, E)\right|^2 \tag{1.58}$$

the differential scattering-cross-section.

At large distances, both the incident and scattered waves are present, and therefore the scattering wave function has the following asymptotic behaviour

$$\psi_{\vec{k}}(\vec{r}\,) \underset{r\to\infty}{\longrightarrow} (2\pi)^{-3/2}\left[e^{i\vec{k}\vec{r}} + \mathsf{f}(\hat{\vec{r}} \leftarrow \hat{\vec{k}}, E)\frac{e^{ikr}}{r}\right]\ . \tag{1.59}$$

If the potential is spherically symmetric, then the scattering amplitude is φ-independent, i.e. depends only on one angle θ and therefore can be expanded in the Legendre polynomials,

$$\mathsf{f}(\hat{\vec{r}} \leftarrow \hat{\vec{k}}, E) = \sum_\ell (2\ell+1) f_\ell(E) P_\ell(\cos\theta)\ , \tag{1.60}$$

where $f_\ell(E)$ is called the partial-wave amplitude. Substituting this expansion together with similar expansion of the plane wave

$$e^{i\vec{k}\vec{r}} = \sum_\ell (2\ell+1) \frac{i^\ell j_\ell(kr)}{kr} P_\ell(\cos\theta) , \tag{1.61}$$

in Eq. (1.59) and using the asymptotic properties of the Riccati–Bessel $j_\ell(z)$ and Riccati–Hankel $h_\ell^{(\pm)}(z)$ functions (see Appendix D), it is easy to obtain the following boundary condition for the radial part of the scattering wave function

$$u_a(r) \underset{r\to\infty}{\longrightarrow} f_\ell^{(\mathrm{in})}(E) h_\ell^{(-)}(kr) + f_\ell^{(\mathrm{out})}(E) h_\ell^{(+)}(kr) , \tag{1.62}$$

where the energy-dependent amplitudes $f_\ell^{(\mathrm{in/out})}$ of the incoming and outgoing waves are called the Jost functions. Actually, these amplitudes are the coefficients C_1 and C_2 of Eq. (1.34). The difference is that C_1 and C_2 are arbitrary (because (1.34) is the general solution) while $f_\ell^{(\mathrm{in/out})}$ have certain values for each collision energy (because (1.62) is the scattering solution that has certain asymptotic behaviour). These values are determined by the character of the interaction forces. In their turn, the Jost functions completely determine what is observed in the scattering process. Indeed, it is not difficult to show that the partial-wave scattering amplitude is related to these functions as follows

$$f_\ell(k) = \frac{f_\ell^{(\mathrm{out})}(E) - f_\ell^{(\mathrm{in})}(E)}{2ikf_\ell^{(\mathrm{in})}(E)} . \tag{1.63}$$

Therefore the scattering problem can be solved if we solve the differential equation (1.30) starting from $r = 0$ with the condition (1.31), to a sufficiently far point $r = R$ and using the asymptotic condition (1.62), somehow manage to extract the Jost functions from the resulting solution.

1.3.3 Resonances

Resonance phenomena are ubiquitous in all branches of physics. Being associated with oscillations and waves, they are characterized by discrete values of resonance frequencies (or wavelengths) at which the energy is accumulated in small regions of space. Another important characteristics of resonances are frequency windows around the discrete values that are called the resonance widths. The accumulated energy leaks out through these windows. Narrow resonances are therefore more pronounced and visible in experiments because for them it is more difficult to leak out as compared with wide resonances.

In quantum physics where the motion of all particles is associated with the de Broglie waves, we encounter the resonance phenomena even more often than in classical physics. Because of the universal relation $E = \hbar\omega$ between the energy and the circular frequency ω, in quantum physics instead of the resonance frequencies, it is more convenient (and logical) to consider the corresponding discrete values E_r of the energy. Naturally, the resonance widths Γ are also measured in the units of energy.

Since E_r assumes discrete values, quantum resonances are similar to bound states, but in contrast to them are not stable. Resonances that are also known as quasi-stationary or quasi-bound states have finite lifetime and stochastically decay according to the radioactivity law

$$n(\vec{r}, t) = n(\vec{r}, 0) \exp(-\Gamma t/\hbar) , \tag{1.64}$$

which should be understood using the quantum ensemble picture with $n(\vec{r}, t)$ being the number of the ensemble particles per unit volume (density of particles) at the point $\vec{r}$ and time t. If $N(t)$ is the total number of particles at time t, then the wave function corresponding to this ensemble can be normalized as

$$\int |\Psi_a(\vec{r}, t)|^2 \, d^3 r = N(t) . \tag{1.65}$$

Then at any point $\vec{r}$ in the configuration space the density of the particles is given by

$$n(\vec{r}, t) = |\Psi_a(\vec{r}, t)|^2 . \tag{1.66}$$

Therefore to get the decay law (1.64), the time-dependent factor in the wave function (1.21) must have $\exp(-\Gamma t/2\hbar)$, which is the case if the Hamiltonian has complex eigenvalues:

$$E_{res} = E_r - i\frac{\Gamma}{2} . \tag{1.67}$$

These eigenvalues are unphysical. Actual resonance phenomena occur at the positive collision energies E_r. The imaginary part of the energy (1.67) just tells us how long the resonance state can live or how wide is the energy interval around E_r, within which the resonance can be formed.

After forming a quasi-stationary state, the quantum system spends some time within a finite volume of space similarly to a truly bound state like, for example, an atom or a nucleus. During that time interval, the particles do many movements and revolutions around each other and "forget" the original direction from which the projectile particle arrived. Finally, when the "departure" time comes, the direction for the decay is chosen at random. Therefore an ensemble of quasi-stationary states decays isotropically (if not polarized by an external field).

This means that the eigenfunctions of H, corresponding to the complex energies (1.67), must have only outgoing waves in their asymptotic behaviour (1.34), i.e.

$$u_a(r) \underset{r\to\infty}{\sim} h_\ell^{(+)}(kr)\,. \tag{1.68}$$

Comparing this condition with Eq. (1.38), we see that bound and resonant states indeed have much in common. The boundary conditions for them are the same. Formally, the only difference is that the bound states have real negative energies while the resonance solutions of the Schrödinger equation can be found at complex energies.

1.4 Semi-classical Wave Function

Similarly to electrodynamics, where for short wavelengths, wave optics is smoothly transformed into geometrical optics, classical mechanics is a limiting case of quantum mechanics when de Broglie wavelength tends to zero. There is an intermediate range of wavelengths for which a physical system is still governed by quantum probability-laws but its wave function can be obtained in a simplified way without solving the Schrödinger equation.

The idea of this so-called semi-classical or WKB[3] approximation as well as the range of its applicability may become clear if we consider a case when the de Broglie wavelength is so short that the potential $V(r) + \ell(\ell+1)/r^2$ (including the centrifugal term) in Eq. (1.30) is practically constant on an interval covering few periods of the wave. On each of such intervals, the two independent solutions of Eq. (1.30) are the exponential functions $\exp(\pm ipr/\hbar)$ where

$$p(r) = \hbar\sqrt{k^2 - V(r) - \ell(\ell+1)/r^2} \tag{1.69}$$

is the classical momentum. This fact suggests that when taking into account slow variations of the potential, the wave function should not be very much different from the exponential one and we could try to find it in the following form

$$u_a(r) = e^{is_a(r)/\hbar}\,, \tag{1.70}$$

[3] WKB stands for the names of G. Wentzel, H. A. Kramers, and L. Brillouin, who introduced it in quantum mechanics. However, similar approximation was known in the theory of ordinary differential equations long before that and is associated with the name of mathematician H. Jeffreys[6]. This is why sometimes this approximation is called JWKB[7].

where $s_a(r)$ is the so-called eikonal.[4] Substituting the ansatz (1.70) into Eq. (1.30) we obtain the following nonlinear equation for the eikonal function

$$s_a'(r) = \pm\sqrt{p^2(r) + i\hbar s_a''(r)}\,, \tag{1.71}$$

which (if the second derivative of the eikonal function is small) can be solved iteratively starting with $s_a'(r) \approx \pm p(r)$. Making the first iteration and assuming that

$$\left|p'(r)/p^2(r)\right| \ll 1\,, \tag{1.72}$$

we arrive at the WKB approximation

$$u_a(r) \approx \frac{1}{\sqrt{p(r)}} \exp\left[\pm\frac{i}{\hbar}\int p(r)dr\right] \tag{1.73}$$

for the two independent solutions of the Schrödinger equation. Condition (1.72) for the validity of this approximation can be formulated in a more clear way if we introduce the local effective wavelength $\lambda(r) = 2\pi\hbar/p(r)$. Indeed, from Eq. (1.72) it follows that this wavelength may vary only slowly, i.e.

$$|\mathrm{d}\lambda/\mathrm{d}r| \ll 1\,. \tag{1.74}$$

In many important cases the accuracy of the WKB approximation is improved if in Eq. (1.69) the product $\ell(\ell+1)$ is replaced by $(\ell+\frac{1}{2})^2$. In particular, the semi-classical quantization condition becomes exact for the hydrogen atom and for the three-dimensional harmonic oscillator. Moreover, the modification $\ell(\ell+1) \to (\ell+\frac{1}{2})^2$ is necessary in order to maintain the regularity condition at the origin when one wishes to include higher order WKB corrections.

1.5 Two-Body Problem

The motion of a particle in a potential which is fixed in space, i.e. one-body problem, is an abstraction that is very seldom encountered in real world. Usually, we have a system of at least two interacting particles whose potential energy $\mathcal{U}$ depends on the particle coordinates $\vec{r}_1$ and $\vec{r}_2$. Both particles can move and the problem might look complicated in the $|\vec{r}_1, \vec{r}_2\rangle$ representation. If, however, we separate the trivial plane wave describing the motion of the system as a whole, it is reduced to an effective one-body problem. This can be done as follows.

[4] The term eikonal comes from the optics where it means the phase of the electromagnetic wave.

The Hamiltonian governing a system of two particles with masses m_1 and m_2 can be written as

$$H = \frac{\vec{p}_1^{\,2}}{2m_1} + \frac{\vec{p}_2^{\,2}}{2m_2} + \mathcal{U}\,, \tag{1.75}$$

where $\vec{p}_1$ and $\vec{p}_2$ are the operators of the particle momenta. Introducing another pair of operators,

$$\vec{P} = \vec{p}_1 + \vec{p}_2 \tag{1.76}$$

and

$$\vec{p} = \frac{m_2\vec{p}_1 - m_1\vec{p}_2}{m_1 + m_2}\,, \tag{1.77}$$

we see that

$$\frac{\vec{p}_1^{\,2}}{2m_1} + \frac{\vec{p}_2^{\,2}}{2m_2} = \frac{\vec{P}^{\,2}}{2M} + \frac{\vec{p}^{\,2}}{2\mu}\,, \tag{1.78}$$

where $M = m_1 + m_2$ is the total mass and

$$\mu = \frac{m_1 m_2}{m_1 + m_2} \tag{1.79}$$

the reduced mass of the system. Let the dot above a vector means its derivative with respect to time, then

$$\vec{P} = m_1\dot{\vec{r}}_1 + m_2\dot{\vec{r}}_2 = M\frac{m_1\dot{\vec{r}}_1 + m_2\dot{\vec{r}}_2}{M} = M\dot{\vec{R}}$$

and

$$\vec{p} = \frac{m_2 m_1\dot{\vec{r}}_1 - m_1 m_2\dot{\vec{r}}_2}{m_1 + m_2} = \frac{m_1 m_2}{m_1 + m_2}(\dot{\vec{r}}_1 - \dot{\vec{r}}_2) = \mu\dot{\vec{r}}\,,$$

where the so-called Jacobi coordinates (see Fig. 1.3)

$$\vec{R} = \frac{m_1\vec{r}_1 + m_2\vec{r}_2}{m_1 + m_2} \tag{1.80}$$

and

$$\vec{r} = \vec{r}_1 - \vec{r}_2 \tag{1.81}$$

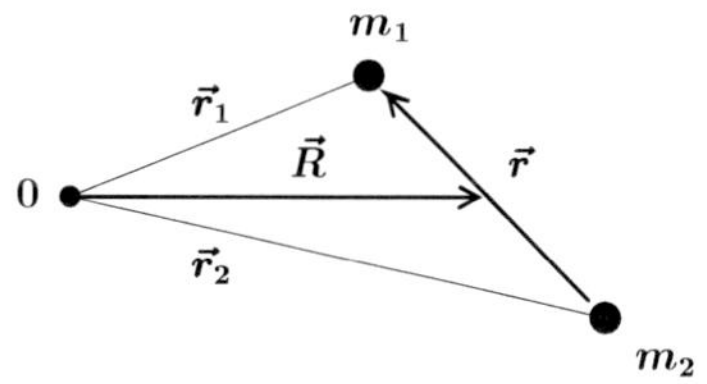

Fig. 1.3 Jacobi coordinates of a two-body system

point to the centre of mass and from particle two to particle one, respectively. They are canonically conjugate to the operators of the total and relative momenta $\vec{P}$ and $\vec{p}$.

If no external fields are present, the space is translationally invariant and the potential (generally, it is non-local),

$$\langle \vec{R}, \vec{r}\,|\mathcal{U}|\vec{R}\,', \vec{r}\,'\rangle = \delta(\vec{R} - \vec{R}\,')U(\vec{r}, \vec{r}\,')$$

does not depend on $\vec{R}$. Then in the Jacobi-vector representation, $|\vec{R}, \vec{r}\,\rangle$, the Schrödinger equation

$$\left(\frac{\hbar^2}{2M}\Delta_{\vec{R}} + \frac{\hbar^2}{2\mu}\Delta_{\vec{r}} - \mathcal{E}\right)\Psi(\vec{R}, \vec{r}\,) = \int \langle\vec{r}|U|\vec{r}\,'\rangle\Psi(\vec{R}, \vec{r}\,')\mathrm{d}^3r\,' \tag{1.82}$$

is separated in the following two:

$$\left(\frac{\hbar^2}{2M}\Delta_{\vec{R}} - E_0\right)\phi(\vec{R}\,) = 0 \tag{1.83}$$

and

$$\left(\frac{\hbar^2}{2\mu}\Delta_{\vec{r}} - E\right)\psi(\vec{r}\,) = \int \langle\vec{r}\,|U|\vec{r}\,'\rangle\psi(\vec{r}\,')\mathrm{d}^3r\,' \tag{1.84}$$

with the total solution being factorized as

$$\Psi(\vec{R}, \vec{r}\,) = \phi(\vec{R}\,)\psi(\vec{r}\,) \tag{1.85}$$

and the total energy being the sum $\mathcal{E} = E_0 + E$. Eq. (1.83) describes the free motion of the system as a whole. Its solution is the plane wave

$$\phi(\vec{R}\,) = \exp\left(i\vec{K}\vec{R}\right), \qquad K^2 = 2ME_0/\hbar^2 . \tag{1.86}$$

Eq. (1.84) describes the relative motion in the reference frame tied to the centre of mass. It is an effective one-body equation for a particle of mass μ moving in the potential U.

Part I

Single-Channel Problems

2 Schrödinger Equation and Its Solutions

From the outset, it should be stated that this book is devoted almost entirely to the so-called binary processes[1] of the type

$$a + b \rightarrow c + d . \tag{2.1}$$

This means that we always deal with only two interacting bodies. These bodies may be composed of many constituent parts, but still in the collision process (2.1) they act as united complexes. We will call the bodies a, b, c, and d particles, but they may be atoms or heavy nuclei. During the interaction, these bodies can be excited or can exchange their constituent parts. Therefore, despite the restriction to the binary processes, the range of applicability of the theory that is presented here is rather wide.

Loosely speaking, each different state of the binary system is a different channel of the interaction process. Examples of such channels are the systems $a + b$ and $a + b^*$, where b^* is an excited state of the particle b. Since the excitation consumes some energy, the channel $a + b^*$ can only be open if the particles a and b collide with a sufficient kinetic energy. The minimal energy that is needed to open a channel is called its threshold.

Apart from the "truly" different channels, there are also the degenerate channels that have the same threshold. They differ by some quantum numbers of the state $a + b$. For example, in the degenerate channels, these particles can be in different spin states or can have different orbital angular momenta.

Traditionally, the multi-channel status is reserved for the problems that have more than one energy threshold. But formally, the number of the channels N is the same as the dimension $(N \times N)$ of the S-matrix that describes the process.

[1] There is only one exception at the end of the book, where many-body systems are considered in Sect. 17.1.

S. A. Rakityansky, *Jost Functions in Quantum Mechanics*,
https://doi.org/10.1007/978-3-031-07761-6_2

In this book, it is used, so to say, "flexible" approach to the terminology: In most cases, the number of channels is associated with the number of different thresholds of the physical system, but not always. For example, the problem of the non-central potential that couples different partial waves (of the same energy) is considered in Chap. 9 as a multi-channel problem. In all cases, it is clear what is meant by the term "channel", and it should not cause any confusion.

* * *

To begin with, let us consider a "truly" single-channel problem: The motion of a structureless and spinless particle of mass μ in a spherically symmetric potential $U(r)$. As an additional condition, we assume that this potential is non-singular, i.e.

$$\lim_{r\to 0} r^2 U(r) = 0\,. \tag{2.2}$$

The potentials that do not obey this condition are considered in Sect. 16.1.

Since the Hamiltonian is rotationally invariant, the angular momentum ℓ and its third component m are conserving. Together with the energy E, they constitute the full set of the quantum numbers that completely define a quantum state $|E,\ell,m\rangle$ of the particle. In order to use the well-known separation of variables of the Laplacian operator in the spherical polar coordinates, the corresponding wave function is factorized into the radial and angular parts,

$$\langle\vec{r}\,|E,\ell,m\rangle = \psi_{E\ell m}(\vec{r}\,) = \frac{u_\ell(E,r)}{r} Y_{\ell m}(\theta,\varphi)\,. \tag{2.3}$$

Its angular dependence given by the spherical harmonics $Y_{\ell m}(\theta,\varphi)$ is known analytically. The radial wave function $u_\ell(E,r)$ obeys the radial Schrödinger equation

$$\left[\frac{d^2}{dr^2} + k^2 - \frac{\ell(\ell+1)}{r^2} - V(r)\right] u_\ell(E,r) = 0\,, \tag{2.4}$$

where $k^2 = 2\mu E/\hbar^2$ is the wave number squared and

$$V(r) = \frac{2\mu}{\hbar^2} U(r) \tag{2.5}$$

is the reduced potential. Since there is no m in this equation, the radial wave function depends only on the angular momentum but not on its third component. A formal proof of the fact that $u_\ell(E,r)$ is m-independent is given further down in Sect. 10.2.1.

2.1 Regular and Irregular Solutions of the Radial Equation

The general properties of equations of the type (2.4) are well-studied and known in the theory of ordinary differential equations (see, for example, book [8]). Since this is an equation of the second order, it has two linearly independent solutions that constitute the basis in the space of all possible solutions. This means that any other solution is a linear combination of the two basis solutions. Of course, the choice of the basis is not unique (like in any space). When $\ell \neq 0$, the presence of the centrifugal potential, $\ell(\ell+1)/r^2$, makes $r = 0$ a point of singularity of Eq. (2.4). This singularity, however, is not very severe (just $\sim 1/r^2$). In the theory of differential equations, such a point is called regular singular point.

Although the singularity at $r = 0$ is not severe, its presence does not remain without consequences: at least one of the two independent solutions is singular (tends to infinity) at $r = 0$. We can choose the basis where one of the solutions is regular, while the other is singular (irregular), or where both are singular.

It is not difficult to establish how the regular and irregular solutions behave when $r \to 0$. Looking for the function $u_\ell(E, r)$ near the point $r = 0$ as

$$u_\ell(E, r) = Ar^a + Br^{a+1} + Cr^{a+2} + \ldots ,$$

and substituting it into the Schrödinger equation (2.4), we find

$$a(a-1)Ar^{a-2} - \ell(\ell+1)Ar^{a-2} \approx 0 ,$$

where we retained only the terms with the lowest power of r. This gives the equation

$$a(a-1) = \ell(\ell+1) ,$$

which has two possible roots:

$$a = \ell + 1 \qquad \text{and} \qquad a = -\ell .$$

Therefore, all possible regular, $u_\ell^{(r)}$, and irregular, $u_\ell^{(i)}$, solutions behave near the origin as

$$u_\ell^{(r)}(E, r) \underset{r\to 0}{\sim} r^{\ell+1} , \qquad u_\ell^{(i)}(E, r) \underset{r\to 0}{\sim} r^{-\ell} \tag{2.6}$$

It is clear that any physical solution must be regular because the wave function (describing the probability density) should be finite everywhere.

There is a bit confusing terminology in scattering theory. The word "regular" is used with two different meanings:

- Any solution that is not singular at $r = 0$ (normalization is not fixed)
- The solution that has specific regular behaviour near $r = 0$ (normalization is fixed)

The reader should be warned that different authors sometimes use different normalizations of that "specific" regular solution.

In this book, the regular solution with the specific normalization is usually denoted as $\phi_\ell(E, r)$ and is defined by the following short-distance behaviour:

$$\phi_\ell(E, r) \underset{r\to 0}{\longrightarrow} j_\ell(kr) \underset{r\to 0}{\longrightarrow} \frac{(kr)^{\ell+1}}{(2\ell+1)!!} \tag{2.7}$$

where j_ℓ is the Riccati–Bessel function (see Appendix D) and $(2\ell+1)!! = 1 \cdot 3 \cdot 5 \cdots (2\ell+1)$. The function $j_\ell(kr)$ is a regular solution of Eq. (2.4) in the absence of the potential, $V(r) \equiv 0$.

It should be clearly understood the difference between the regular solution, $\phi_\ell(E, r)$, and a physical solution, $u_\ell(E, r)$ (which is also regular at $r = 0$). They differ not only in the normalization. There is a principal difference between them.

The regular solution (2.7) is specified by the boundary conditions at a single point ($r = 0$). In fact, Eq. (2.7) specifies both the function itself and its first derivative. In mathematical terminology, this is called Cauchy problem. In contrast to that, a physical solution is specified by the condition $u_\ell(E, 0) = 0$ and by certain behaviour at large distances ($r \to \infty$). For example, for a bound state, the second condition is $u_\ell(E, \infty) = 0$. Mathematicians call it Dirichlet problem (the boundary conditions are imposed at two different points).

The Cauchy problem always has a solution. In contrast to it, the Dirichlet problem may only have solutions with certain values of the parameters in the differential equation. In our case, this means that the regular solution, $\phi_\ell(E, r)$, always exists,[2] while the existence of physical solutions depends on the parameter E (the energy). For example, the bound-state solutions only exist at certain discrete values of the energy. If $u_\ell(E, r)$ exists, then it is proportional to $\phi_\ell(E, r)$.

As it is shown further down in this book, in the simplest single-channel problem, the functions ϕ_ℓ and u_ℓ (if exists) indeed differ only by the normalization. But in a more complicated multi-channel problem, the regular solution is a square matrix, while a physical solution is a single-column-matrix that is constructed as a linear combination of the columns of ϕ. In this case, the principal difference between the regular solution and the physical solution becomes obvious. The columns of ϕ constitute a basis in the space of all possible solutions that are regular at $r = 0$, while the physical solution is a particular vector in that space. In a single-channel problem, that space is one-dimensional, and therefore, any physical vector, u_ℓ, is simply proportional to the single basis vector, ϕ_ℓ.

It may seem that the irregular solutions (behaving as $\sim r^{-\ell}$ at the origin) should be abandoned altogether. As we will see, however, in some cases, it is convenient to use them as the basis and to construct a regular solution as their linear combination. When doing this, the combination coefficients are chosen in such a way that at $r = 0$ the singularities of the irregular solutions cancel each other.

[2] If E is negative but not a bound-state energy, the regular solution $\phi_\ell(E, r)$ exists but does not tend to zero at large distances.

2.2 Finite-Range Potential

A potential that is identically zero for all r beyond some finite value R is said to be of a finite range. Obviously, using such potentials to model physical systems is a simplification of the reality since the interaction fields can never vanish abruptly. However, the approximations based on finite-range potentials are ubiquitous and describe physical phenomena reasonably well.

From mathematical and computational points of view, finite-range potentials are much more simple than those with the tails extending to infinity, since they do not pose the question of convergence when $r \to \infty$. Due to the very same reason, in this section, the Jost functions are introduced in a most clear and simple way. Many of the facts concerning these functions are obtained here without complicated proofs of convergence.

2.2.1 Transformation of Schrödinger Equation

Let us look for the regular solution of the Schrödinger equation

$$\left[\frac{d^2}{dr^2} + k^2 - \frac{\ell(\ell+1)}{r^2}\right]\phi_\ell(E,r) = V(r)\phi_\ell(E,r)\,, \tag{2.8}$$

in a special form that is convenient for constructing the physical wave function, $u_\ell(E,r)$, with a given asymptotic behaviour. To this end, we follow the idea of the variation of parameters method developed long ago in the theory of ordinary differential equations. A simple description of this method can be found, for example, in Sec. 15.2.4 of the book [9].

2.2.1.1 First-Order Differential Equations

As the first step, we consider two independent solutions of the "free" equation

$$\left[\frac{d^2}{dr^2} + k^2 - \frac{\ell(\ell+1)}{r^2}\right]\phi_\ell(E,r) = 0\,, \tag{2.9}$$

which is valid where the potential vanishes. This is the Riccati–Bessel equation whose solutions are well known (see Appendix D). The choice of a pair of its linearly independent solutions is not unique. It can be either the Riccati–Bessel and Riccati–Neumann functions, $j_\ell(kr)$ and $n_\ell(kr)$, or the Riccati–Hankel functions, $h_\ell^{(\pm)}(kr)$. For our purpose, the latter are more appropriate since they represent the incoming and outgoing spherical waves, which we denote as

$$w_\ell^{(\mathrm{in})}(E,r) = h_\ell^{(-)}(kr) \underset{|kr|\to\infty}{\longrightarrow} (+i)^{\ell+1}\exp(-ikr)\,, \tag{2.10}$$

$$w_\ell^{(\mathrm{out})}(E,r) = h_\ell^{(+)}(kr) \underset{|kr|\to\infty}{\longrightarrow} (-i)^{\ell+1}\exp(+ikr)\,. \tag{2.11}$$

The general solution of Eq. (2.9) is a linear combination of them, $W_\ell^{(\text{in})} F_\ell^{(\text{in})} + W_\ell^{(\text{out})} F_\ell^{(\text{out})}$, with some constants (parameters), $F_\ell^{(\text{in/out})}$.

Coming back to Eq. (2.8) with a non-zero potential, we can look for its solution in the form of the same combination but with the parameters considered as unknown functions of the independent variable,

$$\boxed{\phi_\ell(E,r) = W_\ell^{(\text{in})}(E,r)F_\ell^{(\text{in})}(E,r) + W_\ell^{(\text{out})}(E,r)F_\ell^{(\text{out})}(E,r)} \tag{2.12}$$

where $F_\ell^{(\text{in/out})}(E,r)$ are unknown functions. In the theory of ordinary differential equations, this way of finding solution is known as the *variation of parameters method.*

Since instead of one unknown function, ϕ_ℓ, we introduce two unknown functions, $F_\ell^{(\text{in/out})}$, they cannot be independent. We therefore can impose an arbitrary condition that relates them to each other.[3] As such a condition, it is convenient to choose the following equation:

$$W_\ell^{(\text{in})}(E,r)\partial_r F_\ell^{(\text{in})}(E,r) + W_\ell^{(\text{out})}(E,r)\partial_r F_\ell^{(\text{out})}(E,r) = 0\,, \tag{2.13}$$

which is standard in the variation of parameters method and is called the Lagrange condition. It implies that, when calculating the first derivative of $\phi_\ell(E,r)$, we only need to differentiate the spherical waves,

$$\frac{d}{dr}\phi_\ell(E,r) = W_\ell'^{(\text{in})}(E,r)F_\ell^{(\text{in})}(E,r) + W_\ell'^{(\text{out})}(E,r)F_\ell^{(\text{out})}(E,r)\,. \tag{2.14}$$

Let us substitute the ansatz (2.12) into Eq. (2.8) and thus obtain the corresponding equations that determine $F_\ell^{(\text{in})}$ and $F_\ell^{(\text{out})}$. With the Lagrange condition, the second derivative of $\phi_\ell(E,r)$ is

$$\frac{d^2}{dr^2}\phi_\ell = W_\ell''^{(\text{in})}F_\ell^{(\text{in})} + W_\ell''^{(\text{out})}F_\ell^{(\text{out})} + W_\ell'^{(\text{in})}F_\ell'^{(\text{in})} + W_\ell'^{(\text{out})}F_\ell'^{(\text{out})}\,.$$

Therefore, Eq. (2.8) takes the following form:

$$\begin{aligned}
&\left\{\left[\frac{d^2}{dr^2} + k^2 - \frac{\ell(\ell+1)}{r^2}\right]W_\ell^{(\text{in})}(E,r)\right\}F_\ell^{(\text{in})}(E,r)\\
&+\left\{\left[\frac{d^2}{dr^2} + k^2 - \frac{\ell(\ell+1)}{r^2}\right]W_\ell^{(\text{out})}(E,r)\right\}F_\ell^{(\text{out})}(E,r)\\
&+W_\ell'^{(\text{in})}(E,r)F_\ell'^{(\text{in})}(E,r) + W_\ell'^{(\text{out})}(E,r)F_\ell'^{(\text{out})}(E,r) = V(r)\phi_\ell(E,r)\,,
\end{aligned}$$

[3] Of course, with any particular choice of such a condition, the functions $F_\ell^{(\text{in/out})}$ are different, but the function ϕ_ℓ is the same.

where the first two terms (which are in the curly brackets) disappear because both $W_\ell^{(\text{in})}(E,r)$ and $W_\ell^{(\text{out})}(E,r)$ are solutions of Eq. (2.9). This gives

$$W_\ell'^{(\text{in})} F_\ell'^{(\text{in})} + W_\ell'^{(\text{out})} F_\ell'^{(\text{out})} = V\left[W_\ell^{(\text{in})} F_\ell^{(\text{in})} + W_\ell^{(\text{out})} F_\ell^{(\text{out})}\right] . \tag{2.15}$$

From the Lagrange condition (2.13), it follows that

$$F_\ell'^{(\text{out})} = -\left[W_\ell^{(\text{out})}\right]^{-1} W_\ell^{(\text{in})} F_\ell'^{(\text{in})} . \tag{2.16}$$

Substituting Eq. (2.16) into Eq. (2.15) and multiplying the left hand side of the equation by $\left[W_\ell^{(\text{out})}\right]^{-1} W_\ell^{(\text{out})} \equiv 1$ from the left, we obtain

$$\left[W_\ell^{(\text{out})}\right]^{-1} W_\ell^{(\text{out})} \left\{W_\ell'^{(\text{in})} - W_\ell'^{(\text{out})} \left[W_\ell^{(\text{out})}\right]^{-1} W_\ell^{(\text{in})}\right\} F_\ell'^{(\text{in})}$$
$$= V\left[W_\ell^{(\text{in})} F_\ell^{(\text{in})} + W_\ell^{(\text{out})} F_\ell^{(\text{out})}\right] ,$$

which gives

$$F_\ell'^{(\text{in})} = \left[W_\ell^{(\text{out})} W_\ell'^{(\text{in})} - W_\ell'^{(\text{out})} W_\ell^{(\text{in})}\right]^{-1} W_\ell^{(\text{out})} V \left[W_\ell^{(\text{in})} F_\ell^{(\text{in})} + W_\ell^{(\text{out})} F_\ell^{(\text{out})}\right] . \tag{2.17}$$

Similarly, from the Lagrange condition (2.13), it follows that

$$F_\ell'^{(\text{in})} = -\left[W_\ell^{(\text{in})}\right]^{-1} W_\ell^{(\text{out})} F_\ell'^{(\text{out})} , \tag{2.18}$$

and therefore, Eq. (2.15) gives

$$F_\ell'^{(\text{out})} = \left[W_\ell^{(\text{in})} W_\ell'^{(\text{out})} - W_\ell'^{(\text{in})} W_\ell^{(\text{out})}\right]^{-1} W_\ell^{(\text{in})} V \left[W_\ell^{(\text{in})} F_\ell^{(\text{in})} + W_\ell^{(\text{out})} F_\ell^{(\text{out})}\right] . \tag{2.19}$$

The first factors in Eqs. (2.17) and (2.19) differ only by the sign and actually are equal to the inverted Wronskian of the incoming and outgoing spherical waves, i.e. the Riccati–Hankel functions (see Appendix D),

$$h_\ell^{(-)}(kr)\partial_r h_\ell^{(+)}(kr) - h_\ell^{(+)}(kr)\partial_r h_\ell^{(-)}(kr) = 2ik . \tag{2.20}$$

Thus we finally obtain the following coupled system of the first-order differential equations for the new unknown functions:

$$\begin{aligned}
\partial_r F_\ell^{(\mathrm{in})} &= -\frac{h_\ell^{(+)}(kr)}{2ik} V(r) \left[h_\ell^{(-)}(kr) F_\ell^{(\mathrm{in})} + h_\ell^{(+)}(kr) F_\ell^{(\mathrm{out})} \right] \\
\partial_r F_\ell^{(\mathrm{out})} &= \frac{h_\ell^{(-)}(kr)}{2ik} V(r) \left[h_\ell^{(-)}(kr) F_\ell^{(\mathrm{in})} + h_\ell^{(+)}(kr) F_\ell^{(\mathrm{out})} \right]
\end{aligned} \tag{2.21}$$

It should be emphasized that this system is equivalent to the Schrödinger equation (2.8) from which we started. It is, however, more convenient for solving the problem of motion of a particle in a potential. Indeed, when the potential vanishes (for $r > R$), the right hand sides of Eq. (2.21) disappear, and the functions $F_\ell^{(\mathrm{in/out})}(E, r)$ become constants. This means that these unknown functions have a more simple r-dependence than the original unknown solution $\phi_\ell(E, r)$.

The form (2.12) that was chosen for the radial wave function ensures its correct asymptotic behaviour (as a linear combination of the Riccati–Hankel functions) outside the interaction region. Furthermore, this asymptotic behaviour is given analytically, which makes it easy to construct the physical solutions with specific boundary conditions at $r \to \infty$. In a sense, this is a semi-analytical way of solving the Schrödinger problem because the Riccati–Hankel functions that determine the wave function asymptotics are explicitly embedded in the solution (2.12).

What kind of boundary conditions should we impose on the functions $F_\ell^{(\mathrm{in/out})}(E, r)$, when solving Eqs. (2.21)? These conditions should be chosen in such a way that would guarantee the short-distance behaviour of the regular solution, $\phi_\ell(E, r)$, given by Eq. (2.7). Since Eq. (2.21) is of the first order, we only need to specify at $r = 0$ the values of $F_\ell^{(\mathrm{in/out})}(E, r)$.

In the representation (2.12), we have the Riccati–Hankel functions, $h_\ell^{(\pm)}(kr)$, that both are singular at $r = 0$ (see Appendix D),

$$h_\ell^{(\pm)}(kr) = j_\ell(kr) \pm i n_\ell(kr) \underset{r \to 0}{\longrightarrow} \pm i n_\ell(kr) \underset{r \to 0}{\longrightarrow} \mp \frac{i(2\ell - 1)!!}{(kr)^\ell} .$$

Therefore, Eq. (2.7) can only be satisfied if at $r = 0$ both $F_\ell^{(\mathrm{in})}(E, r)$ and $F_\ell^{(\mathrm{out})}(E, r)$ converge to the same constant. Indeed, in such a case, the singular parts of $h_\ell^{(\pm)}(kr)$ cancel each other,

$$h_\ell^{(-)}(kr) + h_\ell^{(+)}(kr) = 2 j_\ell(kr) , \tag{2.22}$$

and

$$\phi_\ell(E, r) \underset{r \to 0}{\longrightarrow} \mathrm{const} \times \left[h_\ell^{(+)}(kr) + h_\ell^{(-)}(kr) \right] = \mathrm{const} \times 2 j_\ell(kr) \sim r^{\ell+1} . \tag{2.23}$$

The regular solution, $\phi_\ell(E,r)$, behaves (by definition) near the origin exactly as $j_\ell(kr)$. From (2.22), it is therefore clear that the boundary conditions for Eq. (2.21) are

$$F_\ell^{(\mathrm{in})}(E,0) = F_\ell^{(\mathrm{out})}(E,0) = \frac{1}{2} \tag{2.24}$$

At this point, it should be already obvious that when solving the physics problems, equations (2.21) have some advantages as compared with the original Schrödinger equation. Indeed, as we have seen, at those energies for which the physical solution, $u_\ell(E,r)$, exists, it differs from $\phi_\ell(E,r)$ only in the normalization factor. It is much easier to deal with the first-order differential equations and with the Cauchy problem rather than with the Dirichlet problem for the second-order equation. As we will see very soon, the asymptotic behaviour of the function $u_\ell(E,r)$ becomes correct automatically if it is constructed according to Eq. (2.12). This is because the Riccati–Hankel functions, which are most important for its behaviour at large distances, are present there explicitly, while the amplitude functions $F_\ell^{(\mathrm{in/out})}(E,r)$ automatically make a correct combination of them for every choice of the energy E. Thanks to the explicit embedding of the Riccati–Hankel functions in the ansatz (2.12), the functions $F_\ell^{(\mathrm{in/out})}(E,r)$ are more smooth. At large distances, they tend to constant values, while all the oscillations or the exponential behaviour (if k is complex) are present in the wave function in an exact way via $h_\ell^{(\pm)}(kr)$.

2.2.1.2 Alternative Form of the Differential Equations

The system of differential equations (2.21) was obtained using the Riccati–Hankel functions, $h_\ell^{(\pm)}(kr)$, as the pair of linearly independent solutions of the "free" equation (2.9). There is another pair of its solutions (the Riccati–Bessel and Riccati–Neumann functions):

$$j_\ell(kr) = \frac{h_\ell^{(+)}(kr) + h_\ell^{(-)}(kr)}{2} \tag{2.25}$$

and

$$n_\ell(kr) = \frac{h_\ell^{(+)}(kr) - h_\ell^{(-)}(kr)}{2i} \tag{2.26}$$

that can be used as well. Instead of embedding in the radial wave function the spherical waves, we may look for $\phi_\ell(E,r)$ in the form

$$\phi_\ell(E,r) = j_\ell(kr)A_\ell(E,r) - n_\ell(kr)B_\ell(E,r) \tag{2.27}$$

which coincides with the ansatz (2.12) if

$$A_\ell(E,r) \equiv F_\ell^{(\mathrm{in})}(E,r) + F_\ell^{(\mathrm{out})}(E,r) \tag{2.28}$$

and

$$B_\ell(E,r) \equiv i\left[F_\ell^{(\mathrm{in})}(E,r) - F_\ell^{(\mathrm{out})}(E,r)\right], \tag{2.29}$$

that is

$$F_\ell^{(\mathrm{in})}(E,r) \equiv \frac{A_\ell(E,r) - iB_\ell(E,r)}{2} \tag{2.30}$$

and

$$F_\ell^{(\mathrm{out})}(E,r) \equiv \frac{A_\ell(E,r) + iB_\ell(E,r)}{2}. \tag{2.31}$$

Apparently, Eqs. (2.28) and (2.29) are valid for the derivatives as well. Therefore, combining Eq. (2.21), i.e. adding them and subtracting as in Eqs. (2.28,2.29), we obtain the following alternative system of differential equations:

$$\boxed{\begin{aligned}
\partial_r A_\ell(E,r) &= -\frac{1}{k}n_\ell(kr)V(r)\left[j_\ell(kr)A_\ell(E,r) - n_\ell(kr)B_\ell(E,r)\right] \\
\partial_r B_\ell(E,r) &= -\frac{1}{k}\, j_\ell(kr)V(r)\left[j_\ell(kr)A_\ell(E,r) - n_\ell(kr)B_\ell(E,r)\right]
\end{aligned}} \tag{2.32}$$

which is equivalent to the system (2.21) as well as to the Schrödinger equation (2.8). The boundary conditions for Eq. (2.32),

$$A_\ell(E,0) = 1 \tag{2.33}$$

$$B_\ell(E,0) = 0 \tag{2.34}$$

follow immediately from (2.24) and the relations (2.28) and (2.29).

In contrast to the ansatz (2.12) where both terms have clear physical meaning at large distances, its alternative form (2.27) mixes the incoming and outgoing waves. Why do we consider it then? There are at least two reasons for doing this:

- Further in the book, it is shown that the representation (2.27) and its generalizations for the multi-channel problems are the most convenient when the analytic structure of the Jost matrices is analysed.

- The second reason is more simple and can be understood right now. It stems from some computational difficulties that may arise, when Eq. (2.21) is solved numerically. As it is explained below, the alternative equations (2.32) allow one to circumvent such numerical difficulties.

We will come back to the first of these reasons in the subsequent sections of the book and now consider the second one.

The representation of $\phi_\ell(E,r)$ in terms of the functions $A_\ell(E,r)$ and $B_\ell(E,r)$ has certain advantages at short distances. When $r \to 0$, the ansatz (2.27) clearly separates the regular and irregular terms, while in (2.12) they are mixed. This becomes important when solving Eq. (2.21) numerically with the boundary conditions (2.24). Indeed, we have to start the integration from $r = 0$. However, for $\ell \neq 0$, both functions $h_\ell^{(\pm)}(kr)$ are irregular, i.e. at the origin they behave as

$$h_\ell^{(\pm)}(kr) \underset{r\to 0}{\longrightarrow} \mp \frac{i(2\ell-1)!!}{(kr)^\ell} . \tag{2.35}$$

From the formal mathematical point of view, this fact does not pose any problem because with $F_\ell^{(\text{in})}(E,0) = F_\ell^{(\text{out})}(E,0)$ the singularities of $h_\ell^{(+)}(kr)$ and $h_\ell^{(-)}(kr)$ cancel each other according to Eq. (2.25). As a result, the solution is regular,

$$\phi_\ell(E,r) \underset{r\to 0}{\longrightarrow} j_\ell(kr) \underset{r\to 0}{\longrightarrow} \frac{(kr)^{\ell+1}}{(2\ell+1)!!} , \tag{2.36}$$

and the right hand sides of Eq. (2.21) are integrable. Indeed, with the condition (2.2), the potential is less singular than $1/r^2$, i.e.

$$V(r) \underset{r\to 0}{\sim} \frac{1}{r^{2-\delta}} , \qquad \delta > 0 , \tag{2.37}$$

which means that on a sufficiently small interval $0 < r < r_{\text{min}}$ near the origin Eq. (2.21) can be written as

$$\partial_r F_\ell^{(\text{in/out})}(E,r) \approx \frac{Q_\ell^{(\text{in/out})}(E)}{r^{1-\delta}} , \tag{2.38}$$

where $Q_\ell^{(\text{in/out})}(E)$ are r-independent coefficients. The integration of Eq. (2.38) can be easily done,

$$F_\ell^{(\text{in/out})}(E,r) \approx \frac{1}{2} + \int_0^r \frac{Q_\ell^{(\text{in/out})}(E)}{\rho^{1-\delta}} d\rho = \frac{1}{2} + Q_\ell^{(\text{in/out})}(E)\frac{r^\delta}{\delta} , \tag{2.39}$$

and shows that $F_\ell^{(\text{in/out})}(E,r)$ are regular as expected.

Although formally there is no problem with Eq. (2.21) at $r = 0$, in numerical calculations, we cannot start from this point but have to shift to a small value $r = r_{\min}$. Besides this, the cancellation of singularities in numerical calculations is always a source of possible errors and instabilities. The errors increase with increasing ℓ since in this case $h_\ell^{(\pm)}(kr)$ become more singular. Therefore, for larger ℓ, the point $r_{\min}$ must be shifted further from the origin. This, in turn, requires an adjustment of the boundary conditions (2.24), which are not accurate at $r = r_{\min}$. The first-order corrections to the boundary conditions can be obtained, for example, using Eq. (2.39). The higher order corrections can be calculated iteratively, as is shown in the next section.

There is another way to avoid the numerical instabilities and the loss of accuracy caused by high values of the angular momentum. Since the representations of $\phi_\ell(E, r)$ in terms of $F_\ell^{(\text{in/out})}$ and the pair A_ℓ, B_ℓ are equivalent, on different segments of the axis $r \in [0, \infty)$, we can use either the differential equations (2.21) or (2.32) depending on whichever of them is more convenient. Apparently, at large distances, it is more natural to represent $\phi_\ell(E, r)$ in the form (2.12), while the representation (2.27) is more convenient near the origin.

Indeed, since in the immediate neighbourhood of the point $r = 0$, the wave function behaves exactly like $j_\ell(kr)$, the second term in the representation (2.27) is negligible as compared to the first term. The greater the ℓ, the wider is the interval $(0, r_{\min})$ on which this is true. This means that the greater the angular momentum, the more stable Eq. (2.32) near the origin is. This situation is just opposite to the one with Eq. (2.21).

Therefore, in numerical calculations, we should solve Eq. (2.32) from $r = r_{\min}$ to an intermediate point $r = r_{\text{int}}$. At this point, using the calculated values of $A_\ell(E, r_{\text{int}})$ and $B_\ell(E, r_{\text{int}})$ and the relations (2.30) and (2.31), we can obtain the values $F_\ell^{(\text{in/out})}(E, r_{\text{int}})$ that are the boundary values for Eq. (2.21) for further integration from $r = r_{\text{int}}$ to $r = R$ (i.e. to the point beyond which $F_\ell^{(\text{in/out})}(E, r)$ do not change).

In most cases, if the starting point $r_{\min}$ is chosen close enough to the origin, the boundary values at this point differ very slightly from those at $r = 0$, i.e. from the values given by Eqs. (2.33, 2.34). Since near the origin Eq. (2.32) is very stable, it is easy to calculate the wave function with a very high accuracy. When an extreme accuracy is required or for some reason $r_{\min}$ is not very close to zero, the boundary values, $A_\ell(E, r_{\min})$ and $B_\ell(E, r_{\min})$, can be iteratively corrected as is described in the next section.

2.2.1.3 Integral Equations

By integrating the left and right hand sides of Eq. (2.21) and taking into account the boundary conditions (2.24), we obtain an equivalent pair of integral Volterra-type equations,

$$
\begin{aligned}
F_\ell^{(\text{in})}(E,r) &= \frac{1}{2} - \frac{1}{2ik}\int_0^r h_\ell^{(+)}(kr')V(r')\Big[h_\ell^{(-)}(kr')F_\ell^{(\text{in})}(E,r') \\
&\qquad\qquad + h_\ell^{(+)}(kr')F_\ell^{(\text{out})}(E,r')\Big]dr' , \\
F_\ell^{(\text{out})}(E,r) &= \frac{1}{2} + \frac{1}{2ik}\int_0^r h_\ell^{(-)}(kr')V(r')\Big[h_\ell^{(-)}(kr')F_\ell^{(\text{in})}(E,r') \\
&\qquad\qquad + h_\ell^{(+)}(kr')F_\ell^{(\text{out})}(E,r')\Big]dr' .
\end{aligned}
\tag{2.40}
$$

These equations enable us to adjust the boundary conditions (2.24) when the starting point is shifted from zero to $r = r_{\text{min}}$. This can be done by solving Eq. (2.40) on the interval $[0, r_{\text{min}}]$ iteratively, namely,

$$
\begin{aligned}
F_\ell^{(\text{in/out})(0)}(E,r) &= \frac{1}{2} , \\
F_\ell^{(\text{in/out})(1)}(E,r) &= \frac{1}{2} \mp \frac{1}{2ik}\int_0^r h_\ell^{(\pm)}(kr')V(r')j_\ell(kr')\,dr' , \\
&\ \ \vdots \\
F_\ell^{(\text{in/out})(\text{N})}(E,r) &= \frac{1}{2} \mp \frac{1}{2ik}\int_0^r h_\ell^{(\pm)}(kr')V(r') \\
&\quad \times\Big[h_\ell^{(-)}(kr')F_\ell^{(\text{in})(N-1)}(E,r') \\
&\quad + h_\ell^{(+)}(kr')F_\ell^{(\text{out})(N-1)}(E,r')\Big]dr' .
\end{aligned}
\tag{2.41}
$$

Similarly to Eq. (2.40), the integral equations that are equivalent to the differential equations (2.32) with the boundary conditions (2.33, 2.34) are obtained by integration of the left and right hand sides of Eq. (2.32). This gives

$$
\begin{aligned}
A_\ell(E,r) &= 1 - \frac{1}{k}\int_0^r n_\ell(kr')V(r')\left[j_\ell(kr')A_\ell(E,r') - n_\ell(kr')B_\ell(E,r')\right]dr' , \\
B_\ell(E,r) &= \quad - \frac{1}{k}\int_0^r j_\ell(kr')V(r')\left[j_\ell(kr')A_\ell(E,r') - n_\ell(kr')B_\ell(E,r')\right]dr' .
\end{aligned}
\tag{2.42}
$$

These Volterra-type equations can be solved iteratively on a small interval $[0, r_{\min}]$ in order to correct the boundary conditions (2.33, 2.34) when the point $r_{\min}$ is not very close to the origin. The iteration procedure can be written as follows:

$$A_\ell^{(0)}(E,r) = 1\,, \qquad B_\ell^{(0)}(E,r) = 0\,, \tag{2.43}$$

$$\begin{aligned} A_\ell^{(\mathrm{N})}(E,r) &= 1 - \frac{1}{k}\int_0^r n_\ell(kr')V(r') \\ &\quad\times \left[j_\ell(kr')A_\ell^{(\mathrm{N}-1)}(E,r') - n_\ell(kr')B_\ell^{(\mathrm{N}-1)}(E,r')\right] dr'\,, \\ B_\ell^{(\mathrm{N})}(E,r) &= \quad -\frac{1}{k}\int_0^r j_\ell(kr')V(r') \\ &\quad\times \left[j_\ell(kr')A_\ell^{(\mathrm{N}-1)}(E,r') - n_\ell(kr')B_\ell^{(\mathrm{N}-1)}(E,r')\right] dr'\,. \end{aligned} \tag{2.44}$$

Certainly, the integral equations (2.40) and (2.42) are useful not only for correcting the boundary conditions. They are equivalent to the corresponding differential equations and therefore can be used instead of them whenever it is appropriate and convenient.

2.2.2 Uniform Bound for the Regular Solution

In this section, it is found a simple function of k and r such that on the whole plane of complex k it is an upper limit for the absolute value of the regular solution $\phi_\ell(E,r)$, i.e. is a uniform majorant for $|\phi_\ell(E,r)|$. This upper bound will be used throughout the book to establish various analytic properties of the Jost functions.

2.2.2.1 Integral Equation for the Regular Solution

Using Eq. (2.40), it is easy to derive an integral equation for the function $\phi_\ell(E,r)$, which is a solution of the Schrödinger equation (2.8) with the boundary conditions (2.7). According to the ansatz (2.12), this wave function is obtained if we multiply the first of Eq. (2.40) by $h_\ell^{(-)}(kr)$ and add it to the second one multiplied by $h_\ell^{(+)}(kr)$. This gives

$$\begin{aligned} &\phi_\ell(E,r) = \\ &= j_\ell(kr) + \frac{1}{2ik}\int_0^r \left[h_\ell^{(+)}(kr)h_\ell^{(-)}(kr') - h_\ell^{(-)}(kr)h_\ell^{(+)}(kr')\right] V(r')\phi_\ell(E,r')\,dr' \\ &= j_\ell(kr) + \frac{1}{k}\int_0^r \left[n_\ell(kr)j_\ell(kr') - j_\ell(kr)n_\ell(kr')\right] V(r')\phi_\ell(E,r')\,dr'\,. \end{aligned} \tag{2.45}$$

With the Green's function defined as[4]

$$g_\ell(E,r,r') = \frac{1}{k}\left[n_\ell(kr)j_\ell(kr') - j_\ell(kr)n_\ell(kr')\right] \tag{2.46}$$

$$= \frac{1}{2ik}\left[h_\ell^{(+)}(kr)h_\ell^{(-)}(kr') - h_\ell^{(-)}(kr)h_\ell^{(+)}(kr')\right] \tag{2.47}$$

$$= \frac{1}{ik}\left[h_\ell^{(+)}(kr)j_\ell(kr') - j_\ell(kr)h_\ell^{(+)}(kr')\right] \tag{2.48}$$

$$= \frac{1}{ik}\left[j_\ell(kr)h_\ell^{(-)}(kr') - h_\ell^{(-)}(kr)j_\ell(kr')\right] \tag{2.49}$$

$$= \frac{1}{k}\left[n_\ell(kr)h_\ell^{(+)}(kr') - h_\ell^{(+)}(kr)n_\ell(kr')\right] \tag{2.50}$$

$$= \frac{1}{k}\left[n_\ell(kr)h_\ell^{(-)}(kr') - h_\ell^{(-)}(kr)n_\ell(kr')\right], \quad \text{for } r' \leqslant r \tag{2.51}$$

$$g_\ell(E,r,r') = 0\,, \quad \text{for } r' > r,$$

this equation reads

$$\boxed{\phi_\ell(E,r) = j_\ell(kr) + \int_0^r g_\ell(E,r,r')V(r')\phi_\ell(E,r')\,dr'} \tag{2.52}$$

This is a Volterra-type equation whose main feature is that the integration runs up to r (not to infinity). Thanks to this feature, the iterative procedure for such equations is always convergent independently of the strength of the potential (see Appendix F).

2.2.2.2 Upper Bound for the Regular Solution

In order to find a majorant function for $|\phi_\ell(E,r)|$, we write the solution of Eq. (2.52) in the form of the iteration series,

$$\phi_\ell(E,r) = \sum_{n=0}^{\infty}\phi_\ell^{(n)}(E,r)\,, \tag{2.53}$$

[4] Equations (2.48–2.51) are obtained by making appropriate transformations with the help of the relations $h_\ell^{(\pm)} = j_\ell \pm i n_\ell$.

where

$$\phi_\ell^{(0)}(E,r) = j_\ell(kr) \tag{2.54}$$

and

$$\phi_\ell^{(n)}(E,r) = \int_0^r g_\ell(E,r,r')V(r')\phi_\ell^{(n-1)}(E,r')\,dr' \tag{2.55}$$

$$= \int_0^r dr_n \int_0^{r_n} dr_{n-1} \cdots \int_0^{r_2} dr_1 \tag{2.56}$$

$$g_\ell(E,r,r_n)V(r_n)g_\ell(E,r_n,r_{n-1})V(r_{n-1})\cdots g_\ell(E,r_2,r_1)V(r_1)\phi_\ell^{(0)}(E,r_1).$$

Looking at this multiple integral, the reader can understand why the series (2.53) is always convergent. This is because among the sequence $r_n, r_{n-1}, r_{n-2}, \dots r_2, r_1$, every subsequent variable runs within a smaller interval. For example, while r_n runs over $[0, r]$, the interval for the variable r_{n-1} is limited to $[0, r_n]$, as is shown in Fig. 2.1. As a result, the two leftmost integrals in Eq. (2.56) cover a triangle region. Using the bounds (D.25, D.29) for the Riccati–Bessel and Riccati–Neumann functions, given in Appendix D, we obtain the following bound for the integrand of Eq. (2.56):

$$\left| g_\ell(E,r,r_n)V(r_n)g_\ell(E,r_n,r_{n-1})V(r_{n-1})\cdots g_\ell(E,r_2,r_1)V(r_1)\phi_\ell^{(0)}(E,r_1)\right|$$

$$\leqslant C|k|^{-n}e^{|\mathrm{Im}\,k|r}\left(\frac{|kr|}{1+|kr|}\right)^{\ell+1}\prod_{m=1}^{n}\frac{|kr_m|}{1+|kr_m|}|V(r_m)|\ ,$$

where the exponential factors $\exp(-|\mathrm{Im}\,k|r_m)$ as well as the higher powers of the ratio $|kr_m|/(1+|kr_m|)$ disappear because in each pair of the subsequent factors $|g_\ell(E,r,r_m)|$ they cancel each other. Therefore,

$$\left|\phi_\ell^{(n)}(E,r)\right| \leqslant Ce^{|\mathrm{Im}\,k|r}\left(\frac{|kr|}{1+|kr|}\right)^{\ell+1}\int_0^r dr_n\int_0^{r_n}dr_{n-1}\cdots\int_0^{r_2}dr_1$$

$$\times\prod_{m=1}^{n}\frac{r_m\,|V(r_m)|}{1+|kr_m|}\ . \tag{2.57}$$

The integrand in this equation is completely symmetric with respect to all possible interchanges of the variables r_m. This means that it does not matter which triangle, lower or upper, to use for integrating over $dr_n dr_{n-1}$ (see Fig. 2.1). In other words,

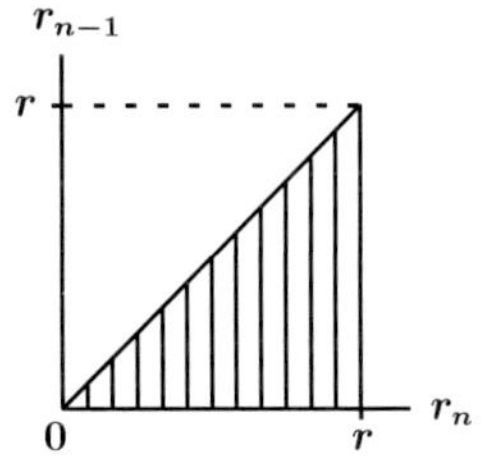

Fig. 2.1 The triangle integration domain for the two leftmost integrals in Eq. (2.56)

we can integrate over the full square and take half of the result,

$$\int_0^r dr_n \int_0^{r_n} dr_{n-1} (\cdots) = \frac{1}{2} \int_0^r \int_0^r dr_n dr_{n-1} (\cdots) .$$

There are $n!$ possible permutations of the product $dr_1 dr_2 \ldots dr_n$. It is easy to see that adding up all the corresponding integrals, we obtain the integral over the whole hyper-cube, in which all r_m run from 0 to r. Therefore,

$$\int_0^r dr_n \int_0^{r_n} dr_{n-1} \cdots \int_0^{r_2} dr_1 (\cdots) = \frac{1}{n!} \int_0^r \int_0^r \ldots \int_0^r dr_1 dr_2 \ldots dr_n (\cdots) ,$$

which implies that

$$\left| \phi_\ell^{(n)}(E, r) \right| \leqslant C \frac{\alpha^n}{n!} e^{|\mathrm{Im}\, kr|} \left(\frac{|kr|}{1 + |kr|} \right)^{\ell+1} , \tag{2.58}$$

where[5]

$$\alpha = \int_0^r \frac{r'}{1 + |kr'|} \left| V(r') \right| dr' . \tag{2.59}$$

When using the bound (2.58) in the series (2.53), we come to the following three conclusions:

- The iteration series (2.53) is uniformly convergent because it is majorated by the absolutely convergent series

$$\sum_{n=0}^{\infty} \frac{\alpha^n}{n!} = e^\alpha ,$$

[5] The reader should have guessed already that we will need this bound for the wave function at large distances, i.e. when $r \to \infty$. It is easy to see that for the potentials that are less singular than r^{-2} and vanishing at infinity faster than r^{-1}, the integral in Eq. (2.59) is always finite.

where α can be made k-independent and r-independent by putting $k = 0$ and $r = \infty$ (which only makes α bigger). This once more corroborates the general theorem stating that the iterative procedure for a Volterra integral equation is always convergent (see Appendix F).

- The function $\phi_\ell(E, r)$ is an analytic function of k on the whole k-plane. This is because it is represented by a uniformly convergent series of analytic functions (2.56), which are analytic because they are integrals of the other analytic functions.
- The function $\phi_\ell(E, r)$ is bound by the following majorant:

$$|\phi_\ell(E, r)| \leqslant C_\ell \left(\frac{|kr|}{1 + |kr|} \right)^{\ell+1} e^{|\mathrm{Im}\, kr|} , \tag{2.60}$$

where C_ℓ is some positive constant.

An integral equation automatically takes into account specific boundary conditions. In this connection, it should be noted that equation (2.52), from which we started the derivations, is written for the regular solution. This is clear from the fact that the inhomogeneous term, $j_\ell(kr)$, in this equation follows from the boundary conditions (2.24) corresponding to the regular solution. However, since in the single-channel problem any physical solution only differs from ϕ_ℓ by its normalization, the estimate (2.60) remains valid for a physical solution as well (with a different constant C_ℓ).

2.2.3 Jost Functions

The function (2.12) with the boundary conditions (2.24), i.e.

$$\phi_\ell(E, r) = W_\ell^{(\mathrm{in})}(E, r) F_\ell^{(\mathrm{in})}(E, r) + W_\ell^{(\mathrm{out})}(E, r) F_\ell^{(\mathrm{out})}(E, r) , \tag{2.61}$$

is the regular solution of the radial Schrödinger equation. Its behaviour at large r is determined by the Riccati–Hankel functions as well as by the asymptotic values of $F_\ell^{(\mathrm{in/out})}(E, r)$, namely,

$$\boxed{\phi_\ell(E, r) \underset{r\to\infty}{\longrightarrow} h_\ell^{(-)}(kr) f_\ell^{(\mathrm{in})}(E) + h_\ell^{(+)}(kr) f_\ell^{(\mathrm{out})}(E)} \tag{2.62}$$

where

$$f_\ell^{(\mathrm{in})}(E) = \lim_{r\to\infty} F_\ell^{(\mathrm{in})}(E, r) \tag{2.63}$$

and

$$f_\ell^{(\mathrm{out})}(E) = \lim_{r\to\infty} F_\ell^{(\mathrm{out})}(E, r) . \tag{2.64}$$

In the case of a finite-range potential, the limits in the above equations are reached at $r = R$ because beyond this point the right hand sides of Eq. (2.21) are zero and therefore the $F_\ell^{(\text{in/out})}(E,r)$ do not change since their derivatives vanish. The energy-dependent functions $f_\ell^{(\text{in/out})}(E)$ are essentially the same that were introduced in 1947 by Res Jost [10]. The only difference is that we normalize them differently.

Letting $r \to \infty$ in the integral equations (2.40) and taking into account the limits (2.63, 2.64) as well as the representation (2.12) of the regular solution, we obtain one more relation between $\phi_\ell(E,r)$ and the Jost functions:

$$f_\ell^{(\text{in/out})}(E) = \frac{1}{2} \pm \frac{i}{2k} \int_0^\infty h_\ell^{(\pm)}(kr) V(r) \phi_\ell(E,r)\, dr \tag{2.65}$$

Nowadays, the Jost functions are introduced practically in all textbooks on quantum mechanics and scattering theory. The choice of their normalization as well as notation varies from book to book. They can be found as $f_\ell^{(\pm)}(E)$, $f_\ell^{(\pm)}(k)$, or $f_\ell(\pm k)$. Some authors use the term Jost function to name the solutions of the Schrödinger equation (2.4) that at large r behave exactly like the incoming and outgoing spherical waves, $\exp(\pm ikr)$. These functions, depending on both the energy and the distance, are also known as the Jost solutions. We consider them further down in Sect. 2.2.4.

Throughout this book, we use the notation $f_\ell^{(\text{in/out})}(E)$ for the Jost functions (and generally Jost matrices), which is different from the traditional notations. There are two reasons for this. First of all, we use the normalization of the so-called regular solution, which is not the same as in some famous textbooks on scattering theory.

Actually, the normalization can be chosen arbitrarily since it does not affect any observable quantities. As a result, both $f^{(\text{in})}(E)$ and $f^{(\text{out})}(E)$ can have an arbitrary common factor. As will be seen in the subsequent chapters, when calculating observable quantities, we either look for zeros of $f_\ell^{(\text{in})}(E)$ or use the ratio of $f_\ell^{(\text{out})}(E)$ and $f_\ell^{(\text{in})}(E)$. In all the cases, their normalization can be cancelled out. Since no observable depends on the Jost function normalization, it can be chosen arbitrarily, for example, as is determined by Eq. (2.24).

The second reason is that the superscripts (in) and (out) are unambiguous since they clearly indicate which of these functions are the asymptotic amplitudes of the incoming and outgoing waves. Thus we avoid possible confusion caused by the existence of notations with opposite signs for the same Jost functions.

In addition to this (and perhaps more importantly than anything else), the notations used here remain unchanged irrespective of how many channels are involved. In multi-channel problems, where it is impossible to introduce a single momentum k (see further chapters), the energy-dependent functions $f^{(\text{in/out})}(E)$ simply become matrices.

2.2.4 Jost Solutions

As was mentioned in Sect. 2.1, both of the two linearly independent solutions of Eq. (2.4) can be chosen irregular. We denote such solutions of the Schrödinger equation

$$\left[\frac{d^2}{dr^2} + k^2 - \frac{\ell(\ell+1)}{r^2} - V(r)\right]\chi_\ell^{(\pm)}(E,r) = 0 \tag{2.66}$$

as $\chi_\ell^{(\pm)}(E,r)$ and define them by the following boundary conditions at infinity:

$$\chi_\ell^{(\pm)}(E,r) \underset{r\to\infty}{\longrightarrow} h_\ell^{(\pm)}(kr) \tag{2.67}$$

$$\underset{r\to\infty}{\longrightarrow} (\mp i)^{\ell+1} e^{\pm ikr} \tag{2.68}$$

These solutions that behave at large distances exactly as the outgoing and incoming spherical waves are called the Jost solutions. The fact that they are irregular can be seen by considering the particular case when $V(r) \equiv 0$ and they coincide with the Riccati–Hankel functions at all distances. More rigorously, the fact that $\chi_\ell^{(\pm)}(E,r)$ are irregular at the origin can be deduced by taking note that ϕ_ℓ and $\chi_\ell^{(\pm)}$ are linearly independent (this is obvious at large distances). Meanwhile, it can exist only one independent regular solution of the radial Schrödinger equation (because it has a regular singular point at $r = 0$).

Similarly to the regular solution (see Eq. (2.12)), we look for the Jost solutions in the form

$$\chi_\ell^{(\pm)}(E,r) = h_\ell^{(-)}(kr)\mathfrak{X}_\ell^{(\pm\text{in})}(E,r) + h_\ell^{(+)}(kr)\mathfrak{X}_\ell^{(\pm\text{out})}(E,r)\,, \tag{2.69}$$

where $\mathfrak{X}_\ell^{(\pm\text{in/out})}(E,r)$ are unknown functions obeying the differential equations

$$\begin{aligned}
\partial_r\mathfrak{X}_\ell^{(\pm\text{in})} &= -\frac{h_\ell^{(+)}(kr)}{2ik}V(r)\left[h_\ell^{(-)}(kr)\mathfrak{X}_\ell^{(\pm\text{in})} + h_\ell^{(+)}(kr)\mathfrak{X}_\ell^{(\pm\text{out})}\right],\\
\partial_r\mathfrak{X}_\ell^{(\pm\text{out})} &= \frac{h_\ell^{(-)}(kr)}{2ik}V(r)\left[h_\ell^{(-)}(kr)\mathfrak{X}_\ell^{(\pm\text{in})} + h_\ell^{(+)}(kr)\mathfrak{X}_\ell^{(\pm\text{out})}\right],
\end{aligned} \tag{2.70}$$

which are the same as Eq. (2.21) but with different boundary conditions. To obtain the Jost solutions, we should start integrating Eq. (2.70) from a faraway point with the boundary values either

$$\mathfrak{X}_\ell^{(+\text{in})}(E,\infty) = 0\,, \quad \mathfrak{X}_\ell^{(+\text{out})}(E,\infty) = 1 \qquad \text{for} \quad \chi_\ell^{(+)}\,, \tag{2.71}$$

or

$$\mathfrak{X}_\ell^{(-\text{in})}(E,\infty)=1\,,\quad \mathfrak{X}_\ell^{(-\text{out})}(E,\infty)=0 \qquad \text{for}\quad \chi_\ell^{(-)}\,, \tag{2.72}$$

and move inwards. Then the function (2.69) would be either $\chi_\ell^{(+)}(E,r)$ or $\chi_\ell^{(-)}(E,r)$, respectively. For a finite-range potential, the faraway point in Eqs. (2.71, 2.72) should be taken at the distance $r=R$ beyond which the potential is identically zero.

Bearing in mind that the regular solution obeys the same differential equation as the Jost solutions, we can expect that there should be some relation between $\phi_\ell(E,r)$ and $\chi_\ell^{(\pm)}(E,r)$. Such a relation can be found using the notion of Wronskian. The definition and some properties of the Wronskian are given in Appendix C, where it is shown that the Wronskian of any two solutions of the radial Schrödinger equation does not depend on r and thus can be calculated at any convenient point within the interval $r\in[0,\infty)$.

Applying this to Eq. (2.66), we can calculate the Wronskian of $\phi_\ell(E,r)$ and $\chi_\ell^{(+)}(E,r)$ using their asymptotics (2.62) and (2.67). Since the Wronskians of the Riccati–Hankel functions are known,

$$W\left[h_\ell^{(-)},h_\ell^{(+)}\right]=2ik \qquad \text{and} \qquad W\left[h_\ell^{(+)},h_\ell^{(+)}\right]=0\,, \tag{2.73}$$

we obtain

$$\begin{aligned} W\left[\phi_\ell,\chi_\ell^{(+)}\right] &= W\left[h_\ell^{(-)},h_\ell^{(+)}\right]f_\ell^{(\text{in})}(E) \\ &= 2ikf_\ell^{(\text{in})}(E)\,. \end{aligned} \tag{2.74}$$

Similarly, one can easily find the relation between $\chi_\ell^{(-)}$ and ϕ_ℓ. Therefore,

$$\boxed{f_\ell^{(\text{in/out})}(E)=\mp\frac{1}{2ik}W\left[\chi_\ell^{(\pm)}(E,r),\phi_\ell(E,r)\right]} \tag{2.75}$$

These relations are very useful in certain derivations and proofs concerning the properties of the Jost functions. In particular, we can express $f_\ell^{(\text{in/out})}$ via the Jost solutions near the origin. In order to do this, we use the fact that Wronskian in Eq. (2.75) does not depend on r and instead of large distances can be calculated near $r=0$, where

$$\phi_\ell(E,r)\underset{r\to 0}{\longrightarrow}\frac{(kr)^{\ell+1}}{(2\ell+1)!!}\,,\qquad \partial_r\phi_\ell(E,r)\underset{r\to 0}{\longrightarrow}\frac{k(\ell+1)(kr)^\ell}{(2\ell+1)!!}\,. \tag{2.76}$$

The Jost solutions are irregular and, according to Eq. (2.6), can be written as

$$\chi_\ell^{(\pm)}(E,r) \underset{r\to 0}{\longrightarrow} \frac{X_\ell^{(\pm)}(E)}{(kr)^\ell}\,, \qquad \partial_r \chi_\ell^{(\pm)}(E,r) \underset{r\to 0}{\longrightarrow} -\frac{k\ell X_\ell^{(\pm)}(E)}{(kr)^{\ell+1}}\,, \tag{2.77}$$

where

$$X_\ell^{(\pm)}(E) = \lim_{r\to 0}(kr)^\ell \chi_\ell^{(\pm)}(E,r)\,.$$

Substituting (2.76) and (2.77) in Eq. (2.75), we find

$$\boxed{f_\ell^{(\text{in/out})}(E) \;=\; \lim_{r\to 0} \frac{\pm i (kr)^\ell}{2(2\ell-1)!!}\chi_\ell^{(\pm)}(E,r)} \tag{2.78}$$

In the particular case $\ell = 0$, the Jost functions are proportional to the values of the Jost solutions at the origin,

$$f_0^{(\text{in/out})}(E) = \frac{\pm i}{2}\chi_0^{(\pm)}(E,0)\,. \tag{2.79}$$

Sometimes the relation (2.78) is used as the definition of the Jost functions.[6] As is seen, it is simple and suitable for numerical calculations only for zero angular momentum. If $\ell \neq 0$, then near the origin, equation (2.78) gives the uncertainty of the type $0\times\infty$ that always leads to numerical instabilities and the loss of accuracy. However, it is still very useful. For example, it is used to define the so-called off-shell Jost functions that generalize the ordinary ones (see Sect. 17.3).

There are several other relations among the regular and Jost solutions. The Jost solutions obey the equations

$$W\left[\chi_\ell^{(-)},\chi_\ell^{(+)}\right] = 2ik \qquad \text{and} \qquad W\left[\chi_\ell^{(+)},\chi_\ell^{(+)}\right] = W\left[\chi_\ell^{(-)},\chi_\ell^{(-)}\right] = 0 \tag{2.80}$$

that follow from (2.73) as well as from the fact that the Wronskian does not depend on r.

Since the Jost solutions constitute a complete basis in the space of all possible solutions of the radial Schrödinger equation, the regular solution is their linear combination,

$$\phi_\ell = \alpha\chi_\ell^{(-)} + \beta\chi_\ell^{(+)}\,,$$

[6] Such a relation is not possible for the Jost matrices in the multi-channel case (see Sect. 9.6.1) and therefore should be avoided as the definition because it cannot be generalized.

with some r-independent coefficients α and β. Apparently, the choice of these coefficients is unique, and we know them at large distances. Indeed, comparing (2.62) with (2.67), we see that

$$\boxed{\phi_\ell(E,r) = \chi_\ell^{(-)}(E,r) f_\ell^{(\mathrm{in})}(E) + \chi_\ell^{(+)}(E,r) f_\ell^{(\mathrm{out})}(E)} \tag{2.81}$$

2.2.4.1 Integral Equations

Similarly to integral equation (2.52) for the regular solution, it is not difficult to derive the integral equations that determine the Jost solutions. To this end, we integrate both sides of the differential equations (2.70) from r to infinity and take into account the boundary conditions (2.71, 2.72). This gives

$$\begin{aligned} \mathfrak{X}_\ell^{(\pm\mathrm{in})}(E,r) &= C_1^{(\pm)} + \frac{1}{2ik}\int_r^\infty h_\ell^{(+)}(kr')V(r')\chi_\ell^{(\pm)}(E,r')\,, \\ \mathfrak{X}_\ell^{(\pm\mathrm{out})}(E,r) &= C_2^{(\pm)} - \frac{1}{2ik}\int_r^\infty h_\ell^{(-)}(kr')V(r')\chi_\ell^{(\pm)}(E,r')\,, \end{aligned} \tag{2.82}$$

where the constants

$$C_1^{(+)} = 0\,, \qquad C_2^{(+)} = 1\,,$$

$$C_1^{(-)} = 1\,, \qquad C_2^{(-)} = 0$$

correspond to the boundary values $\mathfrak{X}_\ell^{(\pm\mathrm{in/out})}(E,\infty)$ given by Eqs. (2.71, 2.72). Now, multiplying the first of Eqs. (2.82) by $h_\ell^{(-)}(kr)$, the second by $h_\ell^{(+)}(kr)$, and adding them, we obtain the integral equations

$$\begin{aligned} \chi_\ell^{(\pm)}(E,r') = h_\ell^{(\pm)}(kr) + \\ &+ \frac{1}{2ik}\int_r^\infty \left[h_\ell^{(-)}(kr)h_\ell^{(+)}(kr') - h_\ell^{(+)}(kr)h_\ell^{(-)}(kr')\right] \\ &\times V(r')\chi_\ell^{(\pm)}(E,r')\,, \end{aligned} \tag{2.83}$$

which we were looking for. It should be noted that under the integral here we have the same expression as in Eq. (2.47) but with the opposite sign. Moreover, in contrast to (2.47), it is non-zero when $r' > r$. In other words, it is mirror-symmetric relative to the point $r' = r$.

2.2.5 Analyticity of the Jost Functions

First of all, we notice that the regular solution, $\phi_\ell(E,r)$, is an analytic function of E in any finite region of complex E-surface, except may be for the point $E=0$. This is proved in Sect. 2.2.2 by analysing the convergence of the successive iterations for the integral equation (2.52) determining $\phi_\ell(E,r)$. The analyticity of $\phi_\ell(E,r)$ also follows from the Poincaré theorem (see Appendix G). In our case, this theorem can be formulated as follows. At every fixed r, a solution $\varphi(\zeta,r)$ of the equation

$$\frac{d^2}{dr^2}\varphi(\zeta,r) + J(\zeta,r)\varphi(\zeta,r) = 0\,, \tag{2.84}$$

where $J(\zeta,r)$ is an analytic function of the parameter ζ, is an entire function of this parameter, if the boundary values $\varphi(\zeta,r_0)$ and $\partial_r\varphi(\zeta,r_0)$ at a regular (or even regular singular) point r_0 are independent of ζ.

Analytic dependence of the Schrödinger equation (2.4) on the parameter E is obvious. The boundary conditions that we impose on $\phi_\ell(E,r)$ are determined by Eqs. (2.23, 2.24). In other words,

$$\phi_\ell(E,r) \underset{r\to 0}{\longrightarrow} j_\ell(kr) \underset{r\to 0}{\longrightarrow} \frac{(kr)^{\ell+1}}{1\cdot 3\cdot 5\cdots(2\ell+1)}\,. \tag{2.85}$$

We impose the boundary conditions at $r_0 = 0$ where there is a singularity because of the term $\ell(\ell+1)/r^2$. However, this singularity is not severe since

$$\left|\lim_{r\to r_0}(r-r_0)^2 J(\zeta,r)\right| < \infty\,.$$

In the theory of differential equations, such points are called regular singular points (see, for example, Sec. 16.1.1 of Ref. [9]). If the boundary conditions are imposed at such a point, then the Poincare theorem requires their independence of the parameter. In order to achieve the energy independence of both $\phi_\ell(E,r)$ and its derivative $\partial_r\phi_\ell(E,r)$ at $r=0$, we can renormalize the wave function by multiplying it by $k^{-(\ell+1)}$. As a result, the function

$$\varphi_\ell(E,r) = \frac{1}{k^{\ell+1}}\phi_\ell(E,r)$$

obeys the same equation and the requirements of the Poincaré theorem. Therefore, the renormalized function $\varphi_\ell(E,r)$ is analytic function of variable E. It is also single-valued function of E because the equation and the boundary conditions are single-valued. The regular solution $\phi_\ell(E,r)$ differs from $\varphi_\ell(E,r)$ by the factor $k^{\ell+1}$ and thus may depend not only on even but also on odd powers of the momentum. This means that $\phi_\ell(E,r)$ is double-valued function of E. Indeed, for a chosen energy E, the momentum has two different values,

$$k = \pm\sqrt{\frac{2\mu}{\hbar^2}E} \ ,$$

depending on the choice of the sign in front of the square root. Therefore, the regular solution (with the normalization chosen in this book) is an analytic single-valued function of k. When considered as a function of E, it is analytic but double-valued.

* * *

Having established that $\phi_\ell(E, r)$ analytically depends on k, we now consider the Jost functions $f_\ell^{(\text{in/out})}(E)$, which can be expressed via the regular solution in the form of the integral (2.65). In order to establish analyticity of $f_\ell^{(\text{in/out})}(E)$, we apply well-known fact from complex analysis (see theorem (B.20) of Appendix B) concerning uniformly convergent integrals. If the integrand analytically depends on a parameter (k in our case), then the resulting function is analytic in that parameter. For a reader who is not inclined to mathematical subtleties, this is pretty obvious since an integration is actually a summation. The integrand in (2.65) is analytic everywhere except may be for the point $E = 0$. Therefore, if we exclude this point, the Jost functions $f_\ell^{(\text{in/out})}(E)$ are analytic functions of k in the domains for which the integrals (2.65) are uniformly convergent. It is clear, however, that if considered as the functions of E, they are double-valued.

For physically reasonable potentials, the internal points of the interval $0 < r < \infty$ do not pose any problem for the convergence. The difficulties may only arise at its end points, namely, at $r = 0$ where the potential may be singular, and at $r \to \infty$ where $V(r)$ may vanish too slowly and one of the Riccati–Hankel functions $h_\ell^{(\pm)}(kr)$ exponentially grows when k is complex.

In this chapter, we consider non-singular potentials obeying the condition (2.2). As we have shown in Sect. 2.2.1.2, this condition guarantees that the integral (2.65) uniformly converges at $r = 0$ (see Eqs. (2.35–2.39)). For the finite-range potentials that we consider in the present section, the integration is cut off at $r = R$, i.e. at the point beyond which $V(r)$ is the identical zero. Therefore, with a non-singular finite-range potential, the integral (2.65) uniformly converges everywhere, which means that the Jost functions are analytic (but double-valued) in any finite domain of the E-surface, except may be for the threshold $E = 0$.

The reader may wonder why do we consider the Jost functions as depending on E instead of k. As the functions of k, they would have simpler analytic properties. In particular, they would be single-valued. This simplicity, however, is only possible in the single-channel problems. When we move to many channels, the energy variable E is still one (common for all the channels), but each channel has its own momentum. To be consistent, it is more logical to treat the Jost functions in the same way as the multi-channel Jost matrices, i.e. as the functions of E.

3 Riemann Surface and the Spectral Points

As we will see shortly (see Sect. 3.3), it is very useful to consider k and therefore E as complex variables. Although complex momenta and energies have no direct physical meaning, an analysis of the analytic properties of the Jost functions can reveal certain properties of the physical system and provide us with valuable information concerning its behaviour at (real) physical energies.

Differential equations (2.21) that determine the functions $F_\ell^{(\mathrm{in/out})}(E,r)$ explicitly involve the parameter k (the momentum). This means that these functions (and therefore the corresponding Jost functions) depend on k (and only through k on the energy E). The fact that for each point E the corresponding momentum

$$k = \pm\sqrt{(2\mu/\hbar^2)E} \tag{3.1}$$

has two different values (with $\pm$) means that $F_\ell^{(\mathrm{in/out})}(E,r)$ are double-valued functions of the energy, i.e. for a given E, there are two different values of $f_\ell^{(\mathrm{in/out})}(E)$ corresponding to the two options of the sign in Eq. (3.1). The only point where this sign is immaterial is the point $E = 0$.

In complex analysis, the multi-valued functions are treated as if they were single-valued but defined on a multi-layered surface consisting of several parallel planes. This complex manifold is called the Riemann surface (see Appendix B). A more detailed and more rigorous introduction to the theory of multi-valued complex functions can be found, for example, in Ref. [11]). If we draw a vertical line through such a surface, the points where this line intersects each of the layers (sheets) correspond to the same value of the independent variable (the energy in our case), as is shown in Fig. 3.1. The values of the function at these points (generally speaking) are different. There are however discrete points, where the function is the same on two or more layers. At these points, the corresponding layers touch each other in such a way that it is possible to go from one of the layers to the other along a continuous path around the points of coincidence. They are called the branch points.

S. A. Rakityansky, *Jost Functions in Quantum Mechanics*,
https://doi.org/10.1007/978-3-031-07761-6_3

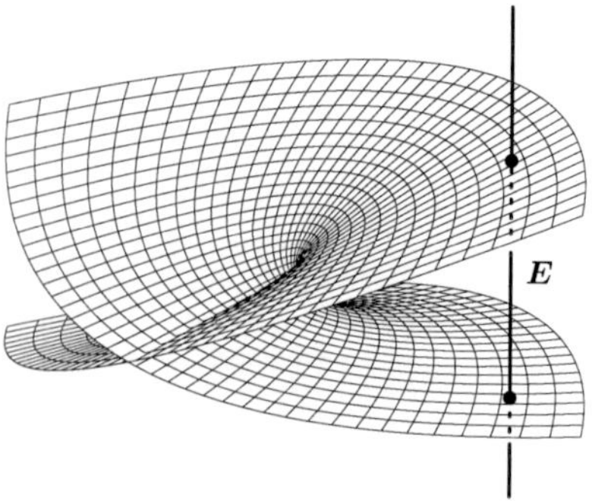

Fig. 3.1 Riemann surface around a square-root branch point. The dots are the points corresponding to the same energy E on different sheets of the surface

As follows from Eq. (3.1), for a single-channel system, the Jost functions are double-valued. They therefore are defined on two sheets of the Riemann surface of complex variable E, where the point $E = 0$ is a square-root branch point. If we make two full circles around this point, the functions $f_\ell^{(\text{in/out})}(E)$ return to the same values from which the circling started (one full circle is not enough). This is because the momentum (which these functions are actually depending on),

$$k = \sqrt{(2\mu/\hbar^2)|E|e^{i\phi}} = \sqrt{(2\mu/\hbar^2)|E|}e^{i\phi/2}\,, \tag{3.2}$$

comes back to its initial value when $\phi \to \phi + 4\pi$, where $|E|$ and ϕ are the polar coordinates of the point E on the complex energy plane relative to the branching point $E = 0$, i.e.

$$E = |E|e^{i\phi}\,. \tag{3.3}$$

In other words, the Jost function has two different values at each point E on this circle. When doing the first circle around the branch point, we are moving on the first sheet and then continue on the second one until coming back to the first sheet after completing the full two circles (see Fig. 3.1). Such continuous transition from one sheet to another is possible if we make a cut from the branch point to infinity and connect opposite rims of the cuts on the two sheets (see Fig. 3.2).

In principle, we can choose the cut arbitrarily, i.e. in any direction. Any particular choice determines the symmetry properties of the Jost functions (with respect to complex conjugation, for example). To be consistent, we follow the standard convention. As is usual in scattering theory, we make straight line cuts on both sheets from the branch point to infinity along the corresponding positive real axes of the energy. Then we connect the upper rim of the first cut with the lower rim of the second cut and vice versa. Via these connections, we can reach any of the two sheets by making an appropriate number of circles around the branch point.

We all used to draw a complex plane in such a way that the real axis is horizontal and its positive direction is to the right (hence, the terms such as "upper half of the plane" and "upper/lower" rim of the cut). In Fig. 3.2, we had to turn the real axis

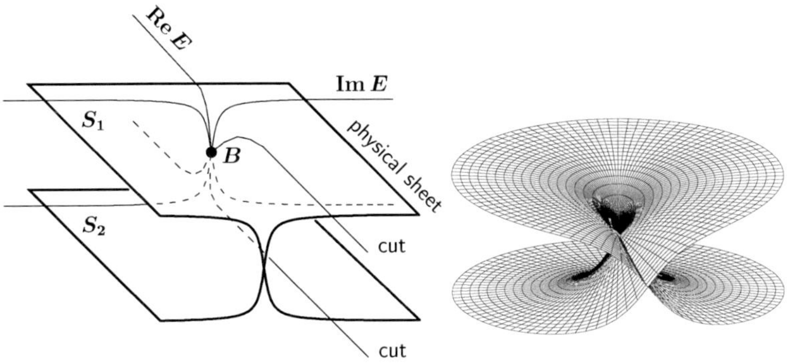

Fig. 3.2 The left part of the figure shows fragments of the physical (S_1) and unphysical (S_2) sheets of the complex energy Riemann surface around the branch point (B) that corresponds to the threshold energy $E = 0$. Each sheet has its own real and imaginary axes, and the point B is their common zero point. Transition from S_1 to S_2 and back is possible through the cuts of both sheets, running from B to infinity along the corresponding real axes. The positive halves of the real axes on the upper and lower sheets are the lines where the rims of the cuts are connected. For the sake of clarity, the right part of the figure shows the corresponding 3D-image of the Riemann sheets and their interconnections

a bit down. This is simply because otherwise it would be difficult to show how the two sheets connect to each other. However, when visualizing the Riemann surface, the reader should mentally turn it back to the traditional position with horizontal real axis. In some cases, it will help to avoid confusion, especially when comparing the E-surface and the momentum plane.

Each sheet has its own real and imaginary axes. At the branch point, all of them intersect. The physical energy (at which the scattering takes place) is on the positive real axis. We choose the cut in such a way that the physical scattering energies lie on the upper rim of the cut. Starting from these physical energies and moving in the anti-clockwise direction around the branch point, we cover the so-called physical sheet of the energy surface. According to Eq. (3.2), all the momenta corresponding to this sheet have positive imaginary parts. When starting the second circle around the branch point, we go down to the sheet that is called unphysical. The momenta corresponding to the energies here all have negative imaginary parts.

The meaning of the names *physical* and *unphysical* will become clear after reading Sects. 3.3.1 and 3.3.2, where it is shown that the bound-state solutions of the Schrödinger equation are only possible for $\text{Im}\, k > 0$ with which the wave function exponentially attenuates when $r \to \infty$ (physical behaviour), while for resonances $\text{Im}\, k < 0$ with which the wave function exponentially grows (unphysical behaviour).

The two full circles around $E = 0$ on the E-surface correspond to one circle on the k-plane as is schematically shown in Fig. 3.3. The reader should take note that in order to simplify the picture, we show in this figure (as well as in some other similar pictures) only one real positive axis of the energy. However, there are two of them: one for the sheet that goes up and the other for the sheet that goes down (as is

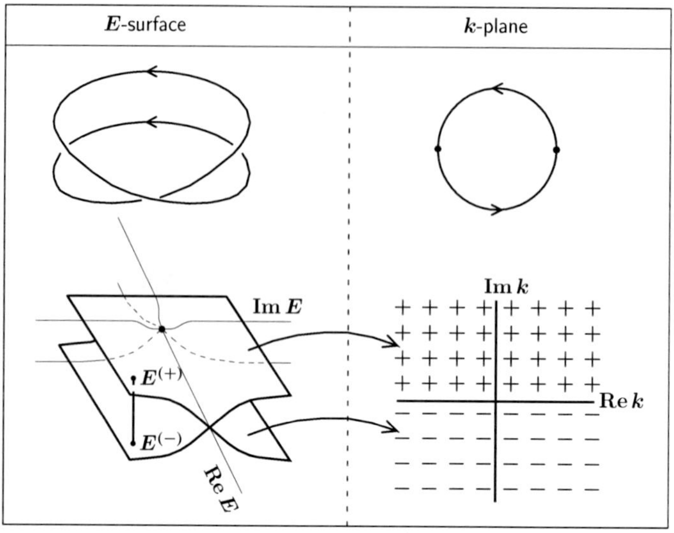

Fig. 3.3 Correspondence between the complex energy surface (left) and the momentum plane (right)

shown in Fig. 3.2). In our drawing, they are simply in the same place and therefore are indistinguishable. This should not cause any confusion. We cannot continuously go over from one of these axes to the other without completing a circle, the same as we cannot jump from negative to positive k.

The whole E-surface is covered when the polar angle ϕ in Eq. (3.2) runs from 0 to 4π. The segment $0 \leqslant \phi < 2\pi$ corresponds to the upper (physical) sheet of the E-surface and to the upper half of the k-plane, where $\text{Im}\, k \geqslant 0$. The other segment $2\pi \leqslant \phi < 4\pi$ sweeps all the points on the unphysical E-sheet and in the lower half of the k-plane, where $\text{Im}\, k \leqslant 0$.

The physical sheet of the E-surface is therefore mapped on the upper half of the k-plane, and its unphysical sheet on the lower half of the k-plane. When it is necessary to indicate which sheet of the Riemann surface an energy point belongs to, we will use the superscript. The points on the upper sheet will be denoted as $E^{(+)}$ and on the lower sheet as $E^{(-)}$. It is clear that

$$E^{(+)} = e^{2\pi i} E^{(-)} . \tag{3.4}$$

Both $E^{(+)}$ and $E^{(-)}$ correspond to the momenta having the same $|k|$, but with opposite signs. As it follows from Eq. (3.2), the replacement $E^{(+)} \rightarrow E^{(-)}$ corresponds to $k \rightarrow -k$. If we look at the two complex numbers $E^{(+)}$ and $E^{(-)}$ represented by their real and imaginary parts, they are identical. The difference shows up when we calculate the momentum as

$$k = \pm\sqrt{(2\mu/\hbar^2)E} \tag{3.5}$$

and choose the sign. It should be chosen differently depending on what kind of energy, $E^{(+)}$ or $E^{(-)}$, is under the square root.[1] However, if instead of the Cartesian representation of the energy point (i.e. as its real and imaginary parts), we use the polar representation (3.3), then there is no problem of choosing the sign. In this case, it comes up automatically.

The Cartesian representation of E may cause confusion in connection with its complex conjugation. When dealing with a single-sheet complex plane, we used to think that complex conjugate points are located symmetrically with respect to the real axis. This is only partially true when the point is on a double-sheet surface. Indeed,

$$\left[\mathrm{Re}\, E^{(+)} + i\,\mathrm{Im}\, E^{(+)}\right]^* \neq \mathrm{Re}\, E^{(+)} - i\,\mathrm{Im}\, E^{(+)}$$

because the polar angle of the complex conjugate point

$$E^* = |E|e^{-i\phi}$$

is counted in the opposite (clockwise) direction as compared to (3.3). As a result, the point E^* is not only on the opposite side of the real axis but also on the other sheet of the Riemann surface. Therefore,

$$\left[\mathrm{Re}\, E^{(+)} + i\,\mathrm{Im}\, E^{(+)}\right]^* = \mathrm{Re}\, E^{(-)} - i\,\mathrm{Im}\, E^{(-)}\,. \tag{3.6}$$

This can also be understood as follows. The complex conjugate energy

$$E^{(+)*} = \frac{\hbar^2}{2\mu}(k^*)^2$$

is calculated as a square of k^*, which has negative imaginary part, hence the superscript $(-)$ on the right hand side of Eq. (3.6).

Perhaps a better understanding of the transformations of the type (3.6) can be gained from Fig. 3.4, where there are shown three different transitions, named as vertical, diagonal, and horizontal, among a set of symmetric points on the Riemann surface.

[1] The reader should be warned that the sign in Eq. (3.5) is not always the same as the superscript of $E^{(\pm)}$. The rule is: if under the square root is $E^{(+)}$, then we choose the sign that makes $\mathrm{Im}\, k \geqslant 0$, and the opposite for $E^{(-)}$. In other words, the sign in the superscript of $E^{(\pm)}$ determines the sign of $\mathrm{Im}\, k$.

vertical = $(k \to -k)$, diagonal = $(k \to k^*)$, horizontal = $(k \to -k^*)$

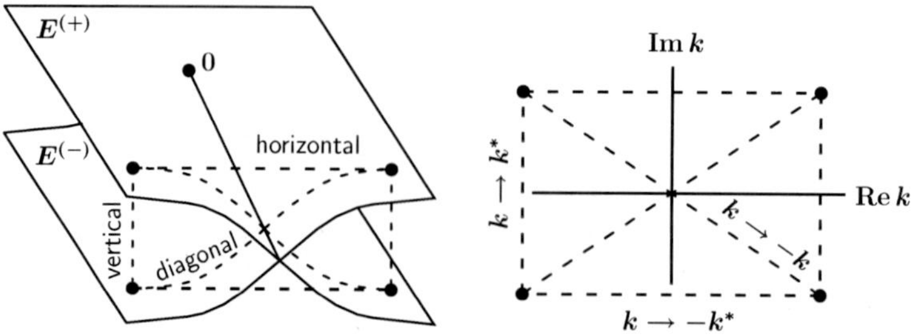

Fig. 3.4 Correspondence of the transitions among the symmetric points on the complex energy surface (left) and in the momentum plane (right). The vertical and diagonal transitions are the sign change and the complex conjugation of the momentum, respectively. The horizontal transition combines both the vertical and diagonal ones

The right hand side of that figure shows the corresponding transitions in the momentum plane. As is seen, the change of the sign of the momentum, $k \to -k$, corresponds to a vertical "jump" from one sheet of the Riemann surface to the other. The complex conjugation, $k \to k^*$, is what is given by Eq. (3.6), i.e. the transition to the diagonal-symmetric point. The horizontal (or mirror) transition moves the point to the other side of the real axis on the same sheet and corresponds to both the conjugation and the sign change, $k \to -k^*$.

3.1 Symmetry Properties of the Jost Functions

The asymptotic coefficients $f_\ell^{(\mathrm{in/out})}$ that determine the long-range behaviour (2.62) of the regular solution, $\phi_\ell(E, r)$, are not independent. They are the amplitudes of the incoming and outgoing waves that swap their roles when the momentum k changes its sign and also under the operation of complex conjugation. These amplitudes should therefore be related to each other at the points shown in Fig. 3.4.

3.1.1 Vertical Symmetry

When we change the sign of the momentum, the value of the energy $E = (\hbar^2/2\mu)k^2$ remains the same, but the point on the E-surface "jumps" up or down between the physical and unphysical sheets in the vertical direction (see Fig. 3.4), because the sign of the imaginary part of the momentum changes,

$$(k \to -k) \quad \leftrightarrow \quad (E \to e^{2i\pi} E)\,. \tag{3.7}$$

As a result of this transformation, the incoming and outgoing waves are interchanged,

$$h_\ell^{(\pm)}(-kr) = (-1)^{\ell+1} h_\ell^{(\mp)}(kr) , \tag{3.8}$$

with a simple sign factor $(-1)^{\ell+1}$ that is the same for both $h_\ell^{(+)}$ and $h_\ell^{(-)}$. When $k \to -k$, the integral equations (2.40) take the form

$$\begin{aligned} F_\ell^{(\text{in})}(e^{2i\pi}E, r) &= \frac{1}{2} + \frac{1}{2ik}\int_0^r h_\ell^{(+)}(-kr')V(r')\Big[h_\ell^{(-)}(-kr')F_\ell^{(\text{in})}(e^{2i\pi}E, r') \\ &\qquad + h_\ell^{(+)}(-kr')F_\ell^{(\text{out})}(e^{2i\pi}E, r')\Big]dr' , \\ F_\ell^{(\text{out})}(e^{2i\pi}E, r) &= \frac{1}{2} - \frac{1}{2ik}\int_0^r h_\ell^{(-)}(-kr')V(r')\Big[h_\ell^{(-)}(-kr')F_\ell^{(\text{in})}(e^{2i\pi}E, r') \\ &\qquad + h_\ell^{(+)}(-kr')F_\ell^{(\text{out})}(e^{2i\pi}E, r')\Big]dr' . \end{aligned}$$

Using Eq. (3.8), they can be re-written as

$$\begin{aligned} F_\ell^{(\text{out})}(e^{2i\pi}E, r) &= \frac{1}{2} - \frac{1}{2ik}\int_0^r h_\ell^{(+)}(kr')V(r')\Big[h_\ell^{(-)}(kr')F_\ell^{(\text{out})}(e^{2i\pi}E, r') \\ &\qquad + h_\ell^{(+)}(kr')F_\ell^{(\text{in})}(e^{2i\pi}E, r')\Big]dr' , \\ F_\ell^{(\text{in})}(e^{2i\pi}E, r) &= \frac{1}{2} + \frac{1}{2ik}\int_0^r h_\ell^{(-)}(kr')V(r')\Big[h_\ell^{(-)}(kr')F_\ell^{(\text{out})}(e^{2i\pi}E, r') \\ &\qquad + h_\ell^{(+)}(kr')F_\ell^{(\text{in})}(e^{2i\pi}E, r')\Big]dr' . \end{aligned}$$

Comparing these equations with (2.40), we see that

$$F_\ell^{(\text{in})}(e^{2i\pi}E, r) \equiv F_\ell^{(\text{out})}(E, r) \tag{3.9}$$

and

$$F_\ell^{(\text{out})}(e^{2i\pi}E, r) \equiv F_\ell^{(\text{in})}(E, r) . \tag{3.10}$$

The functions $F_\ell^{(\text{in/out})}(E, r)$ are the corresponding Jost functions for the potential that is cut off at the point r. Letting $r \to \infty$ in Eqs. (3.9), (3.10), we see that

$$\boxed{f_\ell^{(\text{in})}\left[E^{(+)}\right] = f_\ell^{(\text{out})}\left[E^{(-)}\right]} \tag{3.11}$$

and

$$\boxed{f_\ell^{(\text{out})}\left[E^{(+)}\right] = f_\ell^{(\text{in})}\left[E^{(-)}\right]} \tag{3.12}$$

Therefore, the invariance of the Schrödinger equation with respect to the transformation $k \to -k$ implies that $f_\ell^{(\text{in})}(E)$ and $f_\ell^{(\text{out})}(E)$ are symmetric on the two sheets of the E-surface. The relations (3.11), (3.12) can be called "vertical" symmetry because they relate the values of the Jost functions at the points that are one under the other on a vertical line passing through the Riemann sheets (see Fig. 3.1).

3.1.2 Diagonal Symmetry

If we replace the momentum with its complex conjugate, $k \to k^*$, the energy is also conjugated, $E \to E^*$. Similarly to the preceding section, we apply the operation of complex conjugation to the integral equations (2.40) and use the fact that the Riccati–Hankel functions are related as

$$\left[h_\ell^{(\pm)}(kr)\right]^* = h_\ell^{(\mp)}(k^*r)\,. \tag{3.13}$$

As a result, we obtain

$$F_\ell^{(\text{in/out})}(E,r) \equiv \left[F_\ell^{(\text{out/in})}(E^*,r)\right]^*\,, \tag{3.14}$$

from which it follows immediately that

$$\boxed{f_\ell^{(\text{in/out})}(E) = \left[f_\ell^{(\text{out/in})}(E^*)\right]^*} \tag{3.15}$$

This equation relates the values of the Jost functions at the symmetric points on the diagonal crossing the cut of the Riemann surface (see Fig. 3.4).

3.1.3 Horizontal Symmetry

In the two preceding sections, we have established two types of symmetries among the Jost function values at the points on the energy surface. First, when "jumping" up and down between the sheets, $k \to -k$, we change $f_\ell^{(\text{in})}$ to $f_\ell^{(\text{out})}$ and vice versa.

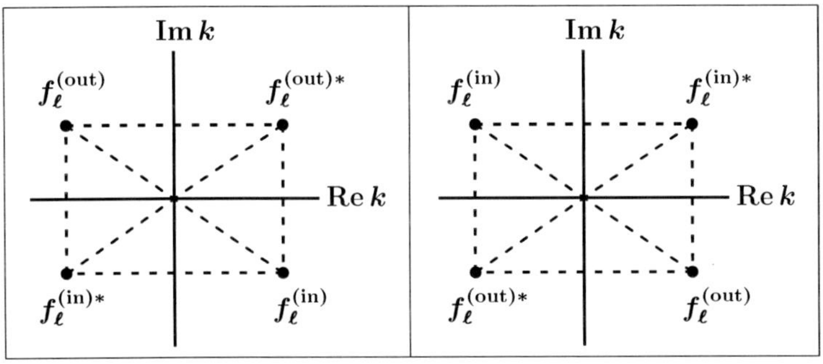

Fig. 3.5 Symmetry properties of the Jost functions on the k-plane. The dashed lines connect the points at which the values indicated next to them are identical. The left and right figures correspond to different choices of the values at the top points

Second, when moving along the diagonal to the other side of the real axis, $k \to k^*$, we change $f_\ell^{(in)}$ to $f_\ell^{(out)*}$ and vice versa.

Now we can combine these two transformations and consider what happens if $k \to -k^*$. This can be done similarly to the following trivial statement: If $a = b$ and $b = c$, then $a = c$. Indeed, the transition from k to $-k^*$ can be considered in the two steps: to the other sheet and then to the other side of the real axis along the diagonal. Therefore, the Jost functions are symmetric with respect to the imaginary axis in the k-plane or, in other words, with respect to the real axis on each sheet of the E-surface, i.e.

$$f_\ell^{(in/out)}(E) = \left[f_\ell^{(in/out)}(e^{2\pi i}E^*)\right]^* \tag{3.16}$$

In connection with this symmetry property, it should be mentioned once again that the operation of complex conjugation moves the point E not only to the other side of the real axis but also to the other sheet of the Riemann surface (see Sect. 3). However, the symmetry (3.16) relates the points on the same sheet. This is why Eq. (3.16) relates the points E and $e^{2\pi i}E^*$, which are on the opposite sides of the real axis but on the same E-sheet. The factor $e^{2\pi i}$ brings the point E^* back to the same sheet where the point E is located.

* * *

The symmetry properties given by Eqs. (3.11), (3.12), (3.15), and (3.16) are graphically summarized in Figs. 3.5 and 3.6, for the k-plane and E-surface, respectively.

It is important to note that the symmetries shown in Figs. 3.5 and 3.6 are valid not only across the cuts of the Riemann surface but also across the negative parts of the real axes. Indeed, since the functions on the left and right hand sides of Eqs. (3.11), (3.12), (3.15), and (3.16) are identical in some non-zero domains of the E-

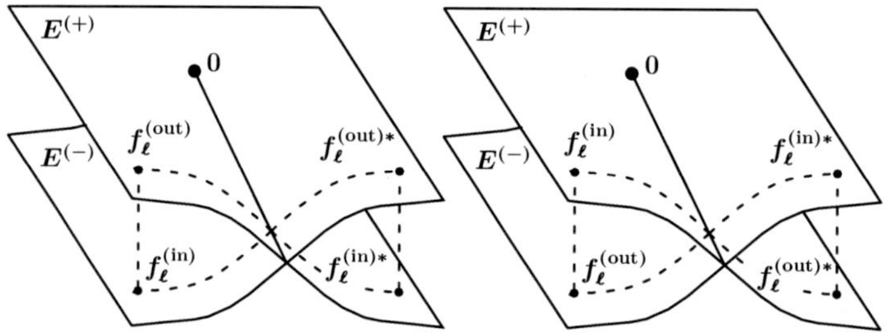

Fig. 3.6 Symmetry properties of the Jost functions on the E-surface. The dashed lines connect the points at which the values indicated next to them are identical. The left and right figures correspond to different choices of the values at the top points

surface, they are identical everywhere.[2] This means that these symmetry properties are not restricted to the positive half of the real axis but can be extended to entire real E-axis.

Using the symmetry properties, we can significantly simplify the task of calculating the Jost functions for complex values of the momentum (energy). Indeed, it is enough to find $f_\ell^{(\text{in})}$ and $f_\ell^{(\text{out})}$ just in one quadrant of the k-plane (i.e. on one half of any sheet of the E-surface), and as a result, we will know them in all four quadrants (i.e. on the whole Riemann surface).

3.2 High-Energy Asymptotics of the Jost Functions

What happens to the Jost functions when $E \to \infty$? To answer this question, consider the integral representation (2.65) of $f_\ell^{(\text{in})}(E)$ in terms of the regular solution,

$$f_\ell^{(\text{in})}(E) = \frac{1}{2} - \frac{1}{2ik}\int_0^\infty h_\ell^{(+)}(kr)V(r)\phi_\ell(E,r)dr\,. \tag{3.17}$$

When $|k| \to \infty$ along the real axis, both $h_\ell^{(+)}(kr)$ and $\phi_\ell(E,r)$ remain bounded (oscillatory), and the Jost function approaches 1/2,

$$f_\ell^{(\text{in})}(E) \underset{|k|\to\infty}{\longrightarrow} \frac{1}{2} + O\left(\frac{1}{k}\right), \qquad \operatorname{Im} k = 0\,. \tag{3.18}$$

In the upper half of the k-plane (on the physical E-sheet), the Riccati–Hankel function $h_\ell^{(+)}(kr)$ is vanishing at large $|k|$, but the wave function is exponentially

[2] This follows from the identity theorem of complex analysis (see statement (B.10) of Appendix B).

growing. Which of them wins? Using the bounds (2.60) and (D.27), we obtain

$$\left|\frac{1}{2ik}\int_0^\infty h_\ell^{(+)}(kr)V(r)\phi_\ell(E,r)dr\right|$$

$$\leqslant \frac{C}{|k|}\int_0^\infty e^{-\operatorname{Im} kr}\left(\frac{|kr|}{1+|kr|}\right)^{-\ell}\left(\frac{|kr|}{1+|kr|}\right)^{\ell+1} e^{|\operatorname{Im} kr|}\,|V(r)|\,dr$$

$$= \frac{C}{|k|}\int_0^\infty \frac{|kr|}{1+|kr|}|V(r)|\,dr \underset{|k|\to\infty}{\longrightarrow} \sim \frac{1}{|k|},$$

which is the same as for real k. Therefore, when $|k| \to \infty$, the Jost function $f_\ell^{(\text{in})}(E)$ approaches 1/2 along any path in the upper half of the k-plane,

$$f_\ell^{(\text{in})}(E) \underset{|k|\to\infty}{\longrightarrow} \frac{1}{2} + O\left(\frac{1}{|k|}\right), \qquad \operatorname{Im} k \geqslant 0. \tag{3.19}$$

In the lower half of the k-plane (on the unphysical E-sheet), the right hand side of Eq. (3.17) does not have a limit when $|E| \to \infty$. Indeed, both $h_\ell^{(+)}(kr)$ and $\phi_\ell(E,r)$ are oscillating, and their absolute values are exponentially growing. As a result, the complex function $f_\ell^{(\text{in})}(E)$ oscillates and therefore may have zeros at any distance down the lower half of the k-plane. The number of such zeros in that region of the momentum plane is thus not limited.

The high-energy behaviour of $f_\ell^{(\text{out})}(E)$ can be deduced from the integral representation similar to (3.17), or simply from the symmetry properties established in Sect. 3.1. It is easy to see that the behaviour of $f_\ell^{(\text{in})}(E)$ and $f_\ell^{(\text{out})}(E)$ is the same only for real k (they both tend to 1/2). For complex k, the function $f_\ell^{(\text{out})}(E)$ behaves in such a way that is exactly opposite to the behaviour of $f_\ell^{(\text{in})}(E)$, namely, it oscillates where $f_\ell^{(\text{in})}$ tends to 1/2 and vice versa.

The behaviour of both $f_\ell^{(\text{in})}$ and $f_\ell^{(\text{out})}$ at large values of $|k|$ is schematically shown in Fig. 3.7. The question marks indicate the domains of the k-plane where

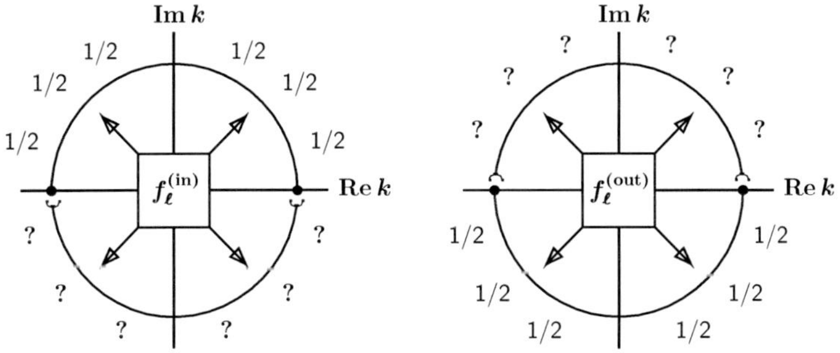

Fig. 3.7 Schematic representation of the asymptotic behaviour of the Jost functions $f_\ell^{(\text{in})}$ (left) and $f_\ell^{(\text{out})}$ (right) when $|k| \to \infty$

Eq. (3.17) does not allow us to establish the high-energy asymptotics of the Jost functions. This does not mean, however, that in these domains they do not have limits. As we will see in Sect. 7.8, with the help of complex rotation, it is possible to analytically continue the function $f_\ell^{(\mathrm{in})}(E)$ to the unphysical sheet of the E-surface and to prove that (in the case of a finite-range potential) it goes to 1/2 in all directions (see Fig. 7.14). Meanwhile, the results obtained here (although for only half of the k-plane) already enable us to establish an important relation, such as Levinson's theorem (see Sect. 4.7).

3.3 Spectral Points

Why do we need the Jost functions? What is the advantage of using them? To answer these questions, we take note that the values of $f_\ell^{(\mathrm{in})}(E)$ and $f_\ell^{(\mathrm{out})}(E)$ determine the asymptotic behaviour of the regular solution, $\phi_\ell(E, r)$, as is given by Eq. (2.62). Actually, the Jost functions are the amplitudes of the incoming and outgoing spherical waves,

$$h_\ell^{(\pm)}(kr) \underset{|kr|\to\infty}{\longrightarrow} \mp i \exp\left[\pm i(kr - \ell\pi/2)\right] . \tag{3.20}$$

It is clear that the ratio of these amplitudes must somehow describe the scattering process (what kind of outgoing wave emerges from a given incoming wave). But not only this can be learned if we know the Jost functions.

The most important information that can be deduced from them comes from the following equations:

$$\boxed{f_\ell^{(\mathrm{in})}(E) = 0} \tag{3.21}$$

and

$$f_\ell^{(\mathrm{out})}(E) = 0 . \tag{3.22}$$

If these equations have solutions, the corresponding roots should form a discrete set of energies $E_1, E_2, E_3, \ldots$. These points must be isolated, i.e. they cannot form a continuous curve; otherwise, the Jost functions would be identical zero in the whole domain of their analyticity.[3] Generally speaking, we should expect that E_n are complex numbers, somehow distributed on the Riemann surface. The roots of Eqs. (3.21) and (3.22) cannot coincide. Indeed, in such a case, the wave function (2.62) would vanish identically, which corresponds to the trivial solution of Eq. (2.4). Therefore on the complex E-surface, there are certain discrete points, at

[3] This is a particular case of a theorem of complex analysis. See, for example, Sect. 2.6 of Ref. [12].

which the function $\phi_\ell(E, r)$ has either only incoming or only outgoing wave in its asymptotics.

As we know, the physical wave function (at those energies where a physical solution exists) is proportional to the regular solution,

$$u_\ell(E_n, r) \sim \phi_\ell(E_n, r) .$$

Clear physical meaning can only be ascribed to those physical wave functions, $u_\ell(E, r)$, that have either both terms on the right hand side of Eq. (2.62) or just the outgoing wave. The points E_n on the energy surface, at which the wave function has only the outgoing wave asymptotics,

$$u_\ell(E_n, r) \underset{r\to\infty}{\sim} h_\ell^{(+)}(k_n r) f_\ell^{(\mathrm{out})}(E_n) , \qquad n = 1, 2, \ldots \tag{3.23}$$

are called *spectral points*. They are determined by Eq. (3.21). As we will see shortly, depending on the location of a spectral point on the Riemann surface, it may correspond to a bound, resonant, or a virtual state solution of the Schrödinger equation.

As was mentioned earlier, none of the observable quantities depends on the normalization of the Jost functions. For example, the distribution of the spectral points on the complex E-surface, determined by Eq. (3.21), obviously does not depend on their normalization.

3.3.1 Bound States

If a particle is bound then, first, it cannot move infinitely far from the centre of attraction, and, second, its energy is below the threshold value (zero for the case under consideration) above which the scattering is possible. This implies that the wave function of a bound state must vanish when $r \to \infty$, and the energy E_n of such a state is negative. Since a bound state is an eigenstate of a Hermitian Hamiltonian, this energy is a real number. The corresponding momentum k_n is therefore pure imaginary,

$$k_n = \pm i\sqrt{(2\mu/\hbar^2)|E_n|} . \tag{3.24}$$

Which sign should we choose for k_n in this equation? To answer this question, we notice that the asymptotics (2.62) for a bound-state solution cannot have both terms. This is because with either choice of the sign, one of the terms is exponentially vanishing, while the other exponentially growing. The energy of a bound state is therefore a root of either Eq. (3.21) or Eq. (3.22). If we choose the "plus" sign in Eq. (3.24), then the bound states have only outgoing waves in their asymptotics, and the corresponding energies are zeros of the function $f_\ell^{(\mathrm{in})}(E)$. The other choice would lead us to the incoming waves and Eq. (3.22).

From the mathematical point of view, both these choices have equal rights. In principle, we could construct a consistent theory with pure incoming waves for the asymptotics of the spectral states. This would simply interchange the physical and unphysical sheets of the Riemann surface. As we have seen in Sect. 3.1, the zeros of $f_\ell^{(\mathrm{in})}(E)$ and $f_\ell^{(\mathrm{out})}(E)$ are symmetric relative to such interchange. Moreover, the values of these functions at all points are symmetric.

The following physical arguments help in breaking the symmetry in making the choice of the sign in Eq. (3.24). It is more logical to assume that a bound particle "tries" to leak out from the "well", and the attractive forces prevent it. In other words, far away from the centre of attraction, the wave function is an outgoing (leaking) wave,

$$u_\ell(E_n, r) \underset{r\to\infty}{\sim} -i \exp\left(-\sqrt{(2\mu/\hbar^2)|E_n|}r - \frac{i\ell\pi}{2}\right) f_\ell^{(\mathrm{out})}(E_n)\,, \tag{3.25}$$

which exponentially attenuates because of the positive sign in Eq. (3.24).

Therefore, the bound states correspond to spectral points on the physical sheet of the energy surface. These points are zeros of the Jost function $f_\ell^{(\mathrm{in})}(E)$, such that $E_n < 0$ and

$$k_n = i\sqrt{(2\mu/\hbar^2)|E_n|}\,. \tag{3.26}$$

3.3.2 Resonances

In contrast to bound states that are defined rigorously and unambiguously, quantum resonances are a source of controversy. Very often, resonances are viewed as "almost bound" states, hence the names such as quasi-stationary or quasi-bound states. Intuitively, they can be visualized as follows (see Fig. 3.8). A particle moving towards a centre of force has such kinetic energy that its "trajectory" around this centre is almost closed. As a result, it does many revolutions around the centre before going away. The more revolutions it does, the longer the resonance lives. From the quantum mechanical point of view, there is no trajectory, of course, and the particle forms a partially localized state that slowly leaks out.

The important concept that lays the foundation for the theory of quantum resonances is that a resonance state (the same way as a bound state) does not "remember" how it was formed, i.e. it "forgets" from which direction the colliding particles arrived. In other words, it can randomly decay with any possible direction for the decay products to move to. This leads to the conclusion that the wave function of a quantum resonance, similarly to the bound states, has only the outgoing wave in its asymptotic behaviour, i.e. a resonance corresponds to a spectral point (see also the qualitative physical consideration given in Sect. 1.3.3).

One may argue that the above assumption is only valid for the long-lived states. Following this argumentation, some authors try to relate resonances to observable

Fig. 3.8 Intuitive visualization of a resonant state

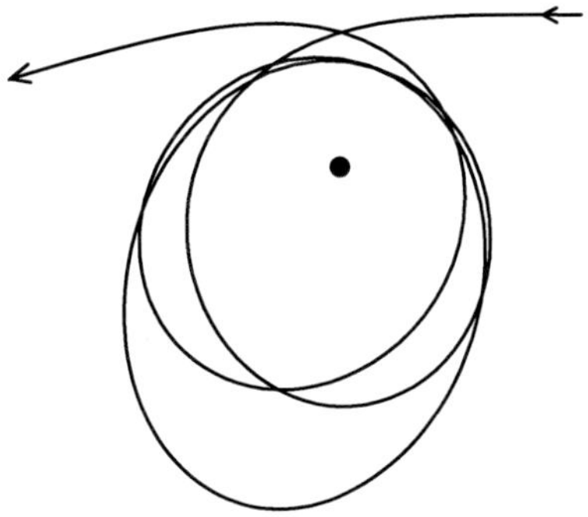

irregularities in the collision process and thus exclude the states with short lifetimes whose contribution to the scattering process is spread over a wide energy range and thus is practically indistinguishable. The possible counterargument is that there is no unambiguous and mathematically reasonable way to separate the long-lived and short-lived states. If we are going to construct a consistent mathematical theory of quantum resonances, the long-lived and short-lived states must be treated equally. When speaking about the bound states, we do not separate well isolated states and the states whose energies are so close to each other that they are practically indistinguishable from the continuum. For example, the highly excited states of the hydrogen atom (near the ionization threshold) are considered on the same mathematical basis as its ground state. Why should we construct the theory of resonances differently?

Thus, in what follows, we assume it as a mathematical axiom that a resonance state solution of the Schrödinger equation corresponds to a spectral point on the energy surface.

We already know that the bound states are the spectral points on the physical sheet of the E-surface, corresponding to real negative values of the energy. Where on this surface are located the resonance spectral points? First of all, we notice that the energy of a resonance state cannot be real. Indeed, in such a case the time dependence of the corresponding quantum state, given by Eq. (1.21), would be oscillatory instead of being exponentially decaying. The attenuating factor $\exp(-\Gamma t/\hbar)$ in the probability density $|\Psi|^2$ appears only if the energy has a negative imaginary part,

$$E_n = E_{\mathrm{r}} - \frac{i}{2}\Gamma\,. \tag{3.27}$$

Which of the two sheets of the Riemann surface should the point E_n belong to? To answer this question, we look at the asymptotic behaviour of the wave function when $r \to \infty$,

$$u_\ell(E_n, r) \underset{r\to\infty}{\sim} -i\exp\left(ik_n r - \frac{i\ell\pi}{2}\right) f_\ell^{(\mathrm{out})}(E_n)\,, \tag{3.28}$$

where

$$k_n = \pm\sqrt{\frac{2\mu}{\hbar^2}\left(E_r - \frac{i}{2}\Gamma\right)} . \tag{3.29}$$

The sign in front of the square root in Eq. (3.29) should be chosen in such a way that $\mathrm{Im}\, k_n < 0$; otherwise, the wave function (3.28) would exponentially vanish at large distances, which is only possible for the bound-state solutions. Therefore, the resonance spectral points are located on the unphysical sheet of the energy surface.

Since $\mathrm{Im}\, k_n < 0$, the resonance solution exponentially grows when $r \to \infty$. Such a behaviour looks unphysical, hence the name for the corresponding sheet of the E-surface. The exponential growth of the resonance wave functions can, however, be logically explained. Indeed, the outgoing spherical wave originates from the source near $r = 0$, where the density of the particles, $|\Psi|^2 \sim \exp(-\Gamma t/\hbar)$, infinitely increases when time goes to the past, $t \to -\infty$. At $t = t_0 > 0$, a given distance r is reached by the particles that left the centre at the earlier time $t = t_0 - r/v$, where v is their average velocity. Therefore, each further layer of the outgoing spherical wave has originated from a more intensive source at the centre, hence the density growth with the radius.

Anyway, in contrast to the bound-state wave functions, a resonance solution of the Schrödinger equation at a complex energy (3.27) has no direct physical meaning. However, the meaning can be given to the energy itself. Although nothing happens at complex energies, both terms of the energy (3.27) determine the physical processes. As we already mentioned, its imaginary part Γ determines how fast the resonance decays. If at $t = 0$ we prepare an ensemble of N_0 identical resonances, their number, $N \sim |\Psi|^2$, will be decreasing with time according to the exponential decay law,

$$N(t) = N_0 e^{-\Gamma t/\hbar} . \tag{3.30}$$

Therefore, the half-life of a resonance,

$$T_{1/2} = \frac{\hbar \ln 2}{\Gamma} , \tag{3.31}$$

is determined by Γ, which is called the resonance width. The term "width" comes from the fact (see further sections) that a resonance can be excited with a reasonable probability, if the energy is within the interval $E_\mathrm{r} - \Gamma/2 < E < E_\mathrm{r} + \Gamma/2$ centred at E_r. Equation (3.31) can also be considered as a form of the uncertainty relation between the energy and time,

$$\Delta E \cdot \Delta t \sim \hbar ,$$

which says that a state with an uncertain energy (smeared over the interval ΔE) cannot exist longer than $\Delta t \sim \hbar/\Delta E$.

The meaning of the real part E_r of the energy (3.27) can be understood if we take notice that the Jost function $f_\ell^{(\text{in})}(E)$ is an analytic function of E (the proof is given in Sect. 2.2.5). Therefore, when we move up from the spectral point $E = E_r - i\Gamma/2$ towards the real positive energy $E = E_r$, the function $f_\ell^{(\text{in})}(E)$ is smoothly transformed from zero to a value $f_\ell^{(\text{in})}(E_r)$. If the spectral point is not far from the real axis (Γ is small), the value $f_\ell^{(\text{in})}(E_r)$ cannot be far from zero. This implies that $f_\ell^{(\text{out})}(E_r)$ is also small; otherwise, the current conservation would be violated. Therefore when passing the point $E = E_r$ along the real axis, we would notice significant variations of both the incoming and outgoing wave amplitudes. This would manifest itself via observable variations of the scattering characteristics (see Sect. 4). In this way, a resonance spectral point influences the scattering process at real collision energies around $E = E_r$. The closer to the real axis the stronger its influence is.

3.3.3 Virtual States and Sub-threshold Resonances

In many books and research papers, one can find the terms *virtual* and *anti-bound* states. Both of them mean the same and are used to name the spectral points located at real negative energies on the unphysical sheet of the E-surface. In other words, they correspond to pure imaginary negative momenta,

$$k_n = -i\sqrt{(2\mu/\hbar^2)|E_n|}\,. \tag{3.32}$$

On the complex k-plane, they are opposite to the bound states (on the other half of the imaginary axis), hence the term "anti-bound".

From the formal mathematical point of view, there is no difference between virtual and resonance spectral points. Indeed, a virtual spectral point is located on the unphysical sheet and corresponds to a wave function with the same exponentially growing asymptotics as any other resonance. Therefore, there is no mathematical reason for introducing a new term, which can only bring up some confusion.

Physically, there is a difference, however. A virtual state is a resonance at a negative collision energy, $E_r < 0$, and with zero width, $\Gamma = 0$. As many numerical calculations show, the string of resonance spectral points usually bends towards the imaginary axis and continues to infinity with $E_r \to -\infty$ and $\Gamma \to \infty$. The points with $E_r < 0$ are called *sub-threshold* resonances. In the collision processes, they are observable insomuch as they influence the values of $f_\ell^{(\text{in/out})}(E)$ on the real axis.

Thus, a virtual (anti-bound) state is a sub-threshold resonance with zero width. How is it possible for a resonance to have $\Gamma = 0$, does this mean that such a resonance never decays? Yes, it never decays simply because it can never be formed in collisions with $E > 0$. In general, the sub-threshold resonance states can be formed when the system is embedded in an external attractive field that restricts their movements. Indeed, with an external binding, the energy of the system becomes negative. The sub-threshold resonances are therefore "would be" resonances, which need external "help" for their formation. For example, some unstable nuclear clusters such as (nn), (nnn), etc. (which are sub-threshold resonance states) can emerge and decay inside heavy nuclei, i.e. within the attractive field collectively generated by the other nucleons.

There is another interpretation of virtual states. When the strength of the attraction of the potential $V(r)$, which is responsible for the existence of bound states, is gradually decreased, the bound states become less and less tightly bound and finally are transformed into resonances. In other words, the bound-state spectral points move towards the threshold $E = 0$ and through it onto the unphysical sheet. It can be shown (see, for example, Chapter 12 of Ref. [13]) that the way they cross the threshold is different for the states with $\ell = 0$ and $\ell \neq 0$. This difference is a consequence of the difference in the energy dependencies of the corresponding functions $f_\ell^{(\mathrm{in})}(E)$ near the threshold spectral point, as is given by Eq. (3.74) further down in this book. The non-zero angular momentum causes the resonances to acquire a positive Γ immediately after crossing the threshold, while the S-wave spectral points remain with $\Gamma = 0$ in a rather broad vicinity of the branching point. A virtual state is therefore only possible for $\ell = 0$. It is a "would be" bound state, which would exist if the attraction were a little bit stronger.

Since virtual state spectral points are located very close to the threshold energy $E = 0$, they must influence the low-energy scattering. It is indeed so. For example, the scattering cross section for the np collision with anti-parallel spins (singlet state) dramatically increases when $E \to 0$. The reason is that in the singlet S-wave state the proton–neutron attraction is almost sufficient for binding. This system has a virtual state at $E \sim -67\,\mathrm{keV}$, which is very small as compared to the depth, $\sim -50\,\mathrm{MeV}$, of np-potential.

3.3.4 Resonance Wave Function Normalization

As we have seen, the resonance corresponds to a spectral point on the unphysical Riemann sheet with the complex momentum

$$k_n = \kappa_n - i\gamma_n\,, \qquad \text{where} \qquad \kappa_n > 0, \quad \gamma_n > 0\,. \tag{3.33}$$

The radial part of its wave function at large distances behaves as is given by Eq. (3.28). With the momentum (3.33), this function oscillates with exponentially growing amplitude,

$$u_\ell(E_n, r) \underset{r\to\infty}{\sim} \exp\left(i\kappa_n r + \gamma_n r\right)\,. \tag{3.34}$$

Therefore, if we try to normalize such a function using the standard recipe, i.e. by integrating its absolute square,

$$u_\ell^*(E_n, r)u_\ell(E_n, r) \underset{r\to\infty}{\sim} \exp\left(2\gamma_n r\right)\,,$$

from zero to infinity, we apparently fail. The situation can be saved, if instead of the absolute square we consider the simple square,

$$u_\ell^2(E_n, r) \underset{r\to\infty}{\sim} \exp\left(i2\kappa_n r + 2\gamma_n r\right) .$$

It is exponentially growing, but in addition to the growth it also oscillates. In the integral, the oscillating "positive" and "negative" fragments of the function partly compensate each other. Therefore, the contribution to the normalization integral,

$$N = \int_0^\infty u_\ell^2(E_n, r)dr , \tag{3.35}$$

from the long-range tail of this function could be zero, and we can hope to obtain a finite value for this integral. However, such an integral of a "wildly" oscillating function cannot be evaluated directly. To find it, we need some regularization trick. There are few of them available. For example, it can be shown that the following limit exists (for the details, see the book [14]):

$$N = \lim_{\varepsilon\to 0}\int_0^\infty u_\ell^2(E_n, r)e^{-\varepsilon r^p}dr , \tag{3.36}$$

where the power p in the regularization factor, $\exp(-\varepsilon r^p)$, can be arbitrarily chosen within the interval $1 < p < \pi/[2\arctan(\gamma_n/\kappa_n)]$. For any p within this interval, the value of the limit (3.36) is the same. Moreover, the same value, N, for the integral (3.35) can be obtained if instead of integrating along the real axis, we do it along the ray $r = \rho e^{i\theta}$ in the complex r-plane (shown in Fig. 3.9),

$$N = \int_0^\infty u_\ell^2(E_n, \rho e^{i\theta})e^{i\theta}d\rho , \tag{3.37}$$

with the rotation angle $\theta > \arctan(\gamma_n/\kappa_n)$. Along such an integration path, we have

$$\mathrm{Im}(k_n r) = \mathrm{Im}\left\{|k_n r|\exp\left[i\left(\theta - \arctan\frac{\gamma_n}{\kappa_n}\right)\right]\right\} > 0 ,$$

which makes the integrand in (3.37) exponentially vanishing at infinity. It can be shown (see Ref. [14]) that the value of the integral (3.37) does not depend on the angle θ when it is greater than $\arctan(\gamma_n/\kappa_n)$.

Therefore, there is a unique number N that can be associated with the resonance wave function and that could be used as its normalization integral. The question arises: on which grounds should we normalize resonances using a recipe that is different from the one for bound states?

To answer such a question, we should recollect a bit of history of quantum mechanics. First of all, we should not forget that neither the Schrödinger equation nor the recipe for calculating the observable quantities from the wave function

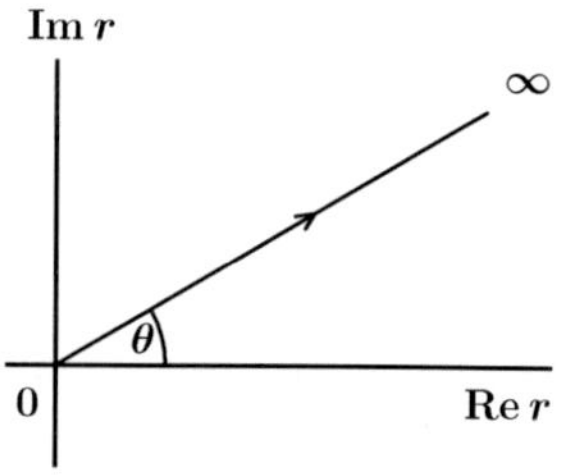

Fig. 3.9 The integration path rotated to the angle θ in the complex r-plane

were rigorously derived from something. Actually, it was not possible. All the basic principles of quantum mechanics emerged as a clever guess. The only substantiation of these principles is that the theory based on them works and is self-consistent. This means that the postulates of quantum mechanics are not "cast in stone". If a more general set of rules is found, which does not break down anything in the existing theory, then it could also be acceptable.

It was suggested by Tore Berggren,[4] to extend the set of the postulates of quantum mechanics by adding the rule for dealing with quantum resonances. His rule consists in using the time-reversed state-vectors as the "bra"-vectors in all the matrix elements of the type $\langle\psi_{n'}|O|\psi_n\rangle$ that involve the resonant states. So, if vector $|\psi_n+\rangle$ describes the state moving forward in time, and $|\psi_n-\rangle$ is the corresponding state moving back in time, then the normalization integral should be taken as

$$N = \langle\psi_n - |\psi_n+\rangle\,. \tag{3.38}$$

The time-reversal operator, $\mathcal{T}$, and its properties are discussed further down in this book in Sect. 10.6. It is shown there that

$$\mathcal{T}\left[\frac{u_\ell(E_n,r)}{r}Y_{\ell,m}(\hat{\vec{r}})\right] = \left[\frac{u_\ell(E_n,r)}{r}Y_{\ell,-m}(\hat{\vec{r}})\right]^*\,, \tag{3.39}$$

i.e. the time reversal results in the complex conjugation of the wave function and inverting of its angular momentum ($m \to -m$). When putting the function (3.39) in the "bra"-position, we do the complex conjugation once more. As a result, in the normalization integral of the type (3.38), the simple square, u_ℓ^2, appears.

Actually, the bound states can also be normalized by integrating the simple square of the radial wave function. Indeed, the radial Schrödinger equation and the corresponding boundary conditions for a bound state are real, which give a real

[4] The references to the original Berggren's works as well as the review of the contributions made to this theory by the other authors can be found in Ref. [14], which is entirely devoted to quantum resonances.

solution. This means that for a bound state, $|u_\ell|^2 = u_\ell^2$. By the way, since a bound state formally exists forever, its time reversal coincides with itself.

It can be shown that different "forward" (+) and "backward" (−) resonance states are mutually orthogonal, i.e.

$$\langle \psi_{n'} - | \psi_n + \rangle \sim \delta_{n'n} .$$

After an appropriate normalization of all the states, it is therefore possible to obtain the following generalization of the closure relation:

$$I = \sum_i |B_i\rangle\langle B_i| + \sum_j |R_j+\rangle\langle R_j - | + \int |C_E\rangle\langle C_E| dE ,$$

where the first sum runs over all the bound states, the second sum takes into account some selected resonances, and the last term is the remaining background integral over the continuum states.

It should be noted that Berggren's theory of resonances is outside the scope of this book and therefore is not considered here any deeper than the short outline given in the present section. The only fact from this theory that we use in the few places of the book is that the resonant states can, in principle, be normalized.

3.3.5 Simplicity of the Bound and Resonant State Zeros

How fast does the Jost function vanish at the spectral points? In other words, which is the first non-zero term in the expansion

$$f_\ell^{(\text{in})}(E) = \dot{f}_\ell^{(\text{in})}(E_n)(E-E_n) + \frac{1}{2}\ddot{f}_\ell^{(\text{in})}(E_n)(E-E_n)^2 + \dots , \tag{3.40}$$

where E_n is a spectral point and the dot over $f_\ell^{(\text{in})}$ means the derivative with respect to E? The answer to this question is: the Jost function vanishes linearly, i.e. the zeros at the spectral points are simple. The proof of this statement is given here by showing that the first derivative of the Jost function is always non-zero at the bound and resonant state spectral points.

* * *

Consider a bound or a resonant state with the energy E_0 that corresponds to a complex momentum k_0. It should be emphasized that here we restrict our consideration to $E_0 \neq 0$. The exotic and extremely rare case of a zero-energy bound state requires a special treatment and is considered separately in Sect. 3.3.6.

At a spectral point, the regular solution ϕ_ℓ and the Jost solution $\chi_\ell^{(+)}$ as well as the physical wave function u_ℓ are proportional to each other, i.e. they only differ by some r-independent coefficients. This can be deduced from Eq. (2.81). In particular, this means that the Jost solution $\chi_\ell^{(+)}(E_0,r)$ is regular at the origin[5] and behaves as

$$\phi_\ell(E_0,r) \underset{r\to 0}{\sim} (k_0 r)^{\ell+1} \qquad \Rightarrow \qquad \chi_\ell^{(+)}(E_0,r) \underset{r\to 0}{\sim} (k_0 r)^{\ell+1} . \tag{3.41}$$

At infinity, the functions $\phi_\ell(E_0,r)$ and $\chi_\ell^{(+)}(E_0,r)$ have the following analytic form:

$$\phi_\ell(E_0,r) \underset{r\to\infty}{\longrightarrow} \frac{e^{ik_0 r}}{i^{\ell+1}} f_\ell^{(\text{out})}(E) , \qquad \chi_\ell^{(+)}(E_0,r) \underset{r\to\infty}{\longrightarrow} \frac{e^{ik_0 r}}{i^{\ell+1}} . \tag{3.42}$$

At the intermediate points, they are also proportional to each other, but their explicit form is, generally speaking, unknown.

According to Eq. (2.75), the first derivative of the Jost function with respect to E can be written as

$$\dot{f}_\ell^{(\text{in})}(E) = \frac{1}{2ik}\left[W(\dot{\phi}_\ell, \chi_\ell^{(+)}) + W(\phi_\ell, \dot{\chi}_\ell^{(+)})\right] + W(\phi_\ell, \chi_\ell^{(+)}) \frac{d}{dE}\left(\frac{1}{2ik}\right) .$$

At $E = E_0$, the last term in this equation vanishes because the functions ϕ_ℓ and $\chi_\ell^{(+)}$ are linearly dependent. Therefore,

$$\dot{f}_\ell^{(\text{in})}(E_0) = \frac{1}{2ik_0}\left\{W\left[\dot{\phi}_\ell(E_0,r), \chi_\ell^{(+)}(E_0,r)\right] + W\left[\phi_\ell(E_0,r), \dot{\chi}_\ell^{(+)}(E_0,r)\right]\right\} . \tag{3.43}$$

We need to prove that the quantity on the left hand side of this equation is non-zero. To this end, let us express the Wronskians on its right hand side in terms of the physical wave function u_ℓ. In order to achieve this, consider the equations

$$\left[\frac{d^2}{dr^2} + \tilde{k}^2 - \frac{\ell(\ell+1)}{r^2} - V(r)\right] \chi_\ell^{(+)}(\tilde{E},r) = 0 ,$$

$$\left[\frac{d^2}{dr^2} + k^2 - \frac{\ell(\ell+1)}{r^2} - V(r)\right] \phi_\ell(E,r) \quad = 0 ,$$

at two different energies, $\tilde{E}$ and E. Multiplying the first of them by $\phi_\ell(E,r)$, the second by $\chi_\ell^{(+)}(\tilde{E},r)$, and subtracting the second from the first, we obtain

$$\phi_\ell(E,r)\chi_\ell^{''(+)}(\tilde{E},r) - \phi_\ell''(E,r)\chi_\ell^{(+)}(\tilde{E},r) + (\tilde{k}^2 - k^2)\phi_\ell(E,r)\chi_\ell^{(+)}(\tilde{E},r) = 0 ,$$

[5] Generally speaking, both Jost solutions are irregular at the origin. However, since the point $r = 0$ is a regular singular point of the radial Schrödinger equation, one of its independent solutions can be regular. When the energy is equal to one of the spectral points, it happens that $\chi_\ell^{(+)}(E_0,r)$ becomes regular, while $\chi_\ell^{(-)}(E_0,r)$ remains irregular.

which means that

$$\frac{d}{dr}W\left[\phi_\ell(E,r),\chi_\ell^{(+)}(\tilde{E},r)\right]=(k^2-\tilde{k}^2)\phi_\ell(E,r)\chi_\ell^{(+)}(\tilde{E},r)\,. \tag{3.44}$$

The differentiation of this equation over the energy E gives

$$\frac{d}{dr}W\left[\dot{\phi}_\ell(E,r),\chi_\ell^{(+)}(\tilde{E},r)\right]=\frac{2\mu}{\hbar^2}\phi_\ell(E,r)\chi_\ell^{(+)}(\tilde{E},r)+(k^2-\tilde{k}^2)\dot{\phi}_\ell(E,r)\chi_\ell^{(+)}(\tilde{E},r)\,.$$

If $E=\tilde{E}=E_0$, then the second term disappears because $k^2-\tilde{k}^2=0$; therefore,

$$\frac{d}{dr}W\left[\dot{\phi}_\ell(E_0,r),\chi_\ell^{(+)}(E_0,r)\right]=\frac{2\mu}{\hbar^2}\phi_\ell(E_0,r)\chi_\ell^{(+)}(E_0,r)\,. \tag{3.45}$$

Similarly, by differentiating Eq. (3.44) over $\tilde{E}$ and putting $E=\tilde{E}=E_0$, we obtain

$$\frac{d}{dr}W\left[\phi_\ell(E_0,r),\dot{\chi}_\ell^{(+)}(E_0,r)\right]=-\frac{2\mu}{\hbar^2}\phi_\ell(E_0,r)\chi_\ell^{(+)}(E_0,r)\,. \tag{3.46}$$

Now we integrate Eqs. (3.45) and (3.46) from 0 to r and from r to ∞, respectively. The important thing is that for a bound state this integration goes along the real axis, but for a resonance state we do the integration in the complex r-plane along the straight line turned to the angle θ, the same as in the integral (3.37). The integrations give

$$\begin{aligned}W\left[\dot{\phi}_\ell(E_0,r),\chi_\ell^{(+)}(E_0,r)\right]&=\frac{2\mu}{\hbar^2}\int_0^r\phi_\ell(E_0,r)\chi_\ell^{(+)}(E_0,r)dr\\&\quad+W\left[\dot{\phi}_\ell(E_0,0),\chi_\ell^{(+)}(E_0,0)\right]\end{aligned} \tag{3.47}$$

and

$$\begin{aligned}W\left[\phi_\ell(E_0,r),\dot{\chi}_\ell^{(+)}(E_0,r)\right]&=\frac{2\mu}{\hbar^2}\int_r^\infty\phi_\ell(E_0,r)\chi_\ell^{(+)}(E_0,r)dr\\&\quad+W\left[\phi_\ell(E_0,\infty),\dot{\chi}_\ell^{(+)}(E_0,\infty)\right]\,.\end{aligned} \tag{3.48}$$

The last terms on the right hand sides of Eqs. (3.47) and (3.48) are zero, i.e.

$$W\left[\dot{\phi}_\ell(E_0,0),\chi_\ell^{(+)}(E_0,0)\right]=W\left[\phi_\ell(E_0,\infty),\dot{\chi}_\ell^{(+)}(E_0,\infty)\right]=0\,. \tag{3.49}$$

This is because in both these Wronskians there are functions that are zero at $r = 0$ as well as when $r \to \infty$ (for a resonance, this limit should be considered along the complex path). Indeed, it is seen from Eqs. (3.41), (3.42) that

$$\dot{\phi}_\ell(E_0, 0) = \chi_\ell^{(+)}(E_0, 0) = 0 \, ,$$

$$\phi_\ell(E_0, \infty) = \dot{\chi}_\ell^{(+)}(E_0, \infty) = \phi'_\ell(E_0, \infty) = \dot{\chi}_\ell^{(+)\prime}(E_0, \infty) = 0 \, .$$

Being substituted into Eq. (3.43), the Wronskians (3.47) and (3.48) give

$$\dot{f}_\ell^{(\text{in})}(E_0) = \frac{\mu}{i\hbar^2 k_0} \int_0^\infty \phi_\ell(E_0, r)\chi_\ell^{(+)}(E_0, r) dr \, . \tag{3.50}$$

Comparing the asymptotics (2.67) and (2.62), where $f_\ell^{(\text{in})}(E_0) = 0$, we see that

$$\phi_\ell(E_0, r) = f_\ell^{(\text{out})}(E_0)\chi_\ell^{(+)}(E_0, r) \, . \tag{3.51}$$

Therefore,

$$\dot{f}_\ell^{(\text{in})}(E_0) = (-1)^{\ell+1} \frac{\mu}{i\hbar^2 k_0} \frac{f_\ell^{(\text{out})}(E_0)}{\mathcal{A}_\ell^2} \int_0^\infty [u_\ell(E_0, r)]^2 \, dr \, . \tag{3.52}$$

At a spectral point, the physical wave function at large distances behaves exponentially,

$$\boxed{u_\ell(E_0, r) \underset{r\to\infty}{\longrightarrow} \mathcal{A}_\ell e^{ik_0 r}} \tag{3.53}$$

where $\mathcal{A}_\ell$ is the asymptotic normalization constant (ANC). Comparing this with (2.68), we find that it is related to the Jost solution as

$$u_\ell(E_0, r) = i^{\ell+1} \mathcal{A}_\ell \, \chi_\ell^{(+)}(E_0, r) \, . \tag{3.54}$$

Therefore,

$$\dot{f}_\ell^{(\text{in})}(E_0) = \frac{\mu}{i\hbar^2 k_0} f_\ell^{(\text{out})}(E_0) \int_0^\infty \left[\chi_\ell^{(+)}(E_0, r)\right]^2 dr \, . \tag{3.55}$$

For a bound state, the Schrödinger equation as well as the boundary conditions are real. The corresponding physical wave function is, therefore, also real. This means that the integral in the last equation is just the normalization integral. As we have seen in the preceding section, the normalization of the wave function of a resonance is also done by integrating its "simple" (not absolute) square. Perhaps, it should be reminded that in the case of a resonance the integration in Eq. (3.55) runs along the complex path shown in Fig. 3.9. In both cases, we can normalize the wave function to unity,

$$\int_0^\infty [u_\ell(E_0, r)]^2\, dr = 1\,,$$

which finally gives

$$\boxed{\dot{f}_\ell^{(\text{in})}(E_0) = (-1)^{\ell+1}\frac{\mu}{i\hbar^2 k_0}\frac{f_\ell^{(\text{out})}(E_0)}{\mathcal{A}_\ell^2}} \tag{3.56}$$

Since $f_\ell^{(\text{out})}(E)$ cannot be zero together with $f_\ell^{(\text{in})}(E)$, the first derivative of the Jost function is not zero, i.e. the first term of the expansion (3.40) is always present. This proves that the Jost function zeros corresponding to bound and resonant states (with $E_0 \neq 0$) are simple, i.e. of the first order.

As far as the virtual states are concerned, the analysis given here as well as the conclusion about the simplicity of the Jost function zeros is not applicable to them. The problem is that the wave function of a virtual state cannot be normalized the same way as it is done for a resonance. For a virtual state, the momentum k_0 is pure imaginary, and therefore, the conditions accompanied the regularization procedures (3.36) and (3.37) cannot be fulfilled. In more simple terms, the wave function of a virtual state exponentially grows without oscillating and, of course, cannot be normalized no matter the absolute or the simple square of it is put under the integral.

Numerical calculations show (see, for example, Ref. [15]) that, indeed, it is possible to have the second- order zeros of the Jost function on the negative imaginary k-axis. This can be found by varying the depth of a potential that supports virtual states and tracing their movements on the imaginary axis of the k-plane. At certain values of the parameters of the potential, two virtual states may fuse, which makes that point a second-order zero. However, any further change of the potential moves these fused virtual states away from each other.

* * *

It should be noted that the simplicity of zeros of the Jost function $f_\ell^{(\text{in})}(E)$ on the energy surface means that the corresponding zeros in the k-plane are simple as well. Indeed, the first term on the right hand side of Eq. (3.40) can be re-written as

$$\dot{f}_\ell^{(\text{in})}(E_n)(E - E_n) \quad = \quad \frac{\hbar^2}{2\mu}\dot{f}_\ell^{(\text{in})}(E_n)(k^2 - k_n^2)$$

$$= \frac{\hbar^2}{2\mu} \overset{\bullet}{f}{}_\ell^{(\mathrm{in})}(E_n)(k+k_n)(k-k_n)$$

$$\underset{k\to k_n}{\longrightarrow} \frac{\hbar^2 k_n}{\mu} \overset{\bullet}{f}{}_\ell^{(\mathrm{in})}(E_n)(k-k_n) ,$$

which is linear in $(k - k_n)$ as well as in $(E - E_n)$. This however is only true if $E_n \neq 0$. As is seen, the vanishing of the Jost function at $E = 0$, which is linear in E, corresponds to a quadratic vanishing in k. The possibility of having a spectral point at $E = 0$ should therefore be handled a bit differently. This is done in the next section.

3.3.6 Spectral Point at Threshold Energy

Although this is very unlikely for a real physical system, it is theoretically possible[6] that $f_\ell^{(\mathrm{in})}(E)$ is zero exactly at $E = 0$. We may encounter such situation, for example, when continuously decrease the depth of a potential generating a bound state. Together with the depth, the binding energy is decreasing and eventually becomes zero. What happens to the state then?

Since the threshold energy is the "border post" separating the regions of bound, resonant, virtual, and scattering states, it equally belongs to each of them. What kind of state represents the zero-energy spectral point? Before trying to answer this question, we need to check if a bound state with zero energy is possible, i.e. if the wave function for $E = 0$ can, in principle, be square integrable.

Consider first the free Schrödinger equation (2.9). Its linearly independent solutions, $j_\ell(kr)$, $n_\ell(kr)$, and $h_\ell^{(\pm)}(kr)$, become meaningless when $E \to 0$. Among them, the function $j_\ell(kr)$ becomes identical zero, while the other three infinite. This does not mean that Eq. (2.9) has no meaningful solutions at zero energy. Apparently, we can multiply the solutions of this linear homogeneous equation by any r-independent factor, and they will remain its solutions. In particular, the functions

$$\tilde{j}_\ell(kr) = k^{-(\ell+1)} j_\ell(kr) , \tag{3.57}$$

$$\tilde{n}_\ell(kr) = k^\ell n_\ell(kr) , \tag{3.58}$$

$$\tilde{h}_\ell^{(\pm)}(kr) = k^\ell h_\ell^{(\pm)}(kr) , \tag{3.59}$$

[6] This is in accordance with the conclusions that follow from Eq. (3.85).

solve Eq. (2.9). In contrast to the "untilded" functions, they remain meaningful even when $k \to 0$. Indeed, as it follows from their series expansions (see Appendix D),

$$\tilde{j}_\ell(kr) \underset{k\to 0}{\longrightarrow} \frac{r^{\ell+1}}{(2\ell+1)!!}\,, \tag{3.60}$$

$$\tilde{n}_\ell(kr) \underset{k\to 0}{\longrightarrow} -\frac{(2\ell-1)!!}{r^\ell}\,, \tag{3.61}$$

$$\tilde{h}_\ell^{(\pm)}(kr) \underset{k\to 0}{\longrightarrow} \mp i\frac{(2\ell-1)!!}{r^\ell}\,. \tag{3.62}$$

When $E = 0$, not all of them remain linearly independent. As we see from Eqs. (3.60–3.62) and as can be checked by the direct substitution, the independent solutions of the free equation,

$$\left[\frac{d^2}{dr^2} - \frac{\ell(\ell+1)}{r^2}\right] u_\ell(0,r) = 0\,, \tag{3.63}$$

are the functions $r^{\ell+1}$ and $r^{-\ell}$. Therefore far away from the interaction region, where the potential vanishes, any solution of the full equation

$$\left[\frac{d^2}{dr^2} - \frac{\ell(\ell+1)}{r^2} - V(r)\right] u_\ell(0,r) = 0 \tag{3.64}$$

has to be a combination of $r^{\ell+1}$ and $r^{-\ell}$. The term $\sim r^{\ell+1}$ is not square integrable with any ℓ, as well as the term $\sim r^{-\ell}$ with $\ell = 0$. However if $\ell \geqslant 1$, the solution $\sim r^{-\ell}$ is square integrable. Thus, a bound state with $E = 0$, in principle, may exist but only with $\ell \geqslant 1$.

If $\ell \neq 0$, then near the point $r = 0$, according to the assumption (2.2), the potential is negligible as compared to the centrifugal term in Eq. (3.64), and thus this equation transforms into (3.63). This means that a physical solution of Eq. (3.64), which must be regular, behaves like $\sim r^{\ell+1}$ when $r \to 0$. Therefore, the wave function of possible zero-energy bound state has the following asymptotic behaviour:

$$u_\ell(0,r) \longrightarrow \begin{cases} \sim r^{\ell+1}\,, & r \to 0\,, \\ \sim r^{-l}\,, & r \to \infty\,, \end{cases} \tag{3.65}$$

i.e. is both regular and square integrable. It is an exotic state, very thinly spread over the space. Its probability density vanishes at infinity not exponentially, but as $r^{-2\ell}$, which means that for $\ell = 1$ the R.M.S. radius,

$$\langle r\rangle_{\rm rms} = \sqrt{\int r^2|\psi(\vec{r}\,)|^2\, d^3r}\,,$$

is infinitely large. However, it is still a bound state. An S-wave bound state at the threshold energy is impossible. Physically, this can be explained by the absence of the centrifugal barrier that helps the states with higher ℓ to be localized.

Thus, we have established (for the class of potentials considered in the present chapter) that bound states at $E = 0$ may only exist with $\ell \geqslant 1$. Do they correspond to zeros of the Jost function? And how to define the Jost function at threshold? The problem is that for $k = 0$, the Riccati–Hankel functions are not appropriate for describing the incoming and outgoing waves (they are infinite for $kr = 0$), and the regular solution $\phi_\ell(E, r)$ with the boundary conditions (2.85) is identically zero for all r. This problem can be circumvented as follows. From the integral equations (2.40) and the definition (2.63), it follows that the Jost function can be expressed in terms of the regular solution via the following integral:

$$
\begin{aligned}
f_\ell^{(\mathrm{in})}(E) &= \frac{1}{2} - \frac{1}{2ik}\int_0^\infty h_\ell^{(+)}(kr)V(r)\phi_\ell(E, r)dr \\
&= \frac{1}{2} - \frac{1}{2i}\int_0^\infty \left[k^\ell h_\ell^{(+)}(kr)\right] V(r) \left[\frac{\phi_\ell(E, r)}{k^{\ell+1}}\right] dr \\
&= \frac{1}{2} - \frac{1}{2i}\int_0^\infty \tilde{h}_\ell^{(+)}(kr)V(r)\tilde{\phi}_\ell(E, r)dr \ ,
\end{aligned}
\tag{3.66}
$$

where $\tilde{\phi}_\ell(E, r) = \phi_\ell(E, r)/k^{\ell+1}$. This "tilded" function is defined for $k = 0$ in a clear way, namely, by the boundary conditions

$$
\tilde{\phi}_\ell(0, r) \underset{r\to 0}{\longrightarrow} \frac{r^{\ell+1}}{(2\ell+1)!!} \ . \tag{3.67}
$$

Apparently, in Eq. (3.66), the integrand exists and does not vanish at $E = 0$, and the integral is finite. Using Eq. (2.74), we can also give an alternative definition of $f_\ell^{(\mathrm{in})}(0)$ as the Wronskian

$$
\begin{aligned}
f_\ell^{(\mathrm{in})}(E) &= \frac{1}{2ik} W\left[\phi_\ell, \chi_\ell^{(+)}\right] \\
&= \frac{1}{2i} W\left[\tilde{\phi}_\ell, \tilde{\chi}_\ell^{(+)}\right] \\
&\underset{E\to 0}{\longrightarrow} -\frac{(2\ell-1)!!}{2} W\left[\tilde{\phi}_\ell(0, r), r^{-\ell}\right] ,
\end{aligned}
\tag{3.68}
$$

where the modified (tilded) Jost solution is

$$\tilde{\chi}_\ell^{(+)}(0, r) \underset{r\to\infty}{\longrightarrow} -i\frac{(2\ell - 1)!!}{r^\ell} . \tag{3.69}$$

Eq. (3.68) shows that at large distances the regular solution $\tilde{\phi}_\ell(0, r)$ of the Schrödinger equation is proportional to $r^{-\ell}$, if and only if $f_\ell^{(\text{in})}(0) = 0$. This proves that for $\ell \geqslant 1$, a zero-energy bound state corresponds to zero of the Jost function at $E = 0$ and vice versa.

What about the S-wave spectral point at zero energy? As we mentioned before, the threshold point is the border separating different types of states and therefore equally belongs to all of them. However, if it happens to be a spectral point, then the "symmetry" is broken, and it must belong to a specific type. Since the threshold S-wave spectral point is not a bound state, it corresponds to a threshold resonance (which can be called zero-energy virtual state as well).

An attentive reader should have noticed that at the point $E = 0$ the Jost functions $f_\ell^{(\text{in})}$ and $f_\ell^{(\text{out})}$ are identical. This follows not only from Fig. 3.5 but can also be seen explicitly, if we write an equation for $f_\ell^{(\text{out})}(0)$ similar to (3.66) and substitute Eq. (3.62) into it. As was said before, the functions $f_\ell^{(\text{in})}$ and $f_\ell^{(\text{out})}$ cannot vanish simultaneously because this would violate the current conservation. But if $f_\ell^{(\text{in})}(0) = 0$, then $f_\ell^{(\text{out})}(0) = 0$ as well. What is wrong with this?

Nothing is wrong. According to Eq. (3.62), the Riccati–Hankel functions are not linearly independent when $k = 0$ (this also follows from the value of their Wronskian (2.20), which vanishes for $k = 0$). Therefore at zero energy, $f_\ell^{(\text{in})}$ and $f_\ell^{(\text{out})}$ cannot be viewed as the amplitudes of the incoming and outgoing waves, and thus their simultaneous vanishing violates nothing. Despite the unclear physical meaning of the Jost functions at threshold, as we have shown the zero of $f_\ell^{(\text{in})}$ at this point corresponds to a bound ($\ell \geqslant 1$) or a virtual ($\ell = 0$) state.

3.3.6.1 Multiplicity of Threshold Zeros

It was shown in Sect. 3.3.5 that for "normal" (non-threshold) bound states, the corresponding zeros of the Jost function are of the first order, i.e. simple. What kind of zero does the Jost function have at the threshold energy?

To find the answer, we consider a bound state with the energy E_0 slightly below zero (for which the analysis given in Sect. 3.3.5 is fully applicable) and assess what happens in the limit $E_0 \to 0$. For this purpose, it is more appropriate to use $\tilde{\phi}_\ell$ and $\tilde{\chi}_\ell^{(+)}$ instead of ϕ_ℓ and $\chi_\ell^{(+)}$. It is also more convenient to consider the derivative with respect to k rather than the energy,

$$\frac{d}{dk} = \frac{\hbar^2 k}{\mu}\frac{d}{dE} .$$

Thus, Eq. (3.50) can be re-written as

$$\frac{d}{dk} f_\ell^{(\text{in})}(E_0) = -i \int_0^\infty \phi_\ell(E_0, r)\chi_\ell^{(+)}(E_0, r)dr \tag{3.70}$$

$$= -ik_0 \int_0^\infty \left[\frac{\phi_\ell(E_0, r)}{k_0^{\ell+1}}\right]\left[k_0^\ell \chi_\ell^{(+)}(E_0, r)\right] dr$$

$$= -ik_0 \int_0^\infty \tilde{\phi}_\ell(E_0, r)\tilde{\chi}_\ell^{(+)}(E_0, r)dr \,. \tag{3.71}$$

For $\ell \geqslant 1$, the last integral is always finite, even when $E_0 \to 0$, because the wave function is square integrable. We thus obtain

$$\frac{d}{dk} f_\ell^{(\text{in})}(E_0) \underset{E_0 \to 0}{\longrightarrow} -ik_0 \int_0^\infty \tilde{\phi}_\ell(0, r)\tilde{\chi}_\ell^{(+)}(0, r)dr \sim k_0 \,, \qquad \ell \geqslant 1 \,,$$

which means that the Jost function vanishes like $\sim k_0^2$, i.e. its zero remains linear in E, like for the "normal" bound states,

$$f_\ell^{(\text{in})}(E_0) \underset{E_0 \to 0}{\longrightarrow} \sim E_0 \,, \qquad \ell \geqslant 1 \,. \tag{3.72}$$

The S-wave case is different. In contrast to a non-zero angular momentum, when $\ell = 0$, the wave function of the zero-energy spectral state is not square integrable. As a result, the integral in Eq. (3.71) diverges when $E_0 \to 0$. So, we have the uncertainty

$$\frac{d}{dk} f_\ell^{(\text{in})}(E_0) \underset{E_0 \to 0}{\longrightarrow} 0 \times \infty \,, \qquad \ell = 0 \,, \tag{3.73}$$

which could mean that this derivative does not vanish when $E_0 \to 0$. We notice that for $\ell = 0$ the functions $\tilde{\chi}_\ell^{(+)}$ and $\chi_\ell^{(+)}$ are the same. Besides that, at a spectral point, the wave function $\tilde{\phi}_\ell$ is proportional to $\tilde{\chi}_\ell^{(+)}$. Therefore (for non-zero E_0), both $\tilde{\chi}_\ell^{(+)}$ and $\tilde{\phi}_\ell$ at large distances behave as $\sim e^{ik_0 r}$.

Let R be a sufficiently large radius, beyond which this asymptotics can be used instead of the exact functions. Splitting up the integral (3.71),

$$\frac{d}{dk} f_0^{(\text{in})}(E_0) = -ik_0 \int_0^R \tilde{\phi}_0(E_0, r)\tilde{\chi}_0^{(+)}(E_0, r)dr$$

$$-ik_0 \int_R^\infty \tilde{\phi}_0(E_0, r)\tilde{\chi}_0^{(+)}(E_0, r)dr \,,$$

we see that the integral over the interval $[0, R]$ is always finite and thus, when multiplied by k_0, vanishes with $E_0 \to 0$. The remaining integral

$$\int_R^\infty \tilde{\phi}_0(E_0, r)\tilde{\chi}_0^{(+)}(E_0, r)dr \sim \int_R^\infty e^{2ik_0 r}dr = -\frac{1}{2ik_0}e^{2ik_0 R} \underset{E_0\to 0}{\sim} \frac{1}{k_0}$$

behaves as $\sim k_0^{-1}$ and, thus when multiplied by k_0, gives a finite and non-zero derivative

$$0 < \left|\frac{d}{dk}f_0^{(\mathrm{in})}(E)\right|_{E=0} < \infty\,.$$

The implication is that the S-wave Jost function vanishes as $\sim k$. Summarizing the conclusions, we obtain

$$f_\ell^{(\mathrm{in})}(E) \underset{E\to 0}{\longrightarrow} \begin{cases} \sim \sqrt{E}\,, & \ell = 0 \\ \sim E\,, & \ell \geqslant 1 \end{cases} \tag{3.74}$$

which is valid for a spectral point at $E = 0$.

3.3.7 Distribution of Spectral Points Over the Riemann Surface

In the preceding sections, we have established that real negative spectral points on the physical sheet of the E-surface are bound states, real negative spectral points on the unphysical sheet are virtual states, and complex spectral points on the lower part of the unphysical sheet are resonances. A typical distribution of the spectral points of all these kinds over the E-surface is shown in Fig. 3.10.

In the simplest case (single-channel, finite-range interaction) that we are considering in this chapter, the Riemann E-surface has simple double-sheet structure

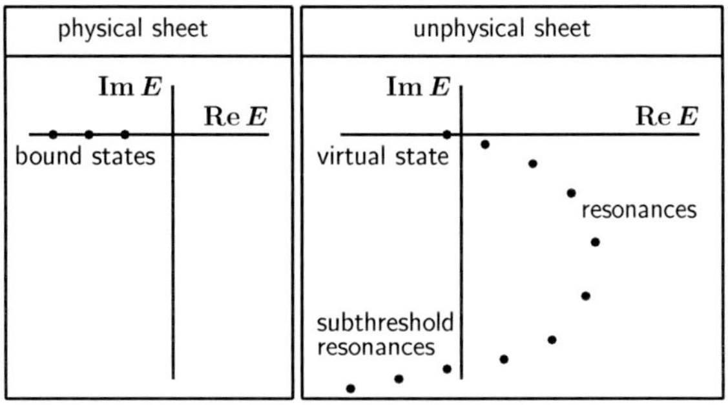

Fig. 3.10 Typical distribution of the spectral points on the physical (left) and unphysical (right) sheets of the complex energy Riemann surface

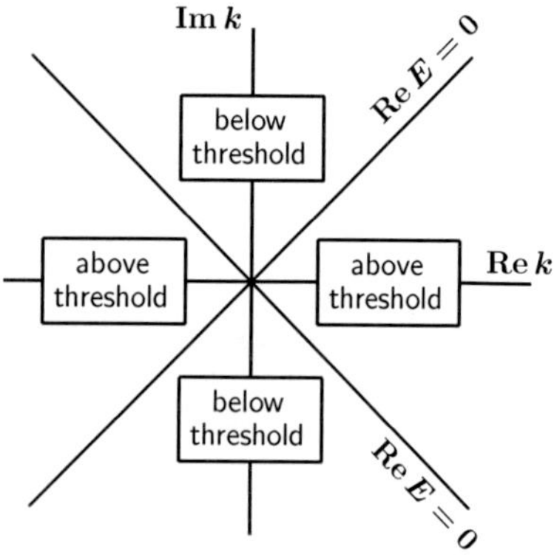

Fig. 3.11 Four domains of the complex momentum plane, where the corresponding energy is above (Re $E > 0$) and below (Re $E < 0$) the threshold

and can be mapped onto a single sheet of the momentum plane. Via the energy–momentum relation,

$$k = \pm\sqrt{(2\mu/\hbar^2)E}\ , \qquad (3.75)$$

the physical sheet of the E-surface is mapped onto the upper half of the k-plane, and the unphysical sheet onto its lower half. Since

$$E = \frac{\hbar^2}{2\mu}\left[(\mathrm{Re}\,k)^2 - (\mathrm{Im}\,k)^2 + i2\,(\mathrm{Re}\,k)\,(\mathrm{Im}\,k)\right]\,,$$

the bisectors of the quadrants of the k-plane, on which $\mathrm{Re}\,k = \mathrm{Im}\,k$, correspond to the upper and lower halves of the line Re $E = 0$ and therefore separate the energies that are above and below the threshold, as is shown in Fig. 3.11.

The distribution of the spectral points on the k-plane that corresponds to their typical distribution on the energy surface (see Fig. 3.10) is shown in Fig. 3.12. Apparently, it is more convenient and easier to deal with a single-sheet k-plane rather than with a double-sheet E-surface. Why do we need and use the E-surface then? The reason is that when we move on to more complicated multi-channel problems with several thresholds, the E-surface (although being multi-sheet) will retain the same clear meaning, while the description using the channel momenta will drastically change and loose the simplicity.

At this point, a logical question may arise: Can the Jost function $f_\ell^{(\mathrm{in})}(E)$ have any other zeros apart from those shown in Fig. 3.10? As far as the physical sheet of the E-surface is concerned, the answer is: No, it cannot; otherwise, the conservation of the probability, i.e. the conservation of the number of particles in a quantum ensemble, would be violated. It can be proved as follows.

First of all, there are no zeros on the positive half of the real axis (i.e. at scattering energies). Indeed, if they were then at such an energy $E_n > 0$, the wave function $\Psi \sim \exp(-iE_nt/\hbar)$ would describe a stationary state with time-

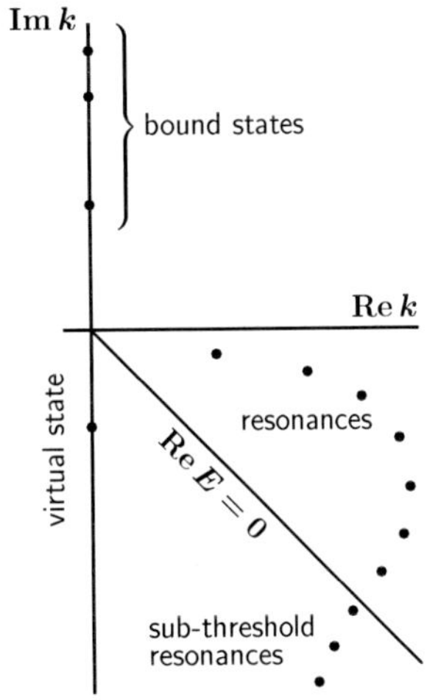

Fig. 3.12 Typical distribution of the spectral points on the complex momentum plane that corresponds to the distribution shown in Fig. 3.10

independent probability density $|\Psi|^2$. The radial part of this function would have the asymptotics

$$u_\ell(E_n, r) \underset{r\to\infty}{\sim} e^{ik_n r} f_\ell^{(\mathrm{out})}(E_n) , \tag{3.76}$$

where

$$k_n = \pm\sqrt{(2\mu/\hbar^2)E_n} . \tag{3.77}$$

Any choice of the sign in the last equation would lead to a contradiction. Assuming k_n to be positive, we conclude that the stationary state Ψ has only outgoing wave when $r \to \infty$. Physically, this would mean that after colliding of just two particles at the energy E_n, we could observe everlasting emission of such particles without any further effort from our side, i.e. a kind of perpetuum mobile. With a negative k_n, we would obtain a pure converging wave, i.e. a kind of "black hole" that completely absorbs the flux of colliding particles.

Suppose now that there is a zero of the Jost function on the physical sheet somewhere away from the real axis. By definition, the physical sheet of the E-surface corresponds to complex k such that $\mathrm{Im}\, k > 0$. Therefore, when $r \to \infty$, the wave function (3.76) is exponentially vanishing, $u_\ell(E_n, r) \sim e^{-(\mathrm{Im}\, k_n)r}$, i.e. is square integrable. This means that the particle is bound. However, since $\mathrm{Im}\, E \neq 0$, its state is not stationary. Its probability density $|\Psi|^2 \sim \exp\{2(\mathrm{Im}\, E)t/\hbar\}$ either grows or vanishes depending on the sign of $\mathrm{Im}\, E$.

Therefore, if the Jost function has zeros on the physical sheet of the E-surface, all of them are on the negative half of the real axis. The above qualitative proof of this statement can be replaced with a more formal quantitative one. In order to prove it more rigorously, consider the time-dependent Schrödinger equation,

$$i\hbar\frac{\partial}{\partial t}\Psi = -\frac{\hbar^2}{2\mu}\Delta\Psi + U\Psi\,, \tag{3.78}$$

together with its complex conjugate,

$$-i\hbar\frac{\partial}{\partial t}\Psi^* = -\frac{\hbar^2}{2\mu}\Delta\Psi^* + U\Psi^*\,. \tag{3.79}$$

Multiplying Eq. (3.78) by Ψ^* and (3.79) by Ψ, and subtracting them, we obtain the continuity equation

$$\frac{\partial}{\partial t}\rho + \vec{\nabla}\cdot\vec{j} = 0\,, \tag{3.80}$$

where $|\Psi|^2 = \rho$ is the particle density and $\vec{j}$ is the current defined by Eq. (1.52). Let v be a spherical volume centred at $r = 0$ and enclosed by the spherical surface s of the radius R. The integration of Eq. (3.80) over v with the help of the divergence theorem gives the integral form of the continuity equation

$$\frac{\partial}{\partial t}\int_v |\Psi|^2\,\mathrm{d}^3r = -\int_s \vec{j}\cdot d\vec{s}\,. \tag{3.81}$$

We can always choose radius of the sphere large enough, such that on the spherical surface the wave function is accurately represented by its asymptotics. Let the energy point

$$E_0 = \frac{\hbar^2k_0^2}{2\mu} = \frac{\hbar^2}{2\mu}(k_r^2 - k_i^2 + 2ik_rk_i)\,,$$

be a zero of the Jost function (here $k_0 = k_r + ik_i$). Then far away from the origin, the wave function is given by

$$\begin{aligned}\Psi_{E_0\ell m}(\vec{r},t) &= \exp\left(-\frac{i}{\hbar}E_0t\right)\frac{u_\ell(E_0,r)}{r}Y_{\ell m}(\theta,\varphi)\\ &\underset{r\to\infty}{\longrightarrow} \exp\left(-\frac{i}{\hbar}E_0t\right)\frac{-ie^{i(k_0r-\ell\pi/2)}}{r}Y_{\ell m}(\theta,\varphi)f_\ell^{(\mathrm{out})}(E_0)\,.\end{aligned} \tag{3.82}$$

The left hand side of Eq. (3.81) is

$$\frac{\partial}{\partial t}\int_v |\Psi|^2 d\vec{r} = \frac{\partial}{\partial t}\exp\left[\frac{i}{\hbar}(E_0^* - E_0)t\right]\int_0^R |u_\ell(E_0, r)|^2\,dr$$

$$= \frac{2\hbar k_r k_i}{\mu}\exp\left(\frac{2\hbar k_r k_i}{\mu}t\right)\int_0^R |u_\ell(E_0, r)|^2\,dr\,. \qquad (3.83)$$

When calculating the right hand side of Eq. (3.81), we notice that among the three components of the current, $\vec{j} = \{j_r, j_\theta, j_\varphi\}$, only the radial component contributes to the surface integral. Bearing in mind that

$$\vec{\nabla} = \left\{\frac{\partial}{\partial r}, \frac{1}{r}\frac{\partial}{\partial \theta}, \frac{1}{r\sin\theta}\frac{\partial}{\partial\varphi}\right\},$$

we find

$$j_r(r,\theta,\varphi) = \exp\left(\frac{2\hbar k_r k_i}{\mu}t\right)\left|f_\ell^{(\mathrm{out})}(E_0)\right|^2 |Y_{\ell m}(\theta,\varphi)|^2\,\frac{\hbar k_r e^{-2k_i r}}{\mu r^2}\,. \qquad (3.84)$$

Since the surface is spherical, the integration

$$\int_s \vec{j}\cdot d\vec{s} = \int_0^\pi d\theta \int_0^{2\pi} d\varphi\,\sin\theta\, j_r(R,\theta,\varphi)$$

just removes the spherical harmonics from (3.84). Equating the left and right hand sides of the continuity equation, we thus obtain

$$2k_r k_i \int_0^R |u_\ell(E_0, r)|^2\,dr = -k_r\left|f_\ell^{(\mathrm{out})}(E_0)\right|^2 \frac{e^{-2k_i R}}{R^2}\,. \qquad (3.85)$$

Because of the minus on the right hand side, this equation can only be fulfilled in the following three cases:

- When $k_r = 0$, which means that the zero k_0 of the Jost function is on the imaginary axis (E_0 is real and negative or zero).
- When $k_i < 0$, i.e. E_0 is on the unphysical sheet of the Riemann surface.
- When $k_r = k_i = 0$, there is a spectral point just at the threshold energy $E_0 = 0$.

This proves that on the physical sheet the only possible spectral points are the bound states. For the unphysical sheet, however, Eq. (3.85) does not give any restrictions on the location of the spectral points.

Examining the symmetry properties summarized in Figs. 3.5 and 3.6, we conclude that every resonance spectral point has its mirror partner that is symmetric relative to the imaginary axis in the k-plane or relative to the real axis on the unphysical sheet of the E-surface (see Figs. 3.13 and 3.14). Indeed, let $k_0 = |k_0|\exp(-i\varphi)$ be a zero of the Jost function $f_\ell^{(\text{in})}$ in the fourth quadrant of the k-plane. Since the zeros of $f_\ell^{(\text{in})}$ and $f_\ell^{(\text{in})*}$ coincide and these two functions are symmetric with respect to the imaginary axis of the k-plane, there should be a mirror zero of the Jost function at $\tilde{k}_0 = |k_0|\exp(i\pi + i\varphi)$. The point k_0 of the k-plane corresponds to the point

$$E_0^{(-)} = \frac{\hbar^2}{2\mu}k_0^2 = \frac{\hbar^2}{2\mu}|k_0|^2\exp(-i2\varphi)$$

on the lower half of the unphysical sheet of the E-surface, while $\tilde{k}_0$ corresponds to the energy

$$E_0^{(-)*} = \frac{\hbar^2}{2\mu}\tilde{k}_0^2 = \frac{\hbar^2}{2\mu}|k_0|^2\exp(i2\pi + i2\varphi) \tag{3.86}$$

on the upper half of that sheet.

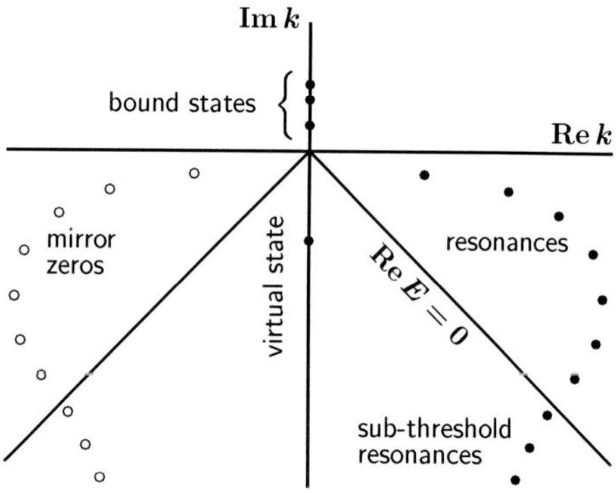

Fig. 3.13 Typical distribution of the spectral points in the complex momentum plane. Each resonance spectral point has a partner that is symmetric relative to the imaginary axis (open circles)

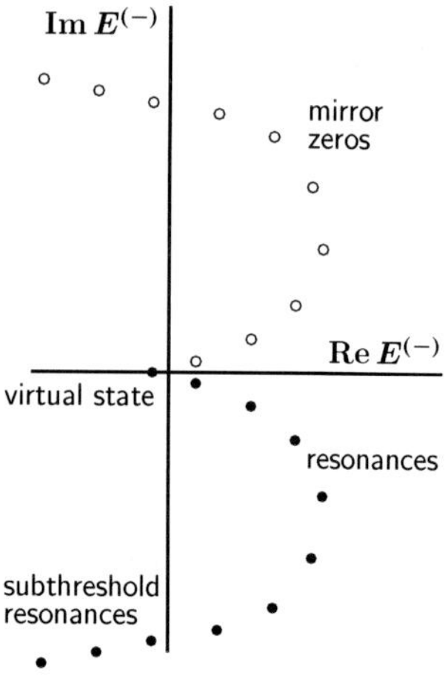

Fig. 3.14 Typical distribution of the spectral points on the unphysical sheet of the energy surface. Each resonance spectral point has a partner that is symmetric relative to the real axis (open circles)

In contrast to the resonance zeros, their mirror partners have no observable effects on the scattering process. This is because the scattering energies are on the upper rim of the cut of the E-surface, and the mirror zeros are far away from this rim. When looking at the E-surfaces depicted earlier, it may seem that the mirror zeros can be very close to the upper rim. This is not true, however. Even if the angle φ in Eq. (3.86) is very small, a continuous path from the point $E_0^{(-)*}$ to the upper rim includes a full circle (2π) around the branch point. When moving along such a path, the Jost function undergoes many changes and "forgets" that it was a zero somewhere behind. On the k-plane, this is even more clear since the mirror zeros are in the third quadrant that is far away from the real positive momenta at which the scattering takes place. The only situation when we can expect some influence from the mirror zeros is when we consider the near-threshold (very low energy) scattering and there is a zero next to the origin. But even in this case the influence of the mirror resonance is overshadowed by a more strong influence of the resonance itself that is always closer to the scattering energies than its mirror partner.

3.3.8 Number of Spectral Points

How many bound states and resonances may a physical system have? In other words, how many zeros may the Jost function have on the physical and unphysical sheets of the energy surface?

As we have shown in Sect. 3.2, on the far periphery of the lower half of the k-plane (i.e. on the unphysical E-sheet, where the resonance spectral points are located), the Jost function oscillates (before approaching a constant value at very large $|k|$). This means that there is no limit on the number of resonance zeros because their location on the k-plane is not restricted to a straight line.

With bound states, the situation is different. On the far periphery of the upper half of the k-plane, the Jost function tends to a constant (see Fig. 3.7),

$$\left| f_\ell^{(\mathrm{in})}(E) - \frac{1}{2} \right| \underset{|k|\to\infty}{\longrightarrow} 0\,, \qquad \mathrm{Im}\, k \geqslant 0\,.$$

This means that for any small $\varepsilon > 0$, we can always find a positive number K such that for all k beyond the semicircle of radius K the value of $f_\ell^{(\mathrm{in})}(E)$ differs from 1/2 less than ε. Therefore, beyond this semicircle the Jost function cannot have zeros.

The zeros of an analytic function are isolated unless this function is the identical zero. In other words, around each zero there is a finite circle within which the analytic function is not zero. This implies that within the domain of analyticity, the zeros cannot condensate, i.e. cannot have a limiting point that is only possible on the border of the region. If there is such a limiting point on the border, then the function is zero at this point.[7] Therefore within any sub-domain that is fully within

[7] This is actually a well-known theorem in complex analysis; see, for example, Sect. 7G of Ref. [16].

the domain of analyticity and does not touch the borders, the analytic function may have only a finite number of zeros.

In the upper half of the k-plane, the function $f_\ell^{(\mathrm{in})}(E)$ may only have zeros on the imaginary axis, in the finite[8] interval $[0, iK)$. The point iK is not on the border of the analyticity domain. It is an internal point, and therefore, the spectral points cannot condensate at $k = iK$. For the point $k = 0$, however, we did not establish analyticity of the Jost function (see Sect. 2.2.5), i.e. it may be a border point. We therefore may expect that $k = 0$ is the point where an infinite set of spectral points condensate, which implies that $f_\ell^{(\mathrm{in})}(0) = 0$. But the zero of the Jost function at $k = 0$ is always simple (see Sect. 3.3.6) and thus is isolated i.e. cannot be a point of condensation.[9] We can finally conclude that the number of bound states with a given angular momentum is always finite.

3.3.9 Integral Equation for a Bound-State Wave Function

The fact that the number of bound states with a given ℓ is limited is very important but is rather vague. We would like to know how many bound states exist or at least to obtain an estimate on this number. An upper bound on the number of possible bound states with a given angular momentum is obtained in the next section. The derivations leading to that upper bound are based on the integral equation for a bound-state wave function, which we are going to derive here.

* * *

The wave function $u_\ell(E_n, r)$ describing a bound state with the energy $E_n = (\hbar^2/2\mu)k_n^2$ and angular momentum $\hbar\ell$ is a solution of the Schrödinger equation

$$\left[D_r - V(r)\right]u_\ell(E_n, r) = 0\,, \tag{3.87}$$

where

$$D_r \equiv \frac{d^2}{dr^2} + k_n^2 - \frac{\ell(\ell+1)}{r^2}\,, \tag{3.88}$$

and obeys the homogeneous boundary conditions,

$$u_\ell(E_n, 0) = u_\ell(E_n, \infty) = 0\,. \tag{3.89}$$

[8] This interval is finite because if a physically meaningful potential supports any bound states, then among them there is always a lowest ground state with a finite binding energy.

[9] It should be emphasized that so far we did not consider the Coulomb and other long-range potentials for which this statement may not be valid. Indeed, as we know the spectrum of hydrogen atom condenses at $E = 0$.

It is possible to replace the differential equation (3.87) with an equivalent integral equation, where the boundary conditions (3.89) are taken into account via its kernel. To achieve this, we introduce the Green's function $g_\ell^{(+)}(r, r')$ obeying the equation[10]

$$D_r g_\ell^{(+)}(r, r') = \delta(r - r') . \tag{3.90}$$

Then the function written as

$$u_\ell(E_n, r) = \int_0^\infty g_\ell^{(+)}(r, r') V(r') u_\ell(E_n, r') dr' \tag{3.91}$$

obeys the Schrödinger equation (3.87). Indeed, this can be checked by the direct substitution,

$$\begin{aligned} D_r u_\ell(E_n, r) &= \int_0^\infty D_r g_\ell^{(+)}(r, r') V(r') u_\ell(E_n, r') dr' \\ &= \int_0^\infty \delta(r - r') V(r') u_\ell(E_n, r') dr' \\ &= V(r) u_\ell(E_n, r) . \end{aligned}$$

When $r \neq r'$, it follows from Eq. (3.90) that the Green's function obeys the "free" Schrödinger equation,

$$D_r g_\ell^{(+)}(r, r') = 0 , \qquad r \neq r' , \tag{3.92}$$

and therefore is a linear combination of any two of its independent solutions, namely, the Riccati functions $j_\ell(k_n r)$, $n_\ell(k_n r)$, $h_\ell^{(\pm)}(k_n r)$. The point $r = r'$ is special because the right hand side of Eq. (3.90) is infinite. This means that on both sides of this point the function $g_\ell^{(+)}(r, r')$ is a superposition of the Riccati functions, but not necessarily the same superposition,

$$g_\ell^{(+)}(r, r') = \begin{cases} j_\ell(k_n r) a(r') + h_\ell^{(+)}(k_n r) b(r') , & r < r' , \\ j_\ell(k_n r) c(r') + h_\ell^{(+)}(k_n r) d(r') , & r > r' . \end{cases} \tag{3.93}$$

In Eq. (3.93), we chose the functions $j_\ell(k_n r)$ and $h_\ell^{(+)}(k_n r)$ as two independent solutions of the "free" equation simply because with this choice it will be easier to satisfy the boundary conditions. Any other choice would be equally legitimate and

[10] The superscript (+) indicates that this Green's function is such that the corresponding wave function asymptotically behaves as the outgoing wave.

would lead to the same result. The combination coefficients a, b, c, and d apparently should depend on r' because there is no other way for $g_\ell^{(+)}(r, r')$ to have the r'-dependence.

As is seen from Eq. (3.91), the boundary conditions (3.89) are satisfied if

$$g_\ell^{(+)}(0, r') \equiv g_\ell^{(+)}(\infty, r') \equiv 0 , \tag{3.94}$$

which implies that $b(r') \equiv 0$ and $c(r') \equiv 0$, i.e. the superpositions (3.93), are reduced to

$$g_\ell^{(+)}(r, r') = \begin{cases} j_\ell(k_n r)a(r') & , r < r' , \\ h_\ell^{(+)}(k_n r)d(r') , & r > r' . \end{cases} \tag{3.95}$$

Consider now the special point $r = r'$. Let ε be a small positive number. Integrating Eq. (3.90) over the interval $[r' - \varepsilon, r' + \varepsilon]$, we obtain

$$\int_{r'-\varepsilon}^{r'+\varepsilon} D_r g_\ell^{(+)}(r, r')dr = 1 .$$

In this integral, the value of r' is fixed, and therefore, $a(r')$ and $d(r')$ are just numbers. Since the function $g_\ell^{(+)}(r, r')$ obeys the second-order differential equation, it must be continuous everywhere, including the point $r = r'$; otherwise, the first derivative would already be infinite. Integral of a continuous function over a shrinking interval tends to zero. Therefore,

$$\lim_{\varepsilon\to 0} \int_{r'-\varepsilon}^{r'+\varepsilon} \left[k_n^2 - \frac{\ell(\ell+1)}{r^2} \right] g_\ell^{(+)}(r, r')dr = 0 ,$$

which implies that

$$\lim_{\varepsilon\to 0} \int_{r'-\varepsilon}^{r'+\varepsilon} \frac{d^2}{dr^2} g_\ell^{(+)}(r, r')dr = 1 . \tag{3.96}$$

This means that the second derivative of $g_\ell^{(+)}(r, r')$ is infinite at the point $r = r'$, which is possible if the first derivative $(d/dr)g_\ell^{(+)}(r, r')$ is discontinuous. This discontinuity is finite; otherwise, the relation

$$\lim_{\varepsilon\to 0} \left\{ \frac{d}{dr} g_\ell^{(+)}(r, r')\bigg|_{r'+\varepsilon} - \frac{d}{dr} g_\ell^{(+)}(r, r')\bigg|_{r'-\varepsilon} \right\} = 1 \tag{3.97}$$

that follows from (3.96) would not be valid. Since the first derivative exists and finite on both sides of r', the function itself is continuous at this point. Therefore, there are two conditions determining the combination coefficients in (3.95), namely, the discontinuity of the first derivative and continuity of $g_\ell^{(+)}(r, r')$ itself,

$$\left[\frac{d}{dr}h_\ell^{(+)}(k_n r)\right]d(r) - \left[\frac{d}{dr}j_\ell(k_n r)\right]a(r) = 1\,, \tag{3.98}$$

$$h_\ell^{(+)}(k_n r)d(r) - j_\ell(k_n r)a(r) = 0\,. \tag{3.99}$$

Using the known Wronskian

$$W\left[j_\ell(kr), h_\ell^{(+)}(kr)\right] = ik\,, \tag{3.100}$$

it is easy to solve the system of equations (3.98), (3.99) and obtain

$$a(r) = \frac{1}{ik_n}h_\ell^{(+)}(k_n r) \tag{3.101}$$

and

$$d(r) = \frac{1}{ik_n}j_\ell(k_n r)\,. \tag{3.102}$$

This gives us the following expression for the Green's function that is part of the kernel of the integral equation (3.91)

$$g_\ell^{(+)}(r, r') = \begin{cases} \dfrac{1}{ik_n}j_\ell(k_n r)h_\ell^{(+)}(k_n r')\,, & r \leqslant r'\,, \\[2ex] \dfrac{1}{ik_n}h_\ell^{(+)}(k_n r)j_\ell(k_n r')\,, & r \geqslant r' \end{cases} \tag{3.103}$$

3.3.10 Bargmann's Inequality

Using the integral equation (3.91) derived in the preceding section, we can now estimate the number of zeros that the Jost function may have on the physical sheet of the E-surface. Let the potential $V(r)$ supports N_ℓ bound states with the angular momentum ℓ. Then the number$\tilde{N}_\ell(\zeta)$ of bound states generated by the scaled potential $\zeta V(r)$, where $0 \leqslant \zeta \leqslant 1$, is such that $\tilde{N}_\ell(0) = 0$ and $\tilde{N}_\ell(1) = N_\ell$. When ζ is decreased from 1 to 0, the bound states become less and less tightly bound and disappear one by one. In other words, the corresponding spectral points move down the positive imaginary axis of the k-plane, and one by one pass through the threshold point $k = 0$. This means that at certain values ζ_n of the scaling factor, there are zero-energy bound states. The number of such values is exactly the same as N_ℓ. So, our task is to estimate how many values ζ_n correspond to threshold bound states.

To this end, we use the integral equation for the wave function of a bound state,

$$u_\ell(E_n, r) = \zeta \int_0^\infty g_\ell^{(+)}(r, r')V(r')u_\ell(E_n, r')dr' , \tag{3.104}$$

where the Green's function is given by Eq. (3.103). Very close to the point $k = 0$, only the leading terms of the series expansions of the Riccati–Bessel and Riccati–Hankel functions can be retained. As a result, the Green's function becomes k-independent,

$$g_\ell^{(+)}(r, r') \underset{k\to 0}{\longrightarrow} \begin{cases} -\dfrac{r^{\ell+1}(r')^{-\ell}}{2\ell+1} , & r \leqslant r' , \\ -\dfrac{r^{-\ell}(r')^{\ell+1}}{2\ell+1} , & r > r' . \end{cases} \tag{3.105}$$

Apparently, if we replace the potential $V(r)$ with pure attractive one $-|V(r)|$, the number of possible bound states can only increase. So any upper bound on N_ℓ, obtained for this pure attractive potential, will be valid for the potential $V(r)$ as well. We therefore replace Eq. (3.104) with

$$u_\ell(E_n, r) = -\zeta \int_0^\infty g_\ell^{(+)}(r, r')|V(r')|u_\ell(E_n, r')dr' , \tag{3.106}$$

where $u_\ell(E_n, r)$ and the set of E_n are, of course, different. We, however, are interested not in the values of E_n, but in their number. Introducing

$$\tilde{g}_\ell(r, r') = -\sqrt{|V(r)V(r')|}g_\ell^{(+)}(r, r') \tag{3.107}$$

and

$$\tilde{u}_\ell(E_n, r) = \sqrt{|V(r)|}u_\ell(E_n, r) ,$$

and putting $\zeta = \zeta_n$, which makes $E_n = 0$, we reduce Eq. (3.106) to

$$\int_0^\infty \tilde{g}_\ell(r, r')\tilde{u}_\ell(0, r')dr' = \zeta_n^{-1}\tilde{u}_\ell(0, r) . \tag{3.108}$$

This equation can be written in the operator form,

$$\tilde{g}_\ell \tilde{u}_\ell^{(n)} = \zeta_n^{-1}\tilde{u}_\ell^{(n)} , \tag{3.109}$$

which is an eigenvalue equation for the operator $\tilde{g}_\ell$. All possible eigenvalues ζ_n^{-1} correspond to zero-energy bound states and thus form the discrete spectrum of $\tilde{g}_\ell$. It is clear that apart from the eigenvalues obeying the constraint $0 \leqslant \zeta_n \leqslant 1$, Eq.

(3.109) may have solutions with $\zeta_n > 1$ (but not with $\zeta_n < 0$ because a negative ζ makes the potential pure repulsive). The corresponding eigenvectors $|\tilde{u}_\ell^{(n)}\rangle$ form a full set and can be normalized to unity. The operator $\tilde{g}_\ell$ can therefore be represented by the following spectral expansion:

$$\tilde{g}_\ell = \sum_n |\tilde{u}_\ell^{(n)}\rangle \zeta_n^{-1} \langle \tilde{u}_\ell^{(n)}| \,,$$

and thus its trace is equal to the sum of its eigenvalues,

$$\operatorname{Sp} \tilde{g}_\ell = \sum_n \langle \tilde{u}_\ell^{(n)} | \tilde{g}_\ell | \tilde{u}_\ell^{(n)} \rangle = \sum_n \zeta_n^{-1} \,.$$

Since the eigenvalues ζ_n that correspond to the N_ℓ bound states of the potential $-|V(r)|$ are less than 1, a sum of the corresponding inverted terms $1/\zeta_n$ is greater than N_ℓ. We therefore can write that

$$N_\ell \leqslant \sum_{n=1}^{N_\ell} \zeta_n^{-1} \leqslant \sum_{n=1}^{\infty} \zeta_n^{-1} = \operatorname{Sp} \tilde{g}_\ell \,.$$

The trace of an operator can be calculated (with the same result) in any representation. In our case, it is easier to use the coordinate representation $|r\rangle$, which gives

$$\operatorname{Sp} \tilde{g}_\ell = \int_0^\infty \tilde{g}_\ell(r, r) dr = \frac{1}{2\ell + 1} \int_0^\infty |V(r)| r \, dr \,,$$

where we used Eqs. (3.105) and (3.107). Finally, we arrive at the following estimate:

$$\boxed{N_\ell \leqslant \frac{1}{2\ell + 1} \int_0^\infty |V(r)| r \, dr} \tag{3.110}$$

This inequality was obtained in Ref. [17] by V. Bargmann and therefore bears his name.

Looking at Eq. (3.110), the reader might be confused by seemingly wrong units on the right hand side. In this case, it should be reminded that $V(r)$ is the reduced potential related to the true potential energy $U(r)$ as

$$V(r) = \frac{2\mu}{\hbar^2} U(r) \,.$$

Bearing this in mind, the reader can easily see that the integral on the right hand side of Eq. (3.110) is dimensionless. Indeed,

$$[V r \, dr] = \frac{[\mu][U][r]^2}{[\hbar^2]} = \frac{[\mu c^2][E][r]^2}{[\hbar c]^2} = \frac{[E]^2[r]^2}{([E][r])^2} = 1 \,.$$

4 Scattering States and the S-Matrix

Quantum scattering is a transformation of an incoming wave into an outgoing wave at real positive energies (when E is above the threshold). The motion of the particle at such energies is not restricted, i.e. it can go infinitely far from the source of the interaction field. The wave function describing it never vanishes even at large r and therefore is not square integrable.

In the time-independent formalism, the scattering is described by the stationary states. Although such states correspond to a single particle, using the statistical ensemble interpretation, they can be related to the probability distributions in the stationary fluxes of particles approaching the interaction region and moving away.

4.1 Partial Waves

If the scattering forces are absent, the incident flux moves undisturbed and is represented by the plane wave $\psi_{\vec{k}}(\vec{r}\,) = (2\pi)^{-3/2}\exp(i\vec{k}\vec{r}\,)$. It is a flux of particles linearly moving in the direction of their incident momentum $\vec{k}$. The plane wave is an extreme and simplest example of a scattering state. We immediately see how it differs from a bound state. Firstly, it is normalized to the δ-function,

$$\int \psi^*_{\vec{k}'}(\vec{r}\,)\psi_{\vec{k}}(\vec{r}\,)\,\mathrm{d}^3r = \delta(\vec{k}\,' - \vec{k}\,)\,, \tag{4.1}$$

and secondly, it depends on two vectors, $\vec{r}$ and $\vec{k}$. The dependence on $\vec{k}$ reflects the fact that, in contrast to a bound state, a scattering state "remembers" how it was formed, i.e. from which direction the particles came.

In the plane-wave flux, the particles move in the same direction in the whole infinite space. There are therefore particles with all possible impact parameters

S. A. Rakityansky, *Jost Functions in Quantum Mechanics*,
https://doi.org/10.1007/978-3-031-07761-6_4

relative to the origin of the coordinate system (where the centre of the potential is supposed to be). Making the partial-wave decomposition (see Appendix A),

$$\psi_{\vec{k}}(\vec{r}\,) = \sqrt{\frac{2}{\pi}}\frac{1}{kr}\sum_{\ell m} i^\ell j_\ell(kr) Y^*_{\ell m}(\hat{\vec{k}}\,) Y_{\ell m}(\hat{\vec{r}}\,)\,, \tag{4.2}$$

$$= \frac{1}{\sqrt{2\pi}kr}\sum_{\ell m} i^\ell \frac{\phi_\ell(E,r)}{f_\ell^{(\text{in})}(E)} Y^*_{\ell m}(\hat{\vec{k}}\,) Y_{\ell m}(\hat{\vec{r}}\,)\,, \tag{4.3}$$

we separate the contributions from the waves with definite values of the angular momentum (impact parameter). In Eq. (4.2) the reader should recognize the regular solution for the "free" motion,

$$\phi_\ell(E,r) = j_\ell(kr) = \frac{1}{2}h_\ell^{(-)}(kr) + \frac{1}{2}h_\ell^{(+)}(kr)\,,$$

with the trivial Jost functions $f_\ell^{(\text{in/out})}(E) = 1/2$. In Eq. (4.3) this solution is divided and multiplied by $f_\ell^{(\text{in})}(E) = 1/2$. The physical wave function, u_ℓ, describing scattering with the angular momentum ℓ, of course, must be regular at $r = 0$. Since the radial Schrödinger equation has two independent solutions and only one of them is regular, the physical solution is proportional to ϕ_ℓ. We can define it as

$$u_\ell(E,r) = \phi_\ell(E,r)\left[f_\ell^{(\text{in})}(E)\right]^{-1}\,, \tag{4.4}$$

then the plane wave is

$$\psi_{\vec{k}}(\vec{r}\,) = \frac{1}{\sqrt{2\pi}kr}\sum_{\ell m} i^\ell u_\ell(E,r) Y^*_{\ell m}(\hat{\vec{k}}\,) Y_{\ell m}(\hat{\vec{r}}\,). \tag{4.5}$$

Taking into account Eq. (A.26) of Appendix A and the relation

$$\int_0^\infty j_\ell(k'r) j_\ell(kr) dr = \frac{\pi}{2}\delta(k'-k)\,, \tag{4.6}$$

it is not difficult to prove that the physical solution, defined by Eq. (4.4), is normalized as

$$\int_0^\infty u_\ell(E',r) u_\ell(E,r) dr = \frac{\pi}{2}\delta(k'-k) \tag{4.7}$$

$$= \frac{\pi\hbar^2 k}{2\mu}\delta(E'-E) \tag{4.8}$$

where we used the relation

$$\delta(E' - E) = \frac{\mu}{\hbar^2 k}\delta(k' - k)\,. \tag{4.9}$$

The normalization of the scattering wave function as well as its partial-wave decomposition should not depend on the choice of the potential. Therefore when the potential is not zero, the function $\psi_{\vec{k}}(\vec{r}\,)$ should obey the same normalization condition (4.1) and should be expressed via the physical radial functions, u_ℓ, in the same way as it is given by Eq. (4.5). In other words, this partial-wave expansion and the normalization (4.7), (4.8) remain valid for any potential.

The reason why we introduced the coefficient $1/f_\ell^{(\mathrm{in})}(E)$ in the physical wave function (4.4) is that the scattering wave function $\psi_{\vec{k}}(\vec{r}\,)$ defined by (4.5) has the correct asymptotic behaviour, i.e. at large r it consists of the properly normalized incident plane wave and a scattered spherical wave. This can be shown as follows.

As we proved in Sect. 3.3.7, there are no spectral points at real positive energies (by the way, this is why we can place $f_\ell^{(\mathrm{in})}$ in the denominator). Therefore the regular solution has both the incoming and outgoing waves at large r,

$$\begin{aligned}\phi_\ell(E,r) &\underset{r\to\infty}{\longrightarrow} h_\ell^{(-)}(kr)f_\ell^{(\mathrm{in})}(E) + h_\ell^{(+)}(kr)f_\ell^{(\mathrm{out})}(E)\\ &\underset{r\to\infty}{\longrightarrow} \left[ie^{-i(kr-\ell\pi/2)}\right]f_\ell^{(\mathrm{in})}(E) + \left[-ie^{+i(kr-\ell\pi/2)}\right]f_\ell^{(\mathrm{out})}(E)\,. \end{aligned} \tag{4.10}$$

Substituting this into Eqs. (4.4, 4.5), we obtain

$$\begin{aligned}
\psi_{\vec{k}}(\vec{r}\,) &\underset{r\to\infty}{\longrightarrow} \frac{1}{\sqrt{2\pi}kr}\sum_{\ell m} i^\ell\left[h_\ell^{(-)}(kr) + h_\ell^{(+)}(kr)\frac{f_\ell^{(\mathrm{out})}(E)}{f_\ell^{(\mathrm{in})}(E)}\right]Y_{\ell m}^*(\hat{\vec{k}}\,)Y_{\ell m}(\hat{\vec{r}}\,)\\
&= \frac{1}{\sqrt{2\pi}kr}\sum_{\ell m} i^\ell\left\{2j_\ell(kr) + h_\ell^{(+)}(kr)\left[\frac{f_\ell^{(\mathrm{out})}(E)}{f_\ell^{(\mathrm{in})}(E)} - 1\right]\right\}Y_{\ell m}^*(\hat{\vec{k}}\,)Y_{\ell m}(\hat{\vec{r}}\,)\\
&= \frac{\exp(i\vec{k}\vec{r}\,)}{(2\pi)^{3/2}} + \frac{2i}{\sqrt{2\pi}r}\sum_{\ell m} i^\ell h_\ell^{(+)}(kr) f_\ell(E)Y_{\ell m}^*(\hat{\vec{k}}\,)Y_{\ell m}(\hat{\vec{r}}\,)\\
&\underset{r\to\infty}{\longrightarrow} \frac{\exp(i\vec{k}\vec{r}\,)}{(2\pi)^{3/2}} + \frac{2i}{\sqrt{2\pi}r}\sum_{\ell m} i^\ell\left[-ie^{+i(kr-\ell\pi/2)}\right]f_\ell(E)Y_{\ell m}^*(\hat{\vec{k}}\,)Y_{\ell m}(\hat{\vec{r}}\,)\\
&= \frac{\exp(i\vec{k}\vec{r}\,)}{(2\pi)^{3/2}} + \sqrt{\frac{2}{\pi}}\frac{e^{ikr}}{r}\sum_{\ell m} f_\ell(E)Y_{\ell m}^*(\hat{\vec{k}}\,)Y_{\ell m}(\hat{\vec{r}}\,)
\end{aligned}$$

$$= \frac{1}{(2\pi)^{3/2}}\left[e^{i\vec{k}\vec{r}} + \frac{e^{ikr}}{r}\sum_{\ell}(2\ell+1)f_\ell(E)P_\ell(\cos\theta)\right]$$

$$= \frac{1}{(2\pi)^{3/2}}\left[e^{i\vec{k}\vec{r}} + \frac{e^{ikr}}{r}\mathsf{f}(\hat{\vec{r}} \leftarrow \hat{\vec{k}}, E)\right], \tag{4.11}$$

where we introduced the so-called partial-wave scattering amplitude

$$\boxed{f_\ell(E) = \frac{f_\ell^{(\mathrm{out})}(E) - f_\ell^{(\mathrm{in})}(E)}{2ikf_\ell^{(\mathrm{in})}(E)}} \tag{4.12}$$

and the full scattering amplitude

$$\mathsf{f}(\hat{\vec{r}} \leftarrow \hat{\vec{k}}, E) = \sum_{\ell}(2\ell+1)f_\ell(E)P_\ell(\cos\theta)\,, \tag{4.13}$$

depending on the direction of vector $\vec{r}$, i.e. on the angle θ between $\vec{k}$ and $\vec{r}$. The first term of Eq. (4.11) represents the undisturbed incident wave, while the second term describes the scattered wave moving in all directions with the amplitude $\mathsf{f}(\hat{\vec{r}} \leftarrow \hat{\vec{k}}, E)$. The value of this amplitude varies with the angle and thus determines the scattering pattern, i.e. how intensive is the scattered flux in various directions. Quantitatively, this is expressed in the form of the so-called differential scattering cross section, which is according to Eq. (1.58) is equal to $|\mathsf{f}(\hat{\vec{r}} \leftarrow \hat{\vec{k}}, E)|^2$.

Such a pattern is a specific characteristic of a potential. It depends on the incident energy and is the main goal of solving the so-called direct scattering problem (when the potential is given). As we see, to solve this problem means to calculate the amplitude which can be obtained if we know the Jost functions.

Substituting $f_\ell^{(\mathrm{in/out})}$ from Eq. (2.65) in Eq. (4.12), we can relate the partial-wave amplitude directly to the physical wave function (4.4),

$$\boxed{f_\ell(E) = -\frac{1}{2k^2}\int_0^\infty j_\ell(kr)V(r)u_\ell(E,r)\,dr} \tag{4.14}$$

* * *

As we see, the differential equations (2.21) or equivalent to them integral equations that determine the functions $F_\ell^{(\mathrm{in/out})}(E,r)$ offer an elegant and universal way of solving all three (seemingly different) types of the problems of quantum mechanics, namely the bound, resonant, and scattering-state problems. If we can solve (in most cases numerically) these equations with complex values of E, then just moving the point E on the Riemann surface to an appropriate region, we can either locate the spectral points or calculate the scattering characteristics. In doing this, we do not need to change the boundary conditions (2.24), which are the same for all three types of problems.

Another important fact is that when locating the spectral states or calculating the scattering amplitude, we also obtain (at the same time) the corresponding functions $F_\ell^{(\text{in/out})}(E,r)$ at all points $r \in [0,\infty)$ and thus through the ansatz (2.12) obtain the wave function itself. This wave function always has the correct asymptotic behaviour given analytically in the form of the Riccati–Hankel functions which are embedded in it from the very beginning. The use of Eqs. (2.21) instead of the equivalent Schrödinger equation is therefore much more convenient and allows us to solve the quantum mechanical problems in a unified way.

Looking at Eqs. (4.11) and (4.12), we see that independently of the boundary conditions, i.e. independently of the value of the constant on the right hand side of Eq. (2.24), the scattering wave function in the form (4.5) is normalized according to Eq. (4.1). Indeed, firstly, any common (normalization) factor of $f_\ell^{(\text{in})}(E)$ and $f_\ell^{(\text{out})}(E)$ cancels out in the ratio (4.12). Secondly, the normalization is the same for any potential and in the case of $V(r) \equiv 0$ the Jost functions are identical $f_\ell^{(\text{in})}(E) \equiv f_\ell^{(\text{out})}(E)$ which means that $f_\ell(E) \equiv 0$ and the second term in Eq. (4.11) vanishes. The remaining first term is apparently normalized correctly. This fact once again shows that the abandoning of the fixed normalization of the Jost functions does not cause any inconveniences.

4.2 S-Matrix

Since the flux of the particles is conserving, the absolute values of the amplitudes $f_\ell^{(\text{in/out})}(E)$ of the incoming and outgoing waves are the same,

$$|f_\ell^{(\text{in})}(E)| = |f_\ell^{(\text{out})}(E)| , \qquad E > 0 . \tag{4.15}$$

This is consistent with the diagonal symmetry property given by Eq. (3.15). The interaction responsible for the scattering can only change the phase of the spherical wave amplitude. In accordance with Eq. (3.15), at each point E on the real positive axis the Jost functions can be written as

$$f_\ell^{(\text{in})}(E) = |f_\ell^{(\text{in})}(E)| e^{-i\delta_\ell(E)} , \tag{4.16}$$

$$f_\ell^{(\text{out})}(E) = |f_\ell^{(\text{out})}(E)| e^{+i\delta_\ell(E)} , \qquad E > 0 \tag{4.17}$$

in terms of the real function $\delta_\ell(E)$, which is called the phase shift.[1] Therefore the incoming wave amplitude is transformed into the outgoing amplitude by a function $S_\ell(E)$, such that

$$f_\ell^{(\text{out})}(E) = S_\ell(E) f_\ell^{(\text{in})}(E) \tag{4.18}$$

and whose absolute value is one,

$$\boxed{S_\ell(E) = e^{i2\delta_\ell(E)}} \qquad E > 0 \tag{4.19}$$

The transformation function $S_\ell(E)$ is called the S-matrix. In the simplest single-channel case this is a trivial 1×1 matrix consisting of just one element, but we will call it matrix anyway, following the tradition and bearing in mind that in more complicated problems (for example, involving particles with spins or non-central potentials) this function becomes a true matrix with the dimensions greater than 1 (see further sections).

Comparing Eqs. (4.12) and (4.18), we see that there is a simple relation between the scattering amplitude and the S-matrix,

$$\boxed{f_\ell(E) = \frac{S_\ell(E) - 1}{2ik}} \tag{4.20}$$

Therefore the differential scattering cross section is given by

$$\begin{aligned} \frac{d\sigma}{d\Omega} &= \left|\sum_\ell (2\ell+1) f_\ell(E) P_\ell(\cos\theta)\right|^2 \\ &= \frac{1}{4k^2}\left|\sum_\ell (2\ell+1)[S_\ell(E) - 1] P_\ell(\cos\theta)\right|^2 . \end{aligned} \tag{4.21}$$

If the differential cross section is integrated over all directions, we obtain the total scattering cross section, $\sigma(E)$, that has the physical meaning of the fraction of the geometrical cross section of the incoming beam, such that all the particles passing through it are disturbed by the potential, i.e. deflected from their initial direction. Using the orthogonality condition for the Legendre polynomials,

$$\int_0^\pi P_\ell(\cos\theta) P_{\ell'}(\cos\theta) \sin\theta \, d\theta = \frac{2}{2\ell+1}\delta_{\ell\ell'} ,$$

[1] The choice of the sign of the phase shift is of course a matter of convention.

when integrating (4.21) over $d\Omega$, we obtain

$$\sigma(E) = \frac{\pi}{k^2} \sum_{\ell} (2\ell + 1) \, |S_\ell(E) - 1|^2 \, . \qquad (4.22)$$

This total cross section is the sum,

$$\sigma(E) = \sum_{\ell} \sigma_\ell(E) \, ,$$

of the partial cross sections

$$\sigma_\ell(E) = \frac{\pi}{k^2} (2\ell + 1) \, |S_\ell(E) - 1|^2 \, . \qquad (4.23)$$

According to (4.19), for real positive energies the S-matrix is on the unitary circle, i.e. $|S_\ell(E)| = 1$. The partial cross section therefore attains its maximum when $S_\ell(E) = -1$. This means that the total cross section in each partial wave has its natural limit,

$$\sigma_\ell(E) \leqslant \frac{4\pi}{k^2} (2\ell + 1) \, . \qquad (4.24)$$

4.3 Phase Shift

The physical meaning of the phase shift introduced via Eqs. (4.16), (4.17) can be clarified if we compare the asymptotics of the incident plane wave and that of the wave function distorted by the potential. In terms of the partial wave expansion (4.5) the radial part of the plane wave behaves as (see Appendix D):

$$u_\ell(E, r) = j_\ell(kr) \underset{r \to \infty}{\longrightarrow} \sin\left(kr - \frac{\ell\pi}{2}\right) , \qquad V(r) \equiv 0 \, . \qquad (4.25)$$

The distorted wave asymptotics is given by Eq. (4.10). Replacing the Jost functions with the expressions (4.16), (4.17), we obtain

$$u_\ell(E, r) \underset{r \to \infty}{\longrightarrow} i |f_\ell^{(\text{in})}(E)| \left[e^{-i(kr - \ell\pi/2 + \delta_\ell)} - e^{+i(kr - \ell\pi/2 + \delta_\ell)} \right]$$

$$\sim \sin\left[kr - \frac{\ell\pi}{2} + \delta_\ell(E)\right] , \qquad V(r) \not\equiv 0 \, . \qquad (4.26)$$

Imagine a situation when you can switch the potential on and off. The effect of turning it on would be the change in the asymptotics of the radial wave function from (4.25) to (4.26), i.e. in the shift of its phase, hence the term phase shift.

It is not difficult to re-write the formulae for the scattering amplitude and cross section in terms of the phase shift:

$$f_\ell(E) = \frac{e^{i\delta_\ell(E)} \sin \delta_\ell(E)}{k} , \tag{4.27}$$

$$\mathsf{f}(\hat{\vec{r}} \leftarrow \hat{\vec{k}}, E) = \frac{1}{k} \sum_\ell (2\ell + 1) e^{i\delta_\ell(k)} \sin \delta_\ell(k) P_\ell(\cos\theta) , \tag{4.28}$$

$$\sigma_\ell(E) = 4\pi(2\ell + 1) \frac{\sin^2 \delta_\ell(E)}{k^2} , \tag{4.29}$$

$$\sigma(E) = \frac{4\pi}{k^2} \sum_\ell (2\ell + 1) \sin^2 \delta_\ell(E) . \tag{4.30}$$

In some cases these formulae are very convenient in derivations and proving certain relations. For example, consider the amplitude (4.28) for forward scattering, i.e. with $\theta = 0$. Since $P_\ell(1) = 1$ for all ℓ, we have

$$\mathsf{f}(\hat{\vec{k}} \leftarrow \hat{\vec{k}}, E) = \frac{1}{k} \sum_\ell (2\ell + 1)(\cos \delta_\ell + i \sin \delta_\ell) \sin \delta_\ell ,$$

hence

$$\mathrm{Im}\, \mathsf{f}(\hat{\vec{k}} \leftarrow \hat{\vec{k}}, E) = \frac{1}{k} \sum_\ell (2\ell + 1) \sin^2 \delta_\ell(E) .$$

Comparing this with Eq. (4.30), we obtain the so-called optical theorem, i.e.

$$\boxed{\sigma(E) = \frac{4\pi}{k} \mathrm{Im}\, \mathsf{f}(\hat{\vec{k}} \leftarrow \hat{\vec{k}}, E)} \tag{4.31}$$

It should be mentioned that the phase shift is not a directly observable quantity. In experiments, we just count the particles scattered in different directions and thus obtain the scattering pattern, i.e. the differential cross section $d\sigma/d\Omega$. In other words, the experimental data can give us only the absolute value of the amplitude $\mathsf{f}(\hat{\vec{r}} \leftarrow \hat{\vec{k}}, E)$. The optical theorem relation (4.31) is the only exception of this rule.

The partial-wave amplitudes and the phase shifts are deduced from experimental data using the so-called phase shift analysis [18]. It is a cumbersome numerical procedure of fitting the cross section measured at a number of angles and energies. However, in spite of this difficulty of extracting them from experimental data, the phase shifts are very convenient instrument of the theoretical analysis.

4.4 Resonant Scattering

Analysis of the behaviour of the cross section near a resonance energy is an example where the use of the phase shifts is very helpful and brings up clarity.

Let $E_0 = E_r - i\Gamma/2$ (where $E_r > 0,\ \Gamma > 0$) be a point on the unphysical sheet of the energy surface not far from the real axis and such that for certain angular momentum ℓ the Jost function at E_0 is zero,

$$f_\ell^{(\mathrm{in})}(E_0) = 0\ .$$

We further assume that this is a simple and well isolated zero, i.e. there is no coincidence or proximity of two or more resonances in the region around E_0. Since the Jost function is analytic, it can be expanded in Taylor series near E_0. The simplicity of the zero means that such an expansion begins from the linear term:[2]

$$f_\ell^{(\mathrm{in})}(E) = \dot{f}_\ell^{(\mathrm{in})}(E_0)(E - E_0) + \frac{1}{2}\ddot{f}_\ell^{(\mathrm{in})}(E_0)(E - E_0)^2 + \dots\ .$$

For the purpose of qualitative estimating, we can retain only the first term in this series,

$$f_\ell^{(\mathrm{in})}(E) \approx \dot{f}_\ell^{(\mathrm{in})}(E_0)(E - E_0)\ . \tag{4.32}$$

On the segment of the real axis near the point E_0, we therefore have

$$|f_\ell^{(\mathrm{in})}(E)|e^{-i\delta_\ell(E)} \approx |\dot{f}_\ell^{(\mathrm{in})}(E_0)|e^{i\arg\dot{f}_\ell^{(\mathrm{in})}}|(E - E_0)|e^{i\arg(E-E_0)}\ ,$$

which means that

$$\delta_\ell(E) \approx -\arg\left[\dot{f}_\ell^{(\mathrm{in})}(E_0)\right] - \arg(E - E_0)$$

$$= \delta_\ell^{\mathrm{bg}} + \delta_\ell^{\mathrm{res}}(E)\ , \tag{4.33}$$

where $\delta_\ell^{\mathrm{bg}} = \mathrm{const}$ is a background phase shift and $\delta_\ell^{\mathrm{res}}(E) = -\arg(E - E_0)$ is a rapidly varying resonant part of the phase shift.

As is shown in Fig. 4.1, the angle $\delta_\ell^{\mathrm{res}}(E)$ is the polar angle of the point E relative to E_0 and taken with the minus sign.

---

[2] The dot over f means the derivative with respect to E.

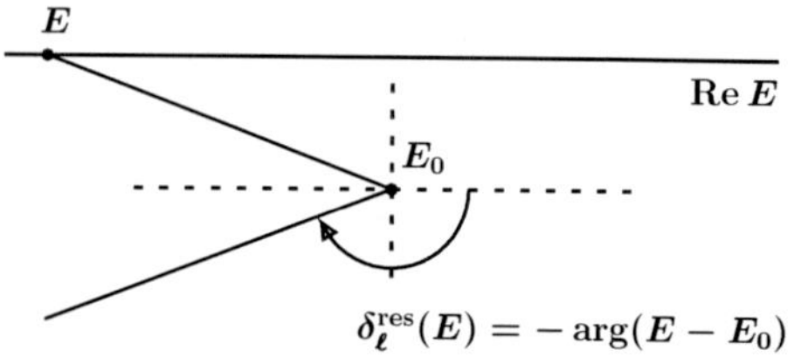

Fig. 4.1 Polar angle of the point E relative to E_0, measured in the clockwise direction (because of the minus sign)

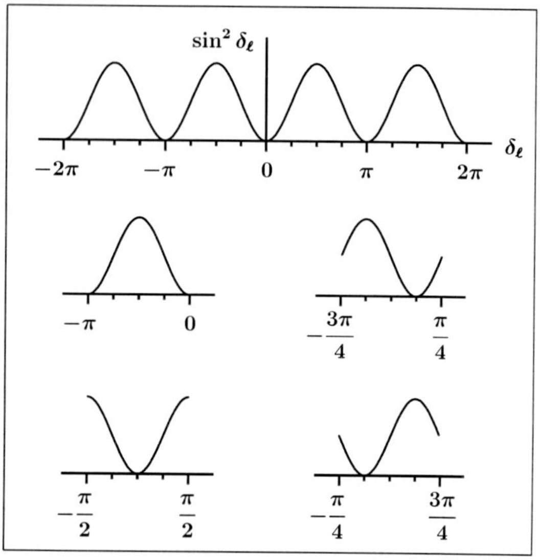

Fig. 4.2 The factor determining the dependence of the cross section σ_ℓ on the phase shift δ_ℓ can be any fragment of the length π that is cut from the curve shown at the top of this figure. Examples of such fragments are given in the middle and at the bottom of the figure

If E_0 is close to the real axis and the point E is far on the left hand side, the angle $\delta_\ell^{\rm res}(E)$ is almost $-\pi$. If the point E moves to the right, this angle increases and almost reaches zero when the point is far to the right. Therefore the resonant part of the phase shift changes from $-\pi$ to zero (i.e. increases by π) when the scattering energy passes by the resonance spectral point. The closer this point is to the real axis, the more abrupt is the change. The energy dependence of the partial wave cross section (4.29) should reflect this change. And if the given partial wave is not overshadowed by the other waves, the effect of the phase shift variation should be seen in the energy dependence of the observable cross section (4.30).

The effect of the increase of δ_ℓ by π depends not only on the rapidity of the change but also on the value of $\delta_\ell^{\rm bg}$. Indeed, the energy dependence of the cross section (4.29) near a resonance is determined by the factor $\sin^2\delta_\ell(E)$. As is seen in Fig. 4.2, starting from different points and cutting from the curve a segment of the length π, we can obtain different types of variations. In Fig. 4.2, there are shown the cuts starting from the angles $-\pi$, $-3\pi/4$, $-\pi/2$, and $-\pi/4$. Of course, all intermediate values are also possible. As is seen, a quantum resonance is not always a peak in the cross section. It can be a deep or a zig-zag. Its experimental manifestation can be even further complicated due to the fact that in the observable cross section (4.30) several partial waves may sum up to a fancy curve.

4.5 Breit-Wigner Resonances

Using the first-order approximation (4.32) for the Jost function near a resonance spectral point $E_0 = E_r - i\Gamma/2$ and the symmetry property, $f_\ell^{(\mathrm{out})} = f_\ell^{(\mathrm{in})*}$, which is valid for real positive energies, we can approximate the S matrix on a segment of the real axis next to E_0, as

$$S_\ell(E) = \frac{f_\ell^{(\mathrm{in})\,*}(E)}{f_\ell^{(\mathrm{in})}(E)}$$

$$\approx \frac{\dot{f}_\ell^{(\mathrm{in})\,*}(E_0)(E - E_0)^*}{\dot{f}_\ell^{(\mathrm{in})}(E_0)(E - E_0)}$$

$$= e^{-2i\delta_\ell^{\mathrm{bg}}} \frac{E - E_r - i\Gamma/2}{E - E_r + i\Gamma/2}\,.$$

Then the partial-wave cross section (4.23) is

$$\sigma_\ell(E) = \frac{\pi}{k^2}(2\ell + 1)\left| e^{-2i\delta_\ell^{\mathrm{bg}}} \frac{E - E_r - i\Gamma/2}{E - E_r + i\Gamma/2} - 1 \right|^2 . \tag{4.34}$$

In the particular case when $\delta_\ell^{\mathrm{bg}} = 0$, this equation is significantly simplified,

$$\boxed{\sigma_\ell(E) = \frac{\pi\hbar^2(2\ell + 1)}{2\mu E} \frac{\Gamma^2}{(E - E_r)^2 + \Gamma^2/4}} \tag{4.35}$$

This is well-known Breit-Wigner formula which was used at the early stages of nuclear physics development (an still is used) to parametrize well pronounced resonance peaks in the cross sections of nuclear reactions. The Breit-Wigner factor

$$W(E) = \frac{\Gamma^2}{(E - E_r)^2 + \Gamma^2/4} \tag{4.36}$$

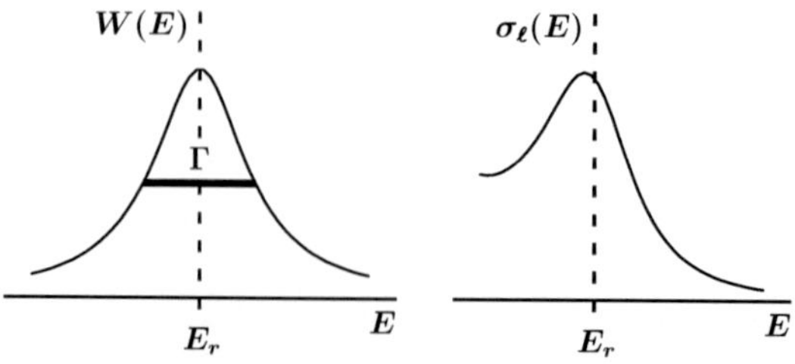

Fig. 4.3 Symmetric (left) Breit-Wigner factor (4.36) and non-symmetric (right) cross section (4.35) skewed due to the factor $1/E$

is a symmetric bell-shaped function with Γ being its width at the half height (see Fig. 4.3). The cross section (4.35), however, is skewed to the left because of the factor $1/E$.

4.6 Analytic Properties of the S-Matrix

When considering the scattering states, we defined the S matrix via Eq. (4.18) for real positive energies. Since both $f_\ell^{(\text{in})}(E)$ and $f_\ell^{(\text{out})}(E)$ are analytic functions, we can extend that definition to the whole Riemann surface of complex energies. In other words, the S-matrix defined as

$$\boxed{S_\ell(E) = f_\ell^{(\text{out})}(E)\left[f_\ell^{(\text{in})}(E)\right]^{-1}} \tag{4.37}$$

can be analytically continued from the real positive axis to all other points of the E-surface, where the Jost functions are analytic. It is clear that the S-matrix is a meromorphic function of E, i.e. analytic everywhere except for isolated poles. The poles are those points where the Jost function $f_\ell^{(\text{in})}(E)$ is zero, i.e. the spectral points. The definition (4.37) is the most general and, as we will see in subsequent chapters, it remains the same for the multi-channel problems.

4.6.1 Symmetry of the S-Matrix

Using the symmetry properties of the Jost functions, discussed in Sect. 3.1 and summarized in Figs. 3.5 and 3.6, we can express the S-matrix in terms of either only $f_\ell^{(\text{in})}(E)$ or only $f_\ell^{(\text{out})}(E)$. For example, Eqs. (3.11) and (3.12) allow us to write

$$S_\ell(E) = f_\ell^{(\text{in})}(e^{2\pi i}E)\left[f_\ell^{(\text{in})}(E)\right]^{-1} . \tag{4.38}$$

In this form,[3] the S-matrix is determined by $f_\ell^{(\mathrm{in})}$ taken at the two opposite points on the two different sheets of the E-surface. It is not difficult to write down some other expressions for the S-matrix in terms of the eight functions, symmetry relations among which are shown in Figs. 3.5 and 3.6. It seems, however, that the clear and unambiguous formula (4.37) is sufficient for all our purposes. When the necessity arises, we can always replace in it either the numerator or the denominator with what is needed and allowed by the symmetry properties.

According to the diagonal symmetry relation (3.15), for real positive energies the Jost functions $f_\ell^{(\mathrm{in})}$ and $f_\ell^{(\mathrm{out})}$ are complex conjugates of each other. Therefore

$$S_\ell^*(E) = S_\ell^{-1}(E)\,, \qquad \text{if} \quad E > 0\,, \tag{4.39}$$

i.e. the S-matrix is unitary at scattering energies. In the simple single-channel case, where no matrix transposition is involved, this means that the absolute value of the S-matrix is 1. As we will see in subsequent chapters, this property is more general and means that

$$S_\ell^\dagger(E)S_\ell(E) = S_\ell(E)S_\ell^\dagger(E) = 1\,, \qquad \text{if} \quad E > 0\,, \tag{4.40}$$

where † denotes the Hermitian conjugation. Physically, the unitarity of the S-matrix reflects the fact that the flux of particles is conserving.

At complex energies, there are also some useful relations that all can be obtained with the help of Figs. 3.5 and 3.6. Applying Eqs. (3.11) and (3.12), we immediately see that "jumping" from one sheet to another results in changing the S_ℓ to S_ℓ^{-1},

$$S_\ell(e^{2\pi i}E) = S_\ell^{-1}(E)\,. \tag{4.41}$$

Analysing the relations (3.16) and (3.15), we conclude that moving to the other side of the real axis on the same sheet of the E-surface results in replacing the S_ℓ with its complex conjugate,

$$S_\ell(e^{2\pi i}E^*) = S_\ell^*(E)\,, \tag{4.42}$$

and the transition from E to E^* makes both the inversion and complex conjugation,

$$S_\ell(E^*) = \left[S_\ell^{-1}(E)\right]^*\,. \tag{4.43}$$

Similarly to Figs. 3.5 and 3.6, the symmetry properties of the S-matrix are summarized in Fig. 4.4. Very often such pictorial representation of the symmetries is more clear and helpful than the corresponding formulae. The same as with the Jost

[3] You may already saw similar formula which in some other books is written as $s_\ell(k) = f_\ell(-k)/f_\ell(k)$.

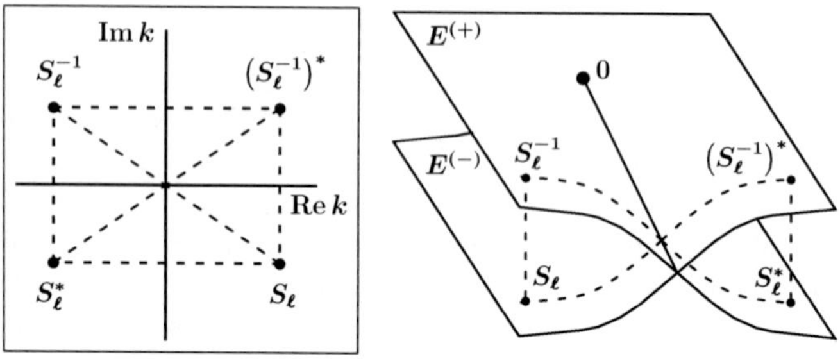

Fig. 4.4 Symmetry properties of the S-matrix on the k-plane (left) and E-surface (right). The dashed lines connect the points at which the values indicated next to them are identical

functions, thanks to the symmetries, the S-matrix can be obtained on the whole k-plane, if we manage to calculate it just in one quadrant.

4.6.2 Spectral Expansion of the S-Matrix

As we have seen, the S-matrix is a meromorphic function of E. Its poles coincide with zeros of the Jost function and thus correspond to spectral points. Let us assume that the spectral points do not coincide with each other and are well isolated. Can we somehow separate the individual contribution of each spectral point to the S-matrix?

Yes, in principle, this can be done on the basis of the Mittag-Leffler theorem[4] that offers a way to decompose the S-matrix in a sum of terms representing all its poles and the background. The theorem itself is rather general. In plain words, it says that a meromorphic function can be expanded in a sum of an entire function and a series of its principal parts at all the poles, i.e.

$$S_\ell(E) = s_\ell(E) + \sum_{n=1}^{\infty} \frac{\text{Res}\,[S_\ell(E_n)]}{E - E_n}\,, \tag{4.44}$$

where $s_\ell(E)$ has no poles (entire function) and E_n are the spectral points.

In practical calculations, we usually need to explicitly separate the contribution not of all the poles but just of few of them. This can be done using the residue theorem. Consider a contour shown in Fig. 4.5, which encloses N poles of the S-matrix. Let E be a point inside the contour (for example, on the real axis). Applying the residue theorem to the function $S_\ell(\zeta)/(\zeta - E)$, we obtain

$$\oint \frac{S_\ell(\zeta)}{\zeta - E} d\zeta = 2\pi i\, S_\ell(E) + 2\pi i \sum_{n=1}^{N} \frac{\text{Res}[S(E_n)]}{E_n - E}\,. \tag{4.45}$$

[4] See, for example, pages 383–386 in the book [19].

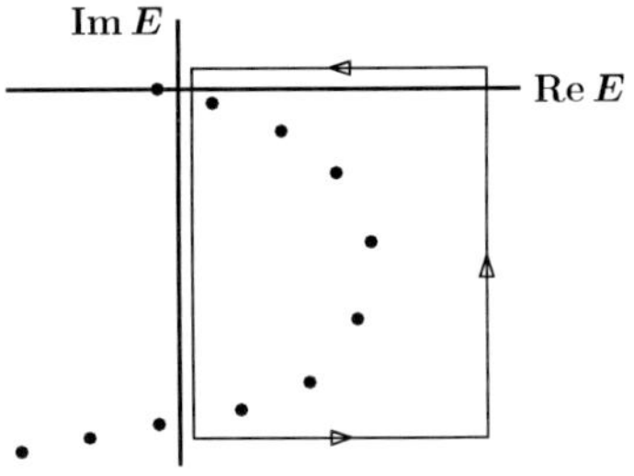

Fig. 4.5 Contour enclosing N selected poles of the S-matrix, used in the expansion (4.46)

Therefore in the expansion that we are looking for,

$$S_\ell(E) = \sum_{n=1}^{N} \frac{\text{Res}[S(E_n)]}{E - E_n} + \frac{1}{2\pi i} \oint \frac{S_\ell(\zeta)}{\zeta - E} d\zeta\ , \tag{4.46}$$

the N selected poles are represented explicitly while the rest of them as well as the background (the Mittag-Leffler's entire function) are taken into account via the contour integral.

Since we know that each spectral point is represented by a singular term in the function (4.44), we can assess its contribution by comparing the observable quantities that are deduced from the full S-matrix and from the truncated one,

$$\tilde{S}_\ell(E) = S_\ell(E) - \frac{\text{Res}\,[S_\ell(E_n)]}{E - E_n}\ , \tag{4.47}$$

from which the corresponding term is excluded.

The use of Eqs. (4.46) and (4.47) in practical calculations would require methods for calculating all the quantities on their right hand sides. As soon as we are able to calculate the S-matrix at complex energies (via the Jost functions, for example) the numerical evaluation of the contour integral is straightforward. The residues can be obtained as is described next.

4.6.3 Residues of the S-Matrix and the ANC

If the spectral points are isolated, then within a small circle around each pole, E_n, the function $S_\ell(E)$ can be expanded in the Laurent series. Since the zeros of the Jost function at these points are simple, this series begins with the term $\sim (E - E_n)^{-1}$,

$$S_\ell(E) = \frac{R_\ell}{E - E_n} + C_0 + C_1(E - E_n) + C_2(E - E_n)^2 + \dots\ , \tag{4.48}$$

where R_ℓ is the residue of the S-matrix at the pole,

$$R_\ell = \text{Res}\,[S_\ell, E_n]\ .$$

The first term of the Laurent series (4.48) is called the principal part of $S_\ell(E)$ at the pole. If E is very close to E_n, we can retain only this leading term,

$$S_\ell(E) \approx \frac{R_\ell}{E - E_n}\ . \tag{4.49}$$

When $f_\ell^{(\text{in})}(E)$ and $f_\ell^{(\text{out})}(E)$ are already obtained, the residue of the S-matrix can be found by calculating the derivative of the Jost function. Indeed, near a spectral point E_n, we have

$$S_\ell(E) = \frac{f_\ell^{(\text{out})}(E)}{f_\ell^{(\text{in})}(E)} \approx \frac{f_\ell^{(\text{out})}(E)}{\dot{f}_\ell^{(\text{in})}(E_n)(E - E_n)}\ , \tag{4.50}$$

from which it follows that

$$\text{Res}\,[S_\ell, E_n] = \frac{f_\ell^{(\text{out})}(E_n)}{\dot{f}_\ell^{(\text{in})}(E_n)}\ . \tag{4.51}$$

In most cases the derivative $\dot{f}_\ell^{(\text{in})}(E_n)$ can be calculated numerically using the central difference approximation

$$\frac{d}{dE} f_\ell^{(\text{in})}(E) \approx \frac{f_\ell^{(\text{in})}(E+\epsilon) - f_\ell^{(\text{in})}(E-\epsilon)}{2\epsilon}\ , \qquad \epsilon \to 0\ . \tag{4.52}$$

If E_n is a bound or resonant state spectral point, we can relate the residue of the S-matrix at this point to the asymptotic normalization constant (ANC), $\mathcal{A}_\ell$, that determines the behaviour of the wave function (3.53) at large distances. Indeed, substituting the derivative (3.56) into Eq. (4.51), we obtain

$$\boxed{\text{Res}\,[S_\ell, E_n] = \frac{(-1)^{\ell+1} i\hbar^2 k_n}{\mu} \mathcal{A}_\ell^2} \tag{4.53}$$

where $k_n = \sqrt{2\mu E_n/\hbar^2}$.

In some cases it is more convenient to consider the S-matrix as a function of the momentum instead of the energy. Near the point k_n in the momentum plane, the principal part of the S-matrix is

$$S_\ell(E) \approx \frac{Q_\ell}{k - k_n} , \tag{4.54}$$

where Q_ℓ is the corresponding residue. It is easy to relate Q_ℓ and R_ℓ. Since

$$\frac{R_\ell}{E - E_n} = \frac{2\mu}{\hbar^2} \frac{R_\ell}{(k - k_n)(k + k_n)} \underset{k \to k_n}{\longrightarrow} \frac{\mu}{\hbar^2 k_n} \cdot \frac{R_\ell}{k - k_n} ,$$

we obtain

$$Q_\ell = \frac{\mu}{\hbar^2 k_n} R_\ell = (-1)^{\ell+1} i \mathcal{A}_\ell^2 . \tag{4.55}$$

4.6.4 Argand Plot

Graphical representation of complex numbers $x + iy = \rho e^{i\phi}$ by points on the complex plane, either in the Cartesian coordinates (x, y) or in the equivalent polar coordinates (ρ, ϕ), is called the Argand diagram. Such a diagram is also useful for depicting a complex valued function depending on a real parameter. For each value of this parameter, the function value corresponds to a point in the complex plane. If the function is continuous, the set of such points forms a curve.

In the scattering theory, the term Argand diagram (or more often Argand plot) has a specific meaning, namely the curve in the complex plane, along which the S-matrix moves when the (real) collision energy increases. Since for real positive energies the S-matrix is unitary,

$$S_\ell(E) = e^{2i\delta_\ell(E)} ,$$

this curve is just a circle of unit radius. Due to the obvious reason, this circle is called the unitary circle.

It is clear that the direction and "speed" with which the point $S_\ell(E)$ moves along the unitary circle, is determined by the character of the energy dependence of the phase shift. When $\delta_\ell(E)$ decreases with the point E moving to the right along the real axis, the S-matrix moves in the clockwise direction along the unitary circle. The increasing $\delta_\ell(E)$ causes the S-matrix to move in the anti-clockwise direction.

This may look trivial and dull until we encounter a resonance. As we have seen in Sect. 4.3, in a vicinity of a resonance energy (i.e. when the point E passes by a resonance pole which is not far from the real axis), the phase shift rapidly increases in π. As a result, the point $S_\ell(E)$ makes a full circle in the anti-clockwise direction.

The sharper the resonance is, the faster the S-matrix moves. Therefore an anti-clockwise circling of the S-matrix is an indication that the scattering is resonant.

This is, of course, an idealized picture for an isolated and narrow resonance. If a resonance pole is not very close to the real axis, the corresponding phase shift "jump" in π is smudged over a wide energy interval. Close to both ends of such an interval the influence of the resonance pole is weak and the S-matrix does not complete a full circle. Besides this, in most cases there is an interference of several resonances. According to Eq. (4.33), near a resonance the phase shift can be split in two terms: The background and resonance terms,

$$\delta_\ell(E) = \delta_\ell^{\mathrm{bg}}(E) + \delta_\ell^{\mathrm{res}}(E) .$$

Their interplay and interference with other resonances may make the movement of the S-matrix along the unitary circle rather complicated. It "speeds up", "slows down", and changes the direction.

The general tendency of the background part of the phase shift is to decrease with the collision energy. Indeed, at high energies, where there is no resonances and only $\delta_\ell^{\mathrm{bg}}$ remains, the movement of the projectile is only weakly affected by the target, and thus the $\delta_\ell(E)$ should be small. Therefore, as a rule of thumb, we can assume the following. In the absence of resonances the S-matrix moves along the unitary circle in the clockwise direction. As soon as the energy approaches the resonance region, the S-matrix "slows down", turns back, and starts moving in the anti-clockwise direction (see Fig. 4.6). Depending on the "sharpness" of the resonance, it can complete a full circle or only part of it before turning back again.

4.6.5 Causality and Analyticity

This section is a bit formal and may be safely skipped if the reader is more inclined towards applications and calculations rather than studying fundamental principles of quantum theory.

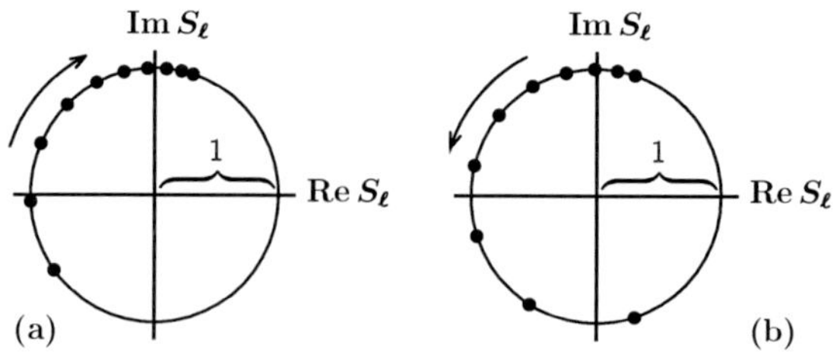

Fig. 4.6 The points shown on the unitary circle, correspond to equidistant values of the collision energy E. (**a**) The clockwise movement of the S-matrix "slows down" when E approaches the left end of a resonance region. (**b**) The anti-clockwise movement of the S-matrix "speeds up" when E approaches centre of a resonance region

In Sects. 2.2.2.2 and 2.2.5, we have established (in two different ways) analyticity of the regular solution, $\phi_\ell(E, r)$, on the whole E-surface. Then, using the integral representation (2.65) of the Jost functions in terms of the function $\phi_\ell(E, r)$, we established analyticity of the functions $f_\ell^{(\text{in})}(E)$ and $f_\ell^{(\text{out})}(E)$. Since the S-matrix is their ratio, we concluded that (see Sect. 4.6) it is a meromorphic function of the energy E. In this section, we are going to prove once more that the function $S_\ell(E)$ is analytic everywhere on the E-surface except for isolated poles. This time, we will establish analyticity of the S-matrix directly from the fundamental physical principle of causality. This principle stating that the cause must proceed the effect should be considered as one of the fundamental axioms which all physical theories are based on.

The reader should bear in mind that, from the point of view of the standards of pure mathematics, the proof given here is not rigorous. Moreover, it is not general because is based on the assumption that the potential is of a finite range, i.e. vanishes beyond some large radius R. Despite this, the proof gives the reader an idea of how certain mathematical properties of quantum theory can be derived from the fundamental principles. A more general proof that the causality implies analyticity of the S-matrix can be found in the original paper [20] by N.G. van Kampen. Since it would be far beyond the scope of this book to reproduce the rigorous proof here, what is given below is a simplified proof exploiting the notion of the Jost function.

* * *

As we know, the functions $f_\ell^{(\text{in})}(E)$ and $f_\ell^{(\text{out})}(E)$ are the amplitudes of the incoming and outgoing spherical waves with the energy E. Real physical systems of colliding particles can never have a definite value of the energy and should be described by the localized wave packets.

Let the potential vanishes beyond the point $r = R$ which is far away from the interaction centre. At this point, we can observe the freely moving incoming and outgoing spherical wave packets,[5]

$$\Phi_\ell^{(\text{in})}(R, t) = \int_0^\infty h_\ell^{(-)}(kR) f_\ell^{(\text{in})}(E) \exp\left(-i\frac{E}{\hbar}t\right) dE \tag{4.56}$$

and

$$\Phi_\ell^{(\text{out})}(R, t) = \int_0^\infty h_\ell^{(+)}(kR) f_\ell^{(\text{out})}(E) \exp\left(-i\frac{E}{\hbar}t\right) dE\,. \tag{4.57}$$

In these integrals the amplitudes $f_\ell^{(\text{in/out})}(E)$ are peaked around an average energy $\langle E \rangle$ and determine how the packets are constructed (i.e. mixed) using the states with definite energies.

[5] Actually, the packets (4.56) and (4.57) are the partial-wave components of the observable packets $\Psi^{(\text{in/out})}(\vec{R}, t)$.

It is clear that in the absence of the incoming packet there is no outgoing packet. In other words they are bounded by the cause and effect relation,

$$\boxed{\begin{array}{c}\Phi_\ell^{(\text{in})}\\ \text{cause}\end{array}} \Rightarrow \boxed{\begin{array}{c}\Phi_\ell^{(\text{out})}\\ \text{effect}\end{array}} \tag{4.58}$$

Time evolution of the system is governed by the linear Schrödinger equation and thus this relation can be represented by a linear operator. Such an operator should take into account the "cause" at all instances of times in the past. From the superposition principle it follows that the "effects" caused by all past instances are summed up. Apparently, the "effect" at time t generated by the "cause" at time t' can only depend on the time elapsed $(t - t')$ but not on absolute values of t and t'. Bearing all this in mind, we conclude that the relation (4.58) is expressed by the following integral

$$\Phi_\ell^{(\text{out})}(R,t) = \int_{-\infty}^{+\infty} G(t-t')\Phi_\ell^{(\text{in})}(R,t')dt' \,, \tag{4.59}$$

where, according to the causality principle, the integration runs only over the past instances, i.e.

$$\boxed{\text{causality}} \Rightarrow \boxed{t' < t} \Rightarrow \boxed{G(\tau) = 0\,, \quad \text{if} \quad \tau < 0} \,. \tag{4.60}$$

Substituting (4.56) into Eq. (4.59), we obtain

$$\begin{aligned}\Phi_\ell^{(\text{out})}(R,t) &= \int_0^\infty dE \int_{-\infty}^{+\infty} dt' G(t-t') h_\ell^{(-)}(kR) f_\ell^{(\text{in})}(E) \exp\left(-i\frac{E}{\hbar}t'\right)\\ &= \int_0^\infty dE\, h_\ell^{(-)}(kR) f_\ell^{(\text{in})}(E) \exp\left(-i\frac{E}{\hbar}t\right) \int_{-\infty}^{+\infty} d\tau\, G(\tau)\exp\left(i\frac{E}{\hbar}\tau\right)\\ &= \int_0^\infty g(E) h_\ell^{(-)}(kR) f_\ell^{(\text{in})}(E) \exp\left(-i\frac{E}{\hbar}t\right) dE \,,\end{aligned}$$

where $g(E)$ is the Fourier transform of $G(\tau)$,

$$g(E) = \int_{-\infty}^{+\infty} G(\tau)\exp\left(i\frac{E}{\hbar}\tau\right) d\tau \,. \tag{4.61}$$

Therefore

$$\int_0^\infty h_\ell^{(+)}(kR) f_\ell^{(\text{out})}(E) \exp\left(-i\frac{E}{\hbar}t\right) dE$$
$$= \int_0^\infty g(E) h_\ell^{(-)}(kR) f_\ell^{(\text{in})}(E) \exp\left(-i\frac{E}{\hbar}t\right) dE\ ,$$

which should be valid for any wave-packet shape. This is only possible if the integrands on both sides of the above equation are the same, i.e.

$$h_\ell^{(+)}(kR) f_\ell^{(\text{out})}(E) = g(E) h_\ell^{(-)}(kR) f_\ell^{(\text{in})}(E)\ ,$$

which means that

$$\begin{aligned} S_\ell(E) &= \frac{h_\ell^{(-)}(kR)}{h_\ell^{(+)}(kR)} g(E) \\ &= \frac{h_\ell^{(-)}(kR)}{h_\ell^{(+)}(kR)} \int_{-\infty}^{+\infty} G(\tau) \exp\left(i\frac{E}{\hbar}\tau\right) d\tau \\ &= \frac{h_\ell^{(-)}(kR)}{h_\ell^{(+)}(kR)} \int_0^\infty G(\tau) \exp\left(i\frac{E}{\hbar}\tau\right) d\tau\ , \end{aligned} \tag{4.62}$$

where the property (4.60) is used to reduce the integration interval. Since time has no special points, the function $G(t)$ must be continuous and cannot have singularities. The S-matrix is therefore analytic at those values of E where the integral (4.62) is convergent. And since the integration runs only over positive τ (an implication of the causality), this integral converges when $\operatorname{Im} E > 0$.

There are two regions of the complex E-surface, where $\operatorname{Im} E > 0$, namely the upper halves of the physical and unphysical sheets. As soon as this is established, we can infer the analyticity of the S-matrix in all other quadrants of the k-plane, using the symmetry properties presented in Fig. 4.4.

For example, in the third quadrant, the S-matrix can only be meromorphic. This is because at any point in this quadrant the value of the S-matrix is the reciprocal of its value at the symmetric point in the first quadrant, and thus it may have poles which are located symmetrically to zeros of the function $S(E)$ in the first quadrant. As we already know, there are indeed poles of the S-matrix in the third quadrant, which are mirror spectral points corresponding to resonances.

4.7 Levinson's Theorem

In Sect. 3.3.10, we obtained an estimate on the number N_ℓ of possible bound states with a given ℓ, the Bargmann's inequality (3.110). It is useful but not satisfactory when one needs the exact number of them. Here, we are going to prove the Levinson's theorem, which relates N_ℓ to the values of the phase shift at threshold and infinitely high energies.

Consider the logarithmic derivative of the Jost function

$$L(E) = \frac{d}{dE} \ln\left[f_\ell^{(\text{in})}(E) \right] = \frac{\dot{f}_\ell^{(\text{in})}(E)}{f_\ell^{(\text{in})}(E)} . \tag{4.63}$$

On the physical E-sheet (i.e. in the upper half of the k-plane), it is an analytic function of E everywhere except for the poles corresponding to the bound-state energies E_n, $n = 1, 2, \dots, N_\ell$. As we have seen in Sect. 3.3.5, these poles are simple, i.e.

$$f_\ell^{(\text{in})}(E) \underset{E \to E_n}{\longrightarrow} \dot{f}_\ell^{(\text{in})}(E_n)(E - E_n) .$$

This implies that the residues of the function $L(E)$ at all the poles are the same and equal to one,

$$\text{Res}\,[L(E), E_n] = 1 , \qquad n = 1, 2, \dots, N_\ell ,$$

hence

$$\sum_{n=1}^{N_\ell} \text{Res}\,[L(E), E_n] = N_\ell . \tag{4.64}$$

Let C be the contour enclosing all the bound-state poles as is shown in Fig. 4.7. Then, according to Cauchy's theorem, we have

$$\oint_C L(E) dE = 2\pi i \sum_{n=1}^{N_\ell} \text{Res}\,[L(E), E_n] = 2\pi i N_\ell . \tag{4.65}$$

This contour consists of a large circle whose radius goes to infinity, the small circle of the radius ε with $\varepsilon \to 0$, and the two straight lines coinciding (when $\varepsilon = 0$) with the upper and lower rims of the cut along the real axis. Since on the physical sheet the Jost function tends to a constant (1/2) when $|E| \to \infty$ (see Sect. 3.2), we have

$$L(E) \underset{|E| \to \infty}{\longrightarrow} 0$$

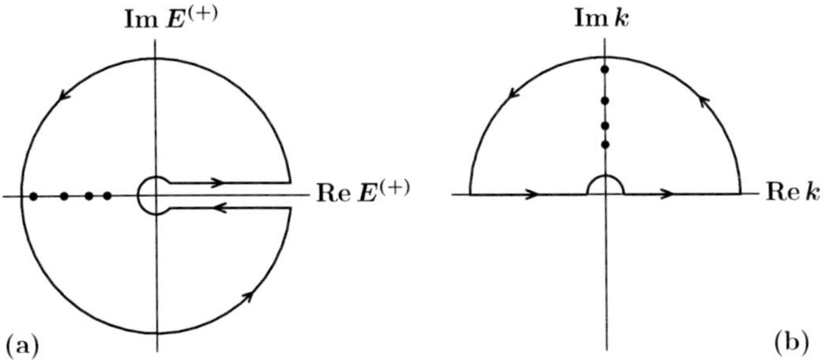

Fig. 4.7 The integration contour used to prove Levinson's theorem. Part (**a**) shows it on the physical E-sheet and part (**b**) shows the equivalent contour in the k-plane

and therefore the integral over the large circle is zero, i.e.

$$\oint_C L(E)dE = \int_{0+i\varepsilon}^{\infty+i\varepsilon} L(E)dE + \int_{\infty-i\varepsilon}^{0-i\varepsilon} L(E)dE + \oint_\varepsilon L(E)dE\,. \tag{4.66}$$

When $\varepsilon \to 0$, the first two integrals on the right hand side of this equation run along two different real axes belonging to the two sheets of the Riemann surface. According to Eq. (4.16), for real energies, we have

$$\ln\left[f_\ell^{(\text{in})}(E)\right] = \ln|f_\ell^{(\text{in})}(E)| - i\delta_\ell(E)\,,$$

and thus Eq. (3.16) taken at real energies implies that

$$\ln\left[f_\ell^{(\text{in})}(E+i\varepsilon)\right] \underset{\varepsilon\to 0}{\longrightarrow} \ln|f_\ell^{(\text{in})}(E)| - i\delta_\ell(E)\,,$$

$$\ln\left[f_\ell^{(\text{in})}(E-i\varepsilon)\right] \underset{\varepsilon\to 0}{\longrightarrow} \ln|f_\ell^{(\text{in})}(E)| + i\delta_\ell(E)\,.$$

Therefore

$$\begin{aligned}\int_{0+i\varepsilon}^{\infty+i\varepsilon} L(E)dE + \int_{\infty-i\varepsilon}^{0-i\varepsilon} L(E)dE &= \ln\left[f_\ell^{(\text{in})}(E)\right]\Big|_{0+i\varepsilon}^{\infty+i\varepsilon} + \ln\left[f_\ell^{(\text{in})}(E)\right]\Big|_{\infty-i\varepsilon}^{0-i\varepsilon} \\ &\underset{\varepsilon\to 0}{\longrightarrow} 2i\left[\delta_\ell(0) - \delta_\ell(\infty)\right]\,. \end{aligned} \tag{4.67}$$

The last integral on the right hand side of Eq. (4.66) is zero unless there is a pole of the integrand at $E = 0$, i.e. there is a threshold spectral point. In such a case, as we know (see Sect. 3.3.6), the Jost function behaves differently depending on ℓ.

According to Eq. (3.74), on the small circle, the Jost function is

$$f_\ell^{(\text{in})}(E) \underset{E\to 0}{\longrightarrow} \begin{cases} a\sqrt{E}\ , & \ell = 0\ , \\ bE\ , & \ell \geqslant 1\ . \end{cases}$$

If we parametrize the points on the circle by the polar angle ϕ, i.e. $E = \varepsilon e^{i\phi}$, then

$$\oint_\varepsilon L(E)dE = i\varepsilon \int_{2\pi}^{0} \frac{\dot{f}_\ell^{(\text{in})}(\varepsilon e^{i\phi})}{f_\ell^{(\text{in})}(\varepsilon e^{i\phi})} e^{i\phi} d\phi = \begin{cases} -i\pi\ , & \ell = 0\ , \\ -2i\pi\ , & \ell \geqslant 1\ . \end{cases} \tag{4.68}$$

Combining Eqs. (4.65)–(4.68), we obtain the Levinson's relation between the phase shifts and the number of bound states,

$$\delta_\ell(0) - \delta_\ell(\infty) = \pi N_\ell\ , \qquad \ell \geqslant 1\ , \tag{4.69}$$

where N_ℓ is the total number of bound states with a given ℓ, including a bound state at $E = 0$, if it exists. For the S-waves, the relation is the same unless there is a spectral point at threshold,

$$\delta_0(0) - \delta_0(\infty) = \begin{cases} \pi N_0\ , & f_0^{(\text{in})}(0) \neq 0\ , \\ \pi\left(N_0 + \dfrac{1}{2}\right)\ , & f_0^{(\text{in})}(0) = 0\ . \end{cases} \tag{4.70}$$

We can use one single formula (4.69) for all cases if assume that an S-wave spectral state at $E = 0$ contributes to the number N_ℓ not 1 but 1/2. Sometimes such spectral states are called semi-bound.

The absolute value of $\delta_\ell(E)$ is irrelevant (we can always add extra 2π without changing anything). What is important is how $\delta_\ell(E)$ changes (shifts) during the scattering process, hence the term phase shift. Usual convention is to assume that

$$\delta_\ell(\infty) = 0\ . \tag{4.71}$$

This is in accordance with the physical reasoning. Indeed, the higher is the collision energy, the less affected is the projectile particle, and at the infinitely high energy it is not affected at all, i.e. the phase does not shift. With the convention (4.71), the curve $\delta_\ell(E)$ begins at $E = 0$ with the value πN_ℓ,

$$\boxed{\delta_\ell(0) = \pi N_\ell} \tag{4.72}$$

Recalling that at the energies corresponding to narrow resonances the phase shift "jumps" by π (see Sect. 4.4), the reader should expect a close relation between bound and resonant states, which is indeed exists as we will see in many places throughout this book. In particular, in Sect. 7.8 it is proved the generalized Levinson's theorem, i.e. the relation similar to (4.72) but including a selected set of the resonance spectral points.

5 Complex Angular Momentum

In quantum mechanics, the allowed physical values of the orbital angular momentum are multiples of the Plank's constant $\ell\hbar$, where $\ell = 0, 1, 2, \dots$ is a non-negative integer number. In the Schrödinger equation, however, the number ℓ is just a parameter. Formally, we can allow it to assume non-integer values, and even consider solutions with complex ℓ. This is similar to what we did to the energy parameter E when considered solutions of the Schrödinger equation with E being a point on the Riemann surface.

The Riccati functions $j_\ell(kr)$, $n_\ell(kr)$, and $h_\ell^{(\pm)}(kr)$ are defined for complex values of ℓ and analytically depend on this parameter. The problem arises only when $\mathrm{Re}\,\ell = -1/2$. This is because at $\ell = -1/2$ the functions $j_\ell(kr)$ and $n_\ell(kr)$ have the same behaviour near the point $r = 0$,

$$
\begin{aligned}
j_\ell(kr) &\underset{r\to 0}{\sim} (kr)^{\ell+1} = (kr)^{1/2}\,, \qquad \ell = -1/2\,, \\
n_\ell(kr) &\underset{r\to 0}{\sim} (kr)^{-\ell} = (kr)^{1/2}\,, \qquad \ell = -1/2\,,
\end{aligned}
$$

and thus become linearly dependent. Actually, the value $\ell = -1/2$ is the border point, to the left and right of which the functions $j_\ell(kr)$ and $n_\ell(kr)$ swap their roles, i.e. swap the character of their short-distance behaviour. It is clear that there must be a way to circumvent this border point and analytically continue any linear combination of $j_\ell(kr)$ and $n_\ell(kr)$ to the other side of it. As we will see, however, it is sufficient to consider only such ℓ for which $\mathrm{Re}\,\ell > -1/2$. With this restriction, no problem arises.

The same as for integer ℓ, the pair of functions $j_\ell(kr)$ and $n_\ell(kr)$ or $h_\ell^{(\pm)}(kr)$ are linearly independent solutions of the "free" Schrödinger equation and therefore we can look for the function $\phi_\ell(E, r)$ in the form of their linear combination with r-dependent coefficients, either as

$$
\phi_\ell(E, r) = h_\ell^{(-)}(kr) F_\ell^{(\mathrm{in})}(E, r) + h_\ell^{(+)}(kr) F_\ell^{(\mathrm{out})}(E, r)\,, \tag{5.1}
$$

S. A. Rakityansky, *Jost Functions in Quantum Mechanics*,
https://doi.org/10.1007/978-3-031-07761-6_5

or as

$$\phi_\ell(E,r) = j_\ell(kr)A_\ell(E,r) - n_\ell(kr)B_\ell(E,r) . \tag{5.2}$$

Despite the fact that ℓ is complex, nothing needs to be changed in the derivations given in sections 2.2.1.1, 2.2.1.2, and 2.2.1.3. Consequently, the functions $F_\ell^{(\text{in/out})}(E,r)$ as well as $A_\ell(E,r)$ and $B_\ell(E,r)$ obey the same differential equations (2.21) and (2.32), with the same boundary conditions (2.24) and (2.33, 2.34). The integral equations of Sect. 2.2.1.3 are also the same.

Therefore for any given (complex) ℓ such that $\operatorname{Re}\ell > -1/2$, we can calculate the Jost functions $f_\ell^{(\text{in/out})}(E)$ by numerically integrating differential equations (2.21) from $r = 0$ to a sufficiently far point $r = R$, where the potential vanishes, i.e. by evaluating the limits

$$f_\ell^{(\text{in/out})}(E) = \lim_{r\to\infty} F_\ell^{(\text{in/out})}(E,r) , \tag{5.3}$$

which, in fact, are reached at $r = R$. The necessary algorithm as well as the corresponding computer routine for calculating the Riccati functions $j_\ell(kr)$, $n_\ell(kr)$, and $h_\ell^{(\pm)}(kr)$ with complex ℓ can be found, for example, in Ref. [21].

Since the boundary conditions (2.24) do not depend on ℓ, it follows from the Poincaré theorem (see Appendix G) that the functions $F_\ell^{(\text{in/out})}(E,r)$ depend on ℓ analytically. For a finite-range potential, this immediately implies that the Jost functions $f_\ell^{(\text{in/out})}(E)$ are analytic functions of ℓ. When the potential extends to infinity, the analyticity of the Jost functions with respect to complex variable ℓ follows from the analyticity of $F_\ell^{(\text{in/out})}(E,r)$ only if the limits (5.3) exist. Such potentials are considered in the subsequent sections. We therefore defer the discussion of the convergence problem till there. For the time being, it should only be noted that the limits (5.3) do exist for any ℓ within the domain $\operatorname{Re}\ell > -1/2$.

5.1 Symmetry Properties of the Jost Functions

As we already know, the asymptotic coefficients $f_\ell^{(\text{in/out})}$ of the regular solution (5.1) are not independent. In Sect. 3.1, we obtained several symmetry relations equating their values at different points of the Riemann surface (see Figs. 3.5 and 3.5). That relations were obtained for an integer angular momentum ℓ and need to be carefully checked and perhaps generalized when ℓ moves into the complex plane.

First of all, we take notice that although the Schrödinger equation is still invariant under the transformation $k \to -k$, it does not lead to the same symmetry properties of the Jost functions (that were derived in Sect. 3.1) when ℓ is complex or even when it is real but not integer. The reason for this can be understood if we look at the corresponding transformations of the Riccati–Hankel functions. These functions can be defined as the solutions of the free radial Schrödinger equation with the following boundary conditions

$$h_\ell^{(\pm)}(kr) \underset{|kr|\to\infty}{\longrightarrow} e^{\pm i\left(kr-\frac{\ell+1}{2}\pi\right)} . \tag{5.4}$$

If k changes its sign, then

$$\begin{aligned} h_\ell^{(+)}(-kr) \underset{|kr|\to\infty}{\longrightarrow} & \exp\left[i\left(-kr-\frac{\ell+1}{2}\pi\right)\right] \\ = & \exp\left[i\left(-kr+\frac{\ell+1}{2}\pi-(\ell+1)\pi\right)\right] \\ = & e^{-i(\ell+1)\pi}\exp\left[-i\left(kr-\frac{\ell+1}{2}\pi\right)\right] , \end{aligned}$$

which means that

$$h_\ell^{(+)}(-kr) = -e^{-i\ell\pi}h_\ell^{(-)}(kr) \quad \text{and} \quad h_\ell^{(-)}(-kr) = -e^{+i\ell\pi}h_\ell^{(+)}(kr) . \tag{5.5}$$

When ℓ is integer, the factor $e^{\pm i\ell\pi}$ is just $+1$ for an even ℓ and is -1 for an odd ℓ. In such a case, the equations of the system (2.21) for the functions $F_\ell^{(\text{in/out})}$ just swap their roles (i.e. "in"↔"out"). This is not true when ℓ is not an integer number, because $e^{+i\ell\pi}$ and $e^{-i\ell\pi}$ are different complex numbers. Generally speaking, this does not allow us to relate $F_\ell^{(\text{in/out})}$ as we did it for an integer ℓ. This means that the so-called vertical symmetry (related to "jumping" up and down between the Riemann sheets) is not valid with complex ℓ.

What about complex conjugation, i.e. the diagonal symmetry? Fortunately, this relation is still valid. To see this, let us apply the operation of complex conjugation to the integral equations (2.40),

$$\begin{aligned} \left[F_\ell^{(\text{in})}(E,r)\right]^* &= \frac{1}{2}+\frac{1}{2ik^*}\int_0^r \left[h_\ell^{(+)}(kr')\right]^* V(r')\left\{\left[h_\ell^{(-)}(kr')\right]^*\left[F_\ell^{(\text{in})}(E,r')\right]^*\right. \\ &\quad \left. + \left[h_\ell^{(+)}(kr')\right]^*\left[F_\ell^{(\text{out})}(E,r')\right]^*\right\}dr' , \\ \left[F_\ell^{(\text{out})}(E,r)\right]^* &= \frac{1}{2}-\frac{1}{2ik^*}\int_0^r \left[h_\ell^{(-)}(kr')\right]^* V(r')\left\{\left[h_\ell^{(-)}(kr')\right]^*\left[F_\ell^{(\text{in})}(E,r')\right]^*\right. \\ &\quad \left. + \left[h_\ell^{(+)}(kr')\right]^*\left[F_\ell^{(\text{out})}(E,r')\right]^*\right\}dr' . \end{aligned}$$

As can be easily verified (see Appendix D)

$$\left[h_\ell^{(\pm)}(kr)\right]^* = h_{\ell^*}^{(\mp)}(k^*r)\,. \tag{5.6}$$

Taking this into account, we obtain

$$\begin{aligned}
\left[F_\ell^{(\text{out})}(E,r)\right]^* &= \frac{1}{2} - \frac{1}{2ik^*}\int_0^r h_{\ell^*}^{(+)}(k^*r')V(r')\left\{h_{\ell^*}^{(-)}(k^*r')\left[F_\ell^{(\text{out})}(E,r')\right]^*\right.\\
&\qquad\left. + h_{\ell^*}^{(+)}(k^*r')\left[F_\ell^{(\text{in})}(E,r')\right]^*\right\}dr'\,,\\
\left[F_\ell^{(\text{in})}(E,r)\right]^* &= \frac{1}{2} + \frac{1}{2ik^*}\int_0^r h_{\ell^*}^{(-)}(k^*r')V(r')\left\{h_{\ell^*}^{(-)}(k^*r')\left[F_\ell^{(\text{out})}(E,r')\right]^*\right.\\
&\qquad\left. + h_{\ell^*}^{(+)}(k^*r')\left[F_\ell^{(\text{in})}(E,r')\right]^*\right\}dr'\,.
\end{aligned}$$

Comparing these equations with (2.40), we see that the functions $\left[F_\ell^{(\text{out})}(E,r)\right]^*$ and $\left[F_\ell^{(\text{in})}(E,r)\right]^*$ obey the same equations as $F_{\ell^*}^{(\text{in})}(E^*,r)$ and $F_{\ell^*}^{(\text{out})}(E^*,r)$, respectively. Therefore

$$\left[F_\ell^{(\text{in/out})}(E,r)\right]^* \equiv F_{\ell^*}^{(\text{out/in})}(E^*,r)\,, \tag{5.7}$$

and consequently

$$\boxed{\left[f_\ell^{(\text{in/out})}(E)\right]^* = f_{\ell^*}^{(\text{out/in})}(E^*)} \tag{5.8}$$

The horizontal symmetry (see Sect. 3.1.3) is a result of the combination of the diagonal and vertical symmetries. Since one of them is broken, the horizontal symmetry is also not valid for complex ℓ. Therefore, among the symmetry relations schematically presented in Figs. 3.5 and 3.6, the only one that remains valid is the diagonal relation. The relations of this type (across the two branch cuts) are schematically shown in Figs. 5.1 and 5.2. Using the symmetry relation (5.8), the S-matrix defined as

$$S_\ell(E) = \frac{f_\ell^{(\text{out})}(E)}{f_\ell^{(\text{in})}(E)} \tag{5.9}$$

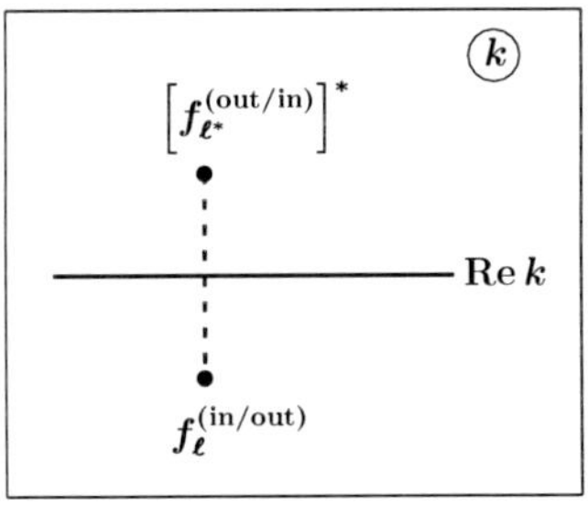

Fig. 5.1 Symmetry of the Jost functions on the k-plane. The dashed line connects the points at which the values indicated next to them are identical

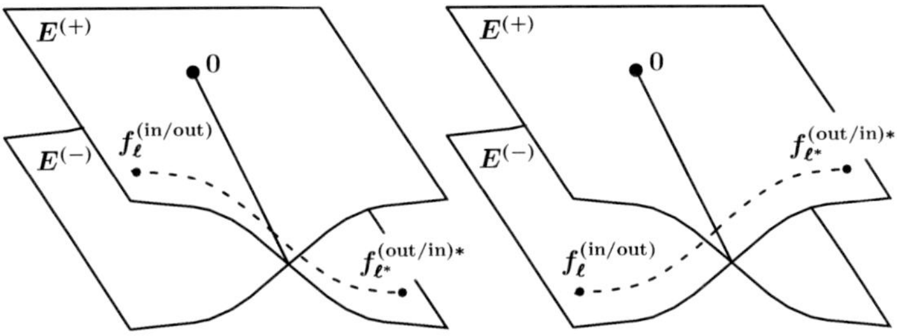

Fig. 5.2 Symmetry of the Jost functions on the E-surface. The dashed line connects the points at which the values indicated next to them are identical. The two parts of the figure show two different diagonal relations

can be re-written in terms of $f_\ell^{(\text{in})}$,

$$S_\ell(E) = \frac{\left[f_{\ell^*}^{(\text{in})}(E^*)\right]^*}{f_\ell^{(\text{in})}(E)} . \tag{5.10}$$

This relation is useful in formal derivations. For example, it enables us to establish the following complex unitarity condition

$$\left[S_{\ell^*}(E^*)\right]^* = \left[S_\ell(E)\right]^{-1} , \tag{5.11}$$

which implies that, if E is real and $S_\ell(E)$ has a pole at $\ell = \ell_0$, then it must have zero at $\ell = \ell_0^*$ and vice versa,

$$S_{\ell_0}(E) = \infty \quad \Longleftrightarrow \quad S_{\ell_0^*}(E) = 0 , \qquad \text{if} \quad \text{Im}\, E = 0 . \tag{5.12}$$

5.2 Regge Poles

The Jost function $f_\ell^{(\mathrm{in})}(E)$ is a function of two complex variables, E and ℓ. It is an analytic function of both of them. In contrast to complex analytic functions of one variable whose zeros are isolated, the zero sets of holomorphic functions of n complex variables are $(n-1)$-dimensional complex manifold. This means that in the case of two variables the zeros belong to continuous curves in the space which is the direct product of individual domains of analyticity for each of the variables.

This can be visualized as follows. Suppose we found a zero of the Jost function at the point (E_0, ℓ_0), i.e. on the E-surface this zero corresponds to $E = E_0$ and in the ℓ-plane to $\ell = \ell_0$. If we move the energy point E from E_0 to E_1 along a continuous curve within the domain of analyticity on the E-surface, the zero of the Jost function can be found with a new value ℓ_1 of variable ℓ, such that the transition from ℓ_0 to ℓ_1 goes along a continuous curve within the domain of analyticity in the complex ℓ-plane (see Fig. 5.3). In other words, when we move either E or ℓ within their respective domains of analyticity, the zeros of the function $f_\ell^{(\mathrm{in})}(E)$ do not disappear (as happens to isolated zeros of a function of one variable) but continuously move along certain curves.

It is easy to understand why zeros of the functions of one and two variables behave so differently. When one of the variables is fixed, say, $\ell = \ell_0$, and the other variable is moved from the value E_0, where the zero was found, to a new value E', the new combination (E', ℓ_0) does not obey the equation

$$f_\ell^{(\mathrm{in})}(E) = 0 \tag{5.13}$$

and the zero disappears. This makes the zeros of $f_\ell^{(\mathrm{in})}(E) = 0$ isolated, when this function is considered as a function of one variable (either E or ℓ) with the other fixed. When both variables are free to move, one of them can "adjust" itself to satisfy Eq. (5.13), if the other variable changes.

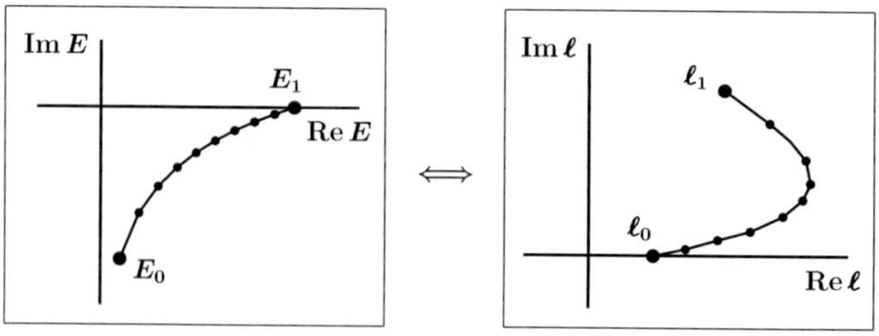

Fig. 5.3 An example of possible correspondence between the movements of complex variables E and ℓ, satisfying Eq. (5.13). When E moves from E_0 to E_1 (left curve), the variable ℓ obeying this equation, moves from ℓ_0 to ℓ_1 (right curve)

As we have already established, when ℓ is a non-negative integer number, the Jost function may have zeros that are the spectral points,

$$E_n = E_{\mathrm{r}} - \frac{i}{2}\Gamma\,, \tag{5.14}$$

corresponding to bound states and resonances (for bound states $\Gamma = 0$). In the complex ℓ-plane, all these zeros are therefore on the real axis at non-negative integer values. What happens to these zeros in the ℓ-plane if we let $\Gamma \to 0$? It is clear that this does not affect the bound-state zeros because for them $\Gamma = 0$ anyway. The resonances, however, move away from the real ℓ-axis. In which direction do they move, up or down?

The answer to this question is "up," i.e. when $\Gamma \to 0$ the Jost function zeros move from the real axis into the first quadrant of the ℓ-plane. This statement can be proved in two different ways. One of them is based on physical argumentation while the other is more formal.

* * *

Physical proof

When ℓ is complex, the centrifugal potential is also complex,

$$\frac{\ell(\ell+1)}{r^2} = \frac{(\mathrm{Re}\,\ell+1)\mathrm{Re}\,\ell - (\mathrm{Im}\,\ell)^2}{r^2} + i\frac{(2\mathrm{Re}\,\ell+1)\mathrm{Im}\,\ell}{r^2}\,. \tag{5.15}$$

Since for spectral points $\mathrm{Re}\,\ell \geqslant 0$, the sign of the imaginary part of this potential is the same as the sign of $\mathrm{Im}\,\ell$. Therefore, when ℓ is below the real axis, the imaginary part of the centrifugal potential is absorptive, and when it is above the real axis the centrifugal potential has an emissive part. Asymptotically, the wave function corresponding to a spectral point, behaves as a pure outgoing wave. With a real positive k (when $\Gamma \to 0$) the state is stationary and the probability flux is directed outwards, as can be seen from Eq. (3.84). The "loss" of the flux can only be compensated if $\mathrm{Im}\,\ell > 0$. This proves that the Jost function cannot have zeros below the real ℓ-axis.

* * *

Formal Proof

Let (E, ℓ) be a resonance spectral point. The corresponding regular solution $\phi_\ell(E, r)$ of the Schrödinger equation

$$\left[\frac{d^2}{dr^2} + k^2 - \frac{\ell(\ell+1)}{r^2} - V(r)\right]\phi_\ell(E, r) = 0 \tag{5.16}$$

has the following asymptotic behaviour

$$\phi_\ell(E,r) \underset{r\to\infty}{\longrightarrow} f_\ell^{(\mathrm{out})}(E)\exp\left[ikr - \frac{i\pi}{2}(\ell+1)\right] . \tag{5.17}$$

When we move E to the real axis ($\Gamma \to 0$), the variable ℓ acquires a non-zero imaginary part $\mathrm{Im}\,\ell$. Our task is to determine the sign of this imaginary part.

Multiplying Eq. (5.16) and its complex conjugate by $\phi_\ell^*(E,r)$ and $\phi_\ell(E,r)$, respectively, and subtracting the second from the first, we obtain

$$2i\,\mathrm{Im}\,[\ell(\ell+1)]\,\frac{|\phi_\ell(E,r)|^2}{r^2} = \frac{d}{dr}\left[\phi_\ell^*(E,r)\phi_\ell'(E,r) - \phi_\ell(E,r)\phi_\ell'^*(E,r)\right] . \tag{5.18}$$

Since we let $\Gamma \to 0$, the energy E in this equation is on the upper rim of the cut, i.e. it is physical scattering energy $E > 0$, which means that the corresponding momentum k is real and positive. With such a momentum, the wave function (5.17) at large distances is pure oscillatory and therefore bounded, $|\phi_\ell(E,r)| < M$. Consequently, the integral

$$0 < \int_0^\infty \frac{|\phi_\ell(E,r)|^2}{r^2}dr < \infty$$

is convergent. Integrating both sides of Eq. (5.18) and taking into account that $\phi_\ell(E,0) = 0$ and $\phi_\ell(E,\infty)$ is determined by Eq. (5.17), we obtain

$$\mathrm{Im}\,[\ell(\ell+1)]\int_0^\infty \frac{|\phi_\ell(E,r)|^2}{r^2}dr = k\left|f_\ell^{(\mathrm{out})}(E)\right|^2 e^{\pi\,\mathrm{Im}\ell} .$$

Since all the other factors in this equation are positive, we have

$$\mathrm{Im}\,[\ell(\ell+1)] > 0 .$$

With the help of Eq. (5.15) where $\mathrm{Re}\,\ell \geqslant 0$, we finally conclude that $\mathrm{Im}\,\ell > 0$.

* * *

Generally speaking, there are infinitely many spectral points at each physical angular momentum $\ell = 0, 1, 2, \ldots$. They may include certain number of bound states and an infinite series of S-wave, P-wave, etc. resonances. In other words, on the real ℓ-axis many spectral points coincide, i.e. all S-wave spectral points correspond to a single value $\ell = 0$, etc.

When $\Gamma \to 0$, the bound-state zeros of the Jost function remain on the real axis of the ℓ-plane while the resonances move upwards. In the upper half of the ℓ-plane, the spectral points do not coincide anymore (there is no reason why, for example,

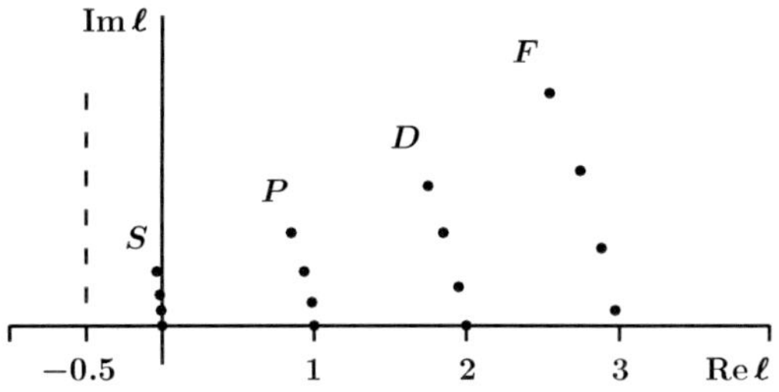

Fig. 5.4 Typical distribution of Regge poles in the complex ℓ-plane

whole series of the S-wave resonances or even any two of such resonances would move from the real axis to a single point in the first quadrant).

The spectral points in the complex ℓ-plane, which can be obtained by letting $\Gamma \to 0$, are usually called the Regge poles, after Tullio Regge who introduced complex angular momentum into quantum mechanics [22]. When speaking about spectral points in the ℓ-plane, it is customary to use the term "poles," rather than "zeros," because at the early stages of the development of this theory the spectral points were associated with singularities of the S-matrix. As we know, there is a one to one correspondence between the S-matrix poles and the Jost function zeros.[1] We therefore will use both terms interchangeably. A typical distribution of Regge poles in the complex ℓ-plane is shown in Fig. 5.4. The series of them denoted as S, P, ... are those that emerge from the points $\ell = 0$, $\ell = 1$, ..., respectively, when $\Gamma \to 0$. Those poles that have greater Γ move to a longer distance from the real axis. This could be expected from a simple intuitive reasoning. Indeed, for wide resonances the transition $\Gamma \to 0$ covers a long distance on the energy surface and the corresponding path on the ℓ-plane should also be long.

5.3 Simplicity of Regge Zeros

At a spectral point, $E_0 = (\hbar k_0)^2/(2\mu)$, on the complex energy surface the Jost function has zero that is simple[2] (see Sect. 3.3.5), i.e. it vanishes as a linear function of the energy,

$$f_\ell^{(\mathrm{in})}(E) \sim (E - E_0)\,.$$

[1] There is certain class of potentials for which the S-matrix may have the so called redundant poles. They appear not because of zeros of the denominator in the expression (5.9) but because of poles of its numerator (see Sect. 7.5). A finite-range potential, however, can never generate redundant poles.

[2] In this statement, we do not take into account the exceptional and exotic case of a threshold spectral point.

It is natural to expect that the corresponding zeros in the ℓ-plane are simple as well, i.e.

$$f_\ell^{(\mathrm{in})}(E_0) \sim (\ell - \ell_0) .$$

We can prove this, in a way that is similar to how it is done in Sect. 3.3.5 for the energy surface.

* * *

To begin with, it should be clearly understood that the fact that at a single (integer) value of $\ell = \ell_0$ several (say, N) bound states are possible does not mean that in such a case the corresponding zero of the function $f_\ell^{(\mathrm{in})}(E)$ at $\ell = \ell_0$ is of the order N. This function depends on the two variables, ℓ and E. These N zeros correspond to different values of the variable E and therefore they do not coincide (despite the fact that on the ℓ-plane they are at the same point). In other words, when considering the zeros of the Jost function in the complex ℓ-plane, we tacitly assume that for each zero the energy variable is fixed. There can be only one Jost function zero in the ℓ-plane for a fixed E. Of course, we can imagine a situation when, say, there are an S-wave and a P-wave bound states at exactly the same energy, but such a coincidence is practically impossible.

For a pair of the parameters, ℓ_0 and E_0, corresponding to a Regge pole, the regular solution $\phi_{\ell_0}(E_0, r)$ is proportional to the outgoing Jost solution, which asymptotically behaves as $\sim \exp(ik_0 r)$, where k_0 is pure imaginary for a bound state and real positive for a resonance.[3]

Recalling Eq. (2.75) that expresses the Jost function in terms of the Wronskian of the regular solution $\phi_\ell(E, r)$ and the Jost solution $\chi_\ell^{(+)}(E, r)$, we find the ℓ-derivative of $f_\ell^{(\mathrm{in})}$ as

$$\frac{\partial}{\partial \ell} f_\ell^{(\mathrm{in})}(E) = \frac{1}{2ik} W\left(\frac{\partial}{\partial \ell}\phi_\ell \,,\, \chi_\ell^{(+)}\right) + \frac{1}{2ik} W\left(\phi_\ell \,,\, \frac{\partial}{\partial \ell}\chi_\ell^{(+)}\right) . \tag{5.19}$$

Our task now is to find the Wronskians on the right hand side of this equation and to show that the first derivative $\partial_\ell f_\ell^{(\mathrm{in})}(E)$ is not zero for a bound or resonant state spectral point. In order to achieve this, consider the following equations

$$\left[\frac{d^2}{dr^2} + k^2 - \frac{\tilde{\ell}(\tilde{\ell}+1)}{r^2} - V(r)\right] \chi_{\tilde{\ell}}^{(+)}(E, r) = 0 ,$$

$$\left[\frac{d^2}{dr^2} + k^2 - \frac{\ell(\ell+1)}{r^2} - V(r)\right] \phi_\ell(E, r) \quad = 0 ,$$

[3] The Regge poles are obtained by letting $\Gamma \to 0$, according to their definition.

at two different angular momenta, $\tilde{\ell}$ and ℓ. Multiplying the first of them by $\phi_\ell(E,r)$, the second by $\chi_{\tilde{\ell}}^{(+)}(E,r)$, and subtracting the second from the first, we obtain

$$\phi_\ell \chi_{\tilde{\ell}}^{''(+)} - \phi_\ell^{''} \chi_{\tilde{\ell}}^{(+)} = \frac{1}{r^2}\left[\tilde{\ell}(\tilde{\ell}+1) - \ell(\ell+1)\right]\phi_\ell \chi_{\tilde{\ell}}^{(+)} ,$$

which means that

$$\frac{d}{dr} W\left(\phi_\ell \,,\, \chi_{\tilde{\ell}}^{(+)}\right) = \frac{1}{r^2}\left[\tilde{\ell}(\tilde{\ell}+1) - \ell(\ell+1)\right]\phi_\ell \chi_{\tilde{\ell}}^{(+)} . \tag{5.20}$$

Differentiating this equation over the variable ℓ and putting $\tilde{\ell} = \ell$, we obtain

$$\frac{d}{dr} W\left(\frac{\partial \phi_\ell}{\partial \ell} \,,\, \chi_\ell^{(+)}\right) = -\frac{2\ell+1}{r^2}\phi_\ell \chi_\ell^{(+)} . \tag{5.21}$$

Similarly, the differentiation of Eq. (5.20) over $\tilde{\ell}$ and putting $\tilde{\ell} = \ell$ gives

$$\frac{d}{dr} W\left(\phi_\ell \,,\, \frac{\partial \chi_\ell^{(+)}}{\partial \ell}\right) = \frac{2\ell+1}{r^2}\phi_\ell \chi_\ell^{(+)} . \tag{5.22}$$

At a Regge pole, (ℓ_0, E_0), the functions ϕ_ℓ and $\chi_\ell^{(+)}$ differ only by a normalization factor (see Eqs. (3.41) and (3.42)). Since here we evaluate the derivatives with respect to ℓ, we should take into account that the relations (3.41) and (3.42) involve some ℓ-dependent coefficients (here $\ell = \ell_0$),

$$\phi_\ell(E_0, r) \underset{r\to 0}{\longrightarrow} a(\ell)(k_0 r)^{\ell+1} , \qquad \chi_\ell^{(+)}(E_0, r) \underset{r\to 0}{\longrightarrow} b(\ell)(k_0 r)^{\ell+1} , \tag{5.23}$$

$$\phi_\ell(E_0, r) \underset{r\to \infty}{\longrightarrow} c(\ell)e^{ik_0 r} , \qquad \chi_\ell^{(+)}(E_0, r) \underset{r\to \infty}{\longrightarrow} d(\ell)e^{ik_0 r} . \tag{5.24}$$

Using these relations, it is easy to see that

$$\phi_\ell(E_0, r) \underset{r\to 0}{\longrightarrow} 0 ,$$

$$\partial_\ell\, \chi_\ell^{(+)}(E_0, r) \underset{r\to 0}{\longrightarrow} [\partial_\ell\, b(\ell)](k_0 r)^{\ell+1} + b(\ell)\ln(k_0 r)(k_0 r)^{\ell+1} \underset{r\to 0}{\longrightarrow} 0 ,$$

$$\left.\begin{array}{r} \partial_\ell\, \phi_\ell(E_0, r) \\ \partial_\ell\, \phi_\ell'(E_0, r) \\ \chi_\ell^{(+)}(E_0, r) \\ \chi_\ell^{'(+)}(E_0, r) \end{array}\right\} \underset{r\to \infty}{\longrightarrow} 0 ,$$

where in the case of a resonance spectral point the limit $r \to \infty$ is taken along the ray in the complex r-plane depicted in Fig. 3.9. Therefore

$$W\left[\partial_\ell \phi_\ell(E_0, r)\,,\; \chi_\ell^{(+)}(E_0, r)\right]_{r=\infty} = W\left[\phi_\ell(E_0, r)\,,\; \partial_\ell\, \chi_\ell^{(+)}(E_0, r)\right]_{r=0} = 0\,.$$

Bearing this in mind and integrating Eq. (5.21) from r to ∞ and Eq. (5.22) from 0 to r, we obtain

$$W\left[\partial_\ell \phi_\ell(E_0, r)\,,\; \chi_\ell^{(+)}(E_0, r)\right] = (2\ell+1)\int_r^\infty \phi_\ell(E_0, r)\chi_\ell^{(+)}(E_0, r)\frac{dr}{r^2}$$

and

$$W\left[\phi_\ell(E_0, r)\,,\; \partial_\ell \chi_\ell^{(+)}(E_0, r)\right] = (2\ell+1)\int_0^r \phi_\ell(E_0, r)\chi_\ell^{(+)}(E_0, r)\frac{dr}{r^2}\,.$$

Substituting these Wronskians into Eq. (5.19), we see that

$$\frac{\partial}{\partial \ell} f_\ell^{(\mathrm{in})}(E_0) = \frac{2\ell+1}{2ik_0}\int_0^\infty \phi_\ell(E_0, r)\chi_\ell^{(+)}(E_0, r)\frac{dr}{r^2}\,. \tag{5.25}$$

Since at a spectral point

$$\phi_\ell(E_0, r) = f_\ell^{(\mathrm{out})}(E_0)\chi_\ell^{(+)}(E_0, r) \tag{5.26}$$

and according to Eq. (3.54), the normalized physical wave function u_ℓ is related to the Jost solution as

$$\chi_\ell^{(+)}(E_0, r) = \frac{u_\ell(E_0, r)}{i^{\ell+1}\mathcal{A}_\ell}\,,$$

we obtain (here $\ell = \ell_0$)

$$\frac{\partial}{\partial \ell} f_\ell^{(\mathrm{in})}(E_0) = \frac{(-1)^{\ell+1}(2\ell+1)f_\ell^{(\mathrm{out})}(E_0)}{2ik_0\mathcal{A}_\ell^2}\int_0^\infty \left[\frac{u_\ell(E_0, r)}{r}\right]^2 dr\,. \tag{5.27}$$

As we have seen in Sect. 3.3.4, for both the bound and resonant states the expectation values can be calculated as the integrals where the simple (not absolute) square of the radial wave function is used. In the case of a resonance, such an integral should be taken along the complex path shown in Fig. 3.9. This means that the integral on the right hand side of Eq. (5.27) is the expectation value of $1/r^2$. Since none of the other factors on the right hand side of this equation can vanish, we conclude that at a bound (or resonant) state spectral point, the first derivative (with respect to ℓ) of the Jost function is always non-zero. This proves that in the complex ℓ-plane, the Jost function zeros are simple, i.e. of the first order.

5.3.1 Asymptotic Normalization Constant (ANC)

As a byproduct of the above proof, Eq. (5.27) allows us to obtain yet another relation between the asymptotic normalization constant, $\mathcal{A}_\ell$, and the residue of the S-matrix, similar to Eq. (4.53). This time the residue is taken in the complex ℓ-plane. As we have proved, the S-matrix has a pole of the first order in the complex ℓ-plane at the point ℓ if there is a bound or a resonant state at the corresponding energy E_0. Therefore the residue of the S-matrix at such a point is

$$\mathrm{Res}\ [S_\ell, \ell] = \frac{f_\ell^{(\mathrm{out})}(E_0)}{\dfrac{\partial}{\partial \ell} f_\ell^{(\mathrm{in})}(E_0)}$$

and thus

$$\boxed{\mathcal{A}_\ell^2 = \frac{(-1)^{\ell+1}(2\ell+1)\langle r^{-2}\rangle}{2ik_0} \mathrm{Res}\ [S_\ell, \ell]} \tag{5.28}$$

The case of a resonance may cause some misunderstanding and therefore needs an additional explanation. The resonance energy, E_0, is complex and the corresponding angular momentum, ℓ, is real and integer (the same as for a bound state). There is a resonance pole of the S-matrix on the real ℓ-axis. This pole corresponds to the complex energy. We can use Eq. (5.28) to relate the residue at this pole to the ANC of the normalized resonance wave function obtained at that complex energy. When $\Gamma \to 0$, the pole moves up to the first quadrant of the complex ℓ-plane and becomes what is called the Regge pole. It is not clear what happens to the ANC, but Eq. (5.28) remains valid and relates "new" ANC to the residue of the S-matrix at the Regge pole.

5.4 Regge Trajectories

Let $E_0 < 0$ be the energy of an S-wave bound state. This state corresponds to the zeros of the Jost function $f_\ell^{(\mathrm{in})}(E)$ at $E = E_0$ on the energy surface and at $\ell = 0$ in the complex ℓ-plane. What happens to the zero in the angular-momentum plane if we gradually increase E by moving this point to the right along the real energy-axis?

To answer this question, let us show that for real energies on the physical sheet, the Jost function $f_\ell^{(\mathrm{in})}(E)$ may have zeros only on the real axis and in the first quadrant of the ℓ-plane, i.e. $\mathrm{Im}\,\ell \geqslant 0$. More specifically,

$$E \leqslant 0 \quad \Rightarrow \quad \mathrm{Im}\,\ell = 0\,, \tag{5.29}$$

$$E > 0 \quad \Rightarrow \quad \mathrm{Im}\,\ell > 0\,. \tag{5.30}$$

Proof

- $\boxed{E \leqslant 0}$

  Apparently, at physical bound-state energies, $E_0 < E_1 < E_2 < \ldots$, the spectral points in the ℓ-plane are at real integer values of ℓ and thus lie on the real axis. The question is: Do they stay on the real axis when E is somewhere in between the points $E_0, E_1, E_2, \ldots$?

  If the energy is negative and $f_\ell^{(\text{in})}(E) = 0$, the corresponding wave function vanishes at $r = 0$ as well as at $r = \infty$, and thus the integration of Eq. (5.18) gives

$$\operatorname{Im}\left[\ell(\ell+1)\right] \int_0^\infty \frac{|\phi_\ell(E,r)|^2}{r^2} dr = 0 ,$$

  from which it follows that $\operatorname{Im} \ell = 0$.

- $\boxed{E > 0}$

  The proof that at physical scattering energies the Regge poles may lie only above real axis of the ℓ-plane was given in Sect. 5.2.

* * *

Now we know that when E increases moving along the real axis from left to right, the Regge zero remains on the real ℓ-axis until E becomes positive, after which it moves upwards. But what does it mean "remains on the real axis"? Apparently, it cannot stay at a fixed point. In which direction does it move? If there are several bound states with $\ell = 0, 1, \ldots$, then there are no other option but to move to the right, from $\ell = 0$ to $\ell = 1$ and so on. It is easy to prove that this is always so, even if there is only one bound state.

Proof Differentiating the Schrödinger equation

$$\left[\frac{d^2}{dr^2} + \frac{2\mu E}{\hbar^2} - \frac{\ell(\ell+1)}{r^2} - V(r)\right] \phi_\ell(E,r) = 0 \tag{5.31}$$

with respect to E, we obtain

$$\left[\frac{d^2}{dr^2} + \frac{2\mu E}{\hbar^2} - \frac{\ell(\ell+1)}{r^2} - V(r)\right] \frac{\partial \phi_\ell}{\partial E} - \left(\frac{2\ell+1}{r^2}\frac{\partial \ell}{\partial E} - \frac{2\mu}{\hbar^2}\right)\phi_\ell = 0 . \tag{5.32}$$

Let us multiply Eq. (5.32) by ϕ_ℓ and subtract from it Eq. (5.31) multiplied by $\partial\phi_\ell/\partial E$. This gives

$$\frac{d}{dr} W\left(\phi_\ell, \frac{\partial \phi_\ell}{\partial E}\right) - \left(\frac{2\ell+1}{r^2}\frac{\partial \ell}{\partial E} - \frac{2\mu}{\hbar^2}\right)\phi_\ell^2 = 0 . \tag{5.33}$$

Since $E < 0$, we have $\phi_\ell(E, 0) = \phi_\ell(E, \infty) = 0$, and thus the Wronskian entering this equation vanishes at $r = 0$ as well as at $r = \infty$. Bearing this in mind and integrating Eq. (5.33) from 0 to ∞, we obtain

$$(2\ell + 1)\frac{\partial \ell}{\partial E}\int_0^\infty \frac{\phi_\ell^2(E, r)}{r^2}dr = \frac{2\mu}{\hbar^2}\int_0^\infty \phi_\ell^2(E, r)dr\ . \tag{5.34}$$

Since ℓ is real and $E < 0$, the boundary conditions determining $\phi_\ell(E, r)$ as well as Eq. (5.31) are real, which implies that $\phi_\ell(E, r)$ is also real. The integrals on both sides of Eq. (5.34) are therefore positive. With the restriction $\ell > -1/2$ this means that

$$\frac{\partial \ell}{\partial E} > 0\ ,$$

i.e. $\ell(E)$ is an ascending function of the energy.

* * *

Therefore the answer to the question asked at the beginning of this section is: while the energy remains negative the Jost function zero moves from left to right along real axis of the ℓ-plane; as soon as E becomes positive, the zero moves upwards into the first quadrant of that plane.

Therefore the Regge zero moves from $\ell = 0$ along the real axis to the right. If there are no P-wave bound states, it does not reach the point $\ell = 1$ before the energy becomes positive. If, however, such a bound state exists, the Regge zero passes through $\ell = 1$ when E becomes equal to the energy of this bound state. The same is applicable to all higher partial-wave bound states, if any. After turning into the first quadrant, the Regge zero passes through the points that correspond to resonances. This happens when $E = E_r$, because these points (Regge poles described in Sect. 5.2) are obtained by letting $\Gamma \to 0$.

The curve along which the Regge zero moves in the complex ℓ-plane when the energy increases from negative to positive values is called the Regge trajectory. Such a trajectory reveals an interesting relation among spectral points. In most cases, each partial wave (S, P, D, etc.) has many spectral points. Say, it can be two different S-wave bound states. Starting from these two states, we obtain two different Regge trajectories. Each of them passes through its own P-wave state (bound or resonant), its own D-wave state (bound or resonant) and so on. Therefore all the spectral points generated by a given potential can be grouped in families, like beads on a necklace. Each family belongs to a different Regge trajectory. Thinking physically, we can view the quantum states of the same family as rotational excitations of the same "basic" S-wave (non-rotating) state. Apparently, it is not necessary that the "basic" state is a bound state. Many potentials do not support bound states at all, but a Regge trajectory always starts from an S-wave state, which can be an S-wave resonance.

A typical example of Regge trajectories is shown in Fig. 5.5. For the sake of clarity, we assumed that the system has only one family of bound states, namely

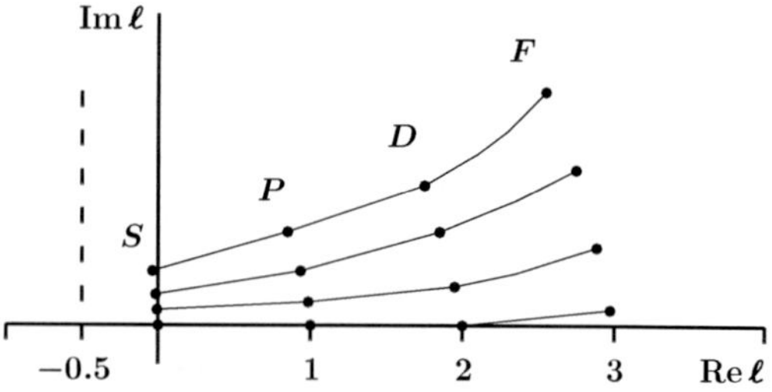

Fig. 5.5 Typical example of Regge trajectories in the complex ℓ-plane

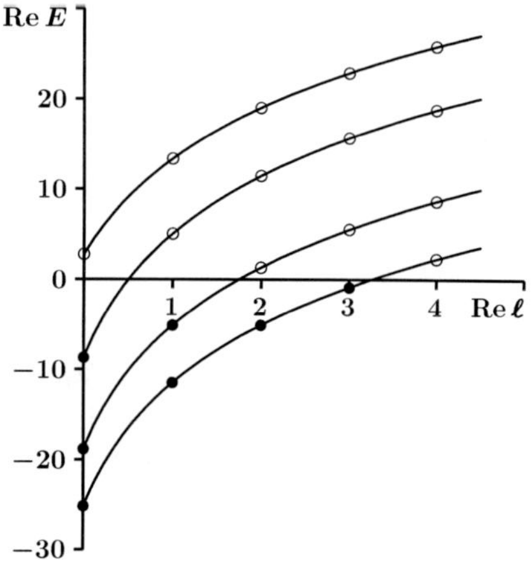

Fig. 5.6 Typical example of the Frautschi curves that group the bound states (filled circles) and resonances (open circles) in families

one S, one P, and one D bound state. When there are several bound states with the same angular momentum (which is very common), the corresponding Regge poles are located at the same points in the ℓ-plane, namely at real $\ell = 0, 1, 2, \ldots$. In such a case several Regge trajectories are indistinguishable while they go along the real axis. It should be emphasized that despite this "coincidence" of the trajectories, they are not identical, because any chosen point on their common path is passed at different values of E.

In order to distinguish different Regge trajectories that relate bound states belonging to different families, they are often depicted as the curves in the plane $\mathrm{Re}\, E$ versus $\mathrm{Re}\, \ell$. Such curves are called the Frautschi curves named after Steven Frautschi who made a significant contribution to the development of the theory of complex angular momenta (see his book [23]). An example of a spectrum whose states are grouped in families with the help of the Frautschi curves is shown in Fig. 5.6. In this example, the system has three S-wave, two P-wave, one D-wave, and one F-wave bound states.

If we assume that every spectral point belongs to a Regge trajectory (which is very natural assumption), then there is the following important implication. Let N_ℓ be a number of bound states with the angular momentum ℓ. It is obvious that none of these bound states is excluded from a family, i.e. belongs to a Regge trajectory, only if

$$N_0 \geqslant N_1 \geqslant N_2 \geqslant \dots \,. \tag{5.35}$$

Indeed, if $N_0 < N_1$ then there are not enough "basic" S-wave states to start N_1 trajectories passing through all the P-wave states, and so on. The inequality relations (5.35) can also be proved more formally and without the assumption that every spectral point belongs to a Regge trajectory.

Proof Repeating the steps that lead to Eq. (5.22), but for $\phi_\ell(E, r)$ and $\phi_\ell^*(E, r)$, we obtain

$$\frac{d}{dr} W\left(\phi_\ell\,,\, \frac{\partial \phi_\ell^*}{\partial \ell}\right) = (2\ell + 1)\frac{|\phi_\ell|^2}{r^2}\,. \tag{5.36}$$

Since $\phi_\ell(E, r)$ is a regular solution of the Schrödinger equation, the Wronskian $W(\phi_\ell\,,\, \partial_\ell \phi_\ell^*)$ vanishes at $r = 0$. Therefore the integration of Eq. (5.36) from 0 to ∞ gives

$$W\left(\phi_\ell\,,\, \frac{\partial \phi_\ell^*}{\partial \ell}\right)\bigg|_{r=\infty} = (2\ell + 1)\int_0^\infty \frac{|\phi_\ell|^2}{r^2} dr > 0\,. \tag{5.37}$$

We are considering real positive (scattering) energies and real (although not necessary integer) angular momenta ℓ. Therefore, according to Eq. (4.15), both $f_\ell^{(\text{in})}(E)$ and $f_\ell^{(\text{out})}(E)$ have the same absolute value, which can be denoted as $|f_\ell(E)|$, i.e.

$$f_\ell^{(\text{in/out})}(E) = |f_\ell(E)| e^{\mp i\delta_\ell(E)}\,. \tag{5.38}$$

When $r \to \infty$ the function $\phi_\ell(E, r)$ takes the following asymptotic form

$$\phi_\ell(E, r) \underset{r\to\infty}{\longrightarrow} |f_\ell| \left[e^{-i\delta_\ell - ikr + i\pi(\ell+1)/2} + e^{+i\delta_\ell + ikr - i\pi(\ell+1)/2}\right]. \tag{5.39}$$

Differentiating this wave function with respect to ℓ and r, it is not difficult to obtain the Wronskian on the left hand side of Eq. (5.37),

$$W\left(\phi_\ell\,,\, \frac{\partial \phi_\ell^*}{\partial \ell}\right)\bigg|_{r=\infty} = 4k|f_\ell|^2\left(\frac{\pi}{2} - \frac{\partial \delta_\ell}{\partial \ell}\right).$$

The right hand side of Eq. (5.37) is positive, hence

$$\frac{\partial \delta_\ell}{\partial \ell} < \frac{\pi}{2}\,. \tag{5.40}$$

Integrating this inequality over ℓ from a value ℓ to $(\ell+1)$, we obtain the following important inequality

$$\delta_{\ell+1}(E) - \delta_\ell(E) < \frac{\pi}{2} , \qquad \text{(for real } \ell \text{ and } E) . \tag{5.41}$$

According to Levinson's theorem $N_\ell = \delta_\ell(0)/\pi$ (see Eq. (4.72)). Dividing Eq. (5.41) by π, we thus conclude that

$$N_{\ell+1} < N_\ell + \frac{1}{2} . \tag{5.42}$$

For integer values of N_ℓ, this means that

$$\boxed{N_{\ell+1} \leqslant N_\ell} \tag{5.43}$$

5.5 Regge Poles and Resonance Parameters

When locating the Jost function zeros $E_{\rm res} = E_{\rm r} - i\Gamma/2$ on the unphysical sheet of the energy surface, we obtain both the collision energy $E_{\rm r}$ at which the resonance is excited, and its lifetime $T_{1/2} = \hbar \ln(2)/\Gamma$. In the ℓ-plane, the Regge poles $\ell_{\rm res}$ corresponding to resonances are also characterized by two real numbers, $\text{Re}\,\ell_{\rm res}$ and $\text{Im}\,\ell_{\rm res}$. It is natural to expect that the resonance parameters $E_{\rm r}$ and Γ should be somehow related to (can be obtained from) $\text{Re}\,\ell_{\rm res}$ and $\text{Im}\,\ell_{\rm res}$.

In order to find such a relation, let us first do some intuitive reasoning. As was discussed in Sec. 3.3.2, the physical reason for the relation connecting the imaginary part of $E_{\rm res}$ and lifetime of the resonant state is the uncertainty relation between the energy and time, $\Delta E \cdot \Delta t \sim \hbar$. Similarly, the uncertainties $\Delta\ell$ and $\Delta\varphi$ of the angular momentum and angular position of a particle are related as[4]

$$\Delta\ell \cdot \Delta\varphi \sim \hbar .$$

Although in quantum mechanics the angular momentum ℓ is a discrete variable, in our qualitative reasoning we can consider it as a continuous classical variable $\ell = k\rho$, determined by the impact parameter ρ. Suppose that there is a resonant state with the angular momentum $\ell_{\rm r}$ and the collision energy $E_{\rm r}$. When a beam of particles with such energies moves toward a scatterer, some of them have the impact

[4] See, for example, Ref. [24].

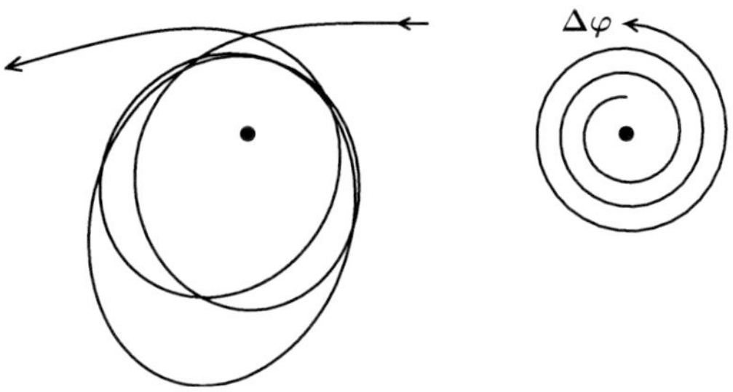

Fig. 5.7 A resonance-scattering orbit (left) and the corresponding angle $\Delta\varphi$ (right) swept by the projectile around the scatterer

parameters within an interval $\Delta\rho$ such that their angular momenta are within the corresponding interval $\Delta\ell$ around ℓ_r. In the quantum world, the resonance can be excited even if ℓ is not exactly equal to ℓ_r. Thus, all the particles whose angular momenta are within $\Delta\ell$ can do it. Of course, the excitation probability diminishes when we shift ℓ either to the left or to the right of ℓ_r. How fast this probability diminishes depends on the sharpness of the resonance.

The uncertainty $\Delta\varphi$ of the angular position can be viewed as the total angle swept by the projectile particle when it revolves around the scatterer before going away, i.e. before the resonance decays (see Fig. 5.7). The value $n = \Delta\varphi/(2\pi)$ is therefore the number of circles the projectile makes around the scatterer. The larger the $\Delta\varphi$ is, the longer the projectile revolves around the scatterer, which means longer lifetime of the resonance.

If we associate $\Delta\ell$ with $\operatorname{Im}\ell_{\rm res}$ (similarly to ΔE and Γ), then

$$\Delta\varphi \sim \frac{\hbar}{\operatorname{Im}\ell_{\rm res}} . \tag{5.44}$$

Therefore, if a Regge pole $\ell_{\rm res}$ is close to the real axis (small $\operatorname{Im}\ell_{\rm res}$) then the resonance has a long lifetime, and vice versa. This is very much the same as we have already learned about the relation between Γ and T. The distance of a Regge pole from the real axis determines the sharpness of the resonance. In the limiting case $\operatorname{Im}\ell_{\rm res} = 0$ the projectile makes an infinite number of circles around the scatterer and thus becomes bound.

Although qualitatively the relation (5.44) between $\operatorname{Im}\ell_{\rm res}$ and $\Delta\varphi$ looks the same as between Γ and T, the exact quantitative relation between $\operatorname{Im}\ell_{\rm res}$ and T is more complicated. In order to derive such a relation, consider a Regge pole at

$$\ell(E_r) = \ell_r + i\ell_i .$$

By definition, this pole corresponds to the Jost function zero on the unphysical sheet of the energy surface at

$$E_{\rm res} = E_r - \frac{i}{2}\Gamma .$$

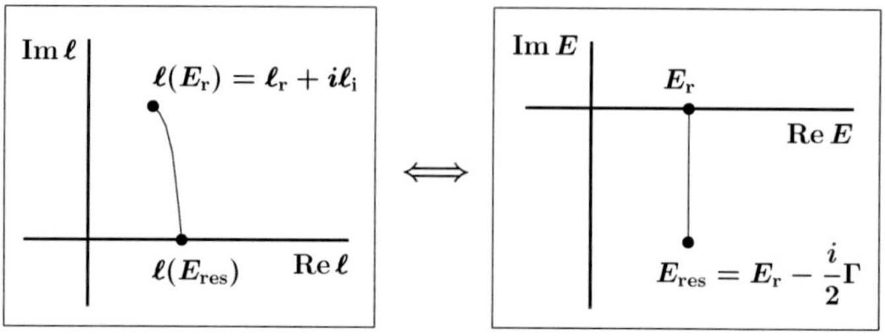

Fig. 5.8 A Regge zero at $\ell = \ell_r + i\ell_i$ (left) and the corresponding Jost function zero E_{res} on the energy surface (right)

If we move the energy from the point E_r down to E_{res}, the Regge pole moves on the ℓ-plane down to an integer number $\ell(E_{res})$, as is schematically shown in Fig. 5.8.

Let the Regge pole be not far from the real axis, i.e. its imaginary part ℓ_i be small. Since the Jost function $f_\ell^{(in)}(E)$ is analytic with respect to both the energy and angular momentum, the corresponding Γ is also small. Then the Taylor series

$$\ell(E) = \ell(E_r) + (E - E_r) \left.\frac{d\ell}{dE}\right|_{E=E_r} + \frac{1}{2}(E - E_r)^2 \left.\frac{d^2\ell}{dE^2}\right|_{E=E_r} + \dots \tag{5.45}$$

converges very fast within the circle $|E - E_r| \sim \Gamma$ that includes the point E_{res}. So, neglecting the higher order derivatives, we have

$$\ell(E_{res}) \approx \ell(E_r) + (E_{res} - E_r) \left.\frac{d\ell}{dE}\right|_{E=E_r} .$$

Therefore

$$\ell(E_{res}) \approx \ell_r + i\ell_i - \frac{i}{2}\Gamma \left.\frac{d\ell}{dE}\right|_{E=E_r} ,$$

which is valid separately for the real and imaginary parts. Equalizing the imaginary parts, we obtain:

$$\boxed{\Gamma \approx \frac{2\ell_i}{\mathrm{Re}\left(\dfrac{d\ell}{dE}\right)_{E=E_r}}} \tag{5.46}$$

As was expected, the resonance width is proportional to the imaginary part ℓ_i of the Regge pole. It should be emphasized, however, that Eq. (5.46) is valid only for narrow resonances. When ℓ_i is not small, the higher order terms of the series (5.45) should be taken into account.

5.6 Watson Transform

The partial-wave analysis allows us to reduce a three-dimensional problem to an infinite set of one-dimensional equations for the partial waves with $\ell = 0, 1, 2, \ldots$. When a bound state or a resonance is considered, the number of the partial waves that we need to take into account is always finite (in many cases, just one). In contrast to it, the scattering process formally involves all possible angular momenta. In other words, the observable cross section

$$\frac{d\sigma}{d\Omega} = \left|\mathsf{f}(\hat{\vec{r}} \leftarrow \hat{\vec{k}}, E)\right|^2 \tag{5.47}$$

that determines the probability of scattering in the direction $\hat{\vec{r}}$ can be obtained by evaluating the infinite sum (see Eq. (4.13)),

$$\mathsf{f}(\hat{\vec{r}} \leftarrow \hat{\vec{k}}, E) = \sum_{\ell=0}^{\infty}(2\ell + 1)f_\ell(E)P_\ell(\cos\theta) , \tag{5.48}$$

involving the partial-wave amplitudes

$$f_\ell(E) = \frac{S_\ell(E) - 1}{2ik} = \frac{1}{2ik}\left[\frac{f_\ell^{(\mathrm{out})}(E)}{f_\ell^{(\mathrm{in})}(E)} - 1\right] . \tag{5.49}$$

At low collision energies, the sum (5.48) converges very fast, but with increasing E the contribution of the higher partial waves becomes more and more significant. This can be intuitively understood using the classical picture where the angular momentum is just the product on the linear momentum, p, and the impact parameter, r. It is clear that the greater is the collision energy (and thus p), the larger is the angular momentum. For the quantum scattering this implies that we have to take into account too many terms in the partial-wave expansion (5.48). This section is devoted to a method that allows one to partly circumvent this difficulty.

* * *

If variable ℓ is allowed to assume complex values, the sum (5.48) can be converted into a contour integral. This is known as the Watson transform. In order to do this, let us consider the following closed contour integral,

$$I = \oint_C \frac{(2\lambda + 1)f_\lambda(E)P_\lambda(-\cos\theta)}{\sin\pi\lambda} d\lambda , \tag{5.50}$$

where the integration path encloses all the non-negative integer values of λ as is shown in Fig. 5.9.

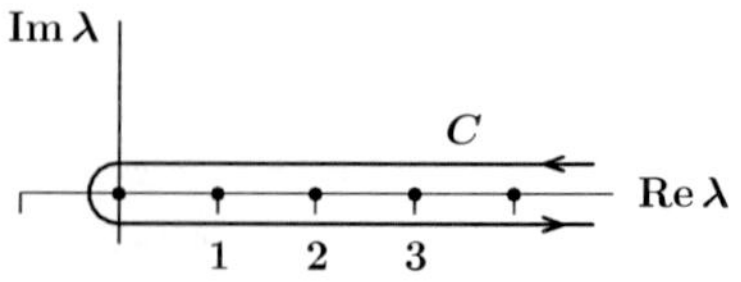

Fig. 5.9 The integration contour C, used in Eq. (5.50), which encloses the non-negative part of the real axis at an infinitesimal distance from it

Apparently, the integrand has a first-order pole at each of these integer values, where

$$\sin \pi \ell = 0 \qquad \text{and} \qquad \left(\frac{d}{d\lambda} \sin \pi \lambda \right)\bigg|_{\lambda=\ell} = (-1)^{\ell} \pi \ , \qquad \ell = 0, 1, 2, \dots \tag{5.51}$$

Therefore (using the residue theory)

$$I = 2\pi i \sum_{\ell=0}^{\infty} \text{Res} \left[\frac{(2\lambda + 1) f_{\lambda}(E) P_{\lambda}(-\cos\theta)}{\sin \pi \lambda} \ , \ \ell \right] , \tag{5.52}$$

where Res $[f(z), z_0]$ stands for the residue of the function $f(z)$ at the point z_0. Since all the poles are of the first order, the residues are easily calculated,

$$\text{Res} \left[\frac{(2\lambda + 1) f_{\lambda}(E) P_{\lambda}(-\cos\theta)}{\sin \pi \lambda} \ , \ \ell \right] = \lim_{\lambda \to \ell} (\lambda - \ell) \frac{(2\lambda + 1) f_{\lambda}(E) P_{\lambda}(-\cos\theta)}{\sin \pi \lambda}$$

$$= (2\ell + 1) f_{\ell}(E) P_{\ell}(-\cos\theta) \lim_{\lambda \to \ell} \frac{\lambda - \ell}{\sin \pi \lambda} = (2\ell + 1) f_{\ell}(E) P_{\ell}(-\cos\theta) \frac{(-1)^{\ell}}{\pi} \ .$$

Taking into account the symmetry property of the Legendre polynomials, namely $P_{\ell}(x) = (-1)^{\ell} P_{\ell}(-x)$, we obtain

$$\mathsf{f}(\hat{\vec{r}} \leftarrow \hat{\vec{k}}, E) = \frac{1}{2i} \oint_C \frac{(2\lambda + 1) f_{\lambda}(E) P_{\lambda}(-\cos\theta)}{\sin \pi \lambda} d\lambda \ , \tag{5.53}$$

which is an alternative to the partial-wave series (5.48). This integral representation of the scattering amplitude is of little use by itself, but it can be further converted into something useful if we recollect that in the first quadrant of the ℓ-plane there are the Regge poles of the function $f_{\lambda}(E)$. Such a transformation is possible thanks to the two facts:

- The integrand in Eq. (5.53) is a meromorphic function of λ, i.e. it is analytic everywhere except for the isolated points (the non-negative integers and the

Regge poles). Thanks to this, the integration contour can be arbitrarily deformed under the condition that it does not cross any of the singularities.

- The amplitude (and thus the whole integrand) tends to zero in all directions on the complex ℓ-plane when $|\ell| \to \infty$. This is because with infinitely large $|\ell|$ the centrifugal potential, $\ell(\ell+1)/r^2$, always dominate over $V(r)$, which means that $V(r)$ can be ignored and the radial Schrödinger equation becomes "free." For the "free" motion the scattering amplitude is zero. Thanks to this, we can ignore any parts of the deformed contour that are at infinity.

Let us deform the integration contour as is shown in steps in Fig. 5.10. As a result, the contour is split into two parts: C_1 that runs down the imaginary axis from $+i\infty$ to $-i\infty$ and circumvents the pole at $\lambda = 0$ on the negative side; and C_2 that encloses all the Regge poles (in the negative direction, i.e. leaving them on the right hand side). Therefore the amplitude (5.53) is also split into two terms,

$$\mathsf{f}(\hat{\vec{r}} \leftarrow \hat{\vec{k}}, E) = \mathsf{f}^{(\mathrm{bg})}(\hat{\vec{r}} \leftarrow \hat{\vec{k}}, E) + \mathsf{f}^{(\mathrm{p})}(\hat{\vec{r}} \leftarrow \hat{\vec{k}}, E) , \tag{5.54}$$

representing the "background" and the "pole" contributions, respectively, i.e.

$$\mathsf{f}^{(\mathrm{bg})}(\hat{\vec{r}} \leftarrow \hat{\vec{k}}, E) = \frac{1}{2i} \int_{+i\infty}^{-i\infty} \frac{(2\lambda+1) f_\lambda(E) P_\lambda(-\cos\theta)}{\sin\pi(\lambda-\epsilon)} d\lambda , \qquad \epsilon \downarrow 0 , \tag{5.55}$$

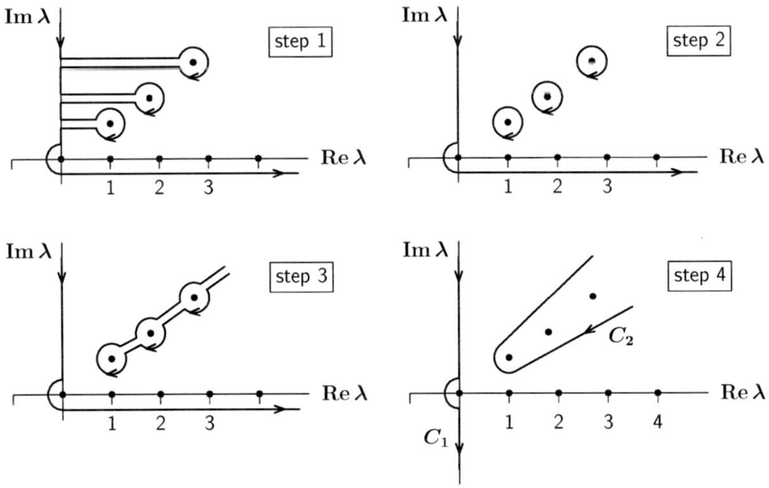

Fig. 5.10 Four steps to deform the integration contour shown in Fig. 5.9. These steps replace the initial contour C, used in Eq. (5.50), with the two contours, C_1 and C_2

$$\mathsf{f}^{(\mathrm{p})}(\hat{\vec{r}} \leftarrow \hat{\vec{k}}, E) = \frac{1}{2i}\oint_{C_2} \frac{(2\lambda+1)f_\lambda(E)P_\lambda(-\cos\theta)}{\sin\pi\lambda} d\lambda \tag{5.56}$$

$$= -\pi \sum_n \frac{2\ell_n + 1}{\sin\pi\ell_n} P_{\ell_n}(-\cos\theta)\mathrm{Res}\left[f_\lambda(E), \ell_n\right] , \tag{5.57}$$

where ℓ_n are the Regge poles.

In Eq. (5.55), we replaced $\sin\pi\lambda$ with $\sin\pi(\lambda - \epsilon)$, where ϵ is an infinitely small real positive number that finally is to be put to zero. This is done because the path C_1 passes by the point $\lambda = 0$ on the left hand side. Replacing the variable,

$$\lambda = -i\zeta ,$$

we obtain

$$\mathsf{f}^{(\mathrm{bg})}(\hat{\vec{r}} \leftarrow \hat{\vec{k}}, E) = \frac{1}{2}\int_{-\infty}^{+\infty} \frac{(1-2i\zeta)f_{-i\zeta}(E)P_{-i\zeta}(-\cos\theta)}{\sin[i\pi(\zeta - i\epsilon)]} d\zeta . \tag{5.58}$$

This integral can be split into three parts as follows

$$\int_{-\infty}^{+\infty} = \int_{-\infty}^{-\delta} + \int_{-\delta}^{+\delta} + \int_{+\delta}^{+\infty} ,$$

where δ is an infinitely small positive number. By doing this splitting, we separated a small interval around the point $\zeta = 0$, where the denominator of the integrand can be replaced with just the first term of its Taylor series, i.e.

$$\sin[i\pi(\zeta - i\epsilon)] \approx i\pi(\zeta - i\epsilon) .$$

Therefore

$$\int_{-\delta}^{+\delta} = \int_{-\delta}^{+\delta} \frac{(1-2i\zeta)f_{-i\zeta}(E)P_{-i\zeta}(-\cos\theta)}{i\pi(\zeta - i\epsilon)} d\zeta .$$

Now we can use well-known Sokhotski-Plemelj theorem,[5]

$$\lim_{\epsilon\to 0^+}\int_a^b \frac{f(x)}{x \pm i\epsilon} dx = \mathcal{P}\int_a^b \frac{f(x)}{x} dx \mp i\pi f(0) , \tag{5.59}$$

[5] A simple derivation of Eq. (5.59) can be found in the Appendix A of the book [25]. A rigorous proof of this relation is given in [26].

where the symbol $\mathcal{P}$ means the Cauchy principal value integration, which we will also denote as $⨍$. Very often this equation is written in the following symbolic form

$$\frac{1}{x \pm i\epsilon} = \mathcal{P}\frac{1}{x} \mp i\pi\delta(x) . \tag{5.60}$$

Applying the relation (5.59), we obtain the following representation of the background part of the full amplitude

$$\mathrm{f}^{(\mathrm{bg})}(\hat{\vec{r}} \leftarrow \hat{\vec{k}}, E) = \frac{1}{2}f_0(E) + ⨍_{-\infty}^{+\infty} \frac{(1-2i\zeta)f_{-i\zeta}(E)P_{-i\zeta}(-\cos\theta)}{2\sin(i\pi\zeta)} d\zeta \, , \tag{5.61}$$

with $f_0(E)$ being the S-wave amplitude. Therefore

$$\begin{aligned} \mathrm{f}(\hat{\vec{r}} \leftarrow \hat{\vec{k}}, E) = \frac{1}{2}f_0(E) \; &+ ⨍_{-\infty}^{+\infty} \frac{(1-2i\zeta)f_{-i\zeta}(E)P_{-i\zeta}(-\cos\theta)}{2\sin(i\pi\zeta)} d\zeta \\ &- \pi \sum_n \frac{2\ell_n + 1}{\sin \pi\ell_n} P_{\ell_n}(-\cos\theta) \mathrm{Res}\left[f_\lambda(E), \ell_n\right] \, , \end{aligned} \tag{5.62}$$

where the sum runs over all the Regge poles that are in the first quadrant of the ℓ-plane.

Therefore instead of summing up too many terms in the series (5.48), we can add together $f_0(E)/2$ and the contributions from few nearest Regge poles. As to the background integral in the representation (5.62), it can be estimated using, for example, the Born approximation for the amplitude $f_{-i\zeta}(E)$. This approximation consists in replacing the scattering wave function $u_\ell(E,r)$ in Eq. (4.14) with the corresponding "free" solution

$$V(r) \equiv 0 \;\Rightarrow \qquad u_\ell(E,r) = 2j_\ell(kr) \, ,$$

which gives

$$\boxed{f_\ell^{(\mathrm{Born})}(E) = -\frac{1}{k^2}\int_0^\infty [j_\ell(kr)]^2 \, V(r) \, dr} \tag{5.63}$$

Green's Functions

6

It should be clearly understood that a quantum mechanical problem can be treated within the two alternative approaches: based on either differential or integral equations. The differential equations require the boundary conditions that are specific for different types of the problems (bound, scattering, or resonant state). The integral equations automatically take the boundary conditions into account. There is only one type of the problem where the boundary conditions are simple. This is the bound-state problem in which the wave function simply vanishes at large distances in all directions. This is why the differential Schrödinger equation is mainly used for bound states. The scattering and resonant states, where the boundary conditions at infinity are more complicated, traditionally are treated within the integral approach.

The differential approach that is developed in this book differs from the traditional one based on the Schrödinger equation. The difference is that instead of the direct calculation of a specific physical wave function, we look for the regular solution, and we look for it in the special semi-analytical form,

$$\phi_\ell(E,r) = h_\ell^{(-)}(kr)F_\ell^{(\mathrm{in})}(E,r) + h_\ell^{(+)}(kr)(E,r)F_\ell^{(\mathrm{out})}(E,r) \,, \tag{6.1}$$

where the correct asymptotic behaviour is explicitly embedded from the outset via the spherical waves $h_\ell^{(\pm)}$. This makes it easy to construct proper physical wave function when the regular solution is found. In contrast to the physical wave function, the boundary conditions for $\phi_\ell(E,r)$ are universal and are imposed at a single point ($r = 0$).

As was stated before, the differential equations (2.21) that determine the functions $F_\ell^{(\mathrm{in/out})}(E,r)$ are sufficient for obtaining the complete solution of the two-body problem with a given interaction potential. In principle, we can develop the whole theory based exclusively on these equations and can avoid using the alternative (traditional) approach with the Green's functions and integral equations. Since, however, there are some useful relations based on the integral equations and involving the Jost functions, we need to define and use the Green's functions.

S. A. Rakityansky, *Jost Functions in Quantum Mechanics*,
https://doi.org/10.1007/978-3-031-07761-6_6

In particular, they are needed to prove the so-called Jost–Pais theorem, which is described further down in Sect. 6.7. This theorem relates the Jost function $f_\ell^{(\mathrm{in})}$ to the Fredholm determinant of the integral equation for the physical wave function of the scattering state.

It should be noted that the Jost–Pais theorem is only used at the end of the book in Sect. 17.2.5, where the non-local potentials are considered. Therefore, this and several sections that follow (up to Sect. 6.7) can be omitted for the first reading.

6.1 Free Green's Function for Scattering Solution

Consider the scattering-state solution, $u_\ell(E,r)$, of the radial Schrödinger equation

$$\big[D_r - V(r)\big]u_\ell(E,r) = 0\,, \tag{6.2}$$

where D_r is the differential operator

$$D_r = \frac{d^2}{dr^2} + k^2 - \frac{\ell(\ell+1)}{r^2}\,, \tag{6.3}$$

which is the "free" part of the radial Hamiltonian. Comparing the second line of the mathematical transformations leading to Eq. (4.11), with Eq. (4.5), we see that the physical solution, $u_\ell(E,r)$, at large distances is a superposition of $j_\ell(kr)$ and $h_\ell^{(+)}(kr)$.

If we introduce the "free" Green's function, $g_\ell^{(+)}(r,r')$, obeying the equation

$$D_r g_\ell^{(+)}(r,r') = \delta(r-r')\,, \tag{6.4}$$

then the function

$$\boxed{u_\ell(E,r) = 2j_\ell(kr) + \int_0^\infty g_\ell^{(+)}(r,r')V(r')u_\ell(E,r')\,dr'} \tag{6.5}$$

solves Eq. (6.2), which is easy to check by the direct substitution.[1] For $r < r'$ as well as for $r > r'$, the right hand side of Eq. (6.4) is zero. This means that for any fixed r' and $r \neq r'$, the Green's function is a solution of the "free" Schrödinger equation, i.e. it is a linear combination of the Riccati–Bessel, Riccati–Neumann, and/or Riccati–Hankel functions, with the r'-dependent combination coefficients. The function $u_\ell(E,r)$ is the physical scattering solution, which must be regular, i.e. $u_\ell(E,0) = 0$. This means that to the left of r' the Green's function can only be proportional to $j_\ell(kr)$. As we established, when $r \to \infty$, the physical solution

[1] The factor 2 in front of j_ℓ in Eq. (6.5) appears because of our choice of the normalization of the Jost functions that are equal to 1/2 when the potential is absent.

should have both $j_\ell(kr)$ and $h_\ell^{(+)}(kr)$. Since $j_\ell(kr)$ is already present in Eq. (6.5), the Green's function to the right of r' can only be proportional to $h_\ell^{(+)}(kr)$. In other words, we come (although from different reasoning) to the same Eq. (3.95) that represents the corresponding Green's function for a bound-state solution. Therefore, following the same steps that take into account the continuity of $g_\ell^{(+)}(r, r')$ and discontinuity of its first derivative at $r = r'$, we obtain the same result, namely,

$$g_\ell^{(+)}(r, r') = \begin{cases} \frac{1}{ik} j_\ell(kr) h_\ell^{(+)}(kr') \,, & r \leqslant r' \\ \frac{1}{ik} h_\ell^{(+)}(kr) j_\ell(kr') \,, & r \geqslant r' \end{cases} \tag{6.6}$$

This is usually written as

$$g_\ell^{(+)}(r, r') = \frac{1}{ik} j_\ell(kr_<) h_\ell^{(+)}(kr_>) \,, \tag{6.7}$$

where $r_<$ and $r_>$ mean the smaller and larger of the two variables, r and r'.

The reader may wonder why the same Green's function corresponds to both the bound-state and scattering solutions. The explanation is that, first, this Green's function is used for such solutions at different (negative and positive) energies, and, second, in contrast to Eq. (3.91), the integral equation for the scattering wave function, Eq. (6.5), has non-zero inhomogeneous term, $2j_\ell(kr)$, which is needed for the scattering-type behaviour of the wave function at large distances.

Let us convince ourselves that the above Green's function gives the correct behaviour of the scattering solution, $u_\ell(E, r)$, both at small and at large distances. Since $g_\ell^{(+)}(0, r') = 0$, its short-range behaviour,

$$u_\ell(E, r) \underset{r \to 0}{\longrightarrow} 2j_\ell(kr) \,,$$

is obviously correct. When $r \to \infty$, the Green's function (6.6) substituted in Eq. (6.5) gives

$$\begin{aligned} u_\ell(E, r) \underset{r \to \infty}{\longrightarrow} \; & 2j_\ell(kr) + \frac{1}{ik} h_\ell^{(+)}(kr) \int_0^\infty j_\ell(kr') V(r') u_\ell(E, r') \, dr' \\ = \; & 2j_\ell(kr) + 2ikh_\ell^{(+)}(kr) f_\ell(E) \qquad (6.8) \\ = \; & 2j_\ell(kr) + 2ikh_\ell^{(+)}(kr) \left[\frac{f_\ell^{(\text{out})}(E)}{f_\ell^{(\text{in})}(E)} - 1 \right] \,, \end{aligned}$$

where we used Eqs. (4.14) and (4.12). Comparing (6.8) with the second line of (4.11), we see that the long-range behaviour of $u_\ell(E, r)$ is also correct.

In exactly the same way as for $g_\ell^{(+)}(r, r')$, we can obtain the Green's function $g_\ell^{(-)}(r, r')$,

$$g_\ell^{(-)}(r, r') = -\frac{1}{ik} j_\ell(kr_<) h_\ell^{(-)}(kr_>) , \tag{6.9}$$

which corresponds the unphysical solution $u_\ell^{(-)}(E, r)$ that is regular at the origin and has incoming (instead of outgoing) spherical waves at large distances.[2]

$$u_\ell^{(-)}(E, r) \underset{r\to\infty}{\longrightarrow} 2j_\ell(kr) - \frac{1}{ik} h_\ell^{(-)}(kr) \int_0^\infty j_\ell(kr') V(r') u_\ell^{(-)}(E, r')\, dr' . \tag{6.10}$$

6.2 Total Green's Function for Scattering Solution

Similarly to the free Green's function, $g_\ell^{(+)}(r, r')$, we can introduce the total Green's function, $G_\ell^{(+)}(r, r')$, obeying the equation

$$[D_r - V(r)]\, G_\ell^{(+)}(r, r') = \delta(r - r') . \tag{6.11}$$

It is also called the resolvent because, if it is known, then the solution of the differential equation can be found by simple integration,[3]

$$\boxed{u_\ell(E, r) = 2j_\ell(kr) + 2\int_0^\infty G_\ell^{(+)}(r, r') V(r') j_\ell(kr')\, dr'} \tag{6.12}$$

This can be checked by direct substitution of (6.12) in the differential equation (6.2).

The way of constructing the total Green's function is the same as for the free one. The only difference is that instead of the functions $j_\ell(kr)$ and $h_\ell^{(+)}(kr)$, which are the solutions of the free radial equation (6.4) for $r < r'$ and for $r > r'$, now we should use the corresponding solutions $\phi_\ell(E, r)$ and $\chi_\ell^{(+)}(E, r)$ of the total radial equation (6.11). This also implies that instead of the Wronskian (3.100) that appears

[2] We do not put the superscript (+) for the physical solution, i.e. $u_\ell^{(+)} \equiv u_\ell$, because the unphysical one is actually never used, and therefore, it is not necessary to complicate the notation for u_ℓ.

[3] The factor 2 in this equation appears as a result of our choice of the normalization of the Jost functions, which are equal to 1/2 for the free motion.

in the denominators of Eqs. (3.101, 3.102), we should put the Wronskian of the regular and the Jost solutions (see Eq. (2.74)):

$$W\left[\phi_\ell(E,r),\, \chi_\ell^{(+)}(E,r)\right] = 2ikf_\ell^{(\mathrm{in})}(E)\,. \tag{6.13}$$

This gives

$$\boxed{G_\ell^{(+)}(r,r') = \frac{\phi_\ell(E,r_<)\chi_\ell^{(+)}(E,r_>)}{2ikf_\ell^{(\mathrm{in})}(E)} = \frac{u_\ell(E,r_<)\chi_\ell^{(+)}(E,r_>)}{2ik}} \tag{6.14}$$

which corresponds and is reduced to (6.7) when the potential completely disappears. And again, similarly to the free Green's function $g_\ell^{(-)}(r,r')$ determining the unphysical solution $u_\ell^{(-)}(E,r)$, it is not difficult to obtain the corresponding total Green's function in the form:

$$G_\ell^{(-)}(r,r') = -\frac{\phi_\ell(E,r_<)\chi_\ell^{(-)}(E,r_>)}{2ikf_\ell^{(\mathrm{out})}(E)}\,. \tag{6.15}$$

The total and free Green's functions can be related to each other via an integral equation. In order to find such an equation, consider the iteration series for Eq. (6.5):

$$\begin{aligned} u_\ell(E,r) &= u_\ell^{(0)}(E,r) + \int_0^\infty g_\ell^{(+)}(r,r')V(r')u_\ell^{(0)}(E,r')\,dr' \\ &+ \iint_0^\infty g_\ell^{(+)}(r,r')V(r')g_\ell^{(+)}(r',r'')V(r'')u_\ell^{(0)}(E,r'')\,dr'dr'' + \cdots\,, \end{aligned}$$

where $u_\ell^{(0)}(E,r) = 2j_\ell(kr)$. Symbolically (in the operator form), this series can be written as

$$\begin{aligned} \langle r|u_\ell\rangle &= \langle r|u_\ell^{(0)}\rangle + \langle r|g_\ell^{(+)}V|u_\ell^{(0)}\rangle + \langle r|g_\ell^{(+)}Vg_\ell^{(+)}V|u_\ell^{(0)}\rangle + \cdots \\ &= \langle r|u_\ell^{(0)}\rangle + \langle r|\left(g_\ell^{(+)} + g_\ell^{(+)}Vg_\ell^{(+)} + g_\ell^{(+)}Vg_\ell^{(+)}Vg_\ell^{(+)} + \cdots\right)V|u_\ell^{(0)}\rangle\,. \end{aligned} \tag{6.16}$$

Comparing the second line of this equation with Eq. (6.12), we see that

$$G_\ell^{(+)} = g_\ell^{(+)} + g_\ell^{(+)}Vg_\ell^{(+)} + g_\ell^{(+)}Vg_\ell^{(+)}Vg_\ell^{(+)} + \cdots\,, \tag{6.17}$$

which can be organized in the following two ways:

$$
\begin{aligned}
G_\ell^{(+)} &= g_\ell^{(+)} + g_\ell^{(+)} V \left(g_\ell^{(+)} + g_\ell^{(+)} V g_\ell^{(+)} + g_\ell^{(+)} V g_\ell^{(+)} V g_\ell^{(+)} + \cdots\right) \\
&= g_\ell^{(+)} + \left(g_\ell^{(+)} + g_\ell^{(+)} V g_\ell^{(+)} + g_\ell^{(+)} V g_\ell^{(+)} V g_\ell^{(+)} + \cdots\right) V g_\ell^{(+)} ,
\end{aligned}
$$

from where it is clear that

$$G_\ell^{(+)} = g_\ell^{(+)} + g_\ell^{(+)} V G_\ell^{(+)} \tag{6.18}$$

$$= g_\ell^{(+)} + G_\ell^{(+)} V g_\ell^{(+)} , \tag{6.19}$$

and similarly

$$G_\ell^{(-)} = g_\ell^{(-)} + g_\ell^{(-)} V G_\ell^{(-)} \tag{6.20}$$

$$= g_\ell^{(-)} + G_\ell^{(-)} V g_\ell^{(-)} . \tag{6.21}$$

In explicit form, these equations read

$$G_\ell^{(\pm)}(r,r') = g_\ell^{(\pm)}(r,r') + \int_0^\infty g_\ell^{(\pm)}(r,r'') V(r'') G_\ell^{(\pm)}(r'',r')\, dr'' \tag{6.22}$$

$$G_\ell^{(\pm)}(r,r') = g_\ell^{(\pm)}(r,r') + \int_0^\infty G_\ell^{(\pm)}(r,r'') V(r'') g_\ell^{(\pm)}(r'',r')\, dr'' \tag{6.23}$$

The expansions of the type (6.17) are usually called Born series. Each term of this series can be symbolically depicted as a Feynman diagram (see Fig. 6.1). In the diagram picture, the Green's functions are called propagators. The free propagator, $g_\ell^{(+)}(r,r')$, "moves" the two-body system without interaction in such a way that the distance between the two particles changes from r' to r. The full propagator, $g_\ell^{(+)}(r,r')$, also changes the distance, but during this change the particles interact with each other (exchange the mediator particle) infinitely many times.

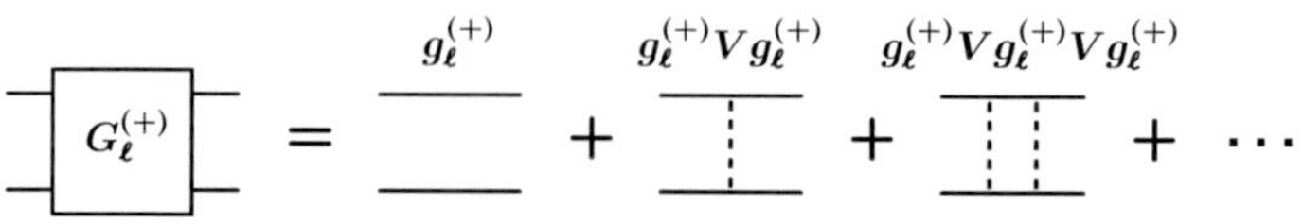

Fig. 6.1 Feynman diagrams representing the Born series (6.17). Horizontal lines correspond to free motion of the two particles, while vertical dashed lines show the action of the potential

6.3 Free and Total Green's Function for the Regular Solution

As we found, the (physical) scattering wave function, $u_\ell(E, r)$, differs from the regular solution, $\phi_\ell(E, r)$, by the normalization (see Eq. (4.4)). However, this is not the only difference. The fundamental difference consists in the different statements of the corresponding differential problems. For the physical solution, the boundary conditions for the Schrödinger equation are imposed at both $r = 0$ and $r \to \infty$ (Dirichlet problem), while for the regular solution they are set at $r = 0$ only (Cauchy problem).

In the simplest single-channel case, this difference is not tangible. This is because here the functional space of regular solutions is one-dimensional, i.e. there is just one basis vector $|\phi_\ell\rangle$ in this space, and any other solution that is also regular is proportional to $|\phi_\ell\rangle$. As we will see further in the book, in the multi-channel problems, the situation drastically changes.

Even in the single-channel case the Cauchy problem (for the regular solution) always (for any energy) has solution, while the Dirichlet problem, for example, at negative energies may have solutions only at certain discrete values of the energy. In this simplest case, the physical solution, if it exists, coincides with the regular solution up to a normalization factor. This means that, when finding the regular solution, $\phi_\ell(E, r)$, we obtain more than is needed because we obtain a solution of the Schrödinger equation even for such energies where the physical solution does not exist.

The fact that the boundary conditions are imposed on $\phi_\ell(E, r)$ at $r = 0$ only is reflected in the corresponding free Green's function $g_\ell(r, r')$ given by Eq. (2.46). This function is zero for $r' > r$ because no conditions are set at infinity, and thus the faraway points should not contribute to the integral part of the equation

$$\phi_\ell(E, r) = j_\ell(kr) + \int_0^\infty g_\ell(E, r, r')V(r')\phi_\ell(E, r')\,dr' . \tag{6.24}$$

In other words, the values of the function $\phi_\ell(E, r)$ on the interval $[0, r]$ cannot depend on its values outside this interval.

In Sect. 2.2.2.1, the free Green's function $g_\ell(r, r')$ is obtained by a simple combination of equations (2.40). However, it can also be derived in the standard way (as it is done in the preceding sections of this book for the physical solution), namely, by considering the differential equation

$$D_r g_\ell(r, r') = \delta(r - r') \tag{6.25}$$

with the condition

$$g_\ell(r, r') = 0 , \qquad \text{for} \quad r' \geqslant r . \tag{6.26}$$

It is seen that the condition (6.26) automatically guaranties fulfilment of the boundary condition for the regular solution, namely,

$$\phi_\ell(E,r) \underset{r\to 0}{\longrightarrow} j_\ell(kr) . \tag{6.27}$$

Indeed, when $r \to 0$, the interval where the integrand in (6.24) is non-zero shrinks to zero.

As is usually done in constructing a Green's function, we consider Eq. (6.25) for r to the left and to the right of r'. In both cases, this is the Riccati–Bessel equation. For $r < r'$, we use its trivial (identical zero) solution, and for $r > r'$, the Green's function is a linear combination of any two of its linearly independent solutions, for example, $j_\ell(kr)$ and $n_\ell(kr)$, where the combination coefficients are functions of r',

$$g_\ell(r,r') = j_\ell(kr)a(r') + n_\ell(kr)b(r') , \qquad r > r' . \tag{6.28}$$

At $r = r'$, the function $g_\ell(r,r')$ must be continuous (otherwise the first derivative in Eq. (6.25) would be infinite), which means that

$$j_\ell(kr)a(r) + n_\ell(kr)b(r) = 0 . \tag{6.29}$$

Since there is the δ-function on the right hand side of Eq. (6.25), the first derivative $\partial_r g_\ell(r,r')$ must have the unit discontinuity (like the Heaviside step-function) at $r = r'$. On the left hand side of r', the derivative is zero; therefore, it must be 1 on the right:

$$\left[\frac{d}{dr} j_\ell(kr)\right] a(r) + \left[\frac{d}{dr} n_\ell(kr)\right] b(r) = 1 . \tag{6.30}$$

Multiplying Eq. (6.29) by $\partial_r n_\ell(kr)$ and Eq. (6.30) by $n_\ell(kr)$, then subtracting the second from the first, we obtain

$$[j_\ell(kr)\partial_r n_\ell(kr) - n_\ell(kr)\partial_r j_\ell(kr)]\, a(r) = -n_\ell(kr) \;\Rightarrow\; a(r) = -\frac{1}{k} n_\ell(kr) ,$$

where we used known value of the Wronskian (see Appendix D.1)

$$W\left[j_\ell(kr), n_\ell(kr)\right] = k . \tag{6.31}$$

Then it follows from Eq. (6.29) that $b(r) = j_\ell(kr)/k$, and therefore,

$$g_\ell(r,r') = \begin{cases} \dfrac{1}{k}\left[n_\ell(kr)j_\ell(kr') - j_\ell(kr)n_\ell(kr')\right] , & r' \leqslant r \\ 0 , & r' \geqslant r \end{cases} \tag{6.32}$$

The other equivalent expressions for $g_\ell(r, r')$ given by Eqs. (2.46)–(2.51) can be obtained either using a different pair of the independent solutions of the Riccati–Bessel equation in (6.28) or simply by the transformations with the help of the relations

$$h_\ell^{(\pm)}(kr) = j_\ell(kr) \pm i n_\ell(kr) \ .$$

Similarly to Eq. (6.12), the total Green's function of the regular solution allows us to obtain $\phi_\ell(E, r)$ via the simple integration:

$$\boxed{\phi_\ell(E, r) = j_\ell(kr) + \int_0^\infty G_\ell(r, r')V(r')j_\ell(kr')\,dr'} \tag{6.33}$$

For that, it should obey the radial equation

$$[D_r - V(r)]\,G_\ell(r, r') = \delta(r - r') \tag{6.34}$$

as well as the condition

$$G_\ell(r, r') \equiv 0 \ , \qquad \text{for} \quad r' \geqslant r \ , \tag{6.35}$$

which guaranties the correct behaviour (6.27) of the regular solution when $r \to 0$. As is seen, everything is the same as for the free Green's function except for the presence of the potential term in the differential equation (6.34).

To construct the total Green's function, $G_\ell(r, r')$, we can follow exactly the same steps as for $g_\ell(r, r')$, but instead of the two independent solutions of the Riccati–Bessel equation, as the building blocks we should now use any two independent solutions of the total radial Schrödinger equation. Let us take them as ϕ_ℓ and $\chi^{(+)}$ (i.e. the regular solution and one of the Jost solutions). Then, the only difference in the derivations is that instead of (6.31), we should use the Wronskian (2.75), i.e.

$$W\left[\phi_\ell, \chi_\ell^{(+)}\right] = 2ikf_\ell^{(\mathrm{in})}(E) \ . \tag{6.36}$$

This gives

$$\boxed{G_\ell(r, r') = \begin{cases} \dfrac{1}{2ik}\left[\chi_\ell^{(+)}(E, r)u_\ell(E, r') - u_\ell(E, r)\chi_\ell^{(+)}(E, r')\right] & , r' \leqslant r \\ 0 & , r' \geqslant r \end{cases}} \tag{6.37}$$

where we replaced the ratio $\phi_\ell / f_\ell^{(\mathrm{in})}$ with the physical solution u_ℓ (see Eq. (4.4).

Using different sets of two independent solutions of the radial Schrödinger equation, we can find alternative (but of course equivalent) representations of the Green's operator G_ℓ. In particular, if we use both Jost solutions whose Wronskian is (see Eq. (2.80)

$$W\left[\chi_\ell^{(-)}, \chi_\ell^{(+)}\right] = 2ik\,, \tag{6.38}$$

then it is easy to find that

$$G_\ell(r, r') = \begin{cases} \dfrac{1}{2ik}\left[\chi_\ell^{(+)}(E, r)\chi_\ell^{(\;)}(E, r') - \chi_\ell^{(-)}(E, r)\chi_\ell^{(+)}(E, r')\right], & r' \leqslant r \\ 0 & , r' \geqslant r \end{cases}. \tag{6.39}$$

The Green's functions $G_\ell(r, r')$ and $g_\ell(r, r')$ are related to each other via an integral equation. It can be found in exactly the same way as we derived Eqs. (6.18, 6.19) relating the total and the free Green's operators $G_\ell^{(+)}$ and $G_\ell^{(+)}$. Iterating Eq. (6.24) and rearranging terms of the iteration series (similarly to what is done in Eq. (6.16)), we obtain the following equations:

$$G_\ell = g_\ell + g_\ell V G_\ell \tag{6.40}$$

$$= g_\ell + G_\ell V g_\ell\,. \tag{6.41}$$

6.4 Free and Total Green's Function for the Jost Solutions

The same as for the regular solution, $\phi_\ell(E, r)$, the differential problem for the Jost solutions, $\chi_\ell^{(\pm)}(E, r)$, is a Cauchy problem because the boundary conditions for them are imposed only at one end of the radial axis, $r \in [0, \infty)$, namely, at $r \to \infty$,

$$\chi_\ell^{(\pm)}(E, r) \underset{r\to\infty}{\longrightarrow} h_\ell^{(\pm)}(kr)\,. \tag{6.42}$$

This means that the corresponding Green's functions (free and total) are non-zero only for $r' > r$. The reason for this is similar to the reason why the Green's functions for the regular solution are non-zero only for $r' < r$. In fact, the way of constructing the Green's functions for $\chi_\ell^{(\pm)}(E, r)$ is the same as for $\phi_\ell(E, r)$. The only difference is that everything is "mirror-reflected" relative to the line $r = r'$ on the rr'-plane.

The free Green's function, $\mathcal{G}_\ell(r, r')$, is common for both $\chi_\ell^{(+)}(E, r)$ and $\chi_\ell^{(-)}(E, r)$ and is defined by the following equations:

$$\chi_\ell^{(\pm)}(E, r) = h_\ell^{(\pm)}(kr) + \int_0^\infty \mathcal{G}_\ell(E, r, r')V(r')\chi_\ell^{(\pm)}(E, r')\,dr'\,, \tag{6.43}$$

$$D_r \mathcal{G}_\ell(r, r') = \delta(r - r') \,, \tag{6.44}$$

$$\mathcal{G}_\ell(r, r') \equiv 0 \,, \qquad \text{for} \quad r' \leqslant r \,. \tag{6.45}$$

Following the same steps as in Sect. 6.3, we obtain

$$\boxed{\mathcal{G}_\ell(r, r') = \begin{cases} 0 & , \; r' \leqslant r \\ -\dfrac{1}{k}\left[n_\ell(kr) j_\ell(kr') - j_\ell(kr) n_\ell(kr')\right] & , \; r' \geqslant r \end{cases}} \tag{6.46}$$

The minus sign in front of the non-zero expression in this equation appears because (in opposite to the case of the regular solution) here the first derivative $\partial_r \mathcal{G}_\ell(r, r')$ becomes zero to the right of r', and therefore, $\partial_r \mathcal{G}_\ell(r, r')$ must be -1 when r approaches r' from the left. This is needed to correctly obtain in Eq. (6.44) the δ-function that is the derivative of the Heaviside step-function. Apart from this sign, this expression is the same as in Eq. (6.32). Therefore, the same as for $g_\ell(r, r')$, we can use for $\mathcal{G}_\ell(r, r')$ any of the equivalent expressions (2.46)–(2.51), but with the opposite sign. In particular, we will need the following:

$$\mathcal{G}_\ell(r, r') = \begin{cases} 0 & , \; r' \leqslant r \\ -\dfrac{1}{ik}\left[h_\ell^{(+)}(kr) j_\ell(kr') - j_\ell(kr) h_\ell^{(+)}(kr')\right] & , \; r' \geqslant r \,. \end{cases} \tag{6.47}$$

In fact, the Green's function $\mathcal{G}_\ell(r, r')$ is the transpose of $g_\ell(r, r')$,

$$\mathcal{G}_\ell(r, r') = g_\ell(r', r) \,. \tag{6.48}$$

When doing the transposition, we automatically acquire the opposite sign, and the condition $r' \leqslant r$ is changed to the opposite as well.

It is easy to see that the total Green's function, $\mathbb{G}_\ell(r, r')$, is also the same for both $\chi_\ell^{(+)}(E, r)$ and $\chi_\ell^{(-)}(E, r)$. It is defined by the equations:

$$\chi_\ell^{(\pm)}(E, r) = h_\ell^{(\pm)}(kr) + \int_0^\infty \mathbb{G}_\ell(E, r, r') V(r') h_\ell^{(\pm)}(E, r') \, dr' \,, \tag{6.49}$$

$$[D_r - V(r)] \, \mathbb{G}_\ell(r, r') = \delta(r - r') \,, \tag{6.50}$$

$$\mathbb{G}_\ell(r, r') \equiv 0 \,, \qquad \text{for} \quad r' \leqslant r \,. \tag{6.51}$$

We can repeat the derivations of the preceding section (with the opposite sign for the discontinuity of the first derivative) and obtain that

$$\mathbb{G}_\ell(r,r') = \begin{cases} 0 & , r' \geqslant r \\ -\dfrac{1}{2ik}\left[\chi_\ell^{(+)}(E,r)u_\ell(E,r') - u_\ell(E,r)\chi_\ell^{(+)}(E,r')\right] & , r' \leqslant r \end{cases} \tag{6.52}$$

Therefore,

$$\mathbb{G}_\ell(r,r') = G_\ell(r',r)\,. \tag{6.53}$$

Iterating Eq. (6.43) and rearranging terms of the iteration series (similarly to what is done in Eq. (6.16)), we obtain the following equations:

$$\mathbb{G}_\ell = \mathcal{G}_\ell + \mathcal{G}_\ell V \mathbb{G}_\ell \tag{6.54}$$

$$= \mathcal{G}_\ell + \mathbb{G}_\ell V \mathcal{G}_\ell\,. \tag{6.55}$$

6.5 Three-Dimensional Free Green's Function

The Green's function (6.7) is associated with the radial (one-dimensional) equation. Actually, it is the partial-wave component of the corresponding full three-dimensional Green's function that is involved in the full ($3D$) integral equation. Further down in the book (see Sect. 17.3), this Green's function is used when the so-called off-shell Jost function is considered. For the sake of completeness, it is therefore introduced here.

Consider the three-dimensional Schrödinger equation for the scattering wave function,

$$\left(\frac{\hbar^2}{2\mu}\Delta + E\right)\psi_{\vec{k}}(\vec{r}\,) = U(\vec{r}\,)\psi_{\vec{k}}(\vec{r}\,)\,. \tag{6.56}$$

Far away from the interaction region, the potential vanishes, and Eq. (6.56) becomes more simple:

$$\left(\frac{\hbar^2}{2\mu}\Delta + E\right)\psi_{\vec{k}}(\vec{r}\,) = 0\,, \qquad r \to \infty\,.$$

This second-order differential equation has two independent solutions:

$$\varphi_{\vec{k}}^{(+)}(\vec{r}\,) = e^{+i\vec{k}\vec{r}} \qquad \text{and} \qquad \varphi_{\vec{k}}^{(-)}(\vec{r}\,) = e^{-i\vec{k}\vec{r}} \,. \tag{6.57}$$

The general solution of this equation is, therefore, their linear combination:

$$\varphi_{\vec{k}}(\vec{r}\,) = C_1\varphi_{\vec{k}}^{(+)}(\vec{r}\,) + C_2\varphi_{\vec{k}}^{(-)}(\vec{r}\,) \,, \tag{6.58}$$

where C_1 and C_2 are arbitrary constants. Formally, the right hand side of Eq. (6.56) can be viewed as the inhomogeneous term. In the theory of differential equations, it is known that the general solution of a linear equation can be constructed as the general solution of the complementary (homogeneous) equation plus any particular solution of the full inhomogeneous equation. Therefore, to find the general solution of Eq. (6.56), we need to add to $\varphi_{\vec{k}}(\vec{r}\,)$ any particular solution of the full equation (6.56). Such a solution can be found using the Green's function defined as

$$\left(\frac{\hbar^2}{2\mu}\Delta + E\right) g(\vec{r}, \vec{r}\,') = \delta(\vec{r} - \vec{r}\,') \,. \tag{6.59}$$

Then the general solution of Eq. (6.56) is

$$\psi_{\vec{k}}(\vec{r}\,) = \varphi_{\vec{k}}(\vec{r}\,) + \int g(\vec{r}, \vec{r}\,')U(\vec{r}\,)\psi_{\vec{k}}(\vec{r}\,')\, d^3r' \,. \tag{6.60}$$

Eq. (6.59) determining the Green's function can be solved in a closed form. To this end, we replace the functions $g(\vec{r}, \vec{r}\,')$ and $\delta(\vec{r} - \vec{r}\,')$ with their Fourier transforms:

$$g(\vec{r}, \vec{r}\,') = \frac{1}{(2\pi)^3} \int e^{i\vec{q}\cdot(\vec{r}-\vec{r}\,')} \gamma(\vec{q}\,)\, d^3q$$

$$\delta(\vec{r} - \vec{r}\,') = \frac{1}{(2\pi)^3} \int e^{i\vec{q}\cdot(\vec{r}-\vec{r}\,')}\, d^3q \,.$$

Substitution of these integrals into Eq. (6.59) gives

$$\frac{\hbar^2}{2\mu}\left(\Delta + k^2\right) \int e^{i\vec{q}\cdot(\vec{r}-\vec{r}\,')} \gamma(\vec{q}\,)\, d^3q = \int e^{i\vec{q}\cdot(\vec{r}-\vec{r}\,')}\, d^3q \,,$$

$$\frac{\hbar^2}{2\mu} \int \left(-q^2 + k^2\right) e^{i\vec{q}\cdot(\vec{r}-\vec{r}\,')} \gamma(\vec{q}\,)\, d^3q = \int e^{i\vec{q}\cdot(\vec{r}-\vec{r}\,')}\, d^3q \,,$$

$$\frac{\hbar^2}{2\mu}\left(-q^2+k^2\right)\gamma(\vec{q}\,)=1\,,$$

$$\gamma(\vec{q}\,)=\frac{2\mu}{\hbar^2}\cdot\frac{1}{k^2-q^2}\,.$$

Therefore,

$$g(\vec{r},\vec{r}\,')=\frac{2\mu}{(2\pi)^3\hbar^2}\int\frac{e^{i\vec{q}\cdot(\vec{r}-\vec{r}\,')}}{k^2-q^2}\,d^3q\,,$$

where the integration over the spherical angles of vector $\vec{q}$ can be easily done:

$$\begin{aligned}g(\vec{r},\vec{r}\,')&=\frac{2\mu}{(2\pi)^2\hbar^2}\int_0^\infty\frac{q^2}{k^2-q^2}\int_0^\pi e^{iq|\vec{r}-\vec{r}\,'|\cos\theta}\sin\theta\,d\theta\,dq\\&=\frac{2\mu}{(2\pi)^2\hbar^2}\int_0^\infty\frac{q^2}{k^2-q^2}\int_{-1}^{+1}e^{iq|\vec{r}-\vec{r}\,'|t}\,dt\,dq\\&=\frac{2\mu}{(2\pi)^2\hbar^2}\int_0^\infty\frac{q^2}{k^2-q^2}\cdot\frac{e^{iq|\vec{r}-\vec{r}\,'|}-e^{-iq|\vec{r}-\vec{r}\,'|}}{iq|\vec{r}-\vec{r}\,'|}\,dq\\&=-\frac{2\mu}{(2\pi)^2i\hbar^2|\vec{r}-\vec{r}\,'|}\int_{-\infty}^{+\infty}\frac{qe^{iq|\vec{r}-\vec{r}\,'|}}{(q-k)(q+k)}\,dq\,.\end{aligned}\tag{6.61}$$

In the last integral, we encounter a problem, namely, at $q=\pm k$ the integrand has poles. To circumvent this problem, we can use the fact that, actually, we need to find any particular solution of Eq. (6.56). This means that we can use any possible trick to avoid the singularities. As such a trick, we can shift the energy variable $E=(\hbar k)^2/(2\mu)$ from the real axis either up or down by adding or subtracting an infinitely small imaginary term:

$$E\longrightarrow E\pm i\varepsilon\,,\qquad \varepsilon>0\,,\tag{6.62}$$

which should be put to zero, $\varepsilon\to 0$, in all final formulae. After such a shift, the poles on the complex momentum plane (q-plane) are also shifted:

$$E+i\varepsilon\Rightarrow\begin{cases}k\to\ \ k+i\epsilon\,,\\-k\to -k-i\epsilon\,,\end{cases}\tag{6.63}$$

$$E-i\varepsilon\Rightarrow\begin{cases}k\to\ \ k-i\epsilon\,,\\-k\to -k+i\epsilon\,,\end{cases}\tag{6.64}$$

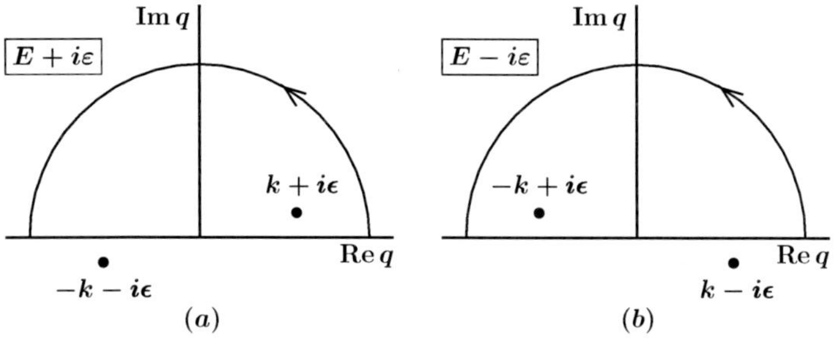

Fig. 6.2 Infinitely small shifts of the poles in Eq. (6.61), corresponding to the shift of the energy either up (*a*) or down (*b*)

where $\epsilon > 0$ is the corresponding infinitely small shift of the momentum. This is illustrated in Fig. 6.2.

In both cases, the integral can be evaluated using the residue theorem that states that a closed contour integral of a meromorphic function $f(z)$ can be found as

$$\oint f(z)\,dz = 2\pi i\,[\mathrm{Res}(f, z_1) + \mathrm{Res}(f, z_2) + \cdots + \mathrm{Res}(f, z_N)]\ ,$$

where $z_1, z_2, \ldots, z_N$ are those poles of the function that are enclosed by the contour, and $\mathrm{Res}(f, z_n)$ is the residue of $f(z)$ at the pole z_n.

For the integral (6.61), the contour can be enclosed with a semicircle of infinitely large radius in the upper half of the complex q-plane because the integrand exponentially tends to zero when the radius of that semicircle increases. Among the two poles of the integrand, only one is inside the contour. In both cases, the pole is of the first order for which

$$\mathrm{Res}(f, z_n) = \lim_{z\to z_n}\,[(z - z_n)f(z)]\ .$$

If we denote the choices of the energy shift up or down in (6.62) by the superscripts (+) or (−), then the residue theorem, applied for the integral (6.61), gives

$$\boxed{g^{(\pm)}(\vec{r}, \vec{r}\,') = -\frac{\mu}{2\pi\hbar^2}\frac{e^{\pm ik|\vec{r}-\vec{r}\,'|}}{|\vec{r} - \vec{r}\,'|}} \tag{6.65}$$

Therefore, the general solution (6.60) of the Schrödinger equation may have the following two different forms:

$$\psi^{(+)}_{\vec{k}}(\vec{r}\,) = C_1 e^{+i\vec{k}\vec{r}} + C_2 e^{-i\vec{k}\vec{r}} - \frac{\mu}{2\pi\hbar^2}\int \frac{e^{+ik|\vec{r}-\vec{r}\,'|}}{|\vec{r}-\vec{r}\,'|} U(\vec{r}\,')\psi^{(+)}_{\vec{k}}(\vec{r}\,')\, d^3r'\,,$$

$$\psi^{(-)}_{\vec{k}}(\vec{r}\,) = C_1 e^{+i\vec{k}\vec{r}} + C_2 e^{-i\vec{k}\vec{r}} - \frac{\mu}{2\pi\hbar^2}\int \frac{e^{-ik|\vec{r}-\vec{r}\,'|}}{|\vec{r}-\vec{r}\,'|} U(\vec{r}\,')\psi^{(-)}_{\vec{k}}(\vec{r}\,')\, d^3r'\,.$$

The choice between the solutions $\psi^{(+)}_{\vec{k}}(\vec{r}\,)$ and $\psi^{(-)}_{\vec{k}}(\vec{r}\,)$ as well as the choice of the coefficients C_1 and C_2 is determined by the boundary conditions, i.e. by the asymptotic behaviour of the scattering wave function. First of all, we see that we must retain only one of the exponentials

$$C_1 e^{+i\vec{k}\vec{r}} + C_2 e^{-i\vec{k}\vec{r}}$$

because they describe the plane waves (incident beams) moving in the opposite directions. We can always choose the coordinate system in such a way that either $C_1 = 0$ or $C_2 = 0$. The plane waves are normalized as is given by Eq. (4.1); therefore, the non-zero coefficient (C_1 or C_2) is equal to $1/(2\pi)^{3/2}$. Without compromising the generality, we can write the two possible solutions as

$$\boxed{\psi^{(\pm)}_{\vec{k}}(\vec{r}\,) = \frac{e^{\pm i\vec{k}\vec{r}}}{(2\pi)^{3/2}} - \frac{\mu}{2\pi\hbar^2}\int d^3r'\,\frac{\exp(\pm ik|\vec{r}-\vec{r}\,'|)}{|\vec{r}-\vec{r}\,'|} U(\vec{r}\,')\psi^{(+)}_{\vec{k}}(\vec{r}\,')} \tag{6.66}$$

which is known as the Lippmann–Schwinger equation.

In order to finally make choice between $\psi^{(+)}_{\vec{k}}(\vec{r}\,)$ and $\psi^{(-)}_{\vec{k}}(\vec{r}\,)$ (which is physical and which is not), we need to examine the asymptotic behaviour of their integral terms.

As the reader should remember, so far we consider finite-range potentials, which are identical zeros beyond certain distance R. This means that the integral over r' in Eq. (6.66) only runs up to that distance, i.e. $r' \leqslant R$. When $r \to \infty$, we can always consider such values of r that $r \gg r'$. Actually, the same can be done in the general case as well, if the potential vanishes at large distances sufficiently fast. In such a case, the contribution into the integral from large values of r' becomes infinitely small, and we can always choose sufficiently large r such that for $r' \sim r$ this contribution is negligible. In other words, the variable r' actually never goes to infinity, and we can always choose such values of r that $r \gg r'$.

Therefore, ignoring all the smaller terms, we can use the following approximation:

$$|\vec{r}-\vec{r}\,'| = \sqrt{r^2 - 2\vec{r}\cdot\vec{r}\,' + r'^2} \underset{r\to\infty}{\approx} r - \frac{\vec{r}}{r}\cdot\vec{r}\,' + \mathcal{O}\left(\frac{r'}{r}\right)^2 .$$

We use this approximation in the exponential function of the integral (6.66), while the denominator can simply be replaced there with r. This gives

$$k|\vec{r}-\vec{r}\,'| \approx kr - k\frac{\vec{r}}{r}\cdot\vec{r}\,' = kr - \vec{k}_r\cdot\vec{r}\,',$$

where

$$\vec{k}_r = k\frac{\vec{r}}{r}$$

is the momentum in the direction of vector $\vec{r}$. Therefore,

$$\psi^{(\pm)}_{\vec{k}}(\vec{r}\,) \underset{r\to\infty}{\longrightarrow} \frac{e^{\pm i\vec{k}\vec{r}}}{(2\pi)^{3/2}} - \frac{\mu}{2\pi\hbar^2}\cdot\frac{e^{\pm ikr}}{r}\int d^3r'\, e^{\mp i\vec{k}_r\vec{r}\,'}U(\vec{r}\,')\psi^{(+)}_{\vec{k}}(\vec{r}\,')$$

$$\underset{r\to\infty}{\longrightarrow} \frac{1}{(2\pi)^{3/2}}\left[e^{\pm i\vec{k}\vec{r}} - \frac{e^{\pm ikr}}{r}\cdot\frac{(2\pi)^2\mu}{\hbar^2}\int d^3r'\,\frac{e^{\mp i\vec{k}_r\vec{r}\,'}}{(2\pi)^{3/2}}U(\vec{r}\,')\psi^{(+)}_{\vec{k}}(\vec{r}\,')\right] . \tag{6.67}$$

Comparing this with Eq. (4.11), we conclude that the correct asymptotic behaviour, i.e. the correct boundary conditions at infinity, is obtained with the choice of the positive sign in Eq. (6.62). In other words, the physical wave function that describes scattering process is $\psi^{(+)}_{\vec{k}}(\vec{r}\,)$. The other solution of the Schrödinger equation, namely, $\psi^{(-)}_{\vec{k}}(\vec{r}\,)$, is unphysical. In this book, we will not use it and therefore (for the sake of notation simplicity) will drop the superscript $(+)$ denoting the physical solution.

The asymptotic behaviour (6.67) of the physical wave function can be written as

$$\psi_{\vec{k}}(\vec{r}\,) \underset{r\to\infty}{\longrightarrow} \frac{1}{(2\pi)^{3/2}}\left[e^{i\vec{k}\vec{r}} + \frac{e^{ikr}}{r}\mathsf{f}(\hat{\vec{r}} \leftarrow \hat{\vec{k}}, E)\right] , \tag{6.68}$$

where the scattering amplitude is

$$\mathsf{f}(\hat{\vec{r}} \leftarrow \hat{\vec{k}}, E) = -\frac{(2\pi)^2\mu}{\hbar^2}\int d^3r'\,\frac{e^{-i\vec{k}_r\vec{r}\,'}}{(2\pi)^{3/2}}U(\vec{r}\,')\psi_{\vec{k}}(\vec{r}\,') , \tag{6.69}$$

i.e.

$$\mathrm{f}(\hat{\vec{r}} \leftarrow \hat{\vec{k}}, E) = -\frac{(2\pi)^2\mu}{\hbar^2}\langle \vec{k}_r\,|U|\psi_{\vec{k}}\rangle \tag{6.70}$$

Here $|\vec{k}_r\,\rangle$ is the state of free motion of the particle with the momentum $|\vec{k}_r\,| = |\vec{k}\,|$ in the direction of vector $\vec{r}$. In simple words, $\vec{k}_r$ is the momentum that the particle has after the scattering.

Substituting in Eq. (6.69), the partial-wave expansions of the plane wave $\langle \vec{k}_r\,|\vec{r}\,'\rangle$ (Eq. (4.2)) as well as of the wave function $\psi_{\vec{k}}(\vec{r}\,')$ (Eq. (4.5)), and performing the angular integration over $\hat{\vec{r}}\,'$ under the assumption that the potential is spherically symmetric, we once again obtain Eqs. (4.13) and (4.14) from a different approach.[4] It would be useful and instructive for the reader to perform these derivations as an exercise.

Coming back to the very first statement of the present section, let us convince ourselves that the radial Green's functions $g_\ell^{(\pm)}(r, r')$ obtained in Sect. 6.1 (see Eqs. (6.7) and (6.9)) are the partial-wave components of the three-dimensional functions (6.65). To this end, we use the formulae of Appendix A. For the sake of clarity, we only do the derivations for $g_\ell^{(+)}(r, r')$. The case of $g_\ell^{(-)}(r, r')$ can be treated in exactly the same way.

As the starting point, we replace all the functions in the Lippmann–Schwinger equation (6.66) with their partial-wave expansions:

$$\begin{aligned}
\frac{1}{\sqrt{2\pi}kr}\sum_{\ell m} i^\ell u_\ell(E,r)Y^*_{\ell m}(\hat{\vec{k}}\,)Y_{\ell m}(\hat{\vec{r}}\,) &= \sqrt{\frac{2}{\pi}}\frac{1}{kr}\sum_{\ell m} i^\ell j_\ell(kr)Y^*_{\ell m}(\hat{\vec{k}}\,)Y_{\ell m}(\hat{\vec{r}}\,) \\
&-\frac{i}{\sqrt{2\pi}k^2 r}\int_0^\infty dr' \sum_{\ell m \ell' m'} j_\ell(kr_<)h_\ell^{(+)}(kr_>)V(r')i^{\ell'}u_{\ell'}(E,r')Y^*_{\ell' m'}(\hat{\vec{k}}\,) \\
&\times \int d\Omega_{\vec{r}\,'}\, Y^*_{\ell m}(\Omega_{\vec{r}_<})Y_{\ell m}(\Omega_{\vec{r}_>})Y_{\ell' m'}(\hat{\vec{r}}\,')\,,
\end{aligned} \tag{6.71}$$

where $V(r) = 2\mu U(r)/\hbar^2$. It is shown in Appendix A that

$$\mathrm{Im}\left[\sum_m Y^*_{\ell m}(\Omega_{\vec{r}_1})Y_{\ell m}(\Omega_{\vec{r}_2})\right] = 0\,,$$

[4] The reader should not forget that in Eq. (4.14) the potential $V = 2\mu U/\hbar^2$ is in the units of $[\text{length}]^{-2}$.

i.e. that such a sum is real. Therefore in the angular integral (6.71), we can do the following replacement:

$$\sum_m Y^*_{\ell m}(\Omega_{\vec{r}_<})Y_{\ell m}(\Omega_{\vec{r}_>}) = \sum_m Y_{\ell m}(\Omega_{\vec{r}_<})Y^*_{\ell m}(\Omega_{\vec{r}_>}) = \sum_m Y_{\ell m}(\Omega_{\vec{r}})Y^*_{\ell m}(\Omega_{\vec{r}\,'}) .$$

Then the integration over the spherical angles of vector $\vec{r}\,'$ gives the Kronecker symbols, $\delta_{\ell\ell'}\delta_{mm'}$, and thus Eq. (6.71) becomes

$$\sum_{\ell m} i^\ell u_\ell(E,r)Y^*_{\ell m}(\hat{\vec{k}})Y_{\ell m}(\hat{\vec{r}}) = 2\sum_{\ell m} i^\ell j_\ell(kr)Y^*_{\ell m}(\hat{\vec{k}})Y_{\ell m}(\hat{\vec{r}})$$

$$+\frac{1}{ik}\int_0^\infty dr' \sum_{\ell m} j_\ell(kr_<)h_\ell^{(+)}(kr_>)V(r')i^\ell u_\ell(E,r')Y^*_{\ell m}(\hat{\vec{k}})Y_{\ell m}(\hat{\vec{r}}) ,$$

from where we obtain the same radial Lippmann–Schwinger equation (6.5):

$$u_\ell(E,r) = 2j_\ell(kr) + \frac{1}{ik}\int_0^\infty dr'\, j_\ell(kr_<)h_\ell^{(+)}(kr_>)V(r')u_\ell(E,r') . \qquad (6.72)$$

Therefore,

$$g^{(+)}(\vec{r},\vec{r}\,') = \langle \vec{r}\,|g^{(+)}|\vec{r}\,'\rangle = \sum_{\ell\ell' mm'} \langle r\ell m|g^{(+)}|r'\ell' m'\rangle Y_{\ell m}(\hat{\vec{r}})Y^*_{\ell' m'}(\hat{\vec{r}}\,') , \qquad (6.73)$$

where

$$\boxed{\langle r\ell m|g^{(+)}|r'\ell' m'\rangle = \delta_{\ell\ell'}\delta_{mm'}\,\frac{2\mu}{\hbar^2}\cdot\frac{g_\ell^{(+)}(r,r')}{rr'}} \qquad (6.74)$$

6.6 Summary

The Green's functions that we obtained in this chapter are schematically depicted in Figs. 6.3, 6.4 and 6.5, where the diagonal line $r = r'$ separates the two sectors, $r' < r$ and $r' > r$. In each product of two functions shown in these figures, the first one depends on r, and the second on r', for example $h_\ell^{(+)} j_\ell = h_\ell^{(+)}(kr)j_\ell(kr')$. On the diagonal line, all the Green's functions are continuous but have a "sharp bend" such that their first derivatives behave as the Heaviside step-function at $r = r'$. This makes the second derivative to be the δ-function.

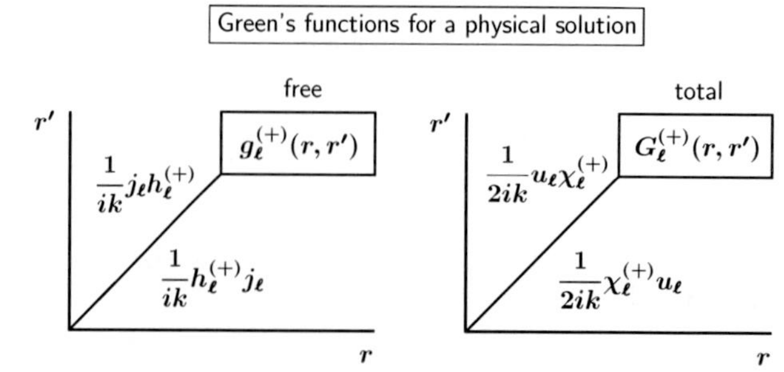

Fig. 6.3 Free (left) and total (right) Green's functions for a physical solution in the two sectors of the rr'-plane

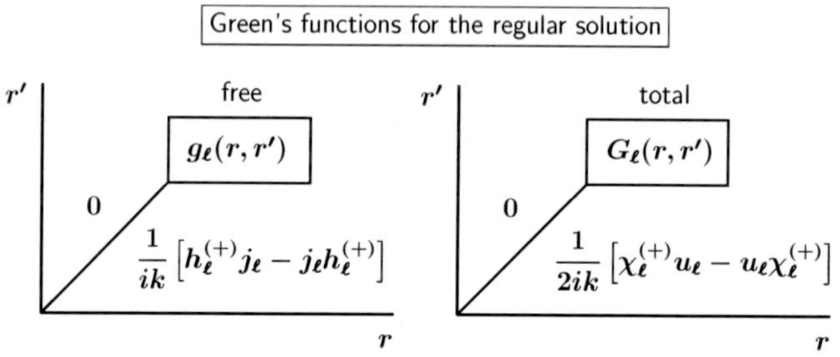

Fig. 6.4 Free (left) and total (right) Green's functions for the regular solution in the two sectors of the rr'-plane

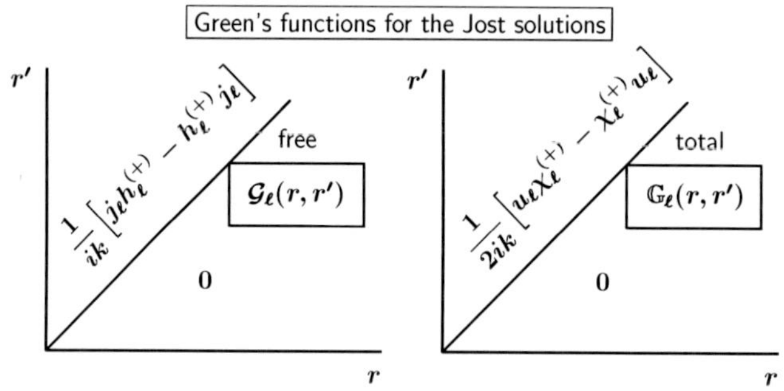

Fig. 6.5 Free (left) and total (right) Green's functions for the Jost solutions in the two sectors of the rr'-plane

It is instructive to compare the corresponding free and total Green's functions. One can see that in the absence of the potential, they become the same because

$$(V \to 0) \;\Rightarrow\; \begin{cases} u_\ell \to 2 j_\ell \,, \\ \\ \chi_\ell^{(+)} \to h_\ell^{(+)} \,. \end{cases} \tag{6.75}$$

It should be noted that Figs. 6.3, 6.4, and 6.5 show only one of several possible (equivalent) representations of each of the Green's functions. The other representations can be obtained using different pairs of linearly independent solutions of the free or the total (including the potential) radial Schrödinger equation. For example, all different representations of $g_\ell(r, r')$ can be found in Sect. 2.2.2.1.

The partial-wave expansions of the free Green's functions for the three-dimensional Scrödinger equation involve the corresponding radial components $g_\ell^{(\pm)}(r, r')$,

$$g^{(\pm)}(\vec{r}, \vec{r}\,') = -\frac{\mu}{2\pi\hbar^2}\frac{e^{\pm ik|\vec{r}-\vec{r}\,'|}}{|\vec{r}-\vec{r}\,'|} = \frac{2\mu}{\hbar^2}\cdot\frac{1}{rr'}\sum_{\ell m} g_\ell^{(\pm)}(r, r')Y_{\ell m}(\hat{\vec{r}})Y^*_{\ell' m'}(\hat{\vec{r}}\,') \,. \tag{6.76}$$

6.7 Jost–Pais Theorem

As is described in Appendix F.3.2, an integral equation of the Fredholm type can be solved using the convergent series known as the Fredholm determinant. When the Fredholm theory is applied to the integral equation (6.5) for the scattering wave function u_ℓ, it can be shown that the Jost function $f_\ell^{(\mathrm{in})}$ is proportional to the Fredholm determinant of its kernel. This finding is known as the Jost–Pais theorem.

In order to prove this theorem, we need to introduce the strength parameter, λ, of the potential, i.e. we replace V with λV. The physical value of this parameter is $\lambda = 1$.

Using the formal notation similar to what was used in Eq. (6.16), the integral equation (6.24) for the regular solution can be written as

$$|\phi_\ell\rangle = |j_\ell\rangle + \lambda g_\ell V|\phi_\ell\rangle \,. \tag{6.77}$$

The only λ-dependent function here is ϕ_ℓ. Taking derivative of this equation over λ, we have

$$|\phi'_\ell\rangle = g_\ell V|\phi_\ell\rangle + \lambda g_\ell V|\phi'_\ell\rangle \,, \tag{6.78}$$

where the prime means derivative over λ. Formal solution of (6.78) with respect to ϕ'_ℓ is

$$|\phi'_\ell\rangle = (1 - \lambda g_\ell V)^{-1} g_\ell V|\phi_\ell\rangle \,.$$

As it follows from Eq. (6.40)

$$(1 - \lambda g_\ell V)^{-1} g_\ell = G_\ell \,. \tag{6.79}$$

Therefore,

$$|\phi'_\ell\rangle = G_\ell V|\phi_\ell\rangle\,, \tag{6.80}$$

i.e.

$$\phi'_\ell(E,r) = \int_0^\infty dr'\, G_\ell(r,r')V(r')\phi_\ell(E,r')\,. \tag{6.81}$$

Now we recall Eq. (2.65) that relates the Jost function to the regular solution:

$$f_\ell^{(\mathrm{in})} = \frac{1}{2} - \frac{\lambda}{2ik}\int_0^\infty dr\, h_\ell^{(+)}(kr)V(r)\phi_\ell(E,r)\,. \tag{6.82}$$

Taking derivative of this equation over λ, we obtain

$$\begin{aligned}\frac{d}{d\lambda}f_\ell^{(\mathrm{in})} &= -\frac{1}{2ik}\int_0^\infty dr\, h_\ell^{(+)}(kr)V(r)\phi_\ell(E,r)\\ &\quad -\frac{\lambda}{2ik}\int_0^\infty dr\, h_\ell^{(+)}(kr)V(r)\phi'_\ell(E,r)\,.\end{aligned}$$

In the second integral on the right hand side of this equation, the derivative ϕ'_ℓ can be replaced by (6.81). This gives

$$\begin{aligned}\frac{d}{d\lambda}f_\ell^{(\mathrm{in})} &= -\frac{1}{2ik}\int_0^\infty dr\, h_\ell^{(+)}(kr)V(r)\phi_\ell(E,r)\\ &\quad -\frac{\lambda}{2ik}\iint_0^\infty dr\,dr'\, h_\ell^{(+)}(kr)V(r)G_\ell(r,r')V(r')\phi_\ell(E,r')\,.\end{aligned} \tag{6.83}$$

At this point, we can use the fact that the transposition of G_ℓ gives the Green's function for the Jost solutions (see Eq. (6.53)). We also divide all terms of (6.83) by the Jost function and take into account that $u_\ell = \phi_\ell/f_\ell^{(\mathrm{in})}$ (see Eq. (4.4)). In this way, we obtain

$$\begin{aligned}\frac{\partial_\lambda f_\ell^{(\mathrm{in})}}{f_\ell^{(\mathrm{in})}} &= -\frac{1}{2ik}\int_0^\infty dr'\left[h_\ell^{(+)}(kr') + \lambda\int_0^\infty dr\, \mathbb{G}(r',r)V(r)h_\ell^{(+)}(kr)\right]\\ &\quad \times V(r')u_\ell(E,r')\,.\end{aligned}$$

According to Eq. (6.49), the expression in the square brackets is equal to the Jost solution. Therefore,

$$\frac{d\ln f_\ell^{(\mathrm{in})}}{d\lambda} = -\frac{1}{2ik}\int_0^\infty \chi_\ell^{(+)}(E,r)V(r)u_\ell(E,r)\,dr\,. \tag{6.84}$$

Taking into account Eq. (6.14), we obtain

$$\frac{d\ \ln f_\ell^{(\mathrm{in})}}{d\lambda} = -\int_0^\infty G_\ell^{(+)}(r,r)V(r)\,dr\ , \tag{6.85}$$

where $G_\ell^{(+)}$ can be replaced with the left hand side of Eq. (6.79), which gives

$$\frac{d\ \ln f_\ell^{(\mathrm{in})}}{d\lambda} = -\int_0^\infty [1-\lambda g_\ell^{(+)}(r,r)V(r)]^{-1} g_\ell^{(+)}(r,r)V(r)\,dr \tag{6.86}$$

$$= -\mathrm{Sp}\left([1-\lambda g_\ell^{(+)}V]^{-1} g_\ell^{(+)}V\right)\ . \tag{6.87}$$

What remains is to integrate both sides of this equation over λ from 0 to 1. In doing this, we recall that our normalization is such that $f_\ell^{(\mathrm{in})} = 1/2$ for $\lambda = 0$, i.e. for free motion. The integration gives

$$f_\ell^{(\mathrm{in})} = \frac{1}{2}\exp\left\{-\int_0^1 \mathrm{Sp}\left[(1-\lambda g_\ell^{(+)}V)^{-1} g_\ell^{(+)}V\right] d\lambda\right\}\ . \tag{6.88}$$

Comparing this with Eq. (F.16) of Appendix F.3.2, we find that the Jost function is proportional to the Fredholm determinant, Δ, of the integral equation with the kernel $K = g_\ell^{(+)}V$, i.e. of the integral equation (6.5) for the scattering wave function $u_\ell(E,r)$:

$$\boxed{f_\ell^{(\mathrm{in})} = \frac{1}{2}\Delta} \tag{6.89}$$

This completes the proof of the Jost–Pais theorem. Additional factor 1/2 in this relation appears due to our choice of the Jost function normalization.

7 Short-Range Potential Extending to Infinity

Although most of the derivations and proofs given in the preceding sections were general, we so far avoided some of the difficulties by considering finite-range potentials, i.e. such potentials that are sharply cut off at a radius R. From now on, we dismiss this restriction and deal with realistic potentials that extend to infinity. Of course, at large distances, the potential vanishes anyway, but not sharply. As we will see, this fact is important and makes changes in the analytic properties of the Jost function as well as in the way it can be calculated.

Since the behaviour of the potential $V(r)$ when $r \to \infty$ (i.e. how fast it vanishes) is important, realistic potentials are usually divided into two classes, the so-called short-range and long-range potentials. Unfortunately, there is no general convention on how to define these classes. Many authors reserve the term short range for the potentials that vanish at infinity exponentially or faster. As we will see, however, there is no principal difference in the way the Jost function can be calculated for an exponentially decaying potential and a potential that vanishes as r^{-n}, if $n > 2$. In this book, we therefore use the term short range for all the potentials that vanish at infinity faster than the centrifugal potential. Mathematically, the condition defining such potentials can be written as

$$\int_R^\infty |V(r)| r \, dr < \infty \, , \tag{7.1}$$

where R is any non-zero radius. It should be reminded that, as before, we restrict our consideration to the potentials that are less singular than $1/r^2$ at the origin. Combining the restrictions imposed on the potential at small and large distances, we have

S. A. Rakityansky, *Jost Functions in Quantum Mechanics*,
https://doi.org/10.1007/978-3-031-07761-6_7

$$\boxed{\int_0^\infty |V(r)| r\, dr < \infty} \tag{7.2}$$

which defines the class of potentials that we consider in this section.

7.1 The Regular Solution

Replacing a finite-range potential with a short-range one, we can repeat without any modification all the derivations given in Sects. 2.2.4 and 2.2.2. The regular solution, $\phi_\ell(E, r)$, of the radial Schrödinger equation can still be found in the form

$$\phi_\ell(E, r) = h_\ell^{(-)}(kr) F_\ell^{(\mathrm{in})}(E, r) + h_\ell^{(+)}(kr) F_\ell^{(\mathrm{out})}(E, r) , \tag{7.3}$$

where the functions $F_\ell^{(\mathrm{in/out})}(E, r)$ obey the same differential and integral equations. The function $\phi_\ell(E, r)$ is analytic on the complex E-surface (k-plane) and is uniformly bound (see Sect. 2.2.2)

$$|\phi_\ell(E, r)| \leqslant C_\ell \left(\frac{|kr|}{1 + |kr|} \right)^{\ell+1} e^{|\mathrm{Im}\, kr|} . \tag{7.4}$$

Dealing with a finite-range potential in the preceding sections, we always knew that beyond certain radius the wave function was a linear combination of the incoming and outgoing spherical waves described by the Riccati–Hankel functions. Intuitively, we expect that the same is valid in the case of a short-range potential as well. This, however, needs to be proved. Such a proof is given next.

7.1.1 Long-Range Asymptotics

Since the regular solution, $\phi_\ell(E, r)$, has the form (7.3), which contains both $h_\ell^{(+)}(kr)$ and $h_\ell^{(-)}(kr)$, we can expect that it behaves asymptotically like their linear combination. As we will see, this is true, but such simple reasoning has a flaw. Indeed, we do not know yet how the functions $F_\ell^{(\mathrm{in/out})}(E, r)$, which are part of the function (7.3), behave at large distances. Actually, this is our task to learn their behaviour, i.e. to establish convergence (or divergence) of the limits

$$f_\ell^{(\mathrm{in/out})}(E) = \lim_{r \to \infty} F_\ell^{(\mathrm{in/out})}(E, r) . \tag{7.5}$$

In a sense, this is a kind of "chicken and egg" problem. Of course, we have the uniform bound (7.4) for the regular solution, which indicates that our guess about exponential behaviour of $\phi_\ell(E, r)$ is most probably correct. This, however, is only an upper bound.

A more accurate estimate of the long-range behaviour of the function $\phi_\ell(E, r)$ can be obtained with the help of the WKB method (see Sect. 1.4).

7.1.1.1 WKB Asymptotic Analysis

Most physicists think that WKB is just an approximate method of quantum mechanics and was proposed at the early stage of its development. However, the first application of the same (or rather similar) approach for finding an approximate solution of a differential equation dates back to 1837 (see Refs. [27] and [28]), that is about a hundred years before G. Wentzel, H. A. Kramers, and L. Brillouin. Actually, the WKB approach is widely used in the theory of ordinary differential equations for deriving the asymptotic behaviour of their solutions.

It can be rigorously proved (see Ref. [29]) that for large values of the argument the two independent solutions $y_1(r)$ and $y_2(r)$ of the differential equation

$$y''(r) + q(r)y(r) = 0$$

have the following form:

$$y_{1,2}(r) = \frac{1}{[q(r)]^{1/4}} \exp\left[\pm i \int_R^r \sqrt{q(t)}dt\right] [1 + \varepsilon_{1,2}(r)] ,$$

where R is an arbitrary number (it determines the normalization) and

$$\lim_{r\to\infty} \varepsilon_{1,2}(r) = 0 .$$

Therefore, in the case of the radial Schrödinger equation,

$$\left[\frac{d^2}{dr^2} + k^2 - \frac{\ell(\ell+1)}{r^2} - V(r)\right] \phi_\ell(E, r) = 0 , \tag{7.6}$$

the leading terms of the asymptotic behaviour of the two linearly independent solutions $\psi_\ell^{(\pm)}(E, r)$ are

$$\psi_\ell^{(\pm)}(E, r) \underset{|kr|\to\infty}{\longrightarrow} \frac{\exp\left[\pm i \int \sqrt{k^2 - V(r) - \frac{\ell(\ell+1)}{r^2}}\, dr\right]}{\left[k^2 - V(r) - \frac{\ell(\ell+1)}{r^2}\right]^{1/4}} , \tag{7.7}$$

where the integral in the exponential function is indefinite (i.e. anti-derivative or primitive function). Since the potential is vanishing,

$$V(r) \underset{r\to\infty}{\longrightarrow} 0\,,$$

we can always choose r to be so large that[1]

$$\sqrt{k^2 - V(r) - \frac{\ell(\ell+1)}{r^2}} = k\sqrt{1 - \frac{V(r)}{k^2} - \frac{\ell(\ell+1)}{(kr)^2}} \approx k - \frac{V(r)}{2k} - \frac{\ell(\ell+1)k}{2(kr)^2}\,.$$

This gives

$$\psi_\ell^{(\pm)}(E,r) \underset{|kr|\to\infty}{\longrightarrow} \frac{1}{\sqrt{k}} \exp\left[\pm ikr \mp \frac{i}{2k}\int V(r)\,dr \pm i\frac{\ell(\ell+1)}{2kr}\right]. \tag{7.8}$$

The potential $V(r)$ obeys the condition (7.1) that is only possible if the primitive function obtained by integrating $V(r)$ is finite at $r = \infty$, i.e.

$$\int V(r)\,dr \underset{r\to\infty}{\longrightarrow} \text{const} < \infty\,.$$

Actually, this constant is zero. Indeed, $V(r)$ vanishes at infinity faster than the centrifugal potential; hence in the "worst case", we have $V(r) \sim 1/r^{2+\varepsilon}$, where $\varepsilon > 0$, which gives

$$\int V(r)\,dr \sim \frac{1}{r^{1+\varepsilon}} \underset{r\to\infty}{\longrightarrow} 0\,. \tag{7.9}$$

Even if it were not zero but a finite constant, this would not change the conclusion that the two linearly independent solutions of Eq. (7.6) asymptotically behave as

$$\psi_\ell^{(\pm)}(E,r) \underset{|kr|\to\infty}{\sim} e^{\pm ikr}\,.$$

Since in the above derivations the momentum was not restricted to any domain of the complex plane, we finally conclude that the general solution of Eq. (7.6) has the following asymptotic behaviour on the whole k-plane:[2]

---

[1] Please take note that $[k] = \text{fm}^{-1}$ and $[V] = \text{fm}^{-2}$. We take k out of the square root in order to leave there the dimensionless quantities and then replace the square root with the first two terms of its Taylor series.

[2] In the above derivations, we tacitly assumed that $k \neq 0$. Apparently near this point, the second and third terms of the exponential function in Eq. (7.8) would depend on the way $k \to 0$. If it goes to zero slowly, i.e. in such a way that $|kr| \to \infty$, then there is no problem. Otherwise, the phase of the wave function is undefined. Therefore, in what follows, it is assumed that the point $k = 0$ is excluded, although we do not always say it explicitly.

$$\phi_\ell(E,r) \underset{|kr|\to\infty}{\longrightarrow} q_\ell^{(-)}(E)e^{-ikr} + q_\ell^{(+)}(E)e^{+ikr}\,, \tag{7.10}$$

where

$$q_\ell^{(\pm)}(E) \sim \frac{\text{const}_\ell^{(\pm)}}{\sqrt{k}} \tag{7.11}$$

are some r-independent quantities (one of which can be zero).

Using the physical reasoning, it is not difficult to understand why the so- called quasi-classical (WKB) approximation becomes very accurate when $r \to \infty$. As we know from quantum mechanics, this approximation is applicable if the wavelength λ is slowly varying function of r (see Sect. 1.4). It is clear that far away from the source of the potential (where it almost vanishes), the particle moves freely, and there are no reasons for varying its effective wavelength that therefore becomes practically constant.

7.2 Jost Functions

Moving smoothly along the same path of reasoning as with a finite-range potential, we stumble at the point where the Jost functions are defined as the limits (2.63) and (2.64). When the potential is cut off at $r = R$, these limits exist for any complex E. In fact, they are reached at $r = R$ because beyond this point the right hand sides of Eq. (2.21) are identical zeros and the functions $F_\ell^{(\text{in/out})}(E,r)$ become constant. What happens when the potential extends to infinity?

Having been armed with the knowledge of the asymptotics (7.10), we can now examine the existence of the limits (7.5) using the integral equations (2.40) written in the form

$$F_\ell^{(\text{in})}(E,r) = \frac{1}{2} - \frac{1}{2ik}\int_0^r h_\ell^{(+)}(kr')V(r')\phi_\ell(E,r')dr'\,, \tag{7.12}$$

$$F_\ell^{(\text{out})}(E,r) = \frac{1}{2} + \frac{1}{2ik}\int_0^r h_\ell^{(-)}(kr')V(r')\phi_\ell(E,r')dr'\,. \tag{7.13}$$

Apparently, the limits (7.5) do exist if the integrals

$$I_\ell^{(\pm)}(k,r) \equiv \int_0^r h_\ell^{(\pm)}(kr')V(r')\phi_\ell(E,r')dr' \tag{7.14}$$

converge[3] when $r \to \infty$. Choosing a large enough radius R and using the asymptotics (7.10) and

$$h_\ell^{(\pm)}(kr) \underset{|kr|\to\infty}{\longrightarrow} (\mp i)^{\ell+1} e^{\pm ikr}\,, \tag{7.15}$$

we obtain

$$\begin{aligned} F_\ell^{(\mathrm{in})}(E,r) &= \mathrm{const} - \frac{1}{2ik}\int_R^r h_\ell^{(+)}(kr')V(r')\phi_\ell(E,r')dr' \\ &\underset{|r|\to\infty}{\longrightarrow} \mathrm{const} + \frac{(-i)^\ell}{2k}\int_R^r V(r')\left[q_\ell^{(-)}(E) + q_\ell^{(+)}(E)e^{2ikr'}\right]dr'\,, \end{aligned} \tag{7.16}$$

and similarly

$$F_\ell^{(\mathrm{out})}(E,r) \underset{|r|\to\infty}{\longrightarrow} \mathrm{const} + \frac{(+i)^\ell}{2k}\int_R^r V(r')\left[q_\ell^{(-)}(E)e^{-2ikr'} + q_\ell^{(+)}(E)\right]dr'\,. \tag{7.17}$$

There are several different cases:

- *Momentum k is real:*
  In this case, both integrals (7.17, 7.17) converge. Indeed, the exponential functions $e^{\pm 2ikr}$ are pure oscillatory, and thus the integrands are bounded by the function $|V(r)|(|q_\ell^{(-)}| + |q_\ell^{(+)}|)$, the integral of which converges according to the condition (7.1).
- *Momentum k is complex:*
  The factors $e^{-2\mathrm{Im}\,(kr')}$ and $e^{+2\mathrm{Im}\,(kr')}$ appearing in the integrands (7.17) and (7.17) play the dominant roles:
  - $\mathrm{Im}\,k > 0$*, but k is not a spectral point:*
    Since k does not coincide with a spectral point k_n, both terms of the asymptotics (7.10) are non-zero. As a result, the function $F_\ell^{(\mathrm{in})}(E,r)$ is convergent, while $F_\ell^{(\mathrm{out})}(E,r)$ diverges unless the potential vanishes so rapidly that the growing factor $e^{+2\mathrm{Im}\,(kr')}$ is compensated. Therefore, in the "worst case scenario", when the potential vanishes at infinity slowly (slower than an exponential function), the integral (7.17) diverges.
  - $\mathrm{Im}\,k < 0$*, but k is not a spectral point:*
    As is easily seen, the situation is just opposite to the one in the previous item, i.e. the function $F_\ell^{(\mathrm{out})}(E,r)$ converges, while $F_\ell^{(\mathrm{in})}(E,r)$ diverges

[3] The convergence of these integrals at the lower limit, i.e. at $r = 0$, is guaranteed by the regularity of the function $\phi_\ell \underset{r\to 0}{\sim} (kr)^{\ell+1}$, and by the condition (7.2).

unless the potential vanishes so rapidly that the growing factor $e^{-2\text{Im}\,(kr')}$ is compensated.

- *Momentum $k = k_n$ is a spectral point:*
  By definition, only the outgoing wave is present in the asymptotics (7.10), i.e. $q_\ell^{(-)}(E_n) = 0$. This means that $F_\ell^{(\text{out})}(E_n, r)$ converges to $f_\ell^{(\text{out})}(E_n)$ irrespective of the sign of $\text{Im}\,k_n$. For the function $F_\ell^{(\text{in})}(E_n, r)$, however, the vanishing of $q_\ell^{(-)}(E_n)$ does not make any difference. The factor $e^{-2\text{Im}\,(kr')}$ remains there, and therefore, $F_\ell^{(\text{in})}(E_n, r)$ converges to its limiting value (zero) only if $\text{Im}\,k_n > 0$.

This can be summarized as follows. The line $\text{Im}\,(kr) = 0$ divides the k-plane into two domains. The function $F_\ell^{(\text{in})}(E, r)$ converges to its limiting value $f_\ell^{(\text{in})}(E)$ if $\text{Im}\,(kr) \geqslant 0$ (on and above the dividing line), while $F_\ell^{(\text{out})}(E, r)$ converges to $f_\ell^{(\text{out})}(E)$ if $\text{Im}\,(kr) \leqslant 0$ (on and below the dividing line) and at all spectral points $k = k_n$ irrespective of the sign of $\text{Im}\,(kr)$,

$$F_\ell^{(\text{in})}(E, r) \underset{r\to\infty}{\longrightarrow} f_\ell^{(\text{in})}(E)\,, \qquad \text{if } \text{Im}\,(kr) \geqslant 0\,, \tag{7.18}$$

$$F_\ell^{(\text{out})}(E, r) \underset{r\to\infty}{\longrightarrow} f_\ell^{(\text{out})}(E)\,, \qquad \text{if } \text{Im}\,(kr) \leqslant 0\,, \tag{7.19}$$

$$F_\ell^{(\text{out})}(E_n, r) \underset{r\to\infty}{\longrightarrow} f_\ell^{(\text{out})}(E_n)\,, \qquad \text{at all spectral points } E_n\,. \tag{7.20}$$

The corresponding domains of the k-plane are shown in Fig. 7.1. On the E-surface, the $F_\ell^{(\text{in})}(E, r)$ converges to $f_\ell^{(\text{in})}(E)$ on the physical sheet, while $F_\ell^{(\text{out})}(E, r)$ to $f_\ell^{(\text{out})}(E)$ on the unphysical sheet. The only common domain of the k-plane, where both limits (7.5) do exist, is the real axis, i.e. both rims of the cut on the E-surface, and the bound state spectral points.

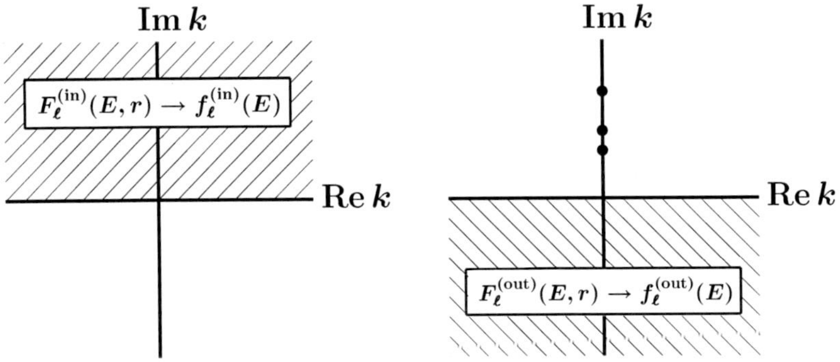

Fig. 7.1 Domains of the k-plane where the functions $F_\ell^{(\text{in})}(E, r)$ (left) and $F_\ell^{(\text{out})}(E, r)$ (right) have finite limits. This is the "worst case scenario", when $V(r)$ vanishes at large distances slower than an exponential function

The conditions (7.18, 7.19, 7.20) can also be obtained using the differential equations (2.21) written in the form

$$\partial_r F_\ell^{(\mathrm{in})}(E,r) = -\frac{h_\ell^{(+)}(kr)}{2ik} V(r)\phi_\ell(E,r)\,, \tag{7.21}$$

$$\partial_r F_\ell^{(\mathrm{out})}(E,r) = \frac{h_\ell^{(-)}(kr)}{2ik} V(r)\phi_\ell(E,r)\,. \tag{7.22}$$

For the reader, it would be pedagogically worthwhile to repeat the analysis using these differential equations. First, because the question at hand is very important, and second, because it gives an insight into the behaviour of these equations, which is helpful in numerical calculations.

Apparently, if the first derivatives of the functions $F_\ell^{(\mathrm{in/out})}(E,r)$ do not vanish at large distances, these functions cannot have finite limits. On the other hand, the mere fact of vanishing of the derivatives is not a sufficient condition for these limits to exist. The derivatives must vanish faster than $\sim r^{-1}$. Indeed, the behaviour

$$\frac{d\varphi(r)}{dr} \underset{r\to\infty}{\sim} \frac{1}{r^{1+\epsilon}}\,, \qquad \epsilon > 0\,, \tag{7.23}$$

of the derivative of a function $\varphi(r)$ implies that

$$\varphi(r) \underset{r\to\infty}{\sim} \int \frac{dr}{r^{1+\epsilon}} = \left(\mathrm{const} - \frac{1}{\epsilon r^\epsilon}\right) \underset{r\to\infty}{\longrightarrow} \mathrm{const}\,,$$

which is divergent unless $\epsilon > 0$. We leave it to the reader (as an exercise) to derive the conditions (7.18, 7.19, 7.20) from differential Eqs. (7.21, 7.22).

7.2.1 Incoming and Outgoing Waves at Complex Momenta

At this point (if not earlier), an attentive reader may suspect a contradiction and inconsistency in our formalism. Indeed, the two functions $F_\ell^{(\mathrm{in/out})}(E,r)$ influence each other since they obey a system of differential (or integral) equations. How can it be that when solving this system, one of the functions is diverging, while the other converges to a finite value (which is correct value of the Jost function)? The correct (converging) function is coupled to a meaningless (diverging) counterpart and is not affected. Doesn't this look suspicious?

Nothing is wrong with the equations. The explanation can be found if we look more closely at the two term decomposition (7.3) of the wave function. In the general case, at all (non-spectral) points above the dividing line, the function $F_\ell^{(\mathrm{out})}(E,r)$ diverges. However, in the representation (7.3), this function is multiplied by the exponentially vanishing factor $h_\ell^{(+)}(kr)$ that partly compensates its growth so that the term $h_\ell^{(-)}F_\ell^{(\mathrm{in})}$ remains dominant, i.e. (see proof below)

$$\frac{h_\ell^{(+)}(kr)F_\ell^{(\mathrm{out})}(E,r)}{h_\ell^{(-)}(kr)F_\ell^{(\mathrm{in})}(E,r)} \underset{r\to\infty}{\longrightarrow} 0\,, \qquad \mathrm{Im}\,(kr) > 0\,. \tag{7.24}$$

And similarly, below the dividing line, the first term of the decomposition (7.3) is negligible (infinitesimal) as compared to the second term,

$$\frac{h_\ell^{(-)}(kr)F_\ell^{(\mathrm{in})}(E,r)}{h_\ell^{(+)}(kr)F_\ell^{(\mathrm{out})}(E,r)} \underset{r\to\infty}{\longrightarrow} 0\,, \qquad \mathrm{Im}\,(kr) < 0\,. \tag{7.25}$$

In other words, despite its divergence, above the dividing line, the second term of the function (7.3) is infinitesimal as compared to the first term. Below the dividing line, the first term becomes negligible (although it diverges there).

Proof If the point k is above the dividing line, then in the asymptotic behaviour (7.17) of the function $F_\ell^{(\mathrm{out})}(E,r)$, all finite terms become negligible as compared to the one involving the exponential function $e^{-2ikr'}$, and thus

$$F_\ell^{(\mathrm{out})}(E,r) \underset{r\to\infty}{\sim} \int_R^r V(r')e^{-2ikr'}dr'\,, \qquad \mathrm{Im}\,(kr) > 0\,.$$

As we have established, in this domain of the k-plane, the function $F_\ell^{(\mathrm{in})}(E,r)$ converges to the Jost function $f_\ell^{(\mathrm{in})}(E)$. Therefore,

$$\begin{aligned}
\frac{h_\ell^{(+)}(kr)F_\ell^{(\mathrm{out})}(E,r)}{h_\ell^{(-)}(kr)F_\ell^{(\mathrm{in})}(E,r)} &\underset{r\to\infty}{\sim} \frac{e^{ikr}\displaystyle\int_R^r V(r')e^{-2ikr'}dr'}{e^{-ikr}f_\ell^{(\mathrm{in})}(E)} \\
&\underset{r\to\infty}{\sim} \int_R^r V(r')e^{2ik(r-r')}dr' \\
&\underset{r\to\infty}{\sim} \int_R^r V(r')e^{2i\mathrm{Re}\,[k(r-r')]-2\mathrm{Im}\,[k(r-r')]}dr'\,,
\end{aligned}$$

and hence,

$$\begin{aligned}
\left|\frac{h_\ell^{(+)}(kr)F_\ell^{(\mathrm{out})}(E,r)}{h_\ell^{(-)}(kr)F_\ell^{(\mathrm{in})}(E,r)}\right| &\leqslant \mathrm{const}\cdot \int_R^r |V(r')|e^{-2\mathrm{Im}\,[k(r-r')]}dr' \\
&= \mathrm{const}\cdot|V(R_0)|\int_R^r e^{-2\mathrm{Im}\,[k(r-r')]}dr'\,,
\end{aligned} \tag{7.26}$$

where, according to the mean-value theorem, the function $|V(r)|$ is taken out of the integral at a middle point $R < R_0 < r$. In the remaining integral, $\mathrm{Im}\,[k(r-r')]$ is always positive, and thus it converges to a finite value,

$$\int_R^r e^{-2\mathrm{Im}\,[k(r-r')]}dr' = \frac{1}{2\mathrm{Im}\,(k)}\left[1-e^{2\mathrm{Im}\,(k)(R-r)}\right] \underset{r\to\infty}{\longrightarrow} \frac{1}{2\mathrm{Im}\,(k)}\,.$$

When $r \to \infty$, the middle point R_0 goes to infinity as well, and since the potential vanishes at large distances, the right hand side of Eq. (7.26) tends to zero. Validity of Eq. (7.25) can be proved in the same way.

* * *

Therefore, when k is above the dividing line, at large distances, the first of Eq. (2.21) decouples from the second one,

$$\begin{aligned}
\partial_r F_\ell^{(\mathrm{in})}(E,r) &\underset{r\to\infty}{\longrightarrow} -\frac{1}{2ik}V(r)h_\ell^{(+)}(kr)h_\ell^{(-)}(kr)F_\ell^{(\mathrm{in})}(E,r)\,, \\
\partial_r F_\ell^{(\mathrm{out})}(E,r) &\underset{r\to\infty}{\longrightarrow} \frac{1}{2ik}V(r)\left[h_\ell^{(-)}(kr)\right]^2 F_\ell^{(\mathrm{in})}(E,r)\,, \qquad \mathrm{Im}\,(kr) > 0\,,
\end{aligned} \tag{7.27}$$

because the outgoing wave component $h_\ell^{(+)}F_\ell^{(\mathrm{out})}$ of the function $\phi_\ell(E,r)$ becomes negligible as compared to its incoming component $h_\ell^{(-)}F_\ell^{(\mathrm{in})}$. As we see, on the right hand side of the differential equation for $F_\ell^{(\mathrm{in})}(E,r)$, the exponential growth of $h_\ell^{(-)}(kr)$ is exactly compensated by the exponential vanishing of $h_\ell^{(+)}(kr)$. Being decoupled, this equation is not affected by the meaningless (diverging) function $F_\ell^{(\mathrm{out})}(E,r)$.

Similarly, below the dividing line, the second of Eq. (2.21) decouples from the first one,

$$\begin{aligned}
\partial_r F_\ell^{(\mathrm{in})}(E,r) &\underset{r\to\infty}{\longrightarrow} -\frac{1}{2ik}V(r)\left[h_\ell^{(+)}(kr)\right]^2 F_\ell^{(\mathrm{out})}(E,r)\,, \\
\partial_r F_\ell^{(\mathrm{out})}(E,r) &\underset{r\to\infty}{\longrightarrow} \frac{1}{2ik}V(r)h_\ell^{(+)}(kr)h_\ell^{(-)}(kr)F_\ell^{(\mathrm{out})}(E,r)\,, \qquad \mathrm{Im}\,(kr) < 0\,.
\end{aligned} \tag{7.28}$$

Thus, the two term decomposition (7.3) has term by term correspondence with the asymptotic form (7.10) of the regular solution only on the dividing line $\mathrm{Im}\,(kr) = 0$ and at all the spectral points. At the other points of the k-plane, only the leading term of (7.3) can be related to the corresponding term of the asymptotics (7.10), while the other one is meaningless (because it is too small and therefore is lost on the background of the large one).

7.2.2 Exponentially Decaying Potentials

As we have mentioned, the conditions (7.18, 7.19, 7.20) for the existence of the limits (7.5) were obtained for the "worst case", i.e. for a potential that vanishes slowly and therefore cannot compensate the exponential growth of the factors $\exp[\pm 2\mathrm{Im}\,(kr')]$ in the integrals (7.17, 7.17). If $V(r)$ vanishes at infinity exponentially, the domains shown in Fig. 7.1 can be widened.

Let the potential $V(r)$ has an exponential tail

$$V(r) \underset{r\to\infty}{\sim} e^{-\beta r}\,, \qquad \beta > 0\,. \tag{7.29}$$

Then the integrals (7.17, 7.17) can be re-written as

$$F_\ell^{(\mathrm{in})}(E,\infty) = \mathrm{const} + \mathrm{const}\cdot\int_R^\infty e^{2i\mathrm{Re}(kr)}e^{-2\mathrm{Im}(kr)-\beta r}dr\,, \tag{7.30}$$

$$F_\ell^{(\mathrm{out})}(E,\infty) = \mathrm{const} + \mathrm{const}\cdot\int_R^\infty e^{-2i\mathrm{Re}(kr)}e^{2\mathrm{Im}(kr)-\beta r}dr\,. \tag{7.31}$$

It is clear that these integrals are convergent if and only if the exponential functions $e^{\mp 2\mathrm{Im}(kr)-\beta r}$ are decaying, i.e.

$$F_\ell^{(\mathrm{in})}(E,r) \underset{r\to\infty}{\longrightarrow} f_\ell^{(\mathrm{in})}(E)\,, \qquad \text{if } \mathrm{Im}\,(k) > -\frac{\beta}{2}\,, \tag{7.32}$$

$$F_\ell^{(\mathrm{out})}(E,r) \underset{r\to\infty}{\longrightarrow} f_\ell^{(\mathrm{out})}(E)\,, \qquad \text{if } \mathrm{Im}\,(k) < +\frac{\beta}{2}\,. \tag{7.33}$$

As compared to the domains shown in Fig. 7.1, the domain of convergence for $F_\ell^{(\mathrm{in})}(E,r)$ is extended downwards, and the corresponding domain for $F_\ell^{(\mathrm{out})}(E,r)$

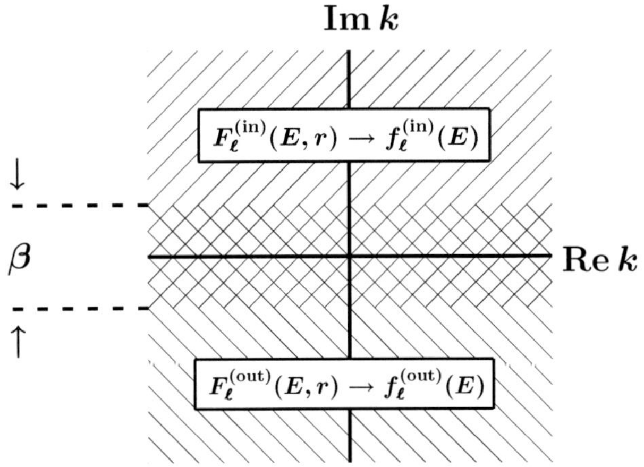

Fig. 7.2 Domains of the k-plane where the functions $F_\ell^{(\mathrm{in/out})}(E,r)$ have finite limits. This is the case when the potential has an exponential tail, $V(r) \underset{r\to\infty}{\sim} e^{-\beta r}$

is symmetrically extended upwards. As a result, their common domain of convergence becomes a band of width β around the real axis (see Fig. 7.2).

It is easy to see that there is a class of potentials for which the limits (7.5) exist on the whole k-plane despite the fact that they have tails extending to infinity. Indeed, if $V(r)$ vanishes faster than the exponential function, it can always "overpower" the growing factor $e^{\mp 2\mathrm{Im}(kr)}$. A simple example is a potential with the following tail:

$$V(r) \underset{r\to\infty}{\sim} e^{-\beta r^2}\,, \qquad \beta > 0\,.$$

It is worthwhile to mention that for the Yukawa potential

$$V(r) = \frac{e^{-\beta r}}{r} \tag{7.34}$$

as well as for any potential of the type

$$V(r) \underset{r\to\infty}{\sim} \frac{e^{-\beta r}}{r^n}$$

the domains of convergence are the same as shown in Fig. 7.2. This is because at large distances the factor r^{-n} cannot make a difference, as compared to $e^{\mp 2\mathrm{Im}(kr)-\beta r}$.

7.3 Analyticity of the Jost Functions

In the domains shown in Figs. 7.1 and 7.2, the Jost functions $f_\ell^{(\mathrm{in/out})}(E)$ are analytic. This can be proved using the theorem of complex analysis (see Appendix B), which states that an integral of a function analytically dependent on a complex parameter is an analytic function of that parameter. The integral must be uniformly convergent, and the integrand as well as its first derivative must be continuous.

For the integrals (7.17, 7.17), all these conditions are apparently satisfied. Indeed, the Riccati–Hankel functions as well as the function $\phi_\ell(E, r)$ are analytic with respect to both k and r and thus have continuous derivatives. As we have shown, the integrals are absolutely convergent and therefore converge uniformly. It may happen that $V(r)$ has a finite number of discontinuities. This obstacle, however, can be easily circumvented by subdivision of the integrals in a sum of integrals over the intervals on which $V(r)$ is continuous.

7.4 Complex Rotation

What we have established so far shows that the differential and integral equations determining the functions $F_\ell^{(\mathrm{in/out})}(E, r)$ have a serious flaw when the potential extends to infinity. Their solutions converge to finite limits ($r \to \infty$) only on a half of the k-plane. In contrast to the case of a finite-range potential, we cannot calculate the Jost function $f_\ell^{(\mathrm{in})}(E)$ on the unphysical sheet of the E-surface (lower half of the k-plane) and therefore cannot locate resonances. This may look frustrating and confusing. How can it be that $f_\ell^{(\text{in})}(E)$ does not exist on the most interesting domain of the energy surface?

Actually, the functions $f_\ell^{(\text{in/out})}(E)$ do exist on the whole E-surface. Otherwise, the regular solution $\phi_\ell(E, r)$ of the Schrödinger equation could not have the asymptotics (7.10). The problem is that in certain domains of the E-surface, we cannot reach the limiting values $f_\ell^{(\text{in/out})}(E)$ by integrating the differential equations (2.21) along the real r-axis. So, this is a flaw of the equations, but not of the regular solution or of its asymptotic amplitudes $f_\ell^{(\text{in/out})}(E)$. It should be noted that at physical (real) energies our equations (2.21) work perfectly. The problem arises when we move away from the real E-axis. To solve the problem, we therefore need to modify (generalize) the equations somehow.

First of all, we notice that the independent variable r in the Schrödinger equation (2.4) can be considered as a complex variable. Why not? If the energy and angular momentum are complex, and the problem arises at complex values of E, it is logical to seek a remedy by moving into complex r-plane.

Second, it can be proved[4] that a solution $\varphi(z)$ of the differential equation

$$\frac{d^2\varphi(z)}{dz^2} + J(z)\varphi(z) = 0$$

is an analytic function of z everywhere in the domain of analyticity of $J(z)$. Therefore, if we assume that the potential $V(r)$ is an analytic function (which is true in many cases), then the regular solution, $\phi_\ell(E, r)$, is also an analytic function of r. Its derivative, $\partial_r\phi_\ell(E, r)$, is therefore also analytic.

From Eqs. (2.12) and (2.14), it follows that

$$F_\ell^{(\text{in/out})}(E, r) = \pm\frac{1}{2ik}\left\{\left[\partial_r h_\ell^{(\pm)}(kr)\right]\phi_\ell(E, r) - h_\ell^{(\pm)}(kr)\left[\partial_r\phi_\ell(E, r)\right]\right\}, \tag{7.35}$$

where on the right hand side all the functions are analytic with respect to variable r. This makes $F_\ell^{(\text{in/out})}(E, r)$ also analytic in r.

For analytic functions, when moving from a point $r = r_1$ to $r = r_2$, we always obtain the same values, $F_\ell^{(\text{in/out})}(E, r_2)$, independently of the path between r_1 and r_2. Thus, if we want to calculate $F_\ell^{(\text{in/out})}(E, r)$ at a point R on the real axis, we can integrate equations (2.21) from $r = 0$ to $r = R$ either directly along the real axis or moving into the complex plane as is shown in Fig. 7.3. If the arc $R'R$ is far away from the origin O, at all its points the potential is negligible as well as the derivatives $\partial_r F_\ell^{(\text{in/out})}$. In other words, the function $\phi_\ell(E, r)$ at $r = R'$ reaches the asymptotics (7.10). Although the value of the function $\phi_\ell(E, r)$ itself is changing

[4] See, for example, Chapter X of the book [8].

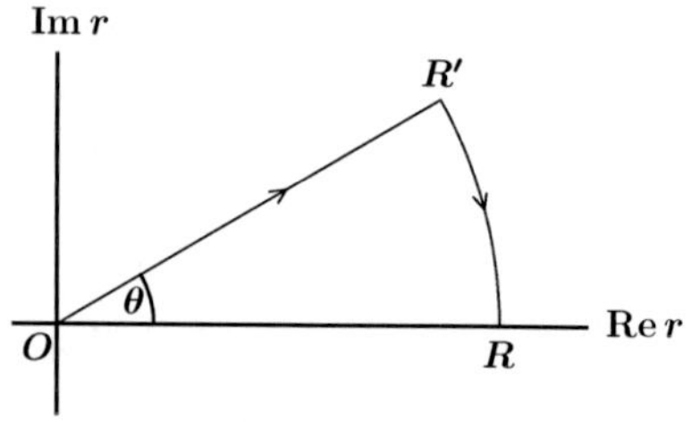

Fig. 7.3 The integration path rotated by the angle θ in the complex r-plane

when we move from R' to R, the asymptotic coefficients $q_\ell^{(-)}(E)$ and $q_\ell^{(+)}(E)$ in Eq. (7.10) are r-independent and thus are the same for all points on the arc $R'R$. Therefore, if we are only interested in these coefficients, which differ from the Jost functions only by constant factors,

$$f_\ell^{(\text{in})}(E) = (-i)^{\ell+1} q_\ell^{(-)}(E) \,,$$

$$f_\ell^{(\text{out})}(E) = (+i)^{\ell+1} q_\ell^{(+)}(E) \,,$$

it is not necessary to go all the way $O \to R' \to R$. It is sufficient to solve Eqs. (2.21) from the point O to R', where these coefficients already attain their asymptotic values.

If we were solving the Schrödinger equation, we would obtain $\phi_\ell(E, R')$ as a single complex number that is difficult to correctly break up into the two terms, $q_\ell^{(-)}(E)e^{-ikR'}$ and $q_\ell^{(+)}(E)e^{+ikR'}$, i.e. it is difficult to extract the Jost functions. This is where the equivalent system of Eq. (2.21) is more convenient because it gives these two terms separately.

Therefore, in order to calculate the Jost functions, we do not have to stick to the real axis of the r-plane. Instead, we can solve the first-order differential equations (2.21) along the turned ray OR',

$$r = \rho e^{i\theta} \,, \qquad \rho = |r| \,. \tag{7.36}$$

The acceptable range for the rotation angle θ is limited by the interval

$$-\pi/2 < \theta < +\pi/2 \,, \tag{7.37}$$

outside of which the incoming and outgoing waves ($\sim e^{\mp ikr}$) swap their roles. This interval can be further narrowed by the properties of the function $V(r)$. The angle should be such that the potential vanishes at the far end of the ray (7.36) faster than $1/|r|^2$, i.e. the condition (7.1) should be satisfied along this ray,

$$\int_{|R'|}^{\infty} |V(\rho e^{i\theta})|\rho \, d\rho < \infty \,; \tag{7.38}$$

otherwise, the functions $F_\ell^{(\text{in/out})}(E, r)$ would not converge to constant values. The Yukawa potential (7.34), for example,

$$V(r) = \frac{e^{-\beta \text{Re}\,(r) - i\beta \text{Im}\,(r)}}{r} ,$$

satisfies this condition within whole interval $\theta \in (-\pi/2, +\pi/2)$. Though the closer θ is to $\pm\pi/2$, the smaller is $\text{Re}\,(r)$ and thus the slower it vanishes, which requires greater $|R'|$ to achieve the convergence.

On the ray (7.36), Eq. (2.21) takes the form:

$$\begin{aligned}
\partial_\rho F_\ell^{(\text{in})}(E, r) &= -\frac{e^{i\theta} h_\ell^{(+)}(kr)}{2ik} V(r) \\
&\quad \times \left[h_\ell^{(-)}(kr) F_\ell^{(\text{in})}(E, r) + h_\ell^{(+)}(kr) F_\ell^{(\text{out})}(E, r) \right] , \\
\partial_\rho F_\ell^{(\text{out})}(E, r) &= \frac{e^{i\theta} h_\ell^{(-)}(kr)}{2ik} V(r) \\
&\quad \times \left[h_\ell^{(-)}(kr) F_\ell^{(\text{in})}(E, r) + h_\ell^{(+)}(kr) F_\ell^{(\text{out})}(E, r) \right] ,
\end{aligned} \tag{7.39}$$

where the derivatives are with respect to the real variable ρ. The boundary conditions for these equations remain the same, namely, given by Eq. (2.24). This means that the integral equations (2.40) are not changed under the rotation (7.36). The only difference is that the integration goes along the ray OR',

$$\begin{aligned}
F_\ell^{(\text{in})}(E, \rho e^{i\theta}) &= \frac{1}{2} - \frac{e^{i\theta}}{2ik} \int_0^\rho h_\ell^{(+)}(kr') V(r') \\
&\quad \times \left[h_\ell^{(-)}(kr') F_\ell^{(\text{in})}(E, r') + h_\ell^{(+)}(kr') F_\ell^{(\text{out})}(E, r') \right] d\rho' , \\
F_\ell^{(\text{out})}(E, \rho e^{i\theta}) &= \frac{1}{2} + \frac{e^{i\theta}}{2ik} \int_0^\rho h_\ell^{(-)}(kr') V(r') \\
&\quad \times \left[h_\ell^{(-)}(kr') F_\ell^{(\text{in})}(E, r') + h_\ell^{(+)}(kr') F_\ell^{(\text{out})}(E, r') \right] d\rho' ,
\end{aligned} \tag{7.40}$$

where $r' = \rho' \exp(i\theta)$.

Equations (7.39) and (7.40) defined on the ray (7.36) are the generalizations of the differential and integral equations determining the functions $F_\ell^{(\text{in/out})}(E, r)$, we were looking for. Do they have any advantages compared with the corresponding equations on the real axis? Yes they do. As we will see shortly, they allow us to calculate the Jost functions in the domains of the k-plane where the limits (7.18, 7.19), taken along the real r-axis, do not converge.

Rewriting Eq. (7.40) in the form

$$F_\ell^{(\text{in})}(E,\rho e^{i\theta}) = \frac{1}{2} - \frac{e^{i\theta}}{2ik}\int_0^\rho h_\ell^{(+)}(k\rho' e^{i\theta})V(\rho' e^{i\theta})\phi_\ell(E,\rho' e^{i\theta})d\rho' \,, \tag{7.41}$$

$$F_\ell^{(\text{out})}(E,\rho e^{i\theta}) = \frac{1}{2} + \frac{e^{i\theta}}{2ik}\int_0^\rho h_\ell^{(-)}(k\rho' e^{i\theta})V(\rho' e^{i\theta})\phi_\ell(E,\rho' e^{i\theta})d\rho' \,, \tag{7.42}$$

and repeating the analysis given in Sect. 7.2, we obtain

$$F_\ell^{(\text{in})}(E,\rho e^{i\theta}) \underset{\rho\to\infty}{\longrightarrow} \text{const} + \tag{7.43}$$

$$+\frac{e^{i(\theta-\ell\pi/2)}}{2k}\int_{|R'|}^{\rho} V(\rho' e^{i\theta})\left[q_\ell^{(-)}(E) + q_\ell^{(+)}(E)e^{2ikr'}\right]d\rho' \,,$$

$$F_\ell^{(\text{out})}(E,\rho e^{i\theta}) \underset{\rho\to\infty}{\longrightarrow} \text{const} + \tag{7.44}$$

$$+\frac{e^{i(\theta+\ell\pi/2)}}{2k}\int_{|R'|}^{\rho} V(\rho' e^{i\theta})\left[q_\ell^{(-)}(E)e^{-2ikr'} + q_\ell^{(+)}(E)\right]d\rho' \,.$$

Similarly to (7.18), (7.19), and (7.20), from these equations, it is easy to derive the conditions for the convergence of the limits

$$f_\ell^{(\text{in/out})}(E) = \lim_{\rho\to\infty} F_\ell^{(\text{in/out})}(E,\rho e^{i\theta}) \,. \tag{7.45}$$

The only difference is that instead of the real k-axis ($\text{Im}\, k = 0$), on which both limits exist, here we have the line $\text{Im}\,(kr) = \rho\text{Im}\,(ke^{i\theta}) = 0$. With the condition (7.38), we thus obtain

$$F_\ell^{(\text{in})}(E,\rho e^{i\theta}) \underset{\rho\to\infty}{\longrightarrow} f_\ell^{(\text{in})}(E) \,, \qquad \text{if } \text{Im}\,(ke^{i\theta}) \geqslant 0 \,, \tag{7.46}$$

$$F_\ell^{(\text{out})}(E,\rho e^{i\theta}) \underset{\rho\to\infty}{\longrightarrow} f_\ell^{(\text{out})}(E) \,, \qquad \text{if } \text{Im}\,(ke^{i\theta}) \leqslant 0 \,, \tag{7.47}$$

$$F_\ell^{(\text{out})}(E_n,\rho e^{i\theta}) \underset{\rho\to\infty}{\longrightarrow} f_\ell^{(\text{out})}(E_n) \,, \qquad \text{at all spectral points } E_n \,. \tag{7.48}$$

When r is real, the sign of $\text{Im}\,(kr)$ is determined solely by the sign of $\text{Im}\,(k)$. On the rotated path, this is not the case anymore. Choosing an appropriate rotation angle, we can always make $\text{Im}\,(ke^{i\theta})$ positive or negative according to our needs.

In the case of no rotation ($\theta = 0$), the dividing line $\text{Im}\,(kr) = 0$ that separates the domains of the complex k-plane where the limits of $F_\ell^{(\text{in})}(E,r)$ and $F_\ell^{(\text{out})}(E,r)$

exist coincides with the real axis (see Fig. 7.1). With a non-zero θ, we have

$$\mathrm{Im}\,(kr) = \rho\,[\mathrm{Re}\,(k)\sin\theta + \mathrm{Im}\,(k)\cos\theta]\ .$$

Therefore, the condition $\mathrm{Im}\,(kr) = 0$ is fulfilled on the line

$$\mathrm{Im}\,(k) = -\tan\theta\cdot\mathrm{Re}\,(k)\ .$$

This can also be seen in the polar coordinates. Let ω be the polar angle of the point k on the complex momentum plane,

$$k = |k|e^{i\omega}\ . \tag{7.49}$$

Then

$$\mathrm{Im}\,(ke^{i\theta}) = |k|\sin(\omega+\theta)\ ,$$

from which it follows that the dividing line in the k-plane corresponds to $\omega = -\theta$. In other words, the dividing line is turned by the same angle θ as the coordinate r, but in the opposite direction. In the "worst case" (slowly vanishing potential), the function $F_\ell^{(\mathrm{in})}(E,r)$ has a finite limit on and above this line, while $F_\ell^{(\mathrm{out})}(E,r)$ on and below it (see Fig. 7.4).

On the energy surface, the dividing line corresponds to the branch cut running from the threshold $E = 0$ to infinity. When the coordinate r is rotated, this dividing line is also turned down (for $\theta > 0$) or up (for $\theta < 0$). Since the energy is a quadratic function of the momentum, $E = \hbar^2k^2/(2\mu)$, the polar angle of a point on the E-surface is 2ω, where ω is defined by Eq. (7.49). Therefore, if the dividing line on the k-plane is turned by the angle $-\theta$, then the corresponding line on the E-surface is turned by the angle -2θ (see Fig. 7.5).

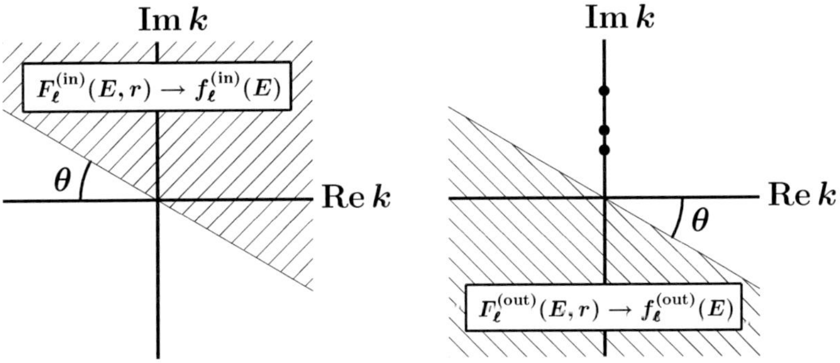

Fig. 7.4 Domains of the k-plane where the functions $F_\ell^{(\mathrm{in})}(E,r)$ (left) and $F_\ell^{(\mathrm{out})}(E,r)$ (right) have finite limits. These domains correspond to a positive rotation angle θ and the "worst case scenario", when $V(r)$ vanishes at large distances slower than an exponential function

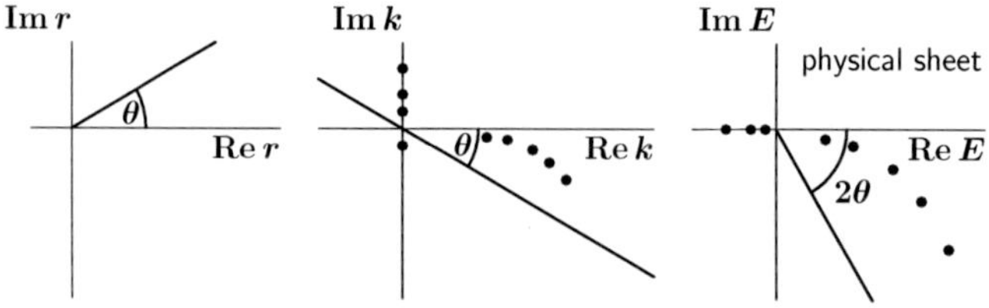

Fig. 7.5 Correspondence among the complex rotation of the coordinate (left), turning of the dividing line in the k-plane (centre), and turning of the corresponding line on the energy surface (right)

By considering complex r, we actually do the analytic continuation of $f_\ell^{(\text{in/out})}(E)$ across the dividing line (the branch cut) to the domains where Eq. (2.21) does not give finite values for these functions.

As we see, the complex rotation solves the problem of calculating the Jost function $f_\ell^{(\text{in})}(E)$ in the fourth quadrant of the k-plane (on the unphysical sheet of the E-surface). We can always choose an appropriate (positive) angle θ that turns the dividing line as is shown in Fig. 7.5 and thus allows us to access the resonance domain. We therefore can locate resonance spectral points as well as calculate the S-matrix on the unphysical sheet. The only problem that remains, is associated with the spectral points corresponding to virtual states. They are on the negative half of the imaginary k-axis, and hence, the required rotation angle is $\theta = \pi/2$, which is not allowed by the condition (7.37).

This difficulty can be circumvented in at least two different ways:

- In many cases, the potentials that support virtual states have exponentially decaying tails (especially in nuclear physics). As we have seen in Sect. 7.2.2, for such potentials, the domain, in which the Jost function $f_\ell^{(\text{in})}(E)$ can be calculated, is extended to a band below the real k-axis. Since virtual states are usually located very close to the threshold, in most cases they are within this band.
- In Sect. 15.1, it is shown that for any short-range potential the Jost function $f_\ell^{(\text{in})}(E)$ can be found in the form of a power series that is convergent within a wide circle around the threshold energy. Using such an expansion, we can locate all the spectral points near $E = 0$, including low-energy resonances, and bound and virtual states.

Suppose we need to calculate the S-matrix at a point k in the fourth quadrant of the k-plane. It may happen that this point is close to the real axis, and the potential has an exponentially decaying tail, such that the point is within the band where both $f_\ell^{(\text{in})}(E)$ and $f_\ell^{(\text{out})}(E)$ can be calculated concurrently, i.e. in one run of integration of Eq. (7.39). In general, however, to calculate the S-matrix, we have to solve these equations twice. First time, we integrate them with $\theta = 0$ and thus obtain $f_\ell^{(\text{out})}(E)$. Second time, we do an appropriate rotation to turn the dividing line below the point

k, and then Eq. (7.39) gives us the Jost function $f_\ell^{(\mathrm{in})}(E)$. The S-matrix is their ratio (4.37).

Numerical examples of using the described approach for the calculations of the Jost functions for several specific potentials can be found in Refs. [30, 31].

7.4.1 Exponentially Decaying Potentials

As we already know, when the potential vanishes at large distances exponentially, the domain of the k-plane where the functions $F_\ell^{(\mathrm{in/out})}(E, r)$ both converge to the corresponding Jost functions is a band along the real axis. In this section, we analyse what happens to this band when the coordinate r is turned into the complex plane.

Earlier in Sect. 7.4, it was shown that the asymptotic behaviour of the functions $F_\ell^{(\mathrm{in/out})}(E, r)$ on the turned ray (7.36) is given by Eqs. (7.43, 7.44), from which it follows that the limits (7.45) do exist if the integrals

$$I^{(\pm)} = \int_{|R'|}^{\rho} V(\rho' e^{i\theta}) \exp\left(\pm 2ik\rho' e^{i\theta}\right) d\rho'$$

converge when $\rho \to \infty$ (the signs "+" and "−" correspond to $F_\ell^{(\mathrm{in})}$ and $F_\ell^{(\mathrm{out})}$, respectively). If the potential has an exponentially vanishing tail (7.29), then

$$I^{(\pm)} \sim \int_{|R'|}^{\rho} e^{\pm 2ikr' - \beta r'} d\rho' ,$$

which are convergent only if

$$\mathrm{Re}\,(\pm 2ikr - \beta r) < 0 .$$

Rewriting this condition as

$$\mp 2\,[\mathrm{Re}\,(k)\mathrm{Im}\,(r) + \mathrm{Im}\,(k)\mathrm{Re}\,(r)] - \beta \mathrm{Re}\,(r) < 0 ,$$

we obtain

$$F_\ell^{(\mathrm{in})}(E, \rho e^{i\theta}) \underset{\rho\to\infty}{\longrightarrow} f_\ell^{(\mathrm{in})}(E) , \qquad \text{if } \mathrm{Im}\,(k) > -\tan(\theta)\mathrm{Re}\,(k) - \frac{\beta}{2} , \tag{7.50}$$

$$F_\ell^{(\mathrm{out})}(E, \rho e^{i\theta}) \underset{\rho\to\infty}{\longrightarrow} f_\ell^{(\mathrm{out})}(E), \qquad \text{if } \mathrm{Im}\,(k) < -\tan(\theta)\mathrm{Re}\,(k) + \frac{\beta}{2} , \tag{7.51}$$

where $\tan(\theta) = \mathrm{Im}\,(r)/\mathrm{Re}\,(r)$ in accordance with Eq. (7.36). Therefore, the function $F_\ell^{(\mathrm{in})}$ converges to a finite limit when k is above the line

$$\mathrm{Im}\,(k) = -\tan(\theta)\mathrm{Re}\,(k) - \frac{\beta}{2} , \tag{7.52}$$

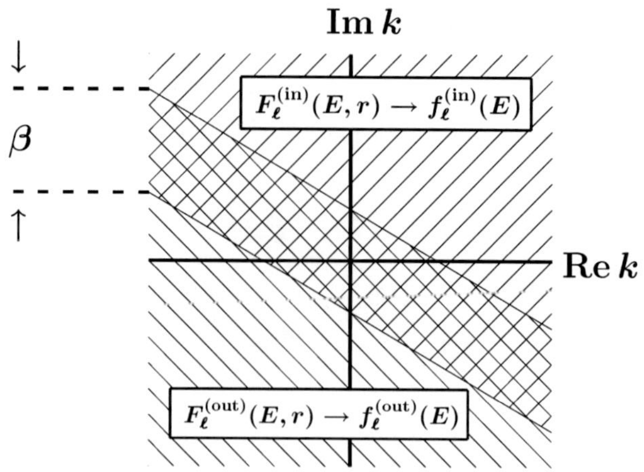

Fig. 7.6 Domains of convergence of the functions $F_\ell^{(in/out)}(E,r)$ for a potential having an exponential tail, $V(r) \underset{r\to\infty}{\sim} e^{-\beta r}$, and with positive rotation angle $\theta = 30°$

while $F_\ell^{(out)}$ is convergent below the line

$$\mathrm{Im}\,(k) = -\tan(\theta)\mathrm{Re}\,(k) + \frac{\beta}{2}\,. \tag{7.53}$$

The corresponding domains are shown in Fig. 7.6. And of course the function $F_\ell^{(out)}$ converges to $F_\ell^{(out)}$ at all spectral points. The common domain of convergence is a band whose vertical width is β. As is easily seen, this is a generalization of the conditions (7.32, 7.33).

7.4.2 Non-analytic Potentials

When considering complex r in the preceding sections, we assumed that the potential could be analytically continued from the real axis into the complex r-plane. This, however, is not always possible. Model potentials used in physics may have a finite number of discontinuities or even be given in the form of a numerical table. For integrating the Schrödinger equation or the equivalent system (2.21) along the real r-axis, this does not pose any problem, but the complex rotation of the coordinate becomes impossible.

Fortunately, in most cases, a model potential has an analytic tail at large distances, although it may have discontinuities at small and intermediate r. Apparently, the convergence or divergence of the functions $F_\ell^{(in/out)}(E,r)$ at large distances is only determined by this tail and does not depend on the behaviour of the potential at short and intermediate r.

Therefore, if the rotation is needed but is not possible for all r, we can do it only for the distances beyond some large radius R_1, where the potential tail is given analytically. As is shown in Fig. 7.7, we can solve the equations along the real axis

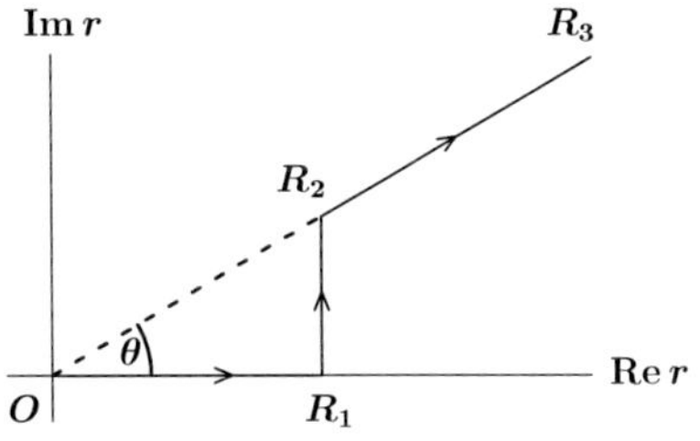

Fig. 7.7 Deformed contour that can be used instead of the one shown in Fig. 7.3, to solve the differential or integral equations determining the functions $F_\ell^{(\text{in/out})}(E, r)$

until a faraway point R_1 is reached. After that, we can turn into the complex plane and via a vertical path $R_1 R_2$ get "back on track" to the ray (7.36) along which the limit can be reached at the point R_3. Of course, the path shown in Fig. 7.7 is not the only possible deformation of the integration contour. It can be deformed as is convenient in a particular problem, but it must be continuous and end up on the far end of the ray (7.36).

If the potential is given as a numerical table, its tail points can be fitted with an appropriate analytic function. One of the main theorems of complex analysis states that two analytic functions coinciding on a segment of a curve coincide everywhere in the domain of analyticity. Therefore, if a numerically given potential is accurately fitted at large distances on the real axis, the fitting function provides its analytic continuation into the complex plane and can be used on the path $R_1 \to R_2 \to R_3$ of Fig. 7.7.

It should be mentioned that the deformed contour of the type shown in Fig. 7.7 is useful not only when dealing with non-analytic potentials. Sometimes, it is advantageous to use such a contour even if the potential allows analytic continuation into the complex plane at all r. For example, if $V(r)$ has an exponential dependence on r at short and intermediate distances, then on the complex path (7.36) it is oscillatory, which may cause the loss of accuracy and numerical instabilities. To avoid possible difficulties of this kind, we can solve the equations along the real axis up to some distance R_1 where the effect of the oscillations is subdued because $V(r)$ begins to vanish. And similarly to the case of a non-analytic potential, after the point R_1, the integration path can be turned into the complex plane and reach the asymptotic point R_3 along the segment $R_2 R_3$ of the ray (7.36).

As we see, the integration path can be divided into several segments that continuously (but not always smoothly) follow after each other. When using such a path in numerical calculations, at each matching point (R_1, R_2, etc.), the values $F_\ell^{(\text{in/out})}(E, R_n)$ obtained by integration of the differential equations (2.21) from R_{n-1} to R_n serve as the boundary (i.e. initial) conditions for the next segment $[R_n, R_{n+1}]$.

This suggests a way of further generalization of the procedure for calculating the Jost functions. Indeed, since the systems of first-order equations (2.21) and the second-order Schrödinger equation (2.4) are equivalent, it does not matter which one of them is used to propagate the solution from R_n to R_{n+1} on an intermediate segment $[R_n, R_{n+1}]$. At the boundary points, the pairs $\{F_\ell^{(\text{in})}(E, R_n), F_\ell^{(\text{out})}(E, R_n)\}$ and $\{\phi_\ell(E, R_n), \phi'_\ell(E, R_n)\}$, constituting the boundary conditions for Eqs. (2.21) and (2.4), can be expressed in terms of each other. Such relations follow from Eqs. (2.12, 2.14, 2.20):

$$\begin{aligned}\phi_\ell(E,R_n) &= h_\ell^{(-)}(kR_n)F_\ell^{(\mathrm{in})}(E,R_n)+h_\ell^{(+)}(kR_n)F_\ell^{(\mathrm{out})}(E,R_n)\,,\\ \phi'_\ell(E,R_n) &= h'^{(-)}_\ell(kR_n)F_\ell^{(\mathrm{in})}(E,R_n)+h'^{(+)}_\ell(kR_n)F_\ell^{(\mathrm{out})}(E,R_n)\,,\end{aligned}\tag{7.54}$$

$$\begin{aligned}F_\ell^{(\mathrm{in})}(E,R_n) &= \frac{1}{2ik}\left[h'^{(+)}_\ell(kR_n)\phi_\ell(E,R_n)-h_\ell^{(+)}(kR_n)\phi'_\ell(E,R_n)\right],\\ F_\ell^{(\mathrm{out})}(E,R_n) &= -\frac{1}{2ik}\left[h'^{(-)}_\ell(kR_n)\phi_\ell(E,R_n)-h_\ell^{(-)}(kR_n)\phi'_\ell(E,R_n)\right],\end{aligned}\tag{7.55}$$

where the "prime" means the derivative with respect to r. Therefore, with the help of Eqs. (7.54, 7.55), we can do the two-way transformations

$$\left\{F_\ell^{(\mathrm{in})}(E,r),\,F_\ell^{(\mathrm{out})}(E,r)\right\}\;\longleftrightarrow\;\left\{\phi_\ell(E,r),\,\phi'_\ell(E,r)\right\}$$

at any point r on the integration path. This gives us the freedom to use the differential equations of either the first or the second order, depending on whichever is more convenient on a particular segment of the path.

It is clear that on the outer segments, i.e. far away from the interaction region, the first-order equations (2.21) have obvious advantages over the Schrödinger equation. This is because at large distances the oscillations of the function $\phi_\ell(E,r)$ are almost completely caused by the Riccati–Hankel functions $h_\ell^{(\pm)}(kr)$ that are explicitly factored out in its representation (7.3). The remaining functions $F_\ell^{(\mathrm{in/out})}(E,r)$ are therefore smooth, and any numerical procedure for their calculation is more stable and accurate than that for the whole regular solution $\phi_\ell(E,r)$. Moreover, as we have seen, the functions $F_\ell^{(\mathrm{in/out})}(E,r)$ in appropriate domains of the k-plane converge to the corresponding Jost functions, while $\phi_\ell(E,r)$ always remains a mixture of the incoming and outgoing waves and thus diverges at all complex k (not coinciding with spectral points). This means that although on the inner segments we are free to use either Eqs. (2.21) or (2.4), on the outer segments it is necessary to switch to the first-order equations for the functions $F_\ell^{(\mathrm{in/out})}(E,r)$.

7.5 Redundant Poles of the S-Matrix

At the so-called spectral points (bound, resonant, and virtual states), the Jost function $f_\ell^{(\mathrm{in})}(E)$ is zero, and thus the S-matrix, defined as

$$S_\ell(E)=\frac{f_\ell^{(\mathrm{out})}(E)}{f_\ell^{(\mathrm{in})}(E)}\,,\tag{7.56}$$

has poles. Can it have poles somewhere else? Generally speaking, the answer is: Yes, it can. It should be emphasized, however, that the Jost function $f_\ell^{(\text{in})}(E)$ cannot vanish anywhere, except for the spectral points. This means that if the S-matrix is singular at a point E_0 that is not a spectral point, then this is only because the function $f_\ell^{(\text{out})}(E)$ is singular at E_0. On the other hand, as we have seen in the preceding sections, for the majority of physically relevant potentials, the Jost function $f_\ell^{(\text{out})}(E)$ is analytic almost everywhere on the k-plane and therefore cannot have singularities. The problem may only arise in the domains where this function cannot be analytically continued to.

If the potential $V(r)$ is an analytic function of r (or at least has an analytic tail) in the sector

$$0 \leqslant |r| < \infty\,, \qquad -\phi < \arg(r) < +\phi\,, \qquad \phi \leqslant \frac{\pi}{2}\,, \tag{7.57}$$

shown in Fig. 7.8a, then with the help of the complex rotation ($-\phi < \theta < +\phi$) the function $f_\ell^{(\text{out})}(E)$ can be analytically continued to the domain shown in Fig. 7.8b. The sector of its possible non-analyticity is centred around the positive imaginary axis of the k-plane. On the E-surface, this corresponds to the sector of the double angle, 2ϕ, around the negative real axis on the physical sheet. The singularities of the function $f_\ell^{(\text{out})}(E)$, if any, can only be in this sector. It should be mentioned that Fig. 7.8b shows the non-analyticity sector for the "worst case" when the potential vanishes at infinity slower than any exponential function. In the case of an exponential tail $\sim e^{-\beta r}$, this sector is lifted up by $\beta/2$ (see Fig. 7.6).

It is clear that the solutions $\sim \phi_\ell(E,r)$ of the Schrödinger equation at the energies where the function $f_\ell^{(\text{out})}(E)$ is singular have no physical meaning since

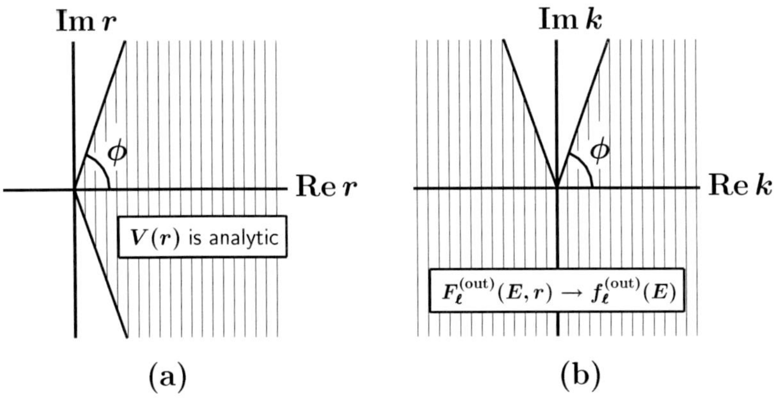

Fig. 7.8 (**a**) Domain of analyticity of the potential, where it obeys the condition (7.38); (**b**) The corresponding domain of analyticity of the Jost function $f_\ell^{(\text{out})}(E)$

they correspond to outgoing waves with infinitely large amplitude. The Jost function $f_\ell^{(\mathrm{in})}(E)$ at such energies is analytic and thus non-singular. As we already mentioned, the singular points of $f_\ell^{(\mathrm{out})}(E)$ cannot coincide with the zeros of $f_\ell^{(\mathrm{in})}(E)$, i.e. with the spectral points. Indeed, in the upper half of the k-plane, the spectral points can be only on the imaginary axis (see Fig. 3.13). These points correspond to bound states. If the function $f_\ell^{(\mathrm{out})}(E)$ were singular at a bound-state spectral point, then as is seen from Eq. (3.25), such a bound state could not be normalized.

Therefore, every spectral point corresponds to a pole of the S-matrix, but the converse is not always true. With some potentials, the S-matrix may have additional poles that have nothing to do with spectral points. They are called the redundant (or false) poles. The reason for such terminology is obvious.

As we know, for a finite-range potential, both $f_\ell^{(\mathrm{in})}(E)$ and $f_\ell^{(\mathrm{out})}(E)$ are analytic functions on the whole k-plane (E-surface). In other words, there is no problem of redundant poles for finite-range potentials. Therefore, if the tail of a potential is cut off at any distance R, the resulting curtailed potential does not generate any redundant poles. They simply disappear as soon as the potential is cut off. On the other hand, the cut-off radius R can be chosen so large that all physical properties of the curtailed and the full potential are identical. This means that even if a given potential generates some redundant poles, they do not affect the physical properties of the system modelled by this potential. In a sense, the redundant poles are spurious phenomena and thus can always be ignored (if avoided when the integration contours are chosen in the k-plane).

When dealing with the S-matrix directly, it is important to distinguish the true and false poles in the upper half of the k-plane. If a pole is found not on the imaginary axis, it can be dismissed right away (because true poles in the upper half of the k-plane correspond exclusively to bound states). If a pole is found on the imaginary axis, a test calculation with a cut-off potential can show if it is a true or false pole (it remains or disappears).

Bearing in mind the possibility of the existence of the redundant poles, we conclude that the definition of the spectral points as zeros of the Jost function rather than poles of the S-matrix is more accurate and free from ambiguities. The Jost function does not have any redundant zeros. Thus, when solving the equation

$$f_\ell^{(\mathrm{in})}(E) = 0\,,$$

we always find only physically relevant states.

An example of a potential that generates a series of redundant poles of the S-matrix on the positive imaginary axis of the k-plane is considered in Sect. 18.1. This is the exponential well

$$V(r) = -V_0 e^{-\beta r}\,, \qquad V_0 > 0\,,$$

for which the Jost functions can be obtained (see Sect. 18.1) in an explicit form, namely,

$$f_0^{(\text{in/out})}(E) = \Gamma\left(1 \mp \frac{2ik}{\beta}\right) J_{\mp 2ik/\beta}\left(\frac{2\sqrt{V_0}}{\beta}\right)\left(\frac{\sqrt{V_0}}{\beta}\right)^{\pm 2ik/\beta} .$$

As is seen, the Jost function $f_0^{(\text{out})}(E) \sim \Gamma(1 + 2ik/\beta)$ has an infinite series of poles at the points where the argument of the gamma-function is a negative integer number, i.e. where

$$k = i\frac{\beta}{2}(n+1)\,, \qquad n = 1, 2, 3, \dots .$$

As was said above, these poles are redundant (spurious) poles of the S-matrix. The Jost function $f_0^{(\text{in})}(E)$ at these poles is finite and analytic. Therefore, if we look for spectral points as zeros of $f_0^{(\text{in})}(E)$, we do not encounter the problem of redundant poles.

7.6 From Finite-Range to Short-Range Potentials

As we have seen, when the potential acquires a tail extending to infinity, the analytic properties of its Jost function become different. The differences, however, are insignificant. Indeed, with the help of the complex rotation, we can analytically continue the functions $f_\ell^{(\text{in/out})}(E)$ almost to the whole E-surface (k-plane). In most cases, the angle ϕ in Eq. (7.57) determining the analyticity domain of the potential $V(r)$ is equal to $\pi/2$. With such potentials, the only domains where these functions are not analytic are the negative real E-axis on the unphysical sheet, for $f_\ell^{(\text{in})}(E)$, and on the physical sheet, for $f_\ell^{(\text{out})}(E)$.

If instead of the E-surface, we consider the k-plane, the function $f_\ell^{(\text{in})}(E)$ is not analytic on the negative imaginary k-axis, while $f_\ell^{(\text{out})}(E)$ on the symmetric domain of the positive imaginary axis (see Fig. 7.9).

Sometimes the corresponding lines of non-analyticity on the E-surface are called "dynamical cuts" (because they are originated from the properties of the potential). The ordinary cuts of the Riemann surface are then called "kinematical cuts" (because they are originated from the kinematical relation between the energy and the momentum). It should be clearly understood that the mathematical nature of these two types of cuts is completely different. The kinematical cut determines the topology of the Riemann surface, while the dynamical cut is just a set of points where the Jost functions are not analytic. In contrast to the kinematical cuts, it is not possible to pass to another Riemann sheet through a dynamical cut.

As soon as the potential tail is cut off at a finite distance R, the possible singularities (i.e. the non-analytic features) of $f_\ell^{(\text{in/out})}(E)$ disappear, no matter how large the cut-off radius R is. This means that the possible non-analytic features of

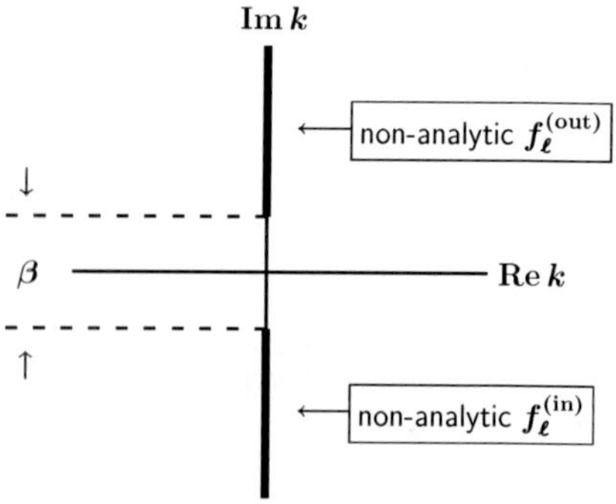

Fig. 7.9 Domains (thick lines) of the k-plane where the functions $f_\ell^{(\mathrm{in/out})}(E)$ are not analytic. This is the case when the potential has an exponential tail, $V(r) \underset{r\to\infty}{\sim} e^{-\beta r}$

$f_\ell^{(\mathrm{in/out})}(E)$ on the imaginary k-axis, caused by the potential tail, cannot manifest themselves through any observable quantities, such as scattering cross section, phase shift, etc., and thus are insignificant.

Therefore, from the formal point of view, the transition from a finite-range to a short-range potential does not change much. It is easy to check that practically everything what was written in the sections devoted to the finite-range potentials remains valid for the potentials with tails falling off faster than $1/r^2$. Looking once more at the derivations and proofs given in this book up to Chap. 6, the reader can see that we almost never used the fact that the potential was assumed to be cut off at a finite radius. So, it is not necessary to repeat the same proofs for the case of a short-range potential. We can therefore take for granted that the functions $F_\ell^{(\mathrm{in/out})}(E, r)$ obey the same differential and integral equations; the Jost functions $f_\ell^{(\mathrm{in/out})}(E)$ have the same symmetry properties (summarized in Figs. 3.5 and 3.6); the spectral points are distributed as is shown in Figs. 3.13 and 3.14; these spectral points have the same multiplicity (i.e. are simple); Levinson's theorem as well as the properties of the Regge poles remains the same, etc.

From the computational point of view, however, the difference between finite-range and short-range potentials is significant. Although the functions $F_\ell^{(\mathrm{in/out})}(E, r)$ obey the same equations, they generally do not converge to $f_\ell^{(\mathrm{in/out})}(E)$ along the real r-axis. When the potential has a tail extending to infinity, we have to follow a contour in the complex r-plane, in order to obtain the solutions of these equations at asymptotically large distances.

7.7 Analytic Structure of the Jost Functions

As we have seen, for a single-channel system, the Jost functions are double-valued functions of the energy and thus are defined on the two-layer Riemann surface schematically shown in Fig. 3.2. In this section, we are going to split these functions into two terms, the first of which is a single-valued analytic function of E, while the second term is a product of another single-valued analytic function of E and the explicit factor $ik^{2\ell+1}$ that is responsible for the branching at $E = 0$.

7.7.1 Factorization

It is clear that $k^2 = 2\mu E/\hbar^2$ as well as all even powers of k are single-valued functions of E. The fact that $f_\ell^{(\text{in/out})}(E)$ are double-valued functions means that they somehow depend not only on the even powers of the momentum, but on its odd powers as well. First, the dependence on the odd powers of k comes from the factor $1/k$ in the differential equations (2.32). Second, the Riccati–Bessel and Riccati–Neumann functions in these equations also involve the odd powers of k. Indeed, their Taylor series can be written as follows (see Appendix D):

$$j_\ell(kr) = \left(\frac{kr}{2}\right)^{\ell+1} \sum_{n=0}^{\infty} \frac{(-1)^n \sqrt{\pi}}{\Gamma(\ell + 3/2 + n)n!} \left(\frac{kr}{2}\right)^{2n} = k^{\ell+1} \tilde{j}_\ell(E, r) , \qquad (7.58)$$

$$n_\ell(kr) = \left(\frac{2}{kr}\right)^{\ell} \sum_{n=0}^{\infty} \frac{(-1)^{n+\ell+1}}{\Gamma(-\ell + 1/2 + n)n!} \left(\frac{kr}{2}\right)^{2n} = k^{-\ell} \tilde{n}_\ell(E, r) , \qquad (7.59)$$

where the tilded functions $\tilde{j}_\ell$ and $\tilde{n}_\ell$ do not involve any odd powers of k and thus are single-valued functions of the energy E. It is known that the Taylor series (7.58, 7.59) are absolutely convergent, and therefore, the tilded functions $\tilde{j}_\ell$ and $\tilde{n}_\ell$ are holomorphic functions of the energy, i.e. they are analytic and single-valued for any complex E, and are defined on the simple single-layered complex plane.[5]

Let us replace the unknown functions $A_\ell(E, r)$ and $B_\ell(E, r)$ in Eq. (2.32) with the corresponding tilded functions defined as follows:[6]

$$A_\ell(E, r) = \tilde{A}_\ell(E, r) , \qquad B_\ell(E, r) = k^{2\ell+1} \tilde{B}_\ell(E, r) . \qquad (7.60)$$

Substituting the factorized representations (7.58, 7.59), and (7.60) into the differential equations (2.32), we have

$$\partial_r \tilde{A}_\ell = -\frac{1}{k} k^{-\ell} \tilde{n}_\ell V \left(k^{\ell+1} \tilde{j}_\ell \tilde{A}_\ell - k^{-\ell} \tilde{n}_\ell k^{2\ell+1} \tilde{B}_\ell\right) ,$$

$$k^{2\ell+1} \partial_r \tilde{B}_\ell = -\frac{1}{k} k^{\ell+1} \tilde{j}_\ell V \left(k^{\ell+1} \tilde{j}_\ell \tilde{A}_\ell - k^{-\ell} \tilde{n}_\ell k^{2\ell+1} \tilde{B}_\ell\right) ,$$

[5] Please take note that in contrast to $n_\ell(kr)$, the tilded function $\tilde{n}_\ell(E, r)$ is not singular at $E = 0$ because the troublesome factor $k^{-\ell}$ has been removed from the Taylor series.

[6] The function $\tilde{A}_\ell$ is not different from A_ℓ, but we still use the tilde over it. This is done to have unified notation in the modified differential equations (7.61, 7.62) as well as to be consistent with the multi-channel case where the tilded $\tilde{A}$ is different from A (see Sect. 13.4).

where all powers of the momentum can be cancelled out. As a result, we obtain the following differential equations for the new (tilded) unknown functions,

$$\partial_r \tilde{A}_\ell(E,r) = -\tilde{n}_\ell(E,r)V(r)\left[\tilde{j}_\ell(E,r)\tilde{A}_\ell(E,r) - \tilde{n}_\ell(E,r)\tilde{B}_\ell(E,r)\right], \quad (7.61)$$

$$\partial_r \tilde{B}_\ell(E,r) = -\tilde{j}_\ell(E,r)V(r)\left[\tilde{j}_\ell(E,r)\tilde{A}_\ell(E,r) - \tilde{n}_\ell(E,r)\tilde{B}_\ell(E,r)\right], \quad (7.62)$$

with the boundary conditions

$$\tilde{A}_\ell(E,0) = 1\,, \qquad \tilde{B}_\ell(E,0) = 0\,. \quad (7.63)$$

The main advantage of these equations is that they do not involve any coefficients or functions depending on odd powers of the momentum k. Since the boundary conditions (7.63) are also k-independent, there is no way for the tilded functions to acquire the dependence on any odd power of k. This means that $\tilde{A}_\ell(E,r)$ and $\tilde{B}_\ell(E,r)$ are single-valued functions of the energy. The multi-valuedness of the original functions, $A_\ell(E,r)$ and $B_\ell(E,r)$, as well as the fact that they are defined on the Riemann surface, is solely due to the momentum factor separated in Eq. (7.60).

Moreover, since for any finite radius r, all the coefficients in differential equations (7.61, 7.62) are holomorphic functions of the parameter E, and the boundary conditions (7.63) do not involve this parameter, according to the Poincaré theorem (see Appendix G), the solutions of these equations, i.e. the functions $\tilde{A}_\ell(E,r)$ and $\tilde{B}_\ell(E,r)$, are also holomorphic functions of the parameter E.

Therefore, for any finite radius r and an arbitrary complex energy E, the functions $F_\ell^{(\mathrm{in/out})}$ have the following analytic structure:

$$F_\ell^{(\mathrm{in/out})}(E,r) = \frac{1}{2}\left[\tilde{A}_\ell(E,r) \mp ik^{2\ell+1}\tilde{B}_\ell(E,r)\right], \quad (7.64)$$

where the double-valued factor $k^{2\ell+1}$ is explicitly isolated. If the potential is cut off at $r = R$, then the corresponding Jost functions (2.63, 2.64) have the same structure for any value of E. If, however, the potential extends to infinity, then taking the limits (2.63, 2.64) is not a trivial procedure. Indeed, at large distances, the tilded Riccati–Bessel and Riccati–Neumann functions,

$$\tilde{j}_\ell = \frac{h_\ell^{(+)} + h_\ell^{(-)}}{2k^{\ell+1}}\,, \qquad \tilde{n}_\ell = \frac{h_\ell^{(+)} - h_\ell^{(-)}}{2ik^{-\ell}}\,,$$

behave as linear combinations of the exponential functions (1.33). There is nothing wrong in this behaviour if k is real. If however the energy is complex and the momentum has a non-zero imaginary part, then $\tilde{j}_\ell(E,r)$ and $\tilde{n}_\ell(E,r)$ involve divergent exponential functions and thus are not holomorphic in E when $r \to \infty$.

Fortunately, these functions in Eqs. (7.61, 7.62) are multiplied by the potential $V(r)$, which tends to zero at large distances. If $V(r)$ vanishes exponentially, as given by Eq. (7.29), then within certain domain $\mathcal{D}$ of complex energies, it can overpower the growth of $\tilde{j}_\ell(E,r)$ and $\tilde{n}_\ell(E,r)$, and thus all the coefficients of these differential equations remain holomorphic functions of E even at $r=\infty$. For an exponentially decaying potential, the domain $\mathcal{D}$ is considered further down in Sect. 7.7.2.

Therefore, we can apply the Poincaré theorem even for $\tilde{A}_\ell(E,\infty)$ and $\tilde{B}_\ell(E,\infty)$, but only within $\mathcal{D}$. This gives

$$\boxed{f_\ell^{(\text{in/out})}(E) = a_\ell(E) \mp ik^{2\ell+1}b_\ell(E)} \qquad E \in \mathcal{D}\,, \tag{7.65}$$

where

$$a_\ell(E) = \tilde{A}_\ell(E,\infty)/2 \qquad \text{and} \qquad b_\ell(E) = \tilde{B}_\ell(E,\infty)/2 \tag{7.66}$$

are holomorphic functions of the energy. The corresponding S-matrix is expressed via these holomorphic functions as

$$\boxed{S_\ell(E) = \frac{a_\ell(E) + ik^{2\ell+1}b_\ell(E)}{a_\ell(E) - ik^{2\ell+1}b_\ell(E)}} \qquad E \in \mathcal{D}\,. \tag{7.67}$$

It should be noted that $a_\ell(E)$ and $b_\ell(E)$ are single-valued functions defined on a simple complex plane without any branch points. As soon as they are found, the Jost functions (7.65) as well as the S-matrix (7.67) can be calculated on any of the two sheets of the Riemann surface by choosing an appropriate sign of the momentum. In Sect. 18.2.1, it is given an exactly solvable example of a single-channel problem, where the representation of the type (7.65) is obtained explicitly.

* * *

Knowledge of the structure (7.65) is very useful in both pure theoretical studies and practical calculations. For example, using these representations of $f_\ell^{(\text{in})}$ and $f_\ell^{(\text{out})}$, we can immediately (and much easier) re-derive all the symmetry properties given in Figs. 3.5 and 3.6:

- **Vertical symmetry:**
  The functions $a_\ell(E)$ and $b_\ell(E)$ do not depend on odd powers of k and therefore are invariant under the transformation $k \to -k$. Therefore, changing of the sign of the momentum in the representations (7.65) results in replacing $f_\ell^{(\text{in})}$ with $f_\ell^{(\text{out})}$, which is what we call vertical symmetry.

- **Diagonal symmetry:**
  At real energies, the functions $a_\ell(E)$ and $b_\ell(E)$ are real because they are asymptotic solutions of Eqs. (7.61, 7.62), which themselves as well as their boundary conditions do not involve any complex numbers (if $\operatorname{Im} E = 0$). Then, from the Schwarz reflection principle (see (B.19) of Appendix B), it follows that they are symmetric relative to the real axis,

$$a_\ell(E^*) = a_\ell^*(E)\,, \qquad b_\ell(E^*) = b_\ell^*(E)\,. \tag{7.68}$$

  Therefore, if we change $k \to k^*$ (which also means $E \to E^*$), then

$$f_\ell^{(\text{in})}(E^*) = a_\ell(E^*) - i\left(k^{2\ell+1}\right)^* b_\ell(E^*) = a_\ell^*(E) - i\left(k^{2\ell+1}\right)^* b_\ell^*(E)$$
$$= \left[a_\ell(E) + ik^{2\ell+1}b_\ell(E)\right]^* = \left[f_\ell^{(\text{out})}(E)\right]^*\,, \tag{7.69}$$

  which relates the values of the Jost functions at the points (E and E^*) located symmetrically across the branch cut on different sheets of the Riemann surface.
- **Horizontal symmetry:**
  Combining the diagonal and vertical symmetries, we relate the Jost functions at the points located opposite to each other across the real axis on the same Riemann sheet:

$$f_\ell^{(\text{in/out})}(E) = \left[f_\ell^{(\text{in/out})}(e^{2\pi i}E^*)\right]^*\,. \tag{7.70}$$

* * *

The fact that the functions $a_\ell(E)$ and $b_\ell(E)$ are holomorphic means that they can be expanded in the Taylor series,

$$a_\ell(E) = \sum_{n=0}^{\infty}\alpha_n(\ell, E_0)(E - E_0)^n\,, \qquad b_\ell(E) = \sum_{n=0}^{\infty}\beta_n(\ell, E_0)(E - E_0)^n\,, \tag{7.71}$$

around an arbitrary point E_0 on the complex energy plane, such that $E_0 \in \mathcal{D}$. The expansion coefficients α_n and β_n depend on the choice of this point. In Sect. 15.1, we will come back to these expansions and derive a set of differential equations that would allow us to calculate α_n and β_n for any given E_0.

With numerically found coefficients α_n and β_n, the truncated series

$$a_\ell(E) \approx \sum_{n=0}^{N}\alpha_n(\ell, E_0)(E - E_0)^n\,, \qquad b_\ell(E) \approx \sum_{n=0}^{N}\beta_n(\ell, E_0)(E - E_0)^n\,, \tag{7.72}$$

substituted in Eqs. (7.65) or (7.67) give approximate semi-analytic expressions for the Jost functions or the S-matrix. These approximate formulae, where the dependencies on E and k are given explicitly, are valid within a circle around E_0 on the complex energy plane. The more terms of the Taylor expansions are retained, the bigger is this circle. Such an expansion can be done, for example, around a point E_0 on the real axis, and then the approximate expression for the Jost function can be used for the nearby complex energies in search for possible resonances.

Alternatively, instead of calculating the expansion coefficients α_n and β_n, they can be treated as unknown fitting parameters. Suppose, for example, that we have experimental values for the scattering cross section (or any other observable quantity) for the (real) energies in the interval $E_1 \leqslant E \leqslant E_2$. We can choose E_0 somewhere in the middle of this interval, write approximate expression for the S-matrix in terms of the unknown α_n and β_n, express the observable quantity via this S-matrix, and fit the data by varying the expansion coefficients.

As soon as the optimal values for α_n and β_n are found, we obtain an analytic expression for the Jost function, which is valid at the nearby complex energies. In this way, we can locate possible resonances, based on given experimental data. The undoubtful advantage of such an approach is that the approximate S-matrix is model-independent and always has proper analytic structure with proper branch point and proper symmetries.

The particular choice of the expansion point at the origin, $E_0 = 0$, gives the so-called effective-range expansion,

$$k^{2\ell+1} \cot \delta_\ell(E) = \sum_{n=0}^{\infty} c_{\ell n} E^n \ . \tag{7.73}$$

A detailed discussion of this expansion can be found in Sect. 15.1.

7.7.2 Domain of Analyticity and Complex Rotation

In Sect. 7.7.1, we introduced the domain $\mathcal{D}$ of the complex energy plane, where the functions $a_\ell(E)$ and $b_\ell(E)$ for the potential with an exponential tail,

$$V(r) \underset{r\to\infty}{\sim} e^{-\beta r} ,$$

are holomorphic. Here we are going to establish how this domain looks like.

* * *

Both functions $\tilde{j}_\ell$ and $\tilde{n}_\ell$ in Eqs. (7.61, 7.62) at large distances are certain linear combinations of the exponential functions $\exp(\pm ikr)$. If E is on the positive half of the real axis, these exponentials do not grow, and therefore, being multiplied by any vanishing potential, they do not cause divergence of the right hand sides of the differential equations when $r \to \infty$. If we move up or down from the real axis, one of the exponential functions

$$e^{\pm ikr} = \exp\left[\pm i\left(\mathrm{Re}\sqrt{\frac{2\mu E}{\hbar^2}}\right)r \mp \left(\mathrm{Im}\sqrt{\frac{2\mu E}{\hbar^2}}\right)r\right]$$

becomes growing. The farther we move from the real axis, the faster it grows. In Eqs. (7.61, 7.62), all such exponentials are squared. Therefore, the potential tail is able to overpower their growth only if

$$2\left|\mathrm{Im}\sqrt{\frac{2\mu E}{\hbar^2}}\right| < \beta\,. \tag{7.74}$$

Writing the complex number $E = |E|\exp(i\varphi)$ in the polar coordinates, we see that

$$\mathrm{Im}\,\sqrt{E} = \mathrm{Im}\left[\sqrt{|E|}\left(\cos\frac{\varphi}{2} + i\sin\frac{\varphi}{2}\right)\right] = \sqrt{|E|}\sin\frac{\varphi}{2}\,.$$

Taking square of both sides of the inequality (7.74), we have

$$\frac{8\mu}{\hbar^2}|E|\sin^2\frac{\varphi}{2} < \beta^2\,,$$

$$\frac{8\mu}{\hbar^2}|E|\frac{1-\cos\varphi}{2} < \beta^2\,,$$

$$\frac{4\mu}{\hbar^2}\left(|E| - |E|\cos\varphi\right) < \beta^2\,,$$

$$\frac{4\mu}{\hbar^2}\left(|E| - \mathrm{Re}\,E\right) < \beta^2\,,$$

$$\frac{4\mu}{\hbar^2}\left[\sqrt{(\mathrm{Re}\,E)^2 + (\mathrm{Im}\,E)^2} - \mathrm{Re}\,E\right] < \beta^2\,.$$

Rearranging terms of the last inequality and taking squares of its sides, we finally arrive at the following condition:

$$\mathcal{D}: \qquad (\mathrm{Im}\,E)^2 < \frac{\hbar^4\beta^4}{16\mu^2} + \frac{\hbar^2\beta^2}{2\mu}\mathrm{Re}\,E\,, \tag{7.75}$$

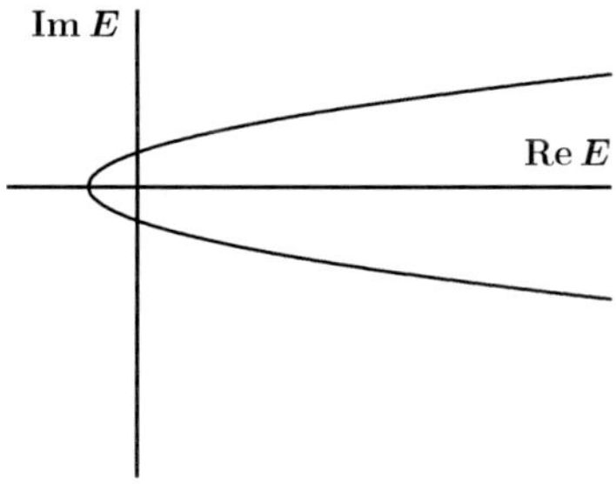

Fig. 7.10 Schematically shown domain $\mathcal{D}$ (within the parabola), where the functions $a_\ell(E)$ and $b_\ell(E)$ are holomorphic, for an exponentially falling potential $V(r)$.

which gives us the domain $\mathcal{D}$, where the functions $a_\ell(E)$ and $b_\ell(E)$ are analytic. It is an area symmetric relative to the positive half of the real axis, and its border is a parabola. The shape of this parabola is determined by the parameters μ and β of a specific problem. A typical example of such a domain is shown in Fig. 7.10.

A question arises: what about the points outside the parabolic domain? The answer is: the functions $a_\ell(E)$ and $b_\ell(E)$ can be analytically continued to most of the complex E-plane with the help of the complex rotation of the coordinate considered in Sect. 7.4. Indeed, if the radius becomes complex, $r = |r|\exp(i\theta)$, then the product

$$e^{\pm 2ikr}e^{-\beta r} = \exp\{i\,[\pm 2\mathrm{Re}\,(kr) - \beta \mathrm{Im}\,r] - [\mp 2\mathrm{Im}\,(kr) + \beta \mathrm{Re}\,r]\}$$

tends to zero at infinity for both choices of the signs $\pm$, only if

$$2\left|\mathrm{Im}\left(\sqrt{\frac{2\mu E}{\hbar^2}}\,r\right)\right| < \beta\,\mathrm{Re}\,r\ , \tag{7.76}$$

which generalizes Eq. (7.74). Taking E and r in the polar coordinates, we have

$$kr = \sqrt{\frac{2\mu E}{\hbar^2}}\,r = \sqrt{\frac{2\mu|E|}{\hbar^2}}\,|r|\exp\left[i\left(\frac{\varphi}{2}+\theta\right)\right]\ ,$$

and thus

$$2\sqrt{\frac{2\mu|E|}{\hbar^2}}\,|r|\sin\left(\frac{\varphi}{2}+\theta\right) < \beta|r|\cos\theta\ ,$$

which gives

$$\mathcal{D}: \qquad \sin^2\left(\frac{\varphi}{2}+\theta\right) < \frac{\hbar^2\beta^2\cos^2\theta}{8\mu|E|}\ . \tag{7.77}$$

In order to understand what shape this domain $\mathcal{D}$ has, we re-write the left hand side of (7.77) as

$$\sin^2\left(\frac{\varphi}{2}+\theta\right)=\frac{1}{2}\left[1-\cos(\varphi+2\theta)\right]=\frac{1}{2}\left(1-\cos\varphi\cos 2\theta-\sin\varphi\sin 2\theta\right) .$$

Then the inequality takes the form

$$|E|-\mathrm{Re}\,(E)\cos 2\theta-\mathrm{Im}\,(E)\sin 2\theta<\frac{\hbar^2\beta^2\cos^2\theta}{4\mu} ,$$

$$(\mathrm{Re}\,E)^2+(\mathrm{Im}\,E)^2<\left[\frac{\hbar^2\beta^2\cos^2\theta}{4\mu}+\mathrm{Re}\,(E)\cos 2\theta+\mathrm{Im}\,(E)\sin 2\theta\right]^2 , \tag{7.78}$$

and finally,

$$(\mathrm{Im}\,E)^2<\frac{\hbar^4\beta^4\cos^4\theta}{16\mu^2}+\frac{\hbar^2\beta^2}{2\mu}\left[\mathrm{Re}\,(E)\cos 2\theta+\mathrm{Im}\,(E)\sin 2\theta\right]\cos^2\theta+ \tag{7.79}$$

$$+|E|^2\sin^2 2\theta+\mathrm{Re}\,(E)\mathrm{Im}\,(E)\sin 4\theta .$$

Apparently, with no rotation ($\theta=0$), this brings us back to the inequality (7.75).

If we turn the energy plane to the angle -2θ (in the opposite direction to the rotation of the coordinate), then the Cartesian coordinates of its points are transformed as follows:

$$\mathrm{Re}\,(E)\rightarrow\mathrm{Re}\,(E)\cos 2\theta-\mathrm{Im}\,(E)\sin 2\theta ,$$

$$\mathrm{Im}\,(E)\rightarrow\mathrm{Re}\,(E)\sin 2\theta+\mathrm{Im}\,(E)\cos 2\theta .$$

Substituting these transformations into Eq. (7.78), we find that on the turned plane the domain $\mathcal{D}$ is given by

$$(\mathrm{Im}\,E)^2<\frac{\hbar^4\beta^4\cos^4\theta}{16\mu^2}+\frac{\hbar^2\beta^2\cos^2\theta}{2\mu}\mathrm{Re}\,E .$$

As we see, this is almost the same parabola as the one represented by the inequality (7.75), but with β replaced with $\beta\cos\theta$ (i.e. with a smaller β). This means that the rotation of the coordinate r in the anti-clockwise direction to the angle θ results in the clockwise rotation of $\mathcal{D}$ to the double angle and in reducing its width by a factor of $\cos\theta$. An example of a rotated domain $\mathcal{D}$ is schematically shown in Fig. 7.11.

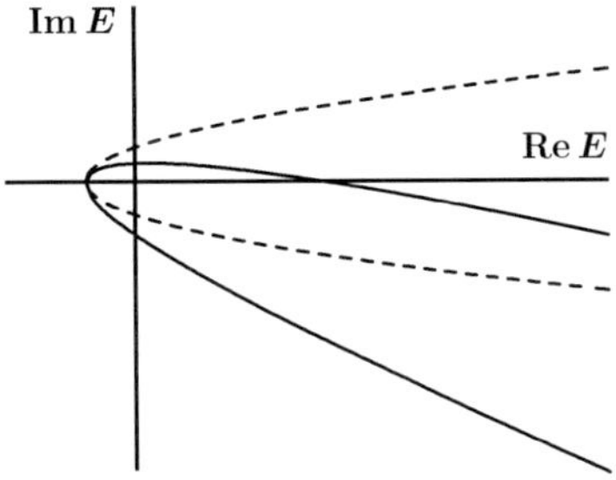

Fig. 7.11 Schematically shown the original analyticity domain $\mathcal{D}$ (within the dashed parabola) and the corresponding rotated domain (within the solid parabola) for the rotation angle $\theta = 0.05\pi$.

By rotating the domain $\mathcal{D}$ clockwise and anti-clockwise, we can cover a significant part of the complex energy plane. In practical applications of the theory, the physically interesting points lie on the real axis (scattering) and below it (resonances). We therefore can say that for most of the practical purposes the structure (7.65) of the Jost functions for a short-range potential is valid everywhere.

7.8 Generalized Levinson's theorem

In Sect. 4.7, it was shown that the number of bound states with a given angular momentum ℓ is related to the difference $[\delta_\ell(0) - \delta_\ell(\infty)]$ between the values of the corresponding phase shift at zero and infinitely high collision energies. This relation, which is known as Levinson's theorem, enables us to predict global behaviour of the function $\delta_\ell(E)$ as soon as we know the number of bound states, and vice versa. Is it possible to derive similar relation that would involve the number of resonances? This is the question that we try to answer in the present section.[7]

* * *

First of all, we need to generalize the definition of the scattering phase shift. In Sect. 4.2, it was defined via Eqs. (4.16, 4.17) only for real positive (i.e. scattering) energies. For a complex energy, the simplest way to define the phase shift is offered by Eq. (4.19), which gives

$$\delta_\ell(E) = \frac{1}{2i} \ln S_\ell(E) . \tag{7.80}$$

Apparently, this is the same function on the real positive axis, but it is, generally speaking, complex when $\operatorname{Im} E \neq 0$.

The main idea of the proof of Levinson's theorem consists in integrating the logarithmic derivative (4.63) of the Jost function along the contour that encloses all

[7] The idea of generalizing Levinson's theorem in the way it is given here was originally presented in Ref. [32].

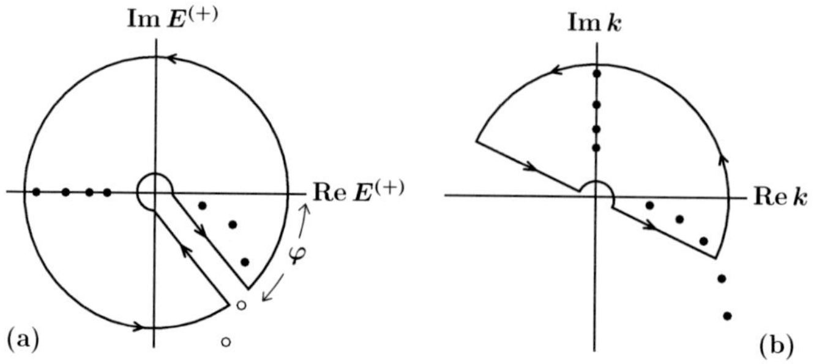

Fig. 7.12 The integration contour used to prove the generalized Levinson's theorem. Part (**a**) shows it on the E-surface (mainly on its physical sheet), and part (**b**) shows the equivalent contour in the k-plane. The angle φ is the angle between the real axis of the E-plane and the straight-line segments of the contour

the bound-state spectral points (see Fig. 4.7). Here, we modify the contour in such a way that it encloses certain (finite) number of resonances as is shown in Fig. 7.12. The greater is the angle, to which the contour is rotated, the more resonances are included. It should be noted that the main part of the contour in Fig. 7.12a is on the upper (physical) sheet of the Riemann surface, and only the part that encloses the resonances is on the lower sheet. This means that the straight-line segments of the contour are on different sheets, i.e. one of the lines is above the other.

When proving Levinson's theorem in Sect. 4.7, we get rid of the integral over the infinitely large circle in the E-plane. This is possible because the Jost function $f_\ell^{(\mathrm{in})}(E)$ tends to a constant value (1/2) at infinitely large k^2 in the upper half of the complex k-plane, which is schematically shown in Fig. 3.7. The corresponding behaviour of this function in the lower half of the k-plane in that figure is undefined (note the question marks). For the purpose of generalizing Levinson's theorem, we need to clarify this behaviour.

To this end, let us consider (as we did it in Sect. 3.2) the integral representations of $f_\ell^{(\mathrm{in})}(E)$ in terms of the regular solution,

$$f_\ell^{(\mathrm{in})}(E) = \frac{1}{2} - \frac{1}{2ik}\int_0^\infty h_\ell^{(+)}(kr)V(r)\phi_\ell(E,r)dr\,. \tag{7.81}$$

Using the uniform bounds (D.27) and (2.60) for the Riccati–Hankel function and for the radial wave function, we obtain the following estimate for the absolute value of the integral term on the right hand side of Eq. (7.81)

$$\left|\frac{1}{2ik}\int_0^\infty h_\ell^{(+)}(kr)V(r)\phi_\ell(E,r)dr\right| \leqslant \frac{\mathrm{const}}{|k|}\int_0^\infty e^{|\mathrm{Im}\,kr|-\mathrm{Im}\,kr}\frac{|kr|}{1+|kr|}|V(r)|\,dr\,. \tag{7.82}$$

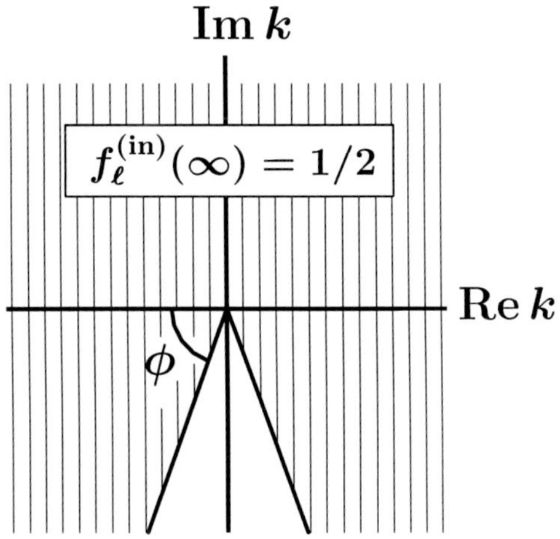

Fig. 7.13 The domain of analyticity of the Jost function $f_\ell^{(\mathrm{in})}(E)$ for a potential with an analytic tail within the sector given by Eq. (7.57) and shown in Fig. 7.8a.

We assume that the potential $V(r)$ is analytic (or at least has an analytic tail) within the sector given by Eq. (7.57) and vanishes at large $|r|$ faster than $1/|r|^2$. Then the integration in (7.82), which goes along the real axis, can be replaced with an equivalent integration along the deformed path in the complex r-plane as is shown in Fig. 7.3 or in Fig. 7.7. If k is within the segment of the complex plane shown in Fig. 7.13,[8] we can always choose the integration path within the sector (7.57) shown in Fig. 7.8a, in such a way that (at least) at large r the product kr remains real, i.e. $\mathrm{Im}\, kr = 0$. Together with our assumption about the potential, this means that the integral on the right hand side of Eq. (7.82) is finite. Moreover, as is seen, if $\mathrm{Im}\, kr = 0$, then this integral becomes k-independent when $|k| \to \infty$. Therefore (being divided by k), the whole right hand side tends to zero in the limit $k \to \infty$ in the lower half of the k-plane, i.e. $f_\ell^{(\mathrm{in})}(\infty) = 1/2$ on the (lower) unphysical sheet of the Riemann surface.

In the majority of practical cases, the analytic tail of the potential is such that the upper limit in Eq. (7.57) is $\phi = \pi/2$, i.e. the maximal possible rotation angle is up to (but not including) $\pi/2$. This means that the only direction where the limit $f_\ell^{(\mathrm{in})}(\infty)$ remains undefined is the direction down the imaginary axis of the k-plane, i.e. the direction along the negative real axis on the unphysical sheet of the E-surface.

The corresponding limits $f_\ell^{(\mathrm{out})}(\infty)$ can be easily obtained using the symmetry properties, summarized in Figs. 3.5 and 3.6. In fact, the functions $f_\ell^{(\mathrm{in})}(E)$ and $f_\ell^{(\mathrm{out})}(E)$ are mirror-symmetric with respect to the point $k = 0$ of the k-plane. Therefore, both $f_\ell^{(\mathrm{in})}(E)$ and $f_\ell^{(\mathrm{out})}(E)$ tend to a constant value (1/2) almost in all directions of the k-plane as it is schematically shown in Fig. 7.14.

[8] The use of the same symbol ϕ for the angle in this figure and for the regular solution should not cause confusion.

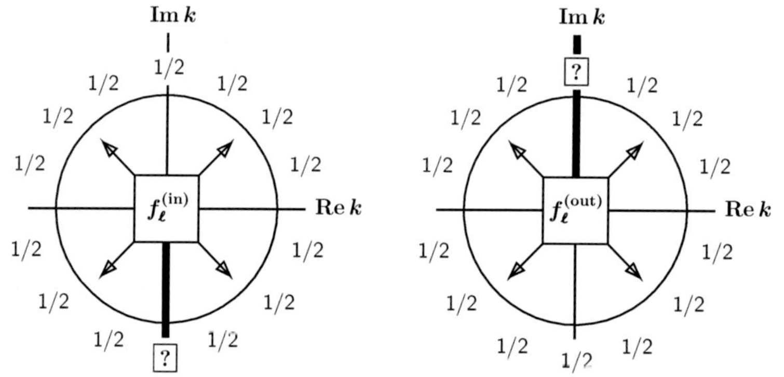

Fig. 7.14 Schematic representation of the asymptotic behaviour of the Jost functions $f_\ell^{(in)}$ (left) and $f_\ell^{(out)}$ (right) when $|k| \to \infty$. The thick lines show the domains where the Jost functions are not analytic. These domains are also shown in Fig. 7.9.

Now we can come back to the integral of the logarithmic derivative of the Jost function over the closed contour shown in Fig. 7.12,

$$\oint L(E)dE = \oint_C L(E)dE + \oint_\varepsilon L(E)dE + \int_A^a L(E)dE + \int_b^B L(E)dE\,, \tag{7.83}$$

where $L(E)$ is defined by Eq. (4.63), C is the infinitely large circle of that contour, and ε is the infinitely small circle around the point $E = 0$. The points a and A,

$$a = \varepsilon e^{2\pi - i\varphi}\,, \qquad A = \infty \cdot e^{2\pi - i\varphi}\,,$$

are the end points of the straight-line segment of the contour on the upper sheet of the Riemann surface, while b and B,

$$b = \varepsilon e^{-i\varphi}\,, \qquad B = \infty \cdot e^{-i\varphi}\,,$$

are the corresponding points on the lower sheet of that surface (exactly under a and A). The integrand,

$$L(E) = \frac{d}{dE} \ln\left[f_\ell^{(in)}(E)\right]\,,$$

at all points of the contour C is zero because the Jost function tends to an E-independent constant, and therefore its derivative is zero on the whole contour C. The integral over the contour ε is exactly the same as is given by Eq. (4.68), i.e. it is zero if there is no bound state at $E = 0$, and it is $-i\pi$ for an S-wave bound state at this point or $-2i\pi$ for a bound state with $\ell > 0$.

The remaining two integrals along the straight lines (the first one is on the upper, and the second on the lower sheet of the Riemann surface) can be found explicitly,

$$\int_A^a L(E)dE + \int_b^B L(E)dE = \ln f_\ell^{(\mathrm{in})}(a) - \ln f_\ell^{(\mathrm{in})}(A) + \ln f_\ell^{(\mathrm{in})}(B) - \ln f_\ell^{(\mathrm{in})}(b) \, .$$

Since the points a and A are exactly above the points b and B, we can use the symmetry properties of the Jost functions (see Fig. 3.6), which give

$$f_\ell^{(\mathrm{in})}(a) = f_\ell^{(\mathrm{out})}(b) \, , \qquad f_\ell^{(\mathrm{in})}(A) = f_\ell^{(\mathrm{out})}(B) \, . \tag{7.84}$$

Therefore,

$$\int_A^a L(E)dE + \int_b^B L(E)dE = \ln \frac{f_\ell^{(\mathrm{out})}(b)}{f_\ell^{(\mathrm{in})}(b)} - \ln \frac{f_\ell^{(\mathrm{out})}(B)}{f_\ell^{(\mathrm{in})}(B)} = \ln S_\ell(b) - \ln S_\ell(B) \, .$$

According to Eq. (7.80), this means that

$$\int_A^a L(E)dE + \int_b^B L(E)dE = 2i \left[\delta_\ell(0) - \delta_\ell(\infty \cdot e^{-i\varphi}) \right] . \tag{7.85}$$

Since all zeros of the Jost function at the spectral points are simple, we can repeat the reasoning given in Sect. 4.7 and find that the closed contour integral shown in Fig. 7.12a is equal to $2\pi i(N_\ell^{(\mathrm{b})} + N_\ell^{(\mathrm{r})})$, where $N_\ell^{(\mathrm{b})}$ is the total number of non-zero bound states with the angular momentum ℓ and $N_\ell^{(\mathrm{r})}$ is the number of the ℓ-wave resonances that are above the straight line $[0, \infty \cdot e^{-i\varphi})$. The integral over the infinitely small circle around the point $E = 0$ is either zero (if no bound state at this point) or $-i\pi$ (for an S-wave bound state) or else $-2i\pi$ (for a bound state with $\ell > 0$).

Therefore, Levinson's theorem has the following general form:

$$\boxed{\delta_\ell(0) - \delta_\ell(\infty \cdot e^{-i\varphi}) = \pi \left[N_\ell^{(\mathrm{b})} + N_\ell^{(\mathrm{r})} \right]} \tag{7.86}$$

where $N_\ell^{(\mathrm{b})}$ includes the zero-energy bound state (if any). If this zero-energy bound state happens to be an S-wave one, its contribution is only $1/2$ instead of normal 1.

Looking at Eq. (7.86), the reader might be confused. Indeed, on the right hand side of this equation, we have a real number (integer, if no S-wave bound state at $E = 0$, and half-integer otherwise), while on the left hand side there is (generally speaking) a complex number. This can be explained as follows. According to Eq. (7.80), at any finite non-real energy, the phase shift $\delta_\ell(E)$ is indeed complex. However, when $E \to \infty$, both $f_\ell^{(\mathrm{in})}(E)$ and $f_\ell^{(\mathrm{out})}(E)$ tend to the same value (1/2), and hence, $S_\ell(E) \to 1$. This does not mean that $\delta_\ell(E) \to 0$. The unit value of S_ℓ

remains unit after any integer number of rotations to the angle 2π around the origin. Actually,

$$S_\ell(\infty \cdot e^{-i\varphi}) = \exp[-2\pi i N_\ell^{(\mathrm{r})}]$$

and thus

$$\delta_\ell(\infty \cdot e^{-i\varphi}) = -\pi N_\ell^{(\mathrm{r})} .$$

These relations follow from the fact that the second term on the right hand side of Eq. (7.86) is due to the second term on its left hand side because for $\varphi = 0$ we have $\delta_\ell(\infty \cdot e^{-i\varphi}) = 0$.

The generalized Levinson's theorem (7.86) can be helpful in practical calculations. Suppose we are looking for resonances of a given potential. How could we be sure that we have located all the resonances with certain ℓ in a given domain of complex energies? Such an assurance can be achieved, if we calculate $\delta_\ell(E)$ along a ray on the unphysical sheet, such that our domain is above this ray. When we move along the ray, the real part of $\delta_\ell(E)$ converges to $-\pi N_\ell^{(\mathrm{r})}$, while its imaginary part converges to zero. In this way, the total number of resonances that must be above the ray can be found.

7.9 Dispersion Relations

Initially, the so-called dispersion relation (Kramers–Kronig relation) was derived in optics. It relates the real and imaginary parts of the refractive index via an integral. The imaginary part describes the absorption of light, while the real part determines the frequency dependence of the speed of light in media and thus its dispersion (hence the name). That integral relation was obtained using the Cauchy theorem of complex analysis (see Appendix B), which states that the value of an analytic function $\omega(z)$ at a point z within its domain of analyticity, $z \in \mathcal{D}$, is equal to the integral

$$\omega(z) = \frac{1}{2\pi i} \oint_C dz' \, \frac{\omega(z')}{z' - z} , \tag{7.87}$$

where the contour C encloses $\mathcal{D}$.

Nowadays, almost any relation based on this theorem is called dispersion relation. There are many of them in quantum scattering theory. Since, however, this book is devoted to the Jost functions, it is only derived here the one that expresses the Jost function via an integral involving the scattering phase shifts. This relation is used in the inverse scattering problem where the unknown two-body potential is reconstructed from the knowledge of the phase shifts and the bound states of the two-body system.

* * *

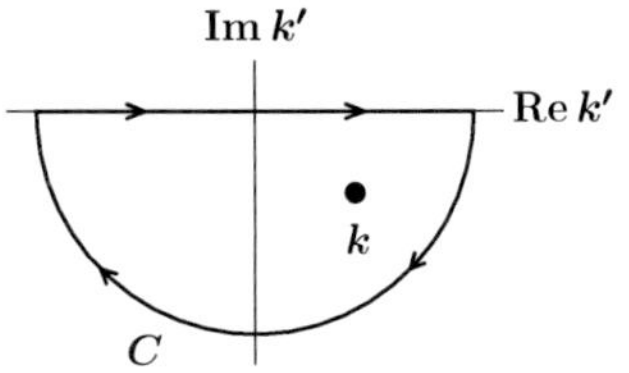

Fig. 7.15 The integration contour used in Eq. (7.90)

Consider the problem with a short-range potential for which the Jost functions tend to constant values as is shown in Fig. 7.14, i.e.

$$f_\ell^{(\text{in/out})}(E) \underset{|E|\to\infty}{\longrightarrow} f_\ell^{(\text{in/out})}(\infty) = \frac{1}{2}\,. \tag{7.88}$$

It should be noted that the limits (7.88) do not exist for $f_\ell^{(\text{in})}$ along the negative imaginary half-axis on the k-plane, and for $f_\ell^{(\text{out})}$ (symmetrically) along the positive imaginary half-axis (which are shown in Fig. 7.14 by the thick lines).[9] On the Riemann surface of the energy, these half-axes correspond to real negative half-axes on the non-physical and physical sheets, respectively.

In principle, the whole derivation of the dispersion relations can be done using the contour integration over the energy variable E on the Riemann surface. It is, however, much easier to depict the contours and to do the derivations if $f_\ell^{(\text{in/out})}(E)$ are considered as the single-valued functions of the momentum k. The functions $f_\ell^{(\text{in})}$ and $f_\ell^{(\text{out})}$ are symmetric relative to the real k-axis (see Fig. 3.5),

$$f_\ell^{(\text{in})*}(E^*) = f_\ell^{(\text{out})}(E)\,. \tag{7.89}$$

It is therefore sufficient to derive the dispersion relation for one of them (it is chosen here the function $f_\ell^{(\text{out})}$), while the corresponding relation for the other function can be obtained by its complex conjugation.

Consider the Jost function $f_\ell^{(\text{out})}(E)$ in the lower half of the k-plane (i.e. on the non-physical Riemann sheet of the energy), where it is analytic at all points. Writing the Cauchy formula (7.87) for

$$\omega(E) = f_\ell^{(\text{out})}(E) - f_\ell^{(\text{out})}(\infty),$$

where the contour C is shown in Fig. 7.15, we have

$$f_\ell^{(\text{out})}(E) - f_\ell^{(\text{out})}(\infty) = -\frac{1}{2\pi i}\oint_C dk'\,\frac{f_\ell^{(\text{out})}(E') - f_\ell^{(\text{out})}(\infty)}{k' - k}\,. \tag{7.90}$$

[9] The fact that the Jost functions cannot be analytic everywhere on the k-plane is also in accordance with Liouville's theorem (see statement (B.9) of Appendix B). Otherwise they would be identically constant, which is the case for the "free" motion.

The negative sign in front of the integral is because the contour runs in the clockwise direction around the complex domain. In this equation, $E = (\hbar k)^2/(2\mu)$ and $E' = (\hbar k')^2/(2\mu)$.

If the radius of the semicircle of the contour C tends to infinity, then thanks to the limit (7.88), the contribution of that semicircle to the integral (7.90) tends to zero, and therefore,

$$f_\ell^{(\text{out})}(E) = f_\ell^{(\text{out})}(\infty) - \frac{1}{2\pi i}\int_{-\infty}^{+\infty} dk' \, \frac{f_\ell^{(\text{out})}(E') - f_\ell^{(\text{out})}(\infty)}{k' - k} \,. \tag{7.91}$$

If we move the point k onto the real axis, the integrand in this equation becomes singular at $k' = k$. The integration over this singularity can be done using the Sokhotski–Plemelj formula (see Eq. (5.59)):

$$\frac{1}{k' - (k - i0)} = \mathcal{P}\frac{1}{k' - k} - i\pi\delta(k' - k) \,, \tag{7.92}$$

which gives

$$f_\ell^{(\text{out})}(E) = f_\ell^{(\text{out})}(\infty) - \frac{1}{2\pi i}\int_{-\infty}^{+\infty} dk' \, \frac{f_\ell^{(\text{out})}(E') - f_\ell^{(\text{out})}(\infty)}{k' - k} +$$

$$+\frac{1}{2}\left[f_\ell^{(\text{out})}(E) - f_\ell^{(\text{out})}(\infty)\right] , \qquad \operatorname{Im} k \uparrow 0 \,,$$

and thus

$$f_\ell^{(\text{out})}(E) = f_\ell^{(\text{out})}(\infty) - \frac{1}{\pi i}\int_{-\infty}^{+\infty} dk' \, \frac{f_\ell^{(\text{out})}(E') - f_\ell^{(\text{out})}(\infty)}{k' - k} \,. \tag{7.93}$$

As is seen in Eq. (7.88), the asymptotic value $f_\ell^{(\text{out})}(\infty)$ is real. Actually, it can always be made real with an appropriate choice of the Jost function normalization (which is arbitrary). Writing Eq. (7.93) separately for the real and imaginary parts of $f_\ell^{(\text{out})}$, we obtain the following dispersion relations:

$$\operatorname{Re}\left[f_\ell^{(\text{out})}(E)\right] = f_\ell^{(\text{out})}(\infty) - \frac{1}{\pi}\int_{-\infty}^{+\infty} dk' \, \frac{\operatorname{Im}\left[f_\ell^{(\text{out})}(E')\right]}{k' - k} \,, \tag{7.94}$$

$$\operatorname{Im}\left[f_\ell^{(\text{out})}(E)\right] = \frac{1}{\pi}\int_{-\infty}^{+\infty} dk' \, \frac{\operatorname{Re}\left[f_\ell^{(\text{out})}(E')\right] - f_\ell^{(\text{out})}(\infty)}{k' - k} \,. \tag{7.95}$$

It follows from the Sokhotski–Plemelj formula (7.92) that

$$\fint_{-\infty}^{+\infty} dk' \, \frac{\mathrm{Im}\left[f_\ell^{(\mathrm{out})}(E')\right]}{k'-k} = \int_{-\infty}^{+\infty} dk' \, \frac{\mathrm{Im}\left[f_\ell^{(\mathrm{out})}(E')\right]}{k'-k+i0} + i\pi\, \mathrm{Im}\left[f_\ell^{(\mathrm{out})}(E)\right] ,$$

which allows us to re-write the relation (7.94) in another form:

$$f_\ell^{(\mathrm{out})}(E) = f_\ell^{(\mathrm{out})}(\infty) - \frac{1}{\pi}\int_{-\infty}^{+\infty} dk' \, \frac{\mathrm{Im}\left[f_\ell^{(\mathrm{out})}(E')\right]}{k'-k+i0} . \tag{7.96}$$

For real energies, we have (see Eq. (4.17))

$$f_\ell^{(\mathrm{out})}(E) = \left|f_\ell^{(\mathrm{out})}(E)\right| e^{i\delta_\ell(E)} ,$$

and thus,

$$\mathrm{Im}\left[f_\ell^{(\mathrm{out})}(E)\right] = \left|f_\ell^{(\mathrm{out})}(E)\right| \sin\delta_\ell(E) = f_\ell^{(\mathrm{out})}(E)\sin\delta_\ell(E) e^{-i\delta_\ell(E)} .$$

Therefore, (7.96) becomes an integral equation for the Jost function,

$$f_\ell^{(\mathrm{out})}(E) = f_\ell^{(\mathrm{out})}(\infty) - \frac{1}{\pi}\int_{-\infty}^{+\infty} dk' \, \frac{\sin\delta_\ell(E')e^{-i\delta_\ell(E')}}{k'-k+i0} f_\ell^{(\mathrm{out})}(E') , \tag{7.97}$$

where the kernel is determined by the phase shifts $\delta_\ell(E)$. What is done further down in this section is finding the solution of this equation under the assumption that in addition to the phase shifts for all energies we are also given the energies of all the bound states for the angular momentum ℓ.

In what follows, we use the logarithm of the function $f_\ell^{(\mathrm{out})}(E)$, which becomes singular at the points

$$E_n = -\frac{(\hbar\kappa_n)^2}{2\mu} , \qquad n = 1, 2, \ldots, N ,$$

where κ_n is the absolute value of the (pure imaginary) momentum corresponding to the n-th bound state. Indeed, the bound states are at the zeros of $f_\ell^{(\mathrm{in})}(E)$, and due to the symmetry between $f_\ell^{(\mathrm{in})}$ and $f_\ell^{(\mathrm{out})}$, they also correspond to zeros of $f_\ell^{(\mathrm{out})}(E)$ at the mirror-symmetric points on the k-plane (see Fig. 3.5),

$$f_\ell^{(\mathrm{in})}(E(i\kappa_n)) = f_\ell^{(\mathrm{out})}(E(-i\kappa_n)) = 0 .$$

In other words, $\ln f_\ell^{(\mathrm{out})}$ is singular at the points $-i\kappa_n$ in the lower half of the k-plane. To get rid of these singularities, we introduce another function:

$$\tilde{f}_\ell^{(\mathrm{out})}(E) = \frac{f_\ell^{(\mathrm{out})}(E)}{f_\ell^{(\mathrm{out})}(\infty)} \prod_{n=1}^{N} \frac{k - i\kappa_n}{k + i\kappa_n} .$$

This function is analytic in the same domain $\mathcal{D}$ as $f_\ell^{(\mathrm{out})}$, does not have zeros in the lower half of the k-plane, and tends to unity when $|k| \to \infty$. Therefore, its logarithm is analytic for $\mathrm{Im}\, k \leqslant 0$ and

$$\ln \tilde{f}_\ell^{(\mathrm{out})}(E) \underset{|k|\to\infty}{\longrightarrow} 0 .$$

In other words, the function $\ln \tilde{f}_\ell^{(\mathrm{out})}(E)$ has the same properties in the lower half of the k-plane as the function $f_\ell^{(\mathrm{out})}(E) - f_\ell^{(\mathrm{out})}(\infty)$. We can therefore repeat all the derivations leading to Eq. (7.93) and obtain

$$\ln \tilde{f}_\ell^{(\mathrm{out})}(E) = -\frac{1}{\pi i} \int_{-\infty}^{+\infty} dk' \, \frac{\ln \tilde{f}_\ell^{(\mathrm{out})}(E')}{k' - k} . \tag{7.98}$$

Now, let $\tilde{\delta}_\ell$ be the polar angle of the complex number $\tilde{f}_\ell^{(\mathrm{out})}$, i.e.

$$\tilde{f}_\ell^{(\mathrm{out})} = \left| \tilde{f}_\ell^{(\mathrm{out})} \right| e^{i\tilde{\delta}_\ell} ,$$

and then

$$\mathrm{Re}\left[\ln \tilde{f}_\ell^{(\mathrm{out})}\right] = \mathrm{Re}\left\{\ln \left|\tilde{f}_\ell^{(\mathrm{out})}\right| + i\tilde{\delta}_\ell\right\} = \ln \left|\tilde{f}_\ell^{(\mathrm{out})}\right| .$$

If k is real, then

$$\mathrm{Im}\, k = 0 \quad \Rightarrow \quad \left|\frac{k - i\kappa_n}{k + i\kappa_n}\right| = 1 \quad \Rightarrow \quad \ln \left|\tilde{f}_\ell^{(\mathrm{out})}(E)\right| = \ln \left|\frac{f_\ell^{(\mathrm{out})}(E)}{f_\ell^{(\mathrm{out})}(\infty)}\right| .$$

For any complex number of unit absolute value, we have

$$|z| \quad \Rightarrow \quad z = e^{i\delta} \quad \Rightarrow \quad \delta = -i \ln z .$$

Therefore,

$$\tilde{\delta}_\ell = \delta_\ell - i \sum_{n=1}^{N} [\ln(k - i\kappa_n) - \ln(k + i\kappa_n)] .$$

Taking real part of Eq. (7.98), we obtain

$$\ln\left|\frac{f_\ell^{(\text{out})}(E)}{f_\ell^{(\text{out})}(\infty)}\right| = -\frac{1}{\pi}\fint_{-\infty}^{+\infty} dk' \, \frac{\text{Im}\left[\ln \tilde{f}_\ell^{(\text{out})}(E')\right]}{k'-k} ,$$

where

$$\text{Im}\left[\ln \tilde{f}_\ell^{(\text{out})}\right] = \delta_\ell - i\sum_{n=1}^{N}\left[\ln(k - i\kappa_n) - \ln(k + i\kappa_n)\right] ,$$

and thus

$$\ln\left|\frac{f_\ell^{(\text{out})}(E)}{f_\ell^{(\text{out})}(\infty)}\right| = -\frac{1}{\pi}\fint_{-\infty}^{+\infty} dk' \, \frac{\delta_\ell(E')}{k'-k} + \frac{i}{\pi}\sum_{n=1}^{N}\fint_{-\infty}^{+\infty} dk' \, \frac{\ln\dfrac{k'-i\kappa_n}{k'+i\kappa_n}}{k'-k} . \tag{7.99}$$

The first integral on the right hand side of this equation we leave as it is, and the second one we can calculate using the residue theory (see Appendix B). To this end, consider (for each term in the sum (7.99)) the closed contour in the upper half of the k-plane, which is shown in Fig. 7.16.

It is clear that the integral over the whole contour is zero because there are no singularities inside,

$$\oint_C dk' \, \frac{\ln\dfrac{k'-i\kappa_n}{k'+i\kappa_n}}{k'-k} = 0 . \tag{7.100}$$

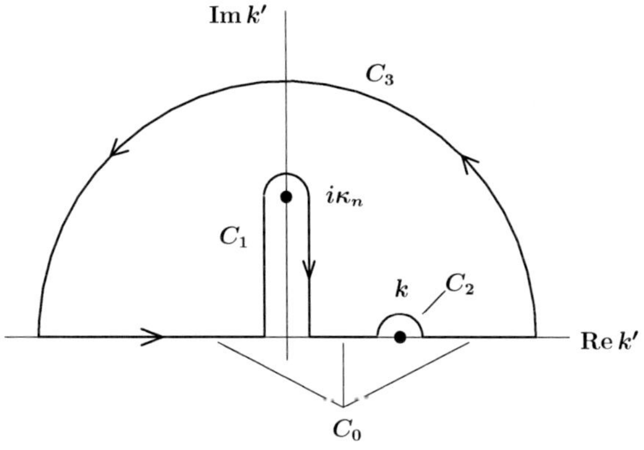

Fig. 7.16 The integration contour $C = C_0 + C_1 + C_2 + C_3$ that is used in Eq. (7.100)

The closed contour C consists of several segments,

$$C = C_0 + C_1 + C_2 + C_3 .$$

The integral over C_0 is what we are looking for,

$$\fint_{-\infty}^{+\infty} dk' \frac{\ln \dfrac{k' - i\kappa_n}{k' + i\kappa_n}}{k' - k} = \int_{C_0} dk' \frac{\ln \dfrac{k' - i\kappa_n}{k' + i\kappa_n}}{k' - k} . \tag{7.101}$$

The contribution from the big semicircle, C_3, when its radius tends to infinity, is zero. The contribution from the small semicircle, C_2, is a real number:

$$\int_{C_2} dk' \frac{\ln \dfrac{k' - i\kappa_n}{k' + i\kappa_n}}{k' - k} = \lim_{\epsilon \to 0} \int_{\pi}^{0} d\varphi \frac{\ln \dfrac{k + \epsilon e^{i\varphi} - i\kappa_n}{k + \epsilon e^{i\varphi} + i\kappa_n}}{\epsilon e^{i\varphi}} i\epsilon e^{i\varphi}$$

$$= i \ln \frac{k - i\kappa_n}{k + i\kappa_n} \int_{\pi}^{0} d\varphi = -i\pi \ln \frac{k - i\kappa_n}{k + i\kappa_n} = \pi \arg \left(\frac{k - i\kappa_n}{k + i\kappa_n} \right) ,$$

where $\arg(z)$ is the polar angle of the complex number z. When calculating the integral over C_1, we should take into account the fact that this contour goes in the clockwise direction around the branch point, $k' = i\kappa_n$, of the logarithm in the integrand. The logarithm is a multi-valued function,

$$\ln z = \ln \left[|z| \exp(i \arg z + 2\pi i m) \right] , \qquad m = 0, \pm 1, \pm 2, \dots .$$

Each time we make a full circle around the branch point in the clockwise direction, the polar angle of z changes from $\arg z$ to $(\arg z - 2\pi i)$. Therefore,

$$\int_{C_1} dk' \frac{\ln \dfrac{k' - i\kappa_n}{k' + i\kappa_n}}{k' - k} = \int_{0}^{i\kappa_n} dk' \frac{\ln \dfrac{k' - i\kappa_n}{k' + i\kappa_n}}{k' - k} + \int_{i\kappa_n}^{0} dk' \frac{\ln \dfrac{k' - i\kappa_n}{k' + i\kappa_n} - 2\pi i}{k' - k} .$$

In this sum, the integrals involving the logarithmic functions cancel each other, and we remain with the simple integrand,

$$\int_{C_1} dk' \frac{\ln \dfrac{k' - i\kappa_n}{k' + i\kappa_n}}{k' - k} = -\int_{i\kappa_n}^{0} dk' \frac{2\pi i}{k' - k} = 2\pi i \ln \frac{k - i\kappa_n}{k}$$

$$= 2\pi i \ln \left[\left| \sqrt{1 + \frac{\kappa_n^2}{k^2}} \right| e^{i \arg(1 - i\kappa_n / k)} \right]$$

$$
= 2\pi i \left[\ln \left| \sqrt{1 + \frac{\kappa_n^2}{k^2}} \right| + i \arg \left(1 - i \frac{\kappa_n}{k} \right) \right]
$$

$$
= \pi i \ln \left(1 + \frac{\kappa_n^2}{k^2} \right) - 2\pi \arg \left(1 - i \frac{\kappa_n}{k} \right) .
$$

It then follows from (7.91) that

$$
\int_{C_0} = - \int_{C_1} - \int_{C_2} = -\pi i \ln \left(1 + \frac{\kappa_n^2}{k^2} \right) + 2\pi \arg \left(1 - i \frac{\kappa_n}{k} \right) - \pi \arg \left(\frac{k - i\kappa_n}{k + i\kappa_n} \right) . \tag{7.102}
$$

Since k is on the real axis, we have

$$
\arg \left(\frac{k - i\kappa_n}{k} \right) = \arg(k - i\kappa_n) - 0 = - \arctan \frac{\kappa_n}{k}
$$

and

$$
\arg \left(\frac{k - i\kappa_n}{k + i\kappa_n} \right) = \arg(k - i\kappa_n) - \arg(k + i\kappa_n) = -2 \arctan \frac{\kappa_n}{k} .
$$

Therefore, the last two terms in Eq. (7.102) cancel each other, and thus, substituting (7.102) in (7.101) and then in Eq. (7.99), we obtain

$$
\ln \left| \frac{f_\ell^{(\text{out})}(E)}{f_\ell^{(\text{out})}(\infty)} \right| = -\frac{1}{\pi} \int_{-\infty}^{+\infty} dk' \, \frac{\delta_\ell(E')}{k' - k} + \ln \left[\prod_{n=1}^{N} \left(1 + \frac{\kappa_n^2}{k^2} \right) \right] . \tag{7.103}
$$

Recollecting that $f_\ell^{(\text{out})}(\infty)$ is real, we can write

$$
\ln \left| \frac{f_\ell^{(\text{out})}(E)}{f_\ell^{(\text{out})}(\infty)} \right| = \ln \left\{ \left| \frac{f_\ell^{(\text{out})}(E)}{f_\ell^{(\text{out})}(\infty)} \right| e^{i\delta_\ell(E)} e^{-i\delta_\ell(E)} \right\}
$$

$$
= \ln \left[\frac{f_\ell^{(\text{out})}(E)}{f_\ell^{(\text{out})}(\infty)} e^{-i\delta_\ell(E)} \right] = \ln \frac{f_\ell^{(\text{out})}(E)}{f_\ell^{(\text{out})}(\infty)} - i\delta_\ell(E) .
$$

Using once more the Sokhotski–Plemelj formula (7.92), we re-write Eq. (7.103) as

$$
\ln \frac{f_\ell^{(\text{out})}(E)}{f_\ell^{(\text{out})}(\infty)} = -\frac{1}{\pi} \int_{-\infty}^{+\infty} dk' \, \frac{\delta_\ell(E')}{k' - k + i0} + \ln \left[\prod_{n=1}^{N} \left(1 + \frac{\kappa_n^2}{k^2} \right) \right] ,
$$

which finally gives

$$f_\ell^{(\text{in/out})}(E) = f_\ell^{(\text{in/out})}(\infty) \prod_{n=1}^{N} \left(1 - \frac{E_n}{E}\right) \exp\left[-\frac{1}{\pi} \int_{-\infty}^{+\infty} dk' \, \frac{\delta_\ell(E')}{k' - k \mp i0}\right] \tag{7.104}$$

where for obtaining the corresponding equation for $f_\ell^{(\text{in})}(E)$, we make use of the relation (7.89). The function $f_\ell^{(\text{out})}(E)$ given by Eq. (7.104) is the solution of the integral equation (7.97). It should not be forgotten that Eq. (7.104) is obtained for and therefore valid for real energies $E > 0$. If one tries to use it for negative energies, then immediately encounters the contradiction, that both $f_\ell^{(\text{in})}(E)$ and $f_\ell^{(\text{out})}(E)$ have zeros at the bound states, $E_n < 0$.

The integration in Eq. (7.104) requires the phase shifts not only at positive but also at negative values of the momentum k. Using Eq. (7.80) and the symmetry property of the Jost functions under the transformation $k \to -k$ (see Sect. 3.1.1), it is easy to show that $\delta_\ell(E(k))$ is an odd function of the momentum. Indeed,

$$\delta_\ell(E(k)) = \frac{1}{2i} \ln S_\ell(E(k)) = \frac{1}{2i} \left[\ln f_\ell^{(\text{out})}(E(k)) - \ln f_\ell^{(\text{in})}(E(k))\right]$$

and

$$\delta_\ell(E(-k)) = \frac{1}{2i} \left[\ln f_\ell^{(\text{in})}(E(k)) - \ln f_\ell^{(\text{out})}(E(k))\right],$$

which gives

$$\delta_\ell(E(-k)) = -\delta_\ell(E(k)) . \tag{7.105}$$

Using this property, we can re-write Eq. (7.104) in another form that involves the integration over the energy variable. To this end, we multiply and divide the integrand by $(k' + k \pm i0)$:

$$\int_{-\infty}^{+\infty} dk' \, \frac{\delta_\ell(E')}{k' - k \mp i0} = \int_{-\infty}^{+\infty} dk' \, \frac{\delta_\ell(E')(k' + k)}{k'^2 - k^2 \mp i0} .$$

Because of the relation (7.105), we have

$$\int_{-\infty}^{+\infty} dk' \, \frac{\delta_\ell(E')k}{k'^2 - k^2 \mp i0} = 0 ,$$

and thus

$$\int_{-\infty}^{+\infty} dk' \, \frac{\delta_\ell(E')}{k' - k \mp i0} = 2 \int_0^{\infty} dk' \, \frac{\delta_\ell(E') k'}{k'^2 - k^2 \mp i0} = \int_0^{\infty} dE' \, \frac{\delta_\ell(E')}{E' - E \mp i0} \, .$$

Therefore,

$$\boxed{f_\ell^{(\text{in/out})}(E) = f_\ell^{(\text{in/out})}(\infty) \prod_{n=1}^{N} \left(1 - \frac{E_n}{E}\right) \exp\left[-\frac{1}{\pi} \int_0^{\infty} dE' \, \frac{\delta_\ell(E')}{E' - E \mp i0}\right]} \tag{7.106}$$

8 Single-Channel Potential with Coulombic Tail

The electric attraction or repulsion between two point-like charges eZ_1 and eZ_2 is one of the few forces in physics that are known exactly. It is described by the Coulomb potential, which is spherically symmetric and has a very simple functional form,

$$U_c(r) = \frac{e^2 Z_1 Z_2}{r} \,. \tag{8.1}$$

Despite its simplicity, this potential causes major mathematical difficulties, when it is present in the Schrödinger equation. The reason for these difficulties is its slow ($\sim r^{-1}$) vanishing at large distances. Formally, the Coulomb force is present at any distance, and therefore, the particles cannot move "freely", i.e. their asymptotic states cannot be viewed as the plane waves. Instead, these states are described by the so-called Coulomb waves, which we consider in the next section.

In real physical problems, very seldom, we deal with pure Coulomb potential. In most cases, some other (short-range) forces are also present. As a result, the total potential has a complicated r-dependence at short distances and asymptotically becomes a Coulomb potential, i.e. has a Coulombic tail. For such a problem, all the mathematical difficulties associated with the r^{-1} dependence remain the same.

It is always convenient to split the total potential into a short-range and the pure Coulomb terms. This can be done by adding and subtracting the pure Coulomb potential having the same tail as the total potential,

$$U(r) = U(r) - U_c(r) + U_c(r) = U_s(r) + U_c(r) \,, \tag{8.2}$$

where $U_s(r)$ vanishes at large distances much faster than $U_c(r)$ (this is illustrated in Fig. 8.1). The fact that the short-range potential $U_s(r) = U(r) - U_c(r)$ acquires the $1/r$ singularity at $r = 0$ does not pose any problem because it is still less singular than the centrifugal potential $\ell(\ell+1)/r^2$ and apparently obeys the condition (2.2).

S. A. Rakityansky, *Jost Functions in Quantum Mechanics*,
https://doi.org/10.1007/978-3-031-07761-6_8

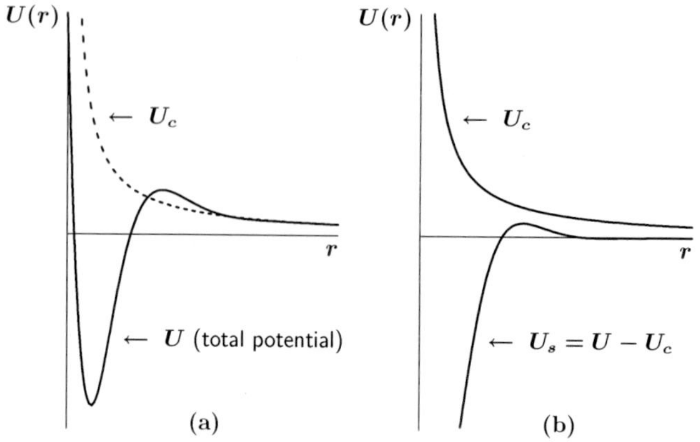

Fig. 8.1 (**a**) Total potential U with a Coulombic tail; (**b**) Splitting of the potential $U = U_s + U_c$ in the short-range and pure Coulomb terms

We therefore assume that whenever the Coulomb forces are present, they always are given in the form of a pure Coulomb potential as an addition to a short-range potential, if any. To begin with, let us consider the problem with a pure Coulomb potential alone.

8.1 Pure Coulomb Potential

8.1.1 Schrödinger Equation

For a pair of point-like particles that interact with each other via only the potential (8.1), the relative motion in the c.m. system is equivalent to the motion of a single particle with the reduced mass μ (see Sect. 1.5), and the corresponding wave function obeys the Schrödinger equation

$$\left(\frac{\hbar^2}{2\mu}\Delta_{\vec{r}} + E - \frac{e^2 Z_1 Z_2}{r}\right)\psi_E(\vec{r}\,) = 0\,. \tag{8.3}$$

After standard separation of the angular dependence, the corresponding radial equation reads

$$\left[\partial_r^2 + k^2 - \frac{\ell(\ell+1)}{r^2} - \frac{2k\eta}{r}\right]u_\ell(E,r) = 0\,, \tag{8.4}$$

where the reduced Coulomb potential is

$$V_c(r) = \frac{2\mu}{\hbar^2}U_c(r) = \frac{2\mu}{\hbar^2}\frac{e^2 Z_1 Z_2}{r} = \frac{2k\eta}{r} \tag{8.5}$$

with

$$\eta = \frac{\mu e^2 Z_1 Z_2}{k\hbar^2} \tag{8.6}$$

being the so-called Sommerfeld parameter. As is easily seen, this parameter is dimensionless. Indeed,

$$[\eta] = \frac{[\mu][e^2]}{[k][\hbar^2]} = \frac{[\mu c^2]}{[k][(\hbar c)]}\left[\frac{e^2}{\hbar c}\right] = \frac{\text{MeV}}{\text{fm}^{-1}\cdot\text{MeV}\cdot\text{fm}} = 1\,,$$

where we use the facts that the product

$$\hbar c \approx 197.3269718\,\text{MeV}\cdot\text{fm} \tag{8.7}$$

has the units [MeV · fm], while the fine-structure constant

$$\alpha = \frac{e^2}{\hbar c} \approx \frac{1}{137.0359895} \tag{8.8}$$

is dimensionless. Although the Sommerfeld parameter has no units, it inversely depends on the momentum and therefore becomes extremely large when the collision energy approaches zero.

8.1.2 Jost Functions

Two linearly independent solutions of Eq. (8.4) are known analytically (see Appendix E). They are the regular, $F_\ell(\eta, kr)$, and irregular, $G_\ell(\eta, r)$, Coulomb functions (the names reflect their behaviour when $r \to 0$). The normalizations of these functions are chosen in such a way that for neutral particles ($\eta \to 0$), i.e. when the Coulomb potential in Eq. (8.4) disappears, they become the Riccati–Bessel and Riccati–Neumann functions, respectively,[1]

$$F_\ell(\eta, kr) \underset{\eta\to 0}{\longrightarrow} j_\ell(kr)\,, \tag{8.9}$$

$$G_\ell(\eta, kr) \underset{\eta\to 0}{\longrightarrow} -n_\ell(kr)\,. \tag{8.10}$$

Similarly to the Riccati–Hankel functions (which are linear combinations of j_ℓ and n_ℓ) that describe free-moving spherical waves,

$$h_\ell^{(\pm)}(kr) = j_\ell(kr) \pm i n_\ell(kr) \underset{r\to\infty}{\longrightarrow} \mp i \exp\left[\pm i\left(kr - \frac{\ell\pi}{2}\right)\right]\,, \tag{8.11}$$

[1] Please note the minus sign on the right hand side of the limit (8.10).

we can construct the following combinations of the Coulomb functions

$$H_\ell^{(\pm)}(\eta, kr) = F_\ell(\eta, kr) \mp i G_\ell(\eta, kr) \tag{8.12}$$

$$\underset{r\to\infty}{\longrightarrow} \mp i \exp\left\{\pm i \left[kr - \frac{\ell\pi}{2} - \eta\ln(2kr) + \sigma_\ell\right]\right\},$$

where the quantity $\sigma_\ell(\eta)$ obeys the following equation:

$$\exp(2i\sigma_\ell) = \frac{\Gamma(\ell+1+i\eta)}{\Gamma(\ell+1-i\eta)}, \tag{8.13}$$

which gives

$$\sigma_\ell(\eta) = \frac{1}{2i}\left[\ln\Gamma(\ell+1+i\eta) - \ln\Gamma(\ell+1-i\eta)\right] \underset{\eta\to 0}{\longrightarrow} 0\,. \tag{8.14}$$

When k is real, $\Gamma(\ell+1-i\eta) = [\Gamma(\ell+1+i\eta)]^*$, and thus

$$\sigma_\ell = \arg\Gamma(\ell+1+i\eta)\,, \qquad \operatorname{Im} k = 0\,. \tag{8.15}$$

Apparently, the functions $H_\ell^{(\pm)}(\eta, kr)$ (which might be called Coulomb–Hankel functions) generalize the spherical waves $h_\ell^{(\pm)}(kr)$, which are their particular case for neutral particles,

$$H_\ell^{(\pm)}(\eta, kr) \underset{\eta\to 0}{\longrightarrow} h_\ell^{(\pm)}(kr)\,. \tag{8.16}$$

These functions should therefore be considered as the spherical waves distorted by the presence of the Coulomb potential. As we see in Eq. (8.12), at large distances, the phase of the spherical wave is distorted by the constant term σ_ℓ (which is called the Coulomb phase shift) and logarithmic term $\eta\ln(2kr)$ that never stops to grow. This logarithm is the most troublesome feature of the Coulomb potential, stemming from the fact that the potential vanishes at infinity too slowly ($\sim 1/r$).

Since $H_\ell^{(\pm)}(\eta, kr)$ are two linearly independent solutions of Eq. (8.4), its general solution is their linear combination

$$u_\ell(E, r) = H_\ell^{(-)}(\eta, kr)c_1 + H_\ell^{(+)}(\eta, kr)c_2\,, \tag{8.17}$$

which is valid for all $r \in [0, \infty)$. We are only interested in physically meaningful solutions, i.e. in those that are regular at $r = 0$. This implies that the constants c_1 and c_2 in Eq. (8.17) must be equal to each other, $c_1 = c_2 = c$. Indeed, in such a case, the singularities (at $r = 0$) of the functions $H_\ell^{(\pm)}$ cancel

$$H_\ell^{(-)}(\eta, kr) + H_\ell^{(+)}(\eta, kr) = 2F_\ell(\eta, kr)\,, \tag{8.18}$$

and the wave function is regular,

$$u_\ell(E,r) = 2cF_\ell(\eta,kr)\,. \tag{8.19}$$

As we know, the normalization of the wave function can be chosen arbitrarily. However when in Sect. 2.2.3 we defined the Jost functions for the short-range potential, we used the radial wave function normalized in such a way that near the point $r = 0$ it behaved exactly as $j_\ell(kr)$, i.e. as the regular solution for completely free motion (no potential is present). Such radial wave function was called the regular solution and was denoted as ϕ_ℓ. To be consistent, we need to find such a value of c that near the origin ($r \to 0$) it would always give $j_\ell(kr)$ on the right hand side of Eq. (8.19).

When $r \to 0$, the leading term determining the behaviour of the regular Coulomb function is

$$F_\ell(\eta,kr) \underset{r\to 0}{\longrightarrow} \frac{C_\ell(\eta)}{(2\ell+1)!!}(kr)^{\ell+1}\,, \tag{8.20}$$

where

$$C_\ell(\eta) = \frac{e^{-\pi\eta/2}}{\Gamma(\ell+1)}\exp\left\{\frac{1}{2}\left[\ln\Gamma(\ell+1+i\eta)+\ln\Gamma(\ell+1-i\eta)\right]\right\} \underset{\eta\to 0}{\longrightarrow} 1 \tag{8.21}$$

is the so-called Coulomb barrier factor.[2] Meanwhile for the Riccati–Bessel function, the corresponding leading term is given by

$$j_\ell(kr) \underset{r\to 0}{\longrightarrow} \frac{(kr)^{\ell+1}}{(2\ell+1)!!}\,. \tag{8.22}$$

Comparing Eqs. (8.20) and (8.22), we see that the constant c in Eq. (8.19) should be chosen as

$$c = \frac{1}{2C_\ell(\eta)}\,.$$

This gives the regular solution in the form:

$$\phi_\ell(E,r) = \frac{1}{C_\ell(\eta)}F_\ell(\eta,kr) \tag{8.23}$$

$$= H_\ell^{(-)}(\eta,kr)\frac{1}{2C_\ell(\eta)} + H_\ell^{(+)}(\eta,kr)\frac{1}{2C_\ell(\eta)}\,. \tag{8.24}$$

---

[2] Please note that in some mathematical texts the barrier factor is defined in such a way that for $\eta = 0$ it becomes $1/(2\ell+1)!!$

The asymptotic forms (8.12) of the functions $H_\ell^{(\pm)}(\eta, kr)$ can be factorized into the r-dependent (oscillating for real energies) factors and the r-independent factors $\exp[\pm i\sigma(\eta)]$ that are determined (via η) by the energy and the strength of the potential. In other words,

$$\phi_\ell(E,r) \underset{r\to\infty}{\longrightarrow} W_\ell^{(\mathrm{in})}(E,r)\frac{e^{-i\sigma_\ell}}{2C_\ell(\eta)} + W_\ell^{(\mathrm{out})}(E,r)\frac{e^{+i\sigma_\ell}}{2C_\ell(\eta)}, \tag{8.25}$$

where

$$W_\ell^{(\mathrm{in/out})}(E,r) = H_\ell^{(\mp)}(\eta, kr)e^{\pm i\sigma_\ell} \tag{8.26}$$

$$\underset{r\to\infty}{\longrightarrow} \pm i\exp\left\{\mp i\left[kr - \frac{\ell\pi}{2} - \eta\ln(2kr)\right]\right\} \tag{8.27}$$

The functions (8.26) are generalizations of the incoming and outgoing spherical waves (2.10, 2.11) defined for the case of short-range interactions in Sect. 2.2.1.1. It is easy to see that in the limit $\eta \to 0$ they coincide with the Riccati–Hankel functions.

As is seen in Eq. (8.27), for real scattering energies, the Coulomb waves are almost harmonic, whose oscillations are slowly but steadily shifted by the r-dependent phase $\eta\ln(2kr)$. This logarithmic shift grows to infinity because the colliding particles feel the presence of each other at any distance. The effect of this infinite phase shift, however, is not observable because the incoming and outgoing waves have the same infinite shift of their phases but with different signs. In other words, before the collision, the incoming wave already has the logarithmic phase shift, which is "undone" during and after the collision. This is of course just intuitive reasoning that explains why the logarithmic phase shift is included in the "free" Coulomb waves. A rigorous mathematical ground for this trick can be found using the time-dependent scattering theory, which is outside the scope of this book.

Following the definition of the Jost functions as the energy-dependent amplitudes of the incoming and outgoing waves, we immediately find from Eq. (8.25) that the Jost functions for a pure Coulomb potential should be given by[3]

$$f_\ell^{\mathrm{c\,(in/out)}}(E) = \frac{e^{\mp i\sigma_\ell}}{2C_\ell(\eta)} \tag{8.28}$$

[3] The reader should understand that this is not a rigorous logical derivation, but rather a "plausible" definition. Indeed, we could have split the factors into the exponential functions (8.12) differently. The only "proof" that the separation of the factors $\exp(\mp i\sigma_\ell)$ from the waves $W_\ell^{(\mathrm{in/out})}$ leads to the correct Jost functions is the fact that they give the correct S-matrix (8.29) and the correct Bohr spectrum (8.33) for the case of an attractive Coulomb force.

Then the pure Coulomb S-matrix is

$$S_\ell^{\rm c}(E) = \frac{f_\ell^{\rm c\,(out)}(E)}{f_\ell^{\rm c\,(in)}(E)} = e^{2i\sigma_\ell} = \frac{\Gamma(\ell+1+i\eta)}{\Gamma(\ell+1-i\eta)}\,. \tag{8.29}$$

Comparing this with Eq. (4.19), we find that $\sigma_\ell(\eta)$ is the pure Coulomb phase shift. Equations (8.13) and (8.21) give

$$e^{\pm i\sigma_\ell} = \sqrt{\frac{\Gamma(\ell+1\pm i\eta)}{\Gamma(\ell+1\mp i\eta)}} \tag{8.30}$$

and

$$C_\ell(\eta) = \frac{e^{-\pi\eta/2}}{\ell!}\sqrt{\Gamma(\ell+1+i\eta)\Gamma(\ell+1-i\eta)}\,. \tag{8.31}$$

This allows us to re-write Eq. (8.28) as follows:

$$f_\ell^{\rm c\,(in/out)}(E) = \frac{e^{\pi\eta/2}\ell!}{2\Gamma(\ell+1\pm i\eta)}\,. \tag{8.32}$$

It is easy to check that for an attractive potential ($Z_1Z_2 < 0$), the zeros of $f_\ell^{\rm c\,(in)}(E)$ corresponding to the poles of the Γ-function in its denominator give the spectrum of a hydrogen-like atom. Indeed, the Γ-function has poles at the zero as well as at all real negative integer numbers, i.e. when

$$\ell+1+i\eta = -n_r\,, \qquad n_r = 0, 1, 2, \ldots$$

This gives pure imaginary discrete values of η,

$$\eta = i(n_r+\ell+1)\,.$$

Taking into account that for an attractive potential

$$\eta = -\frac{\mu e^2|Z_1Z_2|}{k\hbar^2}\,,$$

we obtain pure imaginary values of the wave number (on the physical sheet of the Riemann surface),

$$k = i\,\frac{\mu e^2|Z_1Z_2|}{\hbar^2(n_r+\ell+1)}\,,$$

and finally the Bohr energies of the bound states,

$$E_n = -\frac{\mu e^4 |Z_1 Z_2|^2}{2\hbar^2 n^2}, \tag{8.33}$$

where $n = n_r + \ell + 1 = 1, 2, 3, \ldots$ is the so-called principal quantum number.

8.1.3 Scattering

Coulomb potential is one of the few for which the Schrödinger equation can be solved exactly in the closed form. Its full three-dimensional solution for the scattering state is expressed in terms of the hypergeometric function ${}_1F_1$ as follows:[4]

$$\begin{aligned} \psi^c_{\vec{k}}(\vec{r}\,) &= \frac{e^{i\vec{k}\vec{r}}}{(2\pi)^{3/2}} e^{-\pi\eta/2}\Gamma(1+i\eta)\, {}_1F_1\left[-i\eta; 1; i(kr - \vec{k}\vec{r}\,)\right] \\ &\underset{\eta\to 0}{\longrightarrow} \frac{e^{i\vec{k}\vec{r}}}{(2\pi)^{3/2}} . \end{aligned} \tag{8.34}$$

If the coordinate system is chosen in such a way that the incoming momentum, $\vec{k}$, is directed along the z-axis, then asymptotically this solution behaves as

$$\psi^c_{\vec{k}}(\vec{r}\,) \underset{r\to\infty}{\longrightarrow} \frac{1}{(2\pi)^{3/2}} \left[e^{i[kz+\eta \ln k(r-z)]} + \frac{e^{i(kr-\eta\ln 2kr)}}{r} \mathrm{f}^c(\hat{\vec{r}} \leftarrow \hat{\vec{k}}, E) \right], \tag{8.35}$$

where the pure Coulomb amplitude of scattering to the angle θ is given by

$$\mathrm{f}^c(\hat{\vec{r}} \leftarrow \hat{\vec{k}}, E) = -\frac{\eta}{2k\sin^2(\theta/2)} \exp\left\{2i\sigma_0(\eta) - i\eta \ln\left[\sin^2(\theta/2)\right]\right\} . \tag{8.36}$$

It is instructive to compare Eq. (8.35) with the corresponding asymptotic behaviour of the wave function for neutral particles (see Eq. (4.11)). As is seen, the Coulomb forces distort (by the logarithmic phase) both the incident plane wave and the scattered spherical wave. Despite this, the observable cross section is still found as the square of the amplitude. Indeed, using the same recipe for calculating the currents of the incoming and scattered particles as is described in Sect. 1.3.2.2, we obtain

$$\vec{j}^{\,(\text{in})} = \frac{\hbar\vec{k}}{(2\pi)^3\mu} + \frac{\hbar\eta}{2(2\pi)^3\mu r} \cdot \frac{\hat{\vec{r}} - \hat{\vec{k}}}{\sin^2(\theta/2)}, \tag{8.37}$$

[4] Since this is outside the scope of this book, a derivation of Eq. (8.34) should be looked up elsewhere, for example, in Refs. [13, 25, 33, 34].

$$j_r^{(\mathrm{sc})} r^2 = \frac{|\mathrm{f}(\hat{\vec{r}} \leftarrow \hat{\vec{k}}, E)|^2}{(2\pi)^3} \left(\frac{\hbar k}{\mu} - \frac{\hbar \eta}{\mu r} \right) , \tag{8.38}$$

where in both currents the second terms, appearing due to the logarithmic phases, have the factors $1/r$. Therefore at sufficiently large distances, these terms can be omitted, and the cross section is equal to the square of the amplitude:

$$\frac{d\sigma}{d\Omega} = \frac{j_r^{(\mathrm{sc})} r^2 d\Omega}{|\vec{j}^{\,(\mathrm{in})}|} = \left| \mathrm{f}^c(\hat{\vec{r}} \leftarrow \hat{\vec{k}}, E) \right|^2 . \tag{8.39}$$

This gives famous Rutherford formula:

$$\boxed{\frac{d\sigma}{d\Omega} = \left(\frac{Z_1 Z_2 e^2}{4E \sin^2 \frac{\theta}{2}} \right)^2} \tag{8.40}$$

It can be shown that the partial-wave expansion of the scattering wave function (8.34) has the form:[5]

$$\psi_k^c(\vec{r}\,) = \sqrt{\frac{2}{\pi}} \frac{1}{kr} \sum_{\ell m} i^\ell e^{i\sigma_\ell} F_\ell(\eta, kr) Y_{\ell m}^*(\hat{\vec{k}}\,) Y_{\ell m}(\hat{\vec{r}}\,) , \tag{8.41}$$

and the amplitude (8.36) can be *formally* presented as the series

$$\mathrm{f}^c(\hat{\vec{r}} \leftarrow \hat{\vec{k}}, E) = \frac{1}{2ik} \sum_{\ell=0}^{\infty} (2\ell + 1)(e^{2i\sigma_\ell} - 1) P_\ell(\cos\theta) . \tag{8.42}$$

However, this series does not converge. Mathematical reason for its divergence consists in the fact that σ_ℓ does not vanish when $\ell \to \infty$. Indeed, as it follows from (8.14),

$$\sigma_\ell(\eta) = \arg \Gamma(\ell + 1 + i\eta) . \tag{8.43}$$

Using the property of the Γ-function,

$$\arg \Gamma(z + 1) = \arg \Gamma(z) + \arctan \frac{\mathrm{Im}(z)}{\mathrm{Re}(z)} ,$$

[5] A derivation of Eqs. (8.41, 8.42) can be found in Ref. [34]. A clear presentation of many other relations used in the present chapter is given in Ref. [35].

we therefore have

$$\sigma_\ell = \sigma_{\ell-1} + \arctan\frac{\eta}{\ell} = \sigma_0 + \sum_{\ell'=1}^{\ell} \arctan\frac{\eta}{\ell'}\,,$$

where the sum is diverging when $\ell \to \infty$. From the point of view of physics, the series (8.42) diverges because the Coulomb potential at large distances vanishes too slowly, and therefore, the colliding particles interact at any impact parameter.

Actually, for pure Coulomb potential, we do not need to use the expansion (8.42) because the full amplitude (8.36) is known. The Coulomb partial-wave phase shifts, σ_ℓ, are only used together with the corresponding short-range phase shifts when the potential has both the Coulomb and short-range parts. In such problems, we usually deal with the states (e.g. resonances) that have definite values of the conserving total angular momentum that is built up of the orbital angular momentum and the spin. Such problems involve just few partial waves, and therefore, there is no difficulty of convergence.

8.1.3.1 Gamow Factor

In nuclear physics, we mainly deal with positively charged particles. Due to the electric repulsion, they cannot come close to each other. The lower is the collision energy, the less probable is that they come to the zero distance. To estimate such a probability, P, we should take the absolute square of the wave function (8.34) at $r = 0$. Using the definition of the hypergeometric function ${}_1F_1$,

$${}_1F_1(a;b;z) = 1 + \frac{a}{b}\cdot\frac{z}{1!} + \frac{a(a+1)}{b(b+1)}\cdot\frac{z^2}{2!} + \frac{a(a+1)(a+2)}{b(b+1)(b+2)}\cdot\frac{z^3}{3!} + \cdots\,, \tag{8.44}$$

we see that ${}_1F_1(-i\eta;1;0) = 1$, and thus

$$\left|\psi^c_{\vec{k}}(\vec{r}\,)\right|^2_{r=0} = \frac{e^{-\pi\eta}}{(2\pi)^3}|\Gamma(1+i\eta)|^2\,.$$

The Γ-function in this equation can be replaced with

$$|\Gamma(1+i\eta)|^2 = \frac{\pi\eta}{\sinh(\pi\eta)} = \frac{2\pi\eta}{e^{\pi\eta}-e^{-\pi\eta}}\,.$$

At low collision energies, the Sommerfeld parameter becomes very large, $\eta \to \pm\infty$, where the sign is positive for the repulsive Coulomb force and negative otherwise. Therefore,

$$\left|\psi^c_{\vec{k}}(\vec{r}\,)\right|^2_{r=0} \underset{k\to 0}{\longrightarrow} \begin{cases} \dfrac{\eta}{(2\pi)^2}\cdot e^{-2\pi\eta}\,, & \eta > 0 \text{ repulsive Coulomb force}\,, \\[2ex] -\dfrac{\eta}{(2\pi)^2}\,, & \eta < 0 \text{ attractive Coulomb force}\,. \end{cases} \tag{8.45}$$

As follows from Eq. (8.40), the scattering cross section is the same for the cases of $\eta > 0$ and $\eta < 0$. However, Eq. (8.45) shows that in the repulsive case the probability of finding the particles close to each other is supressed by the factor

$$\boxed{P = e^{-2\pi\eta}} \qquad \eta > 0\,. \tag{8.46}$$

This exponential is known as the Gamow factor that makes it difficult for the particles to approach each other. It can also be viewed as the probability of penetrating through the Coulomb barrier.

8.2 Short-Range Plus Coulomb Potential

Let us consider a more general case, when the interaction potential includes a short-range part $V(r)$. At large distances, where this potential vanishes, the radial Schrödinger equation

$$\left[\partial_r^2 + k^2 - \frac{\ell(\ell+1)}{r^2} - V(r) - \frac{2k\eta}{r}\right] u_\ell(E,r) = 0 \tag{8.47}$$

transforms into the pure Coulomb equation (8.4) whose general solution is a linear combination of the Coulomb functions. This fact gives us a hint how Eq. (8.47) can be replaced with an equivalent set of the first-order differential equations.

8.2.1 First-Order Differential Equations

Generalizing the ansatz (2.27), we look for the Coulomb-modified regular solution of Eq. (8.47) in the form[6]

$$\phi_\ell^c(E,r) = F_\ell(\eta,kr)A_\ell(E,r) + G_\ell(\eta,kr)B_\ell(E,r)\,, \tag{8.48}$$

where the unknown functions $A_\ell(E,r)$ and $B_\ell(E,r)$ depend also on the Sommerfeld parameter η. For the sake of notation simplicity, we suppress this dependence because η is determined by the energy (dependence on which is indicated separately) and by the charge (which, like the mass μ, is constant in a particular problem).

In order to find the equations determining the new unknown functions, we follow the same steps as in Sect. 2.2.1, i.e. we substitute the ansatz (8.48) into Eq. (8.47) and use the Lagrange condition

$$F_\ell(\eta,kr)\partial_r A_\ell(E,r) + G_\ell(\eta,kr)\partial_r B_\ell(E,r) = 0\,, \tag{8.49}$$

[6] The plus sign here is because of the minus sign in Eq. (8.10).

as well as the Wronskian of the Coulomb functions (see Appendix E),

$$G_\ell(\eta, kr)\partial_r F_\ell(\eta, kr) - F_\ell(\eta, kr)\partial_r G_\ell(\eta, kr) = k \,. \tag{8.50}$$

As can be easily checked by the reader, these steps lead to the following set of differential equations for the functions $A_\ell(E, r)$ and $B_\ell(E, r)$

$$\partial_r A_\ell = \frac{1}{k} G_\ell V \left(F_\ell A_\ell + G_\ell B_\ell\right) \tag{8.51}$$

$$\partial_r B_\ell = -\frac{1}{k} F_\ell V \left(F_\ell A_\ell + G_\ell B_\ell\right) \tag{8.52}$$

where we dropped the variable symbols (E, r) and (η, kr) for the sake of clarity. Comparing these equations with the system (2.32), we see that they are almost the same. The only difference is that the regular and irregular Bessel–Neumann functions are replaced with their Coulomb counterparts (please note the different signs of the irregular functions),

$$F_\ell(\eta, kr) \leftrightarrow j_\ell(kr) \,, \qquad G_\ell(\eta, kr) \leftrightarrow -n_\ell(kr) \,. \tag{8.53}$$

We are interested in the regular solutions of Eq. (8.47), which requires $B_\ell(E, 0) = 0$. The value of $A_\ell(E, 0)$ in principle can be any non-zero constant, the choice of which determines the overall normalization of the function (8.48). To be consistent with the zero-charge case, we choose the following boundary conditions for Eqs. (8.51, 8.52):

$$A_\ell(E, 0) = 1 \,, \qquad B_\ell(E, 0) = 0 \tag{8.54}$$

Introducing the linear combinations of the unknown functions A_ℓ and B_ℓ,

$$\mathcal{F}_\ell^{(\mathrm{in/out})}(E, r) = \frac{1}{2}\left[A_\ell(E, r) \mp i B_\ell(E, r)\right] \,, \tag{8.55}$$

and making the same combinations of Eqs. (8.51, 8.52), we can re-write in the alternative form the Coulomb-modified regular solution,

$$\phi_\ell^c(E, r) = H_\ell^{(-)}(\eta, kr)\mathcal{F}_\ell^{(\mathrm{in})}(E, r) + H_\ell^{(+)}(\eta, kr)\mathcal{F}_\ell^{(\mathrm{out})}(E, r) \,, \tag{8.56}$$

as well as the differential equations,

$$\partial_r \mathcal{F}_\ell^{(\text{in})} = -\frac{H_\ell^{(+)}}{2ik} V \left[H_\ell^{(-)} \mathcal{F}_\ell^{(\text{in})} + H_\ell^{(+)} \mathcal{F}_\ell^{(\text{out})} \right] \tag{8.57}$$

$$\partial_r \mathcal{F}_\ell^{(\text{out})} = \frac{H_\ell^{(-)}}{2ik} V \left[H_\ell^{(-)} \mathcal{F}_\ell^{(\text{in})} + H_\ell^{(+)} \mathcal{F}_\ell^{(\text{out})} \right] \tag{8.58}$$

which generalize the corresponding zero-charge equations (2.21). Substituting Eqs. (8.54) into the definitions (8.55), we obtain the boundary conditions for these differential equations,

$$\mathcal{F}_\ell^{(\text{in})}(E, 0) = \mathcal{F}_\ell^{(\text{out})}(E, 0) = \frac{1}{2} \tag{8.59}$$

8.2.2 Integral Equations

Similarly to what was done in Sect. 2.2.1.3 for pure short-range potentials, the pairs of differential equations (8.51, 8.52) and (8.57, 8.58) can be formally integrated from 0 to an arbitrary r. This gives us another formulation of the same problem, in the form of the integral equations, where the corresponding boundary conditions are already incorporated. Therefore, the equations

$$A_\ell(E, r) = 1 + \frac{1}{k} \int_0^r G_\ell(\eta, kr') V(r') \Big[F_\ell(\eta, kr') A_\ell(E, r') + G_\ell(\eta, kr') B_\ell(E, r') \Big] dr' \,, \tag{8.60}$$

$$B_\ell(E, r) = -\frac{1}{k} \int_0^r F_\ell(\eta, kr') V(r') \Big[F_\ell(\eta, kr') A_\ell(E, r') + G_\ell(\eta, kr') B_\ell(E, r') \Big] dr' \tag{8.61}$$

generalize the corresponding zero-charge equations (2.42), while the coupled system

$$\mathcal{F}_\ell^{(\text{in})}(E,r) = \frac{1}{2} - \frac{1}{2ik}\int_0^r H_\ell^{(+)}(\eta,kr')V(r')\Big[H_\ell^{(-)}(\eta,kr')\mathcal{F}_\ell^{(\text{in})}(E,r') +$$

$$+ \; H_\ell^{(+)}(\eta,kr')\mathcal{F}_\ell^{(\text{out})}(E,r')\Big]dr' , \tag{8.62}$$

$$\mathcal{F}_\ell^{(\text{out})}(E,r) = \frac{1}{2} + \frac{1}{2ik}\int_0^r H_\ell^{(-)}(\eta,kr')V(r')\Big[H_\ell^{(-)}(\eta,kr')\mathcal{F}_\ell^{(\text{in})}(E,r') +$$

$$+ \; H_\ell^{(+)}(\eta,kr')\mathcal{F}_\ell^{(\text{out})}(E,r')\Big]dr' \tag{8.63}$$

are the general forms of the integral equations (2.40).

8.2.3 Jost Functions

The differential equations of Sect. 8.2.1 or alternatively the integral equations of Sect. 8.2.2 allow us to find the Coulomb-modified regular solution, $\phi_\ell^c(E,r)$. If we want to obtain the Jost functions, we need to consider the asymptotic behaviour of $\phi_\ell^c(E,r)$ and extract them as the amplitudes of the incoming and outgoing spherical waves (8.26) distorted by the Coulomb potential. In doing this, we should remember that the Jost functions are defined as the amplitudes of the incoming and outgoing waves in the asymptotic behaviour of the solution that has a different normalization, namely, of the so-called regular solution, $\phi_\ell(E,r)$, that is not modified by the Coulomb potential and behaves as

$$\phi_\ell(E,r) \underset{r\to 0}{\longrightarrow} j_\ell(kr) . \tag{8.64}$$

Since

$$\phi_\ell^c(E,r) \underset{r\to 0}{\longrightarrow} F_\ell(\eta,kr) , \tag{8.65}$$

and taking into account Eqs. (8.20) and (8.22), we obtain

$$\phi_\ell(E,r) = \frac{1}{C_\ell(\eta)}\phi_\ell^c(E,r) . \tag{8.66}$$

The Jost functions $f_\ell^{(\text{in/out})}(E)$ are defined via the following asymptotics:

$$\phi_\ell(E,r) \underset{r\to\infty}{\longrightarrow} H_\ell^{(-)}(\eta,kr)e^{i\sigma_\ell}f_\ell^{(\text{in})}(E) + H_\ell^{(+)}(\eta,kr)e^{-i\sigma_\ell}f_\ell^{(\text{out})}(E) . \tag{8.67}$$

Therefore,

$$\begin{aligned}\phi_\ell^c(E,r) &= C_\ell(\eta)\phi_\ell(E,r)\\ &\underset{r\to\infty}{\longrightarrow} H_\ell^{(-)}(\eta,kr)e^{i\sigma_\ell}C_\ell(\eta)f_\ell^{(\text{in})}(E)+ \\ &\quad + H_\ell^{(+)}(\eta,kr)e^{-i\sigma_\ell}C_\ell(\eta)f_\ell^{(\text{out})}(E)\,.\end{aligned} \tag{8.68}$$

Comparing this with Eq. (8.56), we obtain

$$f_\ell^{(\text{in/out})}(E) = \frac{e^{\mp i\sigma_\ell}}{C_\ell(\eta)}\lim_{r\to\infty}\mathcal{F}_\ell^{(\text{in/out})}(E,r)\,. \tag{8.69}$$

Introducing the limits

$$f_\ell^{\text{sc (in/out)}}(E) = 2\lim_{r\to\infty}\mathcal{F}_\ell^{(\text{in/out})}(E,r)\,, \tag{8.70}$$

we see that in the total Jost functions, the pure Coulomb Jost functions (8.28) are factorized,

$$f_\ell^{(\text{in/out})} = f_\ell^{\text{c (in/out)}} f_\ell^{\text{sc (in/out)}}\,. \tag{8.71}$$

It should be emphasized that $f_\ell^{\text{sc (in/out)}}(E)$ are not the pure short-range counterparts of the functions $f_\ell^{\text{c (in/out)}}(E)$. Indeed, the functions $\mathcal{F}_\ell^{(\text{in/out})}(E,r)$ obey Eqs. (8.57, 8.58), where apart from the short-range potential $V(r)$ the Coulomb functions $H_\ell^{(\pm)}(\eta,kr)$ are also involved.

The factorization (8.71) can be intuitively understood as follows. The pure Coulomb factor takes into account the interaction at very large distances, where the short-range potential is absent. The factor $f_\ell^{\text{sc(in/out)}}$ describes the joint effect of both the short-range and Coulomb forces at the short distances.

In nuclear physics, it is customary to calculate separately the phase shifts due to the nuclear and Coulomb forces and then add them together. From the above reasoning, it is clear that such simple addition is an approximation, which is equivalent to replacing in Eqs. (8.57, 8.58) the functions $H_\ell^{(\pm)}(\eta,kr)$ with the corresponding Riccati–Hankel functions $h_\ell^{(\pm)}(kr)$.

If we introduce the pure Coulomb and Coulomb-modified short-range S-matrices in the state with the angular momentum ℓ,

$$S_\ell^{\text{c}} = \frac{f_\ell^{\text{c (out)}}}{f_\ell^{\text{c (in)}}}\,,\qquad S_\ell^{\text{sc}} = \frac{f_\ell^{\text{sc (out)}}}{f_\ell^{\text{sc (in)}}}\,, \tag{8.72}$$

and the corresponding partial-wave amplitudes (see Eq. (4.20)),

$$f_\ell^{\text{c}} = \frac{S_\ell^{\text{c}}-1}{2ik}\,,\qquad f_\ell^{\text{sc}} = \frac{S_\ell^{\text{sc}}-1}{2ik}\,, \tag{8.73}$$

then the total partial-wave amplitude can be written as the following sum:

$$f_\ell = \frac{S^{\rm c}_\ell S^{\rm sc}_\ell - 1}{2ik} = \frac{S^{\rm c}_\ell - 1}{2ik} + \frac{S^{\rm c}_\ell S^{\rm sc}_\ell - 1}{2ik} - \frac{S^{\rm c}_\ell - 1}{2ik}$$

$$= f^{\rm c}_\ell + S^{\rm c}_\ell f^{\rm sc}_\ell \,. \tag{8.74}$$

Again, as is seen, the total amplitude is not a simple sum of the pure Coulomb and pure short-range amplitudes. First, the amplitude $f^{\rm sc}_\ell$ describes the effect of the short-range forces distorted by the Coulomb interaction. Second, this amplitude in Eq. (8.74) is multiplied by the factor $S^{\rm c}_\ell$ that can only be replaced by unity if the Coulomb phase shift σ_ℓ is small.

8.2.4 Jost Solutions

In the case of neutral particles, the so-called Jost solutions, $\chi^{(\pm)}_\ell(E,r)$, of the radial Schrödinger equation are defined by the following boundary conditions at infinity (see Sect. 2.2.4):

$$\chi^{(\pm)}_\ell(E,r) \underset{r\to\infty}{\longrightarrow} h^{(\pm)}_\ell(kr) \underset{r\to\infty}{\longrightarrow} (\mp i)^{\ell+1} e^{\pm ikr} \,, \qquad \eta = 0 \,. \tag{8.75}$$

When we deal with charged particles ($\eta \neq 0$), this definition can be generalized as follows (compared with Eq. (8.27)):

$$\chi^{(\pm)}_\ell(E,r) \underset{r\to\infty}{\longrightarrow} H^{(\pm)}_\ell(\eta,kr) e^{\mp i\sigma_\ell} \underset{r\to\infty}{\longrightarrow} (\mp i)^{\ell+1} e^{\pm i[kr - \eta\ln(2kr)]} \,. \tag{8.76}$$

The regular solution $\phi_\ell(E,r)$ is still defined by the same boundary conditions (8.64) at $r = 0$ as for neutral particles.

The Wronskian of any two solutions of the radial Schrödinger equation (8.47) does not depend on r, i.e. it can be calculated anywhere and, in particular, at infinity. Using the Wronskians (E.14, E.15) (see Appendix E) and the asymptotic form (8.67) of the regular solution, it is straightforward to obtain the relations:

$$\boxed{f^{\rm (in/out)}_\ell = \mp \frac{1}{2ik} W\left[\chi^{(\pm)}_\ell, \phi_\ell\right]} \tag{8.77}$$

In the same way as for the regular solution (see Eq. (8.56)), we look for the Jost solutions in the form

$$\chi^{(\pm)}_\ell(E,r) = H^{(-)}_\ell(kr)\mathfrak{X}^{(\pm{\rm in})}_\ell(E,r) + H^{(+)}_\ell(kr)\mathfrak{X}^{(\pm{\rm out})}_\ell(E,r), \tag{8.78}$$

where $\mathfrak{X}_\ell^{(\pm\text{in/out})}(E,r)$ are unknown functions. Imposing on them the standard Lagrange condition,

$$H_\ell^{(-)}(kr)\partial_r\mathfrak{X}_\ell^{(\pm\text{in})}(E,r) + H_\ell^{(+)}(kr)\partial_r\mathfrak{X}_\ell^{(\pm\text{out})}(E,r) = 0\,,$$

we find that they obey the differential equations:

$$\begin{aligned}
\partial_r\mathfrak{X}_\ell^{(\pm\text{in})} &= -\frac{H_\ell^{(+)}}{2ik}V\left[H_\ell^{(-)}\mathfrak{X}_\ell^{(\pm\text{in})} + H_\ell^{(+)}\mathfrak{X}_\ell^{(\pm\text{out})}\right],\\
\partial_r\mathfrak{X}_\ell^{(\pm\text{out})} &= \frac{H_\ell^{(-)}}{2ik}V\left[H_\ell^{(-)}\mathfrak{X}_\ell^{(\pm\text{in})} + H_\ell^{(+)}\mathfrak{X}_\ell^{(\pm\text{out})}\right].
\end{aligned} \tag{8.79}$$

As it is easy to see, the behaviour (8.76) of the functions $\chi_\ell^{(\pm)}(E,r)$ at infinity is guaranteed if Eqs. (8.79) are solved inwards from a faraway point with the boundary values either

$$\mathfrak{X}_\ell^{(+\text{in})}(E,\infty) = 0\,,\quad \mathfrak{X}_\ell^{(+\text{out})}(E,\infty) = e^{-i\sigma_\ell} \qquad \text{for}\quad \chi_\ell^{(+)}\,, \tag{8.80}$$

or

$$\mathfrak{X}_\ell^{(-\text{in})}(E,\infty) = e^{+i\sigma_\ell}\,,\quad \mathfrak{X}_\ell^{(-\text{out})}(E,\infty) = 0 \qquad \text{for}\quad \chi_\ell^{(-)}\,. \tag{8.81}$$

Comparing Eqs. (8.67) and (8.76), we see that at large distances the regular solution is the following linear combination of the Jost solutions:

$$\phi_\ell(E,r) \underset{r\to\infty}{\longrightarrow} \chi_\ell^{(-)}(E,r)f_\ell^{(\text{in})}(E) + \chi_\ell^{(+)}(E,r)f_\ell^{(\text{out})}(E)\,. \tag{8.82}$$

The Jost solutions $\chi_\ell^{(-)}$ and $\chi_\ell^{(+)}$ constitute a complete basis in the space of all solutions of the radial Schrödinger equation. This means that the linear combination (8.82) is unique and remains valid at any distance, i.e.

$$\boxed{\phi_\ell(E,r) = \chi_\ell^{(-)}(E,r)f_\ell^{(\text{in})}(E) + \chi_\ell^{(+)}(E,r)f_\ell^{(\text{out})}(E)} \qquad \eta \neq 0\,. \tag{8.83}$$

This relation is formally the same as Eq. (2.81) for the neutral particle case.

It is easy to see that Eqs. (2.76, 2.77) as well as the reasoning around them are also valid for the problem of charged particles. This means that in exactly the same way we can derive the following useful relations:

$$f_\ell^{(\text{in/out})}(E) = \lim_{r\to 0} \frac{\pm i(kr)^\ell}{2(2\ell-1)!!}\chi_\ell^{(\pm)}(E,r) \qquad \eta \neq 0\,. \tag{8.84}$$

8.2.5 Analyticity of the Jost Functions

By introducing the Jost functions as the limits (8.69), we tacitly assumed that such limits exist. However, as we saw in Sect. 7.2, the limits of this type may only converge in certain domains of the momentum plane (or the energy surface). Our task now is to establish where these domains are.

First of all, we need to find the long-range asymptotic behaviour of the function $\phi_\ell(E,r)$. This can be done in the same way as we did it for short-range potentials in Sect. 7.1.1, i.e. using the WKB approximation. Actually, all the derivations up to Eq. (7.8) are still valid if the potential has a Coulomb tail,

$$V(r) = V_s(r) + \frac{2k\eta}{r}\,. \tag{8.85}$$

Substituting this potential in the integral of Eq. (7.8), we see that

$$\frac{i}{2k}\int V(r)dr = \frac{i}{2k}\int V_s(r)dr + i\eta\ln(2kr) \underset{|kr|\to\infty}{\longrightarrow} i\eta\ln(2kr)\,,$$

where (as we always did before) we assumed that $k \neq 0$ and used the limit (7.9), which is valid for the short-range part $V_s(r)$ of the potential. Therefore, everywhere on the complex k-plane (with the point $k = 0$ excluded)

$$\phi_\ell(E,r) \underset{|kr|\to\infty}{\longrightarrow} q_\ell^{(-)}(E)e^{-ikr+i\eta\ln(2kr)} + q_\ell^{(+)}(E)e^{+ikr-i\eta\ln(2kr)}\,, \tag{8.86}$$

where $q_\ell^{(\pm)}(E)$ are the r-independent constants (one of which can be zero). This result is not surprising. We could have guessed it without formal derivations. Indeed, at large distances, the Coulomb tail of the potential dominates over $V_s(r)$, and the general solution of the Schrödinger equation is a linear combination of the functions $H^{(\pm)}(\eta, kr)$ that behave as given by Eq. (8.12).

It is obvious that between the two terms

$$kr - \eta\ln(2kr),$$

the linear term, kr, dominates at large distances ($|r| \to \infty$) over the logarithmic term, $\eta\ln(2kr)$, on the whole k-plane (with the point $k = 0$ excluded). Therefore, when analysing the convergence of the limits (8.69), we can simplify the asymptotic behaviour (8.86) of the wave function as well as the behaviour (8.12) of the Coulomb functions $H^{(\pm)}(\eta, kr)$. The simplification consists in dropping all the terms in the exponential functions except the linear ones, i.e. the simplified asymptotics only involve $\exp(\pm ikr)$.

Such a simplification brings us back to Chap. 7. This means that the limits (8.69) exist exactly in the same domains of the k-plane, as were established for a pure short-range potential. These domains could be rotated in exactly the same way, and the Jost functions are analytic in the same domains of the k-plane.

8.2.6 Analytic Structure of the Jost Functions

As is shown in Sect. 7.7, in the simple case of a short-range potential, the Jost function can be split into the two terms as is given by Eq. (7.65), where the factor $k^{2\ell+1}$ responsible for the branching of the Riemann surface of the energy is explicitly separated. When a Coulombic tail is present, a similar splitting of the Jost functions is also possible but is a bit more complicated. The explicit expressions for all the coefficients in such a splitting are what we are going to derive in this section.

* * *

The derivation is based on the following factorized structure of the Coulomb functions (for the details, see Appendix E.2):

$$F_\ell(\eta, kr) = D_\ell(\eta, k)\tilde{F}_\ell(E, r) , \tag{8.87}$$

$$G_\ell(\eta, kr) = M(\eta)D_\ell(\eta, k)\tilde{F}_\ell(E, r) + \frac{k}{D_\ell(\eta, k)}\tilde{G}_\ell(E, r) , \tag{8.88}$$

where $\tilde{F}_\ell(E, r)$ and $\tilde{G}_\ell(E, r)$ are entire (analytic and single-valued) functions of the energy variable E, while the complicated k-dependence, which makes the Coulomb functions multi-valued, is explicitly isolated via the factors

$$D_\ell(\eta, k) = C_\ell(\eta)k^{\ell+1} \qquad \text{and} \qquad M(\eta) = \frac{2\eta h(\eta)}{C_0^2(\eta)} , \tag{8.89}$$

with

$$h(\eta) = \frac{1}{2}\left[\psi(i\eta) + \psi(-i\eta)\right] - \ln\eta , \qquad \psi(z) = \frac{\Gamma'(z)}{\Gamma(z)} . \tag{8.90}$$

The representations (8.87) and (8.88) generalize the more simple formulae (7.58, 7.59) known for the neutral particle case, and are reduced to them when $\eta = 0$.

Using the factorized representation (8.87, 8.88), we can re-write the Coulomb-modified regular solution (8.48) as

$$\phi_\ell^c = F_\ell A_\ell + G_\ell B_\ell = D_\ell \left(\tilde{F}_\ell A_\ell + M \tilde{F}_\ell B_\ell + \frac{k}{D_\ell^2} \tilde{G}_\ell B_\ell \right) = D_\ell (\tilde{F}_\ell \tilde{A}_\ell + \tilde{G}_\ell \tilde{B}_\ell) , \tag{8.91}$$

where the "tilded" unknown functions $\tilde{A}_\ell$ and $\tilde{B}_\ell$ are defined by the following equations:

$$\tilde{A}_\ell(E,r) = A_\ell(E,r) + M(\eta) B_\ell(E,r) , \tag{8.92}$$

$$\tilde{B}_\ell(E,r) = \frac{k}{D_\ell^2(\eta,k)} B_\ell(E,r) . \tag{8.93}$$

With the help of Eqs. (8.87, 8.88) and (8.91), the differential equations (8.51, 8.52) take the form

$$\partial_r A_\ell = \left(\frac{M D_\ell^2}{k} \tilde{F}_\ell + \tilde{G}_\ell \right) V (\tilde{F}_\ell \tilde{A}_\ell + \tilde{G}_\ell \tilde{B}_\ell) , \tag{8.94}$$

$$\partial_r B_\ell = -\frac{D_\ell^2}{k} \tilde{F}_\ell V (\tilde{F}_\ell \tilde{A}_\ell + \tilde{G}_\ell \tilde{B}_\ell) . \tag{8.95}$$

Multiplying Eq. (8.95) by M and adding the result with Eq. (8.94) and then multiplying Eq. (8.95) by k/D_ℓ^2, we obtain

$$\partial_r \tilde{A}_\ell(E,r) = \quad \tilde{G}_\ell(E,r) V(r) \left[\tilde{F}_\ell(E,r) \tilde{A}_\ell(E,r) + \tilde{G}_\ell(E,r) \tilde{B}_\ell(E,r) \right] , \tag{8.96}$$

$$\partial_r \tilde{B}_\ell(E,r) = -\tilde{F}_\ell(E,r) V(r) \left[\tilde{F}_\ell(E,r) \tilde{A}_\ell(E,r) + \tilde{G}_\ell(E,r) \tilde{B}_\ell(E,r) \right] . \tag{8.97}$$

The most important feature of these new differential equations is the fact that all their coefficients are entire functions of the parameter E for any finite value of r. Moreover, the boundary conditions,

$$\tilde{A}_\ell(E,0) = 1 , \qquad \tilde{B}_\ell(E,0) = 0 , \tag{8.98}$$

which immediately follow from (8.54) and (8.92, 8.93), do not depend on E. According to the Poincaré theorem (see Appendix G), the solutions of Eqs. (8.96, 8.97), i.e. the functions $\tilde{A}_\ell(E,r)$ and $\tilde{B}_\ell(E,r)$, are analytic functions of the energy. Furthermore, these functions of E are single-valued because all the coefficients in the differential equations (8.96, 8.97) are single-valued functions of E.

To obtain the Jost functions, we actually need the asymptotic values of $\tilde{A}_\ell(E,r)$ and $\tilde{B}_\ell(E,r)$ at large distances. Combining Eqs. (8.55), (8.69), and (8.92, 8.93), we see that the Jost functions (for a short-range plus a Coulomb potential) have the following structure:

$$f_\ell^{(\mathrm{in/out})}(E) = \frac{e^{\mp i\sigma_\ell}}{2C_\ell(\eta)}\left\{\tilde{A}_\ell(E,\infty) - \left[\frac{2\eta h(\eta)}{C_0^2(\eta)} \pm i\right] C_\ell^2(\eta) k^{2\ell+1}\tilde{B}_\ell(E,\infty)\right\}. \tag{8.99}$$

Similarly to Eq. (7.65) derived for the short-range interaction, we denote $a_\ell(E) = \tilde{A}_\ell(E,\infty)/2$ and $b_\ell(E) = \tilde{B}_\ell(E,\infty)/2$, which gives

$$f_\ell^{(\mathrm{in/out})}(E) = \frac{e^{\mp i\sigma_\ell}}{C_\ell(\eta)}\left\{a_\ell(E) - \left[\frac{2\eta h(\eta)}{C_0^2(\eta)} \pm i\right] C_\ell^2(\eta) k^{2\ell+1} b_\ell(E)\right\} \tag{8.100}$$

Perhaps it needs to be reiterated one more time that in the above the functions $a_\ell(E)$ and $b_\ell(E)$ are analytic single-valued functions of E and are defined on a simple single-sheet complex plain. All the troubles with the branch points of the Riemann surface are caused by the remaining k-dependent coefficients that are given explicitly. In other words, the functions $a_\ell(E)$ and $b_\ell(E)$ are more simple than $f_\ell^{(\mathrm{in/out})}(E)$. In general, $a_\ell(E)$ and $b_\ell(E)$ are complex. However, if E is on the real axis, these functions have real values,

$$\mathrm{Im}\,E = 0 \quad \longrightarrow \quad \mathrm{Im}\,a_\ell(E) = \mathrm{Im}\,b_\ell(E) = 0 \tag{8.101}$$

because neither Eqs. (8.96, 8.97) nor the boundary conditions (8.98) involve any complex numbers, if the energy is real.

Equation (8.100) generalizes the corresponding analytic structure given by Eq. (7.65) for the case of a pure short-range potential. The presence of the Coulomb forces makes the structure more complicated. However, the Jost functions still consist of two terms, and the factor $k^{2\ell+1}$ is still there. In Sect. 18.3.1, it is given an exactly solvable example of a single-channel problem with charged particles, where the representation of the type (8.100) is obtained explicitly.

It is natural to expect that Eq. (8.100) should take the more simple form (7.65) when $\eta \to 0$, i.e. for neutral particles. In order to show this, we notice (see Eqs. (8.13) and (8.21)) that

$$\sigma_\ell(\eta) \underset{\eta\to 0}{\longrightarrow} 0 \qquad \text{and} \qquad C_\ell(\eta) \underset{\eta\to 0}{\longrightarrow} 1\,.$$

Using the recurrency relation $\psi(z) = \psi(z+1) - 1/z$, we find

$$
\begin{aligned}
2\eta h(\eta) &= \eta\,[\psi(i\eta) + \psi(-i\eta)] - 2\eta \ln \eta \\
&= \eta\,[\psi(1+i\eta) + \psi(1-i\eta)] - 2\eta \ln \eta \underset{\eta \to 0}{\longrightarrow} 0
\end{aligned}
$$

because $\psi(1) = -\gamma = -0.5772156649\ldots$ is the Euler's constant and the linear function vanishes faster than the logarithm grows. This gives

$$
f_\ell^{(\text{in/out})}(E) \underset{\eta \to 0}{\longrightarrow} a_\ell(E) \mp i k^{2\ell+1} b_\ell(E)\,,
$$

as it should be.

8.2.7 Domain of Analyticity and Complex Rotation

Taking the limit $r \to \infty$ in Eqs. (8.96, 8.97) is a delicate procedure. When E is a real scattering energy, everything is simple: both the regular and irregular Coulomb functions on the right hand sides of these equations are oscillating with finite amplitudes. Thanks to this, the short-range potential $V(r)$ causes the right hand sides vanish very fast with growing r. In other words, the derivatives $\partial_r \tilde{A}_\ell$ and $\partial_r \tilde{B}_\ell$ tend to zero, and therefore the functions $\tilde{A}_\ell$ and $\tilde{B}_\ell$ become constant, i.e. attain the limits $\tilde{A}_\ell(E,\infty)$ and $\tilde{B}_\ell(E,\infty)$.

However, if E is complex, the Coulomb functions (which can always be written as linear combinations of the functions (8.12) become exponentially growing, which may cause the divergence. This kind of difficulty and a way to go around it were discussed in Sects. 7.7.1 and 7.7.2 for the case of a pure short-range potential.

It was shown that the solutions of differential equations similar to (8.96, 8.97), at $r \to \infty$, remain holomorphic functions of E within a domain $\mathcal{D}$ along the positive real axis. The shape and width of $\mathcal{D}$ are determined by the character of $V(r)$. In particular, for an exponentially falling potential $V(r)$, this domain is of a parabolic type, $\mathrm{Re}E \sim (\mathrm{Im}E)^2$. In any case, the functions $\tilde{A}_\ell(E,\infty)$ and $\tilde{B}_\ell(E,\infty)$ can always be analytically continued outside the domain $\mathcal{D}$ by using complex rotation of the coordinate, $r \to re^{i\theta}$, which causes the $\mathcal{D}$ to turn as is schematically shown in Fig. 7.11.

The reader can easily check that the analysis given in Sect. 7.7.2 is completely applicable to Eqs. (8.96, 8.97). We therefore can conclude that in the analytic structure (8.100), the functions $a_\ell(E)$ and $b_\ell(E)$ are holomorphic (at least) in a wide domain around the positive real axis, which means at all points of practical interest, where Eq. (8.100) could be used to describe scattering processes and resonances.

8.2.8 Short-Range Plus Attractive Coulomb Potential

In all the derivations throughout Chap. 8, we assume that the charges Z_1e and Z_2e have arbitrary signs, i.e. the Coulomb potential can be either attractive or repulsive. A pure Coulomb potential with $Z_1Z_2 < 0$ always generates the hydrogen-like spectrum of the bound-state energies,

$$E_n^c = -\frac{|Z_1Z_2|^2\mu e^4}{2\hbar^2 n^2}\,, \tag{8.102}$$

where the principal quantum number, $n = 1, 2, 3, \ldots$, is composed of the radial quantum number $n_r = 0, 1, 2, \ldots$ and the orbital number $\ell = 0, 1, 2, \ldots$,

$$n = n_r + \ell + 1\,. \tag{8.103}$$

At all these energies, the Sommerfeld parameter is pure imaginary,

$$\eta_n = in\,,$$

which makes singular the denominator on the right hand side of Eq. (8.32) for the Jost function $f_\ell^{\mathrm{c\,(in)}}(E)$,

$$\Gamma(\ell + 1 + i\eta_n) = \Gamma(\ell + 1 - n) = \Gamma(-n_r) = \infty\,. \qquad n_r = 0, 1, 2, \ldots$$

Therefore (as is expected), the spectral points (8.102) correspond to the zeros of $f_\ell^{\mathrm{c\,(in)}}(E)$ and to the poles of the corresponding S-matrix (8.29).

Apparently, if a short-range potential is added to an attractive Coulomb potential, the Coulombic bound-state spectrum is somehow modified. In this connection, the structure of the Jost function $f_\ell^{\mathrm{(in)}}(E)$, given by Eq. (8.100), may cause confusion. Indeed, rewriting this equation as

$$f_\ell^{\mathrm{(in)}}(E) = f_\ell^{\mathrm{c\,(in)}}(E) f_\ell^{\mathrm{sc\,(in)}}(E)\,, \tag{8.104}$$

where

$$f_\ell^{\mathrm{sc\,(in)}}(E) = 2\left\{a_\ell(E) - \left[\frac{2\eta h(\eta)}{C_0^2(\eta)} + i\right] C_\ell^2(\eta) k^{2\ell+1} b_\ell(E)\right\}\,, \tag{8.105}$$

one may erroneously conclude that the Jost function $f_\ell^{\mathrm{(in)}}(E)$ always has zeros at the hydrogen-like energies (8.102), independently of the choice of the short-range potential, because the pure Coulomb Jost function is a factor in (8.104). It is however not difficult to show that the other factor in Eq. (8.104) is singular at the energies (8.102), i.e.

$$f_\ell^{\,\mathrm{sc\,(in)}}(E_n^c) = \infty \,.$$

In order to see this, it is sufficient to notice that the Coulomb barrier factor (8.31) is singular at $E = E_n^c$,

$$C_\ell(\eta)|_{E=E_n^c} = \frac{e^{-\pi\eta/2}}{\ell!}\sqrt{\Gamma(-n_r)\Gamma(2\ell+2+n_r)} = \infty \,, \qquad n_r = 1, 2, 3, \ldots$$

Therefore,

$$f_\ell^{(\mathrm{in})}(E_n^c) = 0 \times \infty \,.$$

Let us show that in this product the zero and the singularity cancel each other in such a way that the Jost function (8.104) is finite and generally non-zero at the Coulomb energies (8.102), i.e.

$$f_\ell^{(\mathrm{in})}(E_n^c) = 0 \times \infty = \mathrm{const}\begin{cases} \neq 0 \\ \neq \infty \end{cases} . \tag{8.106}$$

To this end, we do an infinitely small shift of the energy from a Coulomb spectral point,

$$E = E_n^c + \delta \,, \qquad \delta \to 0 \,,$$

and make use of the following formulae derived in Appendix E.3:

$$f_\ell^{\,\mathrm{c\,(in)}}(E_n^c + \delta) \underset{\delta\to 0}{\sim} \delta \,, \tag{8.107}$$

$$C_\ell(\eta)|_{E=E_n^c+\delta} \underset{\delta\to 0}{\sim} \frac{1}{\sqrt{\delta}} \,, \tag{8.108}$$

$$h(\eta)|_{E=E_n^c+\delta} \underset{\delta\to 0}{\sim} \frac{1}{\delta} \,. \tag{8.109}$$

With such a behaviour of the vanishing and singular functions, the Jost function (8.104) near the Coulomb spectral point behaves as

$$f_\ell^{(\mathrm{in})}(E_n^c + \delta) \underset{\delta\to 0}{\longrightarrow} \delta\left[\mathrm{const} - \left(\mathrm{const}\frac{\delta}{\delta} + i\right)\frac{\mathrm{const}}{\delta}\right] , \tag{8.110}$$

which proves the statement (8.106). Therefore, in general

$$f_\ell^{(\text{in})}(E_n^c) \neq 0\,,$$

i.e. (as it should be) the Coulomb spectral points are not the spectral points anymore when a short-range potential is added. Moreover, the zeros and the singularities exactly cancel each other, and the Jost function $f_\ell^{(\text{in})}$ is regular at E_n^c. Of course, it may happen that $b_\ell(E_n^c) = 0$, and then the corresponding point is a spectral point of the total potential. This, however, would be just a coincidence.

Therefore, the factorized representation (8.100) of the Jost functions is valid even when the Coulombic part of the total potential is attractive. However, when using it near the spectral points of the pure Coulomb potential, one should do it with caution because the zeros and the singularities cancel each other at these points. In formal derivations, this fact does not pose any problem. But in numerical calculations, such a cancellation is always a source of errors. In such a case, it is needed to replace the vanishing and singular functions involved in Eq. (8.100), with their approximate expressions derived in Appendix E.3.

8.2.9 Riemann Surface for a System of Charged Particles

As we have seen in Chap. 3, the Jost functions, $f_\ell^{(\text{in/out})}(E)$, for a short-range potential, as well as the corresponding S-matrix, are double-valued functions of the complex variable E. This is a consequence of the fact that they actually depend on E via the momentum k that (for a fixed E) can have two different values, given by Eq. (3.1). As a result, the Jost functions are defined on a two-sheet Riemann surface with a square-root branch point at $E = 0$ (see Fig. 3.1).

When a Coulomb tail is added to the potential, the situation drastically changes. Equation (3.1) still gives us two different momenta. However, there is an additional source of ambiguity in calculating the Jost functions. Indeed, the Coulombic incoming and outgoing spherical waves (8.26) depend not only on k but also on $\ln k$, which has infinitely many different values (see Appendix B). Being the asymptotic amplitudes of these waves, the Jost functions should also have infinitely many values for any fixed energy E.

This becomes clear from the factorized representations of the Jost functions, given by Eq. (8.100). These representations involve the function $h(\eta)$ that depends on the logarithm of k,

$$h(\eta) = \frac{1}{2}[\psi(i\eta) + \psi(-i\eta)] - \ln\eta = \frac{1}{2}[\psi(i\eta) + \psi(-i\eta)] - \ln\frac{\mu e^2 Z_1 Z_2}{\hbar^2} + \ln k\,. \tag{8.111}$$

The complex function $\ln k$ has infinitely many different values,

$$\ln(k) = \ln\left\{|k|e^{i[\arg(k)+2\pi m]}\right\} = \ln|k| + i\arg(k) + i2\pi m\,, \quad m = 0, \pm 1, \pm 2, \ldots, \tag{8.112}$$

corresponding to different choices of m. It is defined on the spiral Riemann surface with the branch point at $k = 0$ (see Fig. B.9 in the Appendix B). It is customary to define the so-called principal branch of the logarithm as the part of this surface corresponding to $m = 0$ and $-\pi < \arg(k) \leqslant \pi$. The value of the function $\ln(k)$ in Eq. (8.112) corresponds to the energy

$$E = |E|\exp(i\varphi + i4\pi m)\,, \tag{8.113}$$

where $-2\pi < \varphi \leqslant 2\pi$ and $\arg(k) = \varphi/2$. Since the momentum is the square root of the energy, $k = \sqrt{2\mu E/\hbar^2}$, the interval $-2\pi < \varphi \leqslant 0$ covers the unphysical Riemann sheet where $\operatorname{Im} k < 0$, and the interval $0 < \varphi \leqslant 2\pi$ corresponds to the physical sheet with $\operatorname{Im} k > 0$. These two intervals are represented by the two sheets of the Riemann surface stemming from the square-root branch point. We denote the sheets using the symbols $(\pm)_m$, where the sign indicates the sign of $\operatorname{Im} k$ and the subscript m corresponds to the number of 2π in Eq. (8.112). It is logical to call the sheets with $m = 0$ the principal sheets.

The additional arbitrary angle $4\pi m$ in Eq. (8.113) stems from the fact that any number of two full circles around the point $E = 0$ (either in the clockwise or anti-clockwise direction) brings us to the same E and to the same sign of $\operatorname{Im} k$. The additional angle does not play any role until we take the logarithm of k.

Therefore, there are infinitely many different values of the Jost functions (and thus of the S-matrix) for a single E. The presence of a Coulomb tail makes the point $E = 0$ the square root and the logarithm branch point at the same time.

If we start the path on the principal physical sheet, $(+)_0$ and make a full circle around the branch point in the anti-clockwise direction, we cannot go down to the principal unphysical sheet, $(-)_0$, as is shown in Fig. 3.1 for neutral particles. Instead, we go up to another unphysical sheet with $\operatorname{Im} k < 0$ and $m = 1$. Similarly, if we start from the principal unphysical sheet and make a full circle in the clockwise direction, then, instead of going up to the principal physical sheet, we go down to another sheet with $\operatorname{Im} k > 0$ and $m = -1$. This means that (in contrast to the neutral particle case) moving in the same direction (clockwise or anti-clockwise), we can never come back to the same sheet. In other words, the Riemann surface has a "spiral" topology, as is shown in Fig. 8.2. There are infinitely many sheets in this Riemann surface. For each m, we have two sheets: one (unphysical) with $\operatorname{Im} k < 0$ and the other (physical) with $\operatorname{Im} k > 0$. These pairs are stacked one over the other.

All the physically relevant information (about the bound, resonant, and scattering states) is located on the physical and unphysical sheets with $m = 0$ because they are adjacent to the real axis of the collision energies. In practical calculations, we can ignore the sheets that are below or above the principal branch. However, it should

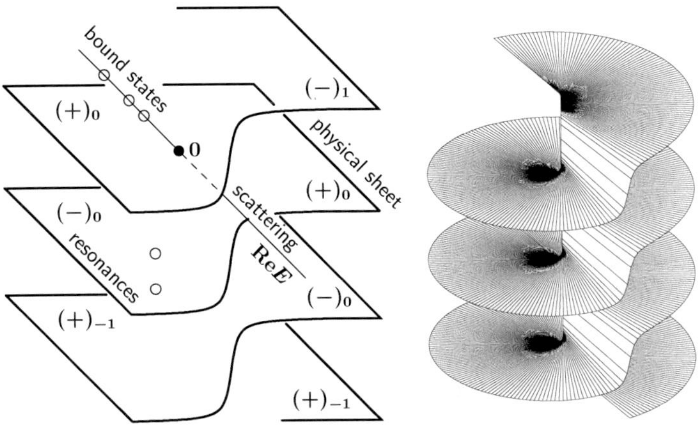

Fig. 8.2 The left part of the figure schematically shows the spiral Riemann surface for a potential with a Coulomb tail. The symbols $(\pm)_m$ label the sheets where $\mathrm{Im}(k)$ is either positive or negative, and the subscript m is the number of $2\pi i$ in Eq. (8.112) stemming from the logarithmic branching. For the sake of clarity, the right part of the figure shows the corresponding 3D-image of the Riemann sheets and their interconnections

not be forgotten that there is a cut along the positive real axis and that (in contrast to the short-range potential) the lower rim of the physical sheet is not connected with the upper rim of the unphysical sheet of the principal branch. In other words, the presence of the Coulomb forces changes the topology of the Riemann surface.

8.2.10 Symmetry Properties of the Jost Functions

As we have seen in Sect. 3.1, the functions $f_\ell^{(\mathrm{in})}$ and $f_\ell^{(\mathrm{out})}$ are related to each other at different points of the Riemann surface, i.e. they obey certain symmetry rules. For the short-range interactions, these rules are graphically depicted in Figs. 3.5 and 3.6. In the present section, it is shown that, when the potential has a Coulomb tail, most of these rules are not valid anymore. The symmetries are broken because the Coulomb potential changes the topology of the Riemann surface (compare Figs. 3.6 and 8.2). This can be seen using the analytic structure (8.100) of the Jost functions. The analysis is similar to the one given in Sect. 7.7.1 for the short-range interaction.

* * *

To begin with, let us consider what happens if we change the sign of the momentum. When k is replaced with $-k$, we "jump" from one sheet to the other along a vertical line (that passes through the points of the same energy on different sheets). Since the unknown functions $a_\ell(E)$ and $b_\ell(E)$ in the semi-analytic representation (8.100) are the same on all the sheets, the possible symmetry relations that could be associated with such a "jump" are determined by the explicitly given factors that undergo certain changes. The "jump" can be either up or down,

$$(+)_m \longleftrightarrow (-)_{m'}\,, \qquad \Delta m = m - m' = 0, \pm 1, \pm 2, \ldots \tag{8.114}$$

It is clear from the definition (8.6) of the Sommerfeld parameter that η changes sign together with k. The exponential factor $e^{-\pi\eta/2}$ in Eq. (8.21) that defines the Coulomb barrier factor is replaced with $e^{+\pi\eta/2}$, i.e. the function $C_\ell(\eta)$ is multiplied by $e^{\pi\eta}$,

$$(+)_m \longleftrightarrow (-)_{m'} \Rightarrow C_\ell(\eta) \rightarrow e^{\pi\eta}C_\ell(\eta)\,. \tag{8.115}$$

As a result, the terms of Eq. (8.100) change differently, and such a change cannot be factorized. This means that the relations (3.11, 3.12) obtained for the neutral particles cannot be valid anymore. In other words, the Coulomb forces break down the "vertical" symmetry.

In the case of neutral particles, one more symmetry property, namely, the "diagonal" symmetry, was derived using the Schwarz reflection principle, graphically illustrated in Fig. B.4 of Appendix B. For the charged particles, this principle can be used as well because on the real axis the functions $a_\ell(E)$ and $b_\ell(E)$ are real (see (8.101)), and therefore,

$$a_\ell(E^*) = a_\ell^*(E) \qquad \text{and} \qquad b_\ell(E^*) = b_\ell^*(E)\,. \tag{8.116}$$

The complex conjugation of the energy implies the conjugation of the momentum as well, i.e. the sign of Im k is changing. In other words, we go from a sheet $(+)_m$ to a sheet $(-)_{m'}$ and vice versa. As is seen from Eq. (8.112), in such a "diagonal" transition, the logarithmic index m is changing to the opposite, $m' = -m$. Applying the operation of complex conjugation to both sides of Eq. (8.100) and using the Schwarz relations (8.116), we obtain

$$f_\ell^{(\text{in/out})}(E_m^*) = \left[f_\ell^{(\text{out/in})}(E_{m'})\right]^*\,, \qquad m' = -m\,, \tag{8.117}$$

where for the sake of clarity we specify the logarithmic index m of the Riemann sheet by the subscript of the energy variable. In other words, the points where we calculate the left and right hand sides of Eq. (8.117) are on the sheets corresponding to different logarithmic indices, $(\pm)_m \leftrightarrow (\mp)_{m'}$.

Three "diagonal" relations that follow from the Schwarz reflection principle are graphically illustrated in Fig. 8.3, where the dashed curve and lines connect the symmetric points corresponding to E and E^* on different sheets. The values indicated next to the connected points are the same. This can be symbolically written as follows:

$$f_\ell^{(\text{in/out})}(d) = \left[f_\ell^{(\text{out/in})}(c)\right]^*\,, \tag{8.118}$$

$$f_\ell^{(\text{in/out})}(a) = \left[f_\ell^{(\text{out/in})}(e)\right]^*\,, \tag{8.119}$$

$$f_\ell^{(\text{in/out})}(g) = \left[f_\ell^{(\text{out/in})}(b)\right]^*\,. \tag{8.120}$$

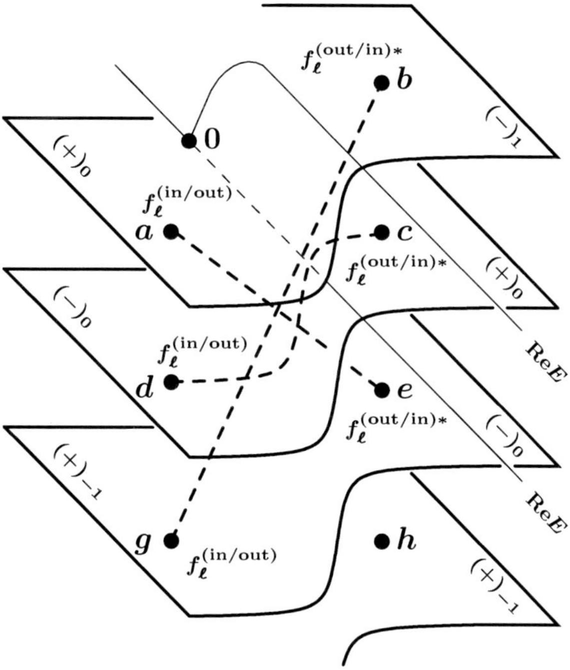

Fig. 8.3 Three "diagonal" relations among the values of the Jost functions for the potential with a Coulomb tail. The values indicated next to the connected points are the same as is given by Eqs. (8.118, 8.119, 8.120)

Of course, there are an infinite number of such relations. However, as was already mentioned, the sheets with $m \neq 0$ do not practically influence the scattering, and therefore, the most important relations are those that are given by Eqs. (8.118) and (8.119), i.e. for the points on the principal sheets. Since the S-matrix is the ratio, $S_\ell = f_\ell^{(\mathrm{out})}/f_\ell^{(\mathrm{in})}$, it follows from these equations that

$$S_\ell(d) = \left[S_\ell^{-1}(c)\right]^* , \tag{8.121}$$

$$S_\ell(a) = \left[S_\ell^{-1}(e)\right]^* . \tag{8.122}$$

In particular, this means that the S-matrix poles correspond to the S-matrix zeros at the "diagonal" points shown in Fig. 8.3.

In the case of neutral particles, there are both the "vertical" and "diagonal" symmetries. A combination of them gives the "horizontal" or "mirror" symmetry. Since the Coulomb forces break the "vertical" symmetry, the "mirror" symmetry is broken as well. In other words, there is no relation between the Jost functions at the points d and e or between a and c shown in Fig. 8.3. An important implication of breaking this symmetry is that, in contrast to the neutral particles, there are no mirror resonances when the Coulomb forces are present.

8.2.11 Asymptotic Normalization Constant (ANC)

At a spectral point, $E_0 = (\hbar k_0)^2/(2\mu)$, the "in"-Jost function is zero, $f_\ell^{(\mathrm{in})}(E_0) = 0$. Therefore, according to Eq. (8.83), the regular and the Jost solutions, ϕ_ℓ and $\chi_\ell^{(+)}$, are proportional to each other and have the following behaviour at the very short and very large distances:

$$\phi_\ell(E_0, r) \underset{r\to 0}{\sim} (k_0 r)^{\ell+1}\,, \qquad \chi_\ell^{(+)}(E_0, r) \underset{r\to 0}{\sim} (k_0 r)^{\ell+1}\,, \tag{8.123}$$

$$\phi_\ell(E_0, r) \underset{r\to\infty}{\longrightarrow} \frac{e^{ik_0 r - i\eta\ln(2k_0 r)}}{i^{\ell+1}} f_\ell^{(\mathrm{out})}(E)\,, \qquad \chi_\ell^{(+)}(E_0, r) \underset{r\to\infty}{\longrightarrow} \frac{e^{ik_0 r - i\eta\ln(2k_0 r)}}{i^{\ell+1}}\,. \tag{8.124}$$

Here, the only difference from the corresponding formulae for the neutral particle case (see Eqs. (3.41, 3.42)) is that at large distances these functions have the additional logarithmic phase term. However, this term does not change the vanishing behaviour of the functions (8.124) and their first derivatives at infinity because the linear term at large r always dominates over the logarithmic one (for the resonances, the asymptotics should be considered along the ray $r = \rho\exp(i\theta)$ depicted in Fig. 3.9),

$$e^{ik_0 r - i\eta\ln(2k_0 r)} \underset{r\to\infty}{\longrightarrow} 0\,, \qquad \mathrm{Im}(k_0 r) > 0\,, \tag{8.125}$$

$$\frac{d}{dr} e^{ik_0 r - i\eta\ln(2k_0 r)} = ik_0\left(1 - \frac{\eta}{k_0 r}\right) e^{ik_0 r - i\eta\ln(2k_0 r)} \underset{r\to\infty}{\longrightarrow} 0\,. \tag{8.126}$$

Taking the derivative of $f_\ell^{(\mathrm{in})}$, given by Eq. (8.77), over the variable E (further down denoted by the dot over the function) and following the same logical steps as in Sect. 3.3.5, we obtain

$$\dot{f}_\ell^{(\mathrm{in})}(E_0) = \frac{\mu}{i\hbar^2 k_0} f_\ell^{(\mathrm{out})}(E_0) \int_0^\infty \left[\chi_\ell^{(+)}(E_0, r)\right]^2 dr\,, \tag{8.127}$$

where the integration goes along the real axis for a bound state and along the ray in the complex r-plane for a resonance.

If we generalize the definition (3.53) of the asymptotic normalization constant (ANC), $\mathcal{A}_\ell$, for the case of the charged particles by inclusion of the logarithmic phase in the behaviour of the normalized physical wave function, u_ℓ, at large distances,

$$\boxed{u_\ell(E_0, r) \underset{r\to\infty}{\longrightarrow} \mathcal{A}_\ell e^{ik_0 r - i\eta\ln(2k_0 r)}} \tag{8.128}$$

then (the same as for neutral particles)

$$u_\ell(E_0, r) = i^{\ell+1} \mathcal{A}_\ell \, \chi_\ell^{(+)}(E_0, r) \, . \tag{8.129}$$

Therefore,

$$\dot{f}_\ell^{(\text{in})}(E_0) = (-1)^{\ell+1} \frac{\mu}{i\hbar^2 k_0} \frac{f_\ell^{(\text{out})}(E_0)}{\mathcal{A}_\ell^2} \int_0^\infty [u_\ell(E_0, r)]^2 \, dr \, . \tag{8.130}$$

There are two important conclusions that can be drawn from this equation. First, all the quantities on its right hand side are non-zero, and thus the first derivative of the Jost function is not zero, which means that all the bound and resonant states correspond to simple (first order) poles of the S-matrix. Second, this equation leads to the relation between the ANC and the residue,

$$\text{Res}\,[S_\ell, E_0] = f_\ell^{(\text{out})}(E_0) \left[\frac{d}{dE} f_\ell^{(\text{in})}(E) \right]_{E=E_0}^{-1} , \tag{8.131}$$

of the S-matrix at a spectral pole, which is the same as for neutral particles, $\eta \to 0$ (see Eq. (4.53)),

$$\boxed{\text{Res}\,[S_\ell, E_0] = i(-1)^{\ell+1} \frac{\hbar^2 k_0}{\mu} \mathcal{A}_\ell^2} \tag{8.132}$$

Part II

Multi-Channel Problems

9 Non-central Potential

When considering quantum objects, very often we have to deal with the forces that depend on the orientation of these objects in space. This is obviously true for molecules, which cannot be spherically symmetric. This is also true for many nuclei, whose shape may significantly deviate from spherical. The elementary particles with non-zero spins (even if we consider them as being structureless) interact with each other differently in the states with different mutual orientation of their intrinsic angular momenta.

In terms of the Schrödinger equation, all these interactions are described by non-central potentials $U(\vec{r})$, whose strength generally depends on the direction $\vec{r}$, from which you approach the source of the force. Such a potential makes the space non-symmetric with respect to rotations. And since the angular-momentum conservation-law follows from the rotational invariance, it is not valid when a non-central potential is present. In other words, the angular momentum before and after the collision can be different.

Mathematically, it is still convenient to expand the wave function in partial waves with definite values of the angular momentum. But now we cannot consider the partial waves independently. The equations for them are coupled to each other. As a result we have to deal with a set of radial Schrödinger equations, which can be written in the matrix form. The Jost function becomes a matrix-function of the energy.

Formally, we can consider a non-symmetric potential whose source is fixed in the space. For a quantum object moving under the action of such forces, the potential is "inherently" non-central, i.e. the isotropy of space is "inherently" broken and the number of coupled partial waves is infinite. In real physical problems, however, we almost never have such a situation that the source of the potential can be considered as immovable. Even if we deal with the electron colliding with a heavy non-symmetric molecule, we can turn the molecule together with the electron in any direction in such a way that the forces acting between them remain the same. This means that for the whole system (molecule plus electron) the total angular

S. A. Rakityansky, *Jost Functions in Quantum Mechanics*,
https://doi.org/10.1007/978-3-031-07761-6_9

momentum J is conserving. That heavy molecule has many rotational states and each of these states has certain angular momentum associated with the rotation (the total spin of the molecule). Adding together (in the quantum way) the molecular spin, the electron spin, and its orbital angular momentum ℓ, we obtain the fixed total angular momentum. It is not difficult to see that for a given J, the number of possible ℓ is always limited.

There are many systems where the non-central part of the potential is solely due to the spin-spin and spin-orbital interaction (nucleon-nucleon forces, for example). In such a case only few partial waves are coupled to each other. Even if we consider an "inherently" non-central potential, we assume that in practical calculations it is always possible to take into account only a finite number of partial waves, only those that are the most important for a particular physical process.

As we will see soon, the Jost function as well as the S-matrix for the problems with non-central potentials are always $(N \times N)$-square matrices, where N is the number of coupled Schrödinger equations for a given J. Despite the fact that $N > 1$, we may consider such a problem as a single-channel one (as was argued at the beginning of Chap. 2). However, the reader should keep in mind that the problem of coupled partial waves is a simplified version of a multi-channel problem. The simplification consists in the fact that all the "channels" (partial waves) have the same threshold energy (degenerate thresholds) and therefore there is only one branch point of the Riemann energy surface, which has only two sheets. Most of the formulae derived here can be used directly or easily generalized for a "true" multi-channel problem (with different thresholds) considered further down in the book.

In all sections of this chapter it is assumed that the potential is non-singular at the origin and of the short-range type at infinity.

9.1 Partial Waves

To make all the formulae more simple and transparent, we consider here a spinless particle moving in an "inherently" non-central potential. The main purpose of this simplified consideration is pedagogical. It allows us to introduce the concept of the Jost matrices that couple different partial-wave states, in a most clear way. Nonetheless, there are some real physical problems, where the potential is "inherently" non-central. As an example of such a problem, we can mention the so-called double-slit experiment, in which the interference of electron-waves is observed behind a macroscopic screen that has two narrow slits. The generalization of all the formulae for more practical and realistic problems that involve spin-dependent potentials is straightforward and is given in Chap. 10.

As we have seen in the case of a spherically symmetric potential, the partial-wave expansions for a discrete spectrum (bound or resonant) states are different from the expansion for a scattering state. The difference stems from the fact that the bound or resonant wave function depends on a single vector variable $\vec{r}$, while the wave function describing a scattering state depends on both the coordinate $\vec{r}$ and the incident momentum $\vec{k}$. However, despite this principal difference, the radial parts of all these wave functions obey the same differential equation.

9.1.1 Discrete States

Since the angular momentum ℓ is not a "good" quantum number in the case of a non-central potential, the wave function of a discrete state,

$$\psi_E(\vec{r}\,) = \langle \vec{r}\,|\psi_E\rangle\ ,$$

is a mixture of all possible partial waves, i.e. its projections on all components of $|\vec{r}\,\rangle$ with definite values of ℓ and m (see Appendix A),

$$|\vec{r}\,\rangle = \sum_{\ell m} |r\ell m\rangle Y^*_{\ell m}(\hat{\vec{r}}\,)\ ,$$

are (generally) non-zero. Introducing the radial components of the wave function for each partial wave (ℓ, m) as

$$\langle r\ell m|\psi_E\rangle = \frac{u_{[\ell]}(E,r)}{r}\ ,$$

we obtain

$$\psi_E(\vec{r}\,) = \sum_{[\ell]} \frac{u_{[\ell]}(E,r)}{r} Y_{[\ell]}(\theta,\varphi)\ , \tag{9.1}$$

where the symbol $[\ell]$ stands for the two subscripts, ℓ and m,

$$[\ell] \equiv \{\ell, m\}\ ,$$

and the sum runs over all possible values of them. Of course the notation $[\ell]$ is not much simpler than ℓm. We use it in order to simplify the generalizations of all the formulae in the subsequent sections of the book. For the particles with spin, for example, the symbol $[\ell]$ will stand for another pair of subscripts, namely ℓ and the total spin s (see Chap. 10). For the few-body systems treated in Sect. 17.1 within the hyperspherical expansion, all the formulae remain the same, but $[\ell]$ is replaced with $[L]$ which is a collection of several quantum numbers.

After substituting the partial-wave expansion (9.1) in the three-dimensional Schrödinger equation (1.28), we multiply this equation from the left by the spherical harmonics $Y^*_{[\ell]}(\theta,\varphi)$ and integrate over the spherical angles. Taking into account the orthogonality of the spherical harmonics, we obtain the following set of coupled radial equations

$$\left[\partial_r + k^2 - \frac{\ell(\ell+1)}{r^2}\right] u_{[\ell]}(E,r) = \sum_{[\ell']} V_{[\ell][\ell']}(r) u_{[\ell']}(E,r) \,, \tag{9.2}$$

where

$$V_{[\ell][\ell']}(r) = \frac{2\mu}{\hbar^2} \iint Y^*_{[\ell]}(\theta,\varphi) U(\vec{r}\,) Y_{[\ell']}(\theta,\varphi) \sin\theta \, d\theta \, d\varphi \tag{9.3}$$

is the matrix of the potential, whose non-diagonal elements (if non-zero) are responsible for the coupling of different partial waves. If the potential is real, $U^* = U$, then (as is seen from the above equation) the matrix V is Hermitian, i.e.

$$V_{[\ell][\ell']}(r) = V^*_{[\ell'][\ell]}(r) \,. \tag{9.4}$$

The spherical harmonics have non-zero imaginary parts because of their dependence on the angle φ in the form $e^{im\varphi}$. If the potential $U(\vec{r}\,)$ is axially symmetric (around the z-axis), i.e. φ-independent, then the integration over φ in (9.3) gives $\delta_{mm'}$ and the matrix

$$V_{[\ell][\ell']}(r) = V_{\ell\ell'}(r)\delta_{mm'} \tag{9.5}$$

becomes real and, according to Eq. (9.4), symmetric, $V = V^T$.

9.1.2 Scattering States

The scattering-state wave function $\psi_{\vec{k}}(\vec{r}\,) = \langle \vec{r}\,|\psi_{\vec{k}}\rangle$ depends on two vectors, namely on the incident momentum $\vec{k}$ and the space coordinate $\vec{r}$, i.e. on their absolute values, k and r, as well as on two different sets of spherical angles denoted as $\hat{\vec{k}}$ and $\hat{\vec{r}}$ or, equivalently, as $\Omega_{\vec{k}}$ and $\Omega_{\vec{r}}$. The expansion of this function is a product of independent partial-wave expansions for the bra- and ket-vectors $\langle \vec{r}\,|$ and $|\psi_{\vec{k}}\rangle$,

$$\langle \vec{r}\,|\psi_{\vec{k}}\rangle = \sum_{[\ell'][\ell]} Y_{[\ell']}(\hat{\vec{r}}\,)\langle r\ell'm'|\psi_{k\ell m}\rangle Y^*_{[\ell]}(\hat{\vec{k}}\,) \,,$$

where the radial part of the wave function, $\langle r\ell'm'|\psi_{k\ell m}\rangle$, is not diagonal, i.e.

$$\langle r\ell'm'|\psi_{k\ell m}\rangle \not\sim \delta_{\ell'\ell}\delta_{m'm} \,,$$

because any incoming partial wave $|k\ell m\rangle$ is transformed by the non-central potential into a mixture of all possible partial waves. This can be more clearly understood in the time-dependent scattering theory where the state $|\psi_{k\ell m}\rangle$ (at $t = 0$) is obtained from the in-asymptote $|k\ell m\rangle$ (at $t = -\infty$) by the action of the Møller operator[1] (in this book, the Møller operators are defined in Sect. 1.3.2),

$$\Omega_+ = \lim_{t\to-\infty} e^{iHt}e^{-iH_0t} ,$$

with H and H_0 being the total and the free Hamiltonians, respectively. Using this operator, we have

$$|\psi_{\vec{k}}\rangle = \Omega_+|\vec{k}\,\rangle = \sum_{\ell m}\Omega_+|k\ell m\rangle Y^*_{\ell m}(\hat{\vec{k}}\,) = \sum_{\ell m}|\psi_{k\ell m}\rangle Y^*_{\ell m}(\hat{\vec{k}}\,) .$$

The operator Ω_+ involves the total Hamiltonian, which includes the non-central potential, and therefore an initial state $|k\ell m\rangle$ with definite values of ℓ and m is distorted when the particle approaches the source of the force. Therefore the resulting state $|\psi_{k\ell m}\rangle$ does not have definite values of the quantum numbers $[\ell]$. The meaning of its subscripts is just the indication of the origin of this state, i.e. they show from which "pure" partial wave this state has evolved. This means that the radial part of the wave function is a matrix,

$$\langle r\ell'm'|\psi_{k\ell m}\rangle = \psi_{[\ell'][\ell]}(E,r) ,$$

in the space of the partial waves $[\ell]$.

As usual, to get rid of the first derivative in the radial Schrödinger equation, we should explicitly separate in this matrix the factor $1/r$. Let us introduce the matrix u, which differs from ψ by this factor,

$$\psi_{[\ell'][\ell]}(E,r) = \frac{u_{[\ell'][\ell]}(E,r)}{r} .$$

This gives

$$\langle\vec{r}\,|\psi_{\vec{k}}\rangle = \psi_{\vec{k}}(\vec{r}\,) = \sum_{[\ell][\ell']} Y_{[\ell]}(\hat{\vec{r}}\,)\frac{u_{[\ell][\ell']}(E,r)}{r}Y^*_{[\ell']}(\hat{\vec{k}}\,) . \tag{9.6}$$

If we substitute the expansion (9.6) into the three-dimensional Schrödinger equation (1.28), multiply this equation from the left by $Y^*_{[\ell]}(\hat{\vec{r}}\,)Y_{[\ell']}(\hat{\vec{k}}\,)$, and then integrate

[1] The most clear presentation of the Møller operators can be found in Ref. [4].

over all four spherical angles, we obtain the set of differential equations

$$\left[\partial_r + k^2 - \frac{\ell(\ell+1)}{r^2}\right] u_{[\ell][\ell']}(E,r) = \sum_{[\ell'']} V_{[\ell][\ell'']}(r) u_{[\ell''][\ell']}(E,r) \,. \tag{9.7}$$

For any fixed $[\ell']$, the function $u_{[\ell][\ell']}(E,r)$ is a column-matrix, which obeys exactly the same Eq. (9.2) as the wave function of a discrete state. All equations of the set (9.7) are homogeneous and thus each column of the matrix $u_{[\ell][\ell']}(E,r)$ can be multiplied by an arbitrary r-independent factor. The factors can be different for different columns. Using this freedom, we can introduce an individual normalization factor $N_{[\ell]}(k)$ for each column (the subscript $[\ell]$ shows which column is multiplied),

$$\psi_{\vec{k}}(\vec{r}) = \sum_{[\ell][\ell']} Y_{[\ell]}(\hat{\vec{r}}) \frac{u_{[\ell][\ell']}(E,r) N_{[\ell']}(k)}{r} Y^*_{[\ell']}(\hat{\vec{k}}) \,. \tag{9.8}$$

Further down, in Sect. 9.7.2, we will find that the correct normalization (to the δ-function) of the scattering wave function can be obtained if

$$N_{[\ell]}(k) = i^\ell/(\sqrt{2\pi}k) \qquad \Longrightarrow \qquad \langle\psi_{\vec{k}}|\psi_{\vec{k}'}\rangle = \delta(\vec{k} - \vec{k}') \,.$$

9.2 Fundamental Matrix of Regular Solutions

In principle the sums in the partial-wave expansions (9.1), (9.8) run to infinity. However, in any physical model we always have reasons to include only a finite number, N, of them. As a result, Eq. (9.2) is a system of N coupled differential equations of the second order. Each solution of such a system is a column-matrix and (since the equations are of the second order) there are $2N$ linearly independent solutions.

In this chapter we consider the case when several angular momenta are coupled to each other. This means that they cannot all be zero and thus the singular term $\ell(\ell+1)/r^2$ is present at least in one of equations (9.2). This term makes the point $r = 0$ a regular singular-point of the whole system of differential equations (9.2). This means that among all its $2N$ solutions one half are regular at $r = 0$, while the other half are irregular,

$$\underbrace{\begin{pmatrix} u_{[\ell_1],1} \\ u_{[\ell_2],1} \\ \vdots \\ u_{[\ell_N],1} \end{pmatrix}, \begin{pmatrix} u_{[\ell_1],2} \\ u_{[\ell_2],2} \\ \vdots \\ u_{[\ell_N],2} \end{pmatrix}, \dots, \begin{pmatrix} u_{[\ell_1],N} \\ u_{[\ell_2],N} \\ \vdots \\ u_{[\ell_N],N} \end{pmatrix}}_{N \text{ regular columns}}, \underbrace{\begin{pmatrix} u_{[\ell_1],N+1} \\ u_{[\ell_2],N+1} \\ \vdots \\ u_{[\ell_N],N+1} \end{pmatrix}, \begin{pmatrix} u_{[\ell_1],N+2} \\ u_{[\ell_2],N+2} \\ \vdots \\ u_{[\ell_N],N+2} \end{pmatrix}, \dots, \begin{pmatrix} u_{[\ell_1],2N} \\ u_{[\ell_2],2N} \\ \vdots \\ u_{[\ell_N],2N} \end{pmatrix}}_{N \text{ irregular columns}} .$$

If we choose only the regular columns (there are N of them), they can be combined in a square matrix, ϕ, of the size $(N \times N)$,

$$\phi_{[\ell][\ell']}(E,r) \underset{r\to 0}{\longrightarrow} 0\,, \qquad \forall\,[\ell][\ell']\,. \tag{9.9}$$

Such a matrix is called the fundamental matrix of regular solutions. Each column of this matrix solves the system (9.2) and is regular at the origin. Moreover all the columns are linearly independent. This means that the columns of ϕ form a basis in the space of the regular solutions and thus any other regular solution is a linear combination of the columns of the fundamental matrix.

As we have seen, a physical solution (which is regular) can be either a single column $u_{[\ell]}$ (for bound and resonant states) or an $(N \times N)$-matrix $u_{[\ell][\ell']}$ (for scattering states). In both cases these solutions are linear combinations of the columns of the fundamental matrix of regular solutions,

$$\text{Bound states and resonanses:} \qquad u_{[\ell]}(E,r) = \sum_{[\ell']} \phi_{[\ell][\ell']}(E,r)c_{[\ell']}\,, \tag{9.10}$$

$$\text{Scattering states:} \qquad u_{[\ell][\ell']}(E,r) = \sum_{[\ell'']} \phi_{[\ell][\ell'']}(E,r)c_{[\ell''][\ell']}\,. \tag{9.11}$$

Here $c_{[\ell']}$ and $c_{[\ell''][\ell']}$ are the corresponding combination coefficients. In Eq. (9.10), each column of the fundamental matrix is present in the combination with its own coefficient. In Eq. (9.11), the same type of combination is done for each column of the scattering solution. Of course the combination coefficients are not arbitrary. The choice of them determines the behaviour of the matrix $u(E,r)$ at large distances. The coefficients should be chosen in such a way as to guarantee the correct asymptotics of the wave function. As we will see, they could be found via the Jost matrix.

When looking for a physical solution of the Schrödinger equation, we not only require its regularity at the origin but also are concerned with its proper behaviour at large distances. In contrast to it, for the fundamental matrix of regular solutions, ϕ, we only have to specify the boundary conditions at $r = 0$. Whatever happens to this matrix at large distances, so be it. The columns of the fundamental matrix are kind of building blocks from which a specific physical solution can be constructed, as is given in Eq. (9.10) or in Eq. (9.11). Each of these blocks is regular and thus the regularity of $u_{[\ell]}(E,r)$ is guaranteed.

In mathematical terminology, when calculating ϕ, we solve the Cauchy problem (boundary conditions are imposed at a single point, at $r = 0$); when finding the physical wave function, we deal with the Dirichlet problem (boundary conditions are imposed at the two end points of the interval $r \in [0,\infty)$). The Cauchy problem has a solution at any energy, while the Dirichlet problem can only be solved at certain set of energies (because it is not always possible to satisfy physical boundary conditions at large distances).

When finding the fundamental matrix ϕ, we actually find more than we need, because the physical solutions form just a small subspace of the whole space of regular solutions spanned by the columns of ϕ. However, there is an advantage in splitting the task in the two steps: at the first step, we satisfy the regularity condition and find ϕ; at the second step, we construct the physical wave functions out of its columns by satisfying the asymptotic boundary conditions.

Such a two-step approach is never used when the set of second-order Schrödinger equations is solved directly. This is because there is no simple way of finding appropriate coefficients $c_{[\ell]}$ in Eqs. (9.10), (9.11). If, however, we transform the second-order equations to the equivalent set of first-order equations (in the same way as we did it for the central potential), then both of the steps become simple and easy.

9.3 Transformation of Schrödinger Equation

In the case of a single radial Schrödinger equation for a central potential, we obtained an equivalent set of two first-order differential equations (see Sect. 2.2.1) whose solutions asymptotically become the Jost functions. A similar transformation is possible for the matrix equation

$$\left[\partial_r + k^2 - \frac{\ell(\ell+1)}{r^2}\right]\phi_{[\ell][\ell']}(E,r) = \sum_{[\ell'']} V_{[\ell][\ell'']}(r)\phi_{[\ell''][\ell']}(E,r) \tag{9.12}$$

that determines the regular basis $\phi_{[\ell][\ell']}(E,r)$. In order to do this, we need to define the "free" solutions, i.e. the incoming and outgoing waves.

9.3.1 Incoming and Outgoing Waves

Our assumption that the potential $U(\vec{r}\,)$ is non-singular and of a short range can be formally expressed as

$$\int_0^\infty |V_{[\ell][\ell']}(r)| r\, dr < \infty\,, \qquad \forall\, [\ell], [\ell']\,, \tag{9.13}$$

which generalizes the condition (7.2). Since the potential matrix $V(r)$ quickly wanes with $r \to \infty$, at large distances the system (9.12) simplifies and all of its equations decouple,

$$\left[\partial_r + k^2 - \frac{\ell(\ell+1)}{r^2}\right]\phi_{[\ell][\ell']}(E,r) \approx 0\,, \qquad r \to \infty\,. \tag{9.14}$$

Despite the decoupling, we may still consider them as a system of differential equations of the second order. Each column of the matrix ϕ solves this system. Actually, this is a set of Riccati–Bessel equations, whose solutions are known, namely one of the four functions $j_\ell(kr)$, $n_\ell(kr)$, $h_\ell^{(\pm)}(kr)$, or the identical zero (trivial solution). Using these possibilities, we need to construct the matrix $\phi(E,r)$. By definition, all columns of $\phi(E,r)$ must be linearly independent. The independence can be achieved if we use the trivial solution everywhere except for the diagonal elements. On the diagonal of the matrix ϕ, we can put $h_\ell^{(\pm)}(kr)$. In other words, the matrix

$$\begin{bmatrix} a_1 h_{\ell_1}^{(-)} + b_1 h_{\ell_1}^{(+)} & 0 & \cdots & 0 \\ 0 & a_2 h_{\ell_2}^{(-)} + b_2 h_{\ell_2}^{(+)} & \cdots & 0 \\ \vdots & \vdots & \vdots & \vdots \\ 0 & 0 & \cdots & a_N h_{\ell_N}^{(-)} + b_N h_{\ell_N}^{(+)} \end{bmatrix} \tag{9.15}$$

is a general solution of Eq. (9.14) and obeys the condition of linear independence of the columns, with any choice of the combination coefficients a_n and b_n.

Looking at the matrix (9.15), we see that it can be written as a combination of the following two matrices representing the incoming and outgoing spherical waves in each of the partial waves

$$W^{(\text{in})}(E,r) = \begin{bmatrix} h_{\ell_1}^{(-)}(kr) & 0 & \cdots & 0 \\ 0 & h_{\ell_2}^{(-)}(kr) & \cdots & 0 \\ \vdots & \vdots & \vdots & \vdots \\ 0 & 0 & \cdots & h_{\ell_N}^{(-)}(kr) \end{bmatrix}, \tag{9.16}$$

$$W^{(\text{out})}(E,r) = \begin{bmatrix} h_{\ell_1}^{(+)}(kr) & 0 & \cdots & 0 \\ 0 & h_{\ell_2}^{(+)}(kr) & \cdots & 0 \\ \vdots & \vdots & \vdots & \vdots \\ 0 & 0 & \cdots & h_{\ell_N}^{(+)}(kr) \end{bmatrix}. \tag{9.17}$$

9.3.2 First-Order Differential Equations

Since asymptotically the fundamental matrix of regular solutions behaves as a combination of the incoming and outgoing waves (9.16), (9.17), let us look for ϕ as the same type of combination on the whole interval $r \in [0,\infty)$ (with unknown r-dependent coefficient-matrices),

$$\boxed{\phi(E,r) = W^{(\text{in})}(E,r)F^{(\text{in})}(E,r) + W^{(\text{out})}(E,r)F^{(\text{out})}(E,r)} \tag{9.18}$$

which generalizes the corresponding representation (2.12). The unknown matrices $F^{(\text{in/out})}(E,r)$ have the same size, $(N \times N)$, as the matrix ϕ.

Since instead of one unknown matrix, ϕ, we introduce two unknown matrices, $F^{(\text{in/out})}$, they cannot be independent. We therefore can impose an arbitrary condition that relates them to each other. As such condition, it is convenient to choose the following equation

$$W^{(\text{in})}(E,r)\partial_r F^{(\text{in})}(E,r) + W^{(\text{out})}(E,r)\partial_r F^{(\text{out})}(E,r) = 0\,, \tag{9.19}$$

which is known as the Lagrange condition.

Substituting the combination (9.18) into the differential equation (9.12), we can repeat all the steps that we did in Sect. 2.2.1.1 and thus can obtain the differential equations for the matrices $F^{(\text{in/out})}$. As a matter of fact, the derivations given in Sect. 2.2.1.1 were deliberately written in the general form, as if all the quantities there were matrices i.e. non-commuting. It is easy to check that the resulting equations are

$$\partial_r F^{(\text{in})} = -\frac{1}{2ik} W^{(\text{out})} V \left[W^{(\text{in})} F^{(\text{in})} + W^{(\text{out})} F^{(\text{out})} \right] \tag{9.20}$$

$$\partial_r F^{(\text{out})} = \frac{1}{2ik} W^{(\text{in})} V \left[W^{(\text{in})} F^{(\text{in})} + W^{(\text{out})} F^{(\text{out})} \right] \tag{9.21}$$

where for the sake of clarity we dropped the variable symbols (E,r). The boundary conditions for these equations generalize the conditions (2.24) and are discussed further down in Sect. 9.3.4.

9.3.3 Alternative Form of the Differential Equations

Similarly to what was done in Sect. 2.2.1.2, we can combine the incoming and outgoing waves in such a way that the resulting matrices,

$$j(E,r) = \frac{W^{(\text{out})}(E,r) + W^{(\text{in})}(E,r)}{2} = \begin{bmatrix} j_{\ell_1}(kr) & 0 & \cdots & 0 \\ 0 & j_{\ell_2}(kr) & \cdots & 0 \\ \vdots & \vdots & \vdots & \vdots \\ 0 & 0 & \cdots & j_{\ell_N}(kr) \end{bmatrix}, \tag{9.22}$$

$$n(E,r)=\frac{W^{(\mathrm{out})}(E,r)-W^{(\mathrm{in})}(E,r)}{2i}=\begin{bmatrix} n_{\ell_1}(kr) & 0 & \cdots & 0 \\ 0 & n_{\ell_2}(kr) & \cdots & 0 \\ \vdots & \vdots & \vdots & \vdots \\ 0 & 0 & \cdots & n_{\ell_N}(kr) \end{bmatrix}, \tag{9.23}$$

can be used for an alternative representation of the fundamental matrix of regular solutions,

$$\boxed{\phi(E,r)=j(E,r)A(E,r)-n(E,r)B(E,r)} \tag{9.24}$$

where

$$A(E,r)=F^{(\mathrm{in})}(E,r)+F^{(\mathrm{out})}(E,r)\,, \tag{9.25}$$

$$B(E,r)=i\left[F^{(\mathrm{in})}(E,r)-F^{(\mathrm{out})}(E,r)\right]\,. \tag{9.26}$$

The same combination of Eqs. (9.20), (9.21) gives the corresponding differential equations for the matrices A and B,

$$\partial_r A=-\frac{1}{k}nV\,(jA-nB) \tag{9.27}$$

$$\partial_r B=-\frac{1}{k}jV\,(jA-nB) \tag{9.28}$$

9.3.4 Boundary Conditions

There are two requirements in the definition of the fundamental matrix, ϕ, of regular solutions of the matrix radial Schrödinger equation (9.12). Firstly, all the elements of this matrix tend to zero when $r\to 0$. Secondly, all the columns of this matrix are linearly independent. Obviously, the boundary condition (9.9) alone is not sufficient because Eq. (9.12) is of the second order and this condition cannot guarantee linear independence of the columns. We therefore need to specify how the elements of ϕ tend to zero (some of them may diminish faster than the others) in order to fulfil the second requirement.

Let us consider Eq. (9.12) near the origin on a small interval $0\leqslant r\leqslant\delta$ with $\delta\to 0$. After multiplying this system of equations by r^2 and using the fact that the potential is non-singular ($r^2V\to 0$), we obtain the decoupled system of equations,

$$\left[r^2\partial_r-\ell(\ell+1)\right]\phi_{[\ell][\ell']}(E,r)\approx 0\,,\qquad r\in[0,\delta]\,. \tag{9.29}$$

Each of these equations has two linearly independent solutions $r^{\ell+1}$ and $r^{-\ell}$ and of course the trivial (identical zero) solution. A regular column can only consist of $r^{\ell+1}$ and zero elements. In order to guarantee their linear independence, we should have only one element equal to $r^{\ell+1}$ and all the others zero in each row of the matrix ϕ, i.e. this matrix should be diagonal as follows

$$\phi(E,r) \underset{r\to 0}{\sim} \begin{pmatrix} r^{\ell_1+1} & 0 & \cdots & 0 \\ 0 & r^{\ell_2+1} & \cdots & 0 \\ \vdots & \vdots & \vdots & \vdots \\ 0 & 0 & \cdots & r^{\ell_N+1} \end{pmatrix} . \tag{9.30}$$

Higher order corrections to Eq. (9.29) give, of course, non-zero off-diagonal elements. However, they should not change the condition (9.30) at very short distances. In other words, to guarantee linear independence of the columns, the off-diagonal elements in each column must diminish faster than the diagonal one in that column,

$$\phi(E,r) \underset{r\to 0}{\sim} \begin{bmatrix} r^{\ell_1+1} & o(r^{\ell_2+1}) & \cdots & o(r^{\ell_N+1}) \\ o(r^{\ell_1+1}) & r^{\ell_2+1} & \cdots & o(r^{\ell_N+1}) \\ \vdots & \vdots & \vdots & \vdots \\ o(r^{\ell_1+1}) & o(r^{\ell_2+1}) & \cdots & r^{\ell_N+1} \end{bmatrix} . \tag{9.31}$$

The inclusion of the higher order terms retains the linear independence of the columns. Indeed,

$$\det \|\phi(E,r)\| \sim \prod_i r^{\ell_i+1} + o\left(\prod_i r^{\ell_i+1}\right) ,$$

because the products involving off-diagonal elements always contain at least one of the factors $o(r^{\ell_1+1}), o(r^{\ell_2+1}), \cdots, o(r^{\ell_N+1})$. Hence

$$\det \|\phi(E,r)\| \underset{r\to 0}{\sim} \prod_i r^{\ell_i+1}$$

and on the interval $[0,\delta]$ we can always find at least one point r_0 where $\det \|\phi(E,r_0)\| \neq 0$. This means that the columns of the matrix (9.31) are linearly independent on the whole interval $[0,\delta]$.

The structure of the matrix (9.31) implies that the behaviour of the regular basis in the immediate vicinity of the point $r = 0$ is such that

$$\lim_{r\to 0} \frac{\phi_{[\ell][\ell']}(E,r)}{r^{\ell'+1}} \sim \delta_{[\ell][\ell']} ,$$

where the angular momentum in the denominator corresponds to the second subscript of the numerator.

Equations of the system (9.29) are actually decoupled, i.e. each element of the matrix $\phi(E, r)$ is a solution of a separate equation. Since all these equations are homogeneous, each element of ϕ can be multiplied by an arbitrary factor and these factors can be different for different elements. We therefore may replace $r^{\ell'+1}$ in the last equation by $(kr)^{\ell'+1}/(2\ell'+1)!!$ to get the boundary condition

$$\lim_{r\to 0} \frac{\phi_{[\ell][\ell']}(E, r)}{j_{\ell'}(kr)} = \delta_{[\ell][\ell']} , \tag{9.32}$$

which can be written in the matrix form as

$$\boxed{\lim_{r\to 0} \phi j^{-1} = I} \tag{9.33}$$

This generalizes the condition (2.36) that we used in the case of a central potential (with a single ℓ).

In order to solve Eqs. (9.20), (9.21) or Eqs. (9.27), (9.28), we need to find the corresponding boundary conditions for the matrices $F^{(\text{in/out})}$ and A, B. Since by definition the matrix $\phi(E, r)$ is regular at $r = 0$, we have

$$h_\ell^{(-)}(kr) F^{(\text{in})}_{[\ell][\ell']}(E, r) + h_\ell^{(+)}(kr) F^{(\text{out})}_{[\ell][\ell']}(E, r) \underset{r\to 0}{\longrightarrow} 0 ,$$

and the Lagrange condition (9.19) gives

$$h_\ell^{(-)}(kr) \partial_r F^{(\text{in})}_{[\ell][\ell']}(E, r) + h_\ell^{(+)}(kr) \partial_r F^{(\text{out})}_{[\ell][\ell']}(E, r) \equiv 0 .$$

On the other hand,

$$\frac{h_\ell^{(+)}(kr)}{h_\ell^{(-)}(kr)} \equiv \frac{j_\ell(kr) + i n_\ell(kr)}{j_\ell(kr) - i n_\ell(kr)} \underset{r\to 0}{\longrightarrow} -1 ,$$

from which we conclude that in the immediate vicinity of the point $r = 0$ the matrices $F^{(\text{in})}(E, r)$ and $F^{(\text{out})}(E, r)$ as well as their first derivatives become identical, i.e.,

$$F^{(\text{in})}_{[\ell][\ell']}(E, r) \underset{r\to 0}{\longrightarrow} F^{(\text{out})}_{[\ell][\ell']}(E, r) , \tag{9.34}$$

$$\partial_r F^{(\text{in})}_{[\ell][\ell']}(E, r) \underset{r\to 0}{\longrightarrow} \partial_r F^{(\text{out})}_{[\ell][\ell']}(E, r) , \tag{9.35}$$

which means that at least the first two leading terms of the power-series expansions of these matrices near $r = 0$ must be the same. Taking into account Eq. (9.25), we see that both $F^{(\text{in})}$ and $F^{(\text{out})}$ tend to $A/2$ and thus

$$\phi_{[\ell][\ell']}(E,r) \underset{r\to 0}{\longrightarrow} \frac{1}{2}\left[h_\ell^{(-)}(kr)+h_\ell^{(+)}(kr)\right] A_{[\ell][\ell']}(E,r) = j_\ell(kr)A_{[\ell][\ell']}(E,r)\,. \tag{9.36}$$

Substituting this in Eq. (9.32), we obtain

$$\lim_{r\to 0} \frac{j_\ell(kr)A_{[\ell][\ell']}(E,r)}{j_{\ell'}(kr)} = \delta_{[\ell][\ell']}\,, \tag{9.37}$$

$$\lim_{r\to 0} \frac{j_\ell(kr)F^{(\text{in/out})}_{[\ell][\ell']}(E,r)}{j_{\ell'}(kr)} = \frac{1}{2}\delta_{[\ell][\ell']}\,, \tag{9.38}$$

which can also be written in matrix form as

$$\boxed{\lim_{r\to 0} jAj^{-1} = I} \tag{9.39}$$

$$\boxed{\lim_{r\to 0} jF^{(\text{in/out})}j^{-1} = I/2} \tag{9.40}$$

where the diagonal matrix j is defined by Eq. (9.22).

Equations (9.34), (9.35) together with Eq. (9.26) tell us that in the matrix $B(E,r)$ the first two leading terms of the power-series expansion of $F^{(\text{in/out})}(E,r)$ cancel each other near $r=0$. The corresponding boundary condition can be derived from the following chain of transformations:

$$nBj^{-1} = in\big[F^{(\text{in})} - F^{(\text{out})}\big]j^{-1} = inj^{-1}jF^{(\text{in})}j^{-1} - inj^{-1}jF^{(\text{out})}j^{-1}$$

$$\underset{r\to 0}{\longrightarrow} inj^{-1}I/2 - inj^{-1}I/2 \underset{r\to 0}{\longrightarrow} 0\,,$$

the result of which can be written as

$$\lim_{r\to 0} \frac{n_\ell(kr)B_{[\ell][\ell']}(E,r)}{j_{\ell'}(kr)} = 0\,. \tag{9.41}$$

Since the ratio $n_\ell(kr)/j_{\ell'}(kr)$ for all ℓ and ℓ' tends to infinity when $r\to 0$, the condition (9.41) means that all elements of matrix B vanish at $r=0$, i.e.

$$nBj^{-1} \underset{r\to 0}{\longrightarrow} 0 \quad\Rightarrow\quad \boxed{B(E,0)=0} \tag{9.42}$$

It may seem that all off-diagonal elements of the matrices A and $F^{(\mathrm{in/out})}$ also vanish at the origin. This, however, is not true. Indeed, unlike (9.32), the conditions (9.37), (9.38) do not demand that all elements of the matrices A and $F^{(\mathrm{in/out})}$ be regular. This can be understood as follows: when $\ell > \ell'$ Eqs. (9.37), (9.38) hold even if $A_{[\ell][\ell']}(E,r) \sim r^{-(\ell-\ell'-\epsilon)}$ and $F^{(\mathrm{in/out})}_{[\ell][\ell']}(E,r) \sim r^{-(\ell-\ell'-\epsilon)}$ with any $\epsilon > 0$. Hence, the left bottom corners of the matrices A and $F^{(\mathrm{in/out})}$ may, in principle, have diverging elements. The regular basis $\phi_{[\ell][\ell']}(E,r)$ itself remains always regular due to the presence of $j_\ell(kr)$ in (9.36) which compensates the diverging terms.

For a better understanding, let us consider a hypothetical example of a quantum state that is a mixture of the three partial waves with $\ell_1 = 1$, $\ell_2 = 3$, and $\ell_3 = 5$. Then the condition (9.37) reads

$$\begin{bmatrix} A_{[\ell_1][\ell_1]} \to 1 & \frac{1}{(kr)^2} A_{[\ell_1][\ell_2]} \to 0 & \frac{1}{(kr)^4} A_{[\ell_1][\ell_3]} \to 0 \\ (kr)^2 A_{[\ell_2][\ell_1]} \to 0 & A_{[\ell_2][\ell_2]} \to 1 & \frac{1}{(kr)^2} A_{[\ell_2][\ell_3]} \to 0 \\ (kr)^4 A_{[\ell_3][\ell_1]} \to 0 & (kr)^2 A_{[\ell_3][\ell_2]} \to 0 & A_{[\ell_3][\ell_3]} \to 1 \end{bmatrix},$$

which means that the following (partly singular) behaviour of the matrix A could be acceptable

$$A(E,r) \underset{r\to 0}{\sim} \begin{bmatrix} 1 & (kr)^3 & (kr)^5 \\ \frac{1}{kr} & 1 & (kr)^3 \\ \frac{1}{(kr)^3} & \frac{1}{kr} & 1 \end{bmatrix}. \tag{9.43}$$

Therefore the formal boundary conditions (9.37), (9.38) cannot be used in a numerical procedure of solving the corresponding differential equations. Indeed, when doing practical (numerical) calculations, we cannot start integration of the differential equations exactly from $r = 0$ (simply because the Riccati–Hankel functions and the Riccati–Neumann functions in Eqs. (9.20), (9.21) and Eqs. (9.27), (9.28) are singular at this point). We have to start at a small $r = \delta$. But what we have for this point is incomplete. If we use again the above hypothetical example, then the starting values of the matrix elements at $r = \delta$ would be

$$A(E,\delta) = 2F^{(\mathrm{in/out})}(E,\delta) = \begin{pmatrix} 1 & 0 & 0 \\ ? & 1 & 0 \\ ? & ? & 1 \end{pmatrix}, \qquad B(E,\delta) = \begin{pmatrix} 0 & 0 & 0 \\ 0 & 0 & 0 \\ 0 & 0 & 0 \end{pmatrix},$$

where the question-marks stand for unknown values. In order to complete the boundary conditions, we need to establish the behaviour (near $r = 0$) of these unknown (possibly singular) matrix elements in an explicit form.

9.3.4.1 Explicit Behaviour of the Regular Solution Near $r = 0$

This and the next two sections describe some recipes that are needed in numerical calculations. It may therefore be skipped for the first reading or omitted completely, if the reader has no intentions to do calculations.

* * *

To obtain the functional dependencies of the matrix elements of $F^{(\mathrm{in/out})}(E, r)$ in an explicit form, we take indefinite integrals on both sides of Eq. (9.20), (9.21). This gives

$$F^{(\mathrm{in/out})}_{[\ell][\ell']}(E,r) = \mathrm{const} \mp \frac{1}{ik} \int h^{(\pm)}_{\ell}(kr) \sum_{[\ell'']} V_{[\ell][\ell'']}(r) \phi_{[\ell''][\ell']}(E,r)\, dr\,. \tag{9.44}$$

The integration constants are fixed for the diagonal and the right-top corner elements by the condition (9.38). Indeed, due to the fact that the potential is less singular than $1/r^2$, the integrals in (9.44) for $\ell \leqslant \ell'$ give (near $r = 0$) the functions

$$\int r^{-\ell} r^{-2+\varepsilon} r^{\ell+1}\, dr = \frac{r^{\varepsilon}}{\varepsilon} \to 0\,, \qquad \ell' = \ell\,, \quad \varepsilon > 0\,,$$

for the diagonal elements, and, according to Eq. (9.31),

$$\int r^{-\ell} r^{-2+\varepsilon} o\left(r^{\ell'+1}\right) dr = o\left(\frac{r^{\ell'-\ell+\varepsilon}}{\ell' - \ell + \varepsilon}\right) \to 0\,, \qquad \ell' > \ell\,, \quad \varepsilon > 0\,,$$

for the elements above the diagonal. As we see, all the integrals for the elements on and above the diagonal vanish at $r = 0$. Hence to fulfil (9.38) we must set $\mathrm{const} = 1/2$ for $\ell = \ell'$ and $\mathrm{const} = 0$ for $\ell < \ell'$.

For the elements below the diagonal, we still have freedom in choosing these constants. If these elements are singular, as illustrated in the example (9.43), then the value of the constant is immaterial ($\mathrm{const} + \infty = \infty$). If they are finite, then the choice of the constants simply determines the normalization of the solutions. The most natural choice would be to put these constants to zero in the same way as for the elements above the diagonal. Thus finally we obtain

$$F^{(\mathrm{in/out})}_{[\ell][\ell']}(E,r) = \frac{1}{2}\delta_{[\ell][\ell']} \mp \frac{1}{ik} \int h^{(\pm)}_{\ell}(kr) \sum_{[\ell'']} V_{[\ell][\ell'']}(r) \phi_{[\ell''][\ell']}(E,r)\, dr\,, \tag{9.45}$$

where the indefinite integrals should be understood as the primitive functions since all arbitrary constants have already been fixed.

From Eq. (9.45) it is clear that the leading terms determining the behaviour of the off-diagonal elements of the matrices $F^{(\text{in/out})}$ (and thus of the matrices A and B) cannot be given in a universal form independent of the potential matrix $V(r)$. In other words, not only the differential equations Eqs. (9.20), (9.21) involve the potential, but the boundary conditions for them depend on the chosen potential as well.

The explicit expressions for the leading terms of all the elements of the matrices $F^{(\text{in/out})}$ on a small interval $r \in (0, \delta]$ can be obtained by replacing all the functions under the indefinite integral in Eq. (9.45) with their power-series expansions near $r = 0$ and by performing the integration. However, we only know the power-series expansions for $h_\ell^{(\pm)}(kr)$ and in principle can obtain them for the elements $V_{[\ell][\ell']}(r)$ of the potential matrix. Meanwhile for $\phi_{[\ell][\ell']}(E, r)$, only the diagonal elements are known,

$$\phi_{[\ell][\ell]}(E, r) \approx j_\ell(kr) , \qquad r \in (0, \delta] .$$

Looking more closely at Eq. (9.45), we see that in fact for each column (with fixed ℓ') there is an independent equation. And according to Eq. (9.31), in each column of the matrix ϕ the diagonal element is dominant. Therefore, as the zero-order approximation for the columns of ϕ, we should take the following

$$\phi^{(0)}_{[\ell][\ell']}(E, r) = j_\ell(kr)\delta_{[\ell][\ell']} , \qquad r \in (0, \delta] , \tag{9.46}$$

which is also consistent with Eq. (9.32). Then, using Eq. (9.36), we can define the corresponding zero order approximation for the matrices $F^{(\text{in/out})}$ as follows

$$F^{(\text{in/out})(0)}_{[\ell][\ell']}(E, r) = \frac{1}{2}\delta_{[\ell][\ell']} , \qquad r \in (0, \delta] . \tag{9.47}$$

Substituting (9.46) into the indefinite integral (9.45) and using the series expansions of $h_\ell^{(\pm)}(kr)$, $j_\ell(kr)$, and $V_{[\ell][\ell']}(r)$, we obtain the first-order approximation $F^{(\text{in/out})(1)}$ for all elements of the matrices $F^{(\text{in/out})}$ and thus we find $\phi^{(1)}$ which includes the next terms of the expansion of the regular solution. Using this iterative procedure we can find as many terms of the expansion as needed by applying the following recurrence formulae

$$F^{(\text{in/out})(n+1)}_{[\ell][\ell']}(E, r) = \frac{1}{2}\delta_{[\ell][\ell']} \mp \frac{1}{ik}\int h_\ell^{(\pm)}(kr) \sum_{[\ell'']} V_{[\ell][\ell'']}(r)\phi^{(n)}_{[\ell''][\ell']}(E, r)\, dr , \tag{9.48}$$

$$\phi^{(n)}_{[\ell][\ell']}(E, r) = h_\ell^{(-)}(kr) F^{(\text{in})(n)}_{[\ell][\ell']}(E, r) + h_\ell^{(+)}(kr) F^{(\text{out})(n)}_{[\ell][\ell']}(E, r) . \tag{9.49}$$

Using Eqs. (9.25), (9.26), it is not difficult to transform this formulae in the form suitable for calculating the corresponding iterations for the matrices A and B,

$$A^{(0)}_{[\ell][\ell']}(E,r) = \delta_{[\ell][\ell']}\,, \qquad B^{(0)}_{[\ell][\ell']}(E,r) = 0\,, \qquad r \in (0,\delta]\,, \tag{9.50}$$

$$A^{(n+1)}_{[\ell][\ell']}(E,r) = \delta_{[\ell][\ell']} - \frac{1}{k}\int n_\ell(kr)\sum_{[\ell'']} V_{[\ell][\ell'']}(r)\phi^{(n)}_{[\ell''][\ell']}(E,r)\,dr\,, \tag{9.51}$$

$$B^{(n+1)}_{[\ell][\ell']}(E,r) = -\frac{1}{k}\int j_\ell(kr)\sum_{[\ell'']} V_{[\ell][\ell'']}(r)\phi^{(n)}_{[\ell''][\ell']}(E,r)\,dr\,, \tag{9.52}$$

$$\phi^{(n)}_{[\ell][\ell']}(E,r) = j_\ell(kr)A^{(n)}_{[\ell][\ell']}(E,r) - n_\ell(kr)B^{(n)}_{[\ell][\ell']}(E,r)\,. \tag{9.53}$$

9.3.4.2 Cancellation of Singularities Near $r = 0$

It should be noted that although the representation (9.18) of the matrix ϕ is suitable for large distances, it may be not good for numerical calculations in the vicinity of $r = 0$. Indeed, near this point the matrices $F^{(\mathrm{in})}$ and $F^{(\mathrm{out})}$ become identical and the singularities of $h^{(+)}_\ell(kr)$ and $h^{(-)}_\ell(kr)$ are cancelled. Although this does not formally cause any problem, in numerical calculations the cancellation of singularities is always a source of possible numerical errors. These errors increase with increasing ℓ since in this case $h^{(\pm)}_\ell(kr)$ becomes more singular. Therefore, for larger ℓ the point $r = \delta$ must be shifted further from the origin. This in turn requires more iterations of Eq. (9.48) to obtain the boundary values $F^{(\mathrm{in/out})}_{[\ell][\ell']}(E,\delta)$ to a required accuracy.

The representations (9.18) and (9.24) of ϕ in terms of $F^{(\mathrm{in/out})}$ and A, B are equivalent. From a practical point of view, however, when calculating the matrix $\phi(E,r)$, it is more convenient to start by integrating equations (9.27), (9.28) from the boundary values $A_{[\ell][\ell']}(E,\delta)$, $B_{[\ell][\ell']}(E,\delta)$ and proceed to some intermediate point r_{int} (far enough from the origin). Thus obtained matrices $A_{[\ell][\ell']}(E,r_{\mathrm{int}})$ and $B_{[\ell][\ell']}(E,r_{\mathrm{int}})$ can be used in the relations

$$F^{(\mathrm{in/out})}(E,r) = \frac{1}{2}\left(A \mp iB\right) \tag{9.54}$$

to find the initial values $F^{(\mathrm{in/out})}_{[\ell][\ell']}(E,r_{\mathrm{int}})$ for further integration of equations (9.20), (9.21) up to a point $r = R$, where the matrices $F^{(\mathrm{in/out})}$ reach their asymptotic values.

It may seem that we can abandon equations (9.20), (9.21) altogether and integrate equations (9.27), (9.28) on the whole interval $[\delta, R]$ instead. However, similarly to the case of a single angular momentum (central potential), the matrices $F^{(\mathrm{in})}$ and $F^{(\mathrm{out})}$ have finite limits ($r \to \infty$) in different domains of the complex k-plane (below and above the dividing line, respectively).[2] The only points where they have limits simultaneously are the spectral points and the dividing line itself. And since A and B involve both $F^{(\text{in/out})}$, they have limits only at these points. Therefore, to obtain the Jost matrix we should start at small r with Eqs. (9.27), (9.28) and finish at large R with Eqs. (9.20), (9.21).

9.3.4.3 Combination of the Second- and the First-Order Equations

There is an alternative and, perhaps, more simple way of avoiding the difficulties with the boundary conditions in numerical calculations. As we have seen, near the origin the fundamental matrix $\phi(E, r)$ is diagonal and behaves as the matrix $j(E, r)$ (see Eq. (9.33)). Matrix ϕ does not involve any diverging elements. We can start the integration of the original second-order equation (9.12) from a small distance $r = \delta$ with the condition

$$\phi(E, \delta) \approx j(E, \delta) \tag{9.55}$$

and proceed up to an intermediate point $r = r_{\text{int}}$. In this way we obtain

$$\phi(E, r_{\text{int}}) \qquad \text{and} \qquad \partial_r \phi(E, r_{\text{int}})\,. \tag{9.56}$$

Then, from this intermediate point, we can start the integration of the first-order equations (9.20), (9.21). The starting values $F^{(\text{in/out})}(E, r_{\text{int}})$ for them is not difficult to find using the regular matrix and its derivative (9.56). Indeed, using Eqs. (9.18) and (9.19), we obtain

$$\phi(E, r_{\text{int}}) = W^{(\text{in})}(E, r_{\text{int}}) F^{(\text{in})}(E, r_{\text{int}}) + W^{(\text{out})}(E, r_{\text{int}}) F^{(\text{out})}(E, r_{\text{int}}) \tag{9.57}$$

and

$$\phi'(E, r_{\text{int}}) = W'^{(\text{in})}(E, r_{\text{int}}) F^{(\text{in})}(E, r_{\text{int}}) + W'^{(\text{out})}(E, r_{\text{int}}) F^{(\text{out})}(E, r_{\text{int}})\,, \tag{9.58}$$

where "prime" means the derivative over r. Multiplying from the left Eq. (9.57) by matrix $W'^{(\text{out})}$ and Eq. (9.58) by matrix $W^{(\text{out})}$ and subtracting the resulting equations, we obtain[3]

$$W'^{(\text{out})}\phi - W^{(\text{out})}\phi' = \left[W'^{(\text{out})} W^{(\text{in})} - W^{(\text{out})} W'^{(\text{in})} \right] F^{(\text{in})}\,.$$

---

[2] This is shown in Sect. 9.5.

[3] When doing all these operations, we use the fact that the matrices $W^{(\text{in/out})}$ and their derivatives are diagonal and therefore commute with each other.

Using known Wronskian of the Riccati–Hankel functions,

$$\left[\partial_r h_\ell^{(+)}(kr)\right] h_\ell^{(-)}(kr) - h_\ell^{(+)}(kr)\left[\partial_r h_\ell^{(-)}(kr)\right] = 2ik\,,$$

we can find $F^{(\text{in})}$. Repeating similar derivations for $F^{(\text{out})}$, we finally obtain

$$\boxed{F^{(\text{in/out})} = \pm\frac{1}{2ik}\left[W'^{(\text{out/in})}\phi - W^{(\text{out/in})}\phi'\right]} \tag{9.59}$$

Equation (9.59) is valid at any r and therefore can be used for transiting from the second-order equation to the system of first-order ones at any point which is convenient for the calculations.

9.4 Asymptotic Behaviour of the Fundamental Matrix

Taking into account Eq. (9.14), we see that asymptotically the fundamental matrix is a combination of the incoming and outgoing waves

$$\phi(E,r) \underset{r\to\infty}{\longrightarrow} W^{(\text{in})}(E,r)C^{(\text{in})} + W^{(\text{out})}(E,r)C^{(\text{out})}\,, \tag{9.60}$$

where $C^{(\text{in/out})}$ are the matrices of the combination coefficients. Replacing the Riccati–Hankel functions with their asymptotic expressions (D.10), (D.9), i.e.

$$W^{(\text{in/out})}(E,r) \underset{r\to\infty}{\longrightarrow} e^{\mp ikr}\begin{bmatrix} e^{\pm i(\ell_1+1)\pi/2} & 0 & \cdots & 0 \\ 0 & e^{\pm i(\ell_2+1)\pi/2} & \cdots & 0 \\ \vdots & \vdots & \vdots & \vdots \\ 0 & 0 & \cdots & e^{\pm i(\ell_N+1)\pi/2}\end{bmatrix}, \tag{9.61}$$

we have

$$\phi(E,r) \underset{r\to\infty}{\longrightarrow} e^{-ikr}\tilde{C}^{(\text{in})} + e^{+ikr}\tilde{C}^{(\text{out})}\,, \tag{9.62}$$

where $\tilde{C}^{(\text{in/out})}$ differ from $C^{(\text{in/out})}$ by the diagonal factors on the right hand side of Eq. (9.61).

If $\operatorname{Im} kr = 0$, then both terms of the last asymptotic relation are oscillating and therefore the regular solution is bounded, i.e. for all $[\ell]$ and $[\ell']$

$$\left|\phi_{[\ell][\ell']}(E,r)\right| < \text{const}\,, \qquad \operatorname{Im} kr = 0\,. \tag{9.63}$$

For $\text{Im}\, kr \neq 0$, the regular solution diverges due to the first or the second term of (9.62). In this case we can keep only the diverging term of it, i.e.

$$\phi(E,r) \underset{r\to\infty}{\longrightarrow} e^{-ikr}\tilde{C}^{(\text{in})}\,, \qquad \text{Im}\, kr > 0\,, \tag{9.64}$$

$$\phi(E,r) \underset{r\to\infty}{\longrightarrow} e^{+ikr}\tilde{C}^{(\text{out})}\,, \qquad \text{Im}\, kr < 0\,. \tag{9.65}$$

There are, however, special discrete points k_n on the complex-momentum plane or, equivalently, the points E_n on the Riemann surface of the energy, called spectral points, whereby definition the regular solution contains only the second term of (9.62), i.e.

$$\phi(E_n,r) \underset{r\to\infty}{\longrightarrow} e^{+ik_n r}\tilde{C}^{(\text{out})}\,. \tag{9.66}$$

The spectral points k_n are sitting on the positive imaginary axis (bound states) and under the real axis (resonances), as is discussed in Sect. 3.3.7.

9.5 Jost Matrices

The natural way to generalize the definition of the Jost functions given in Sect. 2.2.3 is the following. Comparing Eqs. (9.60) and (9.18), we see that the matrices $F^{(\text{in/out})}(E,r)$ at large distances asymptotically become constant (i.e. r-independent). Their energy-dependent limits

$$f^{(\text{in})}(E) = \lim_{r\to\infty} F^{(\text{in})}(E,r)\,, \tag{9.67}$$

$$f^{(\text{out})}(E) = \lim_{r\to\infty} F^{(\text{out})}(E,r) \tag{9.68}$$

are called the Jost matrices for a non-central potential. With these matrices the asymptotic behaviour of the fundamental matrix takes the form

$$\boxed{\phi(E,r) \underset{r\to\infty}{\longrightarrow} W^{(\text{in})}(E,r)f^{(\text{in})}(E) + W^{(\text{out})}(E,r)f^{(\text{out})}(E)} \tag{9.69}$$

Due to exactly the same reasons as were discussed in Sect. 7.2, the limits (9.67), (9.68) cannot be reached along the real r-axis for an arbitrarily chosen point E on the Riemann surface of the energy.

In order to analyse the long-range behaviour of the matrix functions $F^{(\text{in/out})}(E, r)$, we re-write the system of differential equations (9.20), (9.21) as follows

$$\partial_r F^{(\text{in/out})}(E, r) = \mp \frac{1}{2ik} W^{(\text{out/in})}(E, r) V(r) \phi(E, r) \tag{9.70}$$

and make use of the asymptotic expressions (9.63), (9.64), (9.65), (9.66) of the fundamental matrix, obtained in the previous section. If we take into account the fact that the potential matrix V vanishes at infinity faster than the Coulomb potential, i.e. $V(r \to \infty) \propto r^{-(1+\varepsilon)}$, $\varepsilon > 0$, then

$$\partial_r F^{(\text{in})}(E, r) \underset{|r| \to \infty}{\propto} \begin{cases} 1/r^{1+\varepsilon} & , \ \text{Im}\, kr > 0\,, \\ \left[e^{2ikr} \tilde{C}^{(\text{out})} + \tilde{C}^{(\text{in})} \right] / r^{1+\varepsilon} & , \ \text{Im}\, kr = 0\,, \\ e^{2ikr} / r^{1+\varepsilon} & , \ \text{Im}\, kr < 0\,, \end{cases} \tag{9.71}$$

$$\partial_r F^{(\text{out})}(E, r) \underset{|r| \to \infty}{\propto} \begin{cases} e^{-2ikr} / r^{1+\varepsilon} & , \ \text{Im}\, kr > 0\,, \\ \left[\tilde{C}^{(\text{out})} + e^{-2ikr} \tilde{C}^{(\text{in})} \right] / r^{1+\varepsilon} & , \ \text{Im}\, kr = 0\,, \\ 1/r^{1+\varepsilon} & , \ \text{Im}\, kr < 0\,, \end{cases} \tag{9.72}$$

and at all the spectral points

$$\partial_r F^{(\text{in})}(E_n, r) \underset{|r| \to \infty}{\propto} e^{2ik_n r} / r^{1+\varepsilon}\,, \tag{9.73}$$

$$\partial_r F^{(\text{out})}(E_n, r) \underset{|r| \to \infty}{\propto} 1/r^{1+\varepsilon}\,. \tag{9.74}$$

As we see, the matrix $F^{(\text{in})}(E, r)$ does not converge to a finite limit if $\text{Im}\, kr < 0$, while for $F^{(\text{out})}(E, r)$ the same problem takes place when $\text{Im}\, kr > 0$, because their derivatives grow exponentially in these domains. On the real axis, as well as in the mirror-symmetric domains ($\text{Im}\, kr > 0$ for $F^{(\text{in})}$ and $\text{Im}\, kr < 0$ for $F^{(\text{out})}$), these derivatives vanish rapidly enough to guarantee the existence of the limits. Indeed, if we consider a function $\varphi(r)$, whose first derivative asymptotically behaves as

$$\frac{d\varphi}{dr} \underset{r \to \infty}{\sim} \frac{1}{r^{1+\varepsilon}}\,, \qquad \varepsilon > 0\,,$$

then this function has a finite limit,

$$\varphi(r) \underset{r\to\infty}{\sim} \int \frac{1}{r^{1+\varepsilon}} dr = \left(\text{const} - \frac{1}{\varepsilon r^{\varepsilon}}\right) \underset{r\to\infty}{\longrightarrow} \text{const} .$$

Just on the dividing line, Im $kr = 0$, which separates the mirror-symmetric domains, both derivatives oscillate with vanishing amplitudes,

$$\partial_r F^{(\text{in/out})}(E, r) \underset{|r|\to\infty}{\propto} \frac{\exp(\pm 2ikr)\tilde{C}^{(\text{out/in})} + \tilde{C}^{(\text{in/out})}}{r^{1+\varepsilon}} , \tag{9.75}$$

and therefore

$$F^{(\text{in/out})}(E, r) = F^{(\text{in/out})}(E, R) + \text{const} \int_R^r \frac{\exp(\pm 2ik\rho)\tilde{C}^{(\text{out/in})} + \tilde{C}^{(\text{in/out})}}{\rho^{1+\varepsilon}} d\rho , \tag{9.76}$$

where R is a sufficiently large radius beyond which the asymptotic behaviour (9.75) of the derivative can be used. Apparently, the integral in (9.76) converges, because

$$\int_R^r \left| \frac{\exp(\pm 2ik\rho)}{\rho^{1+\varepsilon}} \right| d\rho = \frac{1}{\varepsilon R^{\varepsilon}} - \frac{1}{\varepsilon r^{\varepsilon}} .$$

At the spectral points, the behaviour of the derivatives is given by Eqs. (9.73), (9.74), which implies that $F^{(\text{out})}(E, r)$ has finite limits at all of them. The matrix $F^{(\text{in})}(E, r)$, however, converges only at those points which are above or on the dividing line.

What we found here about the domains where the Jost matrices have finite limits is actually the same as what was established earlier in Sect. 7.2 for the Jost functions associated with spherically symmetric potentials. We therefore can use the same graphical representation of the domains, which is given in Fig. 7.1. In the same way as we did it in Sect. 7.2.2, it is not difficult to show that for an exponentially falling potential, the domains overlap as shown in Fig. 7.2, where β is the smallest of all the decay constants for the matrix elements of the potential (in most cases all the matrix elements have the same β).

9.5.1 Complex Rotation

When considering the single-ℓ case (spherically symmetric potential), in Sect. 7.4 we found a way to analytically continue the Jost function $f_\ell^{(\text{in})}(E)$ to the domain below the dividing line, and the function $f_\ell^{(\text{out})}(E)$ to the symmetric domain above that line. The trick consisted in using the deformed path for integrating the corresponding differential equations. Examples of such a path are shown in Figs. 7.3 and 7.7.

Coming back to the reasoning given in Sect. 7.4, we see that everything remains the same if we just replace the Jost functions $f_\ell^{(\text{in/out})}(E)$ with the corresponding matrices $f^{(\text{in/out})}(E)$. In other words, without repeating all the derivations, we can state here that if the integration of the differential equations (9.20), (9.21) is done along an appropriate path in the complex r-plane, the limits (9.67), (9.68) exist in the domains shown in Figs. 7.4 and 7.5.[4] Furthermore, if the elements of the potential matrix $V(r)$ vanish at infinity exponentially, then the dividing line is replaced with a band of the width β, as is shown in Fig. 7.6. It may happen that the matrix elements of V have different parameters β determining their rates of diminishing at large distances. In such a case, β is the smallest of the decay constants for all matrix elements of V.

The conclusions of Sect. 7.6, summarized in Fig. 7.9, apparently remain valid for the Jost matrices as well. However, not everything that was established for the Jost functions with a single angular momentum can be directly generalized for the Jost matrices corresponding to a non-central interaction. For example, the analytic structure (7.65) of the "simple" Jost functions established in Sect. 7.7 becomes more complicated for the Jost matrices. We will come back to it in Sect. 13.4 (see Eq. (13.26)).

9.6 Jost Solutions

In Sect. 2.2.4, we introduced the so-called Jost solutions, $\chi_\ell^{(\pm)}(E, r)$, of the radial Schrödinger equation by requiring that asymptotically they behave exactly as the Riccati–Hankel functions (see Eq. (2.67)). They are linearly independent and constitute a complete basis in the space of solutions of the Schrödinger equation.

Moving on to the more general case of coupled Schrödinger equations (9.2), we need $2N$ linearly independent column-solutions in order to have a complete basis for this system of differential equations. The solutions can be chosen in such a way that asymptotically they behave as the following columns (there are $2N$ of them):

$$\begin{bmatrix} h_{\ell_1}^{(-)}(kr) \\ 0 \\ \vdots \\ 0 \end{bmatrix}, \begin{bmatrix} 0 \\ h_{\ell_2}^{(-)}(kr) \\ \vdots \\ 0 \end{bmatrix}, \cdots \begin{bmatrix} 0 \\ 0 \\ \vdots \\ h_{\ell_N}^{(-)}(kr) \end{bmatrix}, \tag{9.77}$$

[4] A demonstration of the effectiveness and accuracy of such an approach in numerical calculations can be found in Ref. [36].

and

$$\begin{bmatrix} h_{\ell_1}^{(+)}(kr) \\ 0 \\ \vdots \\ 0 \end{bmatrix}, \begin{bmatrix} 0 \\ h_{\ell_2}^{(+)}(kr) \\ \vdots \\ 0 \end{bmatrix}, \dots, \begin{bmatrix} 0 \\ 0 \\ \vdots \\ h_{\ell_N}^{(+)}(kr) \end{bmatrix}. \tag{9.78}$$

These solutions can be combined in two, $(\pm)$, square matrices with the asymptotic behaviour

$$\chi^{(\pm)}(E,r) \underset{r\to\infty}{\longrightarrow} \begin{bmatrix} h_{\ell_1}^{(\pm)}(kr) & 0 & 0 \cdots & 0 \\ 0 & h_{\ell_2}^{(\pm)}(kr) & 0 \cdots & 0 \\ \vdots & \vdots & \vdots\ \vdots & \vdots \\ 0 & 0 & 0 \cdots & h_{\ell_N}^{(\pm)}(kr) \end{bmatrix}. \tag{9.79}$$

If we look for them as

$$\chi^{(\pm)}(E,r) = W^{(\text{in})}(E,r)\mathfrak{X}^{(\pm\text{in})}(E,r) + W^{(\text{out})}(E,r)\mathfrak{X}^{(\pm\text{out})}(E,r) \tag{9.80}$$

with the Lagrange condition

$$W^{(\text{in})}(E,r)\partial_r\mathfrak{X}^{(\pm\text{in})}(E,r) + W^{(\text{out})}(E,r)\partial_r\mathfrak{X}^{(\pm\text{out})}(E,r) = 0\,, \tag{9.81}$$

then, similarly to the regular solution, the unknown matrix functions $\mathfrak{X}^{(\pm\text{in}/\text{out})}(E,r)$ can be obtained using the differential equations

$$\partial_r\mathfrak{X}^{(\pm\text{in})} = -\frac{1}{2ik}W^{(\text{out})}V\left[W^{(\text{in})}\mathfrak{X}^{(\pm\text{in})} + W^{(\text{out})}\mathfrak{X}^{(\pm\text{out})}\right], \tag{9.82}$$

$$\partial_r\mathfrak{X}^{(\pm\text{out})} = \frac{1}{2ik}W^{(\text{in})}V\left[W^{(\text{in})}\mathfrak{X}^{(\pm\text{in})} + W^{(\text{out})}\mathfrak{X}^{(\pm\text{out})}\right], \tag{9.83}$$

that are the same as Eqs. (9.20), (9.21) but with different boundary conditions, namely with the following conditions at infinity:

$$\mathfrak{X}^{(+\text{in})}(E,r) \underset{r\to\infty}{\longrightarrow} 0\,, \qquad \mathfrak{X}^{(+\text{out})}(E,r) \underset{r\to\infty}{\longrightarrow} I\,,$$

$$\mathfrak{X}^{(-\text{in})}(E,r) \underset{r\to\infty}{\longrightarrow} I\,, \qquad \mathfrak{X}^{(-\text{out})}(E,r) \underset{r\to\infty}{\longrightarrow} 0\,.$$

The notion of Wronskian can be generalized for the case when the functions involved in it are matrices. In such a case the first matrix in the Wronskian should be transposed (see Appendix C). The matrix radial Schrödinger equation, (9.2), for coupled partial waves has the form (C.7) where the only non-diagonal part of the matrix P is the potential which in most cases is symmetric (see Eq. (9.5)). Therefore the Wronskian of any two solutions of this equation (if the potential is symmetric) is r-independent and can be calculated at any convenient point (usually it is taken either at zero or at infinity).

For the diagonal matrices (9.61) the Wronskians are easy to find,

$$W\left[W^{(\mathrm{in})}, W^{(\mathrm{out})}\right] = 2ikI\,, \qquad W\left[W^{(\mathrm{in})}, W^{(\mathrm{in})}\right] = W\left[W^{(\mathrm{out})}, W^{(\mathrm{out})}\right] = 0\,,$$

where I is the unit matrix. Then, using the following Wronskians (which can be calculated at large distances):

$$\begin{aligned} W\left[\chi^{(+)}, \phi\right] &= W\left[W^{(\mathrm{out})}, W^{(\mathrm{in})} f^{(\mathrm{in})} + W^{(\mathrm{out})} f^{(\mathrm{out})}\right] \\ &= W\left[W^{(\mathrm{out})}, W^{(\mathrm{in})}\right] f^{(\mathrm{in})} = -2ikf^{(\mathrm{in})} \end{aligned}$$

and

$$\begin{aligned} W\left[\chi^{(-)}, \phi\right] &= W\left[W^{(\mathrm{in})}, W^{(\mathrm{in})} f^{(\mathrm{in})} + W^{(\mathrm{out})} f^{(\mathrm{out})}\right] \\ &= W\left[W^{(\mathrm{in})}, W^{(\mathrm{out})}\right] f^{(\mathrm{out})} = 2ikf^{(\mathrm{out})}\,, \end{aligned}$$

we can generalize the relation (2.75),

$$\boxed{f^{(\mathrm{in/out})}(E) = \mp\frac{1}{2ik} W\left[\chi^{(\pm)}(E, r), \phi(E, r)\right]} \tag{9.84}$$

It is straightforward to generalize also the relations (2.80) and (2.81), which in the matrix form read

$$W\left[\chi^{(-)}, \chi^{(+)}\right] = 2ikI\,, \quad W\left[\chi^{(+)}, \chi^{(+)}\right] = W\left[\chi^{(-)}, \chi^{(-)}\right] = 0\,, \tag{9.85}$$

and

$$\boxed{\phi(E, r) = \chi^{(-)}(E, r) f^{(\mathrm{in})}(E) + \chi^{(+)}(E, r) f^{(\mathrm{out})}(E)} \tag{9.86}$$

9.6.1 Jost Solutions Near $r = 0$

In the single-channel problems involving either neutral or charged particles, we found the relation allowing to obtain the Jost functions from the behaviour of the Jost solutions near $r = 0$ (see Eqs. (2.78) and (8.84)). Here it is shown that there is no similar simple way to relate the Jost matrices with the behaviour of the Jost solutions near the origin. In other words, the definition of the Jost matrices via Eq. (2.78) is not universal. When the dimension, N, of the Jost matrices is greater than 1, the only clear and unambiguous way of defining them is via the behaviour of the regular solution at large distances (see Eq. (9.69)).

In order to make all the formulae simple, consider the case of $N = 2$, i.e. when only two partial waves are coupled to each other. In general, there are $2N$ linearly independent column-solutions of the radial Schrödinger equation. Half of them can be chosen regular at $r = 0$. However, we can always choose the solutions in such a way that all $2N$ columns are irregular.

The two columns of the matrix ϕ are regular by construction (see Sect. 9.3.4),

$$\phi(E,r) \underset{r\to 0}{\longrightarrow} \begin{bmatrix} \dfrac{(kr)^{\ell_1+1}}{(2\ell_1+1)!!} & o(r^{\ell_2+1}) \\ o(r^{\ell_1+1}) & \dfrac{(kr)^{\ell_2+1}}{(2\ell_2+1)!!} \end{bmatrix} . \tag{9.87}$$

These two columns have exhausted the maximum number, ($N = 2$), of linearly independent regular column-solutions, i.e. there are no other independent regular columns. Since the Wronskians (9.84) are non-zero, the four columns of the matrices $\chi^{(+)}(E,r)$ and $\chi^{(-)}(E,r)$ are linearly independent of the columns of ϕ. This means that all the columns of $\chi^{(\pm)}(E,r)$ are irregular at $r = 0$.

We consider non-singular potentials, which implies that near the origin the centrifugal potential dominates over $V(r)$. As a result the radial differential equations decouple when $r \to 0$. The decoupled equations can be treated separately. As we have established in Sect. 2.1, all the irregular solutions of such equations near the origin are proportional to $1/r^{\ell}$ (see Eq. (2.6)). Therefore

$$\chi^{(\pm)}(E,r) \underset{r\to 0}{\longrightarrow} X^{(\pm)}(E) \begin{bmatrix} (kr)^{-\ell_1} & o(r^{-\ell_2}) \\ o(r^{-\ell_1}) & (kr)^{-\ell_2} \end{bmatrix} , \tag{9.88}$$

where $X^{(\pm)}(E)$ are some (unknown) r-independent diagonal matrices, defined as the limits:

$$X^{(\perp)}(E) = \lim_{r\to 0} \chi^{(\pm)}(E,r) \begin{bmatrix} (kr)^{\ell_1} & 0 \\ 0 & (kr)^{\ell_2} \end{bmatrix} = \begin{bmatrix} X_1^{(+)} & 0 \\ 0 & X_2^{(\pm)} \end{bmatrix} . \tag{9.89}$$

The Wronskian of Eq. (9.84) near the origin can be written as:[5]

$$W\left[\chi^{(\pm)},\phi\right]=\chi^{(\pm)\,T}\partial_r\phi-\left[\partial_r\chi^{(\pm)\,T}\right]\phi$$

$$\underset{r\to 0}{\longrightarrow}\begin{bmatrix}(kr)^{-\ell_1}X_1^{(\pm)} & o(r^{-\ell_1})\\ o(r^{-\ell_2}) & (kr)^{-\ell_2}X_2^{(\pm)}\end{bmatrix}\begin{bmatrix}\dfrac{(\ell_1+1)k(kr)^{\ell_1}}{(2\ell_1+1)!!} & o(r^{\ell_2})\\ o(r^{\ell_1}) & \dfrac{(\ell_2+1)k(kr)^{\ell_2}}{(2\ell_2+1)!!}\end{bmatrix}$$

$$-\begin{bmatrix}-\ell_1k(kr)^{-\ell_1-1}X_1^{(\pm)} & o(r^{-\ell_1-1})\\ o(r^{-\ell_2-1}) & -\ell_2k(kr)^{-\ell_2-1}X_2^{(\pm)}\end{bmatrix}\begin{bmatrix}\dfrac{(kr)^{\ell_1+1}}{(2\ell_1+1)!!} & o(r^{\ell_2+1})\\ o(r^{\ell_1+1}) & \dfrac{(kr)^{\ell_2+1}}{(2\ell_2+1)!!}\end{bmatrix}.$$

As is seen, the matrix multiplications in the last equation produce the unknown terms of the type

$$o(r^{\ell_n})\cdot o(r^{-\ell_{n'}})\underset{r\to 0}{\longrightarrow}?\,,$$

which can be zero or infinite. This demonstrates that for the case $N>1$ it is not possible to generalize the single-channel relation (2.78) and to define the Jost matrices via the behaviour of the Jost solutions at the origin.

9.7 Physical Solutions

When looking for the fundamental matrix of regular solutions, ϕ, we solve the Cauchy problem for the Schrödinger equation. This means that ϕ is determined by the boundary conditions at a single point, namely at $r=0$. In doing this, we do not care what happens to this matrix at infinity. Each column of the matrix ϕ is regular and independent solution of the Schrödinger equation. Considered together, the columns of this matrix constitute a complete basis in the space of the regular solutions. Therefore any physical solution, $u(E,r)$, can be presented as a linear combination of these columns

$$u(E,r)=\phi(E,r)c\,, \tag{9.90}$$

where the matrix of the combination coefficients, c, is to be chosen in such a way that proper behaviour of $u(E,r)$ at infinity is guaranteed.

[5] It should be taken note of the fact that in the Wronskian of matrix functions the first matrix is transposed (see Appendix C).

There are two different types of the physical solutions: the discrete and continuous spectrum solutions. For the discrete spectrum (bound and resonant) states the physical wave function is a single column, each element of which describes the motion in a separate partial wave, $[\ell]$. A single column is obtained from the square matrix ϕ with the help of a single column of the combination coefficients,

$$u_{[\ell]}(E, r) = \sum_{[\ell']} \phi_{[\ell][\ell']}(E, r) c_{[\ell']} \,. \tag{9.91}$$

The physical wave function of the continuous state is a square matrix, where each element, $u_{[\ell][\ell']}$, describes the scattering that begins in the partial wave $[\ell']$ and ends up in the wave $[\ell]$. In order to construct such a wave function, we have to use a square matrix of the combination coefficients,

$$u_{[\ell][\ell']}(E, r) = \sum_{[\ell'']} \phi_{[\ell][\ell'']}(E, r) c_{[\ell''][\ell']} \,. \tag{9.92}$$

9.7.1 Bound States and Resonances

In the case of a central potential the spectral points, E_n, are defined by requiring that at large distances the physical solution at such an energy behaves as pure outgoing spherical wave (see Sect. 3.3). In this simple case the physical solution and the regular one differ by a normalization factor. Therefore to get rid of the incoming wave it is sufficient to require that the incoming Jost function is zero (see Eq. (3.21)).

In a more complicated case of N coupled partial waves, we cannot require that the first term in the asymptotic behaviour (9.69) of the matrix $\phi(E_n, r)$ completely disappears. Indeed, such a requirement would mean that all the elements of the matrix $f^{(\text{in})}(E_n)$ are zero and thus we would have N^2 simultaneous equations determining just one number E_n. Actually, what we need is that the physical solution, $u(E_n, r)$, (but not the matrix ϕ) asymptotically has only outgoing waves. This can be achieved if it is possible to find such a set of the combination coefficients $c_{[\ell]}$ in Eq. (9.91) that the contribution to the asymptotic behaviour of $u(E_n, r)$ from the first term of (9.69) vanishes, i.e.

$$W^{(\text{in})}(E_n, r) f^{(\text{in})}(E_n) c = 0 \,. \tag{9.93}$$

This condition implies that

$$\sum_{[\ell]} f^{(\text{in})}_{[\ell][\ell']}(E_n) c_{[\ell']} = 0 \,, \tag{9.94}$$

which is a system of homogeneous linear equations for the unknown variables $c_{[\ell]}$. It has a non-trivial solution if and only if the determinant of the matrix of its coefficients is zero,

$$\boxed{\det f^{(\mathrm{in})}(E_n) = 0} \tag{9.95}$$

It is clear that this equation generalizes the more simple requirement (3.21). The roots of Eq. (9.95) are the spectral points. Although at such a point the asymptotics (9.69) of the regular solution ϕ has both the incoming and outgoing waves, the Jost matrix consists of such elements that it is possible to construct the physical solution without the incoming waves at large distances.

The bound-state spectral points are on the real negative axis, $E_n < 0$, which is on the physical Riemann sheet, where the corresponding momenta have positive imaginary parts, i.e.

$$k_n = \sqrt{2\mu E_n/\hbar^2} = i\kappa_n\,, \qquad \kappa_n > 0\,.$$

The corresponding wave functions vanish at large distances exponentially,

$$u_{[\ell]}(E_n, r) \underset{r\to\infty}{\longrightarrow} \sum_{[\ell']} \left\{ h_\ell^{(-)}(k_n r) f^{(\mathrm{in})}_{[\ell][\ell']}(E_n) + h_\ell^{(+)}(k_n r) f^{(\mathrm{out})}_{[\ell][\ell']}(E_n) \right\} c_{[\ell']} \tag{9.96}$$

$$\underset{r\to\infty}{\longrightarrow} \mathcal{A}_{[\ell]} e^{-\kappa_n r} \underset{r\to\infty}{\longrightarrow} 0\,,$$

where $\mathcal{A}_{[\ell]}$ is the asymptotic normalization constant in the partial wave $[\ell]$. It is easy to see that

$$\mathcal{A}_{[\ell]} = \frac{1}{i^{\ell+1}} \sum_{[\ell']} f^{(\mathrm{out})}_{[\ell][\ell']}(E_n) c_{[\ell']}\,. \tag{9.97}$$

Therefore, we can locate all possible bound stated by looking for zeros E_n of the Jost matrix determinant on the real negative axis of the physical sheet of the Riemann surface (see Fig. 3.10). For each zero thus found, the linear system (9.94) uniquely determines the coefficients $c_{[\ell]}$ apart from a general normalization factor, which is finally fixed when the physical wave function

$$\psi_{E_n}(\vec{r}\,) = \frac{1}{r} \sum_{[\ell]} Y_{[\ell]}(\theta, \varphi) u_{[\ell]}(E_n, r)\,, \tag{9.98}$$

where

$$u_{[\ell]}(E_n, r) = \sum_{[\ell']} \left\{ h_\ell^{(-)}(k_n r) F^{(\mathrm{in})}_{[\ell][\ell']}(E_n, r) + h_\ell^{(+)}(k_n r) F^{(\mathrm{out})}_{[\ell][\ell']}(E_n, r) \right\} c_{[\ell']}\,, \tag{9.99}$$

is normalized. The normalization integral

$$\int |\psi_{E_n}(\vec{r})|^2 \, d^3r = \int_0^\infty \left\{ |u_{[\ell_1]}|^2 + |u_{[\ell_2]}|^2 + \cdots + |u_{[\ell_N]}|^2 \right\} dr = 1 \tag{9.100}$$

consists of N terms. Each of these terms,

$$P(\ell_i) = \int_0^\infty |u_{[\ell_i]}|^2 dr \,, \qquad i = 1, 2, \ldots, N \,, \tag{9.101}$$

is a quantity that is usually called the percentage of the i-th partial wave. It characterizes the contribution of a particular partial wave to the bound state wave function. Intuitively, this quantity should be understood as the percentage of time that the bound system spends in the state with the angular momentum ℓ_i.

The resonances, like the bound states, are also the spectral points. This means that (by definition) the corresponding physical solutions, $u(E_n, r)$, at large distances are composed of only the outgoing spherical waves $W^{(\mathrm{out})}(E_n, r)$. It is not surprising that this leads us to exactly the same condition (9.94) as for the bound states. And the resonance energies, E_n, are determined by the same equation (9.95). The only difference is that the resonance spectral points are complex and located on the unphysical sheet of the Riemann surface (see Fig. 3.10). The corresponding wave function asymptotically behaves as

$$u_{[\ell]}(E_n, r) \underset{r\to\infty}{\longrightarrow} \mathcal{A}_{[\ell]} e^{ik_n r} \,, \tag{9.102}$$

where $\mathcal{A}_{[\ell]}$ is given by the same Eq. (9.97).

As is discussed in Sect. 3.3.4 the resonance wave function can be normalized to unity. In order to do this, we need to replace the absolute squares of the partial-wave components, $|u_{[\ell_i]}|^2$, in the integral (9.100) with the simple squares, $u_{[\ell_i]}^2$, and the integration should be done along the ray in the complex r-plane depicted in Fig. 3.9. The radial components of the bound-state wave function can always be made real (because the Schrödinger equation as well as the boundary conditions for them are real) and therefore their absolute squares in the normalization integral can, in principle, be replaced with the simple squares as well. In other words, the integration of the simple squares can be considered as a general rule applicable for both the bound and resonant states.

9.7.2 Scattering States

As was mentioned in Sect. 9.1.2, the scattering-state wave function depends on two vectors: $\vec{k}$ that indicates the direction where the projectiles come from, and $\vec{r}$ that points in the direction of observation (the scattering direction). The radial wave function, $u_{[\ell][\ell']}(E, r)$, has therefore two subscripts: the right-hand one corresponds to the incoming partial wave, and the left hand one indicates the quantum numbers of the outgoing partial wave. Since the transitions are possible among all N partial waves $[\ell_1], [\ell_2], \ldots, [\ell_N]$, the radial wave function $u(E, r)$ is not a single column, but an $(N \times N)$-matrix that can give all the transition amplitudes.

Each column of the physical wave function, $u_{[\ell][\ell']}(E, r)$, can be constructed as a linear combination of the columns of the fundamental matrix, ϕ. Of course the sets of the combination coefficients are different for different columns of u. This is why in Eq. (9.11) the combination coefficients $c_{[\ell'][\ell]}$ have the second subscript $[\ell]$ that indicates which column of u is constructed. We therefore need to determine a square matrix of the combination coefficients which build up a physical solution with proper asymptotic behaviour.

* * *

But before doing this, let us determine the normalization constant $N_{[\ell]}(k)$ in Eq. (9.8). We assume (as we do throughout this book) that the total (three-dimensional) wave function is normalized to the δ-function,

$$\langle \psi_{\vec{k}} | \psi_{\vec{k}'} \rangle = \delta(\vec{k} - \vec{k}') . \tag{9.103}$$

It is clear that the coefficient $N_{[\ell]}(k)$ is the same for any choice of the potential. In particular, it is the same if $V \equiv 0$, i.e. when the wave function

$$\langle \vec{r} \, | \psi_{\vec{k}} \rangle = \frac{e^{i\vec{k}\vec{r}}}{(2\pi)^{3/2}}$$

is the plane wave normalized according to (9.103). As is seen from Eq. (9.45), if $V \equiv 0$, then

$$F^{(\text{in/out})}_{[\ell][\ell']}(E, r) \equiv \frac{1}{2}\delta_{[\ell][\ell']} , \qquad \text{no potential} , \tag{9.104}$$

and therefore the regular basis is

$$\phi_{[\ell][\ell']}(E, r) = \frac{1}{2}\left[h^{(-)}_{\ell}(kr) + h^{(+)}_{\ell}(kr)\right]\delta_{[\ell][\ell']} . \tag{9.105}$$

Using Eq. (4.3) for the plane wave,

$$\frac{e^{i\vec{k}\vec{r}}}{(2\pi)^{3/2}} = \frac{1}{\sqrt{2\pi} kr}\sum_{\ell m} Y_{\ell m}(\hat{\vec{r}}\,)\left[h^{(-)}_{\ell}(kr) + h^{(+)}_{\ell}(kr)\right] i^{\ell} Y^{*}_{\ell m}(\hat{\vec{k}}\,) , \tag{9.106}$$

and compare it with Eq. (9.8), we see that for the case of no potential

$$N_{[\ell]}(k) = \frac{i^{\ell}}{\sqrt{2\pi} k} \qquad \text{and} \qquad c_{[\ell'][\ell]} = 2\delta_{[\ell'][\ell]} .$$

As it was said before, the normalization coefficient $N_{[\ell]}(k)$ does not depend on the potential and therefore is the same for any potential. In other words, with such a coefficient the scattering wave function is always normalized to the δ-function as is given by Eq. (9.103).

The combination coefficients $c_{[\ell'][\ell]}$ are of course different for different potentials. Let us choose them as follows

$$c_{[\ell'][\ell]} = \left[f^{(\mathrm{in})}(E)\right]^{-1}_{[\ell'][\ell]} \tag{9.107}$$

and prove that such a choice always gives the correct asymptotic behaviour of the total wave function. In doing this, we follow the same steps as in Sect. 4.1 for the single-ℓ case. To simplify the notation, let us introduce the following diagonal matrix

$$\mathcal{I} = \begin{pmatrix} i^{\ell_1} & 0 & \cdots & 0 \\ 0 & i^{\ell_2} & \cdots & 0 \\ \vdots & \vdots & \vdots & \vdots \\ 0 & 0 & \cdots & i^{\ell_N} \end{pmatrix} \tag{9.108}$$

and the single-row matrix

$$Y(\hat{\vec{r}}) = \left[Y_{[\ell_1]}(\hat{\vec{r}}),\ Y_{[\ell_2]}(\hat{\vec{r}}),\ \cdots,\ Y_{[\ell_N]}(\hat{\vec{r}})\right]. \tag{9.109}$$

Then the scattering wave function (9.8) can be written as

$$\begin{aligned}
\psi_{\vec{k}}(\vec{r}) &= \frac{1}{\sqrt{2\pi}kr} Y(\hat{\vec{r}})u(E,r)\mathcal{I}Y^{\dagger}(\hat{\vec{k}}) = \frac{1}{\sqrt{2\pi}kr} Y(\hat{\vec{r}})\phi(E,r)c(E)\mathcal{I}Y^{\dagger}(\hat{\vec{k}}) \\
&= \frac{1}{\sqrt{2\pi}kr} Y(\hat{\vec{r}})\Big[W^{(\mathrm{in})}(E,r)F^{(\mathrm{in})}(E,r) + \\
&\quad + W^{(\mathrm{out})}(E,r)F^{(\mathrm{out})}(E,r)\Big]\left[f^{(\mathrm{in})}(E)\right]^{-1}\mathcal{I}Y^{\dagger}(\hat{\vec{k}})\,.
\end{aligned}$$

At large distances the functions $F^{(\mathrm{in/out})}(E,r)$ converge to the corresponding Jost matrices.[6] Therefore

$$\begin{aligned}
\psi_{\vec{k}}(\vec{r}) \underset{r\to\infty}{\longrightarrow} \frac{1}{\sqrt{2\pi}kr} Y(\hat{\vec{r}})\Big\{W^{(\mathrm{in})}(E,r) \\
+ W^{(\mathrm{out})}(E,r)f^{(\mathrm{out})}(E)\left[f^{(\mathrm{in})}(E)\right]^{-1}\Big\}\mathcal{I}Y^{\dagger}(\hat{\vec{k}})\,.
\end{aligned}$$

[6] Scattering energies are on the real positive axis and thus the limits (9.67), (9.68) always exist.

If we define the S-matrix as

$$\boxed{S(E) = f^{(\text{out})}(E)\left[f^{(\text{in})}(E)\right]^{-1}} \tag{9.110}$$

then

$$\psi_{\vec{k}}(\vec{r}\,) \underset{r\to\infty}{\longrightarrow} \frac{1}{\sqrt{2\pi k r}} Y(\hat{\vec{r}}\,)\left[W^{(\text{in})}(E,r) + W^{(\text{out})}(E,r)S(E)\right]\mathcal{I}Y^{\dagger}(\hat{\vec{k}}\,)\,. \tag{9.111}$$

Adding and subtracting the matrix $W^{(\text{out})}$, we can split the right hand side of (9.111) in the two terms

$$\psi_{\vec{k}}(\vec{r}\,) \underset{r\to\infty}{\longrightarrow} \frac{1}{\sqrt{2\pi k r}} Y(\hat{\vec{r}}\,)\left[W^{(\text{in})}(E,r) + W^{(\text{out})}(E,r)\right]\mathcal{I}Y^{\dagger}(\hat{\vec{k}}\,)$$
$$+\frac{1}{\sqrt{2\pi k r}} Y(\hat{\vec{r}}\,)W^{(\text{out})}(E,r)\left[S(E)-1\right]\mathcal{I}Y^{\dagger}(\hat{\vec{k}}\,)\,.$$

The first term is the plane wave (9.106). Then the second term should be the scattering part of the wave function. At large distances, we have[7]

$$W^{(\text{out})}(E,r) \underset{r\to\infty}{\longrightarrow} \begin{bmatrix} \frac{e^{i(kr-\frac{\ell_1\pi}{2})}}{i} & 0 & \cdots & 0 \\ 0 & \frac{e^{i(kr-\frac{\ell_2\pi}{2})}}{i} & \cdots & 0 \\ \vdots & \vdots & \vdots & \vdots \\ 0 & 0 & \cdots & \frac{e^{i(kr-\frac{\ell_N\pi}{2})}}{i} \end{bmatrix} = \frac{1}{i}e^{ikr}\mathcal{I}^{\dagger}\,, \tag{9.112}$$

and therefore

$$\psi_{\vec{k}}(\vec{r}\,) \underset{r\to\infty}{\longrightarrow} \frac{e^{i\vec{k}\vec{r}}}{(2\pi)^{3/2}} + \sqrt{\frac{2}{\pi}}\frac{e^{ikr}}{r}\frac{1}{2ik}Y(\hat{\vec{r}}\,)\mathcal{I}^{\dagger}\left[S(E)-1\right]\mathcal{I}Y^{\dagger}(\hat{\vec{k}}\,)\,. \tag{9.113}$$

Generalizing Eq. (4.20), we introduce the matrix of the partial-wave scattering amplitudes

$$f(E) = \frac{1}{2ik}\mathcal{I}^{\dagger}\left[S(E)-1\right]\mathcal{I}\,, \tag{9.114}$$

[7] This is because $i^{\ell} = \exp(i\pi\ell/2)$.

which allows us to re-write the asymptotic form of the wave function as

$$\psi_{\vec{k}}(\vec{r}\,) \underset{r\to\infty}{\longrightarrow} \frac{e^{i\vec{k}\vec{r}}}{(2\pi)^{3/2}} + \sqrt{\frac{2}{\pi}}\frac{e^{ikr}}{r} Y(\hat{\vec{r}}\,) f(E) Y^{\dagger}(\hat{\vec{k}}\,)$$

$$= \frac{1}{(2\pi)^{3/2}}\left[e^{i\vec{k}\vec{r}} + \frac{e^{ikr}}{r} 4\pi Y(\hat{\vec{r}}\,) f(E) Y^{\dagger}(\hat{\vec{k}}\,)\right]$$

$$= \frac{1}{(2\pi)^{3/2}}\left[e^{i\vec{k}\vec{r}} + \frac{e^{ikr}}{r} \mathsf{f}(\hat{\vec{r}} \leftarrow \hat{\vec{k}}, E)\right], \tag{9.115}$$

where the full (not decomposed in the partial waves) amplitude is

$$\mathsf{f}(\hat{\vec{r}} \leftarrow \hat{\vec{k}}, E) = 4\pi Y(\hat{\vec{r}}\,) f(E) Y^{\dagger}(\hat{\vec{k}}\,) = 4\pi \sum_{[\ell'][\ell]} Y_{[\ell']}(\hat{\vec{r}}\,) f_{[\ell'][\ell]}(E) Y^{*}_{[\ell]}(\hat{\vec{k}}\,)\,. \tag{9.116}$$

As is seen, the asymptotic behaviour (9.115) of the scattering wave function is correct. It should be understood that this is because of the choice (9.107) of the combination coefficients.

* * *

Summarizing what was achieved so far: we have established that the combination coefficients (9.107), taken as the elements of the inverted Jost matrix $f^{(\mathrm{in})}(E)$, allow us to construct the wave function

$$\psi_{\vec{k}}(\vec{r}\,) = \frac{1}{\sqrt{2\pi}kr} Y(\hat{\vec{r}}\,)\Big[W^{(\mathrm{in})}(E,r)F^{(\mathrm{in})}(E,r) \tag{9.117}$$

$$+ W^{(\mathrm{out})}(E,r)F^{(\mathrm{out})}(E,r)\Big]\Big[f^{(\mathrm{in})}(E)\Big]^{-1} \mathcal{I} Y^{\dagger}(\hat{\vec{k}}\,)$$

that has proper normalization (9.103) and proper asymptotic behaviour (9.115), namely as the incoming plane wave and the scattered spherical wave. We also found how the scattering amplitude can be calculated when the Jost matrices are known, namely using Eqs. (9.116), (9.114), and (9.110). All the formulae we obtained here are easily transformed to more simple expressions derived in Sect. 4 for the case of a central potential.

* * *

It is shown in Sect. 1.3.2 that the differential cross section can be found as the square of the absolute value of the amplitude,

$$\frac{d\sigma}{d\Omega} = \left|\mathsf{f}(\hat{\vec{r}} \leftarrow \hat{\vec{k}}, E)\right|^2 . \tag{9.118}$$

As is seen in Eq. (9.116), the amplitude f is the product of a row and a column matrices, and thus is just a complex number, i.e. a trivial (1×1)-matrix. For such a matrix, the complex and Hermitian conjugations are equivalent. Therefore

$$\frac{d\sigma}{d\Omega} = \mathsf{f}^*\mathsf{f} = \mathsf{f}^\dagger\mathsf{f} = 16\pi^2 Y(\hat{\vec{k}}) f^\dagger(E) Y^\dagger(\hat{\vec{r}}) Y(\hat{\vec{r}}) f(E) Y^\dagger(\hat{\vec{k}}) . \tag{9.119}$$

Vector $\vec{k}$ points in the direction of the incoming beam of particles. We can always choose this direction as the z-axis of the coordinate system. Then the spherical angle θ of $\vec{k}$ is zero. Taking into account that

$$Y_{\ell m}(0, \varphi) = \delta_{m0}\sqrt{\frac{2\ell+1}{4\pi}} ,$$

we obtain

$$\begin{aligned}\frac{d\sigma}{d\Omega} &= 16\pi^2 \sum_{[\ell_1][\ell_2][\ell_3][\ell_4]} Y_{[\ell_1]}(\hat{\vec{k}}) \left(f^\dagger\right)_{[\ell_1][\ell_2]} \times \\ &\quad \times Y^*_{[\ell_2]}(\hat{\vec{r}}) Y_{[\ell_3]}(\hat{\vec{r}}) f_{[\ell_3][\ell_4]} Y^*_{[\ell_4]}(\hat{\vec{k}}) \\ &= 4\pi \sum_{[\ell_1][\ell_2][\ell_3][\ell_4]} \sqrt{(2\ell_1+1)(2\ell_4+1)}\delta_{m_1,0} \left(f^\dagger\right)_{[\ell_1][\ell_2]} \times \\ &\quad \times Y^*_{[\ell_2]}(\hat{\vec{r}}) Y_{[\ell_3]}(\hat{\vec{r}}) f_{[\ell_3][\ell_4]} \delta_{m_4,0} .\end{aligned} \tag{9.120}$$

Then the total cross section is

$$\sigma = \int \frac{d\sigma}{d\Omega} d\Omega_{\vec{r}} = 4\pi \sum_{[\ell_1][\ell_2]} \sqrt{(2\ell_1+1)(2\ell_2+1)}\delta_{m_1,0} \left(f^\dagger f\right)_{[\ell_1][\ell_2]} \delta_{m_2,0} . \tag{9.121}$$

10 Systems with Non-zero Spins

In all sections of Chap. 9, we considered a problem of a spinless particle moving in an "inherently" non-central potential, whose source is immovable and whose strength depends on the direction. In more practical and realistic problems, we usually deal with a bound state or collision of two quantum bodies (we call them particles, but they may have some internal structure) that interact via an $\vec{r}$-dependent potential, and both may have non-zero spins. In this kind of problems, the rotational invariance of the space around the particles is not broken, and therefore, the total angular momentum J as well as its third component M is conserving. As a result, only few partial waves (which together with the spin form $\vec{J}$) are mixed to each other. If $\vec{s}$ is the total spin of the interacting particles, then

$$\vec{J} = \vec{\ell} + \vec{s} \,, \tag{10.1}$$

and the maximal number of coupled partial waves is determined by the triangle condition

$$|J - s| \leqslant \ell \leqslant |J + s| \,. \tag{10.2}$$

Of course, the total wave function that describes a collision process, in principle, involves all possible values of J and thus consists of an infinite number of terms. However, in most cases, we can find some physical reasons for taking into account only few of them. Moreover, the problem can be solved independently for each value of J.

10.1 Spin-Angular State-Vectors

Before proceeding to the partial wave decomposition, we need to define some eigenvectors of the total spin and the total angular momentum, which are useful in formal derivations.

S. A. Rakityansky, *Jost Functions in Quantum Mechanics*,
https://doi.org/10.1007/978-3-031-07761-6_10

First of all, we define the basis in the spin space of the two interacting quantum objects. The basic vectors $|s\mu\rangle$ are the eigen-vectors of square of the total spin $\vec{s}$ and its third component μ. They are composed of the individual spin states $|s_1\mu_1\rangle$ and $|s_2\mu_2\rangle$ as follows:

$$|s\mu\rangle = \sum_{\mu_1\mu_2} C^{s\mu}_{s_1\mu_1 s_2\mu_2} |s_1\mu_1, s_2\mu_2\rangle \,, \tag{10.3}$$

where

$$C^{s\mu}_{s_1\mu_1 s_2\mu_2} = \langle s_1\mu_1 s_2\mu_2|(s_1 s_2)s\mu\rangle$$

are the Clebsch–Gordan coefficients.[1] When the orbital angular momentum $\vec{\ell}$ is combined with the total spin, we obtain a common eigen-vector of the four operators, namely, ℓ^2, s^2, J^2, and M (the third component of $\vec{J}$),

$$|\mathcal{Y}^{JM}_{[\ell]}\rangle \equiv |(\ell s)JM\rangle = \sum_{m\mu} |\ell m, s\mu\rangle C^{JM}_{\ell m s\mu} \,, \tag{10.4}$$

where the symbol $[\ell]$ stands for the combination of the two quantum numbers, ℓ and s,

$$[\ell] \equiv \{\ell, s\} \,. \tag{10.5}$$

If the eigen-vector $|\vec{r}\,\rangle$ of the coordinate that describes a certain space configuration of the physical system is considered in the spherical system of coordinates, $\{r, \Omega_{\vec{r}}\} \equiv \{r, \hat{\vec{r}}\,\}$, then $|\vec{r}\,\rangle = |r, \hat{\vec{r}}\,\rangle$, and the scalar product

$$\langle \hat{\vec{r}}\,|\ell m\rangle = Y_{\ell m}(\hat{\vec{r}}\,) \tag{10.6}$$

is the standard spherical harmonics (see the Appendix A). Apparently, vectors (10.4) constitute an ortho-normal basis in the spin-angular space. Indeed,

$$\begin{aligned}
\langle \mathcal{Y}^{JM}_{[\ell]}|\mathcal{Y}^{J'M'}_{[\ell']}\rangle &= \int \langle \mathcal{Y}^{JM}_{[\ell]}|\hat{\vec{r}}\,\rangle\langle \hat{\vec{r}}\,|\mathcal{Y}^{J'M'}_{[\ell']}\rangle \, d\hat{\vec{r}} \\
&= \sum_{m\mu m'\mu'} C^{JM}_{\ell m s\mu} C^{J'M'}_{\ell' m' s'\mu'} \langle s\mu|s'\mu'\rangle \int Y^*_{\ell m}(\hat{\vec{r}}\,) Y_{\ell' m'}(\hat{\vec{r}}\,) d\Omega_{\vec{r}} \\
&= \sum_{m\mu m'\mu'} C^{JM}_{\ell m s\mu} C^{J'M'}_{\ell' m' s'\mu'} \delta_{ss'}\delta_{\mu\mu'}\delta_{\ell\ell'}\delta_{mm'} \\
&= \delta_{[\ell][\ell']} \sum_{m\mu} C^{JM}_{\ell m s\mu} C^{J'M'}_{\ell m s\mu} = \delta_{[\ell][\ell']}\delta_{JJ'}\delta_{MM'} \,.
\end{aligned} \tag{10.7}$$

[1] A comprehensive description of all the properties of these coefficients can be found in Ref. [5].

In the spin-angular space, the operator $|\mathcal{Y}^{JM}_{[\ell]}\rangle\langle\mathcal{Y}^{JM}_{[\ell]}|$ makes the projection onto a state with definite values of ℓ, s, J, and M. If we sum up such projection operators with all possible quantum numbers, then we obtain the identity operator, i.e. the closure relation,

$$\sum_{[\ell]JM} |\mathcal{Y}^{JM}_{[\ell]}\rangle\langle\mathcal{Y}^{JM}_{[\ell]}| = I \ . \tag{10.8}$$

It is also useful to define the projection of vector (10.4) onto the angular part of the coordinate eigenstate,

$$|\mathcal{Y}^{JM}_{[\ell]}(\hat{\vec{r}}\,)\rangle = \langle\hat{\vec{r}}\,|\mathcal{Y}^{JM}_{[\ell]}\rangle = \sum_{m\mu} C^{JM}_{\ell m s\mu} Y_{\ell m}(\hat{\vec{r}}\,)|s\mu\rangle = \sum_{\mu} C^{JM}_{\ell,M-\mu,s\mu} Y_{\ell,M-\mu}(\hat{\vec{r}}\,)|s\mu\rangle \ , \tag{10.9}$$

and its projection on both the angular and spin states,

$$Y^{JM}_{\ell s\mu}(\hat{\vec{r}}\,) = \langle\hat{\vec{r}}, s\mu|\mathcal{Y}^{JM}_{[\ell]}\rangle = \sum_{m} C^{JM}_{\ell m s\mu} Y_{\ell m}(\hat{\vec{r}}\,) = C^{JM}_{\ell,M-\mu,s\mu} Y_{\ell,M-\mu}(\hat{\vec{r}}\,) \ . \tag{10.10}$$

The projection (10.9) is a ket-vector in the spin space. For the sake of clarity, let us consider the case when $\ell = 1$, $s = 1$, $J = 1$, and $M = 0$. Taking the numerical values of the corresponding Clebsch–Gordan coefficients, we obtain

$$|\mathcal{Y}^{10}_{[1,1]}(\hat{\vec{r}}\,)\rangle = \frac{1}{\sqrt{2}} Y_{1,1}(\hat{\vec{r}}\,)|1,-1\rangle - \frac{1}{\sqrt{2}} Y_{1,-1}(\hat{\vec{r}}\,)|1,1\rangle \ . \tag{10.11}$$

10.2 Partial-Wave Decomposition for Discrete Spectrum

The discrete states $|\psi_{EJM}\rangle$ (bound states and resonances) have definite values of the energy as well as of the total angular momentum and of its third component. These states are Hilbert vectors in the direct (tensor) product of the configuration and spin spaces. The basis in this combined space can be constructed as the direct product of the states $|\vec{r}\,\rangle$ with definite value of the coordinate, and the basic vectors $|s\mu\rangle$ in the spin space. The set of the projections (scalar products) of the state $|\psi_{EJM}\rangle$ on the basis vectors,

$$\langle\vec{r}, s\mu|\psi_{EJM}\rangle = \psi_{EJM}(\vec{r}, s, \mu) \ , \tag{10.12}$$

is the wave function. Its absolute square is the density of the probability that the physical system can be found at the point $\vec{r}$ with the total spin s and its third component μ.

In order to separate the angular dependence of the wave function, we can expand it over the ortho-normal basis consisting of the eigen-vectors (10.4),

$$\psi_{EJM}(\vec{r}, s, \mu) = \sum_{[\ell]} \langle \vec{r}, s\mu | \mathcal{Y}^{JM}_{[\ell]} \rangle \langle \mathcal{Y}^{JM}_{[\ell]} | \psi_{EJM} \rangle = \sum_{\ell} Y^{JM}_{\ell s\mu}(\hat{\vec{r}}) \langle r, (\ell s)JM | \psi_{EJM} \rangle$$

$$= \sum_{\ell} Y^{JM}_{\ell s\mu}(\hat{\vec{r}}) \frac{u^{J}_{[\ell]}(E, r)}{r} , \tag{10.13}$$

where $u^{J}_{[\ell]}(E, r) = r\langle r, (\ell s)JM | \psi_{EJM} \rangle$ are the radial parts of the wave function for the states with definite values of the quantum numbers ℓ, s, and J. The factor $1/r$ is introduced as usual for getting rid of the first derivative in the corresponding radial Schrödinger equations (see next section). If we only project the state-vector $|\psi_{EJM}\rangle$ onto the eigen-vector of the coordinate,

$$\langle \vec{r} | \psi_{EJM} \rangle = |\psi_{EJM}(\vec{r})\rangle = \sum_{[\ell]} |\mathcal{Y}^{JM}_{[\ell]}(\hat{\vec{r}})\rangle \frac{u^{J}_{[\ell]}(E, r)}{r} , \tag{10.14}$$

then it remains a ket-vector in the spin space. Here the ket-vector $|\mathcal{Y}^{JM}_{[\ell]}(\hat{\vec{r}})\rangle$ is defined by Eq. (10.9).

10.2.1 Radial Schrödinger Equation

Starting with the Schrödinger equation in the operator form,

$$(H - E)|\psi_{EJM}\rangle = 0,$$

we project it onto the basis $\langle \vec{r} |$,

$$\langle \vec{r} | (H - E) | \psi_{EJM} \rangle = 0 .$$

Since the potential U is assumed to be local, this gives[2]

$$\left[-\frac{\hbar^2}{2\mu} \Delta_{\vec{r}} + U(\vec{r}) - E \right] \sum_{[\ell]} |\mathcal{Y}^{JM}_{[\ell]}(\hat{\vec{r}})\rangle \frac{u^{J}_{[\ell]}(E, r)}{r} = 0 .$$

[2] It is assumed that the use of the same symbol for the reduced mass μ in the factor $\hbar^2/2\mu$ and for the third components of the spin will not cause confusion.

Using the fact that $|\mathcal{Y}_{[\ell]}^{JM}(\hat{\vec{r}}\,)\rangle$ is an eigenstate of the operator $\vec{\ell}^{\,2}$, i.e.

$$\vec{\ell}^{\,2}|\mathcal{Y}_{[\ell]}^{JM}(\hat{\vec{r}}\,)\rangle = \hbar^2\ell(\ell+1)|\mathcal{Y}_{[\ell]}^{JM}(\hat{\vec{r}}\,)\rangle\,,$$

we obtain

$$\sum_{[\ell]}\left[\partial_r^2 + k^2 - \frac{\ell(\ell+1)}{r^2} - \frac{2\mu}{\hbar^2}U(\vec{r}\,)\right]|\mathcal{Y}_{[\ell]}^{JM}(\hat{\vec{r}}\,)\rangle u_{[\ell]}^J(E,r) = 0\,. \tag{10.15}$$

Now, we multiply Eq. (10.15) by $\langle\mathcal{Y}_{[\ell]}^{JM}(\hat{\vec{r}}\,)|$ from the left and integrate over the spherical angles $\hat{\vec{r}}$. Using the orthogonality of the spin-angular functions, we obtain exactly the same set of radial equations as we had for the spinless particles (see Sect. 9.1.1),

$$\left[\partial_r + k^2 - \frac{\ell(\ell+1)}{r^2}\right]u_{[\ell]}^J(E,r) = \sum_{[\ell']} V_{[\ell][\ell']}^J(r)u_{[\ell']}^J(E,r)\,. \tag{10.16}$$

In this equation, the matrix elements of the potential,

$$\begin{aligned} V_{[\ell][\ell']}^J(r) &= \frac{2\mu}{\hbar^2}\int\langle\mathcal{Y}_{[\ell]}^{JM}(\hat{\vec{r}}\,)|U(\vec{r}\,)|\mathcal{Y}_{[\ell']}^{JM}(\hat{\vec{r}}\,)\rangle d\hat{\vec{r}} \\ &= \frac{2\mu}{\hbar^2}\sum_{m\mu m'\mu'} C_{\ell m s\mu}^{JM}C_{\ell' m' s'\mu'}^{JM} \\ &\quad\times\iint Y_{\ell m}^*(\theta,\varphi)\langle s\mu|U(\vec{r}\,)|s'\mu'\rangle Y_{\ell' m'}(\theta,\varphi)\sin\theta\,d\theta\,d\varphi\,, \end{aligned} \tag{10.17}$$

do not depend on the third component M of the total angular momentum. Physical explanation of this fact is as follows. The value of M determines orientation of the whole physical system in space. But because of the isotropic invariance, the whole system can be arbitrarily rotated without changing the forces acting between the particles.

Mathematically, the independence of M can be proved either using the Wigner–Eckart theorem or directly (in a more simple way) as follows. The operator $R(\vec{\alpha}\,)$ that performs the rotation of a quantum system around vector $\vec{\alpha}$ to the angle $|\vec{\alpha}|$,

$$R(\vec{\alpha}\,) = e^{-\frac{i}{\hbar}\vec{\alpha}\vec{J}}\,,$$

involves the total angular momentum $\vec{J}$ defined by Eq. (10.1). The isotropic (rotational) invariance means that R commutes with the Hamiltonian and thus with the potential U (which is a part of the Hamiltonian). This also means that U

commutes with $\vec{J}$, i.e. with all three of its components and thus with the ladder operators $J_\pm$,

$$[R, U] = 0 \implies [\vec{J}, U] = 0 \implies [J_\pm, U] = 0\,, \tag{10.18}$$

where

$$J_\pm = J_x \pm i J_y$$

with the properties[3]

$$J_\pm |JM\rangle = \hbar\sqrt{(J \mp M)(J \pm M + 1)}|J, M \pm 1\rangle\,,$$

$$J_\pm^\dagger = J_\mp\,.$$

Taking into account that $[J_x, J_y] = i\hbar J_z$, we obtain

$$J_- J_+ = (J_x - i J_y)(J_x + i J_y) = J^2 - J_z^2 - \hbar J_z\,.$$

The commutation rule (10.18) implies the following sequence of equalities:

$$\langle J, M|J_- U J_+|J, M\rangle = \langle J, M|U J_- J_+|J, M\rangle\,,$$

$$(J_+|J, M\rangle)^\dagger U J_+|J, M\rangle = \langle J, M|U(J^2 - J_z^2 - \hbar J_z)|J, M\rangle\,,$$

$$\begin{aligned}&\hbar^2[J(J+1) - M^2 - M]\langle J, M+1|U|J, M+1\rangle \\ &\quad = \hbar^2[J(J+1) - M^2 - M]\langle J, M|U|J, M\rangle\,,\end{aligned}$$

which means that the matrix element of the potential does not depend on M, i.e.

$$\langle J, M+1|U|J, M+1\rangle = \langle J, M|U|J, M\rangle\,.$$

The difference from the "inherently" non-central potential is that here the total angular momentum J is conserving, and therefore, the matrix V^J is block-diagonal. In other words, for each fixed J, only few equations of the system (10.16) are coupled to each other.

To make it clear, consider a hypothetical example of the system of two particles, one of which has spin $s_1 = 1/2$ and the other $s_2 = 1$. The total spin can be either $s = 1/2$ (doublet state) or $s = 3/2$ (quartet state). Suppose we are interested in the state with $J = 5/2$. If $s = 1/2$, such a state can only be formed in the partial waves

3 See, for example, Ref. [25].

$\ell = 2, 3$. If $s = 3/2$, the required J is possible when the orbital angular momentum is one of the following $\ell = 1, 2, 3, 4$. Therefore, the following six combinations of the quantum numbers $[\ell] \equiv (\ell, s)$ are possible:

$$(\ell, s): \quad \left(2, \frac{1}{2}\right), \left(3, \frac{1}{2}\right), \left(1, \frac{3}{2}\right), \left(2, \frac{3}{2}\right), \left(3, \frac{3}{2}\right), \left(4, \frac{3}{2}\right).$$

There is another conservation law that restricts the possibilities further, namely, the conservation of the parity $\pi = (-1)^\ell$. This law excludes coupling of even and odd orbital angular momenta. Therefore, the above six states are split into the two groups

$$\text{even states:} \quad \left(2, \frac{1}{2}\right), \left(2, \frac{3}{2}\right), \left(4, \frac{3}{2}\right),$$

$$\text{odd states:} \quad \left(3, \frac{1}{2}\right), \left(1, \frac{3}{2}\right), \left(3, \frac{3}{2}\right).$$

If we assign the numbers 1, 2, and 3 to the three even states, and 4, 5, and 6 to the odd states, then the Jost matrices have the following block-diagonal structure:

$$f^{(\text{in/out})J} = \begin{pmatrix} f_{11}^{(\text{in/out})J} & f_{12}^{(\text{in/out})J} & f_{13}^{(\text{in/out})J} & 0 & 0 & 0 \\ f_{21}^{(\text{in/out})J} & f_{22}^{(\text{in/out})J} & f_{23}^{(\text{in/out})J} & 0 & 0 & 0 \\ f_{31}^{(\text{in/out})J} & f_{32}^{(\text{in/out})J} & f_{33}^{(\text{in/out})J} & 0 & 0 & 0 \\ 0 & 0 & 0 & f_{44}^{(\text{in/out})J} & f_{45}^{(\text{in/out})J} & f_{46}^{(\text{in/out})J} \\ 0 & 0 & 0 & f_{54}^{(\text{in/out})J} & f_{55}^{(\text{in/out})J} & f_{56}^{(\text{in/out})J} \\ 0 & 0 & 0 & f_{64}^{(\text{in/out})J} & f_{65}^{(\text{in/out})J} & f_{66}^{(\text{in/out})J} \end{pmatrix}. \tag{10.19}$$

This simple analysis tells us that for $J = 5/2$, we need to solve separately three coupled equations for the even states and three coupled equations for the odd states.[4]

As was already mentioned, Eqs. (10.16) are the same as the corresponding system (9.2) that was obtained in the case of spinless particles. The only difference is in calculating the matrix elements of the potential and in choosing the partial waves that are coupled to each other in a particular discrete spectrum state. Therefore, everything that was established for the spinless particles is valid and applicable here. In particular, the spectrum of the discrete states is determined by Eq. (9.95), i.e. the spectral points are the zeros of the Jost matrix determinant. Of course, one should not forget about the different meaning of the matrix subscript $[\ell]$ here and there.

[4] If the particles composing the system are identical, there is one more restriction on the choice of the coupled partial waves. This is the requirement of certain permutation symmetry (symmetric for bosons, and anti-symmetric for fermions).

10.3 Partial-Wave Decomposition for Scattering States

In order to generalize the partial-wave expansion (9.8) for the case of particles with non-zero spins, consider a quantum state $|\psi_{\vec{k}s\mu}\rangle$, where the subscripts $\vec{k}s\mu$ indicate that this state has evolved from the initial plane wave $|\vec{k}\rangle$ with the spin state $|s\mu\rangle$. Using the Møller operator discussed in Sects. 1.3.2 and 9.1.2, this can be formally written as

$$|\psi_{\vec{k}s\mu}\rangle = \Omega_+|\vec{k}, s\mu\rangle = \sum_{\ell m} \Omega_+|k\ell m, s\mu\rangle Y^*_{\ell m}(\hat{\vec{k}})$$

$$= \sum_{\ell m J M} \Omega_+|k, (\ell s)JM\rangle C^{JM}_{\ell m s\mu} Y^*_{\ell m}(\hat{\vec{k}}) .$$

Introducing the scattering partial-wave state with definite values of the total angular momentum J and its third component M,

$$|\psi^{JM}_{k[\ell]}\rangle = \Omega_+|k, (\ell s)JM\rangle ,$$

we obtain the following partial-wave expansion of the scattering-state- vector (the spin-angular functions used below are defined in Sect. 10.1)

$$|\psi_{\vec{k}s\mu}\rangle = \sum_{JM\ell} |\psi^{JM}_{k[\ell]}\rangle Y^{JM*}_{\ell s\mu}(\hat{\vec{k}}) = \sum_{JM[\ell]} |\psi^{JM}_{k[\ell]}\rangle\langle \mathcal{Y}^{JM}_{[\ell]}(\hat{\vec{k}})|s\mu\rangle . \tag{10.20}$$

The corresponding wave function, i.e. the amplitude of the probability that the system is at the point $\vec{r}$ and in the spin state $|s'\mu'\rangle$, is

$$\psi_{\vec{k}s\mu}(\vec{r}, s', \mu') = \langle \vec{r}, s'\mu'|\psi_{\vec{k}s\mu}\rangle$$

$$= \sum_{JM[\ell'][\ell]} \langle s'\mu'|\mathcal{Y}^{JM}_{[\ell']}(\hat{\vec{r}})\rangle\langle r, (\ell's')JM|\psi^{JM}_{k[\ell]}\rangle\langle \mathcal{Y}^{JM}_{[\ell]}(\hat{\vec{k}})|s\mu\rangle .$$

As usual, we explicitly separate the factor $1/r$ in the radial part of this wave function (which is a matrix),

$$\langle r, (\ell's')JM|\psi^{JM}_{k[\ell]}\rangle = \frac{u^J_{[\ell'][\ell]}(E, r)}{r} N_{[\ell]}(k) ,$$

and thus

$$\psi_{\vec{k}s\mu}(\vec{r}, s', \mu') = \sum_{JM}\sum_{[\ell'][\ell]} \langle s'\mu'|\mathcal{Y}^{JM}_{[\ell']}(\hat{\vec{r}})\rangle \frac{u^J_{[\ell'][\ell]}(E, r)N_{[\ell]}(k)}{r} \langle \mathcal{Y}^{JM}_{[\ell]}(\hat{\vec{k}})|s\mu\rangle , \tag{10.21}$$

where we multiplied each column of the matrix u^J by an appropriate normalization constant $N_{[\ell]}(k)$. This is done for the same reason as in Eq. (9.8), namely, to have proper normalization of the wave function. In Sect. 10.3.2, it will be shown that this constant is exactly the same as for the spinless case, namely, $N_{[\ell]}(k) = i^\ell/(\sqrt{2\pi}k)$.

We can remove in Eq. (10.21) the left and right spin states $\langle s'\mu'|$ and $|s\mu\rangle$, i.e. remove the projections onto specific final and initial spin states. As a result, the wave function becomes an operator in the spin space,

$$\psi_{\vec{k}}(\vec{r}\,) = \sum_{JM}\sum_{[\ell'][\ell]} |\mathcal{Y}^{JM}_{[\ell']}(\hat{\vec{r}}\,)\rangle \frac{u^J_{[\ell'][\ell]}(E,r)N_{[\ell]}(k)}{r} \langle \mathcal{Y}^{JM}_{[\ell]}(\hat{\vec{k}}\,)| \,. \tag{10.22}$$

This representation is useful in formal derivations, while for calculating a specific transition, we always need to sandwich it between the appropriate spin states.

10.3.1 Radial Schrödinger Equation

In order to find the differential equation for the matrix $u^J(E,r)$, we need to substitute the total wave function in the Schrödinger equation and to get rid of its angular part. This equation written in the operator form,

$$(H - E)\,|\psi_{\vec{k}}\rangle = 0\,,$$

after following the same steps as in Sect. 10.2.1, gives

$$\sum_{JM}\sum_{[\ell][\ell']}\left[\partial_r^2 + k^2 - \frac{\ell(\ell+1)}{r^2} - \frac{2\mu}{\hbar^2}U(\vec{r}\,)\right]$$
$$\times\, |\mathcal{Y}^{JM}_{[\ell]}(\hat{\vec{r}}\,)\rangle u^J_{[\ell][\ell']}(E,r)N_{[\ell']}\langle \mathcal{Y}^{JM}_{[\ell']}(\hat{\vec{k}}\,)| = 0\,.$$

Sandwiching this equation between $\langle \mathcal{Y}^{JM}_{[\ell]}(\hat{\vec{r}}\,)|$ and $|\mathcal{Y}^{JM}_{[\ell']}(\hat{\vec{k}}\,)\rangle$, and using the orthogonality condition (10.7), we obtain for each column of the matrix u^J exactly the same set of coupled differential equations as the set (10.16),

$$\left[\partial_r + k^2 - \frac{\ell(\ell+1)}{r^2}\right] u^J_{[\ell][\ell']}(E,r) = \sum_{[\ell'']} V^J_{[\ell][\ell'']}(r) u^J_{[\ell''][\ell']}(E,r)\,, \tag{10.23}$$

where the potential matrix is given by Eq. (10.18). Since this equation is the same as for the spinless case, we can find its fundamental matrix of regular solutions, $\phi^J(E,r)$, in the same way as is described in Sect. 9.2, namely, as the following combination of the incoming and outgoing spherical waves:

$$\boxed{\phi^J(E,r) = W^{(\mathrm{in})}(E,r)F^{(\mathrm{in})J}(E,r) + W^{(\mathrm{out})}(E,r)F^{(\mathrm{out})J}(E,r)} \tag{10.24}$$

where the unknown matrices $F^{(\text{in/out})J}(E,r)$ obey the differential equations

$$\partial_r F^{(\text{in})J} = -\frac{1}{2ik} W^{(\text{out})} V^J \left[W^{(\text{in})} F^{(\text{in})J} + W^{(\text{out})} F^{(\text{out})J} \right] \tag{10.25}$$

$$\partial_r F^{(\text{out})J} = \frac{1}{2ik} W^{(\text{in})} V^J \left[W^{(\text{in})} F^{(\text{in})J} + W^{(\text{out})} F^{(\text{out})J} \right] \tag{10.26}$$

with the boundary conditions determined by[5]

$$\lim_{r\to 0} \frac{j_\ell(kr) F^{(\text{in/out})J}_{[\ell][\ell']}(E,r)}{j_{\ell'}(kr)} = \frac{1}{2}\delta_{[\ell][\ell']} \,. \tag{10.27}$$

Using this fundamental matrix, we then need to construct the physical wave function with proper asymptotic behaviour. This is done by making the combination

$$u^J_{[\ell][\ell']}(E,r) = \sum_{[\ell'']} \phi^J_{[\ell][\ell'']}(E,r) c^J_{[\ell''][\ell']} \tag{10.28}$$

of the columns of $\phi^J(E,r)$. It is not difficult to show that, similarly to Eq. (9.107), the matrix of the combination coefficients here, which gives the correct asymptotics, is the inverted incoming Jost matrix,

$$c^J_{[\ell][\ell']} = \left[f^{(\text{in})J}(E) \right]^{-1}_{[\ell][\ell']} , \tag{10.29}$$

where

$$f^{(\text{in})J}(E) = \lim_{r\to\infty} F^{(\text{in})J}(E,r) \,. \tag{10.30}$$

Before attempting to prove this statement (which will be done in Sect. 10.3.3), we need to find the appropriate normalization factor $N_{[\ell]}(k)$ in the partial-wave expansions (10.21, 10.22). This is done in the next section.

10.3.2 Plane Wave with Non-zero Spin

The normalization factor $N_{[\ell]}(k)$ in the partial-wave expansions (10.21, 10.22) should be the same for any potential. It is easy to determine this factor in the trivial

[5] All the formulae derived in Sect. 9.3.4.1 are applicable here.

case, when the potential is identical zero, i.e. when the incoming plane wave is not distorted at all. This trivial scattering wave function,

$$\psi_{\vec{k}s\mu}(\vec{r},s',\mu') = \langle\vec{r},s'\mu'|\vec{k},s\mu\rangle = \frac{e^{i\vec{k}\vec{r}}}{(2\pi)^{3/2}}\langle s'\mu'|s\mu\rangle\,,$$

can be expanded in the partial waves as follows (see Appendix A):

$$\psi_{\vec{k}s\mu}(\vec{r},s',\mu') = \sum_{\ell'm'\ell m} Y_{\ell'm'}(\hat{\vec{r}})\langle r,\ell'm's'\mu'|k,\ell m s\mu\rangle Y^*_{\ell m}(\hat{\vec{k}})$$

$$= \sum_{JM}\sum_{\ell'm'\ell m} Y_{\ell'm'}(\hat{\vec{r}})C^{JM}_{\ell'm's'\mu'}\langle r,(\ell's')JM|k,(\ell s)JM\rangle C^{JM}_{\ell m s\mu}Y^*_{\ell m}(\hat{\vec{k}})\,.$$

Taking into account Eq. (A.15), we have

$$\langle r,(\ell's')JM|k,(\ell s)JM\rangle = \sqrt{\frac{2}{\pi}}\frac{i^\ell j_\ell(kr)}{kr}\delta_{\ell'\ell}\delta_{s's}$$

$$= \frac{i^\ell}{\sqrt{2\pi}k}\frac{h^{(-)}_\ell(kr)+h^{(+)}_\ell(kr)}{r}\delta_{[\ell'][\ell]}\,,$$

and thus

$$\psi_{\vec{k}s\mu}(\vec{r},s',\mu') = \sum_{JM[\ell'][\ell]}\langle s'\mu'|\mathcal{Y}^{JM}_{[\ell']}(\hat{\vec{r}})\rangle\frac{u^J_{[\ell'][\ell]}(E,r)i^\ell}{\sqrt{2\pi}kr}\langle\mathcal{Y}^{JM}_{[\ell]}(\hat{\vec{k}})|s\mu\rangle\,, \tag{10.31}$$

where

$$u^J_{[\ell'][\ell]}(E,r) = \left[h^{(-)}_\ell(kr)+h^{(+)}_\ell(kr)\right]\delta_{[\ell'][\ell]}\,,$$

with the trivial Jost matrices

$$f^{(\mathrm{in})J}_{[\ell'][\ell]}(E) = f^{(\mathrm{out})J}_{[\ell'][\ell]}(E) = \frac{1}{2}\delta_{[\ell'][\ell]}\,.$$

Comparing (10.31) with (10.21), we see that the normalization constant is

$$N_{[\ell]}(k) - \frac{i^\ell}{\sqrt{2\pi}k}\,. \tag{10.32}$$

10.3.3 Long-Range Asymptotics of the Wave Function

Let us prove that the physical scattering wave function (10.28) constructed as a combination of the columns of the fundamental matrix $\phi^J(E, r)$ with the combination coefficients (10.29) has the correct asymptotic behaviour. In doing this, we need to repeat (in a more general form) the derivations leading to Eq. (9.113) in Sect. 9.7.2. For this purpose, we introduce the row matrices

$$\left|\mathcal{Y}^{JM}(\hat{\vec{r}})\right\rangle = \left\{|\mathcal{Y}^{JM}_{[\ell_1]}(\hat{\vec{r}})\rangle, |\mathcal{Y}^{JM}_{[\ell_2]}(\hat{\vec{r}})\rangle, \dots, |\mathcal{Y}^{JM}_{[\ell_N]}(\hat{\vec{r}})\rangle\right\} , \tag{10.33}$$

$$\left|\mathcal{Y}^{JM}(\hat{\vec{k}})\right\rangle = \left\{|\mathcal{Y}^{JM}_{[\ell_1]}(\hat{\vec{k}})\rangle, |\mathcal{Y}^{JM}_{[\ell_2]}(\hat{\vec{k}})\rangle, \dots, |\mathcal{Y}^{JM}_{[\ell_N]}(\hat{\vec{k}})\rangle\right\} , \tag{10.34}$$

and use the diagonal matrix (9.108), namely,

$$\mathcal{I} = \operatorname{diag}\left\{i^{\ell_1}, i^{\ell_2}, \dots, i^{\ell_N}\right\} . \tag{10.35}$$

The row matrices (10.33, 10.34) are composed of the "ket"-vectors in the spin space. The corresponding "bra"-matrices are obtained by the Hermitian conjugation,

$$\left\langle\mathcal{Y}^{JM}(\hat{\vec{k}})\right| = \left|\mathcal{Y}^{JM}(\hat{\vec{k}})\right\rangle^{\dagger} ,$$

and therefore are the column-matrices composed of the "bra"-vectors.

With the help of these matrices, the scattering wave function (10.22) can be written as

$$\begin{aligned}
\psi_{\vec{k}}(\vec{r}) &= \frac{1}{\sqrt{2\pi}kr}\sum_{JM}|\mathcal{Y}^{JM}(\hat{\vec{r}})\rangle u^J \mathcal{I}\langle\mathcal{Y}^{JM}(\hat{\vec{k}})| \\
&= \frac{1}{\sqrt{2\pi}kr}\sum_{JM}|\mathcal{Y}^{JM}(\hat{\vec{r}})\rangle \phi^J c^J \mathcal{I}\langle\mathcal{Y}^{JM}(\hat{\vec{k}})| \\
&= \frac{1}{\sqrt{2\pi}kr}\sum_{JM}|\mathcal{Y}^{JM}(\hat{\vec{r}})\rangle \left[W^{(\text{in})}F^{(\text{in})J} + W^{(\text{out})}F^{(\text{out})J}\right]\left[f^{(\text{in})J}\right]^{-1} \\
&\quad\times\mathcal{I}\langle\mathcal{Y}^{JM}(\hat{\vec{k}})| .
\end{aligned}$$

At large distances, we have

$$\psi_{\vec{k}}(\vec{r}) \underset{r\to\infty}{\longrightarrow} \frac{1}{\sqrt{2\pi}kr}\sum_{JM}|\mathcal{Y}^{JM}(\hat{\vec{r}})\rangle\left[W^{(\text{in})} + W^{(\text{out})}S^J\right]\mathcal{I}\langle\mathcal{Y}^{JM}(\hat{\vec{k}})| , \tag{10.36}$$

where the S-matrix is the "ratio"

$$S^J(E) = f^{(\text{out})J}(E)\left[f^{(\text{in})J}(E)\right]^{-1} . \tag{10.37}$$

Adding and subtracting the matrix $W^{(\text{out})}$, we obtain

$$\psi_{\vec{k}}(\vec{r}) \underset{r\to\infty}{\longrightarrow} \frac{1}{\sqrt{2\pi}kr}\sum_{JM}|\mathcal{Y}^{JM}(\hat{\vec{r}})\rangle\left[W^{(\text{in})}+W^{(\text{out})}\right]\mathcal{I}\langle\mathcal{Y}^{JM}(\hat{\vec{k}})| +$$

$$+\frac{1}{\sqrt{2\pi}kr}\sum_{JM}|\mathcal{Y}^{JM}(\hat{\vec{r}})\rangle W^{(\text{out})}(S^J-1)\mathcal{I}\langle\mathcal{Y}^{JM}(\hat{\vec{k}})|$$

and then, in exactly the same way as is done in Sect. 9.7.2,

$$\psi_{\vec{k}s\mu}(\vec{r},s',\mu') \underset{r\to\infty}{\longrightarrow} \frac{e^{i\vec{k}\vec{r}}}{(2\pi)^{3/2}}\delta_{s's}\delta_{\mu'\mu}$$

$$+\sqrt{\frac{2}{\pi}}\frac{e^{ikr}}{r}\sum_{JM}\langle s'\mu'|\mathcal{Y}^{JM}(\hat{\vec{r}})\rangle f^J(E)\langle\mathcal{Y}^{JM}(\hat{\vec{k}})|s\mu\rangle ,$$

where the matrix of the partial-wave scattering amplitudes are defined as

$$f^J(E) = \frac{1}{2ik}\mathcal{I}^\dagger\left[S^J(E)-1\right]\mathcal{I} . \tag{10.38}$$

Generalizing Eq. (9.116), we define the full amplitude as

$$\mathsf{f}(\hat{\vec{r}} \leftarrow \hat{\vec{k}}, E) = 4\pi\sum_{JM}|\mathcal{Y}^{JM}(\hat{\vec{r}})\rangle f^J(E)\langle\mathcal{Y}^{JM}(\hat{\vec{k}})| , \tag{10.39}$$

which gives

$$\psi_{\vec{k}s\mu}(\vec{r},s',\mu') \underset{r\to\infty}{\longrightarrow} \frac{1}{(2\pi)^{3/2}}\left[e^{i\vec{k}\vec{r}}\delta_{s's}\delta_{\mu'\mu}+\frac{e^{ikr}}{r}\langle s'\mu'|\mathsf{f}(\hat{\vec{r}} \leftarrow \hat{\vec{k}}, E)|s\mu\rangle\right] . \tag{10.40}$$

This proves that the choice (10.29) of the combination coefficients results in the correct asymptotic behaviour of the physical wave function. The asymptotic relation

(10.40) can also be written for the function (10.22), i.e. for the wave function not sandwiched between the initial and final spin states, as follows:

$$\psi_{\vec{k}}(\vec{r}) \underset{r\to\infty}{\longrightarrow} \frac{1}{(2\pi)^{3/2}}\left[e^{i\vec{k}\vec{r}} + \frac{e^{ikr}}{r}\mathsf{f}(\hat{\vec{r}} \leftarrow \hat{\vec{k}}, E)\right], \tag{10.41}$$

which formally looks exactly like the asymptotic behaviour (4.11) of the simple wave function for the spinless particles. This relation can be written in yet another form if we multiply Eq. (10.41) by the initial spin state $|s\mu\rangle$ from the right,

$$\psi_{\vec{k}s\mu}(\vec{r}) \underset{r\to\infty}{\longrightarrow} \frac{1}{(2\pi)^{3/2}}\left[e^{i\vec{k}\vec{r}}|s\mu\rangle + \frac{e^{ikr}}{r}\sum_{s'\mu'}|s'\mu'\rangle\langle s'\mu'|\mathsf{f}(\hat{\vec{r}} \leftarrow \hat{\vec{k}}, E)|s\mu\rangle\right], \tag{10.42}$$

which is similar to the multi-channel wave function (see Chap. 11 of this book), where different spin states are treated as different channels.

The full amplitude (10.39) is an operator in the spin space. When sandwiched between the spin states,

$$\langle s'\mu'|\mathsf{f}(\hat{\vec{r}} \leftarrow \hat{\vec{k}}, E)|s\mu\rangle = 4\pi\sum_{JM\ell'\ell} Y^{JM}_{\ell's'\mu'}(\hat{\vec{r}})f^{J}_{[\ell'][\ell]}(E)\left[Y^{JM}_{\ell s\mu}(\hat{\vec{k}})\right]^*, \tag{10.43}$$

it becomes just a complex number, which is the amplitude of the probability that the particles colliding with the incoming momentum $\vec{k}$ and the initial spin state $|s\mu\rangle$ can be finally found moving in the direction $\hat{\vec{r}}$ and in the spin state $|s'\mu'\rangle$.[6]

10.3.4 Scattering Observables

The differential scattering cross section is the absolute square of the amplitude (10.43),

$$\left(\frac{d\sigma}{d\Omega}\right)_{s'\mu'\leftarrow s\mu} = \left|\langle s'\mu'|\mathsf{f}(\hat{\vec{r}} \leftarrow \hat{\vec{k}}, E)|s\mu\rangle\right|^2. \tag{10.44}$$

As usual, we can direct the z-axis along the collision line, which means that the spherical angle θ of vector $\vec{k}$ is zero. The spherical harmonics with $\theta = 0$ are very simple,

$$Y_{\ell m}(0,\varphi) = \delta_{m0}\sqrt{\frac{2\ell+1}{4\pi}}. \tag{10.45}$$

[6] Please note that the summation in Eq. (10.43) is over $\ell'\ell$ but not over $[\ell'][\ell]$. This is because the initial and final spins, s and s', are fixed.

Then from Eq. (10.10), it follows that

$$Y_{\ell s\mu}^{JM}(\hat{k}) = \sqrt{\frac{2\ell+1}{4\pi}} C_{\ell 0 s\mu}^{JM}, \qquad \vec{k} \parallel z. \tag{10.46}$$

Since

$$C_{\ell 0 s\mu}^{JM} = \delta_{M\mu} C_{\ell 0 s\mu}^{J\mu},$$

the cross section can be written as

$$\left(\frac{d\sigma}{d\Omega}\right)_{s'\mu'\leftarrow s\mu} = 4\pi \sum_{J_1\ell_1'\ell_1} \sum_{J_2\ell_2'\ell_2} \sqrt{(2\ell_1+1)(2\ell_2+1)} \times \tag{10.47}$$

$$\times f_{[\ell_1'][\ell_1]}^{J_1 *} f_{[\ell_2'][\ell_2]}^{J_2} C_{\ell_1 0 s\mu}^{J_1\mu} C_{\ell_2 0 s\mu}^{J_2\mu} \left[Y_{\ell_1' s'\mu'}^{J_1\mu}(\hat{r})\right]^* Y_{\ell_2' s'\mu'}^{J_2\mu}(\hat{r}).$$

It is not difficult to see that in the spinless case, i.e. with $s = 0$ and therefore with $C_{\ell 000}^{JM} = \delta_{J\ell}\delta_{M0}$, we come back to Eq. (9.120).

Integrating Eq. (10.48) over all possible directions of scattering, we obtain the total cross section for the spin transition $|s\mu\rangle \rightarrow |s'\mu'\rangle$,

$$\sigma_{s'\mu'\leftarrow s\mu} = \int \left(\frac{d\sigma}{d\Omega}\right)_{s'\mu'\leftarrow s\mu} d\hat{r}.$$

In doing such the integration, we only need to integrate the spherical harmonics,

$$\int \left[Y_{\ell_1' s'\mu'}^{J_1\mu}(\hat{r})\right]^* Y_{\ell_2' s'\mu'}^{J_2\mu}(\hat{r}) d\hat{r} =$$

$$= \int \sum_{m_1 m_2} Y_{\ell_1' m_1}^*(\hat{r}) C_{\ell_1' m_1 s'\mu'}^{J_1\mu} Y_{\ell_2' m_2}(\hat{r}) C_{\ell_2' m_2 s'\mu'}^{J_2\mu} d\hat{r}$$

$$= \delta_{\ell_1'\ell_2'} \sum_m C_{\ell_1' m s'\mu'}^{J_1\mu} C_{\ell_1' m s'\mu'}^{J_2\mu} = \delta_{\ell_1'\ell_2'} C_{\ell_1',\mu-\mu',s'\mu'}^{J_1\mu} C_{\ell_1',\mu-\mu',s'\mu'}^{J_2\mu}, \tag{10.48}$$

which gives

$$\sigma_{s'\mu'\leftarrow s\mu} = 4\pi \sum_{J_1\ell_1 J_2\ell_2\ell'} \sqrt{(2\ell_1+1)(2\ell_2+1)} f_{[\ell'][\ell_1]}^{J_1 *} f_{[\ell'][\ell_2]}^{J_2} \times$$

$$\times C_{\ell',\mu-\mu',s'\mu'}^{J_1\mu} C_{\ell',\mu-\mu',s'\mu'}^{J_2\mu} C_{\ell_1 0 s\mu}^{J_1\mu} C_{\ell_2 0 s\mu}^{J_2\mu}, \tag{10.49}$$

where $[\ell_1] = \{\ell_1, s\}$, $[\ell_2] = \{\ell_2, s\}$, and $[\ell'] = \{\ell', s'\}$.

If the incident beam is not polarized (i.e. all possible orientations of the spin have equal probabilities), then we should average over μ,

$$\sigma_{s'\mu'\leftarrow s} = \frac{1}{2s+1}\sum_{\mu}\sigma_{s'\mu'\leftarrow s\mu} \,. \tag{10.50}$$

If the orientation of the final spin is not monitored, we should sum over μ',

$$\sigma_{s'\leftarrow s} = \frac{1}{2s+1}\sum_{\mu'}\sum_{\mu}\sigma_{s'\mu'\leftarrow s\mu} \,. \tag{10.51}$$

When doing the above summations, we notice that the only quantities depending on the third components of the angular momenta and spins in Eq. (10.49) are the Clebsch–Gordan coefficients. The summations over μ' and μ can therefore be done explicitly:

$$\sum_{\mu'\mu} C^{J_1\mu}_{\ell',\mu-\mu',s'\mu'} C^{J_2\mu}_{\ell',\mu-\mu',s'\mu'} C^{J_1\mu}_{\ell_1 0 s\mu} C^{J_2\mu}_{\ell_2 0 s\mu} = \sum_{Mm\mu'\mu} C^{J_1 M}_{\ell',m,s'\mu'} C^{J_2 M}_{\ell',m,s'\mu'} C^{J_1 M}_{\ell_1 0 s\mu} C^{J_2 M}_{\ell_2 0 s\mu} \,,$$

where

$$\sum_{m\mu'} C^{J_1 M}_{\ell' m s'\mu'} C^{J_2 M}_{\ell' m s'\mu'} = \delta_{J_1 J_2}$$

and

$$\sum_{M\mu} C^{JM}_{\ell_1 0 s\mu} C^{JM}_{\ell_2 0 s\mu} = \frac{2J+1}{2\ell_1+1}\delta_{\ell_1\ell_2} \,.$$

This finally gives

$$\sigma_{s'\leftarrow s} = \frac{4\pi}{2s+1}\sum_{J\ell\ell'}(2J+1)\left|f^{J}_{[\ell'][\ell]}\right|^2 \,. \tag{10.52}$$

For an experiment where neither the initial nor the final spin is observed, we need to sum up the above cross section over both s and s':

$$\sigma = 4\pi \sum_{J[\ell][\ell']} \frac{2J+1}{2s+1}\left|f^{J}_{[\ell'][\ell]}\right|^2 \,. \tag{10.53}$$

10.4 Symmetries of the Jost Matrices

As we have seen in the case of a central potential, the Jost functions $f_\ell^{(\mathrm{in})}(E)$ and $f_\ell^{(\mathrm{out})}(E)$ are related to each other at different points of the Riemann surface (see Sect. 3.1). In fact, it is possible to establish for the matrices $f^{(\mathrm{in})J}(E)$ and $f^{(\mathrm{out})J}(E)$ exactly the same relations as are shown in Figs. 3.5 and 3.6. In order to do this, we re-write Eqs. (10.25, 10.26) in the integral form,

$$F^{(\mathrm{in/out})J}_{[\ell][\ell']}(E,r) = \frac{1}{2}\delta_{[\ell][\ell']} \mp \frac{1}{ik}\int_0^r h_\ell^{(\pm)}(kr)\sum_{[\ell'']} V^J_{[\ell][\ell'']}(r)\Big[h_{\ell''}^{(-)}(kr)F^{(\mathrm{in})J}_{[\ell''][\ell']}(E,r)$$

$$+\, h_{\ell''}^{(+)}(kr)F^{(\mathrm{out})J}_{[\ell''][\ell']}(E,r)\Big]\, dr\,. \tag{10.54}$$

Significant difference of these equations from those used in Sect. 3.1 is that they involve the Riccati–Hankel functions with different angular momenta. Therefore, if we change the sign of the momentum $k \to -k$, then

$$h_\ell^{(\pm)}(-kr) = (-1)^{\ell+1}h_\ell^{(\mp)}(kr)\,, \qquad h_{\ell''}^{(\pm)}(-kr) = (-1)^{\ell''+1}h_{\ell''}^{(\mp)}(kr)\,.$$

Generally speaking, the factors $(-1)^{\ell+1}$ and $(-1)^{\ell''+1}$ do not cancel each other because $\ell \neq \ell''$. However, thanks to the parity conservation, the odd and even angular momenta do not mix. As a result, the Jost matrices can always be presented in the block-diagonal form, as was demonstrated in Eq. (10.19), where one non-zero block only involves the even ℓ, while the other only couples the odd ℓ. In other words, due to the parity conservation, Eq. (10.54) can involve either odd or even angular momenta, but never both. In such a case, we always have

$$\Big[\text{parity conservation}\Big] \quad \Rightarrow \quad (-1)^{\ell+1} = (-1)^{\ell''+1}\,. \tag{10.55}$$

After establishing this simple fact, we can repeat the reasoning given in Sect. 3.1 to prove the symmetry properties of the Jost function for a central potential. As is easy to see, exactly the same properties are valid for the matrices $f^{(\mathrm{in/out})J}(E)$:

vertical symmetry	$f^{(\mathrm{in})J}(e^{2i\pi}E) = f^{(\mathrm{out})J}(E)$	(10.56)
diagonal symmetry	$f^{(\mathrm{in/out})J}(E) = \Big[f^{(\mathrm{out/in})J}(E^*)\Big]^*$	(10.57)

horizontal symmetry	$f^{(\text{in/out})J}(E) = \left[f^{(\text{in/out})J}(e^{2\pi i}E^*)\right]^*$

(10.58)

particular case of diagonal symmetry	$\left[f^{(\text{in})J}(E)\right]^* = f^{(\text{out})J}(E)\,, \quad E > 0$

(10.59)

The easiest way to understand the meaning of these properties is to look at Figs. 3.5 and 3.6, where the functions $f_\ell^{(\text{in/out})}$ can be replaced by the corresponding matrices $f^{(\text{in/out})J}$.

10.5 Analytic Structure of the Jost Matrices

In the case of a central potential, it was shown that the Jost functions can be split into two terms as is given by Eq. (7.65), where the functions $a_\ell(E)$ and $b_\ell(E)$ are analytic and single-valued. Such a two term splitting can be obtained for a problem with the non-central forces as well. This is done further down in Sect. 13.4, where such a problem is considered as a particular (simplified) case of a more general multi-channel problem. Each partial wave can formally be treated as a separate channel. These channels are degenerate because they have the same threshold energy.

In Sect. 13.4, it is shown that the Jost matrices for a non-central potential have the following structure:

$$\boxed{f^{(\text{in/out})J}_{[\ell][\ell']}(E) = k^{\ell'-\ell}a^J_{[\ell][\ell']}(E) \mp ik^{\ell+\ell'+1}b^J_{[\ell][\ell']}(E)} \tag{10.60}$$

where the matrices $a^J(E)$ and $b^J(E)$ are the analytic single-valued functions of the complex variable E. The square-root branching of the Riemann surface on which the Jost matrices are defined is due to the factor $k^{\ell+\ell'+1}$ in Eq. (10.60). The other k-dependent factor, $k^{\ell'-\ell}$, is actually E-dependent because the power $\ell' - \ell$ is always even thanks to the parity conservation law, which requires that ℓ and ℓ' are both either even or odd.

It is also shown that on the real axis the matrices $a(E)$ and $b(E)$ are real. Then from the Schwarz reflection principle, it follows that

$$a^J(E^*) = \left[a^J(E)\right]^*\,, \qquad b^J(E^*) = \left[b^J(E)\right]^*\,. \tag{10.61}$$

It is not difficult to show that the symmetry relations (10.56, 10.58, 10.57, 10.59), obtained in the previous section, could also be derived using the structure (10.60) and the properties (10.61). These symmetry relations can be obtained in the same way as it is done for the problem with a central potential.

10.6 Time-Reversal Invariance, Unitarity, and Parity Conservation

There are three important general properties of the S-matrix that are valid for any potential:

- The transposition symmetry that follows from the time-reversal invariance.
- The unitarity that is a consequence of the particle-flux conservation.
- The conservation of parity that implies that odd and even angular momenta cannot mix. This fact reduces the number of non-zero matrix elements of S^J in half.

10.6.1 Time-Reversal Invariance

In classical mechanics, all the processes are symmetric with respect to the time reversal. This means that by changing the direction of time to the opposite, we could (in principle) observe how all the bodies move backwards along the same trajectories. This follows from the fact that the dynamical equations have such a symmetry.

Time-reversal invariance is assumed to be one of the fundamental symmetries of nature, and thus the quantum dynamics must also be time-reversal invariant. There are no trajectories in quantum mechanics. Therefore, instead of trajectories, the time evolution of the wave function should be exactly reversed if the direction of time is changed. In classical mechanics, the backward movement follows the same coordinates $\vec{r}(t)$, but with opposite directions of the momentum, i.e. $-\vec{p}(t)$, and angular momentum, i.e. $-\vec{L}(t)$. Since classical mechanics is a limiting case ($\hbar \to 0$) of quantum mechanics, the corresponding quantum observables should change in the same way. In other words, under the time reversal, all the coordinates remain the same, but linear and angular momenta change their directions to the opposite. The quantum spins should also change direction because they are of the same nature as the orbital angular momenta.

Replacing time variable t with $-t$ in the Schrödinger equation that describes the time evolution of the wave function,

$$i\hbar \frac{\partial}{\partial t} \Psi_a(\vec{r}, t) = (H_0 + V)\Psi_a(\vec{r}, t) , \tag{10.62}$$

and changing the quantum numbers a (linear and angular momenta) to the opposite, we see that the resulting equation is the same as for Ψ^*_{-a}, which could be obtained by complex conjugation of Eq. (10.62) and replacing a with $-a$. Of course, we assume that the potential is real.[7] Therefore, the time-reversed "trajectory" (the backward moving wave function) is Ψ^*_{-a}. In other words,

$$\mathcal{T}\Psi_a = \Psi^*_{-a} , \tag{10.63}$$

where $\mathcal{T}$ is the time-reversal operator. To be more precise, we should write

$$\langle \vec{r}\,|\mathcal{T}|\Psi_a(t)\rangle = \langle \vec{r}\,|\Psi_{-a}(-t)\rangle = \Psi^*_{-a}(\vec{r}, t) . \tag{10.64}$$

Obviously, if wave function is a linear combination of some other functions, then in accordance with Eq. (10.63), the operator $\mathcal{T}$ replaces the combination coefficients with their complex conjugates,

$$\mathcal{T}(c_1\Psi_1 + c_2\Psi_2) = c_1^*\mathcal{T}\Psi_1 + c_2^*\mathcal{T}\Psi_2 . \tag{10.65}$$

In simple words, this means that the operator $\mathcal{T}$ replaces any complex number with its complex conjugate,

$$\mathcal{T}c = c^*\mathcal{T} . \tag{10.66}$$

The operators having the property (10.65) are called anti-linear. It is not difficult to see that in the matrix elements the anti-linear operators cannot be moved to the bra-vectors in the same way as we get used to, i.e.

$$\langle\Psi_1|\mathcal{T}\Psi_2\rangle \neq \langle\mathcal{T}^\dagger\Psi_1|\Psi_2\rangle . \tag{10.67}$$

Indeed, if it were possible, then we would come to the following contradiction:

$$\langle c\Psi_1|\mathcal{T}\Psi_2\rangle = \langle\mathcal{T}^\dagger c\Psi_1|\Psi_2\rangle = \langle c^*\mathcal{T}^\dagger\Psi_1|\Psi_2\rangle = c\langle\mathcal{T}^\dagger\Psi_1|\Psi_2\rangle = c\langle\Psi_1|\mathcal{T}\Psi_2\rangle ,$$

where the complex number c is taken out of the bra-vector and is not conjugated, which contradicts the general rule. Therefore, the rule for moving an anti-linear operator to the bra-vector differs from the corresponding rule for a linear one, and the difference consists in the additional complex conjugation of the matrix element,

$$\langle\Psi_1|\mathcal{T}\Psi_2\rangle = \langle\mathcal{T}^\dagger\Psi_1|\Psi_2\rangle^* . \tag{10.68}$$

This can be seen from comparing the following two equations:

$$\langle c_1\Psi_1|\mathcal{T}c_2\Psi_2\rangle = c_1^*c_2^*\langle\Psi_1|\mathcal{T}\Psi_2\rangle ,$$

[7] If V is complex, the particle-flux is either absorbed or generated, i.e. is not conserved. In such a case, the physical processes cannot be reversed.

$$\langle \mathcal{T}^\dagger c_1\Psi_1|c_2\Psi_2\rangle = c_1c_2\langle \mathcal{T}^\dagger \Psi_1|\Psi_2\rangle ,$$

where c_1 and c_2 are complex numbers.

The invariance means that the system moves backwards along the same "trajectory", i.e. at every instance of time, the distribution of probability in the space is the same as it was for the forward movement. In the case we are considering, the wave function actually is a spinor in the spin space. Therefore, the probability density is given by $\Psi^\dagger\Psi$. The $\mathcal{T}$-invariance thus can be written as

$$\Psi^\dagger\Psi = (\mathcal{T}\Psi)^\dagger \mathcal{T}\Psi = \Psi^\dagger(\mathcal{T}^\dagger\mathcal{T})\Psi . \tag{10.69}$$

Since the state Ψ is arbitrary, this implies that

$$\mathcal{T}^\dagger\mathcal{T} = 1 . \tag{10.70}$$

Anti-linear operators obeying the condition (10.70) are called anti-unitary.

The $\mathcal{T}$-invariance means that the time-reversal operator commutes with the Hamiltonian,

$$\mathcal{T}H = H\mathcal{T} . \tag{10.71}$$

Together with the fact that $\mathcal{T}i = -i\mathcal{T}$, this gives

$$\mathcal{T}\Omega_\pm = \mathcal{T}\left[\lim_{t\to\mp\infty} e^{+iHt}e^{-iH_0t}\right] = \left[\lim_{t\to\mp\infty} e^{-iHt}e^{+iH_0t}\right]\mathcal{T} = \Omega_\mp\mathcal{T} .$$

For the S-matrix defined via the Møller operators (see Eq. (1.43)), we have

$$\mathcal{T}S = \mathcal{T}\Omega_-^\dagger\Omega_+ = \Omega_+^\dagger\mathcal{T}\Omega_+ = \Omega_+^\dagger\Omega_-\mathcal{T} = S^\dagger\mathcal{T}$$

Using this relation in the following chain of equalities:[8]

$$\langle\phi_{a'}|S|\phi_a\rangle = \langle\phi_{a'}|\mathcal{T}^\dagger\mathcal{T}S|\phi_a\rangle = \langle\phi_{a'}|\mathcal{T}^\dagger S^\dagger\mathcal{T}|\phi_a\rangle = \langle\mathcal{T}\phi_{a'}|S^\dagger|\mathcal{T}\phi_a\rangle^*$$

$$= \langle\phi_{-a'}|S^\dagger|\phi_{-a}\rangle^* = \langle\phi_{-a}|S|\phi_{-a'}\rangle \tag{10.72}$$

(where $|\phi_a\rangle$ are the asymptotic time-independent states), we conclude that for the transition $a' \leftarrow a$, i.e. from a set of quantum numbers a to a set a', the probability amplitude is equal to the corresponding amplitude describing the transition $(-a) \leftarrow (-a')$, where $(-a)$ is a set of quantum numbers of a time-reversed state.

The fact that the time reversal changes the directions of all angular momenta (including spins) to the opposite means that their third components change signs, i.e.

$$|JM\rangle \longrightarrow |J, -M\rangle .$$

[8] Take note of Eqs. (10.68) and (10.70).

The partial-wave matrix elements of the S-matrix do not depend on the orientation of any angular momenta (this can be proved in the same way as it is done in Sect. 10.2.1 for the matrix elements of the potential). Therefore, according to Eq. (10.72), it is symmetric,

$$\boxed{S^J_{[\ell'][\ell]}(E) = S^J_{[\ell][\ell']}(E)} \tag{10.73}$$

10.6.2 Unitarity

In the simple case of single-channel scattering caused by a spherically symmetric potential, where the S-matrix has the dimensions (1×1) (i.e. is just a complex number), we have established that for real positive energies its partial-wave components are of unit length (see Sect. 4.2),

$$S_\ell(E) = e^{i2\delta_\ell(E)}\,, \qquad E > 0\,.$$

This is a consequence of the conservation of the flux of particles. In the processes involving non-central interactions (and generally in the multi-channel problems), the total flux (in all open channels) incoming to and outgoing from the sphere of a large radius is also conserving. Let us show that this makes the S-matrix unitary and thus relates (in a certain way) its matrix elements to each other. In order to do this, we re-write Eq. (10.36) in the form

$$\psi_{\vec{k}}(\vec{r}\,) \underset{r\to\infty}{\longrightarrow} \psi^{(\mathrm{in})}_{\vec{k}}(\vec{r}\,) + \psi^{(\mathrm{out})}_{\vec{k}}(\vec{r}\,)\,, \tag{10.74}$$

where the two terms,

$$\psi^{(\mathrm{in})}_{\vec{k}}(\vec{r}\,) = \frac{1}{\sqrt{2\pi}kr}\sum_{JM}|\mathcal{Y}^{JM}(\hat{\vec{r}}\,)\rangle\rangle W^{(\mathrm{in})}\mathcal{I}\langle\mathcal{Y}^{JM}(\hat{\vec{k}}\,)|\,, \tag{10.75}$$

$$\psi^{(\mathrm{out})}_{\vec{k}}(\vec{r}\,) = \frac{1}{\sqrt{2\pi}kr}\sum_{JM}|\mathcal{Y}^{JM}(\hat{\vec{r}}\,)\rangle\rangle W^{(\mathrm{out})}S^J\mathcal{I}\langle\mathcal{Y}^{JM}(\hat{\vec{k}}\,)|\,, \tag{10.76}$$

describe the incoming and outgoing waves. For the sake of clarity, it is worthwhile to recollect that $|\mathcal{Y}^{JM}(\hat{\vec{r}}\,)\rangle\rangle$ and $\langle\mathcal{Y}^{JM}(\hat{\vec{k}}\,)|$ are the "ket"- and "bra"-vectors in the spin space:

$$|\mathcal{Y}^{JM}(\hat{\vec{r}}\,)\rangle\rangle = \left\{|\mathcal{Y}^{JM}_{[\ell_1]}(\hat{\vec{r}}\,)\rangle\rangle, |\mathcal{Y}^{JM}_{[\ell_2]}(\hat{\vec{r}}\,)\rangle\rangle, \dots, |\mathcal{Y}^{JM}_{[\ell_N]}(\hat{\vec{r}}\,)\rangle\rangle\right\}\,,$$

$$\langle\mathcal{Y}^{JM}(\hat{\vec{k}}\,)| = \begin{bmatrix}\langle\mathcal{Y}^{JM}_{[\ell_1]}(\hat{\vec{k}}\,)| \\ \langle\mathcal{Y}^{JM}_{[\ell_2]}(\hat{\vec{k}}\,)| \\ \vdots \\ \langle\mathcal{Y}^{JM}_{[\ell_N]}(\hat{\vec{k}}\,)|\end{bmatrix}\,,$$

where[9]

$$|\mathcal{Y}_{[\ell]}^{JM}(\hat{\vec{r}}\,)\rangle = \sum_{m\mu} Y_{\ell m}(\hat{\vec{r}}\,)|s\mu\rangle C_{\ell m s\mu}^{JM}$$

and

$$\langle\mathcal{Y}_{[\ell]}^{JM}(\hat{\vec{k}}\,)| = \sum_{m\mu} Y_{\ell m}^{*}(\hat{\vec{k}}\,)\langle s\mu|C_{\ell m s\mu}^{JM}\,.$$

The right hand sides of Eqs. (10.76, 10.75) are constructed as the sums of the products of various matrices. The resulting matrices, i.e. $\psi_{\vec{k}}^{(\mathrm{in})}(\vec{r}\,)$ and $\psi_{\vec{k}}^{(\mathrm{out})}(\vec{r}\,)$, have the dimensions (1×1). In other words, they consist of single elements. Indeed, the structure of these products can be symbolically written as (if for a particular J, there are, for example, three partial waves)

$$\Big(|*\rangle,\ |*\rangle,\ |*\rangle\Big)\begin{pmatrix} * & * & * \\ * & * & * \\ * & * & * \end{pmatrix}\begin{pmatrix} \langle *| \\ \langle *| \\ \langle *| \end{pmatrix} = \Big(|*\rangle\langle *| + |*\rangle\langle *| + |*\rangle\langle *|\Big) = (\hat{\bigstar})\,.$$

Despite being represented by the single elements, $(\hat{\bigstar})$, the quantities $\psi_{\vec{k}}^{(\mathrm{in})}(\vec{r}\,)$ and $\psi_{\vec{k}}^{(\mathrm{out})}(\vec{r}\,)$ are not just complex numbers but the operators in the spin space. When sandwiched between the spin states, they become square matrices, $\langle s'\mu'|\psi_{\vec{k}}(\vec{r}\,)|s\mu\rangle$ (see Eq. (10.31)). Each column of such a matrix describes the state that evolves from the incoming spin state $|s\mu\rangle$. Each element in a chosen column is the amplitude of the probability to find the system in the spin state $|s'\mu'\rangle$. The whole square matrix describes the spin transitions from all possible initial spin states to all possible final spin states.

When considering the conservation of the flux, we choose a single initial state and look at its transitions to all possible final states. This means that out of the square matrix, we take only one column that corresponds to a chosen initial state. For our purpose, it does not matter which column to choose because in any case the whole S-matrix is involved in the transformation of the initial to the final state. Therefore, further down in this section, the wave function $\psi_{\vec{k}s\mu}(\vec{r}, s', \mu')$ should be understood as a single column for a fixed initial spin state $|s\mu\rangle$. However, the derivations are done here in the operator form, where the operator $\psi_{\vec{k}}(\vec{r}\,)$ is not sandwiched between the spin states $\langle s'\mu'|$ and $|s\mu\rangle$.

When calculating the particle-fluxes corresponding to $\psi_{\vec{k}}^{(\mathrm{in})}(\vec{r}\,)$ and $\psi_{\vec{k}}^{(\mathrm{out})}(\vec{r}\,)$ that are represented by the column-matrices, we should be careful not to commute

[9] The symbol μ denotes here the third component of the spin and should not be confused with the reduced mass.

the matrices and need to replace the complex conjugation in Eq. (1.52) with the Hermitian conjugation, i.e.

$$\vec{j} = \frac{\hbar}{2i\mu}\left\{\psi^{\dagger}(\vec{r}\,)\,\vec{\nabla}\,\psi(\vec{r}\,) - \left[\vec{\nabla}\,\psi^{\dagger}(\vec{r}\,)\right]\psi(\vec{r}\,)\right\}, \tag{10.77}$$

where ψ can be either $\psi^{(\text{in})}_{\vec{k}}$ or $\psi^{(\text{out})}_{\vec{k}}$. Since we are interested in the fluxes of the particles incoming to and outgoing from a sphere, among the three components of vector $\vec{j}$, we are only interested in the radial flux, which can be found using the first component of the "nabla" operator, $\vec{\nabla} = \{\partial_r,\ (1/r)\partial_\theta,\ (1/r\sin\theta)\partial_\varphi\}$. Since we consider the functions (10.76, 10.75) at large distances, the waves $W^{(\text{in/out})}$ can be replaced with their asymptotic expressions given by Eq. (9.61), namely,

$$W^{(\text{in})}(E,r) \underset{r\to\infty}{\longrightarrow} ie^{-ikr}\mathcal{I}\,, \qquad W^{(\text{out})}(E,r) \underset{r\to\infty}{\longrightarrow} -ie^{+ikr}\mathcal{I}^{\dagger}\,. \tag{10.78}$$

Therefore, the radial dependence of the waves (10.76, 10.75) is only given by the functions $e^{\mp ikr}/r$ and thus

$$\boxed{r\to\infty} \qquad \Rightarrow \qquad \partial_r\psi^{(\text{in/out})}_{\vec{k}}(\vec{r}\,) = \left(-\frac{1}{r} \mp ik\right)\psi^{(\text{in/out})}_{\vec{k}}(\vec{r}\,)\,.$$

Substituting these derivatives in the radial component of Eq. (10.77), we obtain the radial components of the incoming and outgoing fluxes,

$$j_r^{(\text{in/out})} = \mp\frac{\hbar k}{\mu}\psi^{(\text{in/out})\dagger}_{\vec{k}}\,\psi^{(\text{in/out})}_{\vec{k}}\,. \tag{10.79}$$

If this expression is written explicitly (in the matrix form), then it becomes clear that the sum over all possible final spin states is automatically included via the multiplication of the row matrix $\psi^{\dagger}$ and the column-matrix ψ.

The total fluxes that come in and out of a sphere of a large radius can be obtained by integrating these radial fluxes over all the angles. If the particles do not disappear and do not emerge from nowhere, the total incoming and outgoing fluxes must compensate each other, i.e.

$$\int j_r^{(\text{in})}\,d\hat{\vec{r}} + \int j_r^{(\text{out})}\,d\hat{\vec{r}} = \frac{\hbar k}{\mu}\int\left[-\psi^{(\text{in})\dagger}_{\vec{k}}\,\psi^{(\text{in})}_{\vec{k}} + \psi^{(\text{out})\dagger}_{\vec{k}}\,\psi^{(\text{out})}_{\vec{k}}\right]d\hat{\vec{r}} = 0\,. \tag{10.80}$$

When doing the angular integration, we can use the fact that

$$\int \langle\mathcal{Y}^{JM}(\hat{\vec{r}}\,)|\mathcal{Y}^{J'M'}(\hat{\vec{r}}\,)\rangle\,d\hat{\vec{r}} = \delta_{JJ'}\delta_{MM'}\,, \tag{10.81}$$

which follows from the definition (10.33) and the ortho-normal property given by Eq. (10.7). From the definition (10.35), it is also seen that

$$\mathcal{I}^\dagger \mathcal{I} = \mathcal{I}\mathcal{I}^\dagger = 1\,. \tag{10.82}$$

As a result, we obtain

$$\sum_{JM} |\mathcal{Y}^{JM}(\hat{\vec{k}})\rangle \mathcal{I}^\dagger \left[\left(S^J\right)^\dagger S^J - 1\right] \mathcal{I} \langle \mathcal{Y}^{JM}(\hat{\vec{k}})| = 0\,. \tag{10.83}$$

Since the direction of the incident momentum can be chosen arbitrarily, the above equation implies that the S-matrix is unitary (for the real scattering energies),

$$\left(S^J\right)^\dagger S^J = 1\,, \qquad E > 0\,. \tag{10.84}$$

The same result can be obtained in a more simple way by combining the property (10.59) of the Jost matrices with the definition (10.37) of the S-matrix and its symmetry property (10.73) due to the time-reversal invariance. Indeed, since the S-matrix is symmetric, its complex conjugation is equivalent to Hermitian conjugation. This gives[10]

$$\left(S^J\right)^\dagger = \left(S^J\right)^* = f^{(\mathrm{out})J*}\left[f^{(\mathrm{in})J*}\right]^{-1} = f^{(\mathrm{in})J}\left[f^{(\mathrm{out})J}\right]^{-1},$$

and thus

$$\boxed{\left(S^J\right)^\dagger S^J = S^J \left(S^J\right)^\dagger = 1\,, \qquad E > 0} \tag{10.85}$$

After looking at this simple proof, it may seem that the flux conservation is not necessary for the unitarity of the S-matrix. However, we should not forget that the symmetry relation (10.59) was derived from the fact that the Hamiltonian was Hermitian, and its hermicity is only guaranteed if the potential is real, which in turn is equivalent to the conservation of the flux.

10.6.3 Conservation of Parity

Parity is the eigenvalue of the operator $\mathcal{P}$ that does mirror reflection of all the coordinates, i.e. $\vec{r} \to -\vec{r}$. If we apply this operator twice, everything should return to the original state. Therefore, the eigenvalue of $\mathcal{P}$ can only be either $+1$ or -1 (even and odd states).

[10] It should not be forgotten that this is only valid for real positive energies.

Since the momentum is proportional to velocity (time derivative of the coordinate), $\hbar\vec{k} = \mu\,\dot{\vec{r}}$, it also changes its sign ($\vec{k} \;\to\; -\vec{k}$) under the action of $\mathcal{P}$. The angular momentum is the product

$$\vec{\ell} = \hbar[\vec{r} \times \vec{k}],$$

and thus it is not affected by the coordinate inversion. This should be valid for an angular moment of any nature and in particular for spin.

Therefore, the coordinate inversion affects only the spatial part of the wave function. In the partial-wave decomposition of a bound-state wave function (see Eq. (10.13)), the angular dependence of each term is determined by the corresponding spherical harmonics whose parity is the same as the parity of the angular momentum because

$$Y_{\ell m}(-\vec{r}\,) = (-1)^{\ell} Y_{\ell m}(\vec{r}\,)\,.$$

The scattering wave function depends on both $\vec{r}$ and $\vec{k}$ and is expanded in the double partial-wave series (see Eq. (10.21)). The physical meaning of this wave function can be understood as follows. This is the amplitude of the probability that the particle initially moving with the momentum $\vec{k}$ will be found moving in the direction $\vec{r}$ after the collision. Each term of the double partial-wave series (10.21) is the amplitude of the transition from the partial wave ℓ to the partial wave ℓ'. The parity conservation implies that ℓ and ℓ' can be either both even or both odd. In other words, the difference between ℓ and ℓ' is always even, i.e.

$$\ell - \ell' = \pm 2n\,, \qquad n = 0, 1, 2, 3, \ldots \tag{10.86}$$

Apparently, the same is applicable to all the matrices involved in the description of the scattering process. In the potential, the Jost matrices, and the S-matrix, the subscripts ℓ and ℓ' can only run over either even or odd numbers (not mixing them). This actually reduces the dimensions of these matrices by half.

10.6.4 Reciprocity and Detailed Balance

The symmetries we have established manifest themselves in the form of certain relations among the direct and inverse scattering processes. Let us closely look at them, keeping in mind that throughout all our reasoning the parameter E is assumed to be real and positive (scattering energy).

First of all, we see that the partial-wave scattering amplitude defined by Eq. (10.38) is transpose-symmetric in the same way as the corresponding S-matrix is symmetric. Indeed, Eq. (10.38) in explicit form reads

$$f^{J}_{[\ell'][\ell]} = \frac{1}{2ik}\left(i^{-\ell'} S^{J}_{[\ell'][\ell]} i^{\ell} - 1\right)\,, \tag{10.87}$$

where the S-matrix obeys the (time reversal) symmetry relation (10.73) and, according to Eq. (10.86),

$$\ell - \ell' = \pm 2n \qquad \Longrightarrow \qquad i^{\ell-\ell'} = i^{\ell'-\ell} .$$

These facts imply that

$$f^{J}_{[\ell'][\ell]}(E) = f^{J}_{[\ell][\ell']}(E) , \qquad E > 0 . \tag{10.88}$$

Apparently, this symmetry of the partial-wave amplitude should result in some symmetry of the total amplitude (10.43). Let us find this out.

The total amplitude involves the angular functions (10.10), which can be written as

$$Y^{JM}_{\ell s \mu}(\hat{\vec{r}}) = Y_{\ell, M-\mu}(\hat{\vec{r}}) C^{JM}_{\ell, M-\mu, s, \mu} , \tag{10.89}$$

because the Clebsch–Gordan coefficients are zero if $M \neq m + \mu$. Using the properties of the spherical harmonics,

$$Y_{\ell m}(\hat{\vec{r}}) = (-1)^m Y^*_{\ell,-m}(\hat{\vec{r}}) , \qquad Y_{\ell m}(\hat{\vec{r}}) = (-1)^\ell Y_{\ell m}(-\hat{\vec{r}}) ,$$

and of the Clebsch–Gordan coefficients,

$$C^{JM}_{\ell m s \mu} = (-1)^{\ell+s-J} C^{J,-M}_{\ell,-m,s,-\mu} ,$$

we can re-write the angular functions involved in Eq. (10.43), as follows:

$$\begin{aligned} Y^{JM}_{\ell' s' \mu'}(\hat{\vec{r}}) &= Y_{\ell', M-\mu'}(\hat{\vec{r}}) C^{JM}_{\ell', M-\mu', s', \mu'} \\ &= (-1)^{\ell'}(-1)^{M-\mu'}(-1)^{\ell'+s'-J} Y^*_{\ell', \mu'-M}(-\hat{\vec{r}}) C^{J,-M}_{\ell', \mu'-M, s', -\mu'} \\ &= (-1)^{s'-\mu'}(-1)^{J-M}\left[Y^{J,-M}_{\ell', s', -\mu'}(-\hat{\vec{r}})\right]^* , \end{aligned}$$

and similarly,

$$\left[Y^{JM}_{\ell s \mu}(\hat{\vec{k}})\right]^* = (-1)^{s-\mu}(-1)^{J-M} Y^{J,-M}_{\ell, s, -\mu}(-\hat{\vec{k}}) . \tag{10.90}$$

As a result of these transformations, the total angular momentum and spin as well as vectors $\vec{r}$ and $\vec{k}$ have been inverted, and the angular functions have been replaced with their complex conjugates.

Using these transformations and the symmetry property (10.88) in the expression (10.43) for the total amplitude, we obtain

$$\langle s'\mu'|\mathfrak{f}(\hat{\vec{r}} \leftarrow \hat{\vec{k}}, E)|s\mu\rangle$$

$$= 4\pi(-1)^{s'-\mu'}(-1)^{s-\mu} \sum_{JM\ell'\ell} Y^{J,-M}_{\ell,s,-\mu}(-\hat{\vec{k}}) f^{J}_{[\ell][\ell']}(E) \left[Y^{J,-M}_{\ell',s',-\mu'}(-\hat{\vec{r}})\right]^* .$$

Apparently, the summation variable M can be replaced with $-M$. The resulting expression on the right hand side of the above equation gives the amplitude for the reciprocal process, i.e.

$$\boxed{\langle s'\mu'|\mathfrak{f}(\hat{\vec{r}} \leftarrow \hat{\vec{k}}, E)|s\mu\rangle = (-1)^{s'+s-\mu'-\mu}\langle s,-\mu|\mathfrak{f}(-\hat{\vec{k}} \leftarrow -\hat{\vec{r}}, E)|s',-\mu'\rangle} \tag{10.91}$$

This is known as the reciprocity theorem. Equation (10.91) relates the amplitudes for the processes schematically depicted in Fig. 10.1, where the long arrows show velocity directions of the incoming and outgoing particles, and the short arrows indicate the orientations of the spin before and after the collision. The upper and lower halves of the figure correspond to the left and right hand sides of Eq. (10.91).

When we take the absolute squares of both sides of the reciprocity relation (10.91), the phase factors $(-1)^{s'-\mu'}(-1)^{s-\mu}$ disappear. The scattering angle, i.e. the angle between vectors $\vec{k}$ and $\vec{r}$, and between $-\vec{r}$ and $-\vec{k}$, is the same because

$$\vec{k}\cdot\vec{r} = (-\vec{r})\cdot\left(-\vec{k}\right) .$$

Therefore, the reciprocity relation in terms of the differential cross section defined by Eq. (10.44) can be written as follows:

$$\left(\frac{d\sigma}{d\Omega}\right)_{s'\mu'\leftarrow s\mu} = \left(\frac{d\sigma}{d\Omega}\right)_{s,-\mu\leftarrow s',-\mu'} . \tag{10.92}$$

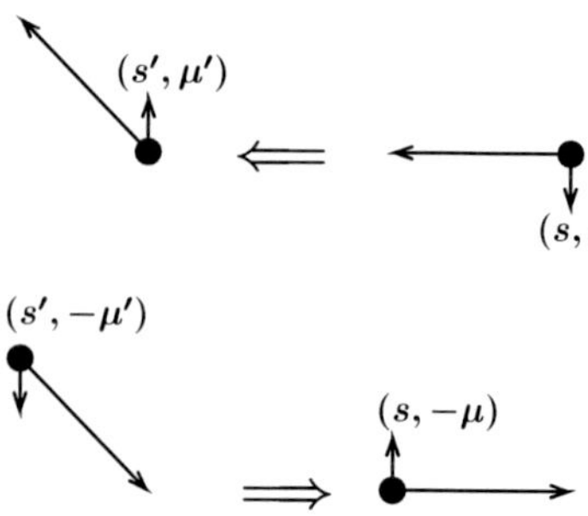

Fig. 10.1 Schematically shown two reciprocal scattering processes, one of which (top) proceeds from right to left and the other (bottom) in the opposite direction. The long arrows indicate the directions of motion at large distances. Short arrows show the directions of the spins

Apparently, the integration over the scattering angle retains this relation for the total cross sections,

$$\sigma_{s'\mu' \leftarrow s\mu} = \sigma_{s,-\mu \leftarrow s',-\mu'} \,. \tag{10.93}$$

Averaging the cross sections over the initial and summing over the final spin orientations, we obtain the total unpolarized cross sections:

$$\sigma_{s' \leftarrow s} = \frac{1}{2s+1} \sum_{\mu\mu'} \sigma_{s'\mu' \leftarrow s\mu} \,,$$

$$\sigma_{s \leftarrow s'} = \frac{1}{2s'+1} \sum_{\mu\mu'} \sigma_{s,-\mu \leftarrow s',-\mu'} \,.$$

Then it follows from Eq. (10.93) that

$$\boxed{(2s+1)\sigma_{s' \leftarrow s} = (2s'+1)\sigma_{s \leftarrow s'}} \tag{10.94}$$

which is known as the principle of detailed balance.

10.7 Simplicity of the Bound and Resonant State Zeros

Let $E_0 = (\hbar k_0)^2/(2\mu)$ be the spectral point corresponding to a bound or a resonant state. The physical (radial) wave function, $u(E, r)$, describing such a state, can be represented by a column-matrix consisting of N elements, where N is the number of the coupled partial waves (see Eq. (9.90)),

$$u(E,r) = \begin{bmatrix} u_{[\ell_1]}(E,r) \\ u_{[\ell_2]}(E,r) \\ \vdots \\ u_{[\ell_N]}(E,r) \end{bmatrix} \underset{r\to\infty}{\longrightarrow} \begin{bmatrix} \mathcal{A}_{[\ell_1]} e^{ik_0 r} \\ \mathcal{A}_{[\ell_2]} e^{ik_0 r} \\ \vdots \\ \mathcal{A}_{[\ell_N]} e^{ik_0 r} \end{bmatrix} . \tag{10.95}$$

We assume that this wave function is normalized to unity, and therefore, the coefficients $\mathcal{A}_{[\ell_n]}$ are the asymptotic normalization constants (ANC). Further down, the symbol $\mathcal{A}$ is used to denote the column-matrix composed of these constants.

It is not difficult to see that the column c of the combination coefficients in Eq. (9.90) should be taken as the following product:

$$c = \left[f^{(\mathrm{out})}(E_0) \right]^{-1} i\mathcal{I}\mathcal{A} \,, \tag{10.96}$$

where the diagonal matrix $\mathcal{I}$ is defined by Eq. (9.108). Indeed, using the representation (9.86) of the regular solution as a combination of the Jost solutions, we have

$$\phi c = \chi^{(-)} S^{-1} i\mathcal{I}\mathcal{A} + \chi^{(+)} i\mathcal{I}\mathcal{A} . \tag{10.97}$$

At a spectral point, the S-matrix is singular and thus

$$S^{-1}(E) \underset{E \to E_0}{\longrightarrow} 0 . \tag{10.98}$$

This removes the first term on the right hand side of Eq. (10.97) and gives the normalized physical solution as

$$u = \chi^{(+)} i\mathcal{I}\mathcal{A} , \tag{10.99}$$

which asymptotically behaves as in (10.95).

In this section, our task is to show that the zero of the Jost matrix determinant (see Eq. (9.95)) at the spectral point E_0 is simple (i.e. of the first order) or, equivalently, that the S-matrix has a simple pole at this point. In order to complete the task, we do matrix generalization of the derivations used in Sect. 3.3.5.

* * *

To begin with, we take Eq. (9.84) representing the Jost matrix in terms of the Wronskian of the Jost and regular solutions and differentiate it with respect to variable E,

$$\dot{f}^{(\text{in})} = -\frac{1}{2ik}\left[W(\dot{\chi}^{(+)}, \phi) + W(\chi^{(+)}, \dot{\phi})\right] - \left(\frac{d}{dE}\frac{1}{2ik}\right) W(\chi^{(+)}, \phi) .$$

Here the dot over the symbols means the derivative over E. Taking into account Eq. (10.98), we see that

$$W(\chi^{(+)}, \phi)\left[f^{(\text{out})}\right]^{-1} = -2ikS^{-1} \underset{E \to E_0}{\longrightarrow} 0 .$$

Therefore at the spectral point, $E = E_0$, we have

$$\dot{f}^{(\text{in})}\left[f^{(\text{out})}\right]^{-1} = -\frac{1}{2ik_0}\left[W(\dot{\chi}^{(+)}, \phi) + W(\chi^{(+)}, \dot{\phi})\right]\left[f^{(\text{out})}\right]^{-1} . \tag{10.100}$$

In order to find the Wronskians that are needed here, consider the radial equations for the transposed Jost solution and for the regular solution at different energies, $\tilde{E} = (\hbar\tilde{k})^2/(2\mu)$ and $E = (\hbar k)^2/(2\mu)$:

$$\left[\chi^{(+)\prime\prime}\right]^T + \left[\chi^{(+)}\right]^T \tilde{k}^2 = \left[\chi^{(+)}\right]^T \mathcal{V} , \tag{10.101}$$

$$\phi'' + k^2\phi = \mathcal{V}\phi , \tag{10.102}$$

where $\mathcal{V}$ is the sum of the centrifugal and the dynamical potentials:

$$\mathcal{V} = \mathrm{diag}\left\{\frac{\ell_1(\ell_1+1)}{r^2}, \frac{\ell_2(\ell_2+1)}{r^2}, \ldots, \frac{\ell_N(\ell_N+1)}{r^2}\right\} + V\,.$$

Multiplying Eq. (10.101) by ϕ from the right and Eq. (10.102) by $\left[\chi^{(+)}\right]^T$ from the left, and subtracting the second from the first, we obtain[11]

$$\partial_r W\left[\chi^{(+)}, \phi\right] = \left[\chi^{(+)}\right]^T (\tilde{k}^2 - k^2)\phi\,. \tag{10.103}$$

In this equation, the matrices $\chi^{(+)}$ and ϕ depend on different energy variables, namely, $\tilde{E}$ and E, respectively. Differentiating Eq. (10.103) over E and then putting $E = \tilde{E}$, we obtain

$$\partial_r W\left[\chi^{(+)}, \dot{\phi}\right] = -\frac{2\mu}{\hbar^2}\left[\chi^{(+)}\right]^T \phi\,. \tag{10.104}$$

Similarly, differentiating with respect to $\tilde{E}$ and putting $\tilde{E} = E$, we get another equation:

$$\partial_r W\left[\dot{\chi}^{(+)}, \phi\right] = +\frac{2\mu}{\hbar^2}\left[\chi^{(+)}\right]^T \phi\,. \tag{10.105}$$

Now we integrate Eq. (10.104) from 0 to r, and Eq. (10.105) from r to infinity. This integration gives us the Wronskians that we need for Eq. (10.100):

$$W\left[\chi^{(+)}, \dot{\phi}\right]_r = W\left[\chi^{(+)}, \dot{\phi}\right]_{r=0} - \frac{2\mu}{\hbar^2}\int_0^r \left[\chi^{(+)}\right]^T \phi\, dr'\,, \tag{10.106}$$

$$W\left[\dot{\chi}^{(+)}, \phi\right]_r = W\left[\dot{\chi}^{(+)}, \phi\right]_{r=\infty} - \frac{2\mu}{\hbar^2}\int_r^\infty \left[\chi^{(+)}\right]^T \phi\, dr'\,. \tag{10.107}$$

Before substituting these Wronskians, we multiply Eq. (10.100) by $\mathcal{A}^T i\mathcal{I}$ from the left and by $i\mathcal{I}\mathcal{A}$ from the right. This gives

$$\mathcal{A}^T i\mathcal{I}\dot{f}^{(\mathrm{in})}\left[f^{(\mathrm{out})}\right]^{-1} i\mathcal{I}\mathcal{A} = \tag{10.108}$$

$$-\frac{1}{2ik_0}\mathcal{A}^T i\mathcal{I}W(\dot{\chi}^{(+)}, \phi)\left[f^{(\mathrm{out})}\right]^{-1} i\mathcal{I}\mathcal{A} - \frac{1}{2ik_0}\mathcal{A}^T i\mathcal{I}W(\chi^{(+)}, \dot{\phi})\left[f^{(\mathrm{out})}\right]^{-1} i\mathcal{I}\mathcal{A}\,.$$

[11] It should be noted that the Wronskian for matrix functions is defined by Eq. (C.5), where the first of the two matrices is transposed.

Now, when substituting (10.106) and (10.107) into Eq. (10.108), we take into account the following obvious relations:

$$\mathcal{A}^T i\mathcal{I}\left[\chi^{(+)}\right]^T = u^T\,, \tag{10.109}$$

$$\phi\left[f^{(\mathrm{out})}\right]^{-1} i\mathcal{I}\mathcal{A} = u\,, \tag{10.110}$$

from which it follows that the Wronskians on the right hand sides of Eqs. (10.106, 10.107) vanish. Indeed,

$$\mathcal{A}^T i\mathcal{I}\, W\left[\chi^{(+)}, \dot{\phi}\right]_{r=0}\left[f^{\,(\mathrm{out})}\right]^{-1} i\mathcal{I}\mathcal{A} = W\left[u, \dot{u}\right]_{r=0} = 0\,,$$

$$\mathcal{A}^T i\mathcal{I}\, W\left[\dot{\chi}^{(+)}, \phi\right]_{r=\infty}\left[f^{\,(\mathrm{out})}\right]^{-1} i\mathcal{I}\mathcal{A} = W\left[\dot{u}, u\right]_{r=\infty} = 0\,,$$

because at $r = 0$ and at $r \to \infty$ the function $u(E, r)$ has very simple dependence on E. The differentiation over the energy at the end points of the interval $r \in [0, \infty)$ leaves u and $r \in [0, \infty)$ linearly dependent and thus their Wronskian is zero. This cannot be said about the intermediate points of the interval $r \in [0, \infty)$.

Therefore, from Eq. (10.108), it follows that

$$\mathcal{A}^T i\mathcal{I}\,\dot{f}^{\,(\mathrm{in})}\left[f^{\,(\mathrm{out})}\right]^{-1} i\mathcal{I}\mathcal{A} = -\frac{i\mu}{\hbar^2 k_0}\int_0^\infty u^T u\, dr\,. \tag{10.111}$$

On the right hand side of this equation, there is the normalization integral,

$$\int_0^\infty u^T u\, dr = \int_0^\infty \left(u_{[\ell_1]}^2 + u_{[\ell_2]}^2 + \cdots + u_{[\ell_N]}^2\right) dr = 1\,. \tag{10.112}$$

As is discussed in Sect. 3.3.4, for a bound state, the radial part of the wave function can always be made real, and therefore, the simple squares under the integral (10.112) are equivalent to the absolute squares. In the case of a resonance, the wave function can also be normalized if we integrate the simple squares along the ray in the complex r-plane, shown in Fig. 3.9.

If c is the column-matrix defined by Eq. (10.96), then Eq. (10.111) can be written as

$$\mathcal{A}^T i\mathcal{I}\,\dot{f}^{\,(\mathrm{in})} c = -\frac{i\mu}{\hbar^2 k_0}\,. \tag{10.113}$$

The set of the combination coefficients c is chosen as the solution of Eq. (9.94), i.e.

$$f^{(\mathrm{in})}(E_0)c = 0\,. \tag{10.114}$$

With such coefficients, we get rid of the incoming waves in the asymptotics of the physical wave function.

Let us choose a point E that is close to E_0 and expand the matrix $f^{(\mathrm{in})}(E)$ in the Taylor series,

$$f^{(\mathrm{in})}(E) = X_0 + (E - E_0)X_1 + (E - E_0)^2 X_2 + \cdots , \tag{10.115}$$

and its inverse matrix in the Laurent series,

$$\left[f^{(\mathrm{in})}(E)\right]^{-1} = \frac{Y_{-n}}{(E - E_0)^n} + \frac{Y_{1-n}}{(E - E_0)^{n-1}} + \frac{Y_{2-n}}{(E - E_0)^{n-2}} + \cdots , \tag{10.116}$$

which begins with certain negative power, n. Our task is to show that $n = 1$, which means that the pole at $E = E_0$ is simple.

In Eqs. (10.115, 10.116), the square matrices X_n and Y_n depend on E_0, but not on E. Therefore, when differentiating (10.115) with respect to variable E, we obtain

$$\dot{f}^{(\mathrm{in})}(E) = X_1 + 2(E - E_0)X_2 + 3(E - E_0)^2 X_3 + \cdots . \tag{10.117}$$

Now, we multiply the series (10.116) by (10.115) from the right and collect in groups all the terms of the same power of $(E - E_0)^m$. Since

$$\left[f^{(\mathrm{in})}(E)\right]^{-1} f^{(\mathrm{in})}(E) = I ,$$

the only non-zero group is the one where the power $m = 0$. The variable E is arbitrary, which implies that

$$m = -n : \quad Y_{-n} X_0 = 0 , \tag{10.118}$$

$$m = 1 - n : \quad Y_{1-n} X_0 + Y_{-n} X_1 = 0 , \tag{10.119}$$

$$m = 2 - n : \quad Y_{2-n} X_0 + Y_{1-n} X_1 + Y_{-n} X_2 = 0 , \tag{10.120}$$

$$\cdots \quad \cdots\cdots\cdots\cdots\cdots\cdots\cdots\cdots\cdots\cdots\cdots ,$$

$$m = 0 : \quad Y_0 X_0 + Y_{-1} X_1 + Y_{-2} X_2 + \cdots + Y_{-n} X_n = 1 . \tag{10.121}$$

It is clear that $f^{(\mathrm{in})}(E_0) = X_0$ from which it follows that $X_0 c = 0$. Therefore, if we multiply Eq. (10.119) by c from the right, then the first term vanishes, and we obtain

$$Y_{-n} X_1 c = 0 .$$

The matrix X_1 is equal to $\dot{f}_\ell^{(\mathrm{in})}(E_0)$, and it follows from Eq. (10.113) that $X_1 c \neq 0$. Therefore, $Y_{-n} = 0$. Now, we multiply Eq. (10.120) by c from the right, which gives

$$Y_{1-n} X_1 c = 0 ,$$

from where we obtain that $Y_{1-n} = 0$. Continue in this way, we obtain that $Y_{-n} = Y_{1-n} = Y_{2-n} = \cdots = 0$ and finally come to Eq. (10.121). Multiplying this equation by c from the right, we find that all the terms vanish except one:

$$Y_{-1}X_1c = 1\ ,$$

from where it follows that $Y_{-1} \neq 0$. In other words, the Laurent series (10.116) begins with the term proportional to $(E - E_0)^{-1}$. This proves that at the spectral points the S-matrix has poles of the first order.

10.8 Asymptotic Normalization Constants

In this section, we are looking for a relation between the residues of the elements of the S-matrix at a spectral point and the asymptotic normalization constants of the corresponding wave function. To find such a relation, we come back to Eq. (10.111), and multiply it by $\mathcal{A}$ from the left and by $\mathcal{A}^T$ from the right. In this way, we obtain the following relation (valid for both the bound and resonant states):

$$\mathcal{A}\mathcal{A}^T\mathcal{I}\dot{f}^{(\mathrm{in})}\left[f^{(\mathrm{out})}\right]^{-1}\mathcal{I}\mathcal{A}\mathcal{A}^T = \frac{i\mu}{\hbar^2 k_0}\mathcal{A}\mathcal{A}^T \tag{10.122}$$

and thus

$$\mathcal{A}\mathcal{A}^T\mathcal{I}\dot{f}^{(\mathrm{in})}\left[f^{(\mathrm{out})}\right]^{-1} = \frac{i\mu}{\hbar^2 k_0}\mathcal{I}^*\ . \tag{10.123}$$

Here, the product $\mathcal{A}\mathcal{A}^T$ is a symmetric and factorized square matrix (which we denote as $\mathbb{A}$),

$$\mathbb{A} = \mathcal{A}\mathcal{A}^T = \begin{pmatrix} \mathcal{A}_1 \\ \mathcal{A}_2 \\ \vdots \\ \mathcal{A}_N \end{pmatrix} (\mathcal{A}_1, \mathcal{A}_2, \cdots, \mathcal{A}_N)$$

$$= \begin{pmatrix} \mathcal{A}_1\mathcal{A}_1 & \mathcal{A}_1\mathcal{A}_2 & \cdots & \mathcal{A}_1\mathcal{A}_N \\ \mathcal{A}_2\mathcal{A}_1 & \mathcal{A}_2\mathcal{A}_2 & \cdots & \mathcal{A}_2\mathcal{A}_N \\ \vdots & \vdots & \vdots & \vdots \\ \mathcal{A}_N\mathcal{A}_1 & \mathcal{A}_N\mathcal{A}_2 & \cdots & \mathcal{A}_N\mathcal{A}_N \end{pmatrix},$$

where for the sake of the notation simplicity, the symbols $\mathcal{A}_{[\ell_n]}$ are replaced with $\mathcal{A}_n$. It is seen that

$$(\mathcal{I}\mathbb{A}\mathcal{I})_{mn} = \sum_{m'n'} \mathcal{I}_{mm'}\mathbb{A}_{m'n'}\mathcal{I}_{n'n} = \sum_{m'n'} i^{\ell_m}\delta_{mm'}\mathbb{A}_{m'n'}i^{\ell_n}\delta_{nn'} = i^{\ell_m}\mathbb{A}_{mn}i^{\ell_n}$$

is also symmetric. Multiplying Eq. (10.123) by $\mathcal{I}$ from the left, we obtain

$$\mathcal{I}\mathbb{A}\mathcal{I}\,\dot{f}^{(\text{in})}\left[f^{(\text{out})}\right]^{-1}=\frac{i\mu}{\hbar^2 k_0}\,. \tag{10.124}$$

If we make an infinitely small shift of the energy from the spectral point (in any direction), i.e. $E=E_0+\epsilon$, then both $\left[f^{(\text{in})}(E)\right]^{-1}$ and $S(E)$ do exist. Inserting in Eq. (10.124) the unit matrix $\left[f^{(\text{in})}\right]^{-1}f^{(\text{in})}$, we have

$$\mathcal{I}\mathbb{A}\mathcal{I}\,\dot{f}^{(\text{in})}\left[f^{(\text{in})}\right]^{-1}f^{(\text{in})}\left[f^{(\text{out})}\right]^{-1}=\frac{i\mu}{\hbar^2 k_0}\,,$$

which can be written as

$$S=-\frac{i\hbar^2 k_0}{\mu}\mathcal{I}\mathbb{A}\mathcal{I}\,\dot{f}^{(\text{in})}\left[f^{(\text{in})}\right]^{-1}\,. \tag{10.125}$$

Let

$$D(E)=\det f^{(\text{in})}(E)\,,$$

and then

$$M(E)=D(E)\left[f^{(\text{in})}(E)\right]^{-1}$$

is the $(N\times N)$-matrix consisting of the transposed minors that are used to construct the inverse matrix. It is clear that $M(E)$ exists even when $\epsilon\to 0$. Multiplying Eq. (10.125) by D and letting $\epsilon\to 0$, we obtain

$$f^{(\text{out})}M=-\frac{i\hbar^2 k_0}{\mu}\mathcal{I}\mathbb{A}\mathcal{I}\,\dot{f}^{(\text{in})}M\,. \tag{10.126}$$

As we have established, the S-matrix is transpose-symmetric due to the time-reversal invariance (see Sect. 10.6.1. This means that the product $f^{(\text{out})}M$ is also symmetric because its matrix elements differ from those of S by the common factor $1/D$. Since the left hand side of Eq. (10.126) is symmetric, its right hand side must be symmetric as well. On the right hand side, we have the product of the symmetric matrix $\mathcal{I}\mathbb{A}\mathcal{I}$ and the matrix

$$\mathbb{B}=\dot{f}^{(\text{in})}M\,. \tag{10.127}$$

The product $(\mathcal{I}\mathbb{A}\mathcal{I})\mathbb{B}$ is symmetric, i.e.

$$(\mathcal{I}\mathbb{A}\mathcal{I}\mathbb{B})_{mn}=(\mathcal{I}\mathbb{A}\mathcal{I}\mathbb{B})_{nm}\,,$$

and therefore,

$$\sum_{n'}(\mathcal{A}_m i^{\ell_m}\mathcal{A}_{n'}i^{\ell_{n'}}\mathbb{B}_{n'n} - \mathcal{A}_n i^{\ell_n}\mathcal{A}_{n'}i^{\ell_{n'}}\mathbb{B}_{n'm}) = 0\,,$$

from which it follows that

$$\sum_{n'}\mathcal{A}_{n'}i^{\ell_{n'}}\mathbb{B}_{n'n} = Q\mathcal{A}_n i^{\ell_n}\,,$$

where Q is the same factor for all n. Therefore, $\mathbb{B}_{mn} = Q\delta_{mn}$, i.e.

$$\mathbb{B} = \begin{pmatrix} Q & 0 & \cdots & 0 \\ 0 & Q & \cdots & 0 \\ \vdots & \vdots & \cdots & \vdots \\ 0 & 0 & \cdots & Q \end{pmatrix}.$$

This means that $Q = \mathrm{Sp}\,\mathbb{B}/N$. On the other hand, according to Eq. (F.13) of Appendix F,

$$\mathrm{Sp}\,\mathbb{B} = \dot{D}\,.$$

Therefore,

$$\mathcal{IAIB} = Q\mathcal{IAI} = \frac{1}{N}\,\dot{D}\,\mathcal{IAI}\,. \tag{10.128}$$

Since the pole of the S-matrix at the spectral point E_0 is of the first order (see Sect. 10.7), the corresponding residue is given by

$$\mathrm{Res}\left(S_{[\ell'][\ell]}, E_0\right) = \frac{1}{\dot{D}}\sum_{[\ell'']} f^{(\mathrm{out})}_{[\ell'][\ell'']} M_{[\ell''][\ell]}\,.$$

Substituting (10.128) into Eq. (10.126), we finally obtain the relation we were looking for

$$\boxed{\mathrm{Res}\left(S_{[\ell'][\ell]}, E_0\right) = -\frac{i\hbar^2 k_0}{\mu N}\mathcal{A}_{[\ell']}\mathcal{A}_{[\ell]}} \tag{10.129}$$

As we see, the residues of the S-matrix at the corresponding poles are factorized, i.e. they are the products of the factors depending on different subscripts.

10.9 Scattering Phase Shifts

For a fixed value of the total angular momentum J, the partial-wave S-matrix $S^J(E)$ has a finite dimension ($N \times N$). Indeed, only few partial waves falling into the limits (10.2) can be involved. Moreover, thanks to the parity conservation, the S-matrix is block-diagonal (similarly, to the Jost matrices in the example (10.19)). The diagonal blocks involve either only even or only odd partial waves and can be considered separately. Therefore, the S-matrix (i.e. either its even or odd block) consists of N^2 complex numbers $S^J_{[\ell'][\ell]}$, i.e. is given by a set of $2N^2$ real parameters, but not all of them are independent. The number of independent parameters is reduced by the facts that the S-matrix is symmetric and unitary.

There are N diagonal elements, and thus there are $N^2 - N$ off-diagonal ones. Half of them, i.e. $(N^2 - N)/2$, are above the diagonal, and they are equal to the symmetric ones that are below the diagonal. This gives us $(N^2 - N)/2$ equations relating the parameters. The unitarity relation (10.84) provides another set of N^2 equations. Therefore, the number M of independent (real) parameters that determine the S-matrix is

$$M = 2N^2 - \frac{N^2 - N}{2} - N^2 = \frac{N(N+1)}{2} .$$

The matrix $S^J_{[\ell'][\ell]}$ is a representation of the S-operator S^J that acts in the N-dimensional space spanned by the ortho-normal basis $\{|[\ell]\rangle\}$ consisting of the states with definite values of the quantum numbers ℓ, s, and J. As we have established, this operator is unitary. Therefore, it can be written in the following form:

$$S^J = \exp(2i\Delta^J) , \tag{10.130}$$

where Δ^J is a Hermitian operator,[12] i.e.

$$\left(S^J\right)^\dagger S^J = 1 \quad \Longrightarrow \quad \left(\Delta^J\right)^\dagger = \Delta^J .$$

The operator Δ^J acts in the same N-dimensional space as the operator S^J does. Let $|n\rangle$ be an eigen-vector of Δ^J and δ^J_n be the corresponding eigenvalue,

$$\Delta^J |n\rangle = \delta^J_n |n\rangle .$$

[12] The factor 2 in the exponential is introduced for consistency with Eq. (4.19) that is valid for a central potential.

Since Δ^J is Hermitian, its eigenvalues are real, and the eigen-vectors can form another ortho-normal basis in the same N-dimensional space. In this basis, the operator Δ^J is diagonal,

$$\Delta^J = \sum_{n=1}^{N} |n\rangle \delta_n^J \langle n| \, .$$

Apparently, in the basis $\{|n\rangle\}$, the operator S^J is also diagonal

$$S^J = \sum_{n=1}^{N} |n\rangle e^{2i\delta_n^J} \langle n| \, . \tag{10.131}$$

The matrix $S^J_{[\ell'][\ell]}$ is obtained by sandwiching Eq. (10.131) between the basis vectors $\langle[\ell']|$ and $|[\ell]\rangle$,

$$S^J_{[\ell'][\ell]} = \sum_{n=1}^{N} \langle[\ell']|n\rangle e^{2i\delta_n^J} \langle n|[\ell]\rangle \, .$$

On the right hand side of this equation, we have the basis-transformation matrix (i.e. the overlap matrix),

$$O_{n,[\ell]} \stackrel{\text{def}}{=} \langle n|[\ell]\rangle \, ,$$

and the corresponding transposed matrix

$$\left(O^T\right)_{[\ell],n} = \langle[\ell]|n\rangle \, .$$

In the N-dimensional space, which we are considering, the basis vectors obey the closure relation, i.e.

$$\sum_{n=1}^{N} |n\rangle\langle n| = 1 \qquad \Longrightarrow \qquad \sum_{n=1}^{N} \langle[\ell']|n\rangle\langle n|[\ell]\rangle = \delta_{[\ell'][\ell]} \, .$$

This means that the basis-transformation matrix O is orthogonal (for which the transposition is equivalent to inversion),

$$O^T O = 1 \, . \tag{10.132}$$

The transposition symmetry of the S-matrix, given by Eq. (10.73), implies that

$$\sum_{n=1}^{N}\langle[\ell']|n\rangle e^{2i\delta_n^J}\langle n|[\ell]\rangle = \sum_{n=1}^{N}\langle[\ell]|n\rangle e^{2i\delta_n^J}\langle n|[\ell']\rangle\,,$$

$$\sum_{n=1}^{N}\langle[\ell]|n\rangle^* e^{2i\delta_n^J}\langle n|[\ell']\rangle^* = \sum_{n=1}^{N}\langle[\ell]|n\rangle e^{2i\delta_n^J}\langle n|[\ell']\rangle\,,$$

which means that

$$\langle[\ell]|n\rangle^* = \langle[\ell]|n\rangle$$

and thus all elements of the matrix O are real.

The set of eigenvalues $\delta_1^J, \delta_2^J, \ldots, \delta_N^J$ can be considered as the first N (out of M) real parameters specifying the S-matrix. The remaining $M - N = N(N-1)/2$ parameters should be chosen among the elements of the matrix O. There are N^2 of such elements. If we choose $N(N-1)/2$ among them, there are still $N^2 - N(N-1)/2 = N(N+1)/2$ elements of the matrix O that remain undefined. For finding them, we have exactly $N(N+1)/2$ equations relating the elements of O to each other. These equations follow from the orthogonality condition (10.132). Indeed, the product $O^T O$ is a symmetric matrix because

$$\left(O^T O\right)^T = O^T O\,,$$

and thus when equating matrix elements of this product with the unit matrix, we only take into account the N diagonal elements and $N(N-1)/2$ elements above (or equivalently below) the diagonal. So, in total, we obtain $N + N(N-1)/2 = N(N+1)/2$ equations.

When parametrizing the matrix O, one should keep in mind that each element of this matrix is an overlap of two unit vectors and therefore its absolute value cannot be greater than 1,

$$-1 \leqslant \langle n|[\ell]\rangle \leqslant +1\,.$$

Due to this reason, the elements of matrix O are usually chosen in the form of sin and cos functions depending on a parameter ε_n. Thus the constructed S-matrix is specified by a set of N eigen-phases δ_n^J and by a set of $N(N-1)/2$ mixing parameters ε_n. Of course, the choice of the parametrization is not unique. It is a matter of convention. Let us consider the most important example of two coupled partial waves.

10.9.1 Example: Two Coupled Partial Waves

The case of two coupled partial waves is very common and thus plays an important role in the scattering theory. When the total spin of the colliding particles is 1, we always have a (2×2) partial-wave S-matrix that couples the states (of the same parity) with $\ell = J \pm 1$. Perhaps, the most familiar example is the nucleon–nucleon scattering in the triplet state (when the spins of the nucleons are parallel). In the nucleon–nucleon system, the total spin determines the permutation symmetry, and therefore the triplet states do not mix with the singlet ones, i.e. the total spin is conserving as well as the total angular momentum J. As a result, the multi-index $[\ell] \equiv \{\ell, s\}$ may only assume the following two values: $\{J-1, 1\}$ and $\{J+1, 1\}$.

As we mentioned above, the choice of the S-matrix parametrization is not unique. In the case of $N = 2$, there are two generally accepted conventions for this. One of them is based on the eigen-phase shifts (it is also known as the Blatt–Biedenharn parametrization[37]), and the other uses the so-called nuclear-bar phase shifts[38]. Let us consider both of them.

10.9.1.1 Eigen-Phase Shifts

This kind of parametrization was considered for the general case of an arbitrary N in Sect. 10.9. Here we just demonstrate how it works for $N = 2$. In this case, the number of independent parameters describing the S-matrix is three. Two of them are the eigen-phases. This means that the matrix O can only have one independent parameter. It is a convention to take it as $\cos\varepsilon$ for the top left corner of the matrix,

$$O = \begin{pmatrix} \cos\varepsilon & x \\ y & z \end{pmatrix} ,$$

where x, y, and z are dependent parameters and should be somehow expressed in terms of the mixing parameter ε. To find x, y, and z, we can use the orthogonality condition,

$$O^T O = \begin{pmatrix} \cos\varepsilon & y \\ x & z \end{pmatrix} \begin{pmatrix} \cos\varepsilon & x \\ y & z \end{pmatrix} = \begin{pmatrix} 1 & 0 \\ 0 & 1 \end{pmatrix} ,$$

which gives us a set of three coupled equations,[13]

$$\cos^2\varepsilon + y^2 = 1 , \tag{10.133}$$

$$x\cos\varepsilon + yz = 0 , \tag{10.134}$$

$$x^2 + z^2 = 1 . \tag{10.135}$$

[13] The possible fourth equation is the same as (10.134).

One of the solutions of this set is $x = \sin\varepsilon$, $y = -\sin\varepsilon$, and $z = \cos\varepsilon$ (the other possible solutions may differ by signs, but this one is commonly used). Therefore[14]

$$S = \begin{pmatrix} \cos\varepsilon & -\sin\varepsilon \\ \sin\varepsilon & \cos\varepsilon \end{pmatrix} \begin{pmatrix} e^{2i\delta_1} & 0 \\ 0 & e^{2i\delta_2} \end{pmatrix} \begin{pmatrix} \cos\varepsilon & \sin\varepsilon \\ -\sin\varepsilon & \cos\varepsilon \end{pmatrix} \tag{10.136}$$

$$= \begin{bmatrix} e^{2i\delta_1}\cos^2\varepsilon + e^{2i\delta_2}\sin^2\varepsilon & (e^{2i\delta_1} - e^{2i\delta_2})\dfrac{\sin 2\varepsilon}{2} \\ (e^{2i\delta_1} - e^{2i\delta_2})\dfrac{\sin 2\varepsilon}{2} & e^{2i\delta_1}\sin^2\varepsilon + e^{2i\delta_2}\cos^2\varepsilon \end{bmatrix} . \tag{10.137}$$

If the S-matrix has been calculated, then the eigen-phase shifts together with the mixing parameter can be found from the set of equations:

$$e^{2i\delta_1}\cos^2\varepsilon + e^{2i\delta_2}\sin^2\varepsilon = S_{11} ,$$

$$(e^{2i\delta_1} - e^{2i\delta_2})\frac{\sin 2\varepsilon}{2} = S_{12} ,$$

$$e^{2i\delta_1}\sin^2\varepsilon + e^{2i\delta_2}\cos^2\varepsilon = S_{22} .$$

10.9.1.2 Nuclear-Bar Phase Shifts

An alternative parametrization is mainly used (as one can guess from the name) in nuclear physics. The main reason for introducing this representation of the S-matrix was the desire to have a convenient way of separating the effects of the strong and Coulomb forces.

In nuclear collisions, we almost always deal with the joint action of a short-range (but very strong) nuclear potential and a long-range (but relatively weak) Coulomb potential. To a very good accuracy, the joint potential coincides with the strong one at short distances and with the Coulomb tail at large distances. This means that the total phase shift is approximately equal to the sum of pure Coulomb one, σ_ℓ, and pure nuclear phase shift, δ_ℓ^N (see the discussion in Sect. 8.2.3 about the nature of such an approximation). When analysing experimental data, in most cases we are interested in obtaining information on pure nuclear processes, i.e. as if the Coulomb forces were switched off. In other words, we look for δ_ℓ^N and try to get rid of such a "nuisance" as the Coulomb phase shift.

The idea of how to "exclude" the pure Coulomb phase shifts can be understood if we come back to Eq. (8.71) that represents the single-channel Jost function as

[14] Please note that we dropped the superscript J to simplify the notation.

the product of the pure Coulomb one and the Coulomb-modified short-range Jost function. The S-matrix can be factorized as well,

$$S_\ell = f_\ell^{(\text{out})}\left[f_\ell^{(\text{in})}\right]^{-1} = f_\ell^{\,\text{c}\,(\text{out})} f_\ell^{\,\text{sc}\,(\text{out})}\left[f_\ell^{\,\text{c}\,(\text{in})} f_\ell^{\,\text{sc}\,(\text{in})}\right]^{-1}$$

$$= f_\ell^{\,\text{c}\,(\text{out})} f_\ell^{\,\text{sc}\,(\text{out})}\left[f_\ell^{\,\text{sc}\,(\text{in})}\right]^{-1}\left[f_\ell^{\,\text{c}\,(\text{in})}\right]^{-1} .$$

Substituting here the explicit expression (8.28) for the pure Coulomb Jost function, we obtain

$$S_\ell = e^{i\sigma_\ell}\,\overline{S}_\ell\,e^{i\sigma_\ell}\,, \tag{10.138}$$

where

$$\overline{S}_\ell = f_\ell^{\,\text{sc}\,(\text{out})}\left[f_\ell^{\,\text{sc}\,(\text{in})}\right]^{-1} = S_\ell^{\text{sc}}$$

is the Coulomb-modified short-range S-matrix. In a sense, it can be viewed as almost the same as the pure short-range S-matrix. If we accept such an approximation then to get rid of the Coulomb effects, we only need to sandwich the total matrix S_ℓ between the factors $e^{-i\sigma_\ell}$ and $e^{-i\sigma_\ell}$.

Equation (10.138) is written for the single-channel case, and therefore, all the quantities in this equation are just complex numbers commuting with each other. In Sect. 14.1, it is shown that a multi-channel S-matrix has a factorized structure similar to the one given by Eq. (10.138), i.e. it is a product of three matrices, where the diagonal matrices consisting of pure Coulomb exponentials $e^{i\sigma_\ell}$ are also on the left and the right sides of the product (see Eq. (14.32). The problem of a non-central potential, which we consider in the present section, is a particular (simplified) case of a multi-channel problem. We can therefore use here the general multi-channel representation (14.32).

After this rather long explanation, it should be clear why the Blatt–Biedenharn representation of the S-matrix in terms of the eigen-phase shifts is not very convenient for nuclear physics applications. Indeed, the matrix of the total (Coulomb plus nuclear) phase shifts is sitting in the middle of the product (10.136) of the three matrices. In such a representation, one cannot get rid of the Coulomb components of the phase shifts by simple multiplication from the left and from the right.

To overcome this difficulty, the following alternative parametrization was suggested, which we consider for a particular (most often used) case of $N = 2$. As we already know, the (2×2) S-matrix is specified by a set of three real parameters. Let us denote two of them as $\bar{\delta}_1$ and $\bar{\delta}_2$, and write the S-matrix in the following symmetric form (which is a particular case of the general representation (14.32)):

$$S = \begin{pmatrix} e^{i\bar{\delta}_1} & 0 \\ 0 & e^{i\bar{\delta}_2} \end{pmatrix} U \begin{pmatrix} e^{i\bar{\delta}_1} & 0 \\ 0 & e^{i\bar{\delta}_2} \end{pmatrix}, \tag{10.139}$$

where U is a unitary and symmetric (2×2)-matrix that depends on the one remaining real parameter. Therefore,

$$U = \begin{pmatrix} x & y \\ y & z \end{pmatrix} \quad \text{and} \quad U^\dagger U = \begin{pmatrix} |x|^2 + |y|^2 & x^* y + y^* z \\ x y^* + y z^* & |y|^2 + |z|^2 \end{pmatrix} = \begin{pmatrix} 1 & 0 \\ 0 & 1 \end{pmatrix} . \tag{10.140}$$

As follows from the Blatt–Biedenharn representation (10.137), the absolute values of the elements of the S-matrix do not exceed 1. Indeed,

$$|S_{11}| \leqslant |e^{2i\delta_1} \cos^2 \varepsilon| + |e^{2i\delta_2} \sin^2 \varepsilon| = 1$$

and similarly for the other matrix elements. The alternative representation (10.139) of the same matrix gives

$$S = \begin{pmatrix} e^{2i\bar{\delta}_1} x & e^{i(\bar{\delta}_1+\bar{\delta}_2)} y \\ e^{i(\bar{\delta}_1+\bar{\delta}_2)} y & e^{2i\bar{\delta}_2} z \end{pmatrix} ,$$

from where it is obvious that

$$|x| \leqslant 1 , \qquad |y| \leqslant 1 , \qquad |z| \leqslant 1 ,$$

and thus the elements of the matrix U can be written as the sin and cos functions of a real parameter. Let us choose $U_{11} = x = \cos 2\bar{\varepsilon}$ and find y and z from the unitarity condition (10.140). From the equations

$$|x|^2 + |y|^2 = 1 \qquad \text{and} \qquad |y|^2 + |z|^2 = 1 ,$$

it follows that $|y| = |\sin 2\bar{\varepsilon}|$ and $|z| = |\cos 2\bar{\varepsilon}|$. At this point, we can choose the signs arbitrarily, which gives different solutions. The convention is to choose $z = \cos 2\bar{\varepsilon}$. Then from the equation

$$x^* y + y^* z = 0,$$

it follows that $y = \pm i \sin 2\bar{\varepsilon}$ where we choose the "+" sign. Finally, we obtain the following "nuclear-bar" parametrization of the S-matrix

$$S = \begin{pmatrix} e^{i\bar{\delta}_1} & 0 \\ 0 & e^{i\bar{\delta}_2} \end{pmatrix} \begin{pmatrix} \cos 2\bar{\varepsilon} & i \sin 2\bar{\varepsilon} \\ i \sin 2\bar{\varepsilon} & \cos 2\bar{\varepsilon} \end{pmatrix} \begin{pmatrix} e^{i\bar{\delta}_1} & 0 \\ 0 & e^{i\bar{\delta}_2} \end{pmatrix} \tag{10.141}$$

$$= \begin{pmatrix} e^{2i\bar{\delta}_1} \cos 2\bar{\varepsilon} & i e^{i(\bar{\delta}_1+\bar{\delta}_2)} \sin 2\bar{\varepsilon} \\ i e^{i(\bar{\delta}_1+\bar{\delta}_2)} \sin 2\bar{\varepsilon} & e^{2i\bar{\delta}_2} \cos 2\bar{\varepsilon} \end{pmatrix} . \tag{10.142}$$

Of course, the nuclear-bar parameters $\bar{\delta}_1$, $\bar{\delta}_2$, and $\bar{\varepsilon}$ can be expressed in terms of the Blatt–Biedenharn parameters δ_1, δ_2, ε and vice versa. Equalizing the matrices (10.137) and (10.142), we obtain the following three equations:

$$e^{2i\bar{\delta}_1}\cos 2\bar{\varepsilon} = e^{2i\delta_1}\cos^2\varepsilon + e^{2i\delta_2}\sin^2\varepsilon\ , \tag{10.143}$$

$$e^{2i\bar{\delta}_2}\cos 2\bar{\varepsilon} = e^{2i\delta_1}\sin^2\varepsilon + e^{2i\delta_2}\cos^2\varepsilon\ , \tag{10.144}$$

$$i e^{i(\bar{\delta}_1+\bar{\delta}_2)}\sin 2\bar{\varepsilon} = (e^{2i\delta_1} - e^{2i\delta_2})\frac{\sin 2\varepsilon}{2}\ . \tag{10.145}$$

After a few algebraic transformations (which include: the division of the real part of Eq. (10.145) by its imaginary part, subtraction of Eq. (10.144) from Eq. (10.143), and the division of the resulting equation by Eq. (10.145)), these relations can be written in a more simple form:

$$\bar{\delta}_1 + \bar{\delta}_2 = \delta_1 + \delta_2\ , \tag{10.146}$$

$$\sin\left(\bar{\delta}_1 - \bar{\delta}_2\right) = \frac{\tan 2\bar{\varepsilon}}{\tan 2\varepsilon}\ , \tag{10.147}$$

$$\sin\left(\delta_1 - \delta_2\right) = \frac{\sin 2\bar{\varepsilon}}{\sin 2\varepsilon}\ . \tag{10.148}$$

For a given set of parameters (δ_1, δ_2, ε), these equations allow us to find the corresponding "bar" values, ($\bar{\delta}_1$, $\bar{\delta}_2$, $\bar{\varepsilon}$), and vice versa.

11 Multi-Channel Schrödinger Equation

So far, we restricted our consideration to either single channel or to degenerate multi-channel problems. However, practically, all real physical processes in the quantum world are "truly" multi-channel. This means that a collision of quantum systems may result in many different final products (channels),

$$A + B \longrightarrow \begin{cases} A + B \,, \\ C + D \,, \\ E + F \,, \\ \text{etc} \,. \end{cases}$$

Even if the particles A and B are not involved in a collision but are bound to each other, they can undergo virtual transitions to other channels (only to those that are allowed by the conservation of the angular momentum, parity, charge, etc.) and back,

$$A + B \longrightarrow C + D \longrightarrow A + B \,.$$

The energy is not conserved in such virtual transitions. However, the deviation ΔE of the energy from its conserving value is only possible for a short time interval Δt in accordance with the uncertainty principle

$$\Delta E \cdot \Delta t \sim \hbar \,. \tag{11.1}$$

Therefore, the more energy is needed for the transition, the less probable it is. The possibility of such virtual transitions means that a bound-state wave function for the particles A and B may have an admixture of the $(C + D)$ and the other

S. A. Rakityansky, *Jost Functions in Quantum Mechanics*,
https://doi.org/10.1007/978-3-031-07761-6_11

allowed configurations (channels). Similarly, the collision process may also go via all possible intermediate channels,

$$A + B \longrightarrow \begin{Bmatrix} A + B \\ C + D \\ E + F \\ \text{etc} \end{Bmatrix} \longrightarrow \begin{cases} A + B\,, \\ C + D\,, \\ E + F\,, \\ \text{etc}\,. \end{cases}$$

From this simple reasoning, it is clear that the description of multi-channel processes is much more complicated than what we had in the single-channel case. It should be emphasized that our further consideration will be restricted to such systems that have only the two-body channels (or can be approximately treated as such), i.e. all the channels only involve two quantum objects.

As we will see, all the quantities describing multi-channel systems, such as the Jost functions, S-matrix, and the wave function, become matrices with the dimensions equal to the number of channels. The Riemann surface of the energy becomes extremely complicated with many sheets and with a branch point at every channel threshold. The threshold for a given channel is defined as the minimal energy (relative to the zero of the chosen energy scale) that is needed for the particles in that channel to move freely.

* * *

In this and in the two chapters that follow, our objective is to introduce the main concepts of the multi-channel processes. We therefore temporarily avoid all unnecessary complications associated with the long-range Coulomb forces (the multi-channel systems of charged particles are considered further down in Chap. 14). In some cases, for the sake of clarity, we limit our consideration to a two-channel problem. It is always clear how to generalize the formulae obtained in this way, for an arbitrary number of channels.

To begin with, let us consider a simplified case, when the type of the particles is the same in all the channels. This means that the different channels correspond to different (excited) internal and spin states of the same interacting bodies, A and B, i.e.

$$A + B \longrightarrow \begin{cases} A + B\,, \\ A^* + B\,, \\ A^* + B^*\,, \\ \text{etc}\,. \end{cases} . \tag{11.2}$$

A way of treating the processes of the type $A+B \to C+D$ is outlined in Sect. 11.1.

There are two different types of the degrees of freedom, which we should keep in mind: the relative motion of the particles associated with the relative AB-coordinate and the internal motion inside each of these particles. As usual, the quantum mechanical two-body problem, after separating the motion of its centre of mass, can be reduced to an effective problem of one body whose dynamics is governed by the Hamiltonian

$$H = H_0 + U + h\,, \tag{11.3}$$

where H_0 and U describe the free relative AB-motion and the forces acting between the two bodies, and h is the Hamiltonian of the internal motion inside the bodies. Since the masses of the bodies A and B can be different for each of their internal states, the reduced masses μ that we obtain for the channels are generally different as well.

Let $|\eta\rangle$ be an eigenstate on the internal Hamiltonian,

$$h|\eta\rangle = E_\eta|\eta\rangle\,. \tag{11.4}$$

Since $h = h_A + h_B$, the total internal state $|\eta\rangle$ is the direct product of the internal states of the particles A and B,

$$|\eta\rangle = |\alpha\rangle \otimes |\beta\rangle = |\alpha, \beta\rangle\,,$$

where

$$h_A|\alpha\rangle = E_\alpha|\alpha\rangle\,, \qquad h_B|\beta\rangle = E_\beta|\beta\rangle\,, \qquad E_\eta = E_\alpha + E_\beta\,.$$

The number of different internal states may be infinite, but in any practical model, we always have some physical reasons to take into account only few of them. Therefore, we assume that there are $\mathcal{N}$ different internal states of the system AB. The internal state-vectors can be chosen in such a way that they form an ortho-normal set,

$$\langle\eta'|\eta\rangle = \delta_{\eta'\eta}\,.$$

When the particles A and B interact with each other, the internal motion inside them is disturbed. Their internal state becomes a mixture of all possible eigenstates (11.4), $|\eta_1\rangle$, $|\eta_2\rangle$, ..., $|\eta_\mathcal{N}\rangle$. Therefore, the total quantum vector $|\Psi\rangle$ in the full Hilbert space (including the internal and relative degrees of freedom) is a superposition of the orthogonal terms corresponding to each of the $\mathcal{N}$ internal states,

$$|\Psi\rangle = |\eta_1\rangle|\psi_{\eta_1}\rangle + |\eta_2\rangle|\psi_{\eta_2}\rangle + \cdots + |\eta_\mathcal{N}\rangle|\psi_\mathcal{N}\rangle\,. \tag{11.5}$$

This is an expansion of the total state-vector over the ortho-normal basis of the internal states. The expansion coefficients describe the relative motion in the corresponding channels (it is used the small letter ψ to distinguish the relative state-vector from the total one denoted by the capital letter Ψ). In this equation, the expansion coefficients $|\psi_\eta\rangle$ are written as the ket-vectors because they become the wave functions of the relative motion only after being projected on the bra-vector of the relative coordinate, $\langle\vec{r}\,|\psi_\eta\rangle$,

$$\langle\vec{r}\,|\Psi\rangle = |\eta_1\rangle\psi_{\eta_1}(\vec{r}\,) + |\eta_2\rangle\psi_{\eta_2}(\vec{r}\,) + \cdots + |\eta_{\mathcal{N}}\rangle\psi_{\mathcal{N}}(\vec{r}\,)\,. \tag{11.6}$$

Very often it is convenient to collect the terms of such an expansion in a column-matrix and represent the total state-vector by that column (similarly to the spinors in the theory of spin),

$$\langle\vec{r}\,|\Psi\rangle = \begin{bmatrix} |\eta_1\rangle\psi_{\eta_1}(\vec{r}\,) \\ |\eta_2\rangle\psi_{\eta_2}(\vec{r}\,) \\ \cdots \\ |\eta_{\mathcal{N}}\rangle\psi_{\mathcal{N}}(\vec{r}\,) \end{bmatrix}\,. \tag{11.7}$$

It should be clearly understood that such a representation is just a convenient way of writing various formulae and nothing more. Actual state-vector (or the corresponding wave function) is always a superposition of the type (11.6). The important thing is that all the terms in the sum (11.6) are orthogonal to each other. It is only possible to represent a state-vector by a column-matrix of the type (11.7), when it is split into mutually orthogonal terms. The reason is that, using the matrix multiplication rule, we obtain the correct scalar products like this:

$$\langle\Psi|\Psi'\rangle = \big(\langle\eta_1,\psi_{\eta_1}|,\langle\eta_2,\psi_{\eta_2}|,\ldots,\langle\eta_{\mathcal{N}},\psi_{\mathcal{N}}|\big)\begin{pmatrix} |\eta_1,\psi'_{\eta_1}\rangle \\ |\eta_2,\psi'_{\eta_2}\rangle \\ \cdots \\ |\eta_{\mathcal{N}},\psi'_{\mathcal{N}}\rangle \end{pmatrix}$$

$$= \langle\psi_{\eta_1}|\psi'_{\eta_1}\rangle + \langle\psi_{\eta_2}|\psi'_{\eta_2}\rangle + \cdots + \langle\psi_{\eta_{\mathcal{N}}}|\psi'_{\eta_{\mathcal{N}}}\rangle\,.$$

Of course, such a splitting is not unique. For example, we can combine the first two terms in the expansion (11.6) in a single term, which is orthogonal to all the other terms. In such a case, the length of the column-matrix representing vector $|\Psi\rangle$ becomes shorter. As an opposite example, we can expand each of the relative-motion wave functions on the right hand side of Eq. (11.6) in the partial waves, which are orthogonal to each other. In such a case, the dimension of the column-matrix becomes greater. In general, the column representation has the dimension equal to the number of the orthogonal terms which the state-vector is divided in, and such a division is a matter of convenience.

When the state-vector is represented by a column-matrix, all the operators become the square matrices of the same dimension. This can be illustrated using a simplified example with only two internal states,

$$|\Psi\rangle = |1\rangle\psi_1 + |2\rangle\psi_2\,.$$

The Schrödinger equation, $H|\Psi\rangle = E|\Psi\rangle$, when multiplied from the left by $\langle 1|$ and $\langle 2|$, gives the following set of equations:

$$\langle 1|H|1\rangle\psi_1 + \langle 1|H|2\rangle\psi_2 = E\psi_1 ,$$

$$\langle 2|H|1\rangle\psi_1 + \langle 2|H|2\rangle\psi_2 = E\psi_2 ,$$

which can be written in the matrix form,

$$\begin{pmatrix} H_{11} & H_{12} \\ H_{21} & H_{22} \end{pmatrix} \begin{pmatrix} \psi_1 \\ \psi_2 \end{pmatrix} = E \begin{pmatrix} \psi_1 \\ \psi_2 \end{pmatrix} .$$

If the interacting particles A and B have non-zero spins, then each channel is sub-divided into the mutually orthogonal spin sub-channels. Suppose that the particles A and B have spins s_α and s_β, respectively. Then these spins can have $(2s_\alpha+1)(2s_\beta+1)$ possible combinations of the values of their z-components, σ_α and σ_β. All these combinations are mutually orthogonal,

$$\langle s_\alpha\sigma_\alpha, s_\beta\sigma_\beta|s_\alpha\sigma'_\alpha, s_\beta\sigma'_\beta\rangle = \delta_{\sigma_\alpha\sigma'_\alpha}\delta_{\sigma_\beta\sigma'_\beta} .$$

Usually, it is more convenient to work within a different basis in the spin space. Using the Clebsch–Gordan coefficients, we can construct the states with definite values of the total two-body spin and its z-component, s and σ,

$$|(s_\alpha s_\beta)s\sigma\rangle = \sum_{\sigma_\alpha\sigma_\beta} C^{s\sigma}_{s_\alpha\sigma_\alpha s_\beta\sigma_\beta}|s_\alpha\sigma_\alpha, s_\beta\sigma_\beta\rangle . \tag{11.8}$$

Of course, the number of mutually orthogonal states $|(s_\alpha s_\beta)s\sigma\rangle$ remains the same, namely, $(2s_\alpha + 1)(2s_\beta + 1)$. Therefore, each term in Eq. (11.5) is sub-divided into mutually orthogonal spin states. Just to simplify the formulae, let us assume that $\mathcal{N} = 2$ (for the two states, η and η'); then

$$|\Psi\rangle = \sum_{i=1}^{N_s} |\eta, s_i\sigma_i\rangle|\psi_\eta^{(i)}\rangle + \sum_{j=1}^{N'_s} |\eta', s'_j\sigma'_j\rangle|\psi_{\eta'}^{(j)}\rangle = \begin{bmatrix} \sum_{i=1}^{N_s} |\eta, s_i\sigma_i\rangle|\psi_\eta^{(i)}\rangle \\ \sum_{j=1}^{N'_s} |\eta', s'_j\sigma'_j\rangle|\psi_{\eta'}^{(j)}\rangle \end{bmatrix} , \tag{11.9}$$

where N_s and N'_s are the numbers of different spin states for the internal states η and η', respectively. Since the spins of the particles A and B in different internal states can be different, these numbers are, generally speaking, not the same, $N_s \neq N'_s$. In Eq. (11.9), the ket-vectors $|\psi_\eta^{(i)}\rangle$ and $|\psi_{\eta'}^{(j)}\rangle$ describe the relative motion of the particles A and B when they are in the corresponding internal and spin states.

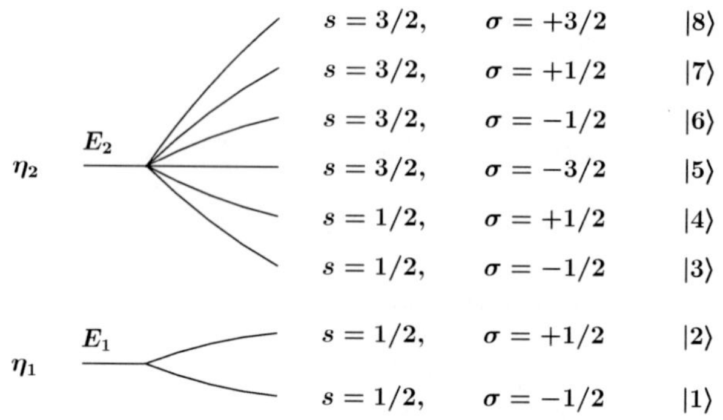

Fig. 11.1 Schematic example of a system with two internal states, η_1 and η_2, corresponding to the (threshold) energies E_1 and E_2. We consider a system such that $s_{\alpha_1} = 1/2$, $s_{\beta_1} = 0$ and $s_{\alpha_2} = 1/2$, $s_{\beta_2} = 1$. Then in the first state, the total spin can only be 1/2, and thus this channel is double-degenerate ($|1\rangle$, $|2\rangle$) with two possible orientations of the spin. In the second state, the spin can be either 1/2 or 3/2, and this channel is split into the six degenerate sub-channels ($|3\rangle$, $|4\rangle$, $|5\rangle$, $|6\rangle$, $|7\rangle$, $|8\rangle$)

Apparently, all the spin states belonging to the same η are degenerate,[1] i.e. they have the same internal energy E_η. An illustrative example given in Fig. 11.1 shows the case when in the state η_1 the total spin is obtained from $s_\alpha = 1/2$ and $s_\beta = 0$, while in the state η_2 it is a result of the spin-coupling of $s_\alpha = 1/2$ and $s_\beta = 1$.

The total number of mutually orthogonal terms in Eq. (11.9) is $N = N_s + N_s'$. This is the total number of the channels (including the degenerate channels). Each element in the column-matrix (11.9) can therefore be expanded in a sub-column,

$$|\Psi\rangle = \begin{bmatrix} |\eta, s_1\sigma_1\rangle|\psi_\eta^{(1)}\rangle \\ |\eta, s_2\sigma_2\rangle|\psi_\eta^{(2)}\rangle \\ \cdots \\ |\eta', s_1'\sigma_1'\rangle|\psi_{\eta'}^{(1)}\rangle \\ |\eta', s_2'\sigma_2'\rangle|\psi_{\eta'}^{(2)}\rangle \\ \cdots \end{bmatrix} \begin{matrix} \left.\vphantom{\begin{matrix}a\\a\\a\end{matrix}}\right\} N_s \text{ degenerate channels} \\ \left.\vphantom{\begin{matrix}a\\a\\a\end{matrix}}\right\} N_{s'} \text{ degenerate channels} \end{matrix} . \tag{11.10}$$

Therefore, N is the dimension of the matrices representing the quantum state and the Hamiltonian. To distinguish all these orthogonal terms, we can use a single label, a sequential number ν that runs from 1 to N. Then we can symbolically denote the

[1] The internal states of the interacting bodies do not depend on the mutual orientation of their spins (and thus on the value of the total spin), but the forces acting between the bodies may strongly depend on how their spins are directed, and therefore, the relative motion, generally speaking, depends on the total spin.

states $|\eta, s\sigma\rangle$ as $|\nu\rangle$, and they are the eigenstates of the internal Hamiltonian,

$$h|\nu\rangle = E_\nu|\nu\rangle\ , \qquad \nu = 1, 2, 3, \ldots, N\ . \tag{11.11}$$

Among the N eigenvalues E_ν, there are repeating values (for the degenerate states). As we will see soon, these eigenvalues are the so-called threshold energies of the corresponding channels.

Using the single sequential label, we can re-write Eq. (11.9) in a more simple and compact form,

$$|\Psi\rangle = |1\rangle|\psi_1\rangle + |2\rangle|\psi_2\rangle + \cdots + |N\rangle|\psi_N\rangle = \begin{pmatrix} |1\rangle|\psi_1\rangle \\ |2\rangle|\psi_2\rangle \\ \cdots \\ |N\rangle|\psi_N\rangle \end{pmatrix} . \tag{11.12}$$

When the ket-vector $|\psi_\nu\rangle$ is projected on the bra-vector of the relative coordinate, $\langle\vec{r}\,|$, it gives the wave function of the relative motion in the channel ν,

$$\langle\vec{r}\,|\psi_\nu\rangle = \psi_\nu(\vec{r}\,)\ . \tag{11.13}$$

It should not be forgotten that the state-vector $|\nu\rangle$ belongs to the Hilbert space that is a direct product of the space of the internal motion and the spin space of the system AB, i.e.

$$|\nu\rangle = |\eta_\nu, s_\nu\sigma_\nu\rangle\ . \tag{11.14}$$

Within this basis, the internal Hamiltonian can be written as a sum of the projection operators,

$$h = \sum_{\nu=1}^{N} |\nu\rangle E_\nu \langle\nu|\ . \tag{11.15}$$

As was already mentioned, there are two types of the degrees of freedom, namely, the relative motion associated with the relative Jacobi coordinate $\vec{r}$ as well as with the relative spin state $|s\sigma\rangle$, and the internal motion inside the interacting bodies that is described by the set of the quantum states $\{|\eta\rangle\}$. Therefore, the set of states

$$|\vec{r}\,\rangle \otimes |s\sigma\rangle \otimes |\eta\rangle = |\vec{r}, \nu\rangle$$

constitute a basis in the full Hilbert space and thus can be used for representing the total quantum state $|\Psi\rangle$ of the whole physical system. These states can be chosen ortho-normal,

$$\langle\vec{r}, \nu|\vec{r}\,', \nu'\rangle = \delta\left(\vec{r} - \vec{r}\,'\right)\delta_{\nu\nu'}\ . \tag{11.16}$$

Using this basis, we can find an appropriate representation of the Schrödinger equation

$$H|\Psi\rangle = E|\Psi\rangle$$

by projecting it onto the set of $\langle\vec{r}, \nu|$,

$$\int \langle\vec{r}, \nu|H|\vec{r}\,'\rangle\langle\vec{r}\,'|\Psi\rangle\, \mathrm{d}^3r' = E\langle\vec{r}, \nu|\Psi\rangle\,. \tag{11.17}$$

If the potential is local (which is our usual assumption), then

$$\langle\vec{r}, \nu|U|\vec{r}\,', \nu'\rangle = \delta\left(\vec{r} - \vec{r}\,'\right) U_{\nu\nu'}(\vec{r}\,)\,, \tag{11.18}$$

and therefore, the total Hamiltonian is diagonal with respect to the relative coordinate,

$$\langle\vec{r}, \nu|H_0 + U + h|\vec{r}\,', \nu'\rangle = \delta\left(\vec{r} - \vec{r}\,'\right)\left[-\frac{\hbar^2}{2\mu_\nu}\Delta_{\vec{r}}\,\delta_{\nu\nu'} + U_{\nu\nu'}(\vec{r}\,) + E_\nu\delta_{\nu\nu'}\right], \tag{11.19}$$

where μ_ν is the reduced mass in the channel ν. Apparently, the projection of $|\Psi\rangle$ on the basis vector $|\vec{r}, \nu\rangle$, i.e.

$$\langle\vec{r}, \nu|\Psi\rangle = \psi_\nu(\vec{r}\,)\,, \tag{11.20}$$

is the wave function describing the relative motion of the interacting bodies that are in the internal state $|\nu\rangle$. In other words, such a projection cuts out a particular term from the expansion (11.12) or selects the corresponding element of the column-matrix. This function only depends on the relative coordinate $\vec{r}$, while all the internal coordinates are folded (integrated) in the scalar product (11.20). The spin variables are folded in this product as well.

Substituting Hamiltonian (11.19) into Eq. (11.17), we obtain the following multi-channel Schrödinger equation:

$$\left[\frac{\hbar^2}{2\mu_\nu}\Delta_{\vec{r}} + (E - E_\nu)\right]\psi_\nu(E, \vec{r}\,) = \sum_{\nu'=1}^{N} U_{\nu\nu'}(\vec{r}\,)\psi_{\nu'}(E, \vec{r}\,)\,. \tag{11.21}$$

This equation can be written in the matrix form. For the sake of clarity, let us assume that there are only two channels ($N = 2$); then

$$\begin{pmatrix} \Delta_{\vec{r}} + k_1^2 & 0 \\ 0 & \Delta_{\vec{r}} + k_2^2 \end{pmatrix}\begin{pmatrix} \psi_1 \\ \psi_2 \end{pmatrix} = \begin{pmatrix} \mathcal{U}_{11} & \mathcal{U}_{12} \\ \mathcal{U}_{21} & \mathcal{U}_{22} \end{pmatrix}\begin{pmatrix} \psi_1 \\ \psi_2 \end{pmatrix}, \tag{11.22}$$

where

$$k_\nu = \sqrt{\frac{2\mu_\nu}{\hbar^2}(E - E_\nu)} \tag{11.23}$$

is the channel momentum (that determines the relative kinetic energy in the channel ν). The energy E_ν of the internal state is the threshold energy for the channel ν because the momentum in that channel can only be real (i.e. the free relative motion is only possible) if $E > E_\nu$. The energy scale is always chosen in such a way that the lowest threshold is zero, $E_1 = 0$.

The matrix $\mathcal{U}$ in Eq. (11.22),

$$\mathcal{U}_{\nu\nu'}(\vec{r}\,) = \frac{2\mu_\nu}{\hbar^2} U_{\nu\nu'}(\vec{r}\,) = \frac{2\mu_\nu}{\hbar^2} \langle \nu | U(\vec{r}\,) | \nu' \rangle \,, \tag{11.24}$$

is the matrix of the reduced (multiplied by the mass factor) potential. It is clear that the channels are coupled to each other because of the off-diagonal elements of the matrix $\mathcal{U}$. If some of these elements are small, then the corresponding couplings are weak.

Intuitively, it is clear that the coupling among different channels should become weak at large distances and should completely disappear when $r \to \infty$. Indeed, when the colliding particles are separated far enough, they cannot affect the internal motion of each other. Formally, this can be seen as follows.

Consider a model in which a bound complex of two particles, a and b, collides with another bound complex consisting of the particles c and d, as is schematically shown in Fig. 11.2. Let the internal coordinates in the complexes be $\vec{r}_1$ and $\vec{r}_2$. In Eq. (11.3), the total Hamiltonian is split into such a way that U only involves the forces acting between the particles belonging to different bodies, while the internal potentials (in our model case, they are V_{ab} and V_{cd}, i.e. the potentials acting in the ab- and cd-pairs, respectively) are included in the internal Hamiltonian h. Therefore, the matrix element of the potential U can be explicitly written as

$$U_{\nu\nu'}(\vec{r}\,) = \int d^3r_1 d^3r_2 \langle \nu | \vec{r}_1, \vec{r}_2 \rangle \left(V_{ac} + V_{ad} + V_{bc} + V_{bd} \right) \langle \vec{r}_1, \vec{r}_2 | \nu' \rangle \,, \tag{11.25}$$

where the subscripts of the potential V indicate which pair of the particles this term belongs to. The pairs (ab) and (cd) are bound, which means that the integration in Eq. (11.25) effectively runs over short values of $\vec{r}_1$ and $\vec{r}_2$ (for which the

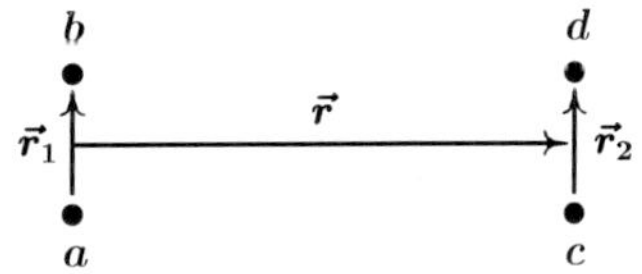

Fig. 11.2 The internal coordinates $\vec{r}_1$ and $\vec{r}_2$ as well as the relative coordinate $\vec{r}$ for a four-body collision $(ab) + (cd)$

corresponding wave functions are non-zero). Each potential term in Eq. (11.25) depends on the corresponding relative coordinate, for example, $V_{ac} = V_{ac}(\vec{r}_{ac})$. When $r \to \infty$, the internal coordinates remain finite, and thus the relative coordinates $\vec{r}_{ac}$, $\vec{r}_{ad}$, $\vec{r}_{bc}$, and $\vec{r}_{bd}$ do not differ from $\vec{r}$,

$$\left.\begin{array}{l}\vec{r}_{ac}\\ \vec{r}_{ad}\\ \vec{r}_{bc}\\ \vec{r}_{bd}\end{array}\right\} \underset{r\to\infty}{\longrightarrow} \vec{r}\,.$$

This allows us to take the potentials (independent of $\vec{r}_1$ and $\vec{r}_2$) out of the integral (11.25) and use the fact that the internal states are ortho-normal,

$$\begin{aligned}U_{\nu\nu'}(\vec{r}\,) &\underset{r\to\infty}{\longrightarrow} \left[V_{ac}(\vec{r}\,) + V_{ad}(\vec{r}\,) + V_{bc}(\vec{r}\,) + V_{bd}(\vec{r}\,)\right] \\ &\quad\times \int d^3r_1 d^3r_2 \langle\nu|\vec{r}_1, \vec{r}_2\rangle\langle\vec{r}_1, \vec{r}_2|\nu'\rangle \\ &\underset{r\to\infty}{\longrightarrow} \left[V_{ac}(\vec{r}\,) + V_{ad}(\vec{r}\,) + V_{bc}(\vec{r}\,) + V_{bd}(\vec{r}\,)\right]\delta_{\nu\nu'}\,. \end{aligned} \tag{11.26}$$

In other words, at large distances, the off-diagonal matrix elements of $U_{\nu\nu'}(\vec{r}\,)$ vanish much faster than the diagonal ones. Asymptotically, this matrix becomes diagonal, and the couplings among different channels disappear.

It should be noted that in this book we do not describe how the matrix elements of the potential $U_{\nu\nu'}(\vec{r}\,)$ should be constructed. We assume that such a matrix is given. Most often, it is obtained phenomenologically by fitting experimental data. In some problems, this matrix can be calculated as the folding integral (11.25). The specific ways of obtaining such a matrix are outside the scope of this book. The only reason why we use here Eq. (11.25) is because it gives a clear understanding of the fact that the off-diagonal elements of the potential must diminish at large distances faster than the diagonal ones.

11.1 Channels with Different Types of Particles

So far we considered a multi-channel system of two interacting bodies A and B, in which neither the rearrangements of their constituents nor the particle transformations were possible. The channels only differed from each other by their internal states (excitations) as was symbolically shown in Eq. (11.2). However, there are many physical processes where the channels involve completely different bodies, as, for example, in the following nuclear reactions,

$$^3\mathrm{He} + {}^2\mathrm{H} \longrightarrow \begin{cases} ^3\mathrm{He} + {}^2\mathrm{H}\,, \\ ^4\mathrm{He} + \;p\,, \end{cases}$$

or in the elementary particle transitions,

$$\pi^- + p \longrightarrow \begin{cases} \pi^- + p \,, \\ \pi^0 + n \,, \\ \Lambda^0 + K^0 \,. \end{cases}$$

As we will see, the channels with different types of particles can be treated within the same multi-channel formalism that we have already formulated.

To simplify the formulae, let us assume that there are only two types of particles, AB and CD, involved in a multi-channel process. When the particles A and B interact, they can transform themselves in another pair of particles, C and D, and vice versa. Of course, each of the particles may have an internal structure.

Let the Hamiltonians describing the internal motion inside the bodies be h_A, h_B, h_C, and h_D. Then the total internal Hamiltonian h of the system can be written as follows:

$$h = |AB\rangle(h_A + h_B)\langle AB| + |CD\rangle(h_C + h_D)\langle CD| \,,$$

where $|AB\rangle\langle AB|$ and $|CD\rangle\langle CD|$ are mutually orthogonal projection operators that specify the types of the interacting bodies. It should be noted that although these operators are orthogonal, i.e. $\langle AB|CD\rangle = 0$, the interaction potential can change the particle types,

$$\langle AB|U|CD\rangle \neq 0 \,. \tag{11.27}$$

A quantum state of the system is a superposition of the two different channels,

$$|\Psi\rangle = |AB\rangle\Psi_{AB} + |CD\rangle\Psi_{CD} \,, \tag{11.28}$$

where the two terms are mutually orthogonal. The Schrödinger equation,

$$H|\Psi\rangle = E|\Psi\rangle \,,$$

multiplied from the left by the channel vectors $\langle AB|$ and $\langle CD|$, gives the following two equations:

$$\langle AB|H|AB\rangle\Psi_{AB} + \langle AB|H|CD\rangle\Psi_{CD} = E\Psi_{AB} \,, \tag{11.29}$$

$$\langle CD|H|AB\rangle\Psi_{AB} + \langle CD|H|CD\rangle\Psi_{CD} = E\Psi_{CD} \,, \tag{11.30}$$

which can be combined in one matrix equation,

$$\begin{bmatrix} \langle AB|H|AB\rangle & \langle AB|H|CD\rangle \\ \langle CD|H|AB\rangle & \langle CD|H|CD\rangle \end{bmatrix} \begin{bmatrix} \Psi_{AB} \\ \Psi_{CD} \end{bmatrix} = E \begin{bmatrix} \Psi_{AB} \\ \Psi_{CD} \end{bmatrix} . \tag{11.31}$$

Apparently, the Hamiltonian (11.3) is a matrix in the space of the channels AB and CD,

$$H = \begin{bmatrix} \langle AB|H_0|AB\rangle & 0 \\ 0 & \langle CD|H_0|CD\rangle \end{bmatrix} + \tag{11.32}$$

$$+ \begin{bmatrix} \langle AB|U|AB\rangle & \langle AB|U|CD\rangle \\ \langle CD|U|AB\rangle & \langle CD|U|CD\rangle \end{bmatrix} + \begin{bmatrix} \langle AB|h|AB\rangle & 0 \\ 0 & \langle CD|h|CD\rangle \end{bmatrix} .$$

It may happen that one or more of the terms in Eq. (11.28) can be further sub-divided into orthogonal terms. Then the dimension of the column-matrix representing the state-vector of the system increases. For example, if the particle A has an excited state, A^*, then the first term of Eq. (11.28) is split into two orthogonal sub-terms,

$$|\Psi\rangle = |AB\rangle\Psi_{AB} + |A^*B\rangle\Psi_{A^*B} + |CD\rangle\Psi_{CD} , \tag{11.33}$$

and the wave function is represented by the column-matrix with three elements,

$$\Psi = \begin{bmatrix} \Psi_{AB} \\ \Psi_{A^*B} \\ \Psi_{CD} \end{bmatrix} .$$

In such a case, the Hamitonian becomes a (3×3)-matrix. The dimensions of the matrices are determined by the number of the channels that are taken into account within a chosen model.

It is obvious that everything that was said in the previous section is applicable here. This means that exactly the same formalism can be used to solve the problems with particle transformations. The only unclear thing is how to obtain the matrix elements of the potential (11.27) for the transitions between the states with different types of the particles. It should be emphasized that constructing the multi-channel potentials is beyond the scope of this book. It is assumed that all necessary potentials are given. The theory developed in this book is focused on the ways of solving the problem with a known potential. In most cases, the potentials of the type (11.27) are phenomenological, i.e. they fit certain experimental data.

11.2 Partial-Wave Decomposition

As we have seen, each solution of the multi-channel Schrödinger equation (11.21) can be represented by a column-matrix, and each element of this matrix, describing the motion in a particular channel, is a function of the three-dimensional vector $\vec{r}$ (relative coordinate). For each of the channels, we can expand this function in a sum of the partial waves, where the dependence on the spherical angles $\hat{\vec{r}}$ is given explicitly via the corresponding spherical harmonics. In this way, the set of the three-dimensional equations (11.21) can be replaced with the corresponding set of one-dimensional radial Schrödinger equations.

We only consider the systems that can be treated as isolated and are not subjected to any external forces. Then the space around them is isotropic, which implies that the total angular momentum is conserving. Therefore, any two channels are coupled to each other only if they have the same total angular momentum, J. In other words, all the channels can be split into groups with a definite J in each one, and for each group, the problem can be solved separately.

In some derivations, we also use the parity conservation. This means that the results of such derivations are not applicable when the weak forces (that may violate the parity conservation) are taken into account. However, most of the formulae derived in this book do not rely on the parity conservation and thus are more general. When the parity conservation is used, it is said explicitly.

In most of the problems, the interacting bodies have non-zero spins. This means that in the partial-wave expansions the orbital angular momentum, ℓ, and the spin, s, should be coupled to form a conserving total angular momentum, J. The corresponding spin-angular functions were introduced in Sect. 10.1. In the multi-channel problems, we need certain matrices composed of such functions. They are introduced in the next section.

For each channel ν, the relative motion is expanded in the partial waves with different values of the orbital angular momentum ℓ. Since the total angular momentum J is conserving and each channel has a definite value s_ν of the total spin, the number of possible partial waves is limited by the triangle inequality,

$$|J - s_\nu| \leqslant \ell \leqslant |J + s_\nu| \,. \tag{11.34}$$

For example, if $s_\nu = 1/2$, there are only two values of ℓ; if $s_\nu = 1$, there are only three of them, etc.

When the angular momenta ℓ and s are coupled, the z-component of the spin does not have definite value anymore, and thus the set of the quantum numbers $\{\eta_\nu, s_\nu\sigma_\nu\}$ as well as the sequential numbers ν introduced for them cannot be used to label different channels. For a fixed J, each channel is specified by the internal state η, the orbital angular momentum ℓ, and the total spin s. We can introduce a new sequence of numbers n that label all possible combinations of these quantum numbers,

$$n :\longrightarrow \{\eta_n, \ell_n, s_n\} \,, \qquad n = 1, 2, 3, \ldots \tag{11.35}$$

After the partial-wave decomposition, some of the channels are degenerate in the same way as before. Indeed, the internal state (and its energy) does not depend on the values of ℓ_n and s_n. Generally speaking, the number of different internal states, $\mathcal{N}$, is less than the total number of the channels, N.

It should be emphasized that after doing the partial-wave decomposition and coupling the orbital and spin-angular momenta, we obtain (generally speaking) a different number of the orthogonal terms than we had in the expansion (11.12). Therefore, the number of the channels, N, becomes different as well.

To understand this, it is sufficient to come back to the example depicted in Fig. 11.1. Let us consider a multi-channel state of that system with the lowest possible value of J, which is $J = 1/2$. For the internal state η_1, this J can be obtained by coupling the spin 1/2 with either $\ell = 0$ or $\ell = 1$. However, these two partial waves have different parities (positive and negative, respectively). If we consider, for example, a state of the system with the positive parity ($\pi = +$), then only the partial wave $\ell = 0$ should be taken into account for the internal state η_1. The same reasoning leads to the conclusion that for the internal state η_2 with $s = 1/2$, the only partial wave that contributes to the state with $J = 1/2$ and $\pi = +$ is $\ell = 0$. From the remaining state $|\eta_2, s = 3/2\rangle$, the only contribution comes from the state with $\ell = 2$. Therefore, for the given example, there are only three channels, namely, $|\eta_1, \ell = 0, s = 1/2\rangle$, $|\eta_2, \ell = 0, s = 1/2\rangle$, and $|\eta_2, \ell = 2, s = 3/2\rangle$ are coupled to each other for $J = 1/2$ and $\pi = +$. Thus, instead of $N = 8$ shown in Fig. 11.1, we obtain $N = 3$. Of course, we have to solve the problem for the other possible values of J, but for each J the problem is solved separately.

11.2.1 Spin-Angular Matrices

In the single-channel case, we used the single-row matrices (10.33) and (10.34). Here we need to generalize that definition for an arbitrary number of channels.

Let the number of different internal states be $\mathcal{N}$. For each internal state η, there are several partial waves of the relative motion. It should be noted that a partial wave is not just a state with certain orbital angular momentum ℓ. For a fixed J, each partial wave is a combination of the two quantum numbers, namely, ℓ and s (see example given at the end of the previous section). In nuclear physics, they are usually denoted using the following symbols:

$$\boxed{{}^{2s+1}\ell_J}\ , \qquad \text{for example,}\quad {}^3S_1,\ {}^3P_1,\ {}^3D_1,\quad \text{etc.}$$

We denote the combinations of the ℓ and s by the multi-index $[\ell] = (\ell, s)$. Let for the internal state η and a fixed J the number of such combinations (partial waves) be N_η. Each partial wave corresponds to a spin-angular ket-vector $|\mathcal{Y}^{JM}_{[\ell]}\rangle$ defined by Eq. (10.4). Combining these spin-angular vectors, we introduce the row-matrix

$$|\mathcal{Y}^{JM}_\eta\rangle = \left\{|\mathcal{Y}^{JM}_{[\ell_1]}\rangle, |\mathcal{Y}^{JM}_{[\ell_2]}\rangle, \ldots, |\mathcal{Y}^{JM}_{[\ell_{N_\eta}]}\rangle\right\} . \tag{11.36}$$

For different internal states, η and η', the number of the elements in this matrix may be different ($N_\eta \neq N_{\eta'}$). The subscripts $[\ell_1]$, $[\ell_2]$,... label the partial waves allowed for a given internal state, and thus they are not the same for a state η' even if $N_\eta = N_{\eta'}$. Making the direct sum of all the matrices (11.36), we introduce the following block-diagonal matrix

$$|\mathcal{Y}^{JM}\rangle = |\mathcal{Y}^{JM}_{\eta_1}\rangle \oplus |\mathcal{Y}^{JM}_{\eta_2}\rangle \oplus \cdots \oplus |\mathcal{Y}^{JM}_{\eta_{\mathcal{N}}}\rangle = \begin{pmatrix} |\mathcal{Y}^{JM}_{\eta_1}\rangle & 0 & 0 & \cdots \\ 0 & |\mathcal{Y}^{JM}_{\eta_2}\rangle & 0 & \cdots \\ \cdots & \cdots & \cdots & \cdots \\ 0 & \cdots & 0 & |\mathcal{Y}^{JM}_{\eta_{\mathcal{N}}}\rangle \end{pmatrix}. \tag{11.37}$$

Generally speaking, this is not a square matrix. Indeed, each block here is a row matrix. The number of lines in the matrix $|\mathcal{Y}^{JM}\rangle$ is equal to the number of different internal states, $\mathcal{N}$. Meanwhile, the number of columns is the sum of the lengths of all the rows $|\mathcal{Y}^{JM}_{\eta}\rangle$. Each partial wave should be treated as a separate channel (some of them are degenerate). Therefore, the number of columns in the matrix (11.37) is equal to the number of the channels, N. In other words, this is a $(\mathcal{N} \times N)$-matrix.

Taking into account the orthonormality of the spin-angular vectors $|\mathcal{Y}^{JM}_{[\ell]}\rangle$ (see Eq. (10.7)), it is not difficult to see that

$$\langle \mathcal{Y}^{JM}|\mathcal{Y}^{J'M'}\rangle = I\delta_{JJ'}\delta_{MM'}\,, \tag{11.38}$$

where I is the unit matrix of the size $(N \times N)$. This can be easily understood, using a simple example with two internal states involving two and three partial waves, respectively. In such a case, $\mathcal{N} = 2$, $N = 5$, and the matrix $|\mathcal{Y}^{JM}\rangle$ has the following structure:

$$|\mathcal{Y}\rangle = \begin{pmatrix} |1\rangle & |2\rangle & 0 & 0 & 0 \\ 0 & 0 & |3\rangle & |4\rangle & |5\rangle \end{pmatrix}. \tag{11.39}$$

For this matrix, the product (11.38) is

$$\langle \mathcal{Y}|\mathcal{Y}\rangle = \begin{pmatrix} \langle 1| & 0 \\ \langle 2| & 0 \\ 0 & \langle 3| \\ 0 & \langle 4| \\ 0 & \langle 5| \end{pmatrix} \begin{pmatrix} |1\rangle & |2\rangle & 0 & 0 & 0 \\ 0 & 0 & |3\rangle & |4\rangle & |5\rangle \end{pmatrix}$$

$$= \begin{pmatrix} \langle 1|1\rangle & \langle 1|2\rangle & 0 & 0 & 0 \\ \langle 2|1\rangle & \langle 2|2\rangle & 0 & 0 & 0 \\ 0 & 0 & \langle 3|3\rangle & \langle 3|4\rangle & \langle 3|5\rangle \\ 0 & 0 & \langle 4|3\rangle & \langle 4|4\rangle & \langle 4|5\rangle \\ 0 & 0 & \langle 5|3\rangle & \langle 5|4\rangle & \langle 5|5\rangle \end{pmatrix} = \begin{pmatrix} 1 & 0 & 0 & 0 & 0 \\ 0 & 1 & 0 & 0 & 0 \\ 0 & 0 & 1 & 0 & 0 \\ 0 & 0 & 0 & 1 & 0 \\ 0 & 0 & 0 & 0 & 1 \end{pmatrix}.$$

The other product of the same matrices,

$$|\mathcal{Y}\rangle\langle\mathcal{Y}| = \begin{pmatrix} |1\rangle & |2\rangle & 0 & 0 & 0 \\ 0 & 0 & |3\rangle & |4\rangle & |5\rangle \end{pmatrix} \begin{pmatrix} \langle 1| & 0 \\ \langle 2| & 0 \\ 0 & \langle 3| \\ 0 & \langle 4| \\ 0 & \langle 5| \end{pmatrix}$$

$$= \begin{pmatrix} |1\rangle\langle 1| + |2\rangle\langle 2| & 0 \\ 0 & |3\rangle\langle 3| + |4\rangle\langle 4| + |5\rangle\langle 5| \end{pmatrix},$$

is a projection operator in the spin-angular space. Therefore, using the closure relation (10.8), we obtain its generalization for a multi-channel problem in the following form:

$$\sum_{JM} |\mathcal{Y}^{JM}\rangle\langle\mathcal{Y}^{JM}| = I\,. \tag{11.40}$$

It should not be forgotten that the matrix $|\mathcal{Y}^{JM}\rangle$ consists of the ket-vectors in the spin-angular space. We can multiply it by the bra-vector $\langle\hat{\vec{r}}\,|$ from the left (compare this with Eq. (10.9)), which gives the corresponding matrix of the ket-vectors in the spin space,

$$|\mathcal{Y}^{JM}(\hat{\vec{r}}\,)\rangle = \langle\hat{\vec{r}}\,|\mathcal{Y}^{JM}\rangle \tag{11.41}$$

$$= \begin{pmatrix} \langle\hat{\vec{r}}\,| & 0 & 0 & \cdots \\ 0 & \langle\hat{\vec{r}}\,| & 0 & \cdots \\ \cdots & \cdots & \cdots & \cdots \\ 0 & \cdots & 0 & \langle\hat{\vec{r}}\,| \end{pmatrix} \begin{pmatrix} |\mathcal{Y}^{JM}_{\eta_1}\rangle & 0 & 0 & \cdots \\ 0 & |\mathcal{Y}^{JM}_{\eta_2}\rangle & 0 & \cdots \\ \cdots & \cdots & \cdots & \cdots \\ 0 & \cdots & 0 & |\mathcal{Y}^{JM}_{\eta_\mathcal{N}}\rangle \end{pmatrix}$$

$$= \begin{pmatrix} |\mathcal{Y}^{JM}_{\eta_1}(\hat{\vec{r}}\,)\rangle & 0 & 0 & \cdots \\ 0 & |\mathcal{Y}^{JM}_{\eta_2}(\hat{\vec{r}}\,)\rangle & 0 & \cdots \\ \cdots & \cdots & \cdots & \cdots \\ 0 & \cdots & 0 & |\mathcal{Y}^{JM}_{\eta_\mathcal{N}}(\hat{\vec{r}}\,)\rangle \end{pmatrix}.$$

Similarly, we can construct the scalar product of the matrix $|\mathcal{Y}^{JM}\rangle$ with the bra-vector of the spherical angles of the momentum,

$$|\mathcal{Y}^{JM}(\hat{\vec{k}}\,)\rangle = \langle\hat{\vec{k}}\,|\mathcal{Y}^{JM}\rangle\,. \tag{11.42}$$

11.2.2 Partial-Wave Decomposition for Discrete Spectrum

Basically, the analysis given here is the same as was done in Sect. 10.2 for a single-channel problem with non-central forces. Of course, there are some differences. First, each channel here may have different thresholds. Second, from channel to channel, the interacting particles may have different compositions, masses, and spins. There are, however, certain conservation laws that impose some restrictions on the quantum numbers for the channels that are coupled to each other.

A multi-channel discrete state, $|\Psi^{JM}\rangle$, has definite values of its energy E as well as of the quantum numbers J and M (to simplify the notation, we drop the label E). It is a sum of the mutually orthogonal terms corresponding to different possible internal states,[2]

$$|\Psi^{JM}\rangle = |\Psi^{JM}_{\eta_1}\rangle + |\Psi^{JM}_{\eta_2}\rangle + \cdots + |\Psi^{JM}_{\eta_{\mathcal{N}}}\rangle\,, \tag{11.43}$$

which can be represented by a column-matrix,

$$|\Psi^{JM}\rangle = \begin{pmatrix} |\Psi^{JM}_{\eta_1}\rangle \\ \\ |\Psi^{JM}_{\eta_2}\rangle \\ \\ \cdots \\ \\ |\Psi^{JM}_{\eta_{\mathcal{N}}}\rangle \end{pmatrix}, \tag{11.44}$$

where the number of elements is $\mathcal{N}$ (the number of different internal states of the system). Introducing the diagonal $(\mathcal{N} \times \mathcal{N})$-matrix

$$|\widehat{\eta}\rangle = \begin{pmatrix} |\eta_1\rangle & 0 & 0 & \cdots \\ \\ 0 & |\eta_2\rangle & 0 & \cdots \\ \\ \cdots & \cdots & \cdots & \cdots \\ \\ 0 & \cdots & 0 & |\eta_{\mathcal{N}}\rangle \end{pmatrix}, \tag{11.45}$$

we can fold (integrate) all the internal coordinates and obtain the corresponding column-matrix of the ket-vectors describing relative motion in all the channels with

---

[2] How can a discrete (bound or resonant) state be a mixture of several channels with different thresholds? This can be understood using, for example, the so-called hypernuclei. A hypernucleus is a bound state of an ordinary nucleus and a hyperon (Λ or Σ particle, for instance). When inside a nucleus, a hyperon can change its type via the virtual processes $\Lambda + n \to \Sigma^- + p$ and vice versa. Therefore, a Λ-hypernucleus always has an admixture of the Σ-component. Meanwhile, the Σ-hyperon is approximately 74 MeV heavier than the Λ-particle, and thus the Σ-nucleus threshold is higher than the Λ-nucleus one. Such a hypernucleus is an example of a two-channel bound state.

different internal states,

$$|\psi^{JM}\rangle = \langle\widehat{\eta}|\Psi^{JM}\rangle = \begin{pmatrix} |\psi^{JM}_{\eta 1}\rangle \\ |\psi^{JM}_{\eta 2}\rangle \\ \cdots \\ |\psi^{JM}_{\eta\mathcal{N}}\rangle \end{pmatrix} . \tag{11.46}$$

It should be noted that the projection $\langle\eta|\Psi^{JM}_{\eta}\rangle$ folds up only the internal coordinates. This scalar product remains a ket-vector in the relative coordinate and in the spin space. Apparently, the unfolded matrix (11.44) can be written in the following factorized form:

$$|\Psi^{JM}\rangle = |\widehat{\eta}\rangle|\psi^{JM}\rangle = \begin{pmatrix} |\eta_1\rangle|\psi^{JM}_{\eta 1}\rangle \\ |\eta_2\rangle|\psi^{JM}_{\eta 2}\rangle \\ \cdots \\ |\eta_{\mathcal{N}}\rangle|\psi^{JM}_{\eta\mathcal{N}}\rangle \end{pmatrix} . \tag{11.47}$$

Now, we make use of the closure relation (11.40),

$$|\psi^{JM}\rangle = |\mathcal{Y}^{JM}\rangle\langle\mathcal{Y}^{JM}| \begin{pmatrix} |\psi^{JM}_{\eta 1}\rangle \\ |\psi^{JM}_{\eta 2}\rangle \\ \cdots \\ |\psi^{JM}_{\eta\mathcal{N}}\rangle \end{pmatrix} = \begin{pmatrix} \sum_{[\ell]} |\mathcal{Y}^{JM}_{[\ell]}\rangle\langle\mathcal{Y}^{JM}_{[\ell]}|\psi^{JM}_{\eta 1}\rangle \\ \sum_{[\ell]} |\mathcal{Y}^{JM}_{[\ell]}\rangle\langle\mathcal{Y}^{JM}_{[\ell]}|\psi^{JM}_{\eta 2}\rangle \\ \cdots \\ \sum_{[\ell]} |\mathcal{Y}^{JM}_{[\ell]}\rangle\langle\mathcal{Y}^{JM}_{[\ell]}|\psi^{JM}_{\eta\mathcal{N}}\rangle \end{pmatrix} . \tag{11.48}$$

Each element of this column-matrix is given as a sum of the allowed partial waves for a particular internal state η. They are the ket-vectors in the relative coordinate and spin space. The coordinate representation of them is obtained by the projection on the bra-vector $\langle\vec{r}\,|$,

$$\langle\vec{r}\,|\psi^{JM}\rangle = |\psi^{JM}(\vec{r}\,)\rangle = \begin{pmatrix} \sum_{[\ell]} |\mathcal{Y}^{JM}_{[\ell]}(\hat{\vec{r}}\,)\rangle\langle r, \mathcal{Y}^{JM}_{[\ell]}|\psi^{JM}_{\eta 1}\rangle \\ \sum_{[\ell]} |\mathcal{Y}^{JM}_{[\ell]}(\hat{\vec{r}}\,)\rangle\langle r, \mathcal{Y}^{JM}_{[\ell]}|\psi^{JM}_{\eta 2}\rangle \\ \cdots \\ \sum_{[\ell]} |\mathcal{Y}^{JM}_{[\ell]}(\hat{\vec{r}}\,)\rangle\langle r, \mathcal{Y}^{JM}_{[\ell]}|\psi^{JM}_{\eta\mathcal{N}}\rangle \end{pmatrix} = \begin{pmatrix} |\psi^{JM}_{\eta 1}(\vec{r}\,)\rangle \\ |\psi^{JM}_{\eta 2}(\vec{r}\,)\rangle \\ \cdots \\ |\psi^{JM}_{\eta\mathcal{N}}(\vec{r}\,)\rangle \end{pmatrix} . \tag{11.49}$$

The elements $|\psi_\eta^{JM}(\vec{r}\,)\rangle$ of this column-matrix are still the ket-vectors in the spin space. In order to obtain the corresponding wave function, we need to project them on the bra-vectors $\langle\widehat{s\sigma}|$ defined as the diagonal matrix,

$$\langle\widehat{s\sigma}| = \begin{pmatrix} \langle s_1\sigma_1| & 0 & 0 & \cdots \\ 0 & \langle s_2\sigma_2| & 0 & \cdots \\ \cdots & \cdots & \cdots & \cdots \\ 0 & \cdots & 0 & \langle s_{\mathcal{N}}\sigma_{\mathcal{N}}| \end{pmatrix} . \tag{11.50}$$

Such a projection gives

$$\begin{aligned} \langle\vec{r}, \widehat{s\sigma}|\psi^{JM}\rangle &= \psi^{JM}(\vec{r}, s_1, \sigma_1, s_2, \sigma_2, \ldots) \\ &= \begin{pmatrix} \sum_\ell Y^{JM}_{\ell s_1\sigma_1}(\hat{\vec{r}}\,)\langle r, \mathcal{Y}^{JM}_{[\ell]}|\psi^{JM}_{\eta_1}\rangle \\ \sum_\ell Y^{JM}_{\ell s_2\sigma_2}(\hat{\vec{r}}\,)\langle r, \mathcal{Y}^{JM}_{[\ell]}|\psi^{JM}_{\eta_2}\rangle \\ \cdots \\ \sum_\ell Y^{JM}_{\ell s_{\mathcal{N}}\sigma_{\mathcal{N}}}(\hat{\vec{r}}\,)\langle r, \mathcal{Y}^{JM}_{[\ell]}|\psi^{JM}_{\eta_{\mathcal{N}}}\rangle \end{pmatrix} \\ &= \begin{bmatrix} \psi^{JM}_{\eta_1}(\vec{r}, s_1, \sigma_1) \\ \psi^{JM}_{\eta_2}(\vec{r}, s_2, \sigma_2) \\ \cdots \\ \psi^{JM}_{\eta_{\mathcal{N}}}(\vec{r}, s_{\mathcal{N}}, \sigma_{\mathcal{N}}) \end{bmatrix} , \end{aligned} \tag{11.51}$$

where the functions $Y^{JM}_{\ell s\sigma}(\hat{\vec{r}}\,)$ are defined by Eq. (10.10). It should be noted that in the above equation the sums over $[\ell] \equiv \{\ell, s\}$ are replaced with the sums over ℓ because the channel spins are fixed by the bra-vector $\langle\vec{r}, s\sigma|$. The physical meaning of this wave function can be understood if we look at each element (row) of the column-matrix (11.51) separately. The absolute square of such an element,

$$\rho_\eta(\vec{r}, s_\eta, \sigma_\eta) = \left|\psi^{JM}_\eta(\vec{r}, s_\eta, \sigma_\eta)\right|^2 ,$$

is the density of the probability that the internal state of the system is η, the relative coordinate is $\vec{r}$, and the spin state is $|s_\eta\sigma_\eta\rangle$. If we sum up these densities for all possible spin states and integrate the sum over the relative coordinate, we obtain the total probability that the system is in the internal state η. And, of course, if we sum up such probabilities for all possible internal states, we should obtain 100%, which is the normalization condition,

$$\sum_{s_1\sigma_1 s_2\sigma_2\ldots} \int d^3r \left[\psi^{JM}(\vec{r}, s_1, \sigma_1, s_2, \sigma_2, \ldots)\right]^{\dagger} \psi^{JM}(\vec{r}, s_1, \sigma_1, s_2, \sigma_2, \ldots) = 1\,. \tag{11.52}$$

The scalar products $\langle r, \mathcal{Y}^{JM}_{[\ell]} | \psi^{JM}_{\eta} \rangle$ in the matrices (11.49, 11.51) are nothing else, but the radial parts of the wave function,

$$\langle r, \mathcal{Y}^{JM}_{[\ell]} | \psi^{JM}_{\eta} \rangle = \langle r, (\ell s) JM | \psi^{JM}_{\eta} \rangle = \frac{u^{J}_{\eta[\ell]}(E, r)}{r}\,, \tag{11.53}$$

where, as usual, the factor $1/r$ is introduced to get rid of the first derivative in the radial Schrödinger equation (see the next section). The independence of the radial wave function of M can be shown in the same way as it was done in Sect. 10.2 for the non-central potential. Therefore, the partial-wave expansion of each row of the matrix (11.51) can be written as

$$\psi^{JM}_{\eta}(\vec{r}, s, \sigma) = \sum_{\ell} Y^{JM}_{\ell s \sigma}(\hat{\vec{r}}\,) \frac{u^{J}_{\eta[\ell]}(E, r)}{r}\,. \tag{11.54}$$

If instead of the matrix (11.51), we consider the matrix (11.49), where we do the projection of the state vector $|\psi^{JM}\rangle$ onto an "incomplete" basis vector $\langle \vec{r}\,|$ (excluding the spin variables), then the resulting scalar product,

$$|\psi^{JM}_{\eta}(\vec{r}\,)\rangle = \sum_{[\ell]} |\mathcal{Y}^{JM}_{[\ell]}(\hat{\vec{r}}\,)\rangle \frac{u^{J}_{\eta[\ell]}(E, r)}{r}\,, \tag{11.55}$$

remains a ket-vector in the spin space. Here, the ket-vector $|\mathcal{Y}^{JM}_{[\ell]}(\hat{\vec{r}}\,)\rangle$ is defined by Eq. (10.9).

At this point, we should recall that each partial wave is treated as a separate channel. Therefore for a given internal state η, there are several sub-channels having the same threshold energy,

$$\eta\,: \qquad \eta[\ell_1],\ \eta[\ell_2],\ \ldots,\ \eta[\ell_{N_\eta}]\,.$$

The terms corresponding to each of these sub-channels in Eq. (11.55) are mutually orthogonal. We can therefore expand each of the elements of the column-matrix (11.49) in a sub-column.

Let us collect the sub-channel radial wave functions in a column-matrix u^J_η and after that collect all such columns in a bigger column-matrix u^J by putting them one under the other:

$$u^J_\eta = \begin{pmatrix} u^J_{\eta[\ell_1]} \\ u^J_{\eta[\ell_2]} \\ \cdots \\ u^J_{\eta[\ell_{N_\eta}]} \end{pmatrix} \quad \Rightarrow \quad u^J = \begin{pmatrix} u^J_{\eta 1} \\ u^J_{\eta 2} \\ \cdots \\ u^J_{\eta \mathcal{N}} \end{pmatrix} = \begin{pmatrix} u^J_1 \\ u^J_2 \\ \cdots \\ u^J_N \end{pmatrix} . \tag{11.56}$$

In the last column-matrix of this equation, we assigned a single sequential subscript n to label the channels in accordance with the re-numbering rule (11.35). The number of the elements in this matrix is equal to the total number N of the channels, including the degenerate sub-channels (partial waves). Using this matrix together with the spin-angular matrix (11.41), we can re-write Eq. (11.55) in the following compact form:

$$|\psi^{JM}(\vec{r}\,)\rangle = |\mathcal{Y}^{JM}(\hat{\vec{r}}\,)\rangle \frac{u^J(E,r)}{r} , \tag{11.57}$$

where on both sides of the equation there are column-matrices of the length $\mathcal{N}$ that is equal to the number of the internal states (the number of truly different channels). Each element of such a matrix is a sum of the allowed partial waves. To get a clear understanding of the product on the right hand side of this equation, it is worthwhile to recall the schematic example given by Eq. (11.39):

$$|\mathcal{Y}\rangle u = \begin{pmatrix} |1\rangle & |2\rangle & 0 & 0 & 0 \\ 0 & 0 & |3\rangle & |4\rangle & |5\rangle \end{pmatrix} \begin{pmatrix} u_1 \\ u_2 \\ u_3 \\ u_4 \\ u_5 \end{pmatrix} = \begin{pmatrix} |1\rangle u_1 + |2\rangle u_2 \\ |3\rangle u_3 + |4\rangle u_4 + |5\rangle u_5 \end{pmatrix} .$$

For the full (unfolded) state-vector, the factorization corresponding to Eq. (11.57) reads

$$|\Psi^{JM}(\vec{r}\,)\rangle = |\widehat{\eta}\,\rangle |\mathcal{Y}^{JM}(\hat{\vec{r}}\,)\rangle \frac{u^J(E,r)}{r} . \tag{11.58}$$

11.2.2.1 Radial Schrödinger Equation

In order to derive the radial equation for the column-matrix $u^J(E,r)$, we begin with the Schrödinger equation in the full space (including the internal motion),

$$(H_0 + U + h - E)|\Psi^{JM}\rangle = 0 , \tag{11.59}$$

and multiply it from the left by the bra-vector $\langle \vec{r}\,|$ of the relative coordinate, which gives

$$\left[H_0(\vec{r}\,) + U(\vec{r}\,) + h - E\right] |\Psi^{JM}(\vec{r}\,)\rangle = 0\,. \tag{11.60}$$

The internal Hamiltonian can be written in the following matrix form:

$$h = |\widehat{\eta}\rangle\, \widehat{E}\, \langle\widehat{\eta}\,|\,, \qquad \widehat{E} = \text{diag}\{E_{\eta_1}, E_{\eta_2}, \ldots, E_{\eta_{\mathcal{N}}}\}\,, \tag{11.61}$$

where $\widehat{E}$ is the diagonal matrix of the threshold energies. Replacing the ket-vector $|\Psi^{JM}(\vec{r}\,)\rangle$ in Eq. (11.60) with the matrix product (11.58) and multiplying this equation from the left by the diagonal matrix $\langle\widehat{\eta}\,|$, we obtain

$$\left[\widehat{H}_0 + \langle\widehat{\eta}\,|U(\vec{r}\,)|\widehat{\eta}\rangle + \widehat{E} - E\right] |\mathcal{Y}^{JM}(\hat{\vec{r}}\,)\rangle \frac{u^J(E,r)}{r} = 0\,, \tag{11.62}$$

where

$$\widehat{H}_0 = \text{diag}\left\{-\frac{\hbar^2}{2\mu_{\eta_1}}\Delta_{\vec{r}}\,, -\frac{\hbar^2}{2\mu_{\eta_2}}\Delta_{\vec{r}}\,, \ldots, -\frac{\hbar^2}{2\mu_{\eta_{\mathcal{N}}}}\Delta_{\vec{r}}\right\}, \tag{11.63}$$

$$U = \begin{pmatrix} |\eta_1\rangle U_{\eta_1\eta_1}\langle\eta_1| & |\eta_1\rangle U_{\eta_1\eta_2}\langle\eta_2| & \cdots & |\eta_1\rangle U_{\eta_1\eta_{\mathcal{N}}}\langle\eta_{\mathcal{N}}| \\ |\eta_2\rangle U_{\eta_2\eta_1}\langle\eta_1| & |\eta_2\rangle U_{\eta_2\eta_2}\langle\eta_2| & \cdots & |\eta_2\rangle U_{\eta_2\eta_{\mathcal{N}}}\langle\eta_{\mathcal{N}}| \\ \cdots & \cdots & \cdots & \cdots \\ |\eta_{\mathcal{N}}\rangle U_{\eta_{\mathcal{N}}\eta_1}\langle\eta_1| & |\eta_{\mathcal{N}}\rangle U_{\eta_{\mathcal{N}}\eta_2}\langle\eta_2| & \cdots & |\eta_{\mathcal{N}}\rangle U_{\eta_{\mathcal{N}}\eta_{\mathcal{N}}}\langle\eta_{\mathcal{N}}| \end{pmatrix}, \tag{11.64}$$

with $U_{\eta'\eta} = \langle\eta'|U|\eta\rangle$. What remains is to multiply Eq. (11.62) by the matrix $\langle\mathcal{Y}^{JM}(\hat{\vec{r}}\,)|$ from the left and integrate over the spherical angles $\hat{\vec{r}}$. In this equation, the matrix in the square brackets has the dimensions $(\mathcal{N} \times \mathcal{N})$. When sandwiched between $\langle\mathcal{Y}^{JM}(\hat{\vec{r}}\,)|$ and $|\mathcal{Y}^{JM}(\hat{\vec{r}}\,)\rangle$, it becomes a matrix of the dimensions $(N \times N)$. This can be easily seen, using once again the schematic example given by Eq. (11.39):

$$\langle\mathcal{Y}|H|\mathcal{Y}\rangle = \begin{pmatrix} \langle 1| & 0 \\ \langle 2| & 0 \\ 0 & \langle 3| \\ 0 & \langle 4| \\ 0 & \langle 5| \end{pmatrix} \begin{pmatrix} H_{11} & H_{12} \\ H_{21} & H_{22} \end{pmatrix} \begin{pmatrix} |1\rangle & |2\rangle & 0 & 0 & 0 \\ 0 & 0 & |3\rangle & |4\rangle & |5\rangle \end{pmatrix}$$

$$= \begin{pmatrix} \langle 1|H_{11}|1\rangle & \langle 1|H_{11}|2\rangle & \langle 1|H_{12}|3\rangle & \langle 1|H_{12}|4\rangle & \langle 1|H_{12}|5\rangle \\ \langle 2|H_{11}|1\rangle & \langle 2|H_{11}|2\rangle & \langle 2|H_{12}|3\rangle & \langle 2|H_{12}|4\rangle & \langle 2|H_{12}|5\rangle \\ \langle 3|H_{21}|1\rangle & \langle 3|H_{21}|2\rangle & \langle 3|H_{22}|3\rangle & \langle 3|H_{22}|4\rangle & \langle 3|H_{22}|5\rangle \\ \langle 4|H_{21}|1\rangle & \langle 4|H_{21}|2\rangle & \langle 4|H_{22}|3\rangle & \langle 4|H_{22}|4\rangle & \langle 4|H_{22}|5\rangle \\ \langle 5|H_{21}|1\rangle & \langle 5|H_{21}|2\rangle & \langle 5|H_{22}|3\rangle & \langle 5|H_{22}|4\rangle & \langle 5|H_{22}|5\rangle \end{pmatrix}.$$

It is clear that all the terms except the potential in the square brackets of Eq. (11.62) result in the diagonal $(N \times N)$-matrices. The potential, $\langle\widehat{\eta}|U(\vec{r})|\widehat{\eta}\rangle$, which may couple all the channels, gives (generally speaking) non-zero values for all the elements of the corresponding $(N \times N)$-matrix.

Finally, we obtain the following matrix equation:

$$\left[\partial_r^2 + K^2 - \frac{\widehat{\ell}(\widehat{\ell}+1)}{r^2}\right] u^J = V^J u^J \,, \tag{11.65}$$

where u^J is the column-matrix defined in Eq. (11.56) with a single sequential subscript. The matrix K is the diagonal matrix of the channel momenta,

$$K = \mathrm{diag}\{k_1, k_2, \ldots, k_N\} \,, \tag{11.66}$$

where some elements are the same (for the channels belonging to the same internal state). The matrix $\widehat{\ell}$ is also diagonal,

$$\widehat{\ell} = \mathrm{diag}\{\ell_1, \ell_2, \ldots, \ell_N\} \,, \tag{11.67}$$

where some elements may also be repeating. This is, first, because for different internal states the allowed angular momenta may be the same, and second, because within a group of degenerate sub-channels a given J may be constructed with different total spins but with the same ℓ. In simple words, the repeating values of k_n and ℓ_n are the consequence of the numbering the channels with the help of a single subscript according to the rule (11.35).

The matrix of the potential is obtained as the integral

$$V^J_{nn'}(r) = \frac{2\mu_n}{\hbar^2} \sum_{mm'} C^{JM}_{\ell m s_n \sigma_n} C^{JM}_{\ell' m' s_{n'} \sigma_{n'}} \int Y^*_{\ell m}(\hat{r}) U_{nn'}(\vec{r}) Y_{\ell' m'}(\hat{r}) d\hat{r} \,, \tag{11.68}$$

where $1 \leqslant n \leqslant N$ is the same sequential subscript that labels different channels (including the degenerate ones). Under this integral, there is another matrix,

$$U_{nn'}(\vec{r}) = \langle \eta_n, s_n\sigma_n | U(\vec{r}) | \eta_{n'}, s_{n'}\sigma_{n'} \rangle \,, \tag{11.69}$$

that couples the channels with different internal and spin states. Here, U is an operator of the total potential acting between the interacting bodies. The fact that V^J does not depend on the quantum number M can be proved in the same way as it is done for a non-central potential (see Sect. 10.2.1).

In the explicit form, the elements of the column-matrix u^J obey the following radial multi-channel Schrödinger equation:

$$\left[\partial_r^2 + k_n^2 - \frac{\ell_n(\ell_n+1)}{r^2}\right] u^J_n(E,r) = \sum_{n'} V^J_{nn'}(r) u^J_{n'}(E,r) \,. \tag{11.70}$$

For the sake of simplicity of notation, the common superscript J in Eq. (11.70) can be dropped, but we should not forget that it is a conserving quantum number and that it restricts the range of possible values of ℓ and thus the range of coupled channels.

At this point, the reader should recall Sect. 10.2 where the partial-wave analysis for the particles with non-zero spins was considered. Apparently, Eq. (11.70) is exactly the same as Eq. (10.16). It should be noted that most of the formulae derived for the non-central potentials can be used here as well.

11.2.3 Partial-Wave Decomposition for Scattering States

This section generalizes the partial-wave analysis that is done in Sect. 10.3 for a non-central potential. In contrast to a discrete (either bound or resonant) state, a scattering state does not have definite values of J and M. It is a mixture (superposition) of the states with different values of these quantum numbers. However, they are still conserving, and thus for each value of the total angular momentum, the problem can be solved separately. When calculating the scattering observables (such as the cross section), the contributions from all possible J should be summed up.

In the single-channel case, the scattering state originates from a plane wave $|\vec{k}\rangle$ that comes from a certain direction. In a muti-channel problem, the channel plane waves $|\vec{k}_\eta, \eta, s\sigma\rangle$, $|\vec{k}_{\eta'}, \eta', s'\sigma'\rangle$, $|\vec{k}_{\eta''}, \eta'', s''\sigma''\rangle, \ldots$ also come from a single direction, but generally speaking in many different channels.[3] The magnitudes of the channel momenta are determined by a common total energy E and by the corresponding threshold energies E_η according to Eq. (11.23). The fact that in all the incoming channels the fluxes of the particles have the same direction means that

$$\vec{k}_\eta \parallel \vec{k}_{\eta'} \parallel \vec{k}_{\eta''} \ldots \quad \Rightarrow \quad \hat{\vec{k}}_\eta = \hat{\vec{k}}_{\eta'} = \hat{\vec{k}}_{\eta''} = \cdots = \hat{\vec{k}}\,, \tag{11.71}$$

where $\hat{\vec{k}}$ are the common spherical angles of the momenta in all the (incoming) channels.

It should be noted that each combination of the quantum numbers (η, s, σ) is a separate channel despite the fact that some of them correspond to the same internal state η and thus to the same energy E_η (degenerate channels). This is schematically illustrated in Fig. 11.1. To label such channels, we introduce a single sequential number (subscript) ν,

$$\nu :\longrightarrow \{\eta_\nu, s_\nu, \sigma_\nu\}\,, \qquad \nu = 1, 2, 3, \ldots \tag{11.72}$$

[3] In practice, we usually use an incoming beam as well as the target in certain internal states, i.e. in one channel. However, the beam-target system may be a mixture of different states of the total spin and its z-component, which formally are different channels.

It can be shown rigorously that the scattering states originated from the incoming asymptotic states in different channels are orthogonal to each other. However, each scattering state originated from a single channel may finally end up in any of the open outgoing channels.[4] Symbolically, this can be depicted as the following diagram:

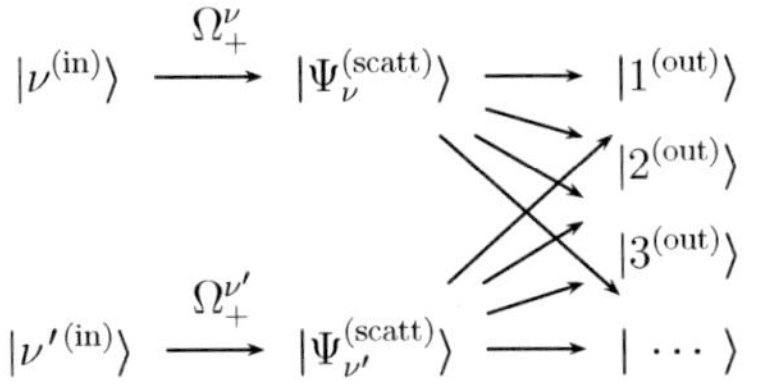

where $\langle\Psi_{\nu'}^{(\mathrm{scatt})}|\Psi_\nu^{(\mathrm{scatt})}\rangle = 0$. Therefore, the scattering state "remembers" from which incoming channel it was originated. This means that a scattering state must have labels (quantum numbers) indicating its origin. These quantum numbers are η_ν, $\vec{k}_\nu$, s_ν, and σ_ν (the internal state, channel momentum, total channel spin, and its z-component). We denote such a scattering state as $|\Psi_\nu(\vec{k}\,)\rangle$, where ν is a sequential number corresponding to a set of the quantum numbers (11.72) of the incoming channel. In order to simplify the notation, we put here the symbol $\vec{k}$ (without the subscript) instead of $\vec{k}_\nu$. Whenever it does not cause confusion, we do the same for the other variables (η, s, σ) as well. The corresponding variables linked to the channel ν' will be denoted with the "primed" symbols.

Since the scattering state originated from the channel ν can end up in any of the exit channels, it is a mixture (superposition) of the states with all possible sets ν' of the quantum numbers $\{\eta', s', \sigma'\}$,

$$|\Psi_\nu(\vec{k}\,)\rangle = \sum_{\nu'=1}^{N} |\Psi_{\nu'\nu}(\vec{k}\,)\rangle \,. \tag{11.73}$$

Similarly to the case of a bound state, the terms in this equation are mutually orthogonal and thus can be collected in a column-matrix,[5]

$$|\Psi_\nu(\vec{k}\,)\rangle = \begin{pmatrix} |\Psi_{1,\nu}(\vec{k}\,)\rangle \\ |\Psi_{2,\nu}(\vec{k}\,)\rangle \\ \cdots \\ |\Psi_{N,\nu}(\vec{k}\,)\rangle \end{pmatrix} \,. \tag{11.74}$$

[4] A proof of this needs the time-dependent consideration of the collision process. It can be found, for example, in Ref. [4].

[5] It is assumed that the reader is not confused by using here the same symbol Ψ with the single and double subscripts. When it is with one subscript, Ψ_ν, it is the whole column-matrix. When we indicate two subscripts, $\Psi_{\nu'\nu}$, it is a single element of that column-matrix, where ν' indicates the position of this element in the column.

The scattering state originated from another incoming state ν forms another column-matrix (11.74). Combining all such columns, we obtain a square matrix,

$$|\Psi(\vec{k}\,)\rangle = \begin{pmatrix} |\Psi_{1,1}(\vec{k}\,)\rangle & |\Psi_{1,2}(\vec{k}\,)\rangle & \cdots & |\Psi_{1,N}(\vec{k}\,)\rangle \\ |\Psi_{2,1}(\vec{k}\,)\rangle & |\Psi_{2,2}(\vec{k}\,)\rangle & \cdots & |\Psi_{2,N}(\vec{k}\,)\rangle \\ \cdots & \cdots & \cdots & \cdots \\ |\Psi_{N,1}(\vec{k}\,)\rangle & |\Psi_{N,2}(\vec{k}\,)\rangle & \cdots & |\Psi_{N,N}(\vec{k}\,)\rangle \end{pmatrix} , \tag{11.75}$$

where each column is originated from a different in-channel ν (the second subscript). The element $|\Psi_{\nu'\nu}(\vec{k}\,)\rangle$ of such a column is the component of the corresponding state-vector in the channel ν'. In simple words, the element $|\Psi_{\nu'\nu}(\vec{k}\,)\rangle$ describes the scattering process that begins in the channel ν and ends up in the channel ν'.

As a way of distinguishing the full state-vector (belonging to the full space of the internal and relative variables) from the relative-motion state-vector, we use the capital and small letters, Ψ and ψ, respectively. The relative-motion state-vector is obtained from the full one by folding the internal variables,

$$|\psi_{\nu'\nu}(\vec{k}\,)\rangle = \langle\eta_{\nu'}|\Psi_{\nu'\nu}(\vec{k}\,)\rangle \,. \tag{11.76}$$

Using these vectors, we can re-write Eq. (11.73) in the following factorized form:

$$|\Psi_{\nu}(\vec{k}\,)\rangle = \sum_{\nu'=1}^{N} |\eta_{\nu'}\rangle|\psi_{\nu',\nu}(\vec{k}\,)\rangle \,. \tag{11.77}$$

All the terms of this sum are mutually orthogonal, and thus it can be represented by a column-matrix,

$$|\Psi_{\nu}(\vec{k}\,)\rangle = \begin{bmatrix} |\eta_1\rangle|\psi_{1,\nu}(\vec{k}\,)\rangle \\ |\eta_2\rangle|\psi_{2,\nu}(\vec{k}\,)\rangle \\ \cdots \\ |\eta_N\rangle|\psi_{N,\nu}(\vec{k}\,)\rangle \end{bmatrix} , \tag{11.78}$$

where among the sequence of the internal states, $|\eta_1\rangle$, $|\eta_1\rangle$, ..., $|\eta_N\rangle$, there are repeating elements (for the degenerate channels). In other words, the sum (11.77) consists of N orthogonal terms, while the number of different internal states, $\mathcal{N}$, is less than N because some of the channels are degenerate. The degenerate terms (sub-channel states) are still mutually orthogonal because they involve different spin states. This is why some of the ket-vectors $|\eta_{\nu'}\rangle$ are repeating in this sum.

Generally speaking, the off-diagonal ($\nu' \neq \nu$) elements of the matrices (11.75), (11.76) are non-zero. This is because the internal and spin states of the colliding bodies are disturbed when they approach each other. Actually, this is the reason why a scattering state originated from a single channel may end up in any of the open channels. The scattering state (11.76) of the relative motion determines the probability that the collision results in the transition $\nu' \leftarrow \nu$.

The scattering state is obtained from the incoming state in the channel ν by the action of the corresponding Møller operator Ω_+^ν,

$$|\Psi_\nu(\vec{k})\rangle = \Omega_+^\nu|\vec{k},\nu\rangle = \sum_{\ell m} \Omega_+^\nu|k\ell m, \eta, s\sigma\rangle Y_{\ell m}^*(\hat{\vec{k}})$$
$$= \sum_{\ell m JM} \Omega_+^\nu|k, \eta, (\ell s)JM\rangle C_{\ell m s\sigma}^{JM} Y_{\ell m}^*(\hat{\vec{k}}) ,$$

where the plane wave is expanded in the partial waves and then the orbital angular momentum and the spin are coupled in the total angular momentum, J. Similarly to what is done in Sect. 10.3 for a non-central potential, we introduce the partial-wave scattering state with definite values of the quantum numbers ℓ, s, J, and M,

$$|\Psi_{\eta[\ell]}^{JM}\rangle = \Omega_+^\nu|k, \eta, (\ell s)JM\rangle ,$$

where $[\ell] = \{\ell, s\}$. In the ket-vector $|\Psi_{\eta[\ell]}^{JM}\rangle$, the channel momentum k is not indicated because it is uniquely determined by the internal state η and the total energy E (common for all the channels). It should be clear that the subscripts $\eta[\ell]$ indicate the quantum numbers of the incoming channel, from which this scattering state evolved. The ket-vector $|\Psi_{\eta[\ell]}^{JM}\rangle$ is a factor in the partial-wave expansion of the state-vector $|\Psi_\nu(\vec{k})\rangle$ (the spin-angular functions used below are defined in Sect. 10.1),

$$|\Psi_\nu(\vec{k})\rangle = \sum_{JM\ell} |\Psi_{\eta[\ell]}^{JM}\rangle Y_{\ell s\sigma}^{JM*}(\hat{\vec{k}}) = \sum_{JM[\ell]} |\Psi_{\eta[\ell]}^{JM}\rangle\langle\mathcal{Y}_{[\ell]}^{JM}(\hat{\vec{k}})|s\sigma\rangle . \tag{11.79}$$

Meanwhile, the ket-vector $|\Psi_\nu(\vec{k})\rangle$ can be split into mutually orthogonal components in all the outgoing channels, as is given by Eq. (11.73). This implies that $|\Psi_{\eta[\ell]}^{JM}\rangle$ is also a sum of the components in all the outgoing channels (the footnote 5 is applicable here as well),

$$|\Psi_{\eta[\ell]}^{JM}\rangle = \sum_{\eta'[\ell']} |\Psi_{\eta'[\ell'],\eta[\ell]}^{JM}\rangle , \tag{11.80}$$

which are mutually orthogonal and thus can be represented by a column-matrix. As the reader should notice, Eq. (11.79) generalizes Eq. (10.20).

Each combination of the subscripts $\eta[\ell]$ corresponds to a sequential number ν. Using the partial-wave scattering states (11.80), we can re-write the expansion (11.73) in the following way:

$$|\Psi_\nu(\vec{k})\rangle = \sum_{\eta'[\ell']} \sum_{JM\ell} |\Psi_{\eta'[\ell'],\eta[\ell]}^{JM}\rangle Y_{\ell s\sigma}^{JM*}(\hat{\vec{k}}) . \tag{11.81}$$

In the column representation, it looks like this:

$$|\Psi_\nu(\vec{k})\rangle = \begin{pmatrix} \sum_{JM\ell} |\Psi^{JM}_{\eta_1[\ell_1],\eta[\ell]}\rangle Y^{JM*}_{\ell s\sigma}(\hat{\vec{k}}) \\ \sum_{JM\ell} |\Psi^{JM}_{\eta_2[\ell_2],\eta[\ell]}\rangle Y^{JM*}_{\ell s\sigma}(\hat{\vec{k}}) \\ \cdots \\ \sum_{JM\ell} |\Psi^{JM}_{\eta_N[\ell_N],\eta[\ell]}\rangle Y^{JM*}_{\ell s\sigma}(\hat{\vec{k}}) \end{pmatrix} . \tag{11.82}$$

Therefore, each element of the column-matrix (11.74) is expanded in the partial waves. Combining such columns for different incoming channels ν, we obtain the corresponding partial-wave expansion of the square matrix (11.75).

The scattering wave function in the space of the relative coordinate and of the total spin can be obtained by projecting the ket-vector (11.79) onto the bra-vector $\langle\vec{r}, \nu'| = \langle\vec{r}, \eta', s'\sigma'|$,

$$\begin{aligned} \psi_{\nu'\nu}(\vec{r}, \vec{k}) &= \langle\vec{r}, \nu'|\Psi_\nu(\vec{k})\rangle \\ &= \sum_{JM\ell'\ell} Y^{JM}_{\ell's'\sigma'}(\hat{\vec{r}})\langle r, \eta', (\ell's')JM|\Psi^{JM}_{\eta[\ell]}\rangle Y^{JM*}_{\ell s\sigma}(\hat{\vec{k}}) \\ &= \sum_{JM[\ell'][\ell]} \langle s'\sigma'|\mathcal{Y}^{JM}_{[\ell']}(\hat{\vec{r}})\rangle\langle r, \eta', (\ell's')JM|\Psi^{JM}_{\eta[\ell]}\rangle\langle\mathcal{Y}^{JM}_{[\ell]}(\hat{\vec{k}})|s\sigma\rangle . \end{aligned} \tag{11.83}$$

Following the same line of thought as in Sect. 10.3, we introduce the matrix u^J of the radial wave functions as

$$\langle r, \eta', (\ell's')JM|\Psi^{JM}_{\eta[\ell]}\rangle = \frac{u^J_{\eta'[\ell'],\eta[\ell]}(E, r)}{r} N_{[\ell]}(k_\nu) .$$

The proof that matrix u^J does not depend on M is the same as the one given in Sect. 10.2.1.

As the reader can see, we multiplied each column of the matrix u^J by an appropriate normalization constant $N_{[\ell]}(k_\nu)$. This is done for the same reason as in Eqs. (9.8) and (10.21), namely, to have proper normalization of the wave function. This constant should be the same for any potential. In particular, it must be the same when the potential is identical zero. In such a case, all the Møller operators are equal to the identity operator, and the scattering states coincide with the corresponding asymptotic states (plane waves). In Sect. 10.3.2, it is shown that

$$N_{[\ell]}(k_\nu) = \frac{i^\ell}{\sqrt{2\pi}k_\nu} . \tag{11.84}$$

Therefore,

$$\psi_{\nu'\nu}(\vec{r},\vec{k}\,) =$$

$$= \sum_{JM}\sum_{\ell'\ell} Y^{JM}_{\ell's'\sigma'}(\hat{\vec{r}}\,)\frac{u^J_{\eta'[\ell'],\eta[\ell]}(E,r)N_{[\ell]}(k_\nu)}{r} Y^{JM*}_{\ell s\sigma}(\hat{\vec{k}}\,)$$

$$= \sum_{JM}\sum_{[\ell'][\ell]} \langle s'\sigma'|\mathcal{Y}^{JM}_{[\ell']}(\hat{\vec{r}}\,)\rangle\frac{u^J_{\eta'[\ell'],\eta[\ell]}(E,r)N_{[\ell]}(k_\nu)}{r}\langle\mathcal{Y}^{JM}_{[\ell]}(\hat{\vec{k}}\,)|s\sigma\rangle$$

$$= \sum_{JM}\sum_{[\ell'][\ell]} \langle s'\sigma'|\mathcal{Y}^{JM}_{[\ell']}(\hat{\vec{r}}\,)\rangle\frac{u^J_{\eta'[\ell'],\eta[\ell]}(E,r)i^\ell}{\sqrt{2\pi}k_\nu r}\langle\mathcal{Y}^{JM}_{[\ell]}(\hat{\vec{k}}\,)|s\sigma\rangle\,. \tag{11.85}$$

It is not difficult to see that

$$\langle\vec{r},\nu'|\Psi^{JM}_{\eta'[\ell'],\eta[\ell]}\rangle = Y^{JM}_{\ell's'\sigma'}(\hat{\vec{r}}\,)\frac{u^J_{\eta'[\ell'],\eta[\ell]}(E,r)}{r}N_{[\ell]}(k_\nu)\,. \tag{11.86}$$

In Eq. (11.85), the left and right spin states, $\langle s'\sigma'|$ and $|s\sigma\rangle$, can be removed. As a result, the wave function becomes an operator in the spin space,

$$\psi_{\eta'\eta}(\vec{r},\vec{k}\,) = \sum_{JM}\sum_{[\ell'][\ell]} |\mathcal{Y}^{JM}_{[\ell']}(\hat{\vec{r}}\,)\rangle\frac{u^J_{\eta'[\ell'],\eta[\ell]}(E,r)i^\ell}{\sqrt{2\pi}k_\nu r}\langle\mathcal{Y}^{JM}_{[\ell]}(\hat{\vec{k}}\,)|\,. \tag{11.87}$$

This operator should be sandwiched between the appropriate spin states when a specific transition is calculated. The operator $\psi_{\eta'\eta}(\vec{r},\vec{k}\,)$ is different from the function $\psi_{\nu'\nu}(\vec{r},\vec{k}\,)$ that is defined in Eq. (11.83) as the scalar product of the bra- and ket-vectors involving the spin states $\langle s'\sigma'|$ and $|s\sigma\rangle$. The same way as in Eq. (11.87), we can remove these spin states and obtain the operator (in the spin space) $\psi_{\eta'\eta}(\vec{r},\vec{k}\,)$ in yet another useful form,

$$\psi_{\eta'\eta}(\vec{r},\vec{k}\,) = \langle\vec{r},\eta'|\Psi_\eta(\vec{k}\,)\rangle\,. \tag{11.88}$$

It should be clearly understood that in Eq. (11.87) for a given J the sum over $[\ell'] = \{\ell',s'\}$ only runs over those quantum numbers ℓ' and s' that are allowed by the internal state $|\eta'\rangle$. Similarly, the sum over $[\ell] = \{\ell,s\}$ covers the combinations of ℓ and s that are allowed by the internal state $|\eta\rangle$. All such combinations can be labeled using a single sequential number n as is schematically illustrated in Table 11.1. In this example, there are three ($\mathcal{N} = 3$) truly different channels (η, η', and η''). They involve two, three, and two degenerate sub-channels, respectively. The total number of channels is seven ($N = 7$).

Table 11.1 Schematic example of assigning a single sequential number n to the spin-angular partial waves in the three internal states, η, η', and η''

Internal state	η		η'			η''	
Allowed combinations (ℓs)	$[\ell_1]$	$[\ell_2]$	$[\ell'_1]$	$[\ell'_2]$	$[\ell'_3]$	$[\ell''_1]$	$[\ell''_2]$
n	1	2	3	4	5	6	7

Using that single subscript, the radial wave function can be written as

$$u^J_{\eta'[\ell'],\eta[\ell]}(E, r) = u^J_{n'n}(E, r)\ .$$

It is an $(N \times N)$ square matrix, where N is the total number of the channels (including all the sub-channels). If we combine the normalization coefficients (11.84) in the diagonal matrix,

$$\widehat{N}(E) = \text{diag}\left\{\frac{i^{\ell_1}}{\sqrt{2\pi}k_1}, \frac{i^{\ell_2}}{\sqrt{2\pi}k_2}, \ldots, \frac{i^{\ell_N}}{\sqrt{2\pi}k_N}\right\} = \frac{1}{\sqrt{2\pi}}K^{-1}\mathcal{I}\ , \tag{11.89}$$

where K is defined by Eq. (11.66) and $\mathcal{I}$ is the diagonal matrix (9.108), then the partial-wave expansion (11.87) of the scattering wave function can be written in the following matrix form:

$$\psi(\vec{r}, \vec{k}) = \sum_{JM} |\mathcal{Y}^{JM}(\hat{\vec{r}}))\frac{u^J(E, r)\widehat{N}(E)}{r}(\mathcal{Y}^{JM}(\hat{\vec{k}})|\ , \tag{11.90}$$

where the $(\mathcal{N} \times N)$-matrices $|\mathcal{Y}^{JM}(\hat{\vec{r}}))$ and $(\mathcal{Y}^{JM}(\hat{\vec{k}})|$ are defined in Sect. 11.2.1. Apparently, $\psi(\vec{r}, \vec{k})$ is a square matrix of the size $(\mathcal{N} \times \mathcal{N})$, i.e. the number of its lines and columns is equal to thc number of the internal states (truly different channels). The motion in each of the truly different channels is represented as a sum of certain partial waves (degenerate sub-channels). This is why the size of the radial square matrix $u^J(E, r)$ is greater, $N > \mathcal{N}$.

Looking at Eq. (11.88), we see that it is useful to introduce the following $(\mathcal{N} \times \mathcal{N})$-matrix in the space of the internal states:

$$\Psi(\vec{r}, \vec{k}) = |\widehat{\eta}\rangle\psi(\vec{r}, \vec{k})\langle\widehat{\eta}| = \begin{pmatrix} |\eta_1\rangle\psi_{\eta_1\eta_1}\langle\eta_1| & |\eta_1\rangle\psi_{\eta_1\eta_2}\langle\eta_2| \\ |\eta_2\rangle\psi_{\eta_2\eta_1}\langle\eta_1| & |\eta_2\rangle\psi_{\eta_2\eta_2}\langle\eta_2| \end{pmatrix}, \tag{11.91}$$

where (for the sake of clarity) it is given explicitly for the case of $\mathcal{N} = 2$. Apparently, this matrix represents an operator in the space of the internal states. Its action in this space is (in a way) similar to the Møller operator action on the asymptotic state. Indeed, using the same simplified example, we see that

$$\begin{pmatrix} |\eta_1\rangle\psi_{\eta_1\eta_1}\langle\eta_1| & |\eta_1\rangle\psi_{\eta_1\eta_2}\langle\eta_2| \\ |\eta_2\rangle\psi_{\eta_2\eta_1}\langle\eta_1| & |\eta_2\rangle\psi_{\eta_2\eta_2}\langle\eta_2| \end{pmatrix} \begin{pmatrix} |\eta_1\rangle \\ 0 \end{pmatrix} = \begin{pmatrix} |\eta_1\rangle\psi_{\eta_1\eta_1} \\ |\eta_2\rangle\psi_{\eta_2\eta_1} \end{pmatrix}, \tag{11.92}$$

which is the scattering state, originated from $|\eta_1\rangle$ (and similarly for the $|\eta_2\rangle$ initial state). It is also clear that

$$\psi = \langle\,\widehat{\eta}\,|\Psi|\,\widehat{\eta}\,\rangle\,. \tag{11.93}$$

The reader should recall the discussion in Sect. 11.1 devoted to the channels with different types of the particles. The representation (11.91) that involves the projection (transition) operators $|\eta'\rangle\langle\eta|$ can be easily generalized for various particle transformations. For that, it is sufficient to consider different types of the particles as their different internal states.

11.2.3.1 Radial Schrödinger Equation

Each column of the matrix (11.91) is a solution of the Schrödinger equation in the full space (involving the internal and relative coordinates), and thus the whole matrix obeys this equation,

$$\left[\widehat{H}_0 + U(\vec{r}\,) + |\widehat{\eta}\rangle\,\widehat{E}\,\langle\widehat{\eta}| - E\right]\Psi(\vec{r},\vec{k}\,) = 0\,, \tag{11.94}$$

where in the square brackets there are the $(\mathcal{N}\times\mathcal{N})$-matrices that are defined by Eqs. (11.61), (11.63), and (11.64). Multiplying Eq. (11.94) from the left by the diagonal matrix $\langle\widehat{\eta}|$ and from the right by its Hermitian conjugate, $|\widehat{\eta}\,\rangle$, we obtain the corresponding equation for the relative-motion matrix (11.90),

$$\left[\widehat{H}_0 + \langle U(\vec{r}\,)\rangle + \widehat{E} - E\right]\psi(\vec{r},\vec{k}\,) = 0\,, \tag{11.95}$$

where $\langle U(\vec{r}\,)\rangle$ is the $(\mathcal{N}\times\mathcal{N})$-matrix consisting of the elements $\langle\eta'|U(\vec{r}\,)|\eta\rangle$.

If now we substitute here the partial-wave expansion (11.90), multiply the whole equation from the left by $\langle\mathcal{Y}^{JM}(\hat{\vec{r}}\,)|$ and from the right by $|\mathcal{Y}^{JM}(\hat{\vec{k}}\,)\rangle$, and then integrate over the spherical angles $\hat{\vec{r}}$ and $\hat{\vec{k}}$, then we obtain exactly the same radial equation as for the case of the discrete states, namely, Eq. (11.65). The only difference is that for the scattering states the matrix $u^J(E,r)$ is not a column but a square matrix of the dimension N. In other words, each column of this matrix obeys the same radial equation as for a bound-state column. By the way, the normalization constants $\widehat{N}$ are the same within each column and thus are cancelled out because the equation is homogeneous.

In the explicit form, the radial Schrödinger equation for the matrix elements $u^J_{nn'}(E,r)$ of the radial wave function has the following form:

$$\left[\partial_r^2 + k_n^2 - \frac{\ell_n(\ell_n+1)}{r^2}\right]u^J_{nn'}(E,r) = \sum_{n''} V^J_{nn''}(r)u^J_{n''n'}(E,r)\,, \tag{11.96}$$

where the subscript n is a sequential number of various combinations (11.35).

Multi-Channel Jost Matrix

12

For the sake of notation clarity, let us consider a two-channel ($N = 2$) case. The generalization of all the formulae for an arbitrary N is straightforward. In this minimal multi-channel case, Eq. (11.70) takes the form

$$\begin{bmatrix} \partial_r^2 + k_1^2 - \dfrac{\ell_1(\ell_1+1)}{r^2} & 0 \\ 0 & \partial_r^2 + k_2^2 - \dfrac{\ell_2(\ell_2+1)}{r^2} \end{bmatrix} \begin{pmatrix} u_1 \\ u_2 \end{pmatrix} = \begin{pmatrix} V_{11} & V_{12} \\ V_{21} & V_{22} \end{pmatrix} \begin{pmatrix} u_1 \\ u_2 \end{pmatrix} . \tag{12.1}$$

We assume that the potential V is not singular. This means that all of its matrix elements are regular or (in the worst case) less singular at $r = 0$ than $\sim r^{-2}$. Therefore the most severe singularity of Eq. (12.1) is due to the centrifugal terms. However if we multiply the equation by r^2, none of its coefficients remain singular at $r = 0$. In the theory of differential equations, such points are called the regular singular points.

A system of N second-order differentia equations has $2N$ linearly independent solutions. At a regular singular point, half of these solutions are regular, while the other half may diverge at such a point. Each solution of Eq. (12.1) is a column. Therefore there are four linearly independent columns that solve this equation,

$$\begin{pmatrix} \phi_1 \\ \phi_2 \end{pmatrix}_1 , \quad \begin{pmatrix} \phi_1 \\ \phi_2 \end{pmatrix}_2 , \quad \begin{pmatrix} \varphi_1 \\ \varphi_2 \end{pmatrix}_1 , \quad \begin{pmatrix} \varphi_1 \\ \varphi_2 \end{pmatrix}_2 ,$$

where the symbols ϕ and φ are used for the regular and irregular solutions, respectively. These columns can be combined in two square matrices,

$$\phi = \begin{pmatrix} \phi_{11} & \phi_{12} \\ \phi_{21} & \phi_{22} \end{pmatrix} , \qquad \varphi = \begin{pmatrix} \varphi_{11} & \varphi_{12} \\ \varphi_{21} & \varphi_{22} \end{pmatrix} ,$$

S. A. Rakityansky, *Jost Functions in Quantum Mechanics*,
https://doi.org/10.1007/978-3-031-07761-6_12

where the second subscript corresponds to the solution number. Since we are only interested in a solution that is relevant to physics, we may forget about the matrix φ and consider the matrix ϕ which is regular at the point $r = 0$. Near this point, the centrifugal potential (behaving as $\sim r^{-2}$) dominates over the potential V (which is non-singular) and therefore the right hand side of Eq. (12.1) as well as the constant term k_n^2 on the left hand side can be omitted, which gives

$$\begin{bmatrix} r^2\partial_r^2 - \ell_1(\ell_1+1) & 0 \\ 0 & r^2\partial_r^2 - \ell_2(\ell_2+1) \end{bmatrix} \begin{pmatrix} \phi_1 \\ \phi_2 \end{pmatrix} \approx 0 \,, \qquad r \to 0 \,. \tag{12.2}$$

It is easy to see that this is the same as Eq. (9.29), where we considered the boundary conditions at $r = 0$ for the regular solution in the case of coupled partial waves. The analysis given there is fully applicable in our present case.[1] In the same way as for coupled partial waves, we can choose the normalization of each element of matrix ϕ arbitrarily. Using this freedom, we can specify the boundary conditions for the matrix ϕ as

$$\lim_{r\to 0} \frac{\phi_{nn'}(E,r)}{j_{\ell_{n'}}(k_{n'}r)} = \delta_{nn'} \,, \tag{12.3}$$

which can be written in the matrix form as

$$\lim_{r\to 0} \phi j^{-1} = I \,, \tag{12.4}$$

where $j = \mathrm{diag}\{j_{\ell_1}(k_1 r), j_{\ell_1}(k_2 r), \ldots, j_{\ell_N}(k_N r)\}$. Equation (12.3) is consistent with the single-channel case (see Eq. (2.36)) as well as with the coupled partial wave case (see Eq. (9.32)). Thus defined columns

$$\begin{pmatrix} \phi_1 \\ \phi_2 \end{pmatrix}_1 \equiv \begin{pmatrix} \phi_{11} \\ \phi_{21} \end{pmatrix} , \qquad \begin{pmatrix} \phi_1 \\ \phi_2 \end{pmatrix}_2 \equiv \begin{pmatrix} \phi_{12} \\ \phi_{22} \end{pmatrix}$$

are linearly independent, i.e. they form a basis in the space of regular solutions. This means that any other regular solution is their linear combination. In particular, a physical solution u is

$$u = \begin{pmatrix} u_1 \\ u_2 \end{pmatrix} = \begin{pmatrix} \phi_1 \\ \phi_2 \end{pmatrix}_1 c_1 + \begin{pmatrix} \phi_1 \\ \phi_2 \end{pmatrix}_2 c_2 = \begin{pmatrix} \phi_{11}c_1 + \phi_{12}c_2 \\ \phi_{21}c_1 + \phi_{22}c_2 \end{pmatrix} = \begin{pmatrix} \phi_{11} & \phi_{12} \\ \phi_{21} & \phi_{22} \end{pmatrix} \begin{pmatrix} c_1 \\ c_2 \end{pmatrix} , \tag{12.5}$$

[1] This is not surprising because different partial waves can always be treated as different channels.

where the combination coefficients c_1 and c_2 should be chosen in such a way that the function u has a correct asymptotic behaviour (when $r \to \infty$). It should be emphasized that the boundary conditions (12.3) uniquely specify the matrix $\phi(E, r)$ on the whole interval $0 \leqslant r < \infty$.[2] Therefore, the choice of the combination coefficients c_n is a separate task, which is needed to guarantee that the physical solution u is meaningful (i.e. behaves correctly at large distances: either as a bound-state solution, or as a resonance, or else as a scattering-state solution).

Since we assumed that the potential is of a short range, at large distances, all its matrix elements vanish, and Eq. (12.1) takes a more simple form,

$$(r \to \infty) \quad \Rightarrow \quad \begin{bmatrix} \partial_r^2 + k_1^2 - \dfrac{\ell_1(\ell_1+1)}{r^2} & 0 \\ 0 & \partial_r^2 + k_2^2 - \dfrac{\ell_2(\ell_2+1)}{r^2} \end{bmatrix} \begin{pmatrix} u_1 \\ u_2 \end{pmatrix} = 0 . \tag{12.6}$$

This is still a system of second-order equations, which has exactly $2N$ linearly independent solutions. And we know these solutions. Of course, there is an infinite number of ways to choose linearly independent solutions. For our purpose, the most convenient choice is to take the following four columns:

$$\begin{bmatrix} h_{\ell_1}^{(-)}(k_1 r) \\ 0 \end{bmatrix}, \quad \begin{bmatrix} 0 \\ h_{\ell_2}^{(-)}(k_2 r) \end{bmatrix}, \quad \begin{bmatrix} h_{\ell_1}^{(+)}(k_1 r) \\ 0 \end{bmatrix}, \quad \begin{bmatrix} 0 \\ h_{\ell_2}^{(+)}(k_2 r) \end{bmatrix} . \tag{12.7}$$

These columns are the solutions of Eq. (12.1) at large distances, where there is no problem of regularity. Therefore, any other solution (does not matter if it is a regular or an irregular one) at large distances can be expanded over the basis (12.7), i.e. can be presented as a linear combination of these $2N$ (four in our simple case) columns. For the regular columns, we thus obtain

$$\begin{pmatrix} \phi_{11} \\ \phi_{21} \end{pmatrix} \underset{r\to\infty}{\longrightarrow}$$

$$\begin{bmatrix} h_{\ell_1}^{(-)}(k_1 r) \\ 0 \end{bmatrix} f_{11}^{(\text{in})} + \begin{bmatrix} 0 \\ h_{\ell_2}^{(-)}(k_2 r) \end{bmatrix} f_{21}^{(\text{in})} + \begin{bmatrix} h_{\ell_1}^{(+)}(k_1 r) \\ 0 \end{bmatrix} f_{11}^{(\text{out})}$$

$$+ \begin{bmatrix} 0 \\ h_{\ell_2}^{(+)}(k_2 r) \end{bmatrix} f_{21}^{(\text{out})} = \begin{bmatrix} h_{\ell_1}^{(-)}(k_1 r) f_{11}^{(\text{in})} \\ h_{\ell_2}^{(-)}(k_2 r) f_{21}^{(\text{in})} \end{bmatrix} + \begin{bmatrix} h_{\ell_1}^{(+)}(k_1 r) f_{11}^{(\text{out})} \\ h_{\ell_2}^{(+)}(k_2 r) f_{21}^{(\text{out})} \end{bmatrix} ,$$

[2] The additional discussion at the end of Sect. 9.3.4 and in whole Sect. 9.3.4.1 is relevant here as well.

$$\begin{pmatrix}\phi_{12}\\ \phi_{22}\end{pmatrix} \underset{r\to\infty}{\longrightarrow}$$

$$\begin{bmatrix}h^{(-)}_{\ell_1}(k_1r)\\ 0\end{bmatrix} f^{(\mathrm{in})}_{12} + \begin{bmatrix}0\\ h^{(-)}_{\ell_2}(k_2r)\end{bmatrix} f^{(\mathrm{in})}_{22} + \begin{bmatrix}h^{(+)}_{\ell_1}(k_1r)\\ 0\end{bmatrix} f^{(\mathrm{out})}_{12}$$

$$+ \begin{bmatrix}0\\ h^{(+)}_{\ell_2}(k_2r)\end{bmatrix} f^{(\mathrm{out})}_{22} = \begin{bmatrix}h^{(-)}_{\ell_1}(k_1r) f^{(\mathrm{in})}_{12}\\ h^{(-)}_{\ell_2}(k_2r) f^{(\mathrm{in})}_{22}\end{bmatrix} + \begin{bmatrix}h^{(+)}_{\ell_1}(k_1r) f^{(\mathrm{out})}_{12}\\ h^{(+)}_{\ell_2}(k_2r) f^{(\mathrm{out})}_{22}\end{bmatrix},$$

where the combination coefficients $f^{(\mathrm{in/out})}_{nn'}$ have two subscripts: the first one shows in which channel the column (12.7) has a non-zero element, and the second subscript indicates which of the columns of the regular matrix is expanded. Combining the regular columns in the square matrix, we have

$$\begin{pmatrix}\phi_{11} & \phi_{12}\\ \phi_{21} & \phi_{22}\end{pmatrix} \underset{r\to\infty}{\longrightarrow} \begin{bmatrix}h^{(-)}_{\ell_1}(k_1r) f^{(\mathrm{in})}_{11} & h^{(-)}_{\ell_1}(k_1r) f^{(\mathrm{in})}_{12}\\ h^{(-)}_{\ell_2}(k_2r) f^{(\mathrm{in})}_{21} & h^{(-)}_{\ell_2}(k_2r) f^{(\mathrm{in})}_{22}\end{bmatrix}$$

$$+ \begin{bmatrix}h^{(+)}_{\ell_1}(k_1r) f^{(\mathrm{out})}_{11} & h^{(+)}_{\ell_1}(k_1r) f^{(\mathrm{out})}_{12}\\ h^{(+)}_{\ell_2}(k_2r) f^{(\mathrm{out})}_{21} & h^{(+)}_{\ell_2}(k_2r) f^{(\mathrm{out})}_{22}\end{bmatrix}$$

$$= \begin{bmatrix}h^{(-)}_{\ell_1}(k_1r) & 0\\ 0 & h^{(-)}_{\ell_2}(k_2r)\end{bmatrix} \begin{bmatrix}f^{(\mathrm{in})}_{11} & f^{(\mathrm{in})}_{12}\\ f^{(\mathrm{in})}_{21} & f^{(\mathrm{in})}_{22}\end{bmatrix} + \begin{bmatrix}h^{(+)}_{\ell_1}(k_1r) & 0\\ 0 & h^{(+)}_{\ell_2}(k_2r)\end{bmatrix}$$

$$\times \begin{bmatrix}f^{(\mathrm{out})}_{11} & f^{(\mathrm{out})}_{12}\\ f^{(\mathrm{out})}_{21} & f^{(\mathrm{out})}_{22}\end{bmatrix}.$$

It is not difficult to see that for an arbitrary N, we have the following asymptotic behaviour of the matrix of regular solutions:

$$\boxed{\phi(E,r) \underset{r\to\infty}{\longrightarrow} W^{(\mathrm{in})}(E,r) f^{(\mathrm{in})}(E) + W^{(\mathrm{out})}(E,r) f^{(\mathrm{out})}(E)} \tag{12.8}$$

where $W^{(\mathrm{in/out})}$ are the diagonal matrices

$$W^{(\mathrm{in/out})}(E,r) = \mathrm{diag}\left\{h^{(\mp)}_{\ell_1}(k_1r), h^{(\mp)}_{\ell_2}(k_2r), \ldots, h^{(\mp)}_{\ell_N}(k_Nr)\right\}, \tag{12.9}$$

describing the incoming and outgoing multi-channel spherical waves. These waves generalize the definitions (9.16) and (9.17). The difference is that here each channel has its own momentum. The energy-dependent square matrices $f^{(\text{in/out})}$ composed of the combination coefficients are the multi-channel Jost matrices.

12.1 First-Order Differential Equations

When considering the coupled partial waves, we derived two different sets of first-order matrix differential equations, which were equivalent to the corresponding Schrödinger equation (see Sects. 9.3.2 and 9.3.3). The reader can easily check that for the multi-channel case all the derivations are exactly the same. Moreover, the resulting equations formally look the same. Let us show this.

Asymptotic behaviour (12.8) of the regular solution ϕ gives us a hint. We look for this solution in the form

$$\boxed{\phi(E,r) = W^{(\text{in})}(E,r)F^{(\text{in})}(E,r) + W^{(\text{out})}(E,r)F^{(\text{out})}(E,r)} \tag{12.10}$$

where $F^{(\text{in/out})}(E,r)$ are unknown matrix-functions obeying the Lagrange condition

$$W^{(\text{in})}(E,r)\partial_r F^{(\text{in})}(E,r) + W^{(\text{out})}(E,r)\partial_r F^{(\text{out})}(E,r) = 0\,. \tag{12.11}$$

The substitution of the ansatz (12.10) into the Schrödinger equation (11.70) and the same tricks used in Sect. 2.2.1.1 give the following set of first-order differential equations:

$$\partial_r F^{(\text{in})} = -\frac{1}{2i}K^{-1}W^{(\text{out})}V\left[W^{(\text{in})}F^{(\text{in})} + W^{(\text{out})}F^{(\text{out})}\right] \tag{12.12}$$

$$\partial_r F^{(\text{out})} = \frac{1}{2i}K^{-1}W^{(\text{in})}V\left[W^{(\text{in})}F^{(\text{in})} + W^{(\text{out})}F^{(\text{out})}\right] \tag{12.13}$$

where the diagonal matrix K consists of the channel momenta. It is defined by Eq. (11.66). The boundary conditions for Eqs. (12.12) and (12.13) can be derived in exactly the same way as it is done in Sect. 9.3.4. Repeating basically the same derivations, we obtain

$$\lim_{r\to 0}\frac{j_{\ell_n}(k_n r)F^{(\text{in/out})}_{nn'}(E,r)}{j_{\ell_{n'}}(k_{n'}r)} = \frac{1}{2}\delta_{nn'}\,. \tag{12.14}$$

The iteration procedure described in Sect. 9.3.4.1 is fully applicable here for obtaining the explicit behaviour of these matrices near $r = 0$.

We can combine the incoming and outgoing waves in the following way:

$$j(E,r) = \frac{W^{(\mathrm{out})}(E,r) + W^{(\mathrm{in})}(E,r)}{2} = \begin{pmatrix} j_{\ell_1}(k_1 r) & 0 & \cdots & 0 \\ 0 & j_{\ell_2}(k_2 r) & \cdots & 0 \\ \vdots & \vdots & \vdots & \vdots \\ 0 & 0 & \cdots & j_{\ell_N}(k_N r) \end{pmatrix}, \tag{12.15}$$

$$n(E,r) = \frac{W^{(\mathrm{out})}(E,r) - W^{(\mathrm{in})}(E,r)}{2i} = \begin{pmatrix} n_{\ell_1}(k_1 r) & 0 & \cdots & 0 \\ 0 & n_{\ell_2}(k_2 r) & \cdots & 0 \\ \vdots & \vdots & \vdots & \vdots \\ 0 & 0 & \cdots & n_{\ell_N}(k_N r) \end{pmatrix}. \tag{12.16}$$

Then introducing the matrices

$$A(E,r) = F^{(\mathrm{in})}(E,r) + F^{(\mathrm{out})}(E,r)\,, \tag{12.17}$$

$$B(E,r) = i\left[F^{(\mathrm{in})}(E,r) - F^{(\mathrm{out})}(E,r)\right], \tag{12.18}$$

we re-write the ansatz (12.10) in the alternative form

$$\boxed{\phi(E,r) = j(E,r)A(E,r) - n(E,r)B(E,r)} \tag{12.19}$$

The corresponding combination of Eqs. (12.12) and (12.13) gives

$$\partial_r A = -K^{-1} n V\,(jA - nB) \tag{12.20}$$

$$\partial_r B = -K^{-1} j V\,(jA - nB) \tag{12.21}$$

with the boundary conditions[3]

$$\lim_{r\to 0} \frac{j_{\ell_n}(k_n r) A_{nn'}(E,r)}{j_{\ell_{n'}}(k_{n'} r)} = \delta_{nn'}\,, \qquad \lim_{r\to 0} \frac{n_{\ell_n}(k_n r) B_{nn'}(E,r)}{j_{\ell_{n'}}(k_{n'} r)} = 0\,, \tag{12.22}$$

[3] These conditions can be derived in the same way as it is done in Sect. 9.3.4. The iteration procedure described in Sect. 9.3.4.1 is also applicable here.

which can also be written in matrix form as

$$\lim_{r\to 0} jAj^{-1} = I\,, \qquad \lim_{r\to 0} nBj^{-1} = 0\,. \tag{12.23}$$

It should be noted that, the same as for the case of non-central potential (see Eq. (9.42)), all the elements of matrix B vanish at $r = 0$,

$$B(E, 0) = 0\,, \tag{12.24}$$

because the ratio $n_{\ell_n}(k_n r)/j_{\ell_{n'}}(k_{n'} r)$ is singular at the origin for any n and n'. In contrast to this, the matrix A does not equal I at $r = 0$. Some of its off-diagonal elements may tend to infinity when $r \to 0$.

12.2 Complex Rotation

Comparing Eqs. (12.8) and (12.10), we should expect that at large distances the matrices $F^{(\text{in/out})}(E, r)$ converge to the corresponding Jost matrices, i.e.

$$\lim_{r\to\infty} F^{(\text{in/out})}(E, r) = f^{(\text{in/out})}(E)\,. \tag{12.25}$$

As we have seen in the simple single-channel case (see Sect. 7.2) as well as in the case of coupled partial waves (see Sect. 9.5), for the potentials diminishing at large distances faster than the Coulomb potential, the limits (12.25) exist and are finite if the energy E is real and positive (a physical collision energy). At complex energies, these limits along the real r-axis (generally speaking) do not exist simultaneously (either $F^{(\text{in})}$ or $F^{(\text{out})}$ diverges).

In the case of a multi-channel system, we encounter the same difficulty at complex energies, and this difficulty can be overcome with the help of the same trick, namely, by using the complex rotation of the coordinate r. The same way as we did it for coupled partial waves in Sect. 9.5, let us re-write Eqs. (12.12) and (12.13) in the form

$$\partial_r F^{(\text{in/out})}(E, r) = \mp\frac{1}{2i} K^{-1} W^{(\text{out/in})}(E, r) V(r)\phi(E, r)\,. \tag{12.26}$$

The limits (12.25) may exist if the above derivatives converge to zero with $r \to \infty$ (i.e. the functions $F^{(\text{in/out})}(E, r)$ become constants) and if this convergence is faster than $1/r$ (see discussion in Sect. 7.2).

In order to make it clear, let us consider once again a two-channel problem ($N = 2$). As can be seen from Eq. (12.6), at large distances, all matrix elements of the $(N \times N)$ matrix $\phi(E, r)$ (which solves this equation) are linear combinations of the Riccat–Hankel functions. If we include all the constant factors in the combination coefficients, then

$$\phi(E, r) \underset{r\to\infty}{\longrightarrow} \begin{pmatrix} e^{-ik_1 r}\alpha_{11} + e^{+ik_1 r}\beta_{11} & e^{-ik_1 r}\alpha_{12} + e^{+ik_1 r}\beta_{12} \\ e^{-ik_2 r}\alpha_{21} + e^{+ik_2 r}\beta_{21} & e^{-ik_2 r}\alpha_{22} + e^{+ik_2 r}\beta_{22} \end{pmatrix}. \tag{12.27}$$

Apparently, for complex values of the channel momenta, there are both exponentially vanishing and exponentially diverging terms in each element of matrix (12.27). This matrix is on the right hand side of Eq. (12.26) and thus may cause the divergence problem. Fortunately, there are also the incoming and outgoing waves in this equation,

$$W^{(\mathrm{in/out})}(E,r) \underset{r\to\infty}{\longrightarrow} \begin{pmatrix} e^{\mp i[k_1 r-(\ell_1+1)\pi/2]} & 0 \\ 0 & e^{\mp i[k_2 r-(\ell_2+1)\pi/2]} \end{pmatrix}, \tag{12.28}$$

which may compensate the diverging terms in certain domains of the complex planes of the channel momenta. This can be seen as follows.

First of all, we notice that according to Eq. (11.26), at large distances, the potential matrix in Eqs. (12.26) is diagonal,

$$V_{nn'}(r) = 0 \qquad \text{for} \qquad n \neq n' , \qquad r \to \infty . \tag{12.29}$$

Therefore, after multiplying all the matrices on the right hand side of Eqs. (12.26) and hiding (for simplicity) all the constants in the r-independent coefficients, we obtain

$$\partial_r F^{(\mathrm{in})}(E,r) \underset{r\to\infty}{\longrightarrow} \begin{pmatrix} V_{11}(r)a_{11} + e^{+2ik_1 r}V_{11}(r)b_{11} & V_{11}(r)a_{12} + e^{+2ik_1 r}V_{11}(r)b_{12} \\ V_{22}(r)a_{21} + e^{+2ik_2 r}V_{22}(r)b_{21} & V_{22}(r)a_{22} + e^{+2ik_2 r}V_{22}(r)b_{22} \end{pmatrix},$$

$$\partial_r F^{(\mathrm{out})}(E,r) \underset{r\to\infty}{\longrightarrow} \begin{pmatrix} V_{11}(r)c_{11} + e^{-2ik_1 r}V_{11}(r)d_{11} & V_{11}(r)c_{12} + e^{-2ik_1 r}V_{11}(r)d_{12} \\ V_{22}(r)c_{21} + e^{-2ik_2 r}V_{22}(r)d_{21} & V_{22}(r)c_{22} + e^{-2ik_2 r}V_{22}(r)d_{22} \end{pmatrix},$$

where $a_{nn'}$, $b_{nn'}$, $c_{nn'}$, and $d_{nn'}$ are the r-independent factors. It is clear that the divergence or convergence of the above matrices is mainly determined by the behaviour of the exponential functions,

$$e^{ikr} \underset{|kr|\to\infty}{\longrightarrow} \begin{cases} 0 \ , & \mathrm{Im}\,(kr) > 0\,, \\ \infty \ , & \mathrm{Im}\,(kr) < 0\,. \end{cases}$$

The potential may to a certain extent save the situation if $V(r)$ at large distances diminishes exponentially, $V \sim e^{-\beta r}$ (see the discussion in Sect. 7.2.2). In such a case, we have

$$e^{ikr-\beta r} \underset{|kr|\to\infty}{\longrightarrow} \begin{cases} 0 \ , & \mathrm{Im}\,(kr) > -\beta\,, \\ \infty \ , & \mathrm{Im}\,(kr) < -\beta\,. \end{cases}$$

The potential is of a short range. This means that the matrix V vanishes at infinity faster than the Coulomb potential, i.e. $V(r \to \infty) \propto r^{-(1+\varepsilon)}\ , \quad \varepsilon > 0$. In the "worst case scenario", all the matrix elements of V have this kind of asymptotic behaviour.

Here we want to answer only one question: what are the conditions for the limits (12.25) to exist? We therefore are not concerned with any constant factors. The order of magnitude of the right hand sides of Eq. (12.26) is the same as for the matrices. Therefore ignoring all exponentially vanishing terms, we obtain

$$\partial_r F^{(\mathrm{in})}_{nn'}(E,r) \underset{|r|\to\infty}{\propto} \begin{cases} 1/r^{1+\varepsilon} & , \ \mathrm{Im}\, k_n r > 0\,, \\ \left(\tilde{a}_{nn'} + e^{2ik_n r}\tilde{b}_{nn'}\right)/r^{1+\varepsilon} & , \ \mathrm{Im}\, k_n r = 0\,, \\ e^{2ik_n r}/r^{1+\varepsilon} & , \ \mathrm{Im}\, k_n r < 0 \end{cases} \tag{12.30}$$

and

$$\partial_r F^{(\mathrm{out})}_{nn'}(E,r) \underset{|r|\to\infty}{\propto} \begin{cases} e^{-2ik_n r}/r^{1+\varepsilon} & , \ \mathrm{Im}\, k_n r > 0\,, \\ \left(\tilde{c}_{nn'} + e^{-2ik_n r}\tilde{d}_{nn'}\right)/r^{1+\varepsilon} & , \ \mathrm{Im}\, k_n r = 0\,, \\ 1/r^{1+\varepsilon} & , \ \mathrm{Im}\, k_n r < 0\,. \end{cases} \tag{12.31}$$

As is shown in Sect. 7.2, the rate $\sim 1/r^{1+\varepsilon}$ of diminishing of the derivative of a function at infinity is a sufficient condition for this function to converge to a constant value. This means that the conditions on $\mathrm{Im}\, k_n r$ on the right hand sides of Eqs. (12.30) and (12.31) determine the domains of the complex k_n-plane where the limits (12.25) exist.

It is not difficult to see that these conditions generalize the corresponding single-channel conditions given by Eqs. (7.18), (7.19), and (7.20) and schematically illustrated in Fig. 7.1.[4] In plain words, the n-th line of the matrix $F^{(\mathrm{in})}$ has a finite limit when k_n is on and above the real axis, while for the corresponding line of $F^{(\mathrm{out})}$ the domain of convergence is on and below that axis. The matrix $F^{(\mathrm{out})}$ also converges at all the spectral points. The last follows from the fact that at such points all the amplitudes $\alpha_{nn'}$ of the incoming waves in Eq. (12.27) are zero by definition.

We therefore encounter the same problem as for the single-channel case: by integrating Eqs. (12.12) and (12.13) along the real r-axis, we can only solve the bound-state and scattering problems. As far as the resonances are concerned, we cannot calculate in this way the Jost matrix $f^{(\mathrm{in})}(E)$ when $\mathrm{Im}\, k_n < 0$ at least for one n.

Fortunately, we can circumvent this difficulty in the same way as for the single-channel case, namely, by integrating the differential equations along the path $|r|\exp(i\theta)$ in the complex r-plane shown in Fig. 7.3. By choosing an appropriate

[4] For an exponentially decaying potential, the common domain of convergence for the matrices $F^{(\mathrm{in})}$ and $F^{(\mathrm{out})}$ is not just the real axis but a band along this axis shown in Fig. 7.2.

rotation angle θ, we can always make $\operatorname{Im} k_n r > 0$ for the channel momentum being in the resonance domain.

For a chosen rotation angle θ, the critical line in the k_n-plane, where $\operatorname{Im} k_n r = 0$, becomes $k = |k_n| \exp(-i\theta)$. The n-th row of the matrix $F^{(\text{in})}$ converges above this line. The corresponding lines on the Riemann energy surface are rotated to the angle 2θ, which follows from squaring Eq. (11.23),

$$|k_n|e^{-i\theta} = \sqrt{\frac{2\mu_n}{\hbar^2}(E - E_n)}\,.$$

This gives

$$E = E_n + \frac{\hbar^2 |k_n|^2}{2\mu_n} e^{-2i\theta}\,. \tag{12.32}$$

Such a rotation "opens up" the resonance domains as is schematically shown in Fig. 12.1. This figure shows only the first two critical lines that begin from the first two thresholds. If there are more than two channels, then all the other critical lines are turned to the same angle.

It should be understood that the N-channel Riemann surface (for a short-range potential) consists of 2^N sheets and has a much more complicated topology as compared with the single-channel one (see Sect. 13). The critical lines shown in Fig. 12.1 are actually turning into different sheets of the Riemann surface. The typical resonance spectral points shown there are also on different sheets. A resonance that can only decay in the first channel (because its energy is not sufficient for producing fragments free moving in the other channels) has negative imaginary part of the first channel momentum, k_1, while the imaginary parts of all the other channel momenta are positive. This is because in the inaccessible (closed) channels the wave function must exponentially vanish at large distances, which reflects the fact that the fragments (corresponding to the closed channels) cannot go away from each other too far. For the resonances that have enough energy to decay in the

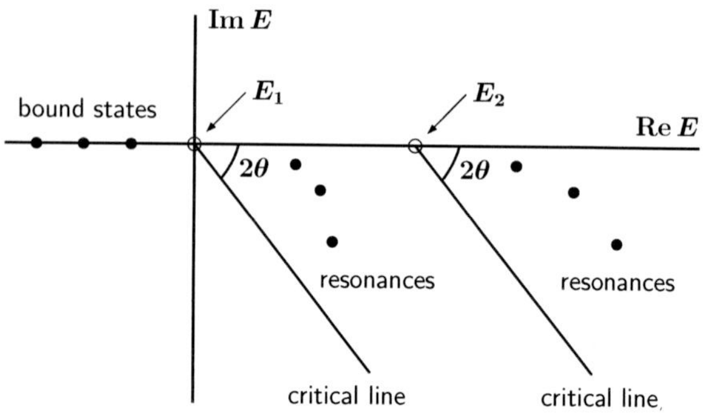

Fig. 12.1 Critical lines on the Riemann surface, above which the matrix $F^{(\text{in})}(E, r)$ is convergent when $r \to \infty$. The critical lines are turning into different sheets of the Riemann surface. The dots schematically show typical distribution of the spectral points, which are also belong to its different sheets.

first two channels, only the first two channel momenta, k_1 and k_2, have negative imaginary parts, and so on. Different combinations of the signs

$$\{\mathrm{sign}(\mathrm{Im}\, k_1), \mathrm{sign}(\mathrm{Im}\, k_2), \ldots, \mathrm{sign}(\mathrm{Im}\, k_N)\}$$

correspond to different sheets of the Riemann surface. This is why the resonances shown in Fig. 12.1 are on the different sheets.

12.3 Jost Solutions

When considering the problem of coupled partial waves, we introduced the Jost solutions of the matrix radial Schrödinger equation (see Sect. 9.6). These solutions are defined by the boundary conditions at infinitely large distances, where they should behave as the incoming and outgoing spherical waves. The problem of coupled partial waves is a particular (degenerate) case of a more general multi-channel problem. It is therefore not difficult to generalize here everything what is described in Sect. 9.6, for the case of coupled channels.

In general, the channels have different reduced masses and the channel momenta. This means that instead of the single momentum k, we have to use the diagonal matrix K, composed of all the channel momenta (see Eq. (11.66)), and the spherical waves, $W^{(\mathrm{in/out})}(E, r)$, should be taken in the form (12.9).

The multi-channel Jost solutions, $\chi^{(\pm)}(E, r)$, are defined by the following boundary conditions at infinity:

$$\chi^{(\pm)}(E, r) \underset{r\to\infty}{\longrightarrow} \begin{bmatrix} h^{(\pm)}_{\ell_1}(k_1 r) & 0 & 0 \cdots & 0 \\ 0 & h^{(\pm)}_{\ell_2}(k_2 r) & 0 \cdots & 0 \\ \vdots & \vdots & \vdots\ \vdots & \vdots \\ 0 & 0 & 0 \cdots & h^{(\pm)}_{\ell_N}(k_N r) \end{bmatrix} . \tag{12.33}$$

If we look for them as

$$\chi^{(\pm)}(E, r) = W^{(\mathrm{in})}(E, r)\mathfrak{X}^{(\pm\mathrm{in})}(E, r) + W^{(\mathrm{out})}(E, r)\mathfrak{X}^{(\pm\mathrm{out})}(E, r) \tag{12.34}$$

with the Lagrange condition

$$W^{(\mathrm{in})}(E, r)\partial_r\mathfrak{X}^{(\pm\mathrm{in})}(E, r) + W^{(\mathrm{out})}(E, r)\partial_r\mathfrak{X}^{(\pm\mathrm{out})}(E, r) = 0\,, \tag{12.35}$$

then, similarly to the regular solution, the unknown matrix-functions $\mathfrak{X}^{(\pm\mathrm{in/out})}(E, r)$ can be found by solving the differential equations

$$\partial_r\mathfrak{X}^{(\pm\mathrm{in})} = -\frac{1}{2i}K^{-1}W^{(\mathrm{out})}V\left[W^{(\mathrm{in})}\mathfrak{X}^{(\pm\mathrm{in})} + W^{(\mathrm{out})}\mathfrak{X}^{(\pm\mathrm{out})}\right], \tag{12.36}$$

$$\partial_r \mathfrak{X}^{(\pm\text{out})} = \frac{1}{2i} K^{-1} W^{(\text{in})} V \left[W^{(\text{in})} \mathfrak{X}^{(\pm\text{in})} + W^{(\text{out})} \mathfrak{X}^{(\pm\text{out})} \right], \tag{12.37}$$

with the boundary conditions

$$\mathfrak{X}^{(+\text{in})}(E,r) \underset{r\to\infty}{\longrightarrow} 0\,, \qquad \mathfrak{X}^{(+\text{out})}(E,r) \underset{r\to\infty}{\longrightarrow} I\,,$$

$$\mathfrak{X}^{(-\text{in})}(E,r) \underset{r\to\infty}{\longrightarrow} I\,, \qquad \mathfrak{X}^{(-\text{out})}(E,r) \underset{r\to\infty}{\longrightarrow} 0\,.$$

Using the following known Wronskians:[5]

$$W\left[W^{(\text{in})}, W^{(\text{out})}\right] = 2iK\,, \qquad W\left[W^{(\text{in})}, W^{(\text{in})}\right] = W\left[W^{(\text{out})}, W^{(\text{out})}\right] = 0\,,$$

and the fact that the Wronskian of any two solutions of the radial Schrödinger equation does not depend on r (it can be calculated at infinity), we find that

$$\begin{aligned} W\left[\chi^{(+)}, \phi\right] &= W\left[W^{(\text{out})}, W^{(\text{in})} f^{(\text{in})} + W^{(\text{out})} f^{(\text{out})}\right] \\ &= W\left[W^{(\text{out})}, W^{(\text{in})}\right] f^{(\text{in})} = -2iK f^{(\text{in})} \end{aligned}$$

and

$$\begin{aligned} W\left[\chi^{(-)}, \phi\right] &= W\left[W^{(\text{in})}, W^{(\text{in})} f^{(\text{in})} + W^{(\text{out})} f^{(\text{out})}\right] \\ &= W\left[W^{(\text{in})}, W^{(\text{out})}\right] f^{(\text{out})} = 2iK f^{(\text{out})}\,. \end{aligned}$$

Therefore

$$\boxed{f^{(\text{in/out})}(E) = \mp \frac{1}{2i} K^{-1} W\left[\chi^{(\pm)}(E,r), \phi(E,r)\right]} \tag{12.38}$$

which generalizes Eq. (9.84), obtained for the problem of coupled partial waves. It is easy to generalize also the relations (9.85) and (9.86), which become

$$W\left[\chi^{(-)}, \chi^{(+)}\right] = 2iK\,, \qquad W\left[\chi^{(+)}, \chi^{(+)}\right] = W\left[\chi^{(-)}, \chi^{(-)}\right] = 0\,, \tag{12.39}$$

[5] The definition of the Wronskian of two matrix-functions involves the transposition of the left one. See Appendix C.

and

$$\boxed{\phi(E,r) = \chi^{(-)}(E,r)f^{(\text{in})}(E) + \chi^{(+)}(E,r)f^{(\text{out})}(E)} \tag{12.40}$$

12.4 Spectral Points

We have introduced the multi-channel Jost matrices as the sets of combination coefficients in the asymptotic behaviour of the regular solutions, as is given by Eq. (12.8). The reader may ask: why did we put the coefficient matrices to the right of the spherical-wave matrices? Do these Jost matrices have the physical meaning, similar to the meaning of the Jost functions in the single-channel case? To answer such questions, in the following sections, we are going to show that our Jost matrices contain information on the spectral points and the scattering in the same way as the single-channel Jost functions do.

First of all, we need to recollect that the threshold energies are the eigenenergies E_n of the internal Hamiltonian h given by Eq. (11.15). As is seen from Eq. (11.23), the momentum k_n in the channel n can be real positive (physical collision momentum) if and only if the total energy E is real positive and is above the energy E_n. If $E > E_n$, the channel n is open, i.e. the particles in this channel are able to move away from each other and their collision is possible. All the channels for which $E < E_n$ are closed channels, i.e. the transitions to them can only be virtual for a short time interval determined by the uncertainty principle (11.1).

12.4.1 Bound States

Bound states are such states of the system that the particles can stay together forever. This means that a bound state only possible at the total energy E_b below all the thresholds,

$$E_b < E_n\,, \qquad \text{for all} \quad n = 1, 2, 3, \ldots$$

The origin of the energy scale can always be chosen arbitrarily. The most logical way is to choose the lowest threshold as zero. Then all the other threshold energies are positive, and any bound state energy is negative (or zero in the extreme case) $E_b \leqslant 0$.

Since the particles cannot escape, all the components of the physical solution u, given by Eq. (12.5), must vanish at large distances,

$$u = \phi c = \begin{pmatrix} \phi_{11} & \phi_{12} \\ \phi_{21} & \phi_{22} \end{pmatrix} \begin{pmatrix} c_1 \\ c_2 \end{pmatrix} \underset{r\to\infty}{\longrightarrow} 0\,,$$

i.e. in all the channels the wave function should be square integrable. Since the energy E is below all the thresholds, all the channel momenta (11.23) are pure imaginary. In principle, we could choose the "+" or "−" sign in front of the square root (11.23). To be consistent with the single-channel case, we choose the "plus" sign (see the discussion in Sect. 3.3.1). In other words, for a multi-channel bound state, all the channel momenta are pure imaginary with positive imaginary parts, i.e.

$$k_n = i|k_n| \, .$$

Similarly to the single-channel case, the behaviour (12.8) of the regular solution at large distances is determined by the Riccati–Hankel functions. As we know,

$$h^{(-)}_{\ell_n}(i|k_n|r) \underset{r\to\infty}{\longrightarrow} \infty \qquad \text{and} \qquad h^{(+)}_{\ell_n}(i|k_n|r) \underset{r\to\infty}{\longrightarrow} 0 \, .$$

This means that the combination coefficients c_n in

$$u \underset{r\to\infty}{\longrightarrow} W^{(\text{in})} f^{(\text{in})} c + W^{(\text{out})} f^{(\text{out})} c \tag{12.41}$$

should be chosen in such a way that the first term on the right hand side (which is diverging) is not there, i.e.

$$W^{(\text{in})}(E, r) f^{(\text{in})}(E) c \equiv 0 \, . \tag{12.42}$$

This is only possible if

$$f^{(\text{in})} c = \begin{bmatrix} f^{(\text{in})}_{11} & f^{(\text{in})}_{12} \\ f^{(\text{in})}_{21} & f^{(\text{in})}_{22} \end{bmatrix} \begin{pmatrix} c_1 \\ c_2 \end{pmatrix} = 0 \, . \tag{12.43}$$

This is a degenerate (homogeneous) system of linear equations for the unknown variables c_n. It has a non-trivial solution if and only if its determinant is zero, i.e.

$$\boxed{\det f^{(\text{in})}(E) = 0} \tag{12.44}$$

This equation generalizes the corresponding Eq. (3.21) of the single-channel case. It may have solutions at certain discrete values of the energy, which are the energies of the bound states. For each root $E = \mathcal{E}$ of Eq. (12.44), the combination coefficients c_n can be found from the linear system

$$\begin{bmatrix} f^{(\text{in})}_{11}(\mathcal{E}) & f^{(\text{in})}_{12}(\mathcal{E}) \\ f^{(\text{in})}_{21}(\mathcal{E}) & f^{(\text{in})}_{22}(\mathcal{E}) \end{bmatrix} \begin{pmatrix} c_1 \\ c_2 \end{pmatrix} = 0 \tag{12.45}$$

with an undetermined common factor, which is finally fixed when the wave function of the bound state u is normalized,

$$\int_0^\infty u^\dagger u\, dr = \int_0^\infty (u_1^*, u_2^*) \begin{pmatrix} u_1 \\ u_2 \end{pmatrix} dr = 1\,.$$

12.4.2 Resonances and Their Partial Decay Widths

In the single-channel case, the resonance spectral points are determined by the same equation as for the bound states, namely, they are the zeros of the Jost function $f_\ell^{(\mathrm{in})}(E)$. It is natural to expect that in the multi-channel problems the equation for the bound and resonance spectral points should be the same as well.

Indeed, following the same logic as for the single-channel resonances (see the discussion in Sect. 3.3.2), we conclude that the resonance solution of the Schrödinger equation should only have the outgoing waves in its long-range asymptotic behaviour. Therefore the coefficients c_n in Eq. (12.41) should be chosen in exactly the same way, i.e. by requiring the condition (12.42), which leads to the same equation (12.44).

The only difference from the bound states is that the resonance spectral points (the roots $\mathcal{E}$ of this equation) should be complex with positive real and negative imaginary parts,

$$\mathcal{E} = \mathcal{E}^{(\mathrm{r})} - \frac{i}{2}\Gamma\,, \qquad \mathcal{E}^{(\mathrm{r})} > 0\,, \qquad \Gamma > 0\,. \tag{12.46}$$

This statement is based on the same reasoning that is given in Sect. 3.3.2 for the single-channel resonances. The quantity Γ in the above equation is the total width of the resonance that is formed at the collision energy $\mathcal{E}^{(\mathrm{r})}$. The total width determines the rate with which the resonance decays. If at the initial instant of time, $t = 0$, the number of generated resonances is $\mathcal{N}(0)$, then it exponentially diminishes as

$$\mathcal{N}(t) = \mathcal{N}(0) \exp\left(-\frac{\Gamma}{\hbar}t\right)\,. \tag{12.47}$$

If the resonance can decay into several channels, then for each channel, n, there is its own width Γ_n, which is called partial width. It is easy to see that the partial widths are fractions of the total width and must add up to Γ. Indeed, for several concurrent decay processes, the function $\mathcal{N}(t)$ is a solution of the following differential equation:

$$\frac{d\mathcal{N}}{dt} = -\frac{\Gamma_1}{\hbar}\mathcal{N} - \frac{\Gamma_2}{\hbar}\mathcal{N} - \cdots = -\frac{1}{\hbar}(\Gamma_1 + \Gamma_2 + \cdots)\mathcal{N}\,, \tag{12.48}$$

from where it follows that

$$\Gamma = \Gamma_1 + \Gamma_2 + \cdots . \tag{12.49}$$

How can we find these partial widths? Before answering this question, let us consider a hypothetical experiment, in which we measure the fluxes of the decay products in two different channels, n and n'. Apparently, the greater is the flux in a particular channel, the greater is the corresponding partial width. Therefore, comparing the intensities of these fluxes, we can determine the branching ratio, i.e. the ratio $\Gamma_n/\Gamma_{n'}$.

Experimental equipment is always a macroscopic object. Typical distances traveled by the decay products before they are detected are many orders of magnitude greater than the size of the quasi-bound state from which they are emitted. It is therefore possible to describe their motion towards the detectors using such notions of classical mechanics as trajectories and velocities.

The counters registering the decay products are at a macroscopic distance r from the source (the emitters). Since the open channels have different thresholds, $E_1 < E_2 < E_3 \ldots$, the kinetic energies, $E_{\rm kin}^{(n)} = E - E_n$, of the decay products are different for the channels. In the channels n and n', the products move towards the counters with the speeds v_n and $v_{n'}$, and thus they need different time to reach them. If $v_n > v_{n'}$, then a pair of simultaneous signals of the detectors (for the channel-products n and n') means that the particle n was emitted from the resonance at a later time than the particle n'. The time difference is

$$\Delta t = \frac{r}{v_{n'}} - \frac{r}{v_n} = \frac{r(v_n - v_{n'})}{v_n v_{n'}} . \tag{12.50}$$

If we deal with short-lived resonances, their number (number of the emitters) during the time interval Δt can be significantly reduced. If the particles n' were emitted at $t = 0$, then the intensity of their flux was determined by the number $\mathcal{N}(0)$ of the emitters, while for the particles n (which arrive to the detectors simultaneously with n') the number of available emitters was smaller, namely,

$$\mathcal{N}(\Delta t) = \mathcal{N}(0) \exp\left(-\frac{\Gamma}{\hbar}\Delta t\right) ,$$

and thus their flux was also reduced. Therefore when trying to obtain the branching ratio $\Gamma_n/\Gamma_{n'}$ by comparing the particle-fluxes in the channels n and n', we should put the corresponding detectors at different distances, r_n and $r_{n'}$, from the emitters. These distances can be found as[6]

$$r_n = \frac{\hbar k_n}{\mu_n} t , \tag{12.51}$$

[6] Of course, for fast-moving particles, the relativistic kinematics should be used.

where k_n and μ_n are the wave number and the reduced mass in the channel n, while t is common time of flight (from the emitter to the detector) for all the channels. In a theoretical analysis, this requirement can be fulfilled if we consider the asymptotic behaviour of the resonance wave function at the corresponding distances (12.51) for different channels. This is what we are going to do.

For the sake of clarity, let us consider a two-channel problem. The generalization for an arbitrary N will be obvious.

After locating a resonance spectral point $\mathcal{E}$, i.e. after finding a complex root of Eq. (12.44), we can solve the homogeneous system of linear equations (12.45) for the unknown coefficients c_1 and c_2. In the two-channel case, this is trivial:

$$c_1 \quad \text{is arbitrary,} \qquad c_2 = -\frac{f_{11}^{(\mathrm{in})}(\mathcal{E})}{f_{12}^{(\mathrm{in})}(\mathcal{E})} c_1 \,,$$

where the arbitrary c_1 should be fixed from the normalization condition. Now, according to Eq. (12.41), the physical wave function asymptotically behaves as

$$\begin{aligned}
u(\mathcal{E}, r) \underset{r\to\infty}{\longrightarrow} \ & W^{(\mathrm{out})}(\mathcal{E}, r) f^{(\mathrm{out})}(\mathcal{E}) c \\
= \ & \begin{bmatrix} h_{\ell_1}^{(+)}(k_1 r) & 0 \\ 0 & h_{\ell_2}^{(+)}(k_2 r) \end{bmatrix} \begin{bmatrix} f_{11}^{(\mathrm{out})}(\mathcal{E}) & f_{12}^{(\mathrm{out})}(\mathcal{E}) \\ f_{21}^{(\mathrm{out})}(\mathcal{E}) & f_{22}^{(\mathrm{out})}(\mathcal{E}) \end{bmatrix} \begin{bmatrix} c_1 \\ -\dfrac{f_{11}^{(\mathrm{in})}(\mathcal{E})}{f_{12}^{(\mathrm{in})}(\mathcal{E})} c_1 \end{bmatrix} \\
= \ & \begin{bmatrix} \mathcal{D}_1 h_{\ell_1}^{(+)}(k_1 r) \\ \mathcal{D}_2 h_{\ell_2}^{(+)}(k_2 r) \end{bmatrix} \underset{r\to\infty}{\longrightarrow} \begin{bmatrix} \mathcal{D}_1 (-i)^{\ell_1+1} e^{ik_1 r} \\ \mathcal{D}_2 (-i)^{\ell_2+1} e^{ik_2 r} \end{bmatrix} ,
\end{aligned} \qquad (12.52)$$

where

$$\mathcal{D}_1 = c_1 \left(f_{11}^{(\mathrm{out})} - \frac{f_{11}^{(\mathrm{in})} f_{12}^{(\mathrm{out})}}{f_{12}^{(\mathrm{in})}} \right) , \qquad (12.53)$$

$$\mathcal{D}_2 = c_1 \left(f_{21}^{(\mathrm{out})} - \frac{f_{11}^{(\mathrm{in})} f_{22}^{(\mathrm{out})}}{f_{12}^{(\mathrm{in})}} \right) . \qquad (12.54)$$

In Eq. (12.52), the Riccati–Hankel functions, $h_{\ell_1}^{(+)}(k_1 r)$ and $h_{\ell_2}^{(+)}(k_2 r)$, are the outgoing spherical waves in the first and the second channels. The coefficients $\mathcal{D}_1$ and $\mathcal{D}_2$ in front of them are proportional to the amplitudes of the probability to find our physical system in the corresponding channels. Apparently, the greater is a particular coefficient, the more probable is the decay into that channel. Any common normalization factor does not change their relative magnitudes.

The partial width Γ_n should be proportional to the corresponding probability $|\mathcal{D}_n|^2$, but it may include additional kinematical factors. A partial width determines the decay rate in a particular channel, and thus it is observable via the corresponding flux of the decay products. Indeed, the total number of particles in the channel n passing through a sphere per unit time (i.e. the integrated flux) is exactly equal to the number of decays in that channel per unit time (i.e. the decay rate).

The flux in the channel n can be found as

$$\vec{\mathcal{J}}_n = \frac{\hbar}{2\mu_n i}\left(\psi_n^* \vec{\nabla}\,\psi_n - \psi_n \vec{\nabla}\,\psi_n^*\right) ,$$

where the channel wave function (concerning the use of a single angular momentum for a channel, see the discussion related to Table 11.1),

$$\psi_n(E,\vec{r}\,) = \frac{u_n(E,r)}{r} Y_{\ell_n m_n}(\hat{\vec{r}}\,) , \tag{12.55}$$

is taken at a large (macroscopic) distance (12.51). We are only interested in the radial component $j_n^{(r)}$ of vector $\vec{\mathcal{J}}_n = \{j_n^{(r)}, j_n^{(\theta)}, j_n^{(\varphi)}\}$, which is normal to a sphere of the radius r. Therefore, out of the three spherical components of the operator $\vec{\nabla} = \{\partial_r, (1/r)\partial_\theta, (1/r\sin\theta)\partial_\varphi\}$, we should use only the radial one. The radial flux for the function (12.55) with u_n given by Eq. (12.52) is

$$j_n^{(r)} = \frac{\hbar\mathrm{Re}(k_n)}{\mu_n r^2}|\mathcal{D}_n|^2 \exp\left[-2\mathrm{Im}(k_n)r\right] Y_{\ell_n m_n}^*(\theta,\varphi) Y_{\ell_n m_n}(\theta,\varphi) .$$

Then the total flux in the n-th channel through a distant sphere of the radius r is the surface integral

$$j_n = \int j_n^{(r)} r^2 \sin\theta\, d\theta\, d\varphi = \frac{\hbar\mathrm{Re}(k_n)}{\mu_n}|\mathcal{D}_n|^2 \exp\left[-2\mathrm{Im}(k_n)r\right] . \tag{12.56}$$

At this point we should recall the discussion about the choice of different asymptotic distances (12.51) for different channels. For a resonance, the momentum $\hbar k_n$ is complex. The speed of the products is determined by its real part, and thus

$$r_n = \frac{\hbar\mathrm{Re}(k_n)}{\mu_n} t . \tag{12.57}$$

At this distance the flux (12.56) is

$$j_n = \frac{\hbar\mathrm{Re}(k_n)}{\mu_n}|\mathcal{D}_n|^2 \exp\left[-2\mathrm{Im}(k_n)\frac{\hbar\mathrm{Re}(k_n)}{\mu_n}t\right] = \frac{\hbar\mathrm{Re}(k_n)}{\mu_n}|\mathcal{D}_n|^2 \exp\left(\frac{\Gamma}{\hbar}t\right) , \tag{12.58}$$

where we used the fact that the total width Γ is expressed via the channel momentum as

$$\Gamma = -\frac{2\hbar^2 \mathrm{Re}(k_n)\mathrm{Im}(k_n)}{\mu_n} \qquad \text{for any} \quad n\,, \tag{12.59}$$

which can be seen if we compare the imaginary parts on both sides of the equation

$$\frac{\hbar^2 k_n^2}{2\mu_n} = \mathcal{E}^{(r)} - \frac{i}{2}\Gamma - E_n\,.$$

For all the channels, the time-dependent factor in Eq. (12.58) is the same. Actually, this is a consequence of the choice of appropriate asymptotic distances (12.57) for different channels. These distances correspond to the same time of flight.

Therefore, in the ratios $j_n/j_{n'}$, the time-dependent factor is always canceled out. Apparently, such a ratio is equal to the ratio of the corresponding decay rates, i.e. it is the branching ratio,

$$\frac{\Gamma_n}{\Gamma_{n'}} = \frac{j_n}{j_{n'}} = \frac{\mu_{n'}\mathrm{Re}(k_n)|\mathcal{D}_n|^2}{\mu_n \mathrm{Re}(k_{n'})|\mathcal{D}_{n'}|^2}\,. \tag{12.60}$$

The fluxes given by Eq. (12.58) originate from the emitter at the same instant of time in all the channels. Therefore the channel flux j_n for a specific n is a fraction f_n of the total flux originated at that time,

$$f_n = \frac{j_n}{j_1 + j_2 + \cdots}\,, \tag{12.61}$$

and the channel width Γ_n is the same fraction of the total width,

$$\Gamma_n = f_n\Gamma = \frac{\mathrm{Re}(k_n)|\mathcal{D}_n|^2\Gamma}{\displaystyle\sum_{n'=1}^{N}\frac{\mu_n}{\mu_{n'}}\mathrm{Re}(k_{n'})|\mathcal{D}_{n'}|^2}\,, \tag{12.62}$$

where N is the number of open channels. In the wave function (12.52), all the amplitudes $\mathcal{D}_n$ have a common normalization factor, which is canceled out in Eq. (12.62). We therefore avoid the problem of "proper" normalization of the resonant wave function. Actually, it does not matter for us what the term "proper" may mean here. Instead of doing proper normalization, we may simply put $c_1 = 1$ in Eqs. (12.53) and (12.54), or alternatively we may require that the sum of absolute squares of the channel amplitudes is one,

$$|\mathcal{D}_1|^2 + |\mathcal{D}_2|^2 + \cdots = 1\,. \tag{12.63}$$

In the two-channel case, such a condition gives

$$c_1 = \left(\left| f_{11}^{(\text{out})} - \frac{f_{11}^{(\text{in})} f_{12}^{(\text{out})}}{f_{12}^{(\text{in})}} \right|^2 + \left| f_{21}^{(\text{out})} - \frac{f_{11}^{(\text{in})} f_{22}^{(\text{out})}}{f_{12}^{(\text{in})}} \right|^2 \right)^{-1/2} .$$

Therefore the amplitudes $\mathcal{D}_1$ and $\mathcal{D}_2$ of the outgoing waves are expressed in terms of the elements of the matrices $f^{(\text{in/out})}(\mathcal{E})$ taken at the complex energy $\mathcal{E}$,

$$\mathcal{D}_1 = \frac{f_{12}^{(\text{in})} f_{11}^{(\text{out})} - f_{11}^{(\text{in})} f_{12}^{(\text{out})}}{\sqrt{\left| f_{12}^{(\text{in})} f_{11}^{(\text{out})} - f_{11}^{(\text{in})} f_{12}^{(\text{out})} \right|^2 + \left| f_{12}^{(\text{in})} f_{21}^{(\text{out})} - f_{11}^{(\text{in})} f_{22}^{(\text{out})} \right|^2}} , \tag{12.64}$$

$$\mathcal{D}_2 = \frac{f_{12}^{(\text{in})} f_{21}^{(\text{out})} - f_{11}^{(\text{in})} f_{22}^{(\text{out})}}{\sqrt{\left| f_{12}^{(\text{in})} f_{11}^{(\text{out})} - f_{11}^{(\text{in})} f_{12}^{(\text{out})} \right|^2 + \left| f_{12}^{(\text{in})} f_{21}^{(\text{out})} - f_{11}^{(\text{in})} f_{22}^{(\text{out})} \right|^2}} . \tag{12.65}$$

When substituted in Eq. (12.62), they allow us to calculate the partial widths. In a similar way the partial widths for any number of channels can be found.

* * *

Numerical calculations that demonstrate using of this method in solving some two-channel problems can be found in Refs. [39, 40].

12.4.3 Simplicity of the Spectral Point Zeros and the ANC

Let E_0 be a bound or a resonant state spectral point corresponding to the set of the channel momenta

$$K_0 = \text{diag}\,\{k_{01}, k_{02}, \ldots, k_{0N}\} \ . \tag{12.66}$$

The physical (radial) wave function, $u(E, r)$, describing such a state can be represented by a column matrix consisting of N elements, where N is the number of the coupled channels (including the degenerate ones that have the same k_{0n} but correspond to different spin-angular states). Asymptotically, this column consists of the outgoing spherical waves,

$$u(E, r) = \begin{bmatrix} u_1(E, r) \\ u_2(E, r) \\ \vdots \\ u_N(E, r) \end{bmatrix} \underset{r \to \infty}{\longrightarrow} \begin{bmatrix} \mathcal{A}_1 e^{ik_{01}r} \\ \mathcal{A}_2 e^{ik_{02}r} \\ \vdots \\ \mathcal{A}_N e^{ik_{0N}r} \end{bmatrix} . \tag{12.67}$$

If the wave function is normalized to unity, then $\mathcal{A}_n$ in Eq. (12.67) are the multichannel asymptotic normalization constants (ANCs). Further down, the symbol $\mathcal{A}$ is used to denote the column-matrix composed of these constants.

The physical solution is regular at $r = 0$. It is therefore a linear combination of the columns of the fundamental matrix of the regular solutions,

$$u(E,r) = \phi(E,r)c\,, \tag{12.68}$$

where c is a column matrix of the combination coefficients. The bound and resonant states are possible at certain discrete energies. At such an energy, $\det f^{(\mathrm{in})}(E_0) = 0$, which implies that it is possible to find such combination coefficients c that completely remove the incoming waves from the asymptotic behaviour of the physical wave function, i.e.

$$u(E_0,r) \underset{r\to\infty}{\longrightarrow} W^{(\mathrm{out})} f^{(\mathrm{out})} c\,. \tag{12.69}$$

It is not difficult to see that the column c of the combination coefficients in Eq. (12.69) should be taken as the following product:

$$c = \left[f^{(\mathrm{out})}(E_0)\right]^{-1} i\mathcal{I}\mathcal{A}\,, \tag{12.70}$$

where the diagonal matrix $\mathcal{I}$ is composed of the ℓ-powers of i and is defined by Eq. (9.108). Indeed, using the representation (12.40) of the regular solution as a combination of the Jost solutions, we have

$$\phi c = \chi^{(-)} S^{-1} i\mathcal{I}\mathcal{A} + \chi^{(+)} i\mathcal{I}\mathcal{A}\,. \tag{12.71}$$

At a spectral point, the S-matrix is singular and thus

$$S^{-1}(E) \underset{E\to E_0}{\longrightarrow} 0\,, \tag{12.72}$$

which gives the physical solution in the form

$$u = \chi^{(+)} i\mathcal{I}\mathcal{A}\,, \tag{12.73}$$

which asymptotically behaves as in (12.67).

In this section, our task is to generalize the derivations of Sects. 10.7 and 10.8. In doing this, we show here that the zeros of the Jost matrix determinant at the spectral points are simple and find the relation between the residues of the S-matrix and the ANC.

* * *

The same as is done in Sect. 10.7, we take Eq. (12.38) representing the Jost matrix in terms of the Wronskian of the Jost and regular solutions and differentiate it with respect to variable E,

$$\dot{f}^{(\mathrm{in})} = -\frac{1}{2i}K^{-1}\left[W(\dot{\chi}^{(+)},\phi) + W(\chi^{(+)},\dot{\phi})\right] - \frac{1}{2i}\left(\frac{d}{dE}K^{-1}\right)W(\chi^{(+)},\phi)\,.$$

Here, as before, the dot over the symbols means the derivative over E. Multiplying this equation by $\left[f^{(\mathrm{out})}\right]^{-1}$ from the right and taking into account Eq. (12.72), we see that at the spectral point the last term vanishes,

$$W(\chi^{(+)},\phi)\left[f^{(\mathrm{out})}\right]^{-1} = -2iKf^{(\mathrm{in})}\left[f^{(\mathrm{out})}\right]^{-1} = -2iKS^{-1} \underset{E\to E_0}{\longrightarrow} 0\,.$$

Therefore, at $E = E_0$, we have

$$\dot{f}^{(\mathrm{in})}\left[f^{(\mathrm{out})}\right]^{-1} = -\frac{1}{2i}K_0^{-1}\left[W(\dot{\chi}^{(+)},\phi) + W(\chi^{(+)},\dot{\phi})\right]\left[f^{(\mathrm{out})}\right]^{-1}\,. \tag{12.74}$$

In order to find the Wronskians that are needed here, consider the radial equations for the transposed Jost solution at the energy $\tilde{E}$ and for the regular solution at the energy E. If $\tilde{K}$ and K are the corresponding matrices of the channel momenta, then

$$\tilde{E}: \qquad \left[\chi^{(+)\prime\prime}\right]^T + \left[\chi^{(+)}\right]^T\tilde{K}^2 = \left[\chi^{(+)}\right]^T\mathcal{V}\,, \tag{12.75}$$

$$E: \qquad \phi'' + K^2\phi = \mathcal{V}\phi\,, \tag{12.76}$$

where $\mathcal{V}$ is the sum of the centrifugal and the dynamical potentials:

$$\mathcal{V} = \mathrm{diag}\left\{\frac{\ell_1(\ell_1+1)}{r^2}, \frac{\ell_2(\ell_2+1)}{r^2}, \ldots, \frac{\ell_N(\ell_N+1)}{r^2}\right\} + V\,.$$

Multiplying Eq. (12.75) by ϕ from the right and Eq. (12.76) by $\left[\chi^{(+)}\right]^T$ from the left and subtracting the second from the first, we obtain[7]

$$\partial_r W\left[\chi^{(+)},\phi\right] = \left[\chi^{(+)}\right]^T(\tilde{K}^2 - K^2)\phi\,. \tag{12.77}$$

This equation looks the same as the corresponding Eq. (10.103) that we derived in the problem of coupled partial waves. The only difference is that instead of simple

[7] It should be noted that the Wronskian for matrix functions is defined by Eq. (C.5), where the first of the two matrices is transposed.

numbers $\tilde{k}$ and k, here we have the diagonal matrices $\tilde{K}$ and K. This seemingly simple difference makes further derivations a bit more complicated.

The same as in the case of coupled partial waves, we differentiate Eq. (12.77) over E and then put $E = \tilde{E}$. This gives

$$\partial_r W\left[\chi^{(+)}, \dot{\phi}\right] = -\left[\chi^{(+)}\right]^T \frac{2\mathcal{M}}{\hbar^2}\phi\,, \tag{12.78}$$

where the diagonal matrix $\mathcal{M}$ is composed of the reduced masses in different channels,

$$\mathcal{M} = \begin{pmatrix} \mu_1 & 0 & \cdots & 0 \\ 0 & \mu_2 & \cdots & 0 \\ \vdots & \vdots & \vdots & \vdots \\ 0 & 0 & \cdots & \mu_N \end{pmatrix}\,. \tag{12.79}$$

Similarly, differentiating with respect to $\tilde{E}$ and putting $\tilde{E} = E$, we get another equation:

$$\partial_r W\left[\dot{\chi}^{(+)}, \phi\right] = +\left[\chi^{(+)}\right]^T \frac{2\mathcal{M}}{\hbar^2}\phi\,. \tag{12.80}$$

Since $\mathcal{M}$ is a matrix, we have to keep it in between $\left[\chi^{(+)}\right]^T$ and ϕ in both equations, (12.78) and (12.80).

Now, we integrate Eq. (12.78) from 0 to r and Eq. (12.80) from r to infinity:

$$W\left[\chi^{(+)}, \dot{\phi}\right]_r = W\left[\chi^{(+)}, \dot{\phi}\right]_{r=0} - \int_0^r \left[\chi^{(+)}\right]^T \frac{2\mathcal{M}}{\hbar^2}\phi\, dr'\,, \tag{12.81}$$

$$W\left[\dot{\chi}^{(+)}, \phi\right]_r = W\left[\dot{\chi}^{(+)}, \phi\right]_{r=\infty} - \int_r^\infty \left[\chi^{(+)}\right]^T \frac{2\mathcal{M}}{\hbar^2}\phi\, dr'\,. \tag{12.82}$$

If multiplied from the left by $\mathcal{A}^T i\mathcal{I}K_0^{-1}$ and from the right by $i\mathcal{I}\mathcal{A}$, the Wronskians on the right hand sides of Eqs. (12.81) and (12.82) vanish. This can be seen in the following way. Firstly, as we established,

$$\phi\left[f^{(\mathrm{out})}\right]^{-1} i\mathcal{I}\mathcal{A} = u\,. \tag{12.83}$$

Now, if we introduce the following modified normalization constants:

$$\tilde{\mathcal{A}} = K_0^{-1}\mathcal{A} = \begin{bmatrix} k_{01}^{-1}\mathcal{A}_1 \\ k_{02}^{-1}\mathcal{A}_2 \\ \vdots \\ k_{0N}^{-1}\mathcal{A}_N \end{bmatrix},$$

then from Eq. (12.73) it follows that

$$\chi^{(+)} i\mathcal{I}\tilde{\mathcal{A}} = \begin{bmatrix} k_{01}^{-1}\mathcal{A}_1 e^{ik_{01}r} \\ k_{02}^{-1}\mathcal{A}_2 e^{ik_{02}r} \\ \vdots \\ k_{0N}^{-1}\mathcal{A}_N e^{ik_{0N}r} \end{bmatrix}, \tag{12.84}$$

i.e.

$$\chi^{(+)} i\mathcal{I}K_0^{-1}\mathcal{A} = K_0^{-1}u \tag{12.85}$$

and thus[8]

$$\mathcal{A}^T i\mathcal{I}K_0^{-1}\left[\chi^{(+)}\right]^T = u^T K_0^{-1}. \tag{12.86}$$

Therefore

$$\mathcal{A}^T i\mathcal{I}K_0^{-1}\, W\left[\chi^{(+)}, \dot{\phi}\right]_{r=0}\left[f^{(\mathrm{out})}\right]^{-1} i\mathcal{I}\mathcal{A} = W\left[K_0^{-1}u, \dot{u}\right]_{r=0} = 0\,, \tag{12.87}$$

$$\mathcal{A}^T i\mathcal{I}K_0^{-1}\, W\left[\dot{\chi}^{(+)}, \phi\right]_{r=\infty}\left[f^{(\mathrm{out})}\right]^{-1} i\mathcal{I}\mathcal{A} = W\left[K_0^{-1}\dot{u}, u\right]_{r=\infty} = 0\,, \tag{12.88}$$

because at $r = 0$ and at $r \to \infty$ the function $u(E, r)$ has very simple dependence on E. The multiplication by the r-independent factor, K_0^{-1}, as well as the differentiation over the energy at the end points of the interval $r \in [0, \infty)$ leaves u and $\dot{u}$ linearly dependent, and thus their Wronskian is zero. This cannot be

[8] Here we take into account the fact that matrices $\mathcal{I}$ and K_0^{-1} are diagonal and therefore commute with each other.

said about the intermediate points of the interval $r \in [0, \infty)$, where u and $\overset{\bullet}{u}$ may be independent.

Let us now substitute Eqs. (12.81) and (12.82) in (12.74) and then multiply the resulting equation from the left by $\mathcal{A}^T i\mathcal{I}$ and from the right by $i\mathcal{I}\mathcal{A}$. Taking into account Eqs. (12.87, 12.88), we obtain

$$\mathcal{A}^T i\mathcal{I}\,\overset{\bullet}{f}^{(\mathrm{in})}\left[f^{(\mathrm{out})}\right]^{-1} i\mathcal{I}\mathcal{A} = -\frac{i}{\hbar}\int_0^\infty u^T\,(\hbar K_0)^{-1}\,\mathcal{M}u\,dr\,. \tag{12.89}$$

The integral on the right hand side of this equation,

$$\begin{aligned}\int_0^\infty u^T\,(\hbar K_0)^{-1}\,\mathcal{M}u\,dr &= \int_0^\infty \left(\frac{\mu_1}{\hbar k_{01}}u_1^2 + \frac{\mu_2}{\hbar k_{02}}u_2^2 + \cdots + \frac{\mu_N}{\hbar k_{0N}}u_N^2\right) dr \\ &= \frac{1}{\langle v\rangle}\,,\end{aligned} \tag{12.90}$$

is the average (expectation value) of the inverted relative velocity, $\langle v\rangle^{-1}$. Indeed, as is discussed in Sect. 3.3.4, for a bound state, the radial part of the wave function can always be made real, and therefore the simple squares under the integral (12.90) are equivalent to the absolute squares. In the case of a resonance, the wave function can also be normalized if we integrate the simple squares along the ray in the complex r-plane, shown in Fig. 3.9. It should be noted that $\langle v\rangle$ is not a real number because for a bound state all the channel momenta are pure imaginary, and for a resonance they are complex.

If c is the column-matrix defined by Eq. (12.70), then Eq. (12.89) can be written as

$$\mathcal{A}^T i\mathcal{I}\,\overset{\bullet}{f}^{(\mathrm{in})} c = -\frac{i}{\hbar\langle v\rangle}\,. \tag{12.91}$$

The corresponding Eq. (10.113) derived in the non-central potential case differs from (12.91) by the right hand side. In the more simple case of coupled partial waves, the channels (partial waves) all have the same relative velocity, which is on the right hand side of Eq. (10.113). In Eq. (12.91), the velocities are averaged over the channels. Otherwise, the mathematical structure of Eqs. (10.113) and (12.91) is the same. It is therefore not necessary to repeat here the derivations of Sects. 10.7 and 10.8 to prove that the spectral points of the multi-channel S-matrix are its poles of the first order and to obtain the factorized representation of its residues in terms of the ANCs:

$$\boxed{\operatorname{Res}\left(S_{n'n}, E_0\right) = -\frac{\hbar\langle v\rangle i^{\ell'+\ell+1}}{N}\mathcal{A}_{n'}\mathcal{A}_n} \tag{12.92}$$

where N is the total number of coupled channels (including the degenerate ones).

12.5 Multi-Channel Scattering

For a scattering state, the wave function has both the incoming and outgoing waves at large r. If $\mathtt{A}_n$ and $\mathtt{B}_n$ are the amplitudes of the incoming and outgoing (scattered) waves in the n-th channel,[9]

$$\mathtt{A} = \begin{pmatrix} \mathtt{A}_1 \\ \mathtt{A}_2 \end{pmatrix}, \qquad \mathtt{B} = \begin{pmatrix} \mathtt{B}_1 \\ \mathtt{B}_2 \end{pmatrix},$$

the scattering boundary condition (at $r \to \infty$) reads

$$u \underset{r\to\infty}{\longrightarrow} W^{(\mathrm{in})}\mathtt{A} + W^{(\mathrm{out})}\mathtt{B}\,. \tag{12.93}$$

Here, $\mathtt{A}$ are given incoming amplitudes that are determined by the conditions of the collision experiment. They describe the intensity of the incoming beam in each of the channels. In most cases, the collision is such that $\mathtt{A}_n$ is non-zero for only one channel. After the scattering, the particles can emerge in any of the open channels. The corresponding intensities of the outgoing particle-flux are described by the amplitudes $\mathtt{B}_n$. These amplitudes are unknown and the task of the theory is to predict them.

On the other hand, we can construct the function u out of the regular basis as

$$u = \phi c \underset{r\to\infty}{\longrightarrow} W^{(\mathrm{in})} f^{(\mathrm{in})} c + W^{(\mathrm{out})} f^{(\mathrm{out})} c\,, \tag{12.94}$$

where the combination coefficients $c = (c_1, c_2)^T$ should be chosen in such a way that they give the same asymptotic behaviour as in Eq. (12.93). This can be achieved if

$$\mathtt{A} = f^{(\mathrm{in})} c \qquad \text{and} \qquad \mathtt{B} = f^{(\mathrm{out})} c\,, \tag{12.95}$$

from which it follows that $\mathtt{B} = f^{(\mathrm{out})}[f^{(\mathrm{in})}]^{-1}\mathtt{A}$, i.e. the outgoing amplitudes $\mathtt{B}$ can be obtained from the given incoming amplitudes $\mathtt{A}$, using the S-matrix,

$$\begin{pmatrix} \mathtt{B}_1 \\ \mathtt{B}_2 \end{pmatrix} = S \begin{pmatrix} \mathtt{A}_1 \\ \mathtt{A}_2 \end{pmatrix},$$

which is constructed out of the Jost matrices in the usual way, namely,

$$S(E) = f^{(\mathrm{out})}(E)\left[f^{(\mathrm{in})}(E)\right]^{-1}. \tag{12.96}$$

[9] Once again, for the sake of clarity, we consider here a two-channel case.

The physical meaning of various matrix elements of the S-matrix can be understood if we consider a two-channel problem,

$$\begin{pmatrix} \mathtt{B}_1 \\ \mathtt{B}_2 \end{pmatrix} = \begin{pmatrix} S_{11}\mathtt{A}_1 + S_{12}\mathtt{A}_2 \\ S_{21}\mathtt{A}_1 + S_{22}\mathtt{A}_2 \end{pmatrix} ,$$

with different choices of the experimental conditions. Suppose that the incident wave is purely in channel 1, i.e. $\mathtt{A}_1 \neq 0$ and $\mathtt{A}_2 = 0$, then

$$\begin{pmatrix} \mathtt{B}_1 \\ \mathtt{B}_2 \end{pmatrix} = \begin{pmatrix} S_{11}\mathtt{A}_1 \\ S_{21}\mathtt{A}_1 \end{pmatrix} \quad \Longrightarrow \quad \begin{pmatrix} S_{11} \\ S_{21} \end{pmatrix} = \begin{pmatrix} \mathtt{B}_1/\mathtt{A}_1 \\ \mathtt{B}_2/\mathtt{A}_1 \end{pmatrix} .$$

The quantities $\mathtt{A}_n$ and $\mathtt{B}_n$ are the amplitudes in the wave function. Therefore their absolute squares are proportional to the probabilities of finding particles in the corresponding channels. Apparently, the ratio of the amplitudes $R_{11} = \mathtt{B}_1/\mathtt{A}_1$ is the amplitude of the probability that the particle coming from channel 1, after the collision remains in the same channel (elastic scattering), while the ratio $T_{2\leftarrow 1} = \mathtt{B}_2/\mathtt{A}_1$ determines the transition probability from channel 1 to channel 2.

Similarly, if we assume that the incident particle is in channel 2, i.e. $\mathtt{A}_1 = 0$ and $\mathtt{A}_2 \neq 0$, we find that $S_{12} = T_{1\leftarrow 2} = \mathtt{B}_1/\mathtt{A}_2$ and $S_{22} = R_{22} = \mathtt{B}_2/\mathtt{A}_2$. The S-matrix is independent of the experimental conditions, and therefore

$$S = \begin{pmatrix} R_{11} & T_{1\leftarrow 2} \\ \\ T_{2\leftarrow 1} & R_{22} \end{pmatrix} .$$

The above explanations can serve as a good basis for an intuitive understanding of the structure of the S-matrix. However, for a practical calculation of the elastic and inelastic cross sections, we need to perform a detailed partial-wave analysis similar to what was done in Sect. 9.7.2 and also in Sect. 10.3.3. For the multi-channel scattering, such an analysis is done in the next section.

12.6 Scattering Observables

All possible measurements are done at macroscopic distances that can always be considered as being infinite as compared to the microscopic volume of space where the actual scattering takes place. Therefore when calculating any measurable scattering quantity (such as the cross section, for example), we need to analyze the asymptotic behaviour of the scattering wave function (11.90). This scattering wave function, $\psi(\vec{r}, \vec{k})$, is a square matrix in the space of $\mathcal{N}$ truly different channels (corresponding to different internal states η). Each element, $\psi_{\eta'\eta}$, of this matrix describes the scattering process that begins in the channel η and ends up in the channel η'. Actually, the matrix element $\psi_{\eta'\eta}$ is an operator in the spin space.

In order to obtain the wave function for the transition between specific spin sub-channels, we need to sandwich it between the corresponding spin states,

$$\psi_{\nu'\nu}(\vec{r},\vec{k}) = \langle s'\sigma'|\psi_{\eta'\eta}(\vec{r},\vec{k})|s\sigma\rangle \,, \tag{12.97}$$

where the subscript ν labels different combinations of the quantum numbers $\{\eta, s\sigma\}$.

The radial matrix $u^J(E,r)$ in that wave function is a physical solution of the radial Schrödinger equation (11.96). Each column of a physical solution is a linear combination of the basis columns, i.e. of the columns of the matrix ϕ defined by the regular boundary conditions (12.3). Asymptotically, the physical solution behaves as is given by Eq. (12.94).

As we will see shortly, the choice of the matrix c of the combination coefficients in that asymptotics, as the inverted Jost matrix,

$$c = \left[f^{(\text{in})}\right]^{-1} \,, \tag{12.98}$$

gives the correct behaviour of the full (three-dimensional) scattering wave function (11.85) at large distances, i.e. as a superposition of the incoming plane wave and a spherical outgoing waves in all open channels. The intensities of these spherical waves in various channels and in various directions will give us the information on how probable is the scattering into each of the channels and in a specific direction.

Substituting $u = \phi c$ into the partial-wave expansion (11.90), with c given by Eq.(12.98) and ϕ in the form (12.10), we obtain (we drop the superscript J as well as the arguments E and r, where possible)

$$\begin{aligned}\psi(\vec{r},\vec{k}) &= \frac{1}{r}\sum_{JM}|\mathcal{Y}^{JM}(\hat{\vec{r}})\rangle\phi\left[f^{(\text{in})}\right]^{-1}\widehat{N}\langle\mathcal{Y}^{JM}(\hat{\vec{k}})| \\ &= \frac{1}{r}\sum_{JM}|\mathcal{Y}^{JM}(\hat{\vec{r}})\rangle\left[W^{(\text{in})}F^{(\text{in})} + W^{(\text{out})}F^{(\text{out})}\right]\left[f^{(\text{in})}\right]^{-1}\widehat{N}\langle\mathcal{Y}^{JM}(\hat{\vec{k}})| \,.\end{aligned} \tag{12.99}$$

At large distances, the matrices $F^{(\text{in/out})}$ converge to the corresponding Jost matrices. Taking into account the definition (12.96) of the S-matrix, we obtain the following asymptotic behaviour of the above wave function:

$$\psi(\vec{r},\vec{k}) \underset{r\to\infty}{\longrightarrow} \frac{1}{r}\sum_{JM}|\mathcal{Y}^{JM}(\hat{\vec{r}})\rangle\left[W^{(\text{in})} + W^{(\text{out})}S\right]\widehat{N}\langle\mathcal{Y}^{JM}(\hat{\vec{k}})| \,. \tag{12.100}$$

Adding and subtracting $W^{(\text{out})}$ in the square brackets of this equation, we can split its right hand side into two terms,

$$\psi(\vec{r}, \vec{k}) \underset{r\to\infty}{\longrightarrow} \frac{1}{r} \sum_{JM} |\mathcal{Y}^{JM}(\hat{\vec{r}})\rangle\!\rangle \left[W^{(\text{in})} + W^{(\text{out})} \right] \widehat{N} \langle\!\langle \mathcal{Y}^{JM}(\hat{\vec{k}})| \quad (12.101)$$

$$+ \frac{1}{r} \sum_{JM} |\mathcal{Y}^{JM}(\hat{\vec{r}})\rangle\!\rangle W^{(\text{out})} (S - 1) \widehat{N} \langle\!\langle \mathcal{Y}^{JM}(\hat{\vec{k}})|$$

$$= \psi^{(\text{pw})}(\vec{r}, \vec{k}) + \psi^{(\text{sc})}(\vec{r}, \vec{k}) .$$

The first of these two terms,

$$\psi^{(\text{pw})}(\vec{r}, \vec{k}) = \frac{1}{r} \sum_{JM} |\mathcal{Y}^{JM}(\hat{\vec{r}})\rangle\!\rangle \left[W^{(\text{in})} + W^{(\text{out})} \right] \widehat{N} \langle\!\langle \mathcal{Y}^{JM}(\hat{\vec{k}})| , \quad (12.102)$$

is the incoming plane wave. It is the diagonal $(\mathcal{N} \times \mathcal{N})$-matrix consisting of the plane waves, $\exp(i\vec{k}_\eta \vec{r})/(2\pi)^{3/2}$ (normalized to the δ-function), in each of the truly different channels. This can be seen, if we compare $\psi^{(\text{pw})}(\vec{r}, \vec{k})$ with the plane-wave expansion (10.31). Therefore

$$\psi^{(\text{pw})}(\vec{r}, \vec{k}) = \frac{||e^{i\vec{k}\vec{r}}||}{(2\pi)^{3/2}} , \quad (12.103)$$

where

$$||e^{i\vec{k}\vec{r}}|| = \begin{pmatrix} e^{i\vec{k}_1\vec{r}} & 0 & \dots & 0 \\ 0 & e^{i\vec{k}_2\vec{r}} & \dots & 0 \\ \dots & \dots & \dots & \dots \\ 0 & 0 & \dots & e^{i\vec{k}_{\mathcal{N}}\vec{r}} \end{pmatrix} . \quad (12.104)$$

The second term in Eq. (12.101) is also an $(\mathcal{N} \times \mathcal{N})$-matrix in the space of the truly different channels. In contrast to $\psi^{(\text{pw})}$, it is not diagonal. Its off-diagonal elements describe various inter-channel transitions. Similarly to Eq. (10.40), this matrix should consist of the spherical outgoing waves in each of the open channels. In order to show this, we need to generalize the logical steps that were used for the non-central potential in Sect. 10.3.3.

The reader should recall that the matrices $|\mathcal{Y}^{JM}(\hat{\vec{r}})\rangle\!\rangle$ and $\langle\!\langle \mathcal{Y}^{JM}(\hat{\vec{k}})|$ are not square; they have the dimensions $(\mathcal{N} \times N)$ and $(N \times \mathcal{N})$, respectively. Meanwhile the matrices $W^{(\text{out})}$ and S are square and of the size $(N \times N)$ because they involve all the channels (including the degenerate ones) explicitly. At large distances, the matrix $W^{(\text{out})}$ consists of the diagonal blocks of the type (9.112) for all the truly different channels,

$$W^{(\mathrm{out})}(E,r) \underset{r\to\infty}{\longrightarrow} \begin{pmatrix} \boxed{e^{ik_1r}} & 0 & \dots & 0 \\ 0 & \boxed{e^{ik_2r}} & \dots & 0 \\ \dots & \dots & \dots & \dots \\ 0 & 0 & \dots & \boxed{e^{ik_{\mathcal{N}}r}} \end{pmatrix} \frac{1}{i}\mathcal{I}^{\dagger}\,, \tag{12.105}$$

where each block,

$$\boxed{e^{ik_\eta r}} = \begin{pmatrix} e^{ik_\eta r} & 0 & \dots & 0 \\ 0 & e^{ik_\eta r} & \dots & 0 \\ \dots & \dots & \dots & \dots \\ 0 & 0 & \dots & e^{ik_\eta r} \end{pmatrix}, \tag{12.106}$$

consists of the identical diagonal elements and has the size equal to the number of different spin sub-channels for a given η. The $(N \times N)$ diagonal matrix $\mathcal{I}$ is defined by Eq. (9.108). Therefore the first factor in the asymptotic behaviour of the matrix $W^{(\mathrm{out})}$ consists of the blocks made of the repeating elements. In the scattering term of Eq. (12.101),

$$\psi^{(\mathrm{sc})}(\vec{r},\vec{k}) = \frac{1}{r}\sum_{JM} |\mathcal{Y}^{JM}(\hat{\vec{r}})\rangle W^{(\mathrm{out})}(S-1)\,\widehat{N}\langle \mathcal{Y}^{JM}(\hat{\vec{k}})|\,, \tag{12.107}$$

the $(N \times N)$-matrix (12.105) with the repeating elements in each block is multiplied from the left by a non-square $(\mathcal{N} \times N)$-matrix $|\mathcal{Y}^{JM}(\hat{\vec{r}})\rangle$. For the sake of clarity, let us once more use the schematic example of two truly different channels with two and three sub-channels in them (see Eq. (11.39)). For that example, the matrix product we are discussing can be symbolically written as follows:

$$\begin{pmatrix} |1\rangle & |2\rangle & 0 & 0 & 0 \\ 0 & 0 & |3\rangle & |4\rangle & |5\rangle \end{pmatrix} \begin{pmatrix} e^{ik_1r} & 0 & 0 & 0 & 0 \\ 0 & e^{ik_1r} & 0 & 0 & 0 \\ 0 & 0 & e^{ik_2r} & 0 & 0 \\ 0 & 0 & 0 & e^{ik_2r} & 0 \\ 0 & 0 & 0 & 0 & e^{ik_2r} \end{pmatrix}$$

$$= \begin{pmatrix} |1\rangle e^{ik_1r} & |2\rangle e^{ik_1r} & 0 & 0 & 0 \\ 0 & 0 & |3\rangle e^{ik_2r} & |4\rangle e^{ik_2r} & |5\rangle e^{ik_2r} \end{pmatrix}$$

$$= \begin{pmatrix} e^{ik_1r} & 0 \\ 0 & e^{ik_2r} \end{pmatrix} \begin{pmatrix} |1\rangle & |2\rangle & 0 & 0 & 0 \\ 0 & 0 & |3\rangle & |4\rangle & |5\rangle \end{pmatrix}.$$

As is seen from this example, the big matrix with repeating elements is moved to the left of the product and each repeating block is replaced with just one element. It is clear that this is valid for any dimensions $\mathcal{N}$ and N because the structure of the matrices is always the same.

If we introduce the diagonal matrix (without repeating elements),

$$||e^{ikr}|| = \begin{pmatrix} e^{ik_1 r} & 0 & \cdots & 0 \\ 0 & e^{ik_2 r} & \cdots & 0 \\ \cdots & \cdots & \cdots & \cdots \\ 0 & 0 & \cdots & e^{ik_{\mathcal{N}} r} \end{pmatrix}, \qquad (12.108)$$

and take into account Eq. (11.89) defining $\widehat{N}$, we can re-write the second (scattering) term of Eq. (12.101) as follows:

$$\psi^{(\mathrm{sc})}(\vec{r}, \vec{k}) = \sqrt{\frac{2}{\pi}} \frac{||e^{ikr}||}{r} \sum_{JM} |\mathcal{Y}^{JM}(\hat{\vec{r}})\rangle \frac{\mathcal{I}^{\dagger}(S-1)\mathcal{I}}{2i} K^{-1} \langle \mathcal{Y}^{JM}(\hat{\vec{k}})| \,. \qquad (12.109)$$

Generalizing the definitions of the partial wave and the full amplitudes (10.38) and (10.39) that are defined for a non-central potential, we introduce the corresponding multi-channel scattering amplitudes as follows:

$$f^{J}(E) = \frac{1}{2i} \mathcal{I}^{\dagger} \left[S^{J}(E) - 1 \right] \mathcal{I} K^{-1} \,, \qquad (N \times N)\text{-matrix} \,, \quad (12.110)$$

$$\mathsf{f}(\hat{\vec{r}} \leftarrow \hat{\vec{k}}, E) = 4\pi \sum_{JM} |\mathcal{Y}^{JM}(\hat{\vec{r}})\rangle f^{J}(E) \langle \mathcal{Y}^{JM}(\hat{\vec{k}})| \,, \qquad (\mathcal{N} \times \mathcal{N})\text{-matrix} \,, \quad (12.111)$$

where the dependencies on the total angular momentum, J, and on the energy, E, are indicated. The matrix elements of the partial-wave amplitude (12.110) have the following explicit form:

$$f^{J}_{n'n} = \frac{S^{J}_{n'n} - \delta_{n'n}}{2ik_n} i^{\ell_n - \ell_{n'}} \,. \qquad (12.112)$$

Using the full scattering amplitude $\mathsf{f}(\hat{\vec{r}} \leftarrow \hat{\vec{k}}, E)$, we can re-write the asymptotic behaviour of the wave function (12.101) in the standard form,

$$\psi(\vec{r}, \vec{k}) \underset{r \to \infty}{\longrightarrow} \frac{1}{(2\pi)^{3/2}} \left[||e^{i\vec{k}\vec{r}}|| + \frac{||e^{ikr}||}{r} \mathsf{f}(\hat{\vec{r}} \leftarrow \hat{\vec{k}}, E) \right] \qquad (12.113)$$

For a particular matrix element (transition between two internal states, η and η'), we obtain

$$\psi_{\eta'\eta}(\vec{r},\vec{k}) \underset{r\to\infty}{\longrightarrow} \frac{1}{(2\pi)^{3/2}}\left[e^{i\vec{k}_\eta\vec{r}}\delta_{\eta'\eta} + \frac{e^{ik_{\eta'}r}}{r}\mathsf{f}_{\eta'\eta}(\hat{\vec{r}} \leftarrow \hat{\vec{k}}, E)\right]. \tag{12.114}$$

In fact, we managed to obtain this (correct) asymptotics thanks to the choice (12.98) of the combination coefficients for constructing the physical solution out of the regular columns. This proves that the choice (12.98) was correct.

It should be clearly understood that each matrix element $f^J_{n'n}(E)$ of the partial-wave amplitude (12.110) is just a complex number, while each matrix element $\mathsf{f}_{\eta'\eta}(\hat{\vec{r}} \leftarrow \hat{\vec{k}}, E)$ of the full amplitude (12.111) is an operator in the spin space. This is because the matrices $|\mathcal{Y}^{JM}(\hat{\vec{r}})\rangle$ and $\langle\mathcal{Y}^{JM}(\hat{\vec{k}})|$ consist of the elements of the type given in the example (10.11). Therefore a particular transition $(\eta', s'\sigma') \leftarrow (\eta, s\sigma)$ is described by the following wave function:

$$\langle s'\sigma'|\psi_{\eta'\eta}(\vec{r},\vec{k})|s\sigma\rangle \tag{12.115}$$

$$\underset{r\to\infty}{\longrightarrow} \frac{1}{(2\pi)^{3/2}}\left[e^{i\vec{k}_\eta\vec{r}}\delta_{\eta'\eta}\delta_{s's}\delta_{\sigma'\sigma} + \frac{e^{ik_{\eta'}r}}{r}\langle s'\sigma'|\mathsf{f}_{\eta'\eta}(\hat{\vec{r}} \leftarrow \hat{\vec{k}}, E)|s\sigma\rangle\right].$$

In this equation, all the factors are not operators or matrices anymore. They are just complex numbers (functions). We can therefore calculate the incoming and scattered particle-fluxes, using the simple formulae derived in the introductory Sect. 1.3.2. As is given by Eqs. (1.54) and (1.56), the incoming (plane wave) current from the state with the quantum numbers (η, s, σ) is

$$\vec{j}^{\,(\mathrm{pw})} = \frac{\hbar\vec{k}_\eta}{(2\pi)^3\mu_\eta}, \tag{12.116}$$

and the radial component of the current of particles in the channel (η', s', σ') can be found as

$$j_r^{(\mathrm{sc})} = \frac{\hbar k_{\eta'}}{(2\pi)^3\mu_{\eta'}r^2}\left|\langle s'\sigma'|\mathsf{f}_{\eta'\eta}(\hat{\vec{r}} \leftarrow \hat{\vec{k}}, E)|s\sigma\rangle\right|^2. \tag{12.117}$$

The ratio of their absolute values is the differential scattering cross section (see Eq. (1.57)):[10]

$$\frac{d\sigma}{d\Omega}(\eta's'\sigma' \leftarrow \eta s\sigma) = \frac{\mu_\eta k_{\eta'}}{\mu_{\eta'}k_\eta}\left|\langle s'\sigma'|\mathsf{f}_{\eta'\eta}(\hat{\vec{r}} \leftarrow \hat{\vec{k}}, E)|s\sigma\rangle\right|^2. \tag{12.118}$$

[10] It is assumed that using the same symbol σ for both the cross section and the z-component of the spin will not cause confusion.

For a particular matrix element of the full amplitude (12.111),

$$\mathsf{f}_{\eta'\eta}(\hat{\vec{r}} \leftarrow \hat{\vec{k}}, E) = 4\pi \sum_{JM} |\mathcal{Y}_{\eta'}^{JM}(\hat{\vec{r}})\rangle f^J(E) \langle \mathcal{Y}_{\eta}^{JM}(\hat{\vec{k}})| \tag{12.119}$$

$$= 4\pi \sum_{JM} \sum_{[\ell'][\ell]} |\mathcal{Y}_{[\ell']}^{JM}(\hat{\vec{r}})\rangle f_{n'n}^J(E) \langle \mathcal{Y}_{[\ell]}^{JM}(\hat{\vec{k}})| \,,$$

the matrices $|\mathcal{Y}_{\eta'}^{JM}(\hat{\vec{r}})\rangle$ and $\langle \mathcal{Y}_{\eta}^{JM}(\hat{\vec{k}})|$ are just a single row and a single column of the partial waves (defined by Eq. (11.36)). The sum over $[\ell'] = \{\ell', s'\}$ and $[\ell] = \{\ell, s\}$ only involves the sub-matrix $f_{n'n}^J$ for the corresponding sub-channels, $n' = \{\eta', \ell', s'\}$ and $n = \{\eta, \ell, s\}$, with fixed η' and η.

It is not difficult to see that the matrix element (12.119) has exactly the same structure as the full amplitude (10.39) for the single-channel case. It is an operator in the spin space. To make it more clear, let us consider an example where both the η' and η channels have two partial-wave sub-channels. In such a case, we have

$$\mathsf{f}_{\eta'\eta} = 4\pi \sum_{JM} \begin{pmatrix} |\mathcal{Y}_{[\ell_1']}^{JM}\rangle & |\mathcal{Y}_{[\ell_2']}^{JM}\rangle \end{pmatrix} \begin{pmatrix} f_{n_1'n_1}^J & f_{n_1'n_2}^J \\ f_{n_2'n_1}^J & f_{n_2'n_2}^J \end{pmatrix} \begin{pmatrix} \langle \mathcal{Y}_{[\ell_1]}^{JM}| \\ \langle \mathcal{Y}_{[\ell_2]}^{JM}| \end{pmatrix} \tag{12.120}$$

$$= 4\pi \sum_{JM} \Big(|\mathcal{Y}_{[\ell_1']}^{JM}\rangle f_{n_1'n_1}^J \langle \mathcal{Y}_{[\ell_1]}^{JM}| + |\mathcal{Y}_{[\ell_2']}^{JM}\rangle f_{n_2'n_1}^J \langle \mathcal{Y}_{[\ell_1]}^{JM}|$$

$$+ |\mathcal{Y}_{[\ell_1']}^{JM}\rangle f_{n_1'n_2}^J \langle \mathcal{Y}_{[\ell_2]}^{JM}| + |\mathcal{Y}_{[\ell_2']}^{JM}\rangle f_{n_2'n_2}^J \langle \mathcal{Y}_{[\ell_2]}^{JM}| \Big) \,.$$

The matrix element of this amplitude over the spin states, $\langle s'\sigma'|\mathsf{f}_{\eta'\eta}|s\sigma\rangle$, can be found using Eq. (10.10).

Since for a pair of fixed incoming and outgoing (truly different) channels, η and η', the full amplitude $\mathsf{f}_{\eta'\eta}(\hat{\vec{r}} \leftarrow \hat{\vec{k}}, E)$ has the same structure as for the single-channel case, all the formulae derived for the differential and total cross sections in Sect. 10.3.4 are applicable here as well. The only difference is the additional kinematical factor that appears in Eq. (12.118) as well as the different meaning of the subscripts of the partial-wave amplitude. For the reader's convenience, these formulae with the appropriate modifications are collected here. The differential cross section is

$$\frac{d\sigma}{d\Omega}(\eta' s'\sigma' \leftarrow \eta s\sigma) = 4\pi \frac{\mu_\eta k_{\eta'}}{\mu_{\eta'} k_\eta} \sum_{J_1 M_1 \ell_1' \ell_1} \sum_{J_2 M_2 \ell_2' \ell_2} \sqrt{(2\ell_1+1)(2\ell_2+1)}$$

$$\times C_{\ell_1 0 s\sigma}^{J_1 M_1} f_{n_1'n_1}^{J_1 *} f_{n_2'n_2}^{J_2} C_{\ell_2 0 s\sigma}^{J_2 M_2} \left[Y_{\ell_1' s'\sigma'}^{J_1 M_1}(\hat{\vec{r}}) \right]^* Y_{\ell_2' s'\sigma'}^{J_2 M_2}(\hat{\vec{r}}) \,, \tag{12.121}$$

where $n_1 = \{\eta, \ell_1, s\}$, $n_2 = \{\eta, \ell_2, s\}$, $n'_1 = \{\eta', \ell'_1, s\}$, and $n'_2 = \{\eta', \ell'_2, s\}$. The total (integrated) cross section can be found as

$$\sigma(\eta' s' \sigma' \leftarrow \eta s \sigma) = 4\pi \frac{\mu_\eta k_{\eta'}}{\mu_{\eta'} k_\eta} \sum_{J_1 M_1 \ell_1 J_2 M_2 \ell_2 \ell' m'} \sqrt{(2\ell_1+1)(2\ell_2+1)} C^{J_1 M_1}_{\ell' m' s' \sigma'} C^{J_1 M_1}_{\ell_1 0 s \sigma}$$

$$\times \left(f^{J_1}\right)^\dagger_{n_1 n'} f^{J_2}_{n' n_2} C^{J_2 M_2}_{\ell_2 0 s \sigma} C^{J_2 M_2}_{\ell' m' s' \sigma'} . \qquad (12.122)$$

If all possible orientations of the incident spin have equal probabilities, then the averaged cross section is

$$\sigma(\eta' s' \sigma' \leftarrow \eta s) = \frac{1}{2s+1} \sum_\sigma \sigma(\eta' s' \sigma' \leftarrow \eta s \sigma) . \qquad (12.123)$$

If the orientation of the final spin is not monitored, then we sum up the cross section over all possible σ'. This enables us to get rid of all the Clebsch–Gordan coefficients:

$$\sum_{M_1 M_2 m' \sigma' \sigma} C^{J_1 M_1}_{\ell' m' s' \sigma'} C^{J_2 M_2}_{\ell' m' s' \sigma'} C^{J_1 M_1}_{\ell_1 0 s \sigma} C^{J_2 M_2}_{\ell_2 0 s \sigma}$$

$$= \sum_{M_1 M_2 \sigma} \delta_{J_1 J_2} \delta_{M_1 M_2} C^{J_1 M_1}_{\ell_1 0 s \sigma} C^{J_1 M_1}_{\ell_2 0 s \sigma} = \delta_{J_1 J_2} \delta_{\ell_1 \ell_2} \frac{2J_1+1}{2\ell_1+1} ,$$

which finally gives

$$\sigma(\eta' s' \leftarrow \eta s) = \frac{4\pi}{2s+1} \frac{\mu_\eta k_{\eta'}}{\mu_{\eta'} k_\eta} \sum_{J \ell \ell'} (2J+1) \left| f^J_{n' n} \right|^2 . \qquad (12.124)$$

If we are not interested in the spin states at all, then we need to sum up the cross sections for all possible initial and final spins,

$$\sigma(\eta' \leftarrow \eta) = 4\pi \frac{\mu_\eta k_{\eta'}}{\mu_{\eta'} k_\eta} \sum_{J[\ell][\ell']} \frac{2J+1}{2s+1} \left| f^J_{n' n} \right|^2 . \qquad (12.125)$$

12.7 Unitarity, Reciprocity, and Detailed Balance

When considering the scattering by a non-central potential, it was shown that the conservation of the particle-flux leads to the unitarity of the S-matrix at real collision energies (see Sect. 10.6.2). The problem of coupled partial waves, considered there, is a particular (simplified) case of a more general multi-channel problem that we are dealing with here. When generalizing the derivations of Sect. 10.6.2, we need to take care of the channel momenta and the reduced masses that are not the same for different channels. Otherwise all the logical steps leading to the conclusion that the S-matrix is unitary are very similar to those used in the non-central potential problem.

* * *

The wave function (12.100) that describes the multi-channel scattering can be split into two terms,

$$\psi(\vec{r},\vec{k}) = \psi^{(\text{in})}(\vec{r},\vec{k}) + \psi^{(\text{out})}(\vec{r},\vec{k})\,, \tag{12.126}$$

$$\psi^{(\text{in})}(\vec{r},\vec{k}) \underset{r\to\infty}{\longrightarrow} \frac{1}{\sqrt{2\pi}r}\sum_{JM}|\mathcal{Y}^{JM}(\hat{\vec{r}})\rangle W^{(\text{in})}K^{-1}\mathcal{I}\langle\mathcal{Y}^{JM}(\hat{\vec{k}})|\,, \tag{12.127}$$

$$\psi^{(\text{out})}(\vec{r},\vec{k}) \underset{r\to\infty}{\longrightarrow} \frac{1}{\sqrt{2\pi}r}\sum_{JM}|\mathcal{Y}^{JM}(\hat{\vec{r}})\rangle W^{(\text{out})}SK^{-1}\mathcal{I}\langle\mathcal{Y}^{JM}(\hat{\vec{k}})|\,. \tag{12.128}$$

These terms correspond to the waves that go into the sphere of the radius r and out of this sphere. The two ("in" and "out") fluxes must compensate each other, and therefore the total flux through the spherical surface of any radius r must be zero.

Since the compensation of the fluxes is taking place at a sphere of any radius, it is convenient to choose the size of such a sphere extremely large. Then to calculate the fluxes, we can use the asymptotic expressions for the wave functions, which are known explicitly.

The same as for the case of coupled partial waves (see Sect. 10.6.2), the wave function (12.126) is an operator in the spin space (this is why the spin-angular factors in it are represented by the "bra" and "ket" vectors). Therefore, to evaluate the "in" and "out" fluxes, we have to use the more general rule:[11]

$$\vec{j} = \frac{\hbar}{2i}\left\{\psi^{\dagger}(\vec{r})M^{-1}\vec{\nabla}\psi(\vec{r}) - \left[M^{-1}\vec{\nabla}\psi(\vec{r})\right]^{\dagger}\psi(\vec{r})\right\}\,, \tag{12.129}$$

where the diagonal matrix M,

[11] Equation (12.129) is derived in the same way as its simple version for a single-channel problem. The only difference is that the kinetic energy operator is given by the diagonal matrix $-(\hbar^2/2)M^{-1}\left(\vec{\nabla}\right)^2$.

$$M = \begin{pmatrix} \boxed{\mu_1} & 0 & \cdots & 0 \\ 0 & \boxed{\mu_2} & \cdots & 0 \\ \cdots & \cdots & \cdots & \cdots \\ 0 & 0 & \cdots & \boxed{\mu_{\mathcal{N}}} \end{pmatrix}, \tag{12.130}$$

consists of the (framed) diagonal blocks of the same reduced masses for the degenerate channels.

We are only interested in the fluxes that are normal to the spherical surface, i.e. we only take the first component of the "nabla" operator, $\vec{\nabla} = \{\partial_r,\ (1/r)\partial_\theta,\ (1/r\sin\theta)\partial_\varphi\}$. This means that we only differentiate the r-dependent factors

$$W^{(\mathrm{in})}(E,r) \underset{r\to\infty}{\longrightarrow} \begin{pmatrix} \boxed{e^{-ik_1 r}} & 0 & \cdots & 0 \\ 0 & \boxed{e^{-ik_2 r}} & \cdots & 0 \\ \cdots & \cdots & \cdots & \cdots \\ 0 & 0 & \cdots & \boxed{e^{-ik_{\mathcal{N}} r}} \end{pmatrix} i\mathcal{I} \tag{12.131}$$

and

$$W^{(\mathrm{out})}(E,r) \underset{r\to\infty}{\longrightarrow} \begin{pmatrix} \boxed{e^{ik_1 r}} & 0 & \cdots & 0 \\ 0 & \boxed{e^{ik_2 r}} & \cdots & 0 \\ \cdots & \cdots & \cdots & \cdots \\ 0 & 0 & \cdots & \boxed{e^{ik_{\mathcal{N}} r}} \end{pmatrix} \frac{1}{i}\mathcal{I}^\dagger , \tag{12.132}$$

where the framed blocks are given by Eq. (12.106) and the diagonal matrix $\mathcal{I}$ consists of the ℓ-powers of the imaginary unit (see Eq. (9.108)). It is not difficult to find that

$$\boxed{r \to \infty} \qquad \Rightarrow \qquad \partial_r \psi^{(\mathrm{in/out})} = \left(-\frac{1}{r} \mp iK\right)\psi^{(\mathrm{in/out})}$$

and thus

$$j_r^{(\mathrm{in/out})} = \mp\hbar\psi^{(\mathrm{in/out})\dagger} M^{-1} K \psi^{(\mathrm{in/out})} ,$$

where K is the diagonal matrix that is similar to (12.130) and consists of the diagonal blocks of the repeating channel momenta for the degenerate channels.

Following the same logics as for Eq. (10.80), we obtain

$$\int \left[\psi^{(\mathrm{out})\dagger} M^{-1} K \psi^{(\mathrm{out})} - \psi^{(\mathrm{in})\dagger} M^{-1} K \psi^{(\mathrm{in})}\right] d\hat{\vec{r}} = 0 , \tag{12.133}$$

which differs from (10.80) by the diagonal matrices of the reduced masses and channel momenta appearing in between the wave functions. The angular integration is done using the orthogonality of the spin-angular components of the wave functions (see Eq. (11.38)). This gives

$$\sum_{JM}\left\{|\mathcal{Y}^{JM}(\hat{\vec{k}})\rangle\mathcal{I}^{\dagger}\left[K^{-1}\left(S^{J}\right)^{\dagger}M^{-1}KS^{J}K^{-1}-K^{-1}M^{-1}\right]\mathcal{I}\langle\mathcal{Y}^{JM}(\hat{\vec{k}})|\right\}=0\,. \tag{12.134}$$

Since this equality must remain valid for any incident momentum $\vec{k}$, we conclude that

$$K^{-1}S^{\dagger}M^{-1}KSK^{-1}-K^{-1}M^{-1}=0\,, \tag{12.135}$$

where in order to simplify notation, we drop the superscript "J". Equation (12.135) can be re-written in the form

$$M^{-\frac{1}{2}}K^{-\frac{1}{2}}\left(M^{\frac{1}{2}}K^{-\frac{1}{2}}S^{\dagger}K^{\frac{1}{2}}M^{-\frac{1}{2}}M^{-\frac{1}{2}}K^{\frac{1}{2}}SK^{-\frac{1}{2}}M^{\frac{1}{2}}-I\right)K^{-\frac{1}{2}}M^{-\frac{1}{2}}=0\,. \tag{12.136}$$

If we introduce the modified (flux adjusted) S-matrix, $\mathbb{S}$,

$$\boxed{\mathbb{S}=M^{-\frac{1}{2}}K^{\frac{1}{2}}SK^{-\frac{1}{2}}M^{\frac{1}{2}}} \tag{12.137}$$

then it follows from Eq. (12.135) that

$$\boxed{\mathbb{S}^{\dagger}\mathbb{S}=I} \tag{12.138}$$

i.e. the new S-matrix is unitary. The physical meaning of the additional factors on both sides of the original S-matrix in the definition (12.137) can be understood, if we take notice of the fact that $\hbar k_n/\mu_n$ is the channel velocity. These factors are needed because the fluxes of the particles in different channels depend on how fast the particles move.

It should be emphasized that the modified matrix $\mathbb{S}$ can only be safely used at real collision energies, where the square roots $\sqrt{k_n}$ have clear meaning. If we try to analytically continue it to complex energies, then these square roots lead to the ambiguity. Indeed, the root of the fourth order, $\sqrt{k_n}\sim\sqrt[4]{E-E_n}$, of a complex number has four different values. For example, $\sqrt[4]{1}$ is equal either to ±1 or to $\pm i$. In other words, each threshold E_n becomes a fourth-order branch point

($\sqrt[4]{E - E_n}$) instead of the second order ($\sqrt[2]{E - E_n}$). As a result, the number of the Riemann sheets is artificially increased. Therefore, if the modified matrix (12.137) is used at complex energies, one should be extremely careful, especially when doing numerical calculations because different compilers work with the complex numbers and with multi-valued functions differently.

* * *

As we have seen, the original S-matrix is not unitary, but it does not depend on the direction of any angular momentum (orbital or spin), and thus (as it follows from to the time-reversal invariance; see Sect. 10.6.1) it is symmetric relative to the swap of the subscripts, $n' \leftrightarrow n$. The only non-symmetric factor in the partial-wave scattering amplitude (12.112) is $1/k_n$. In order to make it symmetric, we need to multiply the amplitude by $\sqrt{k_n/k_{n'}}$. This gives

$$f^J_{n'n} = \sqrt{k_{n'}/k_n}\, \mathfrak{f}^J_{n'n} ,$$

where $\mathfrak{f}^J_{n'n}$ is the symmetric part of the scattering amplitude,

$$\mathfrak{f}^J_{n'n} = i^{\ell_n - \ell_{n'}} \frac{S^J_{n'n} - \delta_{n'n}}{2i\sqrt{k_{n'}k_n}} .$$

Now we recollect the derivations of Sect. 10.6.4 which are applicable here as well. In particular, the same way as obtaining Eq. (10.90), it is not difficult to find that

$$\langle s\mu|\mathcal{Y}^{JM}_{[\ell]}(\hat{\vec{k}})\rangle^\dagger = (-1)^{s-\mu}(-1)^{J-M}\langle\mathcal{Y}^{JM}_{[\ell]}(-\hat{\vec{k}})|s, -\mu\rangle .$$

Using this in Eq. (12.119), we see that

$$\langle s'\mu'|\mathfrak{f}_{\eta'\eta}(\hat{\vec{r}} \leftarrow \hat{\vec{k}}, E)|s\mu\rangle = 4\pi \sum_{JM}\sum_{[\ell'][\ell]} \langle s'\mu'|\mathcal{Y}^{JM}_{[\ell']}(\hat{\vec{r}})\rangle f^J_{n'n}(E)\langle\mathcal{Y}^{JM}_{[\ell]}(\hat{\vec{k}})|s\mu\rangle$$

$$= 4\pi\sqrt{\frac{k_{\eta'}}{k_\eta}}(-1)^{s'-\mu'}(-1)^{s-\mu}\sum_{JM}\sum_{[\ell'][\ell]}\langle s, -\mu|\mathcal{Y}^{JM}_{[\ell]}(-\hat{\vec{k}})\rangle\, \mathfrak{f}^J_{nn'}(E)\langle\mathcal{Y}^{JM}_{[\ell']}(-\hat{\vec{r}})|s', -\mu'\rangle$$

$$= 4\pi\frac{k_{\eta'}}{k_\eta}(-1)^{s'-\mu'}(-1)^{s-\mu}\sum_{JM}\sum_{[\ell'][\ell]}\langle s, -\mu|\mathcal{Y}^{JM}_{[\ell]}(-\hat{\vec{k}})\rangle f^J_{nn'}(E)\langle\mathcal{Y}^{JM}_{[\ell']}(-\hat{\vec{r}})|s', -\mu'\rangle ,$$

from which it follows the multi-channel reciprocity relation:

$$\langle s'\mu'|\mathsf{f}_{\eta'\eta}(\hat{\vec{r}} \leftarrow \hat{\vec{k}}, E)|s\mu\rangle$$
$$= \frac{k_{\eta'}}{k_\eta}(-1)^{s'+s-\mu'-\mu}\langle s, -\mu|\mathsf{f}_{\eta\eta'}(-\hat{\vec{k}} \leftarrow -\hat{\vec{r}}, E)|s', -\mu'\rangle \tag{12.139}$$

It differs from the relation (10.91) by the factor $k_{\eta'}/k_\eta$.

* * *

Taking the absolute squares on both sides of Eq.(12.139) and using the fact that the scattering angle, i.e. the angle between vectors $\vec{k}$ and $\vec{r}$ and between $-\vec{r}$ and $-\vec{k}$, is the same, we find that the differential cross section of the direct and the inverse processes are related as

$$\left(\frac{d\sigma}{d\Omega}\right)_{s'\mu' \leftarrow s\mu} = \left(\frac{k'}{k}\right)^2 \left(\frac{d\sigma}{d\Omega}\right)_{s,-\mu \leftarrow s',-\mu'} . \tag{12.140}$$

This relation remains valid for the total cross sections as well,

$$\sigma_{s'\mu' \leftarrow s\mu} = \left(\frac{k'}{k}\right)^2 \sigma_{s,-\mu \leftarrow s',-\mu'} . \tag{12.141}$$

Averaging the cross sections over the initial and summing over the final spin orientations, we obtain the total unpolarized cross sections:

$$\sigma_{s' \leftarrow s} = \frac{1}{2s+1}\sum_{\mu\mu'} \sigma_{s'\mu' \leftarrow s\mu} ,$$

$$\sigma_{s \leftarrow s'} = \frac{1}{2s'+1}\sum_{\mu\mu'} \sigma_{s,-\mu \leftarrow s',-\mu'} .$$

Then the equality (12.141) implies that

$$(2s+1)k^2\sigma_{s' \leftarrow s} = (2s'+1)k'^2\sigma_{s \leftarrow s'} \tag{12.142}$$

which is known as the principle of detailed balance.

13 Riemann Surfaces for Multi-Channel Systems

In the single-channel case, the Jost functions could be considered either as the functions of the momentum or the energy. Using the momentum (which is actually done in most of the books) is even preferable because there is only one complex k-plane (without branching[1]) in contrast to the two-layer energy surface. However, when the number of channels $N > 1$, the use of the channel momenta leads to the obvious difficulties because the channel momenta are not independent variables. All of them are determined by the single variable E. For a fixed (generally speaking, complex) value of the energy E, each of the N channel momenta can have two different values[2]

$$k_n = \pm\sqrt{\frac{2\mu_n}{\hbar^2}(E - E_n)}\,, \qquad n = 1, 2, \ldots, N\,, \tag{13.1}$$

depending on the choice of the sign in front of the square root. All these momenta are involved as parameters in the Schrödinger equation (11.70) and in the boundary conditions (12.3). The equation itself does not depend on the choice of the signs because the momenta k_n there are squared. The boundary conditions, however, involve the functions $j_{\ell_n}(k_n r)$ that are different for the "+" and "−" signs in Eq. (13.1). The solutions $\phi(E, r)$ are therefore dependent on the choice of the signs in (13.1). And the Jost matrices $f^{(\mathrm{in/out})}(E)$ that are part of the asymptotic expressions (12.8) of these solutions, depend on the choices of that signs as well. This means that the Jost matrices are not single-valued functions of E. At each point E, they have 2^N different values, for all possible combinations of the signs

[1] It should be noted that here we only refer to the potentials without Coulomb tails that cause the logarithmic branching even in the single-channel case.

[2] For the sake of clarity, here we assume that there are no degenerate channels, i.e. that all the channel momenta have different absolute values. The case of degenerate channels is considered further down in Sect. 13.2.

S. A. Rakityansky, *Jost Functions in Quantum Mechanics*,
https://doi.org/10.1007/978-3-031-07761-6_13

of N channel momenta. In other words, the Jost matrices $f^{(\mathrm{in/out})}(E)$ are defined on a Riemann surface of the energy, which consists of 2^N layers (sheets). For each (complex) value of E, there are 2^N possible choices of the set $\{k_1, k_2, \ldots, k_N\}$. Each choice corresponds to a point on a different layer of the Riemann surface. All these points are located one under the other on a vertical line passing through all the layers.

13.1 Cuts and Interconnections

When $E = E_n$, the corresponding channel momentum $k_n = 0$, and therefore the sign in front of the square root (13.1), does not play any role. This means that at the threshold energies two sheets of the Riemann surface (corresponding to $\pm k_n$) coincide, i.e. they are connected to each other at these points.

When we move around any of the threshold point, we go from one layer to another. Indeed, a point on a circle centered at the threshold E_n can be parametrized as $E = E_n + \rho \exp(i\theta)$, where ρ is the distance from E_n and θ is the polar angle. The corresponding channel momentum

$$k_n = \sqrt{\frac{2\mu_n \rho}{\hbar^2}} e^{i\theta/2}$$

changes its sign (which means changing the sheet of the Riemann surface) after one full circle ($\theta = 2\pi$) and comes back to its initial value after two full circles ($\theta = 4\pi$).

Therefore the sheets of the Riemann surface are interconnected and thus form a united multi-layer manifold. The threshold points E_n ($n = 1, 2, \ldots, N$) are the branch points of this manifold. By moving around these points, we can continuously reach any of the 2^N sheets.

In principle, we can construct the Riemann surface rather arbitrarily by making cuts and appropriate connections of the layers. In quantum theory, it is standard that each layer is cut along the real energy axis. The cut begins at the branch point that corresponds to the lowest threshold E_1 and goes to infinity in the positive direction. The edges of the cuts of different layers are interconnected in such a way that the corresponding channel momenta appropriately change their signs when we move around the thresholds.

The simplest two-layer Riemann surface for the single-channel case is easy to visualize (see Fig. 3.2). The two-channel Riemann surface with two branch points and four interconnected layers is much more involved. These connections for the energy intervals $E_1 < E < E_2$ and $E > E_2$ are schematically shown in Figs. 13.1 and 13.2.

In Fig. 13.1, on the left hand side, there are signs labeling different sheets of the manifold. It should be mentioned that these signs are not merely the signs in front of the square root (13.1). These are the signs of the imaginary parts of the

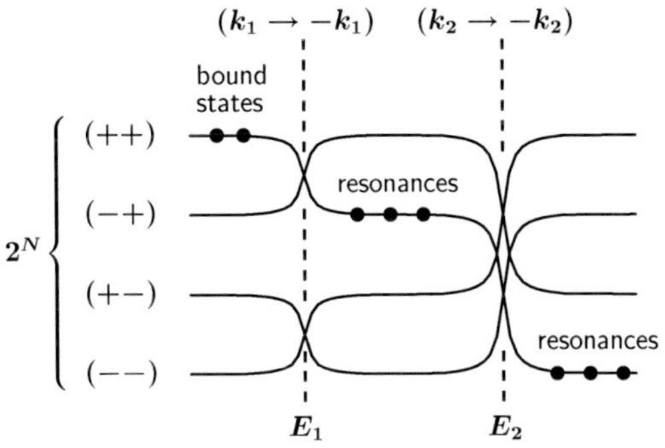

Fig. 13.1 Transitions among the Riemann energy-sheets for a two-channel ($N = 2$) problem. The sheets ($\pm\pm$) correspond to all possible combinations of the signs of the imaginary parts, $\text{Im}\, k_n$, of the channel momenta. Each transition is shown for a full-circle path (2π) around a threshold. On such a path, the corresponding channel momentum changes its sign. The dots schematically show typical distribution of the spectral points

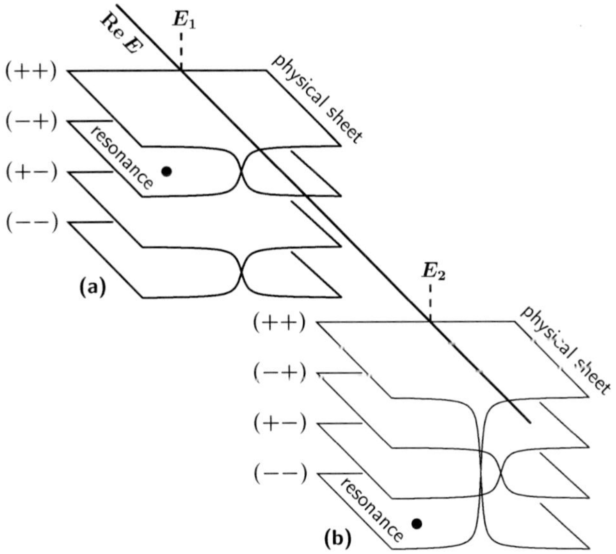

Fig. 13.2 Interconnections of the Riemann energy-sheets for a two-channel ($N = 2$) problem: (**a**) the interconnections between the thresholds E_1 and E_2 and (**b**) the interconnections above the highest threshold. The four sheets ($\pm\pm$) correspond to all possible choices of the signs of $\text{Im}\, k_n$

corresponding channel momenta. Of course, we can always make $\text{Im}\, k_n$ positive or negative by choosing an appropriate sign in Eq. (13.1), but the choice depends also on the value of the energy. In simple terms, this can be explained as follows. In exactly the same way as for the single-channel scattering, each channel momentum k_n covers a whole complex plane. The upper half of such a plane corresponds to a layer of the Riemann surface labeled with "+", and the lower half corresponds to another layer labeled with "−".

Figure 13.1 only shows what happens if we go around the first or the second threshold, i.e. what kind of transition $(\pm\pm)\leftrightarrow(\pm\pm)$ is a result of such a path. Figure 13.2 is more informative. It shows the intricate interconnections of the edges of all four cuts along the real axis. The ways of these connections are different when $\mathrm{Re}\,E$ is between the thresholds and when it is above the highest threshold. The interconnections are chosen in such a way that it corresponds to the sign changes shown in Fig. 13.1. For example, a full circle around E_2 should result in changing the second sign in the label $(\pm\pm)$ of a sheet.

To make it clear, consider, for instance, the resonance shown in the part (b) of Fig. 13.2 on the sheet $(--)$. Let it be our starting point, from which we move around E_2 in the clockwise direction. It should be noted that the sets (stacks) of the energy layers (a) and (b) are the direct continuations of each other. Therefore when passing between E_1 and E_2, we follow the appropriate interconnection shown in the set (a). It is seen that (on our particular path) in between E_1 and E_2, we go up from $(--)$ to $(+-)$ layer. Then we continue encircling the threshold E_2 and find ourselves on the corresponding $(+-)$ layer of the set (b). The full circle has not been completed yet. So, we are to the right of the real axis (i.e. with $\mathrm{Im}\,E > 0$). In order to complete the circle, we need to cross the real axis (i.e. to cross the cut). It is seen that, as a result, we go up to the layer $(-+)$ and finish our "trip" on the $(-+)$ sheet exactly above our starting point. Therefore the final result is that we moved from the $(--)$ to $(-+)$ sheet, i.e. changed the second sign, as was expected according to Fig. 13.1.

In a similar way, it is easy to check that the interconnections shown in Fig. 13.2 always give the correct change of the signs $(\pm\pm)\leftrightarrow(\pm\pm)$ for all possible paths around the thresholds. It is also seen that a single circle (actually, any closed loop) around all the thresholds changes all the signs at once, i.e. it results, for example, in the transition $(++)\leftrightarrow(--)$. And indeed, it should be like this because in the degenerate case ($E_1 = E_2 = \cdots = E_N$) there is only one channel momentum which must change sign when we go around the common threshold energy.

In general, for an N-channel problem the Riemann surface consists of 2^N sheets. All of them are cut along the real axis above the lowest threshold, E_1. The rule for their interconnections is simple: between the thresholds E_1 and E_2, a continuous path crossing the real axis results in changing of the first sign in the sheet label; the continuous path across the interconnection between E_2 and E_3 changes the first two signs, etc.,

$$
\begin{aligned}
&E_1 < (\text{crossing Re}E) < E_2 \Rightarrow (\text{sign}_1, \text{sign}_2, \dots) \to (-\text{sign}_1, \text{sign}_2, \dots)\,,\\
&E_2 < (\text{crossing Re}E) < E_3 \Rightarrow (\text{sign}_1, \text{sign}_2, \dots) \to (-\text{sign}_1, -\text{sign}_2, \dots)\,,\\
&\text{etc.}
\end{aligned}
\tag{13.2}
$$

13.2 Degenerate Channels

What happens to the Riemann surface when two or more thresholds are the same, i.e. when we deal with degenerate channels? Such a situation is not uncommon as it may seem. For example, in the problem of coupled partial waves for a non-central potential, the channels with different ℓ have exactly the same thresholds.

Consider a degenerate two-channel problem where $E_1 = E_2$. The corresponding channel momenta,

$$k_1 = \pm\sqrt{\frac{2\mu_1}{\hbar^2}(E - E_1)} \qquad \text{and} \qquad k_2 = \pm\sqrt{\frac{2\mu_2}{\hbar^2}(E - E_2)}\,, \tag{13.3}$$

in most cases have the same absolute values, $|k_1| = |k_2|$, because in most cases (like coupled partial waves) $\mu_1 = \mu_2$.[3] However, we still can choose the signs in front of the square roots (13.3) for k_1 and k_2 independently. This means that formally we still have four sheets in the Riemann surface.

Looking at Fig. 13.1, we see that, when $E_2 \to E_1$, the fragment of that schematic picture between the thresholds collapses and disappears. Making a full circle around the common threshold, we find that both channel momenta change their signs at the same time. In other words, we can only have the transitions

$$(++) \leftrightarrow (--) \qquad \text{and} \qquad (+-) \leftrightarrow (-+)$$

and nothing else. None of the transitions where only one sign is changing, such as $(++) \leftrightarrow (+-)$, is possible. This means that the four sheets of the Riemann surface are decoupled in pairs, as is shown in Fig. 13.3. There is no way of getting from one pair of the sheets to the other. Since the physical states are associated with the sheets $(++)$ and $(--)$, we can ignore the pair $(+-)$, $(-+)$.

It is clear that the same reasoning can be used for any number of degenerate channels. For example, in the case of $E_1 = E_2 = E_3$, we should only consider the sheets $(+\,+\,+)$ and $(-\,-\,-)$ connected to each other. In other words, for the description of all the degenerate channels, we only need to consider just two sheets of the Riemann surface. This is true even in the most general case when the reduced masses in the channels are different. In such a case, the absolute values of the channel momenta are different, but the signs in front of the square roots (13.3) should be taken the same anyway.

[3] An example of the opposite case is the motion of an electron through a one-dimensional semiconductor nanostructure, where its effective mass may depend on the chemical composition of a particular layer of the structure. The nanostructure crystal can be considered as an "infinite reservoir" of the energy that can absorb and give it and thus can compensate the change of the mass. The reduction of a one-dimensional problem to a two-channel one is considered in Sect. 16.2.

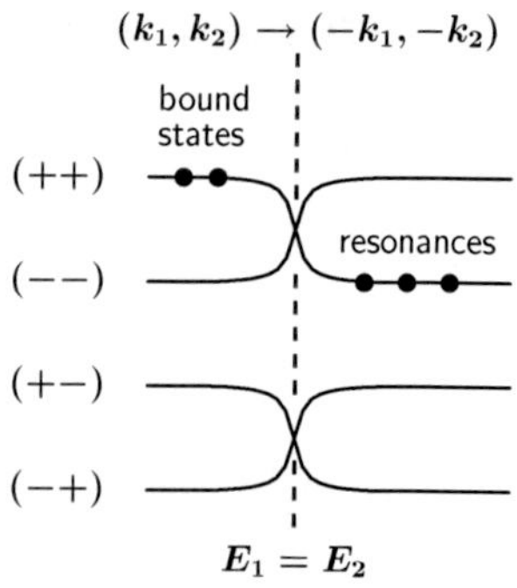

Fig. 13.3 Transitions among the Riemann energy-sheets for a degenerate two-channel problem. The dots schematically show typical distribution of the spectral points

13.3 Distribution of the Spectral Points

Concerning the distribution of the spectral points on the Riemann surface, most of the facts that were proved for the single-channel case in Sect. 3.3 are valid for the multi-channel Riemann surface as well. A rigorous mathematical analysis of such a distribution is far beyond the scope of the present book.

Instead, we can get a qualitative understanding if we look at the asymptotic behaviour of the physical wave function. At a spectral point, by definition, i.e. because of the condition (12.42), the wave function has only the outgoing waves,

$$u \underset{r\to\infty}{\longrightarrow} W^{(\text{out})} f^{(\text{out})} c \underset{r\to\infty}{\longrightarrow} -i \begin{pmatrix} e^{ik_1 r - \frac{i\ell_1\pi}{2}} & 0 & \cdots & 0 \\ 0 & e^{ik_2 r - \frac{i\ell_2\pi}{2}} & \cdots & 0 \\ \vdots & \vdots & \vdots & \vdots \\ 0 & 0 & \cdots & e^{ik_N r - \frac{i\ell_N\pi}{2}} \end{pmatrix} f^{(\text{out})} c\,. \tag{13.4}$$

Depending on the sign of $\operatorname{Im} k_n$, the corresponding component of the wave function is either exponentially growing ($\operatorname{Im} k_n < 0$) or exponentially vanishing ($\operatorname{Im} k_n > 0$). If a channel is open for a resonance to decay, then the wave function should grow (see the explanation in Sect. 3.3.2), otherwise it should vanish.

Based on this intuitive understanding, we see that the physically meaningful solutions of the Schrödinger equation can only correspond to those spectral points that are shown in Figs. 13.1 and 13.2. The bound states are on the negative part of the real axis on the physical sheet $(++)$, where the wave function exponentially vanishes at large distances in all the channels. The resonances that can decay in all the channel are on the unphysical sheet with respect to all the channels, i.e. $(--)$. The resonances that can only decay into a limited number of open channels are on the sheet which is unphysical with respect to the open channels and is physical with respect to closed channels. Of course all the resonances are not on the real axis but rather shifted down to the energies with negative imaginary parts (negative imaginary part is needed to have a probability density that diminishes with time),

$$\mathcal{E} = \mathcal{E}^{(\mathrm{r})} - \frac{i}{2}\Gamma\,.$$

In our two-channel example in Figs. 13.1 and 13.2, these are the resonances that can only decay in the first channel (because their energies $\mathcal{E}^{(\mathrm{r})} < E_2$). They are shown on the sheet $(-+)$. The corresponding wave function grows in the open channel and diminishes in the closed channel.

There is another intuitive explanation of the pattern of the spectral points shown in Fig. 13.1. As we know from the single-channel case, when the attraction of the potential becomes stronger, the resonances move towards the threshold and after passing through it become bound states. In order to have something similar in the multi-channel problem, the resonances must be on the same "curve" shown in Fig. 13.1. Otherwise they would not be able to go to an appropriate Riemann sheet after passing through a threshold. Such movements of the spectral points are demonstrated using several exactly solvable two-channel models in Ref. [15].

As will be shown further down in Sect. 13.5, there are other points on the Riemann surface where the determinant of $f^{(\mathrm{in})}$ is zero. They are the so-called "mirror" zeros, which appear due to the symmetry of the Jost matrix with respect to complex conjugation of the energy. These "mirror" zeros are however too far from the upper edge of the cut of the real axis on the physical sheet. This upper edge is where the scattering happens. Being too far away from it, the "mirror" zeros do not have any influence on the observable cross section.

We cannot prove that there are not some other zeros on the other sheets of the Riemann surface (in "wrong" places). For example, in the two-channel case, the "wrong" places are the sheets $(--)$ and $(+-)$ in the part (a) of Fig. 13.2 and the sheets $(-+)$ and $(+-)$ in the part (b). We cannot exclude the situation when, in addition to the bound-state and resonance spectral points, the multi-channel S-matrix may have poles at "wrong" locations on the Riemann surface, i.e. where the corresponding wave function asymptotically does not behave "correctly".[4] Such solutions of the Schrödinger equation have no physical meaning. However, mathematically these "wrong" poles may influence the behaviour of the S-matrix at the physical scattering energies if they are close enough to the real axis passing through the physical sheet. Such singular points are called the "shadow" poles of the S-matrix.[5] Apparently, if such a "shadow" pole has an influence on the observable quantities, it cannot be ignored, but (in contrast to the resonances) the imaginary part of the corresponding energy cannot be interpreted as the width of a decay process (because the corresponding wave function has wrong asymptotic behaviour).

[4] The existence of such "wrong" poles is explicitly demonstrated in Ref. [15] at least in one exactly solvable two-channel model.

[5] See, for example, Ref. [41].

As was already mentioned, due to certain symmetries, the Jost matrices have equal values at different points on the Riemann surface. In the single-channel case, we established such symmetries. They are summarized in Fig. 3.6. Since a single-channel problem is a particular case ($N = 1$) of a multi-channel one, we should expect that a similar (perhaps a more general) set of the symmetry relations is valid for the multi-channel problems as well. And indeed, further down in Sect. 13.5, it will be established for $N > 1$ the relations of the type given in Fig. 3.6.

13.4 Analytic Structure of the Jost Matrices

In Sect. 7.7, we showed that in the simple single-channel case the Jost function can be written as a sum of two terms as is given by Eq. (7.65), where the factor $k^{2\ell+1}$ is responsible for the branching of the Riemann E-surface. When we deal with a multi-channel problem, a similar two-term splitting of the Jost matrices is also possible. This is what we are going to consider in the current section.

* * *

According to Eqs. (7.58) and (7.59), each diagonal element of the matrices (12.15) and (12.16) can be written as the product of an integer power of a channel momentum and an entire function of the energy,

$$j = \begin{pmatrix} k_1^{\ell_1+1} & 0 & \cdots & 0 \\ 0 & k_2^{\ell_2+1} & \cdots & 0 \\ \vdots & \vdots & \vdots & \vdots \\ 0 & 0 & \vdots & k_N^{\ell_N+1} \end{pmatrix} \tilde{j}\,, \qquad n = \begin{pmatrix} k_1^{-\ell_1} & 0 & \cdots & 0 \\ 0 & k_2^{-\ell_2} & \cdots & 0 \\ \vdots & \vdots & \vdots & \vdots \\ 0 & 0 & \vdots & k_N^{-\ell_N} \end{pmatrix} \tilde{n}\,. \tag{13.5}$$

Here $\tilde{j}$ and $\tilde{n}$ denote the diagonal matrices,

$$\tilde{j} = \mathrm{diag}\left\{\tilde{j}_{\ell_1}(E,r), \tilde{j}_{\ell_2}(E,r), \ldots, \tilde{j}_{\ell_N}(E,r)\right\}\,, \tag{13.6}$$

$$\tilde{n} = \mathrm{diag}\left\{\tilde{n}_{\ell_1}(E,r), \tilde{n}_{\ell_2}(E,r), \ldots, \tilde{n}_{\ell_N}(E,r)\right\}\,, \tag{13.7}$$

with the "tilded" functions defined by Eqs. (7.58) and (7.59). In such a representation, all possible odd powers of the channel momenta are factorized. The remaining "tilded" functions are analytic single-valued functions of the energy. These factorized representations can be written in the compact form,

$$j = D\tilde{j}\,, \qquad n = KD^{-1}\tilde{n}\,, \tag{13.8}$$

where

$$D = \mathrm{diag}\left\{k_1^{\ell_1+1}, k_2^{\ell_2+1}, \ldots, k_N^{\ell_N+1}\right\}\,. \tag{13.9}$$

Let us look for the matrices A and B that are involved in the regular solution (12.19) of the multi-channel Schrödinger equation, in the form

$$A = D^{-1}\tilde{A}D\,, \qquad B = DK^{-1}\tilde{B}D\,. \tag{13.10}$$

Substituting them in the differential equations (12.20) and (12.21), we obtain

$$D^{-1}\partial_r\tilde{A}D = -D^{-1}\tilde{n}V\left(D\tilde{j}D^{-1}\tilde{A}D - KD^{-1}\tilde{n}DK^{-1}\tilde{B}D\right),$$

$$DK^{-1}\partial_r\tilde{B}D = -K^{-1}D\tilde{j}V\left(D\tilde{j}D^{-1}\tilde{A}D - KD^{-1}\tilde{n}DK^{-1}\tilde{B}D\right).$$

In order to get rid of all the momentum-dependent coefficients, the first of these equations should be multiplied from the left by D, the second by KD^{-1} also from the left, and both by D^{-1} from the right. Taking into account the fact that all the matrices K, D, $\tilde{j}$, and $\tilde{n}$ are diagonal and therefore commute with one another, we arrive at the following differential equations for the "tilded" functions

$$\partial_r\tilde{A} = -\tilde{n}V\left(\tilde{j}\tilde{A} - \tilde{n}\tilde{B}\right), \tag{13.11}$$

$$\partial_r\tilde{B} = -\tilde{j}V\left(\tilde{j}\tilde{A} - \tilde{n}\tilde{B}\right), \tag{13.12}$$

which are the matrix generalization of the corresponding single-channel equations (7.61) and (7.62). All the quantities in these equations are the $(N \times N)$-matrices. The potential V depends on the radius r, while all the other matrices are the functions of both r and E. The boundary conditions for Eqs. (13.11) and (13.12) can be obtained from the corresponding conditions (12.23) using the relations (13.8) and (13.10) and from the fact that the diagonal matrices D, K, j, $\tilde{j}$, n, and $\tilde{n}$ commute with one another. Simple matrix algebra gives

$$\tilde{j}\,\tilde{A}\,\tilde{j}^{-1}\underset{r\to 0}{\longrightarrow} I\,, \qquad \tilde{n}\,\tilde{B}\,\tilde{j}^{-1}\underset{r\to 0}{\longrightarrow} 0\,. \tag{13.13}$$

The important thing is that neither the coefficients in Eqs. (13.11) and (13.12) nor the boundary conditions (13.13) involve odd powers of the momenta. This implies that $\tilde{A}(E,r)$ and $\tilde{B}(E,r)$ are single-valued functions of E.

The boundary conditions (13.13) are energy independent (i.e. the limits in them are the same for all values of E) and the coefficients in Eqs. (13.11, 13.12) are analytic functions of E. In principle, in the same way as we did it in the single-channel case, we could use here the Poincaré theorem, which would imply that for any finite value of r the matrices $\tilde{A}(E,r)$ and $\tilde{B}(E,r)$ are analytic single-valued (holomorphic) functions of E. However, such an application of the Poincaré theorem might look sloppy. Indeed, the condition (13.13) completely determines only the matrix $\tilde{B}$, whose elements are all zero at $r = 0$,

$$\tilde{n}\,\tilde{B}\,\tilde{j}^{-1}\underset{r\to 0}{\longrightarrow} 0 \qquad \Rightarrow \qquad \tilde{B}(E,0) = 0\,, \tag{13.14}$$

while for $\tilde{A}$ this condition is imposed not on the function itself but on its product with the other functions. As was discussed in Sect. 9.3.4, in a sense, such a condition is incomplete because it does not specify the values of all the matrix elements of $\tilde{A}$ (some of which may tend to infinity at $r = 0$).

This vague point can be avoided if we prove the analyticity of the matrices $\tilde{A}(E, r)$ and $\tilde{B}(E, r)$ in a different way. Following the same steps that were used in Sect. 9.3.4.1, we can transform differential equations (13.11) and (13.12) in the equivalent integral equations:

$$\partial_r \tilde{A} = I - \int \tilde{n} V \left(\tilde{j} \tilde{A} - \tilde{n} \tilde{B} \right) dr \,, \tag{13.15}$$

$$\partial_r \tilde{B} = \quad - \int \tilde{j} V \left(\tilde{j} \tilde{A} - \tilde{n} \tilde{B} \right) dr \,, \tag{13.16}$$

where the indefinite integrals should be understood as the primitive functions. The point $r = 0$ is a singular point of Eqs. (13.11) and (13.12) because they involve the Riccati–Neumann functions. In order to establish the behaviour of the matrices $\tilde{A}(E, r)$ and $\tilde{B}(E, r)$ at small distances away from the point $r = 0$, we can solve the corresponding integral equations by iterations with the zero-order iteration

$$\tilde{A}^{(0)} = I \,, \qquad \tilde{B}^{(0)} = 0 \,.$$

Then, at a small δ, the leading terms, determining the behaviour of $\tilde{A}(E, r)$ and $\tilde{B}(E, r)$, are given by the first-order iteration:

$$\tilde{A}(E, \delta) \approx \tilde{A}^{(1)} = I - \left(\int \tilde{n} V \tilde{j} dr \right)_{r=\delta} \,, \tag{13.17}$$

$$\tilde{B}(E, \delta) \approx \tilde{B}^{(1)} = - \left(\int \tilde{j} V \tilde{j} dr \right)_{r=\delta} \,. \tag{13.18}$$

The functions under the integrals (13.17) and (13.18) are analytic and single-valued with respect to the variable E. It is clear that the integration over r produces the functions that remain analytic and single-valued with respect to E. After finding these functions, we can impose the boundary conditions (13.17, 13.18) for the differential equations (13.11, 13.12) at the point $r = \delta$ instead of $r = 0$. These new boundary conditions analytically depend on E. In this way we avoid dealing with the singular point ($r = 0$) of these differential equations. Since the point $r = \delta$ is regular, we do not need the Poincaré theorem and can prove the analyticity of $\tilde{A}(E, r)$ and $\tilde{B}(E, r)$ at any finite r as is described in Appendix G.

Therefore, the matrices $\tilde{A}(E, r)$ and $\tilde{B}(E, r)$ are holomorphic with respect to E, i.e. analytic and defined on a simple complex energy plane without any branch points. All the complications associated with the multi-valuedness and Riemann surface for the matrix-functions $A(E, r)$ and $B(E, r)$ stem from the matrices D and

K (which involve the channel momenta) in Eqs. (13.10). In the explicit form, these equations read

$$A_{mn}(E,r) = k_m^{-\ell_m-1}\tilde{A}_{mn}(E,r)k_n^{\ell_n+1}\,, \qquad B_{mn}(E,r) = k_m^{\ell_m}\tilde{B}_{mn}(E,r)k_n^{\ell_n+1}\,. \tag{13.19}$$

It should also be noted that on the real axis the matrices $\tilde{A}$ and $\tilde{B}$ are real, because for $\mathrm{Im}\,E = 0$ the differential equations (13.11) and (13.12) as well as the boundary conditions (13.13) are real.

Combining Eqs. (12.17) and (12.18), it is easy to obtain the corresponding expressions for the matrices $F^{(\mathrm{in/out})}$, namely,

$$F^{(\mathrm{in/out})} = \frac{1}{2}(A \mp iB) = \frac{1}{2}D^{-1}\left(\tilde{A} \mp iD^2K^{-1}\tilde{B}\right)D\,, \tag{13.20}$$

or in the explicit form:

$$F_{mn}^{(\mathrm{in})} = \frac{k_n^{\ell_n+1}}{2k_m^{\ell_m+1}}\tilde{A}_{mn} - \frac{ik_m^{\ell_m}k_n^{\ell_n+1}}{2}\tilde{B}_{mn}\,, \tag{13.21}$$

$$F_{mn}^{(\mathrm{out})} = \frac{k_n^{\ell_n+1}}{2k_m^{\ell_m+1}}\tilde{A}_{mn} + \frac{ik_m^{\ell_m}k_n^{\ell_n+1}}{2}\tilde{B}_{mn}\,. \tag{13.22}$$

In order to find the corresponding Jost matrices, we have to consider the limit (12.25) when $r \to \infty$,

$$f^{(\mathrm{in/out})} - D^{-1}aD \mp iDK^{-1}bD\,, \tag{13.23}$$

$$\boxed{f_{mn}^{(\mathrm{in/out})}(E) = \frac{k_n^{\ell_n+1}}{k_m^{\ell_m+1}}a_{mn}(E) \mp ik_m^{\ell_m}k_n^{\ell_n+1}b_{mn}(E)} \tag{13.24}$$

where

$$a_{mn}(E) = \frac{1}{2}\tilde{A}_{mn}(E,\infty)\,, \qquad b_{mn}(E) = \frac{1}{2}\tilde{B}_{mn}(E,\infty)\,. \tag{13.25}$$

In this connection, it should be noted that everything what was said in Sect. 7.7 about the single-channel functions $\tilde{A}_\ell(E,\infty)$ and $\tilde{B}_\ell(E,\infty)$ remains valid for the matrices $\tilde{A}(E,\infty)$ and $\tilde{B}(E,\infty)$ as well. They are analytic single-valued functions of E at least along the real positive axis within the domain $\mathcal{D}$ shown in Fig. 7.10. Using the complex rotation of the coordinate r (as is explained in Sect. 7.7.2), this domain can be made much wider. An example of a rotated domain $\mathcal{D}$ is schematically shown in Fig. 7.11. In Sect. 18.4.1, it is given an exactly solvable example of a two-channel problem, where the representation of the type (13.24) is obtained explicitly.

As a final note in this section, we should mention that Eq. (13.24) is also applicable in the single-channel problem with a non-central potential (see Sect. 9). Such a potential couples different partial waves. Formally, we can treat partial waves with different angular momenta as different channels. These channels are degenerate, i.e. their thresholds are the same, and therefore there is only one channel momentum, k, and Eq. (13.24) takes the form

$$f^{(\text{in/out})}_{[\ell][\ell']}(E) = k^{\ell'-\ell} a_{[\ell][\ell']}(E) \mp i k^{\ell+\ell'+1} b_{[\ell][\ell']}(E)\,, \tag{13.26}$$

where

$$a_{[\ell][\ell']}(E) = \frac{1}{2}\tilde{A}_{[\ell][\ell']}(E,\infty)\,, \qquad b_{[\ell][\ell']}(E) = \frac{1}{2}\tilde{B}_{[\ell][\ell']}(E,\infty)\,. \tag{13.27}$$

13.5 Symmetry Properties of the Jost Matrices

In Sect. 3.1, we have established the symmetry among the values of the single-channel Jost functions at different points of the Riemann surface. These properties were derived from the analysis of the integral equations determining the Jost functions. They are summarized in Figs. 3.5 and 3.6. When the number of channels is more than one, the Riemann surface is much more complicated and a similar type of symmetries is more difficult to establish using the corresponding integral equations. Instead, we can use the factorized representations (13.24), which are very helpful for this task.

If we change the signs of all the channel momenta to the opposite, the matrices $a(E)$ and $b(E)$ in Eq. (13.24) remain unchanged, while the factorized momenta generate a common factor $(-1)^{\ell_m+\ell_n} = (-1)^{\ell_m-\ell_n}$ and the sign between the two terms is also changed to the opposite. In other words,

$$f^{(\text{in})}_{mn}(-k_1,-k_2,\ldots,-k_N) = (-1)^{\ell_m+\ell_n} f^{(\text{out})}_{mn}(k_1,k_2,\ldots,k_N)\,, \tag{13.28}$$

where for the sake of clarity, we replaced the single independent variable E with the set of channel momenta,

$$(E) \longrightarrow (k_1,k_2,\ldots,k_N)\,.$$

The set $(k_1,k_2,\ldots,k_N)$ specifies not only the energy but also the sheet of the Riemann surface where the point E is located. Apparently, the sets $(k_1,k_2,\ldots,k_N)$ and $(-k_1,-k_2,\ldots,-k_N)$ correspond to the same energy E, but to different points on the Riemann surface. These two points are on a vertical line passing through the Riemann sheets. Eq. (13.28) implies that the two Jost matrices $f^{(\text{in})}$ and $f^{(\text{out})}$ are

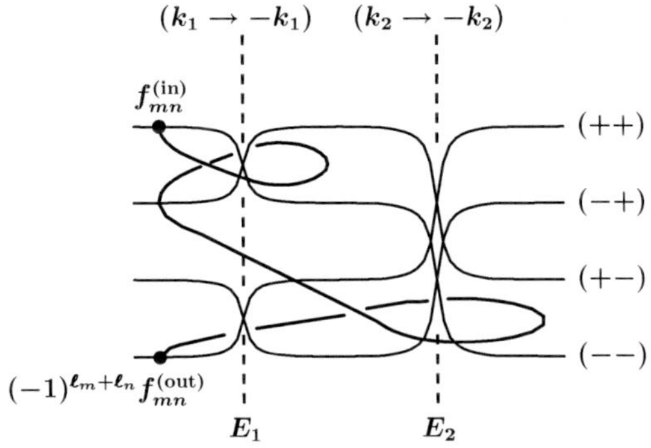

Fig. 13.4 One of the possible paths (thick looped curve) between the physical and the unphysical sheet of the Riemann surface, $(++) \leftrightarrow (--)$, that corresponds to changing the signs of both channel momenta in a two-channel problem. The quantities indicated next to the initial and final points (thick dots) on the sheets are equal to each other

not completely independent. The values of their matrix elements at different points on the Riemann surface are the same. Therefore the S matrix can be written in terms of only $f^{(\mathrm{in})}$,

$$S_{mn} = \sum_{n'} (-1)^{\ell_m+\ell_{n'}} f_{mn'}^{(\mathrm{in})}(-k_1, -k_2, \dots, -k_N) \left[f^{(\mathrm{in})}(k_1, k_2, \dots, k_N) \right]_{n'n}^{-1} . \tag{13.29}$$

When the sign of a channel momentum k_n is changed, the corresponding point E on the Riemann surface makes one full 2π-turn (circle) around the branch point at the threshold energy E_n (see Eq. (3.7)). As a result, the point E finds itself just above or below its initial location, i.e. on another sheet of the Riemann surface, that is connected to the previous sheet via the so-called kinematical cut along the real axis. Therefore when all N channel momenta change their signs, the point E makes N successive turns (one by one) around all the thresholds. This is schematically illustrated in Fig. 13.4 for the transition between the $(++)$ and $(--)$ sheets in a two-channel problem. A similar two-loop path can be drawn for the transition between the sheets $(-+)$ and $(+-)$, where Eq. (13.28) is applicable as well. Alternatively, for changing the signs of all the channel momenta, we can make one full circle around all the thresholds.

It should be noted that in general Eq. (13.28) does not imply that whole matrices $f^{(\mathrm{in})}(E)$ and $f^{(\mathrm{out})}(E)$ are equal to each other at the corresponding points on the Riemann surface. The complete equality is only possible when all the angular momenta have the same parity, i.e. all are even or all are odd. In such a case, $(-1)^{\ell_m+\ell_n} = 1$ for all matrix elements,

$$\text{if } (-1)^{\ell_m+\ell_n} = 1 \qquad \forall\, m, n\,, \quad \text{then}$$

$$\boxed{f^{(\mathrm{in/out})}(-k_1, -k_2, \dots, -k_N) = f^{(\mathrm{out/in})}(k_1, k_2, \dots, k_N)} \tag{13.30}$$

This is the case when the product of the internal parities of the interacting bodies remains the same in all the channels. Indeed, if we exclude the processes where the weak interaction plays an important role, the conserving total parity $\mathcal{P}$ for the reaction $a + b \to c + d$ is given by

$$\mathcal{P} = \mathcal{P}_a \mathcal{P}_b (-1)^{\ell_{ab}} = \mathcal{P}_c \mathcal{P}_d (-1)^{\ell_{cd}},$$

and thus if $\mathcal{P}_a \mathcal{P}_b = \mathcal{P}_c \mathcal{P}_d$, then the orbital angular momenta ℓ_{ab} and ℓ_{cd} must be of the same parity. For the obvious reason, the relation (13.30) can be called the "vertical" symmetry.

Another symmetry property can be derived using the Schwartz reflection principle [42] (see statement (B.19) and Fig. B.4 in Appendix B). It is applicable to the functions (matrices) $a(E) = \tilde{A}(E, \infty)/2$ and $b(E) = \tilde{B}(E, \infty)/2$, which are real on the real axis. This is because at real energies the differential equations (13.11) and (13.12) determining them, as well as the corresponding boundary conditions (13.13), are completely real. Therefore there are the following Schwartz symmetries with respect to the real axis:

$$a(E^*) = a^*(E)\,, \qquad b(E^*) = b^*(E)\,. \tag{13.31}$$

When the energy point E is changed to E^*, all the channel momenta are replaced with their complex conjugate partners. After doing this in Eq. (13.24), we obtain the following:

$$\begin{aligned} f_{mn}^{(\mathrm{in})}(E^*) &= \left(\frac{k_n^{\ell_n+1}}{k_m^{\ell_m+1}}\right)^* a_{mn}(E^*) - i\left(k_m^{\ell_m} k_n^{\ell_n+1}\right)^* b_{mn}(E^*) \\ &= \left(\frac{k_n^{\ell_n+1}}{k_m^{\ell_m+1}}\right)^* a_{mn}^*(E) - i\left(k_m^{\ell_m} k_n^{\ell_n+1}\right)^* b_{mn}^*(E) \\ &= \left[\frac{k_n^{\ell_n+1}}{k_m^{\ell_m+1}} a_{mn}(E) + i k_m^{\ell_m} k_n^{\ell_n+1} b_{mn}(E)\right]^* = \left[f_{mn}^{(\mathrm{out})}(E)\right]^*. \end{aligned}$$

Since this is valid for all matrix elements, we can write this property for the entire matrices,

$$\boxed{f^{(\mathrm{in/out})}(E^*) = \left[f^{(\mathrm{out/in})}(E)\right]^*} \tag{13.32}$$

This is a multi-channel generalization of the cross-axis symmetry shown in Fig. 3.6. From that Figure, it is clear that the relation (13.32) can be called the "diagonal" symmetry.

Apparently, when replacing the energy variable E with its complex conjugate partner, E^*, all the channel momenta, k_n, are automatically replaced with k_n^* as well (this is clear from Eq. (13.1)). This means that the imaginary parts of all the channel momenta change their signs. As a result, the point E^* is on a different sheet of the Riemann surface, which can symbolically be written as follows:

$$E \to E^* \quad \Rightarrow \quad (\text{sign}_1, \text{sign}_2, \ldots) \to (-\text{sign}_1, -\text{sign}_2, \ldots)\,. \tag{13.33}$$

For example, in a two-channel case, a point E on the sheet $(+-)$ after the complex conjugation moves to the sheet $(-+)$. The symmetry (13.32) is valid for each pair of the sheets having the opposite signatures as in Eq. (13.33).

We can combine the change of the signs and complex conjugation of all the channel momenta, i.e. the symmetries given by Eqs. (13.28) and (13.32). Such a combination results in the following symmetry of the elements of the Jost matrices:

$$f_{mn}^{(\text{in})}(k_1, k_2, \ldots, k_N) = (-1)^{\ell_m+\ell_n}\left[f_{mn}^{(\text{in})}(-k_1^*, -k_2^*, \ldots, -k_N^*)\right]^*\,, \tag{13.34}$$

$$f_{mn}^{(\text{out})}(k_1, k_2, \ldots, k_N) = (-1)^{\ell_m+\ell_n}\left[f_{mn}^{(\text{out})}(-k_1^*, -k_2^*, \ldots, -k_N^*)\right]^*\,. \tag{13.35}$$

This becomes an entire matrix symmetry when the parities of all the angular momenta are the same (which happens most often),

$$\boxed{f^{(\text{in/out})}(E) = \left[f^{(\text{in/out})}(e^{i2\pi}E^*)\right]^*} \qquad \text{if } (-1)^{\ell_m+\ell_n} = 1\ \forall\, m, n\,. \tag{13.36}$$

This relation can be called the "horizontal" symmetry, because it relates the points on the same sheet of the Riemann surface, located symmetrically on the opposite sides of the cut.

All three types of the symmetry relations, namely, the "vertical", "diagonal", and 'horizontal", are schematically depicted in Fig. 13.5 for a two-channel problem. The points labeled by the odd numbers are on the same vertical line passing through all the sheets. These points correspond to the same energy, E. The even numbers label the symmetric points corresponding to the complex conjugate energy, E^*. The vertical, diagonal, and horizontal dashed lines connect the points at which the Jost matrices obey the corresponding relations. For example,

$$f^{(\text{in})}(7) = \left[f^{(\text{out})}(2)\right]^*\,,$$

$$f^{(\text{in})}(7) = \left[f^{(\text{in})}(8)\right]^*\,.$$

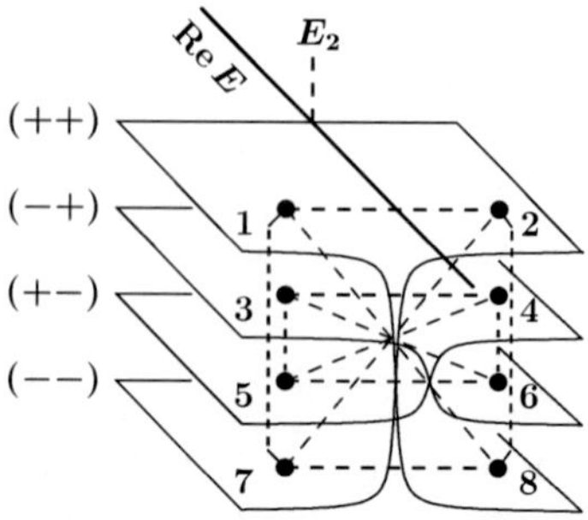

Fig. 13.5 Fragment of the Riemann surface shown in Fig. 13.2b. The points with the odd numbers correspond to the same energy E, while the points with the even numbers correspond to the complex conjugate energy E^*. The dashed lines connect the points at which the values of the Jost matrices are related according to Eqs. (13.30), (13.32), and (13.36)

It is easy to find the corresponding relations for the S-matrix; for example,

$$S(2) = f^{(\text{out})}(2)\left[f^{(\text{in})}(2)\right]^{-1} = f^{(\text{in})*}(7)\left[f^{(\text{out})*}(7)\right]^{-1}$$

$$= \left\{f^{(\text{out})*}(7)\left[f^{(\text{in})*}(7)\right]^{-1}\right\}^{-1} = \left[S^*(7)\right]^{-1} .$$

Therefore if the point 7 is a resonance, where the S-matrix is singular, then the "diagonal" point 2 is a zero of the S-matrix determinant.

In the single-channel case, the symmetry of the "horizontal" type results in the existence of the so-called "mirror" zeros of the Jost function (or equivalently, the "mirror" poles of the S-matrix). Indeed, if $f_\ell^{(\text{in})}(E) = 0$, then $f_\ell^{(\text{in})}(e^{i2\pi}E^*)$ is also zero simply because $\left[f_\ell^{(\text{in})}(e^{i2\pi}E^*)\right]^* = f_\ell^{(\text{in})}(E)$. Schematically, the single-channel "mirror" zeros on the k-plane are shown in Fig. 3.13, and the corresponding distribution of them on the unphysical sheet of the Riemann surface is depicted in Fig. 3.14.

In a multi-channel problem, the "mirror" poles are only possible if $(-1)^{\ell_m+\ell_n} = 1$ for all m and n. Actually, they are the zeros of the "(in)" Jost matrix determinant. In the two-channel case, they can be between the thresholds E_1 and E_2 on the sheet $(-+)$ on the other side of the real axis, and for possible resonances above the threshold E_2, they are on the sheet $(--)$ also on the opposite side of the real axis. For example, if the point 7 in Fig. 13.5 is a resonance, then point 8 is its "mirror" partner.

In order to reach the real axis of the physical sheet $(++)$ from such a "mirror" resonance, we have to go around two branch points. This means that it is far away from the scattering energies and thus its influence on the observable quantities is very weak, if any.

It should be noted the following. The "vertical" and "diagonal" symmetries are derived independently of each other, while the "horizontal" symmetry results from their combination. When considering the problems with the Coulomb forces, we found (see Sect. 8.2.10) that the "vertical" symmetry is always broken by the Coulomb tail of the potential. As a result, the "horizontal" symmetry is broken as well. This means that there are no "mirror" resonance poles of the S-matrix when the particles have non-zero electric charges. This is considered in more detail in Sect. 14.9.

Multi-channel Problems of Charged Particles 14

In the majority of real experiments, we deal with charged particles, and thus the Coulomb tail of the potential is present at least in one of the channels.[1] By adding and subtracting the pure Coulomb potential with the same tail, it is always possible to split the potential into a short-range and the pure Coulomb terms, as is done in Eq. (8.2).

As it follows from Eq. (11.26), a long-range tail can only be present in the diagonal elements of the potential. This means that in order to generalize the multi-channel radial Schrödinger equation (11.70), we only need to add the Coulomb potential in its left hand side (which is diagonal),

$$\left[\partial_r^2 + k_n^2 - \frac{\ell_n(\ell_n+1)}{r^2} - \frac{2k_n\eta_n}{r}\right] u_n^J(E,r) = \sum_{n'} V_{nn'}^J(r) u_{n'}^J(E,r)\,, \tag{14.1}$$

where η_n is the Sommerfeld parameter for the channel n (for its definition, see Eq. (8.6)). The radial part, $u_n^J(E,r)$, of the wave function and the short-range part, $V_{nn'}^J(r)$, of the potential are defined here in exactly the same way as for neutral particles (see Sect. 11.2.2). Actually, most of the derivations given so far for the short-range multi-channel interactions are still valid here (with an appropriate replacement of the Riccati–Bessel–Neumann–Hankel functions by the corresponding Coulomb functions). We therefore skip some detailed explanations that can be found in the preceding sections of the book.

[1] It should be mentioned once more that we only consider the binary channels, i.e. the channels involving two interacting quantum objects.

S. A. Rakityansky, *Jost Functions in Quantum Mechanics*,
https://doi.org/10.1007/978-3-031-07761-6_14

14.1 Jost Matrix

Equation (14.1) has $2N$ linearly independent column-solutions, and only half of them can be regular at $r = 0$ (because it has a regular singular point at $r = 0$). Combining the regular columns (which we denote by the symbol φ) in a square matrix (where the second subscript labels different independent solutions),

$$\varphi(E,r) = \begin{pmatrix} \varphi_{11} \\ \varphi_{21} \\ \vdots \\ \varphi_{N1} \end{pmatrix} \oplus \begin{pmatrix} \varphi_{12} \\ \varphi_{22} \\ \vdots \\ \varphi_{N2} \end{pmatrix} \oplus \cdots \oplus \begin{pmatrix} \varphi_{1N} \\ \varphi_{2N} \\ \vdots \\ \varphi_{NN} \end{pmatrix} = \begin{pmatrix} \varphi_{11} & \varphi_{12} & \cdots & \varphi_{1N} \\ \varphi_{21} & \varphi_{22} & \cdots & \varphi_{2N} \\ \vdots & \vdots & \cdots & \vdots \\ \varphi_{N1} & \varphi_{N2} & \cdots & \varphi_{NN} \end{pmatrix}, \tag{14.2}$$

we obtain the so-called fundamental matrix of regular solutions (the basis). Any physical solution is a linear combination of the columns (14.2).

Since the potential $V^J_{nn'}(r)$ on the right hand side of Eq. (14.1) is of a short range, at large distances it vanishes, and the "free" equation takes the following decoupled form:

$$\left[\partial_r^2 + k_n^2 - \frac{\ell_n(\ell_n+1)}{r^2} - \frac{2k_n\eta_n}{r}\right] u^J_n(E,r) = 0\,, \qquad r \to \infty\,. \tag{14.3}$$

Two linearly independent solutions of this equation are well known. They are described in Appendix E as well as in Sect. 8.1.2. These linearly independent solutions can be any two of the four functions: $F_{\ell_n}(\eta_n, k_n r)$, $G_{\ell_n}(\eta_n, k_n r)$, $H^{(+)}_{\ell_n}(\eta_n, k_n r)$, and $H^{(-)}_{\ell_n}(\eta_n, k_n r)$. They generalize the corresponding Riccati–Bessel, Riccati–Neumann, and Riccati–Hankel functions,[2] $j_{\ell_n}(k_n r)$, $n_{\ell_n}(k_n r)$, $h^{(+)}_{\ell_n}(k_n r)$, and $h^{(+)}_{\ell_n}(k_n r)$, that are used in the absence of the Coulomb forces.

Despite being decoupled, the system (14.3) formally consists of N differential equations ($n = 1, 2, \ldots, N$) of the second order. A solution of this system is a column $(u_1, u_2, \ldots, u_N)^{\mathrm{T}}$, each entry of which describes the motion in an individual channel (we drop the superscript J for the sake of notation simplicity). There are $2N$ linearly independent solutions of this kind. These $2N$ independent columns form the basis, such that any other solution is their linear combination. It should be noted that, in contrast to Eq. (14.2), this basis includes all the $2N$ columns. The difference stems from the fact that at large distances (where Eq. (14.3) is valid) there is no requirement of regularity, and therefore we cannot exclude any of the columns.

[2] Please note the minus sign in Eq. (8.10).

The choice of the $(2N)$-basis is rather arbitrary. There are two convenient choices. One of them is the set of the columns of the following two diagonal matrices:

$$F(E,r) = \mathrm{diag}\left\{F_{\ell_1}(\eta_1,k_1r),\, F_{\ell_2}(\eta_2,k_2r),\,\dots,\, F_{\ell_N}(\eta_N,k_Nr)\right\}\ , \tag{14.4}$$

$$G(E,r) = \mathrm{diag}\left\{G_{\ell_1}(\eta_1,k_1r),\, G_{\ell_2}(\eta_2,k_2r),\,\dots,\, G_{\ell_N}(\eta_N,k_Nr)\right\}\ . \tag{14.5}$$

The other convenient choice is the set of the columns of the diagonal matrices $H^{(+)}$ and $H^{(-)}$ that are defined as

$$H^{(\pm)}(E,r) = F(E,r) \mp iG(E,r) \tag{14.6}$$

and represent the Coulomb spherical waves.

Since the set of $2N$ columns of the square matrices $F(E,r)$ and $G(E,r)$ constitutes a complete basis in the space of the solutions at large distances, the regular solutions behave at these distances as their linear combinations,

$$\varphi(E,r) \underset{r\to\infty}{\longrightarrow} F(E,r)\mathbb{A} + G(E,r)\mathbb{B}\ , \tag{14.7}$$

where $\mathbb{A}$ and $\mathbb{B}$ are square matrices composed of the combination constants. This asymptotic behaviour gives us a hint of the most convenient way of finding the regular solutions. We can look for them in the form of (14.7) at all distances, but of course with some r-dependent matrices instead of the constants $\mathbb{A}$ and $\mathbb{B}$.

Therefore, similarly to Eq. (8.48), we look for the fundamental matrix (14.2) in the form

$$\varphi(E,r) = F(E,r)A(E,r) + G(E,r)B(E,r) \tag{14.8}$$

with the Lagrange condition

$$F(E,r)\partial_r A(E,r) + G(E,r)\partial_r B(E,r) = 0 \tag{14.9}$$

and, substituting it into the Schrödinger equation (14.1), obtain the following differential equations for the unknown square matrices A and B (the derivations go via the same steps as in Sect. 12.1):

$$\partial_r A = \quad K^{-1}GV\,(FA + GB)\ , \tag{14.10}$$

$$\partial_r B = -K^{-1}FV\,(FA + GB)\ , \tag{14.11}$$

where K is the diagonal matrix of the channel momenta,

$$K = \mathrm{diag}\,\{k_1, k_2, \dots, k_N\}\ , \tag{14.12}$$

and V is the potential matrix that appears on the right hand side of Eq. (14.1). Using the same reasoning as in Sect. 9.3.4, it is not difficult to show that the regularity of $\varphi(E,r)$ at $r=0$ and linear independence of its columns can be achieved with the following boundary conditions:

$$\lim_{r\to 0}\frac{F_{\ell_n}(\eta_n,k_nr)A_{nn'}(E,r)}{F_{\ell_{n'}}(\eta_{n'},k_{n'}r)}=\delta_{nn'}\,,\qquad \lim_{r\to 0}\frac{G_{\ell_n}(\eta_n,k_nr)B_{nn'}(E,r)}{F_{\ell_{n'}}(\eta_{n'},k_{n'}r)}=0\,,\tag{14.13}$$

which can also be written in matrix form as

$$\lim_{r\to 0} FAF^{-1}=I\,,\qquad \lim_{r\to 0} GBF^{-1}=0\,.\tag{14.14}$$

Since the ratio $G_{\ell_n}(\eta_n,k_nr)/F_{\ell_{n'}}(\eta_{n'},k_{n'}r)$ is singular at $r=0$ for any n and n', we conclude that all elements of the matrix B vanish at the origin, i.e.

$$\lim_{r\to 0} GBF^{-1}=0\qquad\Rightarrow\qquad B(E,0)=0\,.\tag{14.15}$$

The conditions (14.14) generalize the corresponding conditions (12.22, 12.23) for the multi-channel problem of neutral particles.

With the conditions (14.14), the fundamental matrix obeys the condition

$$\lim_{r\to 0}\varphi F^{-1}=I\tag{14.16}$$

and therefore behaves near the origin as

$$\begin{aligned}\varphi(E,r)&\underset{r\to 0}{\longrightarrow} F(E,r)\\&\underset{r\to 0}{\longrightarrow} CL^{-1}\operatorname{diag}\left\{(k_1r)^{\ell_1+1},(k_2r)^{\ell_2+1},\dots,(k_Nr)^{\ell_N+1}\right\}\,,\end{aligned}\tag{14.17}$$

where C and L are the following diagonal matrices:

$$C=\operatorname{diag}\left\{C_{\ell_1}(\eta_1),C_{\ell_2}(\eta_2),\dots,C_{\ell_N}(\eta_N)\right\}\,,\tag{14.18}$$

$$L=\operatorname{diag}\left\{(2\ell_1+1)!!,(2\ell_2+1)!!,\dots,(2\ell_N+1)!!\right\}\,.\tag{14.19}$$

Similarly to the single-channel case (see Eq. (8.66)), we consider another fundamental matrix of regular solutions, which differs from φ by a constant factor,

$$\phi(E,r)\underset{r\to 0}{\longrightarrow}\operatorname{diag}\left\{j_{\ell_1}(k_1r),j_{\ell_2}(k_2r),\dots,j_{\ell_N}(k_Nr)\right\}\,.\tag{14.20}$$

Matrix ϕ is the one, which is used to define the Jost matrices. The reason is that these Jost matrices describe the total (combined) effect of both the short-range and

Coulomb potentials, and therefore this effect should be viewed as the "distortion" of the completely "free" solution (which is the diagonal matrix composed of the Riccat–Bessel functions).

To find out what is the constant factor relating the matrices $\varphi(E, r)$ and $\phi(E, r)$, we recall Eqs. (8.20) and (8.22), from which it follows that

$$F \underset{r\to 0}{\longrightarrow} jC \,. \tag{14.21}$$

Substituting this in Eq. (14.16), we obtain the boundary conditions

$$\lim_{r\to 0} \varphi C^{-1} j^{-1} = I \,. \tag{14.22}$$

Comparing this with the boundary conditions for the regular solution ϕ (see Eq. (12.4)), namely, $\lim_{r\to 0} \phi j^{-1} = I$, we find that

$$\phi = \varphi C^{-1} \,. \tag{14.23}$$

Each column of ϕ is a solution of the original system of differential equations (14.1). Therefore, at large distances, it can be written as a linear combination of $2N$ columns of the basis, i.e. the columns of the matrices $H^{(-)}$ and $H^{(+)}$. The combination coefficients can be combined in two square matrices $f^{(\mathrm{in})}$ and $f^{(\mathrm{out})}$, which gives

$$\phi(E, r) \underset{r\to\infty}{\longrightarrow} H^{(-)}(E, r)\sigma^{(+)} f^{(\mathrm{in})}(E) + H^{(+)}(E, r)\sigma^{(-)} f^{(\mathrm{out})}(E) \,, \tag{14.24}$$

with

$$\upsilon^{(+)} = \mathrm{diag}\left\{e^{\pm i\sigma_{\ell_1}}, e^{\pm i\sigma_{\ell_2}}, \ldots, e^{\pm i\sigma_{\ell_N}}\right\} \,. \tag{14.25}$$

These $f^{(\mathrm{in/out})}$-matrices are called the Jost matrices. Each element of these matrices has two subscripts: one to show which column of ϕ is expanded and the other to indicate which column of the basis is multiplied and added to the combination. On the other hand, the elements of the Jost matrices can be viewed as the amplitudes of the incoming and outgoing spherical waves for each column of ϕ in each channel. These spherical waves are "distorted" by the logarithmic phase shifts (due to the Coulomb forces), but the pure Coulomb phase shifts, σ_ℓ, are not included in them. This is why in Eq. (14.24) defining the Jost matrices, there are the factors $\sigma^{(\pm)}$ which remove σ_ℓ from $H^{(\mp)}$. For more detailed explanations, the reader should go back to Sect. 8.1.2.

By analogy with Eq. (8.55), we introduce the linear combinations of the unknown matrices A and B,

$$\mathcal{F}^{(\mathrm{in/out})}(E, r) = \frac{1}{2}\left[A(E, r) \mp i B(E, r)\right] \,, \tag{14.26}$$

which gives

$$\begin{aligned} u(E,r) &= H^{(-)}(E,r)\mathcal{F}^{(\text{in})}(E,r) + H^{(+)}(R,r)\mathcal{F}^{(\text{out})}(E,r) \\ &\underset{r\to\infty}{\longrightarrow} H^{(-)}(E,r)\mathcal{F}^{(\text{in})}(E,\infty) + H^{(+)}(E,r)\mathcal{F}^{(\text{out})}(E,\infty), \end{aligned} \tag{14.27}$$

and therefore

$$\boxed{f^{(\text{in/out})}(E) = \lim_{r\to\infty} \sigma^{(\mp)}\mathcal{F}^{(\text{in/out})}(E,r)C^{-1}} \tag{14.28}$$

which generalizes Eq. (8.69).

Similarly to the single-channel case (see Eq. (8.71)), we can factorize the total Jost matrices (14.28) in the pure Coulomb and the Coulomb-modified short-range ones,

$$f^{(\text{in/out})} = f^{\text{c (in/out)}} f^{\text{sc (in/out)}}, \tag{14.29}$$

where

$$f^{\text{c (in/out)}} = \frac{1}{2}\sigma^{(\mp)}C^{-1} \tag{14.30}$$

and

$$f^{\text{sc (in/out)}} = 2C \lim_{r\to\infty} \mathcal{F}^{(\text{in/out})}(E,r)C^{-1}. \tag{14.31}$$

The same as in the single-channel case, the matrices $f^{\text{sc (in/out)}}$ are not pure short-range counterparts of the pure Coulomb ones. The reasoning given after Eq. (8.71) is fully applicable here as well. Using the representations (14.29) of the Jost matrices, the S-matrix can also be factorized,

$$S = \sigma^{(+)}C^{-1}S^{\text{sc}}C\sigma^{(+)}, \tag{14.32}$$

where

$$S^{\text{sc}} = f^{\text{sc (out)}}\left[f^{\text{sc (in)}}\right]^{-1}. \tag{14.33}$$

As a final note for this section, it is worthwhile to mention that the combinations of the form (14.26) transform differential equations (14.10), (14.11) into another form:

$$\partial_r \mathcal{F}^{(\text{in})} = -\frac{1}{2i}K^{-1}H^{(+)}V\left[H^{(-)}\mathcal{F}^{(\text{in})} + H^{(+)}\mathcal{F}^{(\text{out})}\right], \tag{14.34}$$

$$\partial_r \mathcal{F}^{(\mathrm{out})} = \frac{1}{2i} K^{-1} H^{(-)} V \left[H^{(-)} \mathcal{F}^{(\mathrm{in})} + H^{(+)} \mathcal{F}^{(\mathrm{out})} \right] , \tag{14.35}$$

which generalize the corresponding single-channel equations (8.57), (8.58). The boundary conditions for these equations follow from the conditions (14.14) and Eq. (14.26), which give

$$F \mathcal{F}^{(\mathrm{in/out})} F^{-1} = \frac{1}{2} (F A F^{-1} \mp i F G^{-1} G B F^{-1}) .$$

And since FG^{-1} and GBF^{-1} vanish at $r = 0$, we obtain

$$\lim_{r \to 0} F \mathcal{F}^{(\mathrm{in/out})} F^{-1} = \frac{1}{2} I . \tag{14.36}$$

By formal integration, both sets of differential equations, (14.10), (14.11) and (14.34), (14.35), are transformed into the corresponding Volterra-type integral equations:

$$A = I + K^{-1} \int_0^r G V (F A + G B) \, dr' , \tag{14.37}$$

$$B = \ - K^{-1} \int_0^r F V (F A + G B) \, dr' , \tag{14.38}$$

and

$$\mathcal{F}^{(\mathrm{in})} = \frac{I}{2} - \frac{K^{-1}}{2i} \int_0^r H^{(+)} V \left[H^{(-)} \mathcal{F}^{(\mathrm{in})} + H^{(+)} \mathcal{F}^{(\mathrm{out})} \right] dr' , \tag{14.39}$$

$$\mathcal{F}^{(\mathrm{out})} = \frac{I}{2} + \frac{K^{-1}}{2i} \int_0^r H^{(-)} V \left[H^{(-)} \mathcal{F}^{(\mathrm{in})} + H^{(+)} \mathcal{F}^{(\mathrm{out})} \right] dr' . \tag{14.40}$$

14.2 Jost Solutions

In the presence of the Coulomb forces, the multi-channel Jost solutions, $\chi^{(\pm)}(E, r)$, are defined by the following boundary conditions at infinity:

$$\chi^{(\pm)}(E, r) \underset{r \to \infty}{\longrightarrow} H^{(\pm)}(E, r) \sigma^{(\mp)} . \tag{14.41}$$

Comparing these definition with the asymptotic behaviour of the fundamental matrix of the regular solutions (14.24), we obtain the relation

$$\boxed{\phi(E, r) = \chi^{(-)}(E, r) f^{(\mathrm{in})}(E) + \chi^{(+)}(E, r) f^{(\mathrm{out})}(E)} \tag{14.42}$$

which is formally the same as for neutral particles (see Eq. (12.40)). Using the following known Wronskian (see Appendix E),

$$W\left[H^{(-)}(E,r), H^{(+)}(E,r)\right] = 2iK\ ,$$

we find that

$$\boxed{f^{(\text{in/out})}(E) = \mp\frac{1}{2i}K^{-1}W\left[\chi^{(\pm)}(E,r), \phi(E,r)\right]} \tag{14.43}$$

Similarly to the regular solution (14.27), we can look for the Jost solutions in the form:

$$\chi^{(\pm)}(E,r) = H^{(-)}(E,r)\mathfrak{X}^{(\pm\text{in})}(E,r) + H^{(+)}(E,r)\mathfrak{X}^{(\pm\text{out})}(E,r) \tag{14.44}$$

with the Lagrange condition

$$H^{(-)}(E,r)\partial_r\mathfrak{X}^{(\pm\text{in})}(E,r) + H^{(+)}(E,r)\partial_r\mathfrak{X}^{(\pm\text{out})}(E,r) = 0\ . \tag{14.45}$$

Then the unknown matrix-functions $\mathfrak{X}^{(\pm\text{in}/out)}(E,r)$ can be found by solving the differential equations

$$\partial_r\mathfrak{X}^{(\pm\text{in})} = -\frac{1}{2i}K^{-1}H^{(+)}V\left[H^{(-)}\mathfrak{X}^{(\pm\text{in})} + H^{(+)}\mathfrak{X}^{(\pm\text{out})}\right]\ , \tag{14.46}$$

$$\partial_r\mathfrak{X}^{(\pm\text{out})} = \frac{1}{2i}K^{-1}H^{(-)}V\left[H^{(-)}\mathfrak{X}^{(\pm\text{in})} + H^{(+)}\mathfrak{X}^{(\pm\text{out})}\right]\ , \tag{14.47}$$

with the boundary conditions

$$\mathfrak{X}^{(+\text{in})}(E,r) \underset{r\to\infty}{\longrightarrow} 0\ , \qquad \mathfrak{X}^{(+\text{out})}(E,r) \underset{r\to\infty}{\longrightarrow} \sigma^{(-)}\ ,$$

$$\mathfrak{X}^{(-\text{in})}(E,r) \underset{r\to\infty}{\longrightarrow} \sigma^{(+)}\ , \qquad \mathfrak{X}^{(-\text{out})}(E,r) \underset{r\to\infty}{\longrightarrow} 0\ .$$

14.3 Complex Rotation

When considering the single-channel problem with a short-range plus Coulomb potential, it was shown in Sect. 8.2.5 that the presence of a Coulomb tail does not change the character of the asymptotic behaviour (vanishing or diverging) of the wave function at complex values of the momentum k. This is because between the two terms, $kr - \eta\ln(2kr)$, the linear term always dominates over the logarithmic one.

This implies that the analysis done in Sect. 12.2 for a multi-channel problem with a short-range interaction is completely applicable when a Coulomb tail is present. In other words, in order to calculate the Jost matrices in the domain of resonances, we have to rotate the coordinate, $r = |r|\exp(i\theta)$, to an appropriate angle θ such that opens up the corresponding domain as is schematically shown in Fig. 12.1.

14.4 Simplicity of the Spectral Point Zeros and the ANC

In the definition of the asymptotic normalization constants, $\mathcal{A}_n$, for the problem of the charged particles, we include the logarithmic phase in the asymptotic behaviour of the normalized wave function, u, describing a bound or a resonant state at $E = E_0$ (compare this with Eqs. (12.67) and (8.128)):

$$u(E_0, r) = \begin{bmatrix} u_1(E_0, r) \\ u_2(E_0, r) \\ \vdots \\ u_N(E_0, r) \end{bmatrix} \underset{r\to\infty}{\longrightarrow} \begin{pmatrix} \mathcal{A}_1 \exp[ik_{01}r - i\eta\ln(2k_{01}r)] \\ \mathcal{A}_2 \exp[ik_{02}r - i\eta\ln(2k_{02}r)] \\ \vdots \\ \mathcal{A}_N \exp[ik_{0N}r - i\eta\ln(2k_{0N}r)] \end{pmatrix} . \tag{14.48}$$

The Jost solutions for the case of charged particles are defined in such a way (see Sect. 14.2) that their relation to the Jost matrices via the Wronskian with the regular solution (see Eq. (14.43)) is exactly the same as for the neutral particles. We can therefore repeat all the derivations of Sect. 12.4.3 and prove that all the poles of the S-matrix, corresponding to the bound and resonant states, are simple, i.e. of the first order. The derivations are exactly the same and there is no need to give them here again.

And without any change in the derivations, we can obtain that the residues of the S-matrix at the spectral points are related to the asymptotic normalization constants in exactly the same way as for the neutral particles:

$$\boxed{\mathrm{Res}\,(S_{n'n}, E_0) = -\frac{\hbar\langle v\rangle i^{\ell'+\ell+1}}{N}\mathcal{A}_{n'}\mathcal{A}_n} \qquad \eta \neq 0\,, \tag{14.49}$$

where $\langle v\rangle$ is the average relative velocity in the N-channel state,

$$\langle v\rangle = \int_0^\infty \left(\frac{\hbar k_{01}}{\mu_1}u_1^2 + \frac{\hbar k_{02}}{\mu_2}u_2^2 + \cdots + \frac{\hbar k_{0N}}{\mu_N}u_N^2\right) dr\,, \tag{14.50}$$

and N is the total number of coupled channels (including the degenerate ones). In the case of a resonance, the integration here should be done along the complex path shown in Fig. 3.9.

14.5 Scattering Observables

When doing measurements of the scattering cross sections, we always put the equipment at macroscopic distances which are practically infinite as compared to the typical size of the quantum objects. Therefore, in the derivations of the formulae that relate the Jost matrices with the cross sections or with the partial decay widths of the resonant states, we take the particle fluxes at $r \to \infty$.

The troublesome feature of the radial part of the wave function that describes a system of charged particles is the ever-growing logarithmic phase,

$$u_n(E,r) \underset{r\to\infty}{\sim} e^{ik_nr - i\eta\ln(2k_nr)} ,$$

which hinders most of the derivations. It can be suggested at least two ways out of such a difficulty:

- When calculating the particle fluxes, we need the radial derivatives of the type

$$\frac{d}{dr}e^{ik_nr-i\eta\ln(2k_nr)} = ik_n\left(1-\frac{\eta}{k_nr}\right)e^{ik_nr-i\eta\ln(2k_nr)} ,$$

  where we can simply assume that the second term at the macroscopic distances vanishes[3],

$$1-\frac{\eta}{k_nr} \underset{r\to\infty}{\longrightarrow} 1 .$$

- Alternatively, we can replace the Coulomb potential with either a screened potential,

$$\frac{e^2Z_1Z_2}{r} \longrightarrow \frac{e^2Z_1Z_2}{r}\exp(-\epsilon r) ,$$

  or a potential that is cut off at a large radius R. In both cases, the formulae relating the Jost matrices and the observable quantities are the same as for a short-range interaction. The important fact is that none of these formulae explicitly depends on either ϵ or R. Therefore if we let $\epsilon \to 0$ or $R \to \infty$, these formulae remain the same, but the Jost matrices involved in them become the correct ones that take into account the Coulomb forces.[4]

This means that all the difficulties associated with the Coulomb forces are only giving troubles when we calculate the Jost matrices, but after that all the formulae that relate the Jost matrices to the observables are the same as for the neutral particles.

---

[3] See, for example, Sec. 1.3.1 of the book [43].

[4] For more details, see Refs. [4, 44].

14.6 Analytic Structure of the Jost Matrices

In Sect. 8.2.6, we use the regular and irregular Coulomb functions written in a special factorized form given by Eqs. (8.87), (8.88, (8.89), (8.90), where all the factors causing the square root and the logarithmic branching are given explicitly. Using the same representation of the Coulomb functions, we can factorize the diagonal Coulomb matrices (14.4), (14.5) as follows:

$$F = D\tilde{F}\,, \tag{14.51}$$

$$G = MD\tilde{F} + KD^{-1}\tilde{G}\,, \tag{14.52}$$

where

$$D = \text{diag}\left\{C_{\ell_1}(\eta_1)k_1^{\ell_1+1}, C_{\ell_2}(\eta_2)k_2^{\ell_2+1}, \ldots, C_{\ell_N}(\eta_N)k_N^{\ell_N+1}\right\}\,, \tag{14.53}$$

$$M = \text{diag}\left\{\frac{2\eta_1 h(\eta_1)}{C_0^2(\eta_1)}, \frac{2\eta_2 h(\eta_2)}{C_0^2(\eta_2)}, \ldots, \frac{2\eta_N h(\eta_N)}{C_0^2(\eta_N)}\right\}\,, \tag{14.54}$$

and the diagonal matrices $\tilde{F}(E,r)$ and $\tilde{G}(E,r)$ are holomorphic functions of E. It should be noted that in further derivations, we use the fact that the above auxiliary matrices (D, M, $\tilde{F}$, and $\tilde{G}$) are diagonal and therefore commute with each other. Using simple matrix algebra, the transformations (14.51), (14.52) can be inverted,

$$\tilde{F} = D^{-1}F\,, \tag{14.55}$$

$$\tilde{G} = -K^{-1}MDF + K^{-1}DG\,. \tag{14.56}$$

Let us define the following combinations of the unknown matrices A and B:

$$\tilde{A} = DAD^{-1} + MDBD^{-1}\,, \tag{14.57}$$

$$\tilde{B} = KD^{-1}BD^{-1}\,, \tag{14.58}$$

which generalizes the corresponding single-channel transformation (8.92, 8.93). Substituting the factorized representations (14.51) and (14.52) in Eq. (14.8) and using the fact that the diagonal matrices commute with each other, we obtain

$$\begin{aligned}\varphi &= FA + GB = D\tilde{F}A + MD\tilde{F}B + KD^{-1}\tilde{G}B \\ &= \left(D\tilde{F}AD^{-1} + MD\tilde{F}BD^{-1} + KD^{-1}\tilde{G}BD^{-1}\right)D\end{aligned}$$

$$= \left(\tilde{F} D A D^{-1} + \tilde{F} M D B D^{-1} + \tilde{G} K D^{-1} B D^{-1}\right) D$$

$$= \left(\tilde{F}\tilde{A} + \tilde{G}\tilde{B}\right) D .$$

Therefore, Eqs. (14.10), (14.11) can be written as

$$\partial_r A = \quad K^{-1} G V (\tilde{F}\tilde{A} + \tilde{G}\tilde{B}) D , \tag{14.59}$$

$$\partial_r B = -K^{-1} F V (\tilde{F}\tilde{A} + \tilde{G}\tilde{B}) D . \tag{14.60}$$

It is easy to obtain the transformations inverse to (14.57), (14.58),

$$A = D^{-1}\tilde{A} D - M K^{-1} D \tilde{B} D , \tag{14.61}$$

$$B = K^{-1} D \tilde{B} D . \tag{14.62}$$

Substituting them on the left hand sides of Eqs. (14.59), (14.60), we have

$$D^{-1}\partial_r \tilde{A} - M K^{-1} D \partial_r \tilde{B} = \quad K^{-1} G V (\tilde{F}\tilde{A} + \tilde{G}\tilde{B}) , \tag{14.63}$$

$$K^{-1} D \partial_r \tilde{B} = -K^{-1} F V (\tilde{F}\tilde{A} + \tilde{G}\tilde{B}) . \tag{14.64}$$

Multiplying the second of these equations by M from the left and adding to the first equation, we obtain

$$\partial_r \tilde{A} = D (K^{-1} G - M K^{-1} F) V (\tilde{F}\tilde{A} + \tilde{G}\tilde{B}) ,$$

$$\partial_r \tilde{B} = -D^{-1} F V (\tilde{F}\tilde{A} + \tilde{G}\tilde{B}) .$$

Finally, Eqs. (14.55), (14.56) allow us to get rid of all the "troublesome" (non-holomorphic) factors in the differential equations for the "tilded" functions:

$$\partial_r \tilde{A}(E,r) = \quad \tilde{G}(E,r) V(r) \left[\tilde{F}(E,r)\tilde{A}(E,r) + \tilde{G}(E,r)\tilde{B}(E,r)\right] , \tag{14.65}$$

$$\partial_r \tilde{B}(E,r) = -\tilde{F}(E,r) V(r) \left[\tilde{F}(E,r)\tilde{A}(E,r) + \tilde{G}(E,r)\tilde{B}(E,r)\right] . \tag{14.66}$$

This set of equations is the matrix generalization of the single-channel equations (8.96), (8.97). The boundary conditions for them can be derived from the corresponding conditions (14.14) as follows. Using the fact that the diagonal matrices, M, D, K, F, and G commute with one another, we obtain

$$\tilde{F}\tilde{A}\tilde{F}^{-1} = F D^{-1} (D A D^{-1} + M D B D^{-1}) D F^{-1} = F A F^{-1} + M F G^{-1} G B F^{-1}$$

and similarly

$$\tilde{G}\tilde{B}\tilde{F}^{-1} = (K^{-1}DG - K^{-1}MDF)\tilde{B}DF^{-1} = GBF^{-1} + MFG^{-1}GBF^{-1} .$$

Then, since both FG^{-1} and GBF^{-1} vanish at $r = 0$, we come to the following boundary conditions:

$$\begin{aligned} &\lim_{r\to 0} \tilde{F}\tilde{A}\tilde{F}^{-1} = I , \\ &\lim_{r\to 0} \tilde{G}\tilde{B}\tilde{F}^{-1} = 0 \qquad \Rightarrow \qquad \tilde{B}(E,0) = 0 , \end{aligned} \tag{14.67}$$

which remain simple and energy independent. It is not difficult to see that $\tilde{A}(E,\infty)$ and $\tilde{B}(E,\infty)$ are single-valued functions of the energy variable and that their analyticity with respect to E can be proven in exactly the same way as it is done in Sec. 13.4 for the multi-channel problem with short-range potentials. The domain $\mathcal{D}$ where their asymptotic values $\tilde{A}(E,\infty)$ and $\tilde{B}(E,\infty)$ are analytic is also the same.

Using Eqs. (14.28) and (14.26), where the matrices A and B are replaced with their factorized representations (14.61) and (14.62), we obtain

$$\begin{aligned} f^{(\text{in/out})}(E) &= \frac{1}{2}\sigma^{(\mp)} \lim_{r\to\infty} \left[A(E,r) \mp iB(E,r)\right] C^{-1} \\ &= \frac{1}{2}\sigma^{(\mp)} \lim_{r\to\infty} \left[D^{-1}\tilde{A}D - MK^{-1}D\tilde{B}D \mp iK^{-1}D\tilde{B}D\right] C^{-1} \\ &= \frac{1}{2}\sigma^{(\mp)} \lim_{r\to\infty} \left[D^{-1}\tilde{A}D - (M \pm i)K^{-1}D\tilde{B}D\right] C^{-1} \\ &= \sigma^{(\mp)} \left[D^{-1}aD - (M \pm i)K^{-1}DbD\right] C^{-1} , \end{aligned} \tag{14.68}$$

where

$$a(E) = \frac{1}{2}\tilde{A}(E,\infty) , \qquad b(E) = \frac{1}{2}\tilde{B}(E,\infty) . \tag{14.69}$$

In the explicit form for the elements of the Jost matrices, this reads $(m, n = 1, 2, \ldots, N)$:

$$\boxed{\begin{aligned} f^{(\text{in/out})}_{mn}(E) = \frac{\exp(\mp i\sigma_{\ell_m})}{C_{\ell_m}(\eta_m)} \Bigg\{ &\frac{k_n^{\ell_n+1}}{k_m^{\ell_m+1}} a_{mn}(E) \\ &- \left[\frac{2\eta_m h(\eta_m)}{C_0^2(\eta_m)} \pm i\right] C^2_{\ell_m}(\eta_m) k_m^{\ell_m} k_n^{\ell_n+1} b_{mn}(E) \Bigg\} \end{aligned}} \tag{14.70}$$

The corresponding S-matrix is the "ratio":

$$S = f^{(\mathrm{out})}\left[f^{(\mathrm{in})}\right]^{-1} \tag{14.71}$$

$$= \sigma^{(+)}\left[D^{-1}aD - (M - i)K^{-1}DbD\right]\left[D^{-1}aD - (M + i)K^{-1}DbD\right]^{-1}\sigma^{(+)}$$

$$= \sigma^{(+)}\left[D^{-1}a - (M - i)K^{-1}Db\right]\left[D^{-1}a - (M + i)K^{-1}Db\right]^{-1}\sigma^{(+)} ,$$

where we used the fact that $[\sigma^{(-)}]^{-1} = \sigma^{(+)}$. In Sect. 18.5.1, the reader can find an exactly solvable example of a two-channel problem with charged particles, where the representation of the type (14.70) is obtained explicitly.

It is easy to see that in the zero-charge limit (i.e. when all $\eta_n \to 0$), the factorized representations of the Jost matrices, (14.70), are reduced to the more simple representations (13.24). This is because

$$\sigma_\ell(\eta) \underset{\eta\to 0}{\longrightarrow} 0 , \qquad C_\ell(\eta) \underset{\eta\to 0}{\longrightarrow} 1 , \qquad \eta h(\eta) \underset{\eta\to 0}{\longrightarrow} 0 . \tag{14.72}$$

It is important to note that for real energies, the differential equations (14.65) and (14.66) as well as the boundary conditions (14.67) do not involve any complex functions or numbers. This means that the solutions of these equations and therefore the matrices $a(E)$ and $b(E)$ are real when $\operatorname{Im} E = 0$. Then it follows from the Schwarz reflection principle, graphically illustrated in Fig. B.4 of Appendix B, that

$$a(E^*) = [a(E)]^* \qquad \text{and} \qquad b(E^*) = [b(E)]^* . \tag{14.73}$$

14.7 Attractive Coulomb Forces

Using the definition (8.28) of the pure Coulomb Jost matrices, we can re-write the factorized representation (14.70) as

$$f_{mn}^{(\mathrm{in/out})}(E) = f_{\ell_m}^{\mathrm{c}\,(\mathrm{in/out})}(E) f_{mn}^{\mathrm{sc}\,(\mathrm{in/out})}(E) , \tag{14.74}$$

where

$$f_{mn}^{\mathrm{sc}\,(\mathrm{in/out})} = 2\left\{\frac{k_n^{\ell_n+1}}{k_m^{\ell_m+1}}a_{mn}(E) - \left[\frac{2\eta_m h(\eta_m)}{C_0^2(\eta_m)} \pm i\right] C_{\ell_m}^2(\eta_m) k_m^{\ell_m} k_n^{\ell_n+1} b_{mn}(E)\right\} . \tag{14.75}$$

Therefore,

$$\det f^{(\mathrm{in})}(E) = f_{\ell_1}^{\mathrm{c}\,(\mathrm{in})} f_{\ell_2}^{\mathrm{c}\,(\mathrm{in})} \cdots f_{\ell_N}^{\mathrm{c}\,(\mathrm{in})} \cdot \det f^{\mathrm{sc}\,(\mathrm{in})}(E) . \tag{14.76}$$

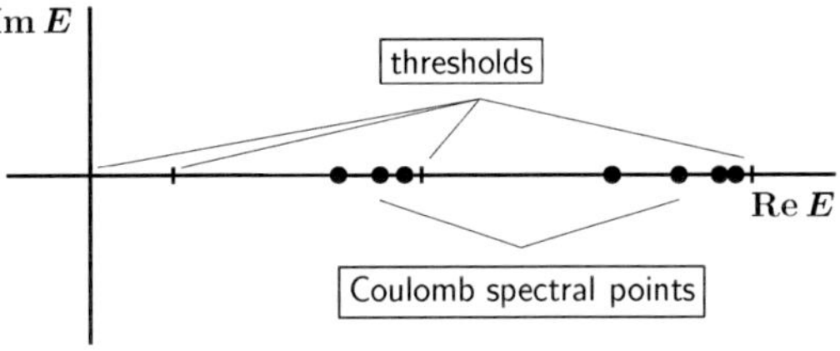

Fig. 14.1 Schematically shown pure Coulomb spectral points in the third and fourth channels (below the corresponding thresholds)

In the case of an attractive Coulomb potential (at least in one of the channels), this factorization of the determinant may cause confusion. Indeed, it looks like we should always have some spectral points on the real axis because the attractive Coulomb forces always generate hydrogen-like bound states. At the corresponding spectral energies, given by Eq. (8.102), one of the Coulomb factors in Eq. (14.76) is zero and thus

$$\det f^{(\mathrm{in})} \sim f_{\ell_1}^{\mathrm{c}\,(\mathrm{in})} f_{\ell_2}^{\mathrm{c}\,(\mathrm{in})} \cdots f_{\ell_N}^{\mathrm{c}\,(\mathrm{in})} = 0\,. \tag{14.77}$$

It looks like the right hand side of Eq. (14.76) is zero at every Coulombic spectral point in all the channels, no matter what kind of the short-range potential we have. To resolve this puzzle, we, first of all, notice that if there are several channels with attractive Coulomb forces, their pure Coulombic spectral points, generally speaking, do not coincide because the corresponding binding energies are counted from different thresholds, and the charges as well as the reduced masses are (generally speaking) different in different channels. An example, when the Coulomb attraction is present in the third and the fourth channels, is schematically shown in Fig. 14.1. Of course, some coincidences may happen (when the channel thresholds are too close to each other), but they do not affect the reasoning given below.

If the complex variable E is not close to one of the pure Coulombic spectral points, the factorized representations (14.74) can be used as they are, without any difficulties. If, however, the energy approaches one of these spectral points, the representations (14.74) lead to the same uncertainty of the type $(0 \times \infty)$ as is discussed for the single-channel problem in Sect. 8.2.8.

Suppose the energy E is close to a Coulombic spectral point in the channel m. If the Coulombic spectral points in different channels do not coincide, then only one row (m) of the matrix $f_{mn}^{\mathrm{sc}\,(\mathrm{in})}$ is multiplied by zero, i.e. by $f_{\ell_m}^{\mathrm{c}\,(\mathrm{in})} = 0$. However all the elements of the matrix $f_{mn}^{\mathrm{sc}\,(\mathrm{in/out})}$ in this row are infinite. This is because it involves the square, $C^2_{\ell_m}(\eta_m)$, of the Coulomb barrier factor (see Eq. (14.70)), which is singular at the Coulomb spectral points. In Appendix E.3, it is shown that the Coulomb barrier factor has the following behaviour near a spectral point E^c:

$$C_\ell(\eta)|_{E=E^c+\delta} \underset{\delta\to 0}{\sim} \frac{1}{\sqrt{\delta}} \tag{14.78}$$

and thus

$$f_{mn}^{\mathrm{sc\,(in)}}(E^c+\delta) \underset{\delta\to 0}{\sim} 1/\delta\,. \tag{14.79}$$

This singularity is exactly compensated by the vanishing factor (see Appendix E.3):

$$f_{\ell_j}^{\mathrm{c\,(in)}}(E^c+\delta) \underset{\delta\to 0}{\sim} \delta\,. \tag{14.80}$$

Therefore,

$$f_{mn}^{\mathrm{(in)}}(E^c+\delta) = f_{\ell_m}^{\mathrm{c\,(in)}}(E^c+\delta) f_{mn}^{\mathrm{sc\,(in)}}(E^c+\delta) \underset{\delta\to 0}{\sim} \mathrm{const}\,, \tag{14.81}$$

where const $\neq 0$ and has a finite value. This means that similarly to the single-channel case, the pure Coulomb spectral points (generally speaking) are not the spectral points of the total multi-channel Hamiltonian.

14.8 Riemann Surfaces

When considering the single-channel problem of charged particles (see Sect. 8.2.9), it was established that the presence of the Coulomb potential drastically changes the topology of the Riemann surface of the energy. The branch point at $E = 0$ becomes not only of the square root type but of the logarithmic type as well. As a result, the Riemann surface becomes spiral with an infinite number of sheets, as is shown in Fig. 8.2. Apparently, one should expect similar changes to the multi-channel Riemann surface, when the Coulomb interaction is present at least in one of the channels.

The logarithmic branching stems from the fact that the Coulomb incoming and outgoing spherical waves,

$$\begin{aligned} W^{\mathrm{(in/out)}}(E,r) &= H^{(\mp)}(E,r)\sigma^{(\pm)} \\ &\underset{r\to\infty}{\longrightarrow} \begin{pmatrix} \pm e^{\mp i[k_1 r-\ell_1\pi/2-\eta_1\ln(2k_1 r)]} & 0 \\ 0 & \pm e^{\mp i[k_2 r-\ell_2\pi/2-\eta_2\ln(2k_2 r)]} \end{pmatrix}, \end{aligned} \tag{14.82}$$

(where for the sake of notation clarity the number of channels is $N = 2$) depend on the logarithms of the channel momenta, $\ln(k_n)$. For each channel with non-zero charges, this logarithm has an infinite number of values that differ by the term $i2\pi m$,

$$\ln(k_n) = \ln|k_n| + i\arg(k_n) + i2\pi m_n\,, \qquad m_n = 0, \pm 1, \pm 2, \dots\,, \tag{14.83}$$

where $-\pi < \arg(k_n) \leqslant \pi$ (this angle covers the principal branch of the logarithm function corresponding to $m_n = 0$).

The Jost matrices are the amplitudes of these waves. Since the waves are multi-valued, the Jost matrices (and thus the S-matrix) have many (infinite number of) values for a given energy E. In other words, they are defined on a Riemann surface with an infinite number of parallel sheets. These sheets are cut along their real axes and the rims of the cuts are interconnected in such a way that provides the correct transitions when we go around the branch points (which are the thresholds).

The factorized representation (14.70) of the Jost matrices gives their dependencies on $\ln(k_n)$ in an explicit form. The logarithm appears in the functions $h(\eta_n)$ for each of the channels, where the Coulomb forces are present (see also Eq. (8.111)).

A sheet of the Riemann surface is specified by the signs of $\mathrm{Im}\, k_n$ $(n = 1, 2, \ldots, N)$ and by the numbers $m_1, m_2, \ldots, m_N$ which are the logarithmic indices of Eq. (14.83) for each of the channels with non-zero charges. We symbolically denote the sheets by giving (in brackets) the signs of all the channel momenta and the logarithmic indices for all the channels with non-zero charges (as the subscripts). For example, $(++)_{00}$ symbolizes the Riemann sheet of a two-channel problem, where both channel momenta have positive imaginary parts and both logarithmic indices are zero.

Figure 14.2 shows an example of the Riemann surface for a two-channel system, where the Coulomb potential is present in both channels. The sheets in the part (b) of this figure are the continuations of the corresponding sheets (with the same labels) of the part (a). Each sheet has its own real axis. It is shown only one of them, which is on the physical sheet. The E-values on this axis correspond to the observable scattering energies $(E > E_1)$ and to the bound states $(E < E_1)$. The dots show the locations of the resonances, when one or two channels are open.

The parts (a) and (b) show how we go from one sheet to another sheet when crossing the real axis either between E_1 and E_2 (part (a)) or above E_2 (part (b)). When doing one full circle around a single threshold, we only change the corresponding sign in the label $(+\pm)_{m_1 m_2}$, while the logarithmic index remains unchanged. This index changes after two full circles, in accordance with Eq. (8.113). When going around both thresholds, we change both signs. For example, we can go around both thresholds (E_1 and E_2) as is shown in Fig. 14.3. If we start from the sheet $(+-)_{00}$ and move in the anti-clockwise direction, then the first circle brings us to the sheet $(-+)_{00}$ and the second circle reverses the sign changes but increases both logarithmic indices. As a result, we find ourselves on the sheet $(+-)_{11}$.

The general rule for making the interconnections of the cuts is the same as for the neutral particles (see Eq. (13.2)). They should be connected in such a way that provides the following: when crossing an interconnection between the thresholds E_1 and E_2, we must find ourselves on a sheet whose label only differs in the first sign; when doing the same between E_2 and E_3 (or just above E_2 for a two-channel problem), we find that only the first two signs are changing to the opposite; i.e. between E_n and E_{n+1}, the first n signs are changing. As to the logarithmic indices, we should count the number of loops around the thresholds that are needed to reach a particular sheet from the principal branch (where $m_1 = m_2 = \cdots = m_N = 0$).

It should be noted that Fig. 14.2 is not complete. It only shows the principal sheets $(m_1 = m_2 = 0)$ and the fragments of the spiral sequences that can be reached when

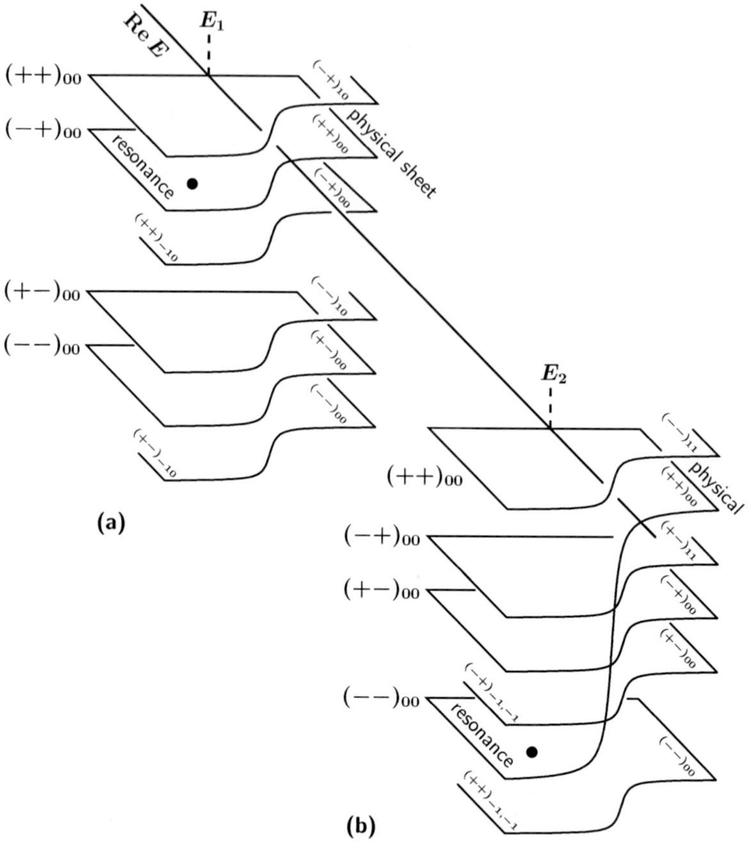

Fig. 14.2 Interconnections of the Riemann sheets for the two-channel problem, where the Coulomb potential is present in both channels: (**a**) the interconnections between the thresholds E_1 and E_2; (**b**) the interconnections above the highest threshold. The symbols $(\pm\pm)_{m_1 m_2}$ label the sheets where $\text{Im}(k_1)$ and $\text{Im}(k_2)$ are either positive or negative, and the subscripts $m_1 m_2$ are the numbers of $i2\pi$ in Eq. (14.83) for each of the channels

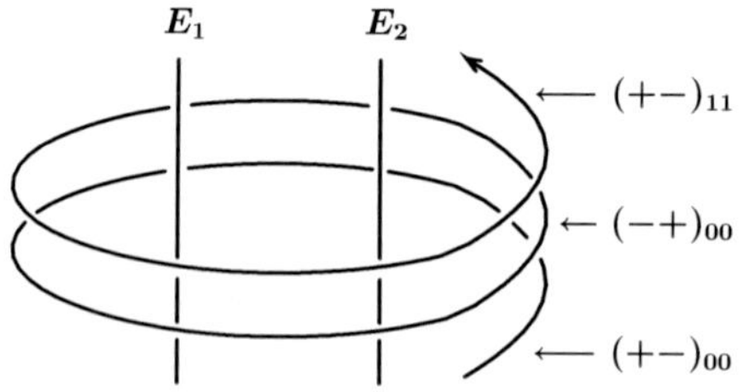

Fig. 14.3 Two-circle path around both thresholds (E_1 and E_2) on the Riemann surface of a two-channel problem with the Coulomb forces in both channels. If the starting point is on the sheet $(+-)_{00}$, then the final point is on the sheet $(+-)_{11}$

we go either around the first threshold or around both thresholds. There are many other spiral sequences, such, for example, as $(\pm\pm)_{01}$, $(\pm\pm)_{02}, \ldots$, etc. They can be reached via the paths that are somewhat more complicated than the one shown in Fig. 14.3. Such a path is a combination of circles around one and two thresholds.

It is instructive to compare Fig. 14.2 with the corresponding surface for the neutral particles, depicted in Fig. 13.2. It is seen that the Coulomb forces change the topology of the surface. The main feature is that, when moving around a threshold in one direction (clockwise or anti-clockwise), we can never come back to any of the previous sheets. In contrast to the neutral particle case, such an encircling of a branch point leads us down or up the spiral sequence of the sheets. It should be noted that all the sheets with the non-zero logarithmic indices, $m_n \neq 0$, are far away from the real axis where the scattering energies belong to. This means that the values of the Jost matrices on these sheets cannot influence any observable quantities. We may therefore safely ignore these ($m_n \neq 0$) sheets and only consider the principal branch of the Riemann surface, where $m_1 = m_2 = \cdots = m_N = 0$. However, we should not forget that the topology (set of the interconnections) of these principal sheets is different as compared with the neutral particle case.

If the Coulomb forces are present not in all the channels, the Riemann surface looks a bit more simple. As an example, Fig. 14.4 shows the surface for the problem where these forces only act in the second channel (for instance, the two-channel nuclear reaction $n\alpha \to dt$). The sheets of this Riemann surface are labeled by a single logarithmic index m. Below the second threshold, the principal branch of this surface looks exactly as for neutral particles. However, we should not forget that the spiral sheets of the part (b) of this figure continue in all directions. They are also present above and below the principal sheets of the part (a). In order to reach them from these principal sheets, we have to go around the second threshold.

14.9 Symmetry Properties of the Jost Matrices

The matrices $f^{(\mathrm{in})}$ and $f^{(\mathrm{out})}$ are related to each other at different points of the Riemann surface, i.e. they obey certain symmetry rules. Since the unknown matrices $a(E)$ and $b(E)$ in the semi-analytic representation (14.70) are the same on all the sheets, the symmetry relations among the Jost matrices are determined by the explicitly given factors that undergo certain changes when we replace k_n with $-k_n$ or with k_n^*.

In Sect. 13.5, the symmetry relations of this kind are obtained for the multi-channel problem of neutral particles, namely, the so-called "vertical" symmetry (13.30), "diagonal" symmetry (13.32), and "mirror" ("horizontal") symmetry given by Eq. (13.36). For a two-channel problem, they are schematically depicted in Fig. 13.5. In the present section, it is shown that the Coulomb forces break some of these symmetries down. This is not surprising because in the single-channel problem of charged particles, we have already seen that the only symmetry that survives is the "diagonal" one (see Sect. 8.2.10).

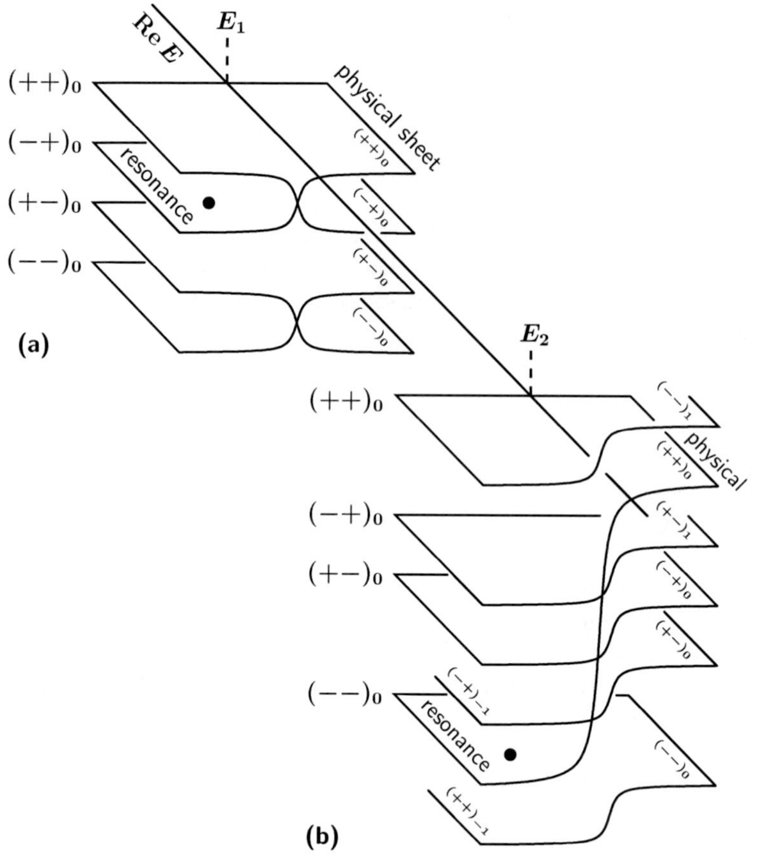

Fig. 14.4 Interconnections of the Riemann sheets for the two-channel problem, where the Coulomb potential is only present in the second channel: (**a**) the interconnections between the thresholds E_1 and E_2 and (**b**) the interconnections above the highest threshold. The symbols $(\pm\pm)_{m_2}$ label the sheets where $\mathrm{Im}(k_1)$ and $\mathrm{Im}(k_2)$ are either positive or negative, and the subscript m_2 is the number of $i2\pi$ in Eq. (14.83) for the second channel

If the multi-channel potential has a Coulomb tail at least in one of the channels, the Jost matrices have the structure given by Eq. (14.68) or, equivalently, by Eq. (14.70). This representation, where the matrices $C, \sigma^{(\pm)}, D$, and M are known in the explicit form, can be used for establishing the changes caused by the "diagonal", "vertical", and "horizontal" shifts of the point on the Riemann surface.

* * *

First of all, let us consider what happens if we change the sign of a channel momentum k_n ("vertical" shift). The sign can be changed by the exponential factor

$$k_n \rightarrow -k_n : \qquad e^{i\pi(2j+1)}k_n , \qquad j = \pm 1, \pm 2, \ldots \tag{14.84}$$

For the channel n without a Coulomb potential, all values of j in such a factor are equivalent. However, if there are Coulomb forces in the channel n, then the functions involved in the matrices of Eq. (14.70) undergo the following changes:

$$k_n \rightarrow e^{i\pi(2j+1)}k_n \Rightarrow \begin{cases} \exp(\pm i\sigma_{\ell_n}) \rightarrow \exp(\mp i\sigma_{\ell_n}) , \\ C_{\ell_n}(\eta_n) \quad \rightarrow e^{\pi\eta_n} C_{\ell_n}(\eta_n) , \\ h(\eta_n) \quad \rightarrow h(\eta_n) + i\pi(2j+1) . \end{cases} \tag{14.85}$$

These changes follow from Eqs. (8.30), (8.31), and (8.111), where the Sommerfeld parameter, η_n, changes sign together with k_n. Apparently, after the change of the signs of all the channel momenta, these functions acquire different values. The elements of the Jost matrices change incoherently, i.e. these changes cannot be factorized, and thus the relation of the type (13.28), generally speaking, cannot be valid anymore. In other words, the "vertical" symmetry is broken by the Coulomb forces.

* * *

Let us turn to the "diagonal" symmetry, which for neutral particles follows from the Schwarz reflection principle. Does this symmetry remain valid in the presence of the Coulomb potential?

The "diagonal" transition $E \rightarrow E^*$ moves the point to the other side of the real axis. Both the original point E and the new point E^* can be on any of the Riemann sheets. It is clear that the complex conjugation of the energy results in the complex conjugation of all the channel momenta and thus of all the Sommerfeld parameters,

$$E \rightarrow E^* \Rightarrow \quad k_n \rightarrow k_n^* , \quad \eta_n \rightarrow \eta_n^* , \qquad n = 1, 2, 3, \ldots, N .$$

The transition $k_n \rightarrow k_n^*$ changes the sign of $\mathrm{Im}\, k_n$. This means that the point E^* is on a different sheet of the Riemann surface. Actually, all the signs in the sheet label $(\pm, \pm, \ldots)_{m_1 m_2, \ldots}$ are changing to the opposite. Apparently, the logarithmic indices, $m_1 m_2, \ldots$, change to the opposite as well. This is clear from Eq. (14.83) that defines them.

Looking at Eqs. (8.30) and (8.31) for the Coulomb phase shift and the barrier factor, we see that

$$\exp\left[\pm i\sigma_\ell(\eta^*)\right] = \exp\left[\mp i\sigma_\ell(\eta)\right]^* \qquad \text{and} \qquad C_\ell(\eta^*) = \left[C_\ell(\eta)\right]^* . \tag{14.86}$$

For the explicitly given diagonal matrices involved in Eq. (14.68), this implies that

$$C(E^*) = [C(E)]^* \ , \tag{14.87}$$

$$\sigma^{(\pm)}(E^*) = \left[\sigma^{(\mp)}(E)\right]^* \ , \tag{14.88}$$

$$D(E^*) = [D(E)]^* \ , \tag{14.89}$$

$$M(E^*) = [M(E)]^* \ . \tag{14.90}$$

Taking into account the Schwarz relations (14.73), we therefore arrive at the same symmetry property as for the single-channel case, given by Eq. (8.117),

$$f^{(\text{in/out})}(E^*_{m_1 m_2,\dots}) = \left[f^{(\text{out/in})}(E_{-m_1,-m_2,\dots})\right]^* \ , \tag{14.91}$$

where the subscripts (logarithmic indices $m_1, m_2, \dots$) of the energy variable specify the sheet of the Riemann surface. The corresponding relations for the S-matrix can be obtained as follows:

$$\begin{aligned} S(E^*_{m_1 m_2,\dots}) &= f^{(\text{out})}(E^*_{m_1 m_2,\dots})\left[f^{(\text{in})}(E^*_{m_1 m_2,\dots})\right]^{-1} \\ &= f^{(\text{in})*}(E_{-m_1,-m_2,\dots})\left[f^{(\text{out})*}(E_{-m_1,-m_2,\dots})\right]^{-1} \\ &= \left\{f^{(\text{out})*}(E_{-m_1,-m_2,\dots})\left[f^{(\text{in})*}(E_{-m_1,-m_2,\dots})\right]^{-1}\right\}^{-1} \\ &= \left[S^*(E_{-m_1,-m_2,\dots})\right]^{-1} \ . \end{aligned} \tag{14.92}$$

It should not be forgotten that the transition $E^*_{m_1 m_2,\dots} \to E_{-m_1,-m_2,\dots}$ moves the point to a different sheet of the Riemann surface, where all the labels are changing to the opposite, for example,

$$(++-)_{1,0,-1} \ \to \ (--+)_{-1,0,1} \ .$$

In particular, the relation (14.92) means that a resonance pole of the S matrix corresponds to a zero of its determinant at the symmetric point on the physical sheet of the Riemann surface.

For a two-channel problem, the diagonal relations (14.91) are illustrated in Fig. 14.5. Of course, there are much more (actually infinite number of) "diagonal" relations of the type (14.91). Figure 14.5 only shows the symmetric points on the principal branch of the Riemann surface. The other sheets (with non-zero

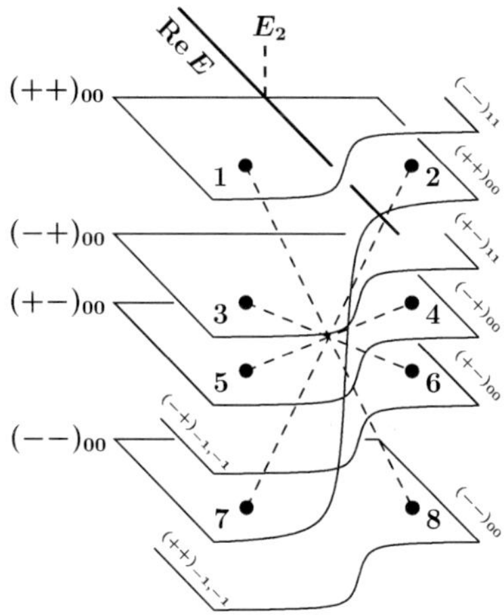

Fig. 14.5 Fragment of the Riemann surface shown in Fig. 14.2b. The points with the odd numbers correspond to the same energy E, while the points with the even numbers correspond to the complex conjugate energy E^*. The four dashed lines connect the points at which the values of the Jost matrices are related according to Eq. (14.91)

logarithmic indices) have no practical influence on the observables and thus can be safely ignored. In this figure, the points labeled by the odd numbers are on the same vertical line and thus correspond to the same energy E. The even numbers label the symmetric complex conjugate energies, E^*. The diagonal dashed lines correspond to the following four relations:

$$f^{(\text{in/out})}(1) = \left[f^{(\text{out/in})}(8)\right]^* ,$$

$$f^{(\text{in/out})}(3) = \left[f^{(\text{out/in})}(6)\right]^* ,$$

$$f^{(\text{in/out})}(5) = \left[f^{(\text{out/in})}(4)\right]^* ,$$

$$f^{(\text{in/out})}(7) = \left[f^{(\text{out/in})}(2)\right]^* .$$

It is instructive to compare the relations depicted in Fig. 14.5 with the corresponding relations for the neutral particles given in Fig. 13.5.

* * *

Since there are no "vertical" symmetries for the charged particles, the "horizontal" symmetries (which result from the combination of the "diagonal" and "vertical" ones) are also broken. In particular, this means that there are no "mirror" poles of the S-matrix, if the charges are non-zero. For example, if the point 7 in Fig. 14.5 is a resonance, then the point 8 could be its "mirror" partner, but in contrast to the neutral particle case (see Fig. 13.5), the points 7 and 8 are not related by a "horizontal" symmetry.

15 Effective-Range Expansion and Its Generalizations

In all branches of science, which are based on mathematical equations, it is very common to approximate the unknown solutions by a finite number of terms of the Taylor series. In quantum scattering theory, the most famous approximation of this type is called the effective-range expansion. It was introduced in nuclear physics for parametrizing the observables at low collision energies.[1] For that purpose, a certain function of the momentum, $\mathcal{K}(k)$, was expanded in the Maclaurin series of the energy variable E,

$$\mathcal{K}(k) = \lambda_0 + \lambda_1 E + \lambda_2 E^2 + \cdots , \tag{15.1}$$

where the coefficients λ_n can either be used as the free parameters to fit experimental data or be calculated within a model. In such an approach, the main task is to find such a function $\mathcal{K}(k)$, which is a single-valued function of the energy, i.e. its power series expansion does not involve the odd powers of the momentum. In the simplest case of a single-channel short-range interaction and zero angular momentum, it was established that the left hand side of Eq. (15.1) is

$$\mathcal{K}(k) = k \cot \delta(k) .$$

The present chapter is devoted to the derivation of this (standard) single-channel effective-range expansions for a short-range potential and for a potential with a

[1] One of the first publications on this subject was the paper by Hans Bethe [45]. A short historical review can be found in the book [46]. Nowadays the effective-range expansions are still frequently used not only in nuclear physics but also for parametrizing the low-energy collisions between atoms and molecules (see, for example, Refs. [35, 47–56]).

S. A. Rakityansky, *Jost Functions in Quantum Mechanics*,
https://doi.org/10.1007/978-3-031-07761-6_15

Coulomb tail. Here, it is also constructed a more general power-series expansion that can be done around an arbitrary complex energy, E_0,

$$\lambda_0 + \lambda_1(E - E_0) + \lambda_2(E - E_0)^2 + \cdots .$$

Then such an expansion is further generalized for an arbitrary number of channels. All these derivations are based on the semi-analytic representations (7.65), (8.100), (13.24), and (14.70) of the corresponding Jost functions and Jost matrices, where the factors responsible for the branching of the Riemann surface are given explicitly.

The analytic structures of the Jost functions for a short-range potential and for a potential with the Coulomb tail are different. Of course, in the limit when the Sommerfeld parameter vanishes, $\eta \to 0$, the more complicated Coulomb case gives the corresponding simplified formulae for the short-range potential. In principle, we could derive the most complicated formulae for the coupled-channel problem with charged particles and then from it could obtain all the other more simple equations. However, it is still worthwhile to begin with considering the simple case separately. This gives the reader a clear idea of how the power-series expansion can be done in all the complicated cases.

15.1 Single-Channel Short-Range Potential

If the potential is of a short range, the Jost functions have the following factorized structure (see Eq. (7.65)):

$$f_\ell^{(\mathrm{in/out})}(E) = a_\ell(E) \mp ik^{2\ell+1} b_\ell(E) , \tag{15.2}$$

where the explicitly given factor $k^{2\ell+1}$ is responsible for the square-root branching of the Riemann surface, while the unknown functions $a_\ell(E)$ and $b_\ell(E)$ are single-valued and analytic. Since they are analytic, we can expand these functions in the Taylor series around an arbitrary point E_0 on the complex energy plane (within the domain $\mathcal{D}$, which is defined in Sect. 7.7.2),

$$a_\ell(E) = \sum_{n=0}^{\infty} \alpha_n(\ell, E_0)(E - E_0)^n , \qquad b_\ell(E) = \sum_{n=0}^{\infty} \beta_n(\ell, E_0)(E - E_0)^n . \tag{15.3}$$

Then for the S-matrix (via which any observable quantity can be expressed), the following ratio of the two series is obtained:

$$S_\ell(E) = \frac{f_\ell^{(\mathrm{out})}(E)}{f_\ell^{(\mathrm{in})}(E)} = \frac{\displaystyle\sum_{n=0}^{\infty} (\alpha_n + ik^{2\ell+1}\beta_n)(E - E_0)^n}{\displaystyle\sum_{n=0}^{\infty} (\alpha_n - ik^{2\ell+1}\beta_n)(E - E_0)^n} . \tag{15.4}$$

If the sums in this equation are truncated, $\sum_{n=0}^{\infty} \approx \sum_{n=0}^{M}$, we obtain an approximation of the S-matrix (and thus of any observable).

In numerical analysis, the approximation of a function in the form of the ratio of two polynomials is called the Padè approximation. It is known that the Padè approximation is much more accurate than any polynomial one.[2] The reason is that it takes into account possible poles of the function at complex values of the independent variable. These poles correspond to the zeros of the polynomial which is in the denominator.

When the summations in Eq. (15.4) are truncated, the numerator and denominator are not the polynomials of the variable E because they involve the odd powers of $\sqrt{E} \sim k$. However, the singularities caused by the zeros of the denominator are taken into account anyway. If we consider the numerator and the denominator as the functions of the momentum, then they are polynomials, and for a finite number of terms, the ratio (15.4) becomes an approximation of the Padè type. This means that Eq. (15.4) with the truncated sums has all the properties of the Padè approximation. There is another advantage of such an approximation, namely, it has the correct analytic structure. Indeed, no matter how many terms are taken into account in the sums, the factor $k^{2\ell+1}$ is always there. This factor guarantees the proper change of the Jost functions (and thus of the S-matrix) when we go around the threshold energy. In other words, the proper topology of the Riemann surface is built in the approximation.

15.1.1 Effective-Range Expansion

Consider the case when the expansions (15.3) are done around the threshold energy, i.e. $E_0 = 0$. The S-matrix takes the form:

$$S_\ell(E) = \frac{\displaystyle\sum_{n=0}^{\infty}(\alpha_n + ik^{2\ell+1}\beta_n)E^n}{\displaystyle\sum_{n=0}^{\infty}(\alpha_n - ik^{2\ell+1}\beta_n)E^n} \,. \tag{15.5}$$

We have established (see Sect. 7.7.1) that for real scattering energies, the functions $a_\ell(E)$ and $b_\ell(E)$ in Eq. (15.2) are real. This implies that for real E_0, all the expansion coefficients are real as well,

$$\operatorname{Im} E_0 = 0 \qquad \Rightarrow \qquad \begin{cases} \operatorname{Im}\alpha_n(\ell, E_0) = 0\,, \\ \operatorname{Im}\beta_n(\ell, E_0) = 0\,, \qquad \forall \quad n, \ell\,. \end{cases} \tag{15.6}$$

This fact is also proved in a different way in the next section.

[2] See, for example, book [106].

The S-matrix at real collision energies is a complex number of unit length. It can be expressed in terms of the scattering phase shift,

$$S_\ell(E) = e^{2i\delta_\ell} = \frac{f_\ell^{(\text{out})}(E)}{f_\ell^{(\text{in})}(E)} \,. \tag{15.7}$$

Since α_n and β_n are real, the numerator and denominator in Eq. (15.5) are complex conjugate of each other at such energies (see also Fig. 3.6), i.e.

$$f_\ell^{(\text{out})}(E) = f_\ell^{(\text{in})*}(E) \,, \qquad \text{Im}\, E = 0 \,. \tag{15.8}$$

As it has been mentioned many times, the normalization of the Jost functions is arbitrary because the normalization factor always cancelled out in the calculation of any observable. This means that the numerator and denominator in the ratio (15.7) may have a common factor that makes their absolute values different from unity. Therefore

$$\sum_{n=0}^{\infty} \alpha_n E^n + ik^{2\ell+1} \sum_{n=0}^{\infty} \beta_n E^n = |f_\ell^{(\text{out})}| e^{i\delta_\ell} \,. \tag{15.9}$$

Equalizing the real and imaginary parts in this equation, we have

$$\sum_{n=0}^{\infty} \alpha_n E^n = |f_\ell^{(\text{out})}| \cos\delta_\ell \,, \tag{15.10}$$

$$k^{2\ell+1} \sum_{n=0}^{\infty} \beta_n E^n = |f_\ell^{(\text{out})}| \sin\delta_\ell \,. \tag{15.11}$$

Dividing the first of these equations by the second one, we obtain

$$\boxed{k^{2\ell+1} \cot\delta_\ell = \frac{\alpha_0 + \alpha_1 E + \alpha_2 E^2 + \cdots}{\beta_0 + \beta_1 E + \beta_2 E^2 + \cdots}} \tag{15.12}$$

In order to derive the standard effective-range expansion of the type (15.1), we need to perform the division of the series in Eq. (15.12). This can be done by expanding the function

$$\mathcal{K}(E) = \frac{\alpha_0 + \alpha_1 E + \alpha_2 E^2 + \cdots}{\beta_0 + \beta_1 E + \beta_2 E^2 + \cdots} \tag{15.13}$$

in the Maclaurin series,

$$\mathcal{K}(E) = \mathcal{K}(0) + \mathcal{K}'(0)E + \mathcal{K}''(0)\frac{E^2}{2} + \mathcal{K}'''(0)\frac{E^3}{6} + \mathcal{K}^{(4)}(0)\frac{E^4}{24} + \cdots \tag{15.14}$$

The lowest order derivatives of the function (15.13) can be easily found,

$$\mathcal{K}(0) = \frac{\alpha_0}{\beta_0}, \qquad \mathcal{K}'(0) = \frac{\alpha_1}{\beta_0} - \frac{\alpha_0\beta_1}{\beta_0^2}, \tag{15.15}$$

$$\mathcal{K}''(0) = \frac{2\alpha_2}{\beta_0} + \frac{2\alpha_0\beta_1^2}{\beta_0^3} - \frac{2\alpha_0\beta_2}{\beta_0^2} - \frac{2\alpha_1\beta_1}{\beta_0^2}, \tag{15.16}$$

$$\mathcal{K}'''(0) = \frac{6\alpha_3}{\beta_0} + \frac{6\alpha_1\beta_1^2}{\beta_0^3} - \frac{6\alpha_0\beta_1^3}{\beta_0^4} - \frac{6\alpha_0\beta_3}{\beta_0^2} - \frac{6\alpha_1\beta_2}{\beta_0^2} - \frac{6\alpha_2\beta_1}{\beta_0^2} + \frac{12\alpha_0\beta_1\beta_2}{\beta_0^3}, \tag{15.17}$$

$$\mathcal{K}^{(4)}(0) = \frac{24\alpha_4}{\beta_0} + \frac{24\alpha_0\beta_2^2}{\beta_0^3} + \frac{24\alpha_2\beta_1^2}{\beta_0^3} - \frac{24\alpha_1\beta_1^3}{\beta_0^4} + \frac{24\alpha_0\beta_1^4}{\beta_0^5} - \frac{24\alpha_0\beta_4}{\beta_0^2} - \frac{24\alpha_1\beta_3}{\beta_0^2}$$
$$- \frac{24\alpha_2\beta_2}{\beta_0^2} - \frac{24\alpha_3\beta_1}{\beta_0^2} + \frac{48\alpha_0\beta_1\beta_3}{\beta_0^3} + \frac{48\alpha_1\beta_1\beta_2}{\beta_0^3} - \frac{72\alpha_0\beta_1^2\beta_2}{\beta_0^4}. \tag{15.18}$$

For the historical reasons, the expansion (15.14) is usually written in the following traditional form:

$$k^{2\ell+1}\cot\delta_\ell = -\frac{1}{\mathsf{a}_\ell} + \frac{\mathsf{r}_\ell}{2}k^2 - \mathsf{P}_\ell\mathsf{r}_\ell^3k^4 + \mathsf{Q}_\ell\mathsf{r}_\ell^5k^6 - \mathsf{L}_\ell\mathsf{r}_\ell^7k^8 + \cdots \tag{15.19}$$

In the case of zero angular momentum, $\ell = 0$, the first two parameters, namely, a_0 and r_0 are called the scattering length and the effective radius (or the effective range, hence the name of the expansion).

Using the relation $E = (\hbar k)^2/2\mu$ and simple algebra, it is not difficult to find the following relations:

$$\mathsf{a}_\ell = -1/\mathcal{K}(0), \qquad \mathsf{r}_\ell = \mathcal{K}'(0)\hbar^2/\mu, \tag{15.20}$$

$$\mathsf{P}_\ell = -\frac{\mu\mathcal{K}''(0)}{8\hbar^2[\mathcal{K}'(0)]^3}, \qquad \mathsf{Q}_\ell = \frac{\mu^2\mathcal{K}'''(0)}{48\hbar^4[\mathcal{K}'(0)]^5}, \qquad \mathsf{L}_\ell = -\frac{\mu^3\mathcal{K}^{(4)}(0)}{384\hbar^6[\mathcal{K}'(0)]^7}. \tag{15.21}$$

Therefore, if the coefficients α_n and β_n in the Maclaurin expansions of the two terms, (15.3), of the Jost functions are known, then the above formulae allow us to calculate any number of the effective-range parameters (15.19).

The next section is devoted to a method for calculating the coefficients α_n and β_n, when the interaction potential is known. Alternatively, they can be used as the free parameters for fitting any available experimental data. In this connection, it should be noted that, for such a fitting, the representation (15.12) is more suitable than the traditional effective-range parametrization (15.19). The reason is that the representation (15.12) is of the Padè type, which is always more accurate than a simple power series.

15.1.2 Expansion Coefficients: Calculation

The holomorphic functions of the energy $a_\ell(E)$ and $b_\ell(E)$ are the asymptotic values of the corresponding r-dependent functions (see Eq. (7.66)),

$$a_\ell(E) = \frac{1}{2} \lim_{r\to\infty} \tilde{A}_\ell(E,r) \qquad \text{and} \qquad b_\ell(E) = \frac{1}{2} \lim_{r\to\infty} \tilde{B}_\ell(E,r)\,, \tag{15.22}$$

which are the solutions of the system of differential equations,

$$\partial_r \tilde{A}_\ell = -\tilde{n}_\ell V \left(\tilde{j}_\ell \tilde{A}_\ell - \tilde{n}_\ell \tilde{B}_\ell \right), \tag{15.23}$$

$$\partial_r \tilde{B}_\ell = -\tilde{j}_\ell V \left(\tilde{j}_\ell \tilde{A}_\ell - \tilde{n}_\ell \tilde{B}_\ell \right), \tag{15.24}$$

with the boundary conditions

$$\tilde{A}_\ell(E,0) = 1\,, \qquad \tilde{B}_\ell(E,0) = 0\,. \tag{15.25}$$

These differential equations involve the holomorphic parts of the Riccati–Bessel and Riccati–Neumann functions,

$$\tilde{j}_\ell(E,r) = j_\ell(kr)/k^{\ell+1} \qquad \text{and} \qquad \tilde{n}_\ell(E,r) = n_\ell(kr)k^\ell\,, \tag{15.26}$$

defined by Eqs. (7.58) and (7.59) (see also Eqs. (D.32) and (D.33) of the Appendix D.2.1).

As is seen from Eqs. (15.23) and (15.24), if the potential is identical zero beyond a finite value of the radius, r, then the limits (15.22) are achieved at that radius. In other words, for any finite r, the functions $\tilde{A}_\ell(E,r)$ and $\tilde{B}_\ell(E,r)$ determine the corresponding components $a_\ell(E)$ and $b_\ell(E)$ in the analytic structure (15.2) of the Jost functions for the potential that is cut off at r (in the spirit of the so-called variable phase approach). We therefore can expand (at a fixed r) the functions $\tilde{A}_\ell(E,r)$ and $\tilde{B}_\ell(E,r)$ in the Taylor series in the same way as is given by Eq. (15.3),

$$\tilde{A}_\ell(E,r) = \sum_{n=0}^{\infty} \tilde{\alpha}_n(\ell, E_0, r)(E - E_0)^n , \tag{15.27}$$

$$\tilde{B}_\ell(E,r) = \sum_{n=0}^{\infty} \tilde{\beta}_n(\ell, E_0, r)(E - E_0)^n . \tag{15.28}$$

In other words, we look for the solutions of differential equations (15.23) and (15.24) in the form of the power series with the unknown coefficients (r-dependent functions) $\tilde{\alpha}_n$ and $\tilde{\beta}_n$. The holomorphic functions $\tilde{j}_\ell(E,r)$ and $\tilde{n}_\ell(E,r)$, involved in those equations, can be expanded in the Taylor series as well,

$$\tilde{j}_\ell(E,r) = \sum_{n=0}^{\infty} \varphi_n(\ell, E_0, r)(E - E_0)^n , \tag{15.29}$$

$$\tilde{n}_\ell(E,r) = \sum_{n=0}^{\infty} \gamma_n(\ell, E_0, r)(E - E_0)^n . \tag{15.30}$$

In the particular case $E_0 = 0$ (Maclaurin expansion), the expansion coefficients $\varphi_n(\ell, 0, r)$ and $\gamma_n(\ell, 0, r)$ are given explicitly by the terms of the series (D.32) and (D.33) in Appendix D.2.1. For the general case ($E_0 \neq 0$), these coefficients can be found as follows (these relations are derived in Appendix D.2.1):

$$\varphi_n(\ell, E_0, r) = \frac{1}{n!}\left[\left(\frac{\mu r}{\hbar^2}\right)^n \frac{j_{\ell+n}(kr)}{k^{\ell+n+1}}\right]_{E=E_0} , \tag{15.31}$$

$$\gamma_n(\ell, E_0, r) = \frac{1}{n!}\left(\frac{\mu r}{\hbar^2}\right)^n \left[k^{\ell-n} n_{\ell-n}(kr)\right]_{E=E_0} , \tag{15.32}$$

where the variable k should be expressed via the corresponding energy, $k = \pm\sqrt{2\mu E/\hbar^2}$. It is not difficult to see that in the ratio $j_\ell(kr)/k^{\ell+1}$ as well as in the product $k^\ell n_\ell(kr)$, the choice of the sign in front of this square root is immaterial.

Substituting the series expansions for the functions $\tilde{A}_\ell$, $\tilde{B}_\ell$, $\tilde{j}_\ell$, and $\tilde{n}_\ell$ into the differential equations (15.23) and (15.24), we have

$$\sum_n \tilde{\alpha}'_n (E - E_0)^n = -\sum_{ijk} \gamma_i V(\varphi_j \tilde{\alpha}_k - \gamma_j \tilde{\beta}_k)(E - E_0)^{i+j+k} ,$$

$$\sum_n \tilde{\beta}'_n (E - E_0)^n = -\sum_{ijk} \varphi_i V(\varphi_j \tilde{\alpha}_k - \gamma_j \tilde{\beta}_k)(E - E_0)^{i+j+k} .$$

Since the variable E is arbitrary, the coefficients of each power of $(E - E_0)$ on the left and right hand sides of these equations must be the same. This gives us the following differential equations for the unknown coefficients:

$$\boxed{\begin{aligned} \partial_r \tilde{\alpha}_n &= - \sum_{i+j+k=n} \gamma_i V (\varphi_j \tilde{\alpha}_k - \gamma_j \tilde{\beta}_k) \\ \partial_r \tilde{\beta}_n &= - \sum_{i+j+k=n} \varphi_i V (\varphi_j \tilde{\alpha}_k - \gamma_j \tilde{\beta}_k) \end{aligned}} \tag{15.33}$$

The boundary conditions (15.25) are independent of E. This implies that at $r = 0$ the only non-vanishing coefficient is $\tilde{\alpha}_0$,

$$\tilde{\alpha}_n(\ell, E_0, 0) = \delta_{n0} , \qquad \tilde{\beta}_n(\ell, E_0, 0) = 0 , \qquad \forall n . \tag{15.34}$$

It should be noted an important fact: if the central point E_0 of the expansion is real, then all the coefficients φ_n and γ_n are also real, and therefore the solutions of the differential equations (15.33) with the real boundary conditions (15.34) are real as well.

Another important note is about the necessity of using the complex rotation (described in Sect. 7.4), i.e. the need of going along the path given in Fig. 7.3 when numerically solving Eqs. (15.33) with a complex central point E_0. If we move E_0 into the complex plane, the solutions of these differential equations diverge at large distances. The reason for this divergence lies in the fact that one of the functions $h_\ell^{(\pm)}(kr)$ is always divergent if $\mathrm{Im}\, k \neq 0$,

$$h_\ell^{(+)}(kr) \underset{r\to\infty}{\longrightarrow} \infty , \quad \text{for} \quad \mathrm{Im}\, k < 0 , \tag{15.35}$$

$$h_\ell^{(-)}(kr) \underset{r\to\infty}{\longrightarrow} \infty , \quad \text{for} \quad \mathrm{Im}\, k > 0 . \tag{15.36}$$

Both $j_\ell(kr)$ and $n_\ell(kr)$ are linear combinations of $h_\ell^{(\pm)}(kr)$ and therefore also exponentially diverge when $\mathrm{Im}\, k \neq 0$, which makes divergent the expansion coefficients (15.31, 15.32) as well as the solutions of Eqs. (15.33). This technical difficulty can be overcome using the same (complex rotation) trick as is described in Sect. 7.4.

In principle, the power-series expansions (15.27), (15.28) include an infinite number of terms. In practice, however, we may take into account just a few terms ($\sum_{n=0}^{\infty} \approx \sum_{n=0}^{M}$) and therefore have to solve a certain number of equations in the infinite system (15.33). It should be emphasized that not all equations of this system are coupled to one another. Indeed, due to the condition $i + j + k = n$ on their right hand sides, the equations for $\tilde{\alpha}_M$ and $\tilde{\beta}_M$ (for any $M \geqslant 0$) are only linked to the corresponding equations with $n < M$. For example, the first pair of equations,

$$\begin{aligned} \partial_r \tilde{\alpha}_0 &= -\gamma_0 V(\varphi_0 \tilde{\alpha}_0 - \gamma_0 \tilde{\beta}_0) \,, \\ \partial_r \tilde{\beta}_0 &= -\varphi_0 V(\varphi_0 \tilde{\alpha}_0 - \gamma_0 \tilde{\beta}_0) \,, \end{aligned} \tag{15.37}$$

is self-contained and is not linked to any other equation of the system. The second pair is linked only to the first one, and so on. This means that by considering a finite number of equations of this system, we do not introduce an additional truncation error.

15.1.3 Constructing Potentials with Given Properties

It often happens that we need to construct a potential that generates certain spectral points. The method of the power-series expansion can be used for adjusting the potential in such a way that it generates a pole of the S-matrix at a given energy E_0. For that, we have to modify the potential until it gives

$$f_\ell^{(\text{in})}(E_0) = 0 \,.$$

Taking the Jost function in the form (15.2), where the functions $a_\ell(E)$ and $b_\ell(E)$ are represented by the series (15.3), we see that $f_\ell^{(\text{in})}(E)$ has a root at E_0 if

$$\alpha_0(\ell, E_0) - i k_0^{2\ell+1} \beta_0(\ell, E_0) = 0 \,, \tag{15.38}$$

with $k_0 = \sqrt{2\mu E_0/\hbar^2}$. The sign in front of this square root should be chosen in such a way that the point E_0 is on the appropriate sheet of the Riemann surface. Since

$$\alpha_0(\ell, E_0) = \frac{1}{2}\tilde{\alpha}_0(\ell, E_0, \infty) \qquad\qquad \beta_0(\ell, E_0) = \frac{1}{2}\tilde{\beta}_0(\ell, E_0, \infty) \,,$$

we only need to solve the pair of differential equations (15.37) and vary the parameters of the potential until the condition (15.38) is achieved. Of course, for a resonance spectral point, Eqs. (15.37) should be solved along the complex path as is described in the previous section.

15.1.4 Expansion Coefficients: Fitting the Data

As we have established (see Sect. 15.1), the S-matrix can be approximated by the following ratio:

$$S_\ell(E) \approx \frac{\sum\limits_{n=0}^{M}(\alpha_n + ik^{2\ell+1}\beta_n)(E-E_0)^n}{\sum\limits_{n=0}^{M}(\alpha_n - ik^{2\ell+1}\beta_n)(E-E_0)^n}\,, \tag{15.39}$$

where α_n and β_n are unknown parameters. This approximation is constructed in such a way that the proper analytic structure of the S-matrix is automatically guaranteed no matter how many terms are taken into account in the expansions of the numerator and the denominator. It was also found that, if the central point, E_0, of the expansion is taken on the real axis, then all the unknown parameters, α_n and β_n, are real as well.

The S-matrix can be used for calculating any experimental observable. In this way, that observable becomes the function of the unknown parameters, which can be adjusted to fit the data. Suppose, for example, that we have sets of experimental cross sections for several partial waves:

$$\begin{array}{lll} \sigma_\ell\left(E_i^{(\ell)}\right) \pm \delta_i^{(\ell)}\,, & & i = 1, 2, 3, \dots, N^{(\ell)}\,, \\ \sigma_{\ell'}\left(E_j^{(\ell')}\right) \pm \delta_j^{(\ell')}\,, & & j = 1, 2, 3, \dots, N^{(\ell')}\,, \\ \dots\dots\dots\dots & \text{etc.} & \dots\dots \end{array}$$

Here σ_ℓ and $\delta^{(\ell)}$ are the cross section in the partial wave ℓ and the corresponding standard deviation, while the energy interval covered by the points $E_i^{(\ell)}$ is around the energy E_0 where we expect to find a resonance. In order to parametrize the S-matrix, we construct the χ^2 function

$$\chi^2 = \sum_{i=1}^{N^{(\ell)}}\left[\frac{\sigma_\ell(E_i^{(\ell)}) - \sigma_\ell^{\text{fit}}(E_i^{(\ell)})}{\delta_i^{(\ell)}}\right]^2 \tag{15.40}$$

$$+\sum_{j=1}^{N^{(\ell')}}\left[\frac{\sigma_{\ell'}(E_j^{(\ell')}) - \sigma_{\ell'}^{\text{fit}}(E_j^{(\ell')})}{\delta_j^{(\ell')}}\right]^2 + \cdots \text{ etc.}\,,$$

where the fitting cross section for the partial wave ℓ (see Eq. (4.23)),

$$\sigma_\ell^{\text{fit}}(E) = \frac{\pi}{k^2}(2\ell+1)\left|S_\ell(E) - 1\right|^2\,, \tag{15.41}$$

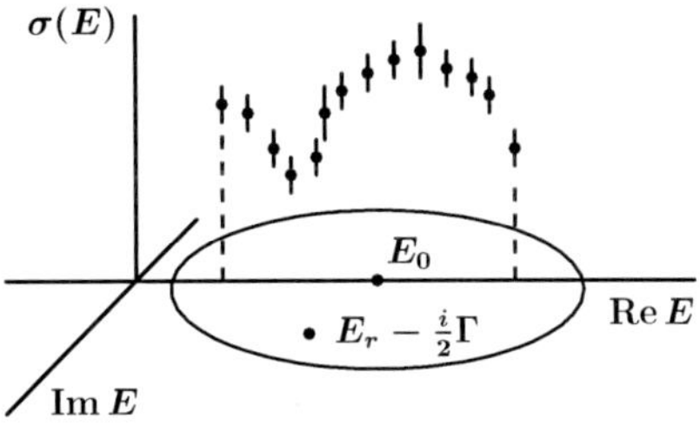

Fig. 15.1 The data points fitted within an interval centred at E_0 on the real axis give us the S-matrix valid within a circle of the adjacent complex domain, where the resonance singularities can be found

depends on the expansion coefficients α_n and β_n in Eq. (15.39). These coefficients therefore serve as the fitting parameters.

After finding the optimal parameters, we obtain analytic expressions for the Jost functions and the S-matrix, Eq. (15.39), valid within a circle around E_0 on all the sheets of the Riemann surface, which is schematically illustrated in Fig. 15.1. Using these expressions, we should be able not only to locate the nearest poles (resonances) of the S-matrix but also to calculate any other observable for which we do not have experimental data.

There is a possibility that some of the poles we locate are spurious, i.e. appearing as a result of the truncation of the Taylor series in the numerator and the denominator of Eq. (15.39). The reason for possible appearance of the spurious poles can be understood using the following simple example. Consider the exponential function e^z that does not have zeros anywhere on the complex z-plane. However, if it is approximated by a finite number of the Taylor terms, it becomes a polynomial that has M zeros, where M is its highest power. Apparently, all such zeros are spurious.

In our case, each S-matrix pole corresponds to a zero of the Jost function in the denominator of Eq. (15.39). Of course this is not a polynomial of the variable E. However, if considered as a function of the momentum k, it becomes a polynomial. This is why the spurious poles may appear. Intuitively, it is clear that such spurious zeros of the Jost function should be somewhere outside the circle, within which the Taylor expansion is sufficiently accurate. However, when extracting the resonance parameters from experimental data in this way, one should be careful and do an additional check.

As with any spurious solutions, there is a simple recipe to identify them: they are unstable and drastically change as a result of any small changes in the parameters of the problem. In our problem, there is a parameter whose choice is rather arbitrary and the results should not depend on such a choice. This is the center E_0 of the Taylor expansion. This expansion is valid within a circle around E_0, and therefore it should be chosen somewhere not far from the energy where it is expected to find resonances. Of course, for different choices of E_0, the fitting parameters α_n and β_n are different, but the non-spurious zeros of the Jost function must be the same. Therefore, in order to exclude any spurious poles, we only need to perform the fit with two different values of E_0 and take only those poles that are stable against the change of E_0.

15.2 Single-Channel Coulomb-Tailed Potential

If the potential has a Coulomb tail, the Jost functions have the following factorized structure (see Eq. 8.100):

$$f_\ell^{(\text{in/out})}(E) = \frac{e^{\mp i\sigma_\ell}}{C_\ell(\eta)} \left\{ a_\ell(E) - \left[\frac{2\eta h(\eta)}{C_0^2(\eta)} \pm i \right] C_\ell^2(\eta) k^{2\ell+1} b_\ell(E) \right\}, \tag{15.42}$$

where the explicitly given factors are responsible for the square-root and logarithmic branching of the Riemann surface, while the unknown functions $a_\ell(E)$ and $b_\ell(E)$ are single-valued and analytic. The corresponding S-matrix is given by

$$S_\ell = e^{2i\sigma_\ell} \frac{a_\ell - \left(\dfrac{2\eta h}{C_0^2} - i \right) C_\ell^2 k^{2\ell+1} b_\ell}{a_\ell - \left(\dfrac{2\eta h}{C_0^2} + i \right) C_\ell^2 k^{2\ell+1} b_\ell}. \tag{15.43}$$

If the energy is real, all the functions of E in this equation are real. This means that the numerator and the denominator are complex conjugate of each other, i.e. their ratio has unit absolute value and thus (see also Eqs. (8.71) and (8.72))

$$S_\ell = e^{2i\sigma_\ell} e^{2i\delta_\ell^{\text{sc}}}, \tag{15.44}$$

where

$$\delta_\ell^{\text{sc}} = \arg \left[a_\ell - \left(\frac{2\eta h}{C_0^2} - i \right) C_\ell^2 k^{2\ell+1} b_\ell \right] \tag{15.45}$$

is the Coulomb-modified short-range part of the phase shift. For any complex number z, we have

$$z = |z| e^{i \arg z} = |z| \cos(\arg z) + i |z| \sin(\arg z) \qquad \Rightarrow \qquad \cot(\arg z) = \frac{\operatorname{Re} z}{\operatorname{Im} z}.$$

Therefore

$$\cot \delta_\ell^{\text{sc}} = \frac{a_\ell - \dfrac{2\eta h}{C_0^2} C_\ell^2 k^{2\ell+1} b_\ell}{C_\ell^2 k^{2\ell+1} b_\ell} = \frac{a_\ell}{C_\ell^2 k^{2\ell+1} b_\ell} - \frac{2\eta h}{C_0^2}.$$

This gives

$$\left(\cot \delta_\ell^{\text{sc}} + \frac{2\eta h}{C_0^2} \right) C_\ell^2 k^{2\ell+1} = \frac{a_\ell(E)}{b_\ell(E)}. \tag{15.46}$$

Since the functions $a_\ell(E)$ and $b_\ell(E)$ are holomorphic (within the domain $\mathcal{D}$), they can be expanded in the Taylor series around an arbitrary point E_0,

$$a_\ell(E) = \sum_{n=0}^{\infty} \alpha_n(\ell, E_0)(E - E_0)^n \,, \qquad b_\ell(E) = \sum_{n=0}^{\infty} \beta_n(\ell, E_0)(E - E_0)^n \,, \tag{15.47}$$

and thus

$$\left(\cot\delta_\ell^{\mathrm{sc}} + \frac{2\eta h}{C_0^2}\right) C_\ell^2 k^{2\ell+1} = \frac{\alpha_0 + \alpha_1(E - E_0) + \alpha_2(E - E_0)^2 + \cdots}{\beta_0 + \beta_1(E - E_0) + \beta_2(E - E_0)^2 + \cdots} \tag{15.48}$$

which generalizes the corresponding equation (15.12) obtained previously for a short-range potential. The standard effective-range expansion for a Coulomb-tailed potential can be obtained from Eq. (15.48) by putting $E_0 = 0$ and using the same formulae (15.14)–(15.21).

15.2.1 Expansion Coefficients

In exactly the same way as for the neutral particles, the holomorphic functions $a_\ell(E)$ and $b_\ell(E)$ in Eq. (15.42) are the asymptotic values of the corresponding r-dependent functions (see Eq. (8.99)),

$$a_\ell(E) = \frac{1}{2}\lim_{r\to\infty} \tilde{A}_\ell(E, r) \qquad \text{and} \qquad b_\ell(E) = \frac{1}{2}\lim_{r\to\infty} \tilde{B}_\ell(E, r) \,, \tag{15.49}$$

which are the solutions of the system of differential equations,

$$\partial_r \tilde{A}_\ell = \tilde{G}_\ell V \left(\tilde{F}_\ell \tilde{A}_\ell + \tilde{G}_\ell \tilde{B}_\ell\right) \,, \tag{15.50}$$

$$\partial_r \tilde{B}_\ell = -\tilde{F}_\ell V \left(\tilde{F}_\ell \tilde{A}_\ell + \tilde{G}_\ell \tilde{B}_\ell\right) \,, \tag{15.51}$$

with the boundary conditions

$$\tilde{A}_\ell(E, 0) = 1 \,, \qquad \tilde{B}_\ell(E, 0) = 0 \,. \tag{15.52}$$

These differential equations involve the holomorphic parts, $\tilde{F}_\ell(E, r)$ and $\tilde{G}_\ell(E, r)$, of the regular and irregular Coulomb functions, defined by Eqs. (8.87) and (8.88).

Being holomorphic, they can be expanded in the Taylor series around an arbitrary complex point E_0,

$$\tilde{F}_\ell(E,r) = \sum_{n=0}^{\infty} \varphi_n(\ell, E_0, r)(E - E_0)^n \,, \tag{15.53}$$

$$\tilde{G}_\ell(E,r) = \sum_{n=0}^{\infty} \gamma_n(\ell, E_0, r)(E - E_0)^n \,. \tag{15.54}$$

Possible ways of finding the expansion coefficients, $\varphi_n(\ell, E_0, r)$ and $\gamma_n(\ell, E_0, r)$, for these series, are described in Appendix E. It can be shown that they are real for real E_0. Similarly to the neutral particle case, we can look for the solutions of the differential equations (15.50) and (15.51) in the form of the power series,

$$\tilde{A}_\ell(E,r) = \sum_{n=0}^{\infty} \tilde{\alpha}_n(\ell, E_0, r)(E - E_0)^n \,, \tag{15.55}$$

$$\tilde{B}_\ell(E,r) = \sum_{n=0}^{\infty} \tilde{\beta}_n(\ell, E_0, r)(E - E_0)^n \,. \tag{15.56}$$

Substituting the series (15.53–15.56) into these differential equations and equalizing the coefficients of each power of $(E - E_0)$, we obtain the following equations:

$$\boxed{\begin{aligned} \partial_r \tilde{\alpha}_n &= \sum_{i+j+k=n} \gamma_i V (\varphi_j \tilde{\alpha}_k + \gamma_j \tilde{\beta}_k) \\ \partial_r \tilde{\beta}_n &= -\sum_{i+j+k=n} \varphi_i V (\varphi_j \tilde{\alpha}_k + \gamma_j \tilde{\beta}_k) \end{aligned}} \tag{15.57}$$

with the boundary conditions

$$\tilde{\alpha}_n(\ell, E_0, 0) = \delta_{n0} \,, \qquad \tilde{\beta}_n(\ell, E_0, 0) = 0 \,, \qquad \forall\, n \,. \tag{15.58}$$

Formally, these differential equations look almost as Eqs. (15.33) for the neutral particle case. The only difference is in the signs on the right hand side, which stems from the definition of the standard irregular Coulomb function (see Eq. (8.53)). However, the functions $\varphi_n(\ell, E_0, r)$ and $\gamma_n(\ell, E_0, r)$ are different here, and the formulae (15.42) for constructing the Jost functions are more involved. It should also be noted that (in the same way as for the neutral particles) not all the differential equations (15.57) are coupled to one another. Indeed, because of the summation condition $i + j + k = n$, the first pair of the equations couples only $\tilde{\alpha}_0$ and $\tilde{\beta}_0$, the second pair couples $\tilde{\alpha}_0$, $\tilde{\beta}_0$, $\tilde{\alpha}_1$, and $\tilde{\beta}_1$, and so on. For the same reason as for the neutral particles, the integration path for the differential equations (15.57) should be turned into the complex plane (complex rotation) when the energy E_0 is chosen to be complex.

Substituting the expansions (15.47) into the denominator of the S-matrix (15.43), it is easy to see that, if the potential has a Coulomb tail, the condition (15.38) for placing a spectral point at E_0 should be replaced with

$$\alpha_0(\ell, E_0) - \left[\frac{2\eta_0 h(\eta_0)}{C_0^2(\eta_0)} + i\right] C_\ell^2(\eta_0) k_0^{2\ell+1} \beta_0(\ell, E_0) = 0\,, \tag{15.59}$$

where $\eta_0 = (\mu e^2 Z_1 Z_2)/(k_0 \hbar^2)$.

* * *

If instead of calculating the expansion coefficients α_n and β_n, we want to use them as the free parameters to fit a set of experimental data, then the procedure is the same as described in Sect. 15.1.4. The only difference is that the parametrized expression (15.39) for the S-matrix should be replaced with the more general one, namely,

$$S_\ell(E) \approx \exp(2i\sigma_\ell) \frac{\displaystyle\sum_{n=0}^{M} \left[\alpha_n - \left(\frac{2\eta h}{C_0^2} - i\right) C_\ell^2 k^{2\ell+1} \beta_n\right] (E - E_0)^n}{\displaystyle\sum_{n=0}^{M} \left[\alpha_n - \left(\frac{2\eta h}{C_0^2} + i\right) C_\ell^2 k^{2\ell+1} \beta_n\right] (E - E_0)^n}\,. \tag{15.60}$$

It is convenient to choose the point E_0 on the real axis, as schematically shown in Fig. 15.1. Then the parameters α_n and β_n are real. Otherwise (if, for example, E_0 is shifted to complex energies towards a resonance) they are complex, which means that the number of the free parameters is doubled.

15.3 Multi-channel short-Range Potential

As it was established in Sec. 13.4, the multi-channel Jost matrices have the following structure:

$$f^{(\text{in/out})}(E) = D^{-1} a(E) D \mp i K^{-1} D b(E) D\,, \tag{15.61}$$

where the diagonal matrices K and D involve all the channel momenta, k_n,

$$K = \text{diag}\{k_1, k_2, \ldots, k_N\}\,, \tag{15.62}$$

$$D = \text{diag}\left\{k_1^{\ell_1+1}, k_2^{\ell_2+1}, \ldots, k_N^{\ell_N+1}\right\}\,. \tag{15.63}$$

The matrix K and (generally speaking) the matrix D are multi-valued functions of the energy E. This is because K and (generally speaking) D depend on the odd powers of the channel momenta. In contrast to them, the matrices $a(E)$ and $b(E)$ are single-valued. This means that all the trouble with the branching of the Riemann surface, on which the Jost matrices are defined, is due to the matrices K and D in Eq. (15.61).

Using the Jost matrices (15.61), we obtain the following S-matrix, structured in the explicitly given factors that are responsible for the branching of the Riemann surface and the holomorphic matrix-functions of the variable E:

$$
\begin{aligned}
S(E) &= \left[D^{-1}a(E)D + iK^{-1}Db(E)D\right]\left[D^{-1}a(E)D - iK^{-1}Db(E)D\right]^{-1} \\
&= \left[D^{-1}a(E) + iK^{-1}Db(E)\right]\left[D^{-1}a(E) - iK^{-1}Db(E)\right]^{-1} . \qquad (15.64)
\end{aligned}
$$

The analytic matrix-functions $a(E)$ and $b(E)$ can be expanded in the Taylor series around an arbitrary point E_0 (within the domain $\mathcal{D}$):

$$
a(E) = \sum_{n=0}^{\infty} \alpha_n(E_0)(E - E_0)^n , \qquad b(E) = \sum_{n=0}^{\infty} \beta_n(E_0)(E - E_0)^n , \qquad (15.65)
$$

where the expansion coefficients, α_n and β_n, are the square matrices of the same dimension as the matrices a and b. Apparently, these expansion coefficients depend on the choice of the central point E_0. If we truncate the sums in Eqs. (15.65) at a term number M, the S-matrix can be approximated by the following "ratio":

$$
\begin{aligned}
S(E) \approx & \left[\sum_{n=0}^{M}\left(D^{-1}\alpha_n + iK^{-1}D\beta_n\right)(E - E_0)^n\right] \qquad (15.66) \\
& \times \left[\sum_{n=0}^{M}\left(D^{-1}\alpha_n - iK^{-1}D\beta_n\right)(E - E_0)^n\right]^{-1} .
\end{aligned}
$$

In principle, it is possible to introduce the channel phase shifts, $\delta_{nn'}$, and then to recast the approximation (15.66) in a form involving the channel momenta and $\cot\delta_{nn'}$, i.e. in the form resembling the single-channel effective-range approximation (15.19). However, for an arbitrary number of channels, such a transformation would be extremely cumbersome. Moreover, the parametrization of $\cot\delta_{nn'}$ would not have any advantage over the parametrization of the S-matrix itself, given by Eq. (15.66). It is therefore not worthwhile to perform such a transformation. Instead, we can consider the approximation (15.66) as the multi-channel analog of the effective-range approximation.

The "ratio" (15.66) has the correct analytic structure, i.e. correctly describes the branching of the Riemann surface at all the thresholds, no matter how many terms of the Taylor series are taken into account. We can therefore use it to fit experimental data in the same way as is described in Sect. 15.1.4. In such a fit, the elements of the matrices α_n and β_n serve as the adjustable parameters.

When these parameters are determined, we obtain an analytic expression (15.66) for the S-matrix, which can be calculated anywhere on the Riemann surface. In this way, we can easily locate possible spectral points. The choice of the Riemann sheet, for which the calculation is done, is determined by the choice of the signs in front of the square roots that relate the energy and the channel momenta. These signs should be chosen in such a way that the imaginary parts of the channel momenta have the appropriate signs.

Any possible spurious spectral points can be excluded in the same way as is described for the single-channel case, i.e. by repeating the fit with two different central points E_0.

In Sec. 13.4, it was shown that on the real axis, the matrices $a(E)$ and $b(E)$ are real. This means that if the central point, E_0, is chosen on the real axis, as it is illustrated in Fig. 15.1, then all the expansion parameters in Eqs. (15.65) are real as well. Therefore it is convenient to take E_0 in Eq. (15.66) as a real number when doing the fit of experimental data. In this way the number of the free parameters is reduced by the factor of two.

* * *

The matrices $a(E)$ and $b(E)$ are the asymptotic values of the r-dependent matrix-functions $\tilde{A}(E,r)$ and $\tilde{B}(E,r)$ (see Sec. 13.4),

$$a(E) = \frac{1}{2}\lim_{r\to\infty}\tilde{A}(E,r) \qquad \text{and} \qquad b(E) = \frac{1}{2}\lim_{r\to\infty}\tilde{B}(E,r)\,, \tag{15.67}$$

obeying the system of differential equations,

$$\partial_r\tilde{A} = -\tilde{n}V\left(\tilde{j}\tilde{A} - \tilde{n}\tilde{B}\right)\,, \tag{15.68}$$

$$\partial_r\tilde{B} = -\tilde{j}V\left(\tilde{j}\tilde{A} - \tilde{n}\tilde{B}\right)\,, \tag{15.69}$$

with the E-independent boundary conditions,

$$\begin{aligned} &\tilde{j}(E,r)\tilde{A}(E,r)\big[\tilde{j}(E,r)\big]^{-1} \underset{r\to 0}{\longrightarrow} I\,, \\ &\tilde{n}(E,r)\tilde{B}(E,r)\big[\tilde{j}(E,r)\big]^{-1} \underset{r\to 0}{\longrightarrow} 0 \qquad \Rightarrow \qquad \tilde{B}(E,0) = 0\,. \end{aligned} \tag{15.70}$$

The diagonal matrices

$$\tilde{j}(E,r) = \mathrm{diag}\left\{\tilde{j}_{\ell_1}(E,r), \tilde{j}_{\ell_2}(E,r), \ldots, \tilde{j}_{\ell_N}(E,r)\right\} \tag{15.71}$$

and

$$\tilde{n}(E,r) = \mathrm{diag}\left\{\tilde{n}_{\ell_1}(E,r), \tilde{n}_{\ell_2}(E,r), \ldots, \tilde{n}_{\ell_N}(E,r)\right\} \tag{15.72}$$

are composed of the holomorphic parts of the Riccati–Bessel and the Riccati–Neumann functions. Since they are holomorphic, we can expand them in the Taylor series around an arbitrary complex point E_0,

$$\tilde{j}(E,r) = \sum_{n=0}^{\infty} \varphi_n(E_0,r)(E-E_0)^n \,, \tag{15.73}$$

$$\tilde{n}(E,r) = \sum_{n=0}^{\infty} \gamma_n(E_0,r)(E-E_0)^n \,, \tag{15.74}$$

where the diagonal matrices

$$\varphi_n(E_0,r) = \mathrm{diag}\left\{\varphi_n(\ell_1,E_0,r), \varphi_n(\ell_2,E_0,r), \ldots, \varphi_n(\ell_N,E_0,r)\right\} \tag{15.75}$$

and

$$\gamma_n(E_0,r) = \mathrm{diag}\left\{\gamma_n(\ell_1,E_0,r), \gamma_n(\ell_2,E_0,r), \ldots, \gamma_n(\ell_N,E_0,r)\right\} \tag{15.76}$$

are made of the expansion coefficients (15.31) and (15.32).

Generalizing the derivations of Sect. 15.1.2, we look for the solutions of Eqs. (15.68) and (15.69) in the form of the power series,

$$\tilde{A}(E,r) = \sum_{n=0}^{\infty} \tilde{\alpha}_n(E_0,r)(E-E_0)^n \,, \tag{15.77}$$

$$\tilde{B}(E,r) = \sum_{n=0}^{\infty} \tilde{\beta}_n(E_0,r)(E-E_0)^n \,, \tag{15.78}$$

where $\tilde{\alpha}_n$ and $\tilde{\beta}_n$ are the r-dependent matrices. Substituting these series into Eqs. (15.68, 15.69) and replacing there the matrices $\tilde{j}$ and $\tilde{n}$ with the corresponding series (15.73, 15.74), we obtain a set of equations where the coefficients of the same powers of $(E - E_0)$ on the left and right hand sides should be the same. This gives

$$\boxed{\begin{aligned} \partial_r \tilde{\alpha}_n &= -\sum_{i+j+k=n} \gamma_i V(\varphi_j \tilde{\alpha}_k - \gamma_j \tilde{\beta}_k) \\ \partial_r \tilde{\beta}_n &= -\sum_{i+j+k=n} \varphi_i V(\varphi_j \tilde{\alpha}_k - \gamma_j \tilde{\beta}_k) \end{aligned}} \tag{15.79}$$

The boundary conditions for these equations follow from the fact that Eqs. (15.70) are independent of E, which implies that at $r = 0$ the only non-vanishing coefficient is $\tilde{\alpha}_0$, i.e.

$$\lim_{r\to 0} \tilde{j}(E,r)\tilde{\alpha}_n(E_0,r)\big[\tilde{j}(E,r)\big]^{-1} = \delta_{n0} I\,, \qquad \tilde{\beta}_n(E_0,0) = 0\,, \qquad \forall\, n\,. \tag{15.80}$$

The differential equations (15.79) look exactly like the corresponding equations (15.33) for the single-channel problem. However, here the quantities $\tilde{\alpha}_n$ and $\tilde{\beta}_n$ are not just the functions but are square matrices. In the same way as for the single-channel case, the summation condition $i + j + k = n$ decouples the first M pairs of the equations from any equations with $n > M$.

It is clear that the expansion coefficients α_n and β_n in the series (15.65) are the asymptotic values of the corresponding "tilded" coefficients,

$$\alpha_n(E_0) = \frac{1}{2}\lim_{r\to\infty} \tilde{\alpha}_n(E_0,r) \qquad \text{and} \qquad \beta_n(E_0) = \frac{1}{2}\lim_{r\to\infty} \tilde{\beta}_n(E_0,r)\,. \tag{15.81}$$

Therefore, in order to find α_n and β_n, we have to start with the boundary values (15.80) and (numerically) integrate Eqs. (15.79) up to a faraway point where the potential vanishes and the limits (15.81) are reached (within a required accuracy). The reader can easily check that, similarly to the single-channel case, such a procedure is converging if the energy E_0 is real. When it becomes complex, the integration path has to be turned into the complex r-plane as is shown in Fig. 7.3.

In the single-channel case, it was shown that in order to have a spectral point at a given energy E_0, we need to adjust the potential until the condition (15.38) is fulfilled. In the multi-channel problem, the steps towards constructing the potential with a given spectral point are similar. Indeed, as is seen from Eq. (15.64), the energy E_0 is a spectral point if

$$\det\Big[D^{-1}a(E_0) - iK^{-1}Db(E_0)\Big] = \det\Big\{D^{-1}\Big[a(E_0) - iK^{-1}D^2 b(E_0)\Big]\Big\} = 0\,.$$

Since the determinant of a product of matrices is equal to the product of their determinants, we have the following condition:

$$\det\Big[a(E_0) - iK^{-1}D^2 b(E_0)\Big] = 0\,, \tag{15.82}$$

where the matrices K and D involve the channel momenta corresponding to the energy E_0, and the signs in calculating these channel momenta are chosen such that put the point E_0 on the appropriate sheet of the Riemann surface. If we use the expansions (15.65), then

$$a(E_0) = \alpha_0(E_0) \qquad \text{and} \qquad b(E_0) = \beta_0(E_0)\,.$$

This gives

$$\det\Big[\alpha_0(E_0) - iK^{-1}D^2\beta_0(E_0)\Big] = 0\,, \tag{15.83}$$

which generalizes the corresponding condition (15.38).

15.4 Multi-channel Coulomb-Tailed Potential

If at least in one of the channels the potential has a Coulomb tail, then the Jost matrices involve the "troublesome" Coulomb-related functions (see Eq. (14.68)),

$$f^{(\mathrm{in/out})}(E) = \sigma^{(\mp)}\Big[D^{-1}a - (M \pm i)K^{-1}Db\Big]DC^{-1}\,, \tag{15.84}$$

where the "troubles" come from the energy-dependent multi-valued matrices K, C, $\sigma^{(\mp)}$, D, and M, which are defined by Eqs. (14.12), (14.18), (14.25), (14.53), and (14.54), respectively. These explicitly given multi-valued matrices are responsible for the branching of the Riemann surface, while the unknown matrices $a(E)$ and $b(E)$ are analytic and single-valued, i.e. are defined on a simple energy plane without any branch points.

It is not difficult to repeat all the reasoning given in the previous section (Sect. 15.3) for a short-range multi-channel potential and to obtain the Coulomb generalization of all the formulae derived there. For the reader's convenience, these generalized formulae are given here (without derivations, which can be easily done).

The Jost matrices (15.84) give the following S-matrix:

$$S(E) = \sigma^{(+)}\Big[D^{-1}a - (M - i)K^{-1}Db\Big]\Big[D^{-1}a - (M + i)K^{-1}Db\Big]^{-1}\sigma^{(+)}\,, \tag{15.85}$$

where the analytic matrix-functions $a(E)$ and $b(E)$ can be expanded in the Taylor series around an arbitrary point E_0 (within the domain $\mathcal{D}$):

$$a(E) = \sum_{n=0}^{\infty}\alpha_n(E_0)(E - E_0)^n\,, \qquad b(E) = \sum_{n=0}^{\infty}\beta_n(E_0)(E - E_0)^n\,. \tag{15.86}$$

Using the truncated sums, we can approximate the S-matrix as follows:

$$S(E) \approx \sigma^{(+)}\left\{\sum_{n=0}^{M}\Big[D^{-1}\alpha_n - (M - i)K^{-1}D\beta_n\Big](E - E_0)^n\right\} \tag{15.87}$$

$$\times\left\{\sum_{n=0}^{M}\Big[D^{-1}\alpha_n - (M + i)K^{-1}D\beta_n\Big](E - E_0)^n\right\}^{-1}\sigma^{(+)}\,,$$

which can be considered as the multi-channel analog of the effective-range approximation for a Coulomb-tailed potential. Eq. (15.87) correctly describes the branching of the Riemann surface at all the thresholds, no matter how many terms in the expansions are taken into account. It can be used to fit experimental data in the same way as is described in the preceding sections. In such a fit, the elements of the matrices α_n and β_n serve as the adjustable parameters. If the central point, E_0, is chosen on the real axis, then all these parameters are real as well.

Alternatively (to fitting the data), the expansion coefficients α_n and β_n can be calculated, if the potential is known. They are the asymptotic values of the corresponding r-dependent "tilded" coefficients,

$$\alpha_n(E_0) = \frac{1}{2}\lim_{r\to\infty}\tilde{\alpha}_n(E_0, r) \qquad \text{and} \qquad \beta_n(E_0) = \frac{1}{2}\lim_{r\to\infty}\tilde{\beta}_n(E_0, r)\,, \tag{15.88}$$

which obey the system of differential equations (which can be derived in exactly the same way as it is done in the preceding section for the case of neutral particles),

$$\boxed{\begin{aligned} \partial_r\tilde{\alpha}_n &= \sum_{i+j+k=n} \gamma_i V(\varphi_j\tilde{\alpha}_k + \gamma_j\tilde{\beta}_k) \\ \partial_r\tilde{\beta}_n &= -\sum_{i+j+k=n} \varphi_i V(\varphi_j\tilde{\alpha}_k + \gamma_j\tilde{\beta}_k) \end{aligned}} \tag{15.89}$$

with the following simple boundary conditions:

$$\lim_{r\to 0}\tilde{F}(E, r)\tilde{\alpha}_n(E_0, r)\left[\tilde{F}(E, r)\right]^{-1} = \delta_{n0}I\,, \qquad \tilde{\beta}_n(E_0, 0) = 0\,, \qquad \forall\, n\,. \tag{15.90}$$

In Eqs. (15.89), the diagonal matrices

$$\varphi_n(E_0, r) = \text{diag}\,\{\varphi_n(\ell_1, E_0, r), \varphi_n(\ell_2, E_0, r), \ldots, \varphi_n(\ell_N, E_0, r)\} \tag{15.91}$$

and

$$\gamma_n(E_0, r) = \text{diag}\,\{\gamma_n(\ell_1, E_0, r), \gamma_n(\ell_2, E_0, r), \ldots, \gamma_n(\ell_N, E_0, r)\} \tag{15.92}$$

are made of the corresponding expansion coefficients in the Taylor series (15.53) and (15.54) representing the holomorphic parts of the regular and irregular Coulomb functions. A way of finding these coefficients is described in Appendix E.

In order to find α_n and β_n, the (numerical) integration of the differential Eqs. (15.89) should start with the boundary values (15.90) at $r = 0$ and should proceed up to a faraway point where the short-range part, V, of the potential vanishes. Such a procedure is converging if the energy E_0 is real. When it becomes complex, the integration path has to be turned into the complex r-plane as is shown in Fig. 7.3.

In order to adjust the potential for putting a spectral point at the energy E_0, we only need to solve just the first pair ($n = 0$) of the differential equations (15.89) and check if the following condition is fulfilled:

$$\det\left[\alpha_0(E_0) - (M + i)K^{-1}D^2\beta_0(E_0)\right] = 0\,, \tag{15.93}$$

which generalizes the corresponding short-range-potential condition (15.83).

* * *

Examples of numerical calculations where the methods described in this chapter are used in some model and real physical problems can be found in Refs. [58–63].

Part III

Special Issues

This part of the book is mainly devoted to some special problems where it is not possible to directly use the methods described in the preceding sections. It is shown here how such problems can be approached on the same basis (i.e. using the Jost matrices) but with the necessary modifications and with the help of some special tricks.

In addition to that this part includes Chap. 18 (the last one) that illustrates some of the properties of the Jost functions and matrices using several exactly solvable models. Within these models the Jost matrices are obtained in the explicit form which demonstrates their analytic structure established in the preceding sections for the general case.

Singular and Low-Dimensional Potentials 16

16.1 Singular Potential

In all preceding chapters, we always assumed that the potential was less singular than $1/r^2$ at the origin (see Eq. (2.2)). Therefore, the methods for calculating the Jost functions developed so far cannot be directly applied to potentials, which are more singular than the centrifugal barrier. A significant number of practical problems, however, where the Jost function could be very useful, involve such potentials.

For example, interatomic and intermolecular forces at short distances are strongly repulsive due to the overlap of the electron clouds, and thus they are usually represented by repulsive singular potentials such as the Lennard–Jones-(12,6) one, which has $\sim 1/r^{12}$ behaviour.[1] It is, therefore, desirable to extend the Jost function method for potentials of this kind.

It is well known that the attractive singular potentials do not admit physically meaningful solutions with the usual boundary conditions [67]. Intuitively, it is easy to understand that a too strong attraction would cause unphysical fall of the particle into the "black hole" at the origin. This can be seen as follows. Suppose a potential behaves near the centre like

$$U(r) \underset{r\to 0}{\longrightarrow} \frac{\beta}{r^s} . \tag{16.1}$$

Consider a wave function that is finite in a small sphere of a radius r_0 around the centre and is zero outside this sphere. This means that we somehow measured the

[1] See, for example, Refs. [64–66].

S. A. Rakityansky, *Jost Functions in Quantum Mechanics*,
https://doi.org/10.1007/978-3-031-07761-6_16

position of the particle and found that it is located near the point $r = 0$ with the uncertainty

$$\Delta r \sim r_0 ,$$

which gives the uncertainty of the momentum

$$\Delta p \sim \frac{\hbar}{r_0} .$$

These uncertainties correspond to the kinetic energy

$$E_{\text{kin}} \sim \frac{(\Delta p)^2}{2\mu} \sim \frac{\hbar^2}{2\mu r_0^2}$$

and to the potential energy

$$E_{\text{pot}} \sim \frac{\beta}{(\Delta r)^s} \sim \frac{\beta}{r_0^s} .$$

The total energy

$$E = E_{\text{kin}} + E_{\text{pot}} \sim \frac{\hbar^2}{2\mu r_0^2} + \frac{\beta}{r_0^s}$$

is a sum of two terms. Depending on which of them dominates, the total energy may go to plus or minus infinity when $r_0 \to 0$. Apparently, for an attractive singular potential ($\beta < 0$ and $s > 2$), the total energy goes to $-\infty$, which means that the particle can be bound with infinitely large binding energy (i.e. this corresponds to physically meaningless fall to the centre). For a repulsive potential ($\beta > 0$) with any value of s, the total energy goes to $+\infty$, i.e. the particle can only reach the point $r = 0$ if it approaches it with infinitely high collision energy (which is physically meaningful). We therefore exclude attractive singular potentials from our further consideration.

Although the repulsive singular potentials do not pose any problem regarding physical interpretation, the integration of the Schrödinger equation as well as that of the relevant equations for the Jost functions has inherent mathematical difficulties resulting from the fact that the singularity of the potential makes the point $r = 0$ an irregular singular point of the equation. In particular, the regular solution cannot be defined by universal boundary conditions independent of the potential. This drawback, stemming from the extremely strong repulsion near the origin, can fortunately be tackled by the WKB approximation, which provides the correct radial behaviour of the wave function in the neighbourhood of the point $r = 0$.

In this way, using the WKB approximation, one can find the regular solution of the Schrödinger equation on a small interval from $r = 0$ to some intermediate point

$r_{\rm int}$. This gives the values of the wave function and its first derivative at $r_{\rm int}$, i.e. the boundary conditions for finding the solution on the whole interval $[r_{\rm int}, \infty)$. Using these boundary conditions, the equations for the Jost functions can be integrated, from $r_{\rm int}$ outwards, the same way as for the non-singular potential.

16.1.1 Boundary Conditions

Let $U(r)$ be a repulsive singular potential. The particle moving under the influence of such a force (with strong repulsion at the origin) can only approach the point $r = 0$ with a very low probability. This means that the wave function diminishes very fast when $r \to 0$, and therefore, we should expect that the usual behaviour of the regular solution given by (see Eq. (2.36))

$$\lim_{r\to 0} \frac{\phi_\ell(E,r)}{j_\ell(kr)} = 1 \tag{16.2}$$

is not valid anymore. The function $\phi_\ell(E,r)$ tends to zero much faster (depending on the behaviour of the potential) than $j_\ell(kr)$, i.e.

$$\phi_\ell(E,r) = o\left[j_\ell(kr)\right] , \qquad r \to 0 .$$

This however does not give us a recipe for imposing meaningful boundary conditions on $\phi_\ell(E,r)$ near $r = 0$.

In order to find such a recipe, let us look for the solution of the radial Schrödinger equation using the WKB approximation (see Sect. 1.4). Formal criterion for applicability of this approximation is given by Eq. (1.74). Physical interpretation of this condition can be deduced from an alternative name for the WKB approximation, namely, the semi-classical or short-wave approximation. It is applicable when the local wavelength is so short that the potential is practically constant on an interval covering many periods of the wave, i.e. the wave function oscillates more fast than the potential changes, and thus the wavelength is locally almost constant, which means that the derivative of the wavelength is almost zero.

The radial Schrödinger equation can be written in the form

$$\left[\frac{d^2}{dr^2} + K^2(r)\right]\phi_\ell(E,r) = 0 , \tag{16.3}$$

where

$$K(r) = \sqrt{k^2 - V(r) - \ell(\ell+1)/r^2} \tag{16.4}$$

is a function of r that (if the WKB approximation is applicable) varies slowly as compared to oscillations of $\phi_\ell(E,r)$. Apparently, if $K(r)$ is almost constant, then

Eq. (16.3) has a regular solution $\sim \sin(Kr)$, and therefore, K is the local wave number that corresponds to the local wavelength

$$\lambda(r) = \frac{2\pi}{K(r)} . \tag{16.5}$$

The derivative of this wavelength is

$$\frac{d\lambda}{dr} = \frac{\pi \left[\frac{dV}{dr} - \frac{2\ell(\ell+1)}{r^3} \right]}{[k^2 - V(r) - \ell(\ell+1)/r^2]^{3/2}} . \tag{16.6}$$

If the potential is given by Eq. (16.1) with $s > 2$, then at very short distances the constant k^2 as well as the centrifugal term can be neglected because

$$|V(r)| \gg \ell(\ell+1)/r^2 \quad \text{and} \quad \left|\frac{dV}{dr}\right| \gg 2\ell(\ell+1)/r^3 , \qquad r \to 0 .$$

For the derivative (16.6), this gives

$$|d\lambda/dr| \underset{r\to 0}{\longrightarrow} \pi \left| \frac{dV}{dr} V^{-3/2} \right| = \frac{\pi s r^{(s-2)/2}}{\sqrt{2\mu\beta/\hbar^2}} \underset{r\to 0}{\longrightarrow} 0,$$

and therefore, the WKB approximation is valid near the point $r = 0$. Using this approximation on the interval $(0, r_{\text{int}}]$, we obtain[2]

$$\phi_\ell(E, r) = \frac{1}{\sqrt{K(r)}} \exp\left[i \int_r^a K(r')dr' \right] , \qquad r \in (0, r_{\text{int}}] , \tag{16.7}$$

where a is an arbitrary radius such that $a > r_{\text{int}}$. It can be taken, for example, at the classical turning point. It is obvious that an additional integration caused by increasing of a can only change the overall normalization of the solution, which is not of our concern.

Looking at Eq. (16.4), we see that at very small distances the potential under the square root dominates, and $K(r)$ becomes pure imaginary

$$K(r) \underset{r\to 0}{\longrightarrow} \sqrt{-V(r)} = i\sqrt{|V(r)|} ,$$

which makes the wave function (16.7) exponentially diminishing when r approaches zero (as we expected).

[2] As was mentioned at the end of Sect. 1.4, in all the formulae related to the WKB approximation for the radial Schrödinger equation, the product $\ell(\ell+1)$ should be replaced with $(\ell+\frac{1}{2})^2$, which is known as the Langer's correction. The details can be found in Ref. [7].

The derivative of the wave function can be easily found by explicit differentiation of Eq. (16.7). This gives

$$\partial_r \phi_\ell(E,r) = \left(-\frac{K'}{2K^{3/2}} - iK^{1/2}\right) \exp\left[i\int_r^a K(r')dr'\right]$$

$$= \left[\frac{\partial_r V}{4K^{5/2}} - \frac{\ell(\ell+1)}{2K^{5/2}r^3} - iK^{1/2}\right] \exp\left[i\int_r^a K(r')dr'\right] . \quad (16.8)$$

The wave function can be expressed in terms of the functions $F_\ell^{(\mathrm{in/out})}(E,r)$ in the usual way (2.12),

$$\phi_\ell(E,r) = h_\ell^{(-)}(kr) F_\ell^{(\mathrm{in})}(E,r) + h_\ell^{(+)}(kr) F_\ell^{(\mathrm{out})}(E,r) ,$$

and with the help of the Lagrange condition (2.13), for its derivative, we obtain

$$\partial_r \phi_\ell(E,r) = \left[\partial_r h_\ell^{(-)}(kr)\right] F_\ell^{(\mathrm{in})}(E,r) + \left[\partial_r h_\ell^{(+)}(kr)\right] F_\ell^{(\mathrm{out})}(E,r) .$$

Combining the last two equations, we have

$$\left[\partial_r h_\ell^{(\pm)}(kr)\right] \phi_\ell(E,r) - h_\ell^{(\pm)}(kr) \partial_r \phi_\ell(E,r) \quad (16.9)$$

$$= \left\{\left[\partial_r h_\ell^{(\pm)}(kr)\right] h_\ell^{(\mp)}(kr) - h_\ell^{(\pm)}(kr) \left[\partial_r h_\ell^{(\mp)}(kr)\right]\right\} F_\ell^{(\mathrm{in/out})}(E,r) ,$$

where on the right hand side there is the Wronskian

$$\left[\partial_r h_\ell^{(+)}(kr)\right] h_\ell^{(-)}(kr) - h_\ell^{(+)}(kr) \left[\partial_r h_\ell^{(-)}(kr)\right] = 2ik .$$

From Eq. (16.9), we finally obtain the boundary conditions for the functions $F_\ell^{(\mathrm{in/out})}(E,r)$,

$$F_\ell^{(\mathrm{in})}(E,r_{\mathrm{int}}) \quad (16.10)$$

$$= \frac{1}{2ik} \left\{\left[\partial_r h_\ell^{(+)}(kr_{\mathrm{int}})\right] \phi_\ell(E,r_{\mathrm{int}}) - h_\ell^{(+)}(kr_{\mathrm{int}}) \partial_r \phi_\ell(E,r_{\mathrm{int}})\right\} ,$$

$$F_\ell^{(\mathrm{out})}(E,r_{\mathrm{int}}) \quad (16.11)$$

$$= -\frac{1}{2ik} \left\{\left[\partial_r h_\ell^{(-)}(kr_{\mathrm{int}})\right] \phi_\ell(E,r_{\mathrm{int}}) - h_\ell^{(-)}(kr_{\mathrm{int}}) \partial_r \phi_\ell(E,r_{\mathrm{int}})\right\} ,$$

where $\phi_\ell(E,r_{\mathrm{int}})$ and $\partial_r \phi_\ell(E,r_{\mathrm{int}})$ are given by Eqs. (16.7) and (16.8).

In summary: when dealing with a repulsive singular potential, we choose a small enough radius r_{int} such that on the interval $(0, r_{\text{int}}]$ the WKB approximation is sufficiently accurate,[3] and using the boundary conditions (16.10), (16.11), solve the differential equations for the functions $F_\ell^{(\text{in/out})}(E, r)$ on the interval $[r_{\text{int}}, R_{\max}]$ in the usual way. If the rotation is needed, then the integration should follow the deformed path shown in Fig. 7.7, where $R_1 \geqslant r_{\text{int}}$. An example of using the described approach in the analysis of the interatomic He–He interaction can be found in Ref. [68].

As is discussed in Sect. 2.2.1.2, in numerical calculations, there is a technical problem of cancellation of singularities of the Riccati–Hankel functions near the point $r = 0$. This difficulty becomes significant for large ℓ but can be avoided if near the origin the equations for $F_\ell^{(\text{in/out})}(E, r)$ are replaced by their linear combinations whose solutions are the functions $A_\ell(E, r)$ and $B_\ell(E, r)$ defined by Eqs. (2.28, 2.29). Making the corresponding combinations of Eqs. (16.10), (16.11), we find that the boundary conditions for the functions $A_\ell(E, r)$ and $B_\ell(E, r)$ are expressed via the following Wronskians:

$$A_\ell(E, r_{\text{int}}) = \frac{1}{k} W\big[\phi_\ell(E, r), n_\ell(kr)\big]\bigg|_{r=r_{\text{int}}}, \tag{16.12}$$

$$B_\ell(E, r_{\text{int}}) = -\frac{1}{k} W\big[\phi_\ell(E, r), j_\ell(kr)\big]\bigg|_{r=r_{\text{int}}}. \tag{16.13}$$

In this form, it is clear that for a non-singular potential, when near the origin the regular solution, $\phi_\ell(E, r)$, behaves as $j_\ell(kr)$, these boundary conditions assume the standard form given by Eqs. (2.33), (2.34). The derivatives of the functions $n_\ell(kr)$ and $j_\ell(kr)$, involved in Eqs. (16.12), (16.13), can be found using the relations (D.23), (D.24) given in Appendix D.1.

16.2 One-Dimensional Problems

Practically, all textbooks on quantum mechanics consider one-dimensional ($1D$) problems as simplified examples of quantum systems. As a result, most of us have a perception that one-dimensional problems are just simple "quantum toys" having no relation to reality. This, however, is wrong.

First, many practical problems of modern semiconductor technology, in fact, can be considered as the one-dimensional scattering problems.[4] The artificially grown semiconductor crystals are composed of alternating layers of different materials with nanometre thickness. For the electrons and holes moving perpendicular to the

[3] The accuracy can be checked by repeating the calculations with different values of r_{int}.

[4] See, for example, Refs. [69–71].

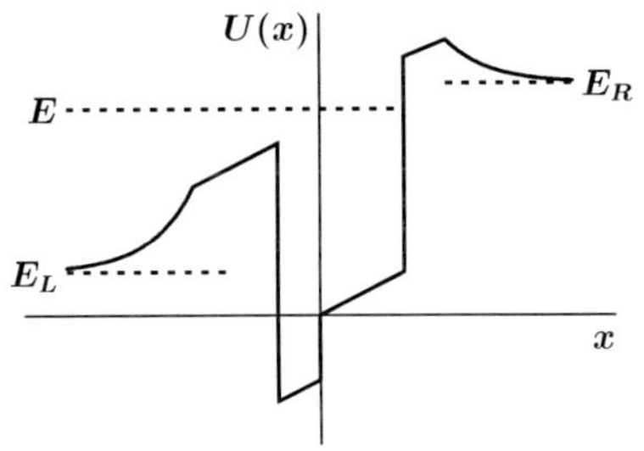

Fig. 16.1 An example of a biased one-dimensional potential, whose asymptotic values E_L and E_R are the threshold energies for the left and right channels. With the collision energy $E_L < E < E_R$ (shown by the dashed line), the only possible process is the elastic scattering in the left channel

layers, these layers represent a one-dimensional alternating sequence of potential wells and barriers. By adjusting the chemical composition and thickness of the layers, it is possible, in principle, to construct a device with a given potential profile.

And, second, the one-dimensional problems are not simple as it may seem. Suffice it to say that they are inherently two-channel problems (otherwise the corresponding S-matrix could not have the dimension 2×2).

The two channels involved in a $1D$-problem are the motion on the left and right halves, $x \in (-\infty, 0]$ and $x \in [0, +\infty)$, of the line. The left and right channels are opened at the corresponding threshold energies E_L and E_R (see Fig. 16.1). Elastic scattering, i.e. pure reflection, is only possible when one of the channels is closed (e.g. when $E < E_R$). The transmission is the inelastic process of the transition between the channels, $L \leftrightarrow R$. This is only possible when the collision energy E is above both thresholds.

When the left and right thresholds are different, $E_L \neq E_R$, the system is called biased. In semiconductor nano-structures, such a bias is usually achieved by applying an electric field perpendicular to the layers, i.e. in parallel with the motion of the electrons.

16.2.1 Schrödinger Equation for a $1D$ System

The one-dimensional problem can be reformulated in such a way that it becomes analogous to the three-dimensional one with zero angular momentum [72]. After that, almost everything described in the previous sections of this book becomes applicable to it.

In order to make things similar to the three-dimensional multi-channel theory, let us replace the "double-infinite" coordinate, $x \in (-\infty, +\infty)$, with the positive coordinate, $r \in [0, +\infty)$, defined as $r = |x|$,

$$r = \begin{cases} -x \ , \ \text{for } L\text{-channel} \ , \\ \\ +x \ , \ \text{for } R\text{-channel} \ , \end{cases} \tag{16.14}$$

and introduce two different potentials:

$$U_L(r) = U(-r) - E_L \tag{16.15}$$

and

$$U_R(r) = U(+r) - E_R \ , \tag{16.16}$$

acting in the L and R channels, respectively. The introduction of r can be visualized as folding of the left half of the xU-plane (see Fig. 16.1) around vertical axis as shown in Fig. 16.2, i.e. like folding of a sheet of paper.

When moving between $r = 0$ and $r = \infty$, a particle actually moves between $x = 0$ and $x = \pm\infty$, depending on the channel which it is in, and therefore passes through the corresponding potential $U_n(r)$ (where $n = L, R$) with the corresponding collision energy $E - E_n$. This is exactly what we have when deal with the three-dimensional multi-channel systems. For example, the dynamic variable r (the distance) characterizing motion of an electron that collides with an atom is the same no matter which state the atom was excided to, but the electron–atom interaction as well as the collision energy is different in different channels.

When the projectile particle passes the origin ($x = 0$), it transits to another channel. In the picture with the single variable r, this means that it moves backwards (scattered) in another channel. If $E_L \neq E_R$, the particle loses or gains the amount of energy $\Delta E = |E_L - E_R|$. Therefore, such a transition can formally be considered as an excitation (or de-excitation) of some internal degrees of freedom of the system (the target) with the excitation energy ΔE. For example, in the case of a semiconductor nano-structure, the solid body (the crystal), inside which the electron

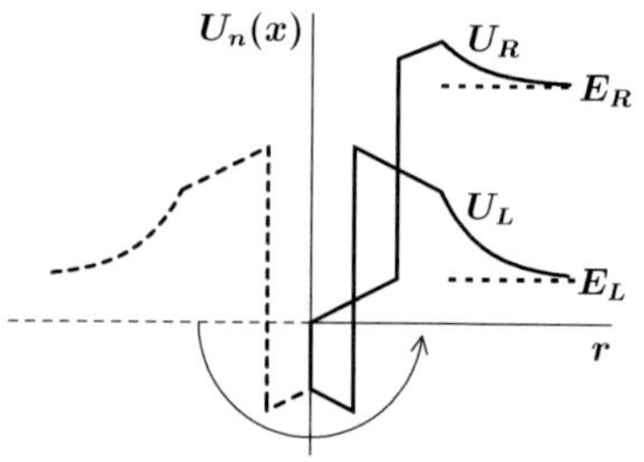

Fig. 16.2 Mapping of the "double-infinite" line $x \in (-\infty, +\infty)$ onto a positive line $r \in [0, +\infty)$ with the potentials $U_L(r)$ and $U_R(r)$ acting in the left and right channels

is moving, serves as that "infinitely large" energy reservoir that takes or gives the energy difference ΔE.

We are not interested in the physical nature of these internal degrees of freedom, and just for the sake of formal derivations, we introduce the "internal hamiltonian" h associated with them and having two eigenstates

$$h|\eta_n\rangle = E_n|\eta_n\rangle \ , \qquad n = L, R \ , \tag{16.17}$$

which are ortho-normal, i.e. $\langle\eta_n|\eta_{n'}\rangle = \delta_{nn'}$. Therefore, $|\eta_n\rangle\langle\eta_n|$ is the operator projecting onto the channel n, and the Hamiltonian of the whole system can be split into the three terms

$$H = H_0 + U + h \ , \tag{16.18}$$

where H_0 describes the free motion of the projectile with the mass μ_n along the r-axis in the channels L and R, i.e.

$$\langle r|H_0|\Psi\rangle = -\frac{\hbar^2}{2}\left[\frac{|\eta_L\rangle\langle\eta_L|}{\mu_L} + \frac{|\eta_R\rangle\langle\eta_R|}{\mu_R}\right]\frac{d^2}{dr^2}\Psi(r) \ , \tag{16.19}$$

and the operator U is the potential

$$U = U_L|\eta_L\rangle\langle\eta_L| + U_R|\eta_R\rangle\langle\eta_R| \ , \tag{16.20}$$

with U_L and U_R being the operators in the r-space, such that

$$\langle r|U_n|\Psi\rangle = U_n(r)\Psi(r) \ , \qquad n = L, R \ . \tag{16.21}$$

The same way as we did it for the $3D$ multi-channel problem (see Eq. (11.5)), let us look for solution of the Schrödinger equation

$$H|\Psi\rangle = E|\Psi\rangle \tag{16.22}$$

in the form of the expansion over the target states, i.e.

$$|\Psi\rangle = |\eta_L\rangle|\phi_L\rangle + |\eta_R\rangle|\phi_R\rangle \ . \tag{16.23}$$

Then, multiplying Eq. (16.22) from the left by $\langle r|\langle\eta_n|$, we obtain Schrödinger equation for wave function of the projectile in the channel n,

$$\left[-\frac{\hbar^2}{2\mu_n}\frac{d^2}{dr^2} + U_n(r) + E_n\right]\phi_n(E,r) = E\phi_n(E,r) \ . \tag{16.24}$$

Introducing the channel momenta,

$$k_n = \sqrt{\frac{2\mu_n}{\hbar^2}(E - E_n)}\,, \tag{16.25}$$

we re-write it as

$$\left[\partial_r^2 + k_n^2\right]\phi_n(E, r) = V_n(r)\phi_n(E, r)\,, \tag{16.26}$$

where

$$V_n(r) = \frac{2\mu_n}{\hbar^2}U_n(r)\,, \qquad n = L, R\,. \tag{16.27}$$

Combining the channel wave functions in a column

$$\psi(E, r) = \begin{bmatrix} \phi_L(E, r) \\ \\ \phi_R(E, r) \end{bmatrix}, \tag{16.28}$$

two equations (16.26) can be written as a single matrix equation,

$$\left[\partial_r^2 + K^2\right]\psi(E, r) = V(r)\psi(E, r)\,, \tag{16.29}$$

where

$$K = \begin{pmatrix} k_L & 0 \\ \\ 0 & k_R \end{pmatrix} \tag{16.30}$$

and

$$V(r) = \begin{bmatrix} V_L(r) & 0 \\ \\ 0 & V_R(r) \end{bmatrix}. \tag{16.31}$$

Eq. (16.29) is a particular case (with zero angular momenta) of the radial multi-channel equation (11.65).

In general, a matrix differential equation of the second order has $2N$ linearly independent column-solutions, if the length of the column is N. This is because for each element of a column, we have to specify two boundary conditions (e.g. the values of this element and of its derivative at certain point), i.e. there are $2N$ boundary conditions in total. At the same time, the number of boundary conditions is always equal to the number of independent solutions because these conditions should uniquely specify the coefficients in the linear combination of the independent solutions, when a general solution is expanded over them.

When considering the radial equation of the N-channel three-dimensional problem, we only use half of the $2N$ independent solutions because the other half is singular at $r = 0$. For the $1D$-problem, there is no requirement of regularity at $r = 0$ because here we do not separate the factor $1/r$ when deriving Eq. (16.29). Despite this, we only use half of the four independent solutions of this equation as well. The reason is that in each independent column-solution, the boundary conditions are common for the left and right channels,

$$\phi_L(E,0) = \phi_R(E,0) \quad \text{and} \quad \phi_L'(E,0) = \phi_R'(E,0)\,, \tag{16.32}$$

because on the x-axis these two functions are the continuations of each other, and therefore, they must smoothly match at the origin. This means that we only have two boundary conditions and thus two independent solutions. By the way, this same matching (16.32) is the only reason why the L and R channels are coupled to each other since the potential (16.31) is diagonal.

Therefore, we have two linearly independent column-solutions, which we distinguish by the subscripts 1 and 2. We can combine them in a square matrix,

$$\begin{bmatrix} \phi_L(E,r) \\ \\ \phi_R(E,r) \end{bmatrix}_1 \oplus \begin{bmatrix} \phi_L(E,r) \\ \\ \phi_R(E,r) \end{bmatrix}_2 \Rightarrow \begin{bmatrix} \phi_{L1}(E,r) & \phi_{L2}(E,r) \\ \\ \phi_{R1}(E,r) & \phi_{R2}(E,r) \end{bmatrix} \equiv \phi(E,r)\,. \tag{16.33}$$

This is the so-called fundamental solution matrix such that any particular solution is a linear combination of its columns.

When $r \to \infty$, all four elements of matrix (16.33) behave like linear combinations of the incoming and outgoing waves in the corresponding channels. If neither of $V_n(r)$ has a long-range tail, then all these waves $W_n^{(\text{in/out})}(r)$ are plane, i.e.

$$W_n^{(\text{in})}(E,r) = h_0^{(-)}(k_n r)\,, \tag{16.34}$$

$$W_n^{(\text{out})}(E,r) = h_0^{(+)}(k_n r)\,, \tag{16.35}$$

where

$$h_0^{(\pm)}(z) = \mp i e^{\pm iz} \tag{16.36}$$

are the Riccati–Hankel functions.

In the theory of one-dimensional processes, the plane waves are usually written as the simple exponentials, $\exp(\pm ikx)$. They differ from the functions (16.36) by the constant phase factors $(\mp i)$ that even do not change their normalization. It is therefore possible to define the plane waves as in Eqs. (16.34) and (16.35). Such a definition makes the $1D$-theory a particular case of the corresponding $3D$-theory with zero angular momentum and thus allows us to use all the formulae derived for the $3D$ multi-channel processes.

If the long-range Coulomb tails are present, the plane waves should be replaced with the Coulombic waves $H_0^{(\pm)}(\eta, kr) = F_0(\eta, kr) \mp iG_0(\eta, kr)$ made of the standard regular F_0 and irregular G_0 Coulomb wave functions, similarly to what was done in Sect. 14 (see Eq. (14.6)). However, to avoid unnecessary complications, in what follows, we assume that the potential is of a short range.

16.2.2 Boundary Conditions

Since in the one-dimensional problem the physical wave function, $\psi^{(\text{phys})}(E, r)$, at $r = 0$ may have any value (not necessarily zero), the boundary conditions for Eq. (16.26) differ from those used for the radial Schrödinger equation in the $3D$-problem.

The function $\psi^{(\text{phys})}(E, r)$ is constructed as a linear combination of the columns of the fundamental solution,[5]

$$\psi^{(\text{phys})}(E, r) = \phi(E, r)\begin{pmatrix} c_1 \\ c_2 \end{pmatrix} = \begin{bmatrix} \phi_{11}(E, r) \\ \phi_{21}(E, r) \end{bmatrix} c_1 + \begin{bmatrix} \phi_{12}(E, r) \\ \phi_{22}(E, r) \end{bmatrix} c_2 \,. \qquad (16.37)$$

To be able to obtain $\psi^{(\text{phys})}(E, 0) \neq 0$, we have to have at least one of the columns of the matrix ϕ non-zero at $r = 0$. The derivative

$$\partial_r \psi^{(\text{phys})}(E, r) = \begin{bmatrix} \phi'_{11}(E, r) \\ \phi'_{21}(E, r) \end{bmatrix} c_1 + \begin{bmatrix} \phi'_{12}(E, r) \\ \phi'_{22}(E, r) \end{bmatrix} c_2 \qquad (16.38)$$

at the origin may also assume any value. Moreover, due to the continuity conditions (16.32), the top and bottom elements (as well as their derivatives) of the matrix ϕ at $r = 0$ must be the same,

$$\phi_{11}(E, 0) = \phi_{21}(E, 0) \quad , \quad \phi_{12}(E, 0) = \phi_{22}(E, 0) \,,$$

$$\phi'_{11}(E, 0) = \phi'_{21}(E, 0) \quad , \quad \phi'_{12}(E, 0) = \phi'_{22}(E, 0) \,.$$

We should not forget one more requirement: the values of ϕ and ϕ' at the origin should guarantee linear independence of the columns of $\phi(E, r)$.

---

[5] For the sake of simplicity, we use here the subscripts 1 and 2 instead L and R, which should not cause confusion.

Any given values of $\psi^{(\text{phys})}(E, 0)$ and $\partial_r \psi^{(\text{phys})}(E, 0)$ can be obtained with an appropriate choice of the coefficients c_1 and c_2, if at $r = 0$ one of the columns of ϕ is zero and the corresponding column of ϕ' is non-zero, while for the other column, ϕ is non-zero and ϕ' is zero. It is clear that these non-zero column values can be any (it is matter of convenience) because the given values of $\psi^{(\text{phys})}(E, 0)$ and $\partial_r \psi^{(\text{phys})}(E, 0)$ are obtained by the choice of c_1 and c_2. The zero and non-zero values of the columns automatically make them linearly independent.

Therefore, it is possible to choose the boundary conditions for the fundamental solution matrix in an infinite number of ways, all within the above requirements. One of the most simple choices is the following:

$$\phi(E, 0) = \begin{pmatrix} 1 & 0 \\ 1 & 0 \end{pmatrix} \qquad \text{and} \qquad \partial_r \phi(E, r)|_{r=0} = \begin{pmatrix} 0 & 1 \\ 0 & 1 \end{pmatrix} . \tag{16.39}$$

With these conditions, the values of the functions (16.37) and (16.38) at $r = 0$ are equal to c_1 and c_2, respectively.

16.2.3 Jost Matrices

The asymptotic behaviour of the fundamental solution $\phi(E, r)$ defined by Eq. (16.33) looks the same as for the multi-channel problem (with $N = 2$ and $\ell_1 = \ell_2 = 0$), discussed in Chap. 12,

$$\phi(E, r) \underset{r \to \infty}{\longrightarrow} W^{(\text{in})}(E, r) f^{(\text{in})}(E) + W^{(\text{out})}(E, r) f^{(\text{out})}(E) , \tag{16.40}$$

where

$$W^{(\text{in/out})}(E, r) = \begin{bmatrix} W_L^{(\text{in/out})}(E, r) & 0 \\ 0 & W_R^{(\text{in/out})}(E, r) \end{bmatrix} \tag{16.41}$$

and $f^{(\text{in/out})}(E)$ are the Jost matrices. The same way as in Sect. 12.1, the fundamental solution can be expressed via the unknown matrix-functions $F^{(\text{in/out})}(E, r)$,

$$\phi(E, r) = W^{(\text{in})}(E, r) F^{(\text{in})}(E, r) + W^{(\text{out})}(E, r) F^{(\text{out})}(E, r) , \tag{16.42}$$

obeying the Lagrange condition,

$$W^{(\text{in})}(E, r) \partial_r F^{(\text{in})}(E, r) + W^{(\text{out})}(E, r) \partial_r F^{(\text{out})}(E, r) = 0 , \tag{16.43}$$

and the following differential equations:

$$\partial_r F^{(\text{in})} = -\frac{1}{2i} K^{-1} W^{(\text{out})} V \left[W^{(\text{in})} F^{(\text{in})} + W^{(\text{out})} F^{(\text{out})} \right] , \tag{16.44}$$

$$\partial_r F^{(\text{out})} = \frac{1}{2i} K^{-1} W^{(\text{in})} V \left[W^{(\text{in})} F^{(\text{in})} + W^{(\text{out})} F^{(\text{out})} \right] . \tag{16.45}$$

When we deal with a $1D$-problem, the boundary conditions for these equations are different from those given by Eq. (12.14). These conditions can be derived from (16.39) in the following way. Similarly to Eqs. (16.10), (16.11), it is easy to express the matrix-functions $F^{(\text{in/out})}(E, r)$ in terms of $\phi(E, r)$ and $\partial_r \phi(E, r)$:

$$F^{(\text{in})} = \frac{1}{2i} K^{-1} \left\{ \left[\partial_r W^{(\text{out})} \right] \phi - W^{(\text{out})} \partial_r \phi \right\} , \tag{16.46}$$

$$F^{(\text{out})} = -\frac{1}{2i} K^{-1} \left\{ \left[\partial_r W^{(\text{in})} \right] \phi - W^{(\text{in})} \partial_r \phi \right\} . \tag{16.47}$$

Taking into account that

$$W^{(\text{in/out})}(E, 0) = \begin{pmatrix} \pm i & 0 \\ 0 & \pm i \end{pmatrix} , \qquad \partial_r W^{(\text{in/out})}(E, r) \Big|_{r=0} = K , \tag{16.48}$$

and using the values (16.39), we obtain the following boundary conditions for the differential equations (12.12, 12.13):

$$F^{(\text{in/out})}(E, 0) = \frac{1}{2} \begin{pmatrix} \mp i & k_1 \\ & \\ \mp i & k_2 \end{pmatrix} . \tag{16.49}$$

As is seen, in contrast to the $3D$-problem, these boundary conditions are non-diagonal, different for $F^{(\text{in})}$ and $F^{(\text{out})}$, and depend on the channel momenta.

Since we reformulated the $1D$-problem as a particular case of a two-channel $3D$-problem, most of what is described in Chap. 11 is applicable here as well. In particular, the Jost matrices are obtained as the limits (12.25). For a complex energy, these limits can be reached when the differential equations (12.12), (12.13) are solved along a complex path, as explained in Sect. 12.2.

16.2.4 Riemann Surface

Since in the $1D$-problem there are two channels and we consider short-range potentials, the Riemann surface of the energy, which the Jost matrices are defined on, consists of four sheets. In general, this surface has the same topology as illustrated by Figs. 13.1 and 13.2. In the particular case of degenerate channels (unbiased potential), i.e. when $E_L = E_R$, the topology of the Riemann surface is more simple. As it is shown in Sect. 13.2, in such a case, we can ignore the Riemann sheets $(+-)$ and $(-+)$ and should choose the same signs in front of the square roots (16.25) for k_1 and k_2.

As to the symmetry relations that we established in the $3D$-problem for values of the Jost matrices at different points of the Riemann surface, they are not valid in the $1D$-problem. The reason is that, in contrast to the $3D$-case, the boundary conditions at $r = 0$ for the matrices $F^{(\mathrm{in})}(E,r)$ and $F^{(\mathrm{out})}(E,r)$ are different. The differential equations (12.12, 12.13) under the sign change or complex conjugation of the channel momenta transform into each other in the same way as in the $3D$-problem. However, the difference between the corresponding boundary conditions does not allow us to relate $F^{(\mathrm{in})}(E,r)$ and $F^{(\mathrm{out})}(E,r)$ to each other. The conclusion is that in the $1D$-problem there are no any symmetry relations depicted in Fig. 13.5.

In addition to destroying the symmetry relations, the complicated and k-dependent boundary conditions (16.49) do not allow us to derive the semi-analytic representations of the Jost matrices analogous to what we obtained in the $3D$-case (see Eq. (13.24)). Despite our effort to reformulate the $1D$-problem in the same way as the $3D$-one, it seems impossible to achieve the complete correspondence.

16.2.5 Spectral Points

The bound states and resonances correspond to the roots of Eq. (12.44) on the appropriate sheets of the Riemann surface. The bound states are located on the sheet $(++)$. The resonances that can only decay into the left channel (which we label with the subscript 1) are on the sheet $(-+)$, and those resonances that can decay into both channels are on the sheet $(--)$. Their locations on the Riemann surface are schematically shown in Figs. 13.1 and 13.2.

For the resonances, the partial widths for decaying into the left and right channels can be found as is described in Sect. 12.4.2. It is not difficult to check that despite the fact that for the $1D$-problem Eq. (12.55) does not involve the factor $Y_{\ell m}/r$, the flux in the channel n still is

$$j_n = \frac{\hbar \mathrm{Re}(k_n)}{\mu_n} |\mathcal{D}_n|^2 \exp[-2\mathrm{Im}(k_n)r] \,, \tag{16.50}$$

and the channel width Γ_n can be found in the same way,

$$\Gamma_n = \frac{\mathrm{Re}(k_n)|\mathcal{D}_n|^2\Gamma}{\dfrac{\mu_n}{\mu_1}\mathrm{Re}(k_1)|\mathcal{D}_1|^2 + \dfrac{\mu_n}{\mu_2}\mathrm{Re}(k_2)|\mathcal{D}_2|^2} \,, \tag{16.51}$$

where the amplitudes $\mathcal{D}_1$ and $\mathcal{D}_2$ of the outgoing waves are expressed in terms of the elements of the Jost matrices as is given by Eqs. (12.64), (12.65).

16.2.6 Scattering

Going from the discrete (bound and resonant) states to the scattering states, we should recollect the reasoning that leads to Eq. (11.75). In contrast to a discrete state, which is represented by a single column, a scattering-state wave function is represented by a square matrix. This is because a scattering state "remembers" which channel it is originated from as well as "knows" the probability of going to different outgoing channels. For the $1D$ two-channel problem, the scattering wave function is a (2×2) matrix,

$$\psi^{(\mathrm{sc})} = \begin{bmatrix} \psi_{11}^{(\mathrm{sc})} & \psi_{12}^{(\mathrm{sc})} \\ \psi_{21}^{(\mathrm{sc})} & \psi_{22}^{(\mathrm{sc})} \end{bmatrix}, \tag{16.52}$$

where an element $\psi_{n'n}^{(\mathrm{sc})}$ describes the scattering process that begins in channel n and ends up in channel n'. In other words, each column of matrix (16.52) is the scattering wave function for the process started either in the first (left) or the second (right) channel. Each column is a separate physical wave function that can be constructed using the fundamental solution as is given by Eq. (16.37). Therefore for making two columns, we need two sets of the combination coefficients c. This means that instead of a single column, like in Eq. (16.37), these coefficients form a square matrix,

$$\psi^{(\mathrm{sc})}(E,r) = \phi(E,r)\begin{pmatrix} c_{11} & c_{12} \\ c_{21} & c_{22} \end{pmatrix}. \tag{16.53}$$

How to find appropriate coefficients? The answer is: they must generate correct asymptotic behaviour of $\psi^{(\mathrm{sc})}(E,r)$ when $r \to \infty$. At large distances, the scattering wave function should consist of two terms: in each channel, we must have the original incoming wave and a distorted outgoing wave. This can be achieved if we choose

$$\begin{pmatrix} c_{11} & c_{12} \\ c_{21} & c_{22} \end{pmatrix} = \left[f^{(\mathrm{in})}(E)\right]^{-1}. \tag{16.54}$$

Indeed, using Eq. (16.40) with these coefficients, we obtain

$$\psi^{(\mathrm{sc})}(E,r) \underset{r\to\infty}{\longrightarrow} W^{(\mathrm{in})}(E,r) + W^{(\mathrm{out})}(E,r)S(E), \tag{16.55}$$

where the distortion (change of the amplitude) is due to the multiplication by the S-matrix.

For the $1D$-scattering states, the analysis given in Sect. 12.5 is fully applicable. As is shown there, the S-matrix for a real collision energy is composed of the reflection and transmission amplitudes, $R_{n\leftarrow n}$ and $T_{n'\leftarrow n}$,

$$S(E) = f^{(\text{out})}(E)\left[f^{(\text{in})}(E)\right]^{-1} = \begin{bmatrix} R_{L\leftarrow L}(E) \ , & T_{L\leftarrow R}(E) \\ T_{R\leftarrow L}(E) \ , & R_{R\leftarrow R}(E) \end{bmatrix} . \tag{16.56}$$

These amplitudes are not directly observable. What we can (in principle) measure are the fluxes (currents) of the incoming and outgoing particles. In the same spirit as in Sect. 1.3.2.2, we can introduce the scattering cross section, $\sigma_{n'n}$, which characterizes the transition of the flux of particles from the original channel n to another (or the same) channel n'. In any space, a cross section has a smaller dimension (less one). In a $3D$-problem, the cross section has the units of area, while for a $2D$-scattering the cross section has the units of length. In a $1D$-problem, "less one" means zero. Therefore, the cross section $\sigma_{n'n}$ that we introduce here is dimensionless. It is just a fraction of the original flux that goes in the channel n'.

Writing Eq. (16.55) in the explicit form,

$$\psi^{(\text{sc})}_{n'n}(E,r) \underset{r\to\infty}{\longrightarrow} h_0^{(-)}(k_n r)\delta_{n'n} + h_0^{(+)}(k_{n'} r) S_{n'n}(E) \ , \tag{16.57}$$

and calculating the fluxes (as is prescribed by Eq. (1.52)) for the incoming and outgoing waves,

$$j^{(\text{in})}_n = -\frac{\hbar k_n}{\mu_n} \quad \text{and} \quad j^{(\text{out})}_{n'} = \frac{\hbar k_{n'}}{\mu_{n'}} |S_{n'n}|^2 \ , \tag{16.58}$$

we find that

$$\boxed{\sigma_{n'n}(E) = \left| \frac{j^{(\text{out})}_{n'}}{j^{(\text{in})}_n} \right| = \frac{\mu_n k_{n'}}{\mu_{n'} k_n} |S_{n'n}(E)|^2} \tag{16.59}$$

16.3 Two-Dimensional Problems

The same as for the $1D$-problems the motion of a particle in two dimensions (on a plane) requires a special treatment if we want to formulate it in the framework of the Jost functions. First of all, we need to perform the partial-wave expansion of the wave function and to obtain the radial Schrödinger equation. The partial-wave decomposition of the wave function, scattering amplitude, and cross section for a particle moving on a plane is done using the cylindrical coordinates where the z-axis (perpendicular to the plane) is needed to define the orbital angular momentum.

All the steps of such a decomposition are similar to the three-dimensional case, but the resulting formulae are not obvious and cannot be easily obtained from the corresponding $3D$-analysis.

16.3.1 Partial-Wave Decomposition

Consider a particle of mass μ, moving on a plane and being affected by a force that is described by a potential $U(\vec{r}\,)$, which is assumed to be of short range and circularly symmetric,

$$U(\vec{r}\,) = U(|\vec{r}\,|)\,.$$

To make the derivations simple, we assume that our particle does not have spin (this restriction can be easily revoked later). In the coordinate representation, the Hamiltonian,

$$H = -\frac{\hbar^2}{2\mu}\Delta + U\,,$$

of such a particle is most conveniently expressed using the polar coordinates. In these coordinates, the Laplacian has the following form:

$$\Delta = \frac{1}{r}\frac{\partial}{\partial r}\left(r\frac{\partial}{\partial r}\right) + \frac{1}{\hbar^2 r^2}\mathcal{L}^2 \tag{16.60}$$

with

$$\mathcal{L}^2 = \hbar^2\frac{\partial^2}{\partial\varphi^2}\,, \tag{16.61}$$

which can be considered as the two-dimensional operator of the square of the angular momentum. Its eigenfunctions $\mathcal{Y}_m(\varphi)$ obeying the equation

$$\mathcal{L}^2\mathcal{Y}_m(\varphi) = -\hbar^2 m^2\mathcal{Y}_m(\varphi) \tag{16.62}$$

and normalized as

$$\int_0^{2\pi}\mathcal{Y}_m^*(\varphi)\mathcal{Y}_{m'}(\varphi)d\varphi = \delta_{mm'} \tag{16.63}$$

are easy to find

$$\mathcal{Y}_m(\varphi) = \frac{1}{\sqrt{2\pi}}e^{im\varphi}\,, \qquad m = 0, \pm 1, \pm 2, \ldots \tag{16.64}$$

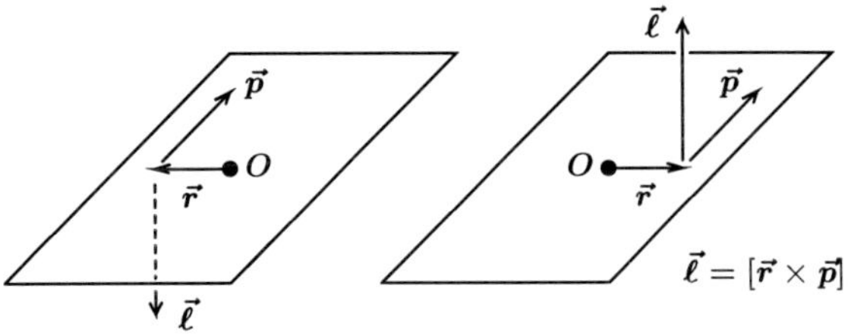

Fig. 16.3 Two possible directions (down on the left and up on the right) of the angular momentum, $\vec{\ell}$, for a particle moving on a plane

It is well known that these functions constitute the ortho-normal basis of complex Fourier series on the interval $[0, 2\pi]$. The completeness of this basis implies that

$$\sum_{m=-\infty}^{+\infty} \mathcal{Y}_m(\varphi)\mathcal{Y}_m^*(\varphi') = \delta(\varphi - \varphi') \,. \tag{16.65}$$

Each value, m, of the angular momentum (except for zero) is represented twice: with two opposite signs. Classically, these two states correspond to the motion of the particle at the same distance r from the centre and with the same velocity, but at different sides of the centre (see Fig. 16.3).

The quantum number of the angular momentum $\ell = |m|$ is always non-negative, and irrespective of its magnitude (if $\ell \neq 0$), the vector $\vec{\ell}$ can have two (only two) directions: up or down (like the spin 1/2). The quantum number $m = \pm\ell$ is its z-component. In principle, we can use the same notation for the eigenfunctions of the operator $\mathcal{L}^2$ as in the three-dimensional case, namely, $Y_{\ell m}$ with two subscripts. However, because of the relation $m = \pm\ell$, the subscript ℓ is redundant.

This can be formulated in a different way. The functions (16.64) form a complete ortho-normal set on the interval $\varphi \in [0, 2\pi]$. This means that any (reasonable) function $f(\varphi)$ defined on this interval can be written as their linear combination, and such a combination can be written in the following two (equivalent) ways:

$$f(\varphi) = \sum_{m=-\infty}^{+\infty} a_m \mathcal{Y}_m(\varphi) = \sum_{\ell=0}^{\infty} \sum_{m=\pm\ell} \tilde{a}_{\ell m} \mathcal{Y}_m(\varphi) \,, \qquad a_m = \tilde{a}_{|m| m} \,.$$

In other words, the following summations are equivalent

$$\sum_{m=-\infty}^{+\infty} \longleftrightarrow \sum_{\ell=0}^{\infty} \sum_{m=\pm\ell} \,.$$

The operator corresponding to the quantum number m is obtained as follows. The gradient operator in the cylindrical coordinates is

$$\vec{\nabla} = \hat{r}\frac{\partial}{\partial r} + \hat{\varphi}\frac{1}{r}\frac{\partial}{\partial \varphi} + \hat{z}\frac{\partial}{\partial z} \,, \tag{16.66}$$

where

$$\hat{\vec{r}} = [\hat{\vec{\varphi}} \times \hat{\vec{z}}] \,, \quad \hat{\vec{\varphi}} = [\hat{\vec{z}} \times \hat{\vec{r}}] \,, \quad \hat{\vec{z}} = [\hat{\vec{r}} \times \hat{\vec{\varphi}}]$$

are the corresponding unit vectors. Then

$$[\vec{r} \times \vec{p}] = r\hat{\vec{r}} \times \frac{\hbar}{i} \vec{\nabla} = \frac{\hbar r}{i} \left(\hat{\vec{z}} \frac{1}{r} \frac{\partial}{\partial \varphi} - \hat{\vec{\varphi}} \frac{\partial}{\partial z} \right)$$

and thus

$$\ell_z = \frac{\hbar}{i} \frac{\partial}{\partial \varphi} \,. \tag{16.67}$$

Apparently, both $\mathcal{L}^2$ and ℓ_z commute with the Hamiltonian. The quantum numbers ℓ and m are therefore conserving. When specifying m, we implicitly specify the quantum number ℓ as well. This means that the quantum state of the particle is determined by two conserving quantum numbers, namely, the energy E and the z-component of the angular momentum m. The corresponding wave function, obeying the Schrödinger equation,

$$H\psi_{Em}(\vec{r}) = E\psi_{Em}(\vec{r}) \,,$$

can be factorized in the radial and angular parts

$$\psi_{Em}(\vec{r}) = \frac{u_m(E, r)}{\sqrt{r}} \mathcal{Y}_m(\varphi) \,, \tag{16.68}$$

where $\sqrt{r}$ in the denominator is introduced to obtain the radial equation without the first derivative. Substituting this factorized form into the Schrödinger equation, we obtain

$$\left[\frac{d^2}{dr^2} + k^2 - \frac{m^2 - 1/4}{r^2} - V(r) \right] u_m(E, r) = 0 \,, \tag{16.69}$$

where k is the wave number (linear momentum) defined by

$$k^2 = \frac{2\mu}{\hbar^2} E \tag{16.70}$$

and $V(r)$ is the reduced (in the units of $[\text{length}]^{-2}$) potential

$$V(r) = \frac{2\mu}{\hbar^2} U(r) \,.$$

Noting that Eq. (16.69) is exactly the same for both choices of the sign for m, we conclude that $u_m(E, r)$ actually depends on ℓ but not on m. The radial equation can therefore be re-written in the way we used to see it in the three-dimensional problems:

$$\left[\frac{d^2}{dr^2} + k^2 - \frac{\lambda(\lambda+1)}{r^2} - V(r)\right] u_\ell(E, r) = 0 \,, \tag{16.71}$$

where we introduced

$$\lambda = \ell - \frac{1}{2} \tag{16.72}$$

and did the replacement

$$\ell^2 - \frac{1}{4} = \left(\ell - \frac{1}{2}\right)\left(\ell + \frac{1}{2}\right) = \lambda(\lambda+1)\,.$$

Formally, Eq. (16.71) looks exactly like the radial equation of the three-dimensional problem. The only difference is that λ is not an integer number

$$\lambda = -\frac{1}{2},\ \frac{1}{2},\ \frac{3}{2},\ \frac{5}{2},\ \ldots$$

This simple fact makes a huge difference: as we will see, it changes the analytic properties of the Jost function and thus the S-matrix because the Riccati–Neumann function $n_\lambda(kr)$ with a half-integer λ has a logarithmic branch point on the Riemann surface of the energy.

16.3.2 Jost Functions

As was already mentioned, to avoid some mathematical complications, we assume that the potential $V(r)$ is of short range, i.e. it vanishes at large distances faster than any power of $1/r^2$ (e.g. exponentially). Then at large distances, where $V(r) \to 0$, the radial equation (16.71) simplifies and has two linearly independent solutions, namely the Riccati–Hankel functions:

$$h^{(\pm)}_{\ell-1/2}(kr) = \sqrt{\frac{\pi kr}{2}}\, H^{(\pm)}_\ell(kr) \underset{|z|\to\infty}{\longrightarrow} e^{\pm i(kr-\ell\pi/2-\pi/4)} = \mp i e^{\pm i(kr-\lambda\pi/2)}\,, \tag{16.73}$$

where $H^{(\pm)}_\ell(z)$ are the cylindrical Hankel functions. The functions

$$W^{(\mathrm{in/out})}_\ell(E, r) = h^{(\mp)}_{\ell-1/2}(kr) \tag{16.74}$$

represent the incoming and outgoing circular waves on the plane. Therefore, by analogy with the $3D$-problem, the amplitudes, $f_\ell^{(\text{in/out})}(E)$, of these waves in the asymptotic form of the regular solution,[6] $\phi_\ell(E,r)$, of Eq. (16.71),

$$\phi_\ell(E,r) \underset{r\to\infty}{\longrightarrow} W_\ell^{(\text{in})}(E,r)f_\ell^{(\text{in})}(E) + W_\ell^{(\text{out})}(E,r)f_\ell^{(\text{out})}(E)\,, \tag{16.75}$$

should be considered as the Jost functions. It is clear that they have the same physical meaning as their $3D$-counterparts. Indeed, the discrete energies $\mathcal{E}_n$ of possible bound states and resonances that should only have the outgoing waves at large distances can be found in the same way as in the $3D$-problem, namely, as the roots of the equation

$$f_\ell^{(\text{in})}(\mathcal{E}_n) = 0\,, \tag{16.76}$$

while the ratio of the amplitudes of the outgoing and incoming waves,

$$S_\ell(E) = \frac{f_\ell^{(\text{out})}(E)}{f_\ell^{(\text{in})}(E)}\,, \tag{16.77}$$

is the S-matrix that describes the $2D$-scattering.

It is easy to see that most of the reasoning and derivations used in Chap. 2 for the $3D$-problem are applicable here as well. Indeed, since in that chapter we considered even complex values of ℓ, the equations obtained there should be applicable for the half-integer values of angular momentum.

Therefore, the $2D$ regular solution $\phi_\ell(E,r)$ can be expressed via the unknown functions $F_\ell^{(\text{in/out})}(E,r)$ as is given by Eq. (2.12), or alternatively via the functions $A_\ell(E,r)$ and $B_\ell(E,r)$ as

$$\phi_\ell(E,r) = j_\lambda(kr)A_\ell(E,r) - n_\lambda(kr)B_\ell(E,r)\,, \tag{16.78}$$

i.e. similarly to Eq. (2.27). These unknown functions obey the same differential equations (2.21) and (2.32) with the same boundary conditions (2.24) and (2.33, 2.34). The only difference is that the Riccati–Bessel, Riccati–Neumann, and Riccati–Hankel functions in all these equations should be taken with the half-integer index (16.72).

---

[6] It should be noted that in a single-channel problem (which we consider here) the regular solution, ϕ_ℓ, and a physical solution, u_ℓ, differ by the normalization. The normalization of the Jost functions depends on how the regular solution is normalized and actually does not play any role in their properties as well as in the observable quantities. We therefore may, in principle, forget here about the difference between the functions ϕ_ℓ and u_ℓ. However, to be consistent with the rest of the book, we take the following normalization of the regular solution: $\phi_\ell(E,r) \underset{r\to 0}{\longrightarrow} j_\lambda(kr)$.

The same as in the $3D$-problem, the $2D$ Jost functions are the asymptotic values of $F_\ell^{(\mathrm{in/out})}(E,r)$,

$$f_\ell^{(\mathrm{in/out})}(E) = \lim_{r\to\infty} F_\ell^{(\mathrm{in/out})}(E,r)\,, \tag{16.79}$$

which can be found by integrating the differential equations (2.21) along (generally speaking) a complex path in the r-plane as is described in Sect. 7.4.

A question arises: is it then any difference between the $2D$- and $3D$- problems? The answer is "yes". The differences appear when we consider the analytic properties of the $2D$ Jost functions, the Riemann surface that they are defined on, as well as their relations to the observable quantities. All these are described further down.

16.3.3 Analytic Properties of the Jost Functions

It is obvious that, like in the $3D$-problem, the $2D$ Jost functions are not single-valued functions of the energy variable. Let us try and find such representations for them where all the factors responsible for branching of the corresponding Riemann surface are given explicitly (similar to what we obtained in the $3D$-case).

First of all, we note that the analytic structure of the Riccati–Bessel and Riccati–Neumann functions with half-integer indices is more complicated than what is given by Eqs. (7.58), (7.59) for an integer ℓ. As is explained in Appendix D.2.2, for a half-integer λ, these functions can be written in the following form:

$$j_\lambda(kr) = k^{\lambda+1}\tilde{j}_\lambda(E,r)\,, \tag{16.80}$$

$$n_\lambda(kr) = k^{-\lambda}\tilde{n}_\lambda(E,r) + k^{\lambda+1}\mathfrak{h}(k)\tilde{j}_\lambda(E,r)\,, \tag{16.81}$$

where the tilded functions are analytic single-valued functions of variable E, and

$$\mathfrak{h}(k) = \frac{2}{\pi}\ln\left(\frac{kR}{2}\right) \tag{16.82}$$

with R being an arbitrary distance (e.g. the unit length). Since in the Riccati–Neumann function the dependence on the logarithm of k is present, we may expect that the Jost functions are defined on a spiral Riemann surface with an infinite number of sheets (similarly to the $3D$-problem of charged particles). Let us check if this is true.

To establish what kind of consequences for the Jost functions follow from the above representations of $j_\lambda(kr)$ and $n_\lambda(kr)$, let us consider the differential equations

$$\begin{aligned}\partial_r A_\ell(E,r) &= -\frac{1}{k} n_\lambda(kr) V(r) \left[j_\lambda(kr) A_\ell(E,r) - n_\lambda(kr) B_\ell(E,r) \right] , \\ \partial_r B_\ell(E,r) &= -\frac{1}{k} j_\lambda(kr) V(r) \left[j_\lambda(kr) A_\ell(E,r) - n_\lambda(kr) B_\ell(E,r) \right] ,\end{aligned} \tag{16.83}$$

which are the same as Eq. (2.32), but with the half-integer subscript, λ, in the Riccati–Bessel and Riccati–Neumann functions. For our purpose, we need their linear combinations:

$$\partial_r (A_\ell - \mathfrak{h} B_\ell) = -\frac{1}{k} (n_\lambda - \mathfrak{h} j_\lambda) V (j_\lambda A_\ell - n_\lambda B_\ell) , \tag{16.84}$$

$$\partial_r k^{-(2\lambda+1)} B_\ell = -k^{-2(\lambda+1)} j_\lambda V (j_\lambda A_\ell - n_\lambda B_\ell) . \tag{16.85}$$

Taking into account (16.80), (16.81), we see that

$$n_\lambda - \mathfrak{h} j_\lambda = k^{-\lambda} \tilde{n}_\lambda + k^{\lambda+1} \mathfrak{h} \tilde{j}_\lambda - k^{\lambda+1} \mathfrak{h} \tilde{j}_\lambda = k^{-\lambda} \tilde{n}_\lambda$$

and

$$\begin{aligned} j_\lambda A_\ell - n_\lambda B_\ell &= k^{\lambda+1} \tilde{j}_\lambda A_\ell - k^{-\lambda} \tilde{n}_\lambda B_\ell - k^{\lambda+1} \mathfrak{h} \tilde{j}_\lambda B_\ell \\ &= k^{\lambda+1} \tilde{j}_\lambda (A_\ell - \mathfrak{h} B_\ell) - k^{-\lambda} \tilde{n}_\lambda B_\ell . \end{aligned}$$

Substitution of these expressions in Eqs. (16.84), (16.85) gives

$$\partial_r (A_\ell - \mathfrak{h} B_\ell) = -k^{-(\lambda+1)} \tilde{n}_\lambda V \left[k^{\lambda+1} \tilde{j}_\lambda (A_\ell - \mathfrak{h} B_\ell) - k^{-\lambda} \tilde{n}_\lambda B_\ell \right] , \tag{16.86}$$

$$\partial_r k^{-(2\lambda+1)} B_\ell = -k^{-(\lambda+1)} \tilde{j}_\lambda V \left[k^{\lambda+1} \tilde{j}_\lambda (A_\ell - \mathfrak{h} B_\ell) - k^{-\lambda} \tilde{n}_\lambda B_\ell \right] . \tag{16.87}$$

Introducing the following tilded functions

$$\tilde{A}_\ell(E,r) \equiv A_\ell(E,r) - \mathfrak{h}(k) B_\ell(E,k) , \tag{16.88}$$

$$\tilde{B}_\ell(E,r) \equiv k^{-(2\lambda+1)} B_\ell(E,r) , \tag{16.89}$$

we re-write Eqs. (16.86), (16.87) in the form:

$$\partial_r \tilde{A}_\ell = -\tilde{n}_\lambda V \left(\tilde{j}_\lambda \tilde{A}_\ell - \tilde{n}_\lambda \tilde{B}_\ell \right) , \tag{16.90}$$

$$\partial_r \tilde{B}_\ell = -\tilde{j}_\lambda V \left(\tilde{j}_\lambda \tilde{A}_\ell - \tilde{n}_\lambda \tilde{B}_\ell \right) . \tag{16.91}$$

The boundary conditions for these equations follow from (2.33), (2.34) and read

$$\tilde{A}_\ell(E,0) = 1 , \qquad \tilde{B}_\ell(E,0) = 0 . \tag{16.92}$$

For any finite r, all the coefficient functions in equations (16.90), (16.91) are entire functions of the parameter E and the boundary conditions are E-independent. According to the Poincaré theorem (see Appendix G), the solutions of these equations, i.e. the functions $\tilde{A}_\ell(E,r)$ and $\tilde{B}_\ell(E,r)$, are entire (analytic single-valued) functions of the complex variable E. An important property of these functions is that they are real for real E. Such a conclusion follows from the fact that neither Eqs. (16.90), (16.91) nor the boundary conditions (16.92) involve any complex numbers if E is real.

Using the functions $\tilde{A}_\ell$ and $\tilde{B}_\ell$, we obtain the following representations for the functions A_ℓ and B_ℓ:

$$A_\ell(E,r) = \tilde{A}_\ell(E,r) + k^{2\lambda+1}\mathfrak{h}(k)\tilde{B}_\ell(E,r) , \tag{16.93}$$

$$B_\ell(E,r) = k^{2\lambda+1}\tilde{B}_\ell(E,r) . \tag{16.94}$$

Apart from the single-valued parts, Eqs. (16.93), (16.94) involve the factors $k^{2\lambda+1}$ and $\mathfrak{h}(k)$. Since λ is a half-integer, the power $(2\lambda + 1)$ is always even, and thus $k^{2\lambda+1}$ is also a single-valued function of E, but $\mathfrak{h}(k)$ has a logarithmic branch point at $E = 0$.

The functions $F_\ell^{(\text{in/out})}$ have similar structure:

$$F_\ell^{(\text{in})}(E,r) = \frac{1}{2}(A_\ell - iB_\ell) = \frac{1}{2}\left\{\tilde{A}_\ell(E,r) + k^{2\lambda+1}[\mathfrak{h}(k) - i]\tilde{B}_\ell(E,r)\right\} , \tag{16.95}$$

$$F_\ell^{(\text{out})}(E,r) = \frac{1}{2}(A_\ell + iB_\ell) = \frac{1}{2}\left\{\tilde{A}_\ell(E,r) + k^{2\lambda+1}[\mathfrak{h}(k) + i]\tilde{B}_\ell(E,r)\right\} . \tag{16.96}$$

For obtaining the Jost functions as the limits (16.79), we need the corresponding limits of the functions $\tilde{A}_\ell(E,r)$ and $\tilde{B}_\ell(E,r)$. It is not obvious, however, that they remain holomorphic in E, when $r \to \infty$. Strictly speaking, this question requires a careful consideration and proof. Fortunately, we can avoid this and base our conclusion on a similar consideration of Sect. 7.7.2, where it was shown that the asymptotic values of the functions $\tilde{A}_\ell(E, r)$ and $\tilde{B}_\ell(E, r)$ are holomorphic practically on whole complex plane of variable E. It is easy to see that the proof given in Sect. 7.7.2 is fully applicable here.

Denoting the asymptotic values of the functions $\tilde{A}_\ell(E, r)$ and $\tilde{B}_\ell(E, r)$ as

$$a_\ell(E) = \lim_{r\to\infty} \tilde{A}_\ell(E, r)/2 , \qquad b_\ell(E) = \lim_{r\to\infty} \tilde{B}_\ell(E, r)/2 , \tag{16.97}$$

we obtain the representations of the Jost functions that we were looking for

$$f_\ell^{(\mathrm{in})}(E) = a_\ell(E) + k^{2\lambda+1}[\mathfrak{h}(k) - i]b_\ell(E) \tag{16.98}$$

$$f_\ell^{(\mathrm{out})}(E) = a_\ell(E) + k^{2\lambda+1}[\mathfrak{h}(k) + i]b_\ell(E) \tag{16.99}$$

where $a_\ell(E)$ and $b_\ell(E)$ are single-valued analytic functions of E within the complex domain $\mathcal{D}$ (practically covering whole complex plane) that is defined in Sect. 7.7.2. For the S-matrix, the representations (16.98), (16.99) give

$$S_\ell(E) = \frac{a_\ell(E) + k^{2\lambda+1}[\mathfrak{h}(k) + i]b_\ell(E)}{a_\ell(E) + k^{2\lambda+1}[\mathfrak{h}(k) - i]b_\ell(E)} \tag{16.100}$$

As was already mentioned, for a half-integer λ, the power $2\lambda + 1$ of the momentum is always even. This means that the only "troublesome" quantity in the Jost functions (16.98), (16.99) is the logarithmic function $\mathfrak{h}(k)$. Due to its presence, the Jost functions have an infinite number of values for a single value of E. It is not difficult to see that the reasoning given in Sect. 8.2.9, is fully applicable here. Therefore, the $2D$ Jost functions (16.98), (16.99) as well as the corresponding S-matrix are defined on the spiral Riemann surface shown in Fig. 8.2.

When it comes to the symmetry properties of the $2D$ Jost functions on the Riemann surface, we see that these properties are the same as for the $3D$-problem with charged particles. Indeed, there is no the so-called vertical symmetry (with respect to the change $k \leftrightarrow -k$) because

$$\ln(-k) = \ln\left[ke^{\pm i\pi(1\pm 2m)}\right] = \ln(k) \pm i\pi \pm 2im , \qquad m = 0, \pm 1, \pm 2, \ldots ,$$

which means that as a result of a "vertical jump" the function $\mathfrak{h}(k)$ is changing, while all the other quantities in (16.98), (16.99) remain intact. This does not lead to any relation between $f_\ell^{(\mathrm{in})}$ and $f_\ell^{(\mathrm{out})}$. The diagonal symmetry, however, is valid.

The same as for the $3D$-problem, it follows from the Schwarz reflection principle. The fact that $a_\ell(E)$ and $b_\ell(E)$ are real on the real axis implies

$$a_\ell(E^*) = a_\ell^*(E) \qquad \text{and} \qquad b_\ell(E^*) = b_\ell^*(E)\,. \tag{16.101}$$

Together with the simple relation

$$\ln(k^*) = \ln\left\{|k|e^{-i[\arg(k)+2\pi m]}\right\} = \ln|k| - i\arg(k) - i2\pi m = [\ln(k)]^*\,,$$

the Schwarz reflections (16.101) lead to the same symmetry (8.117) schematically depicted in Fig. 8.3.

16.3.4 Power-Series Expansions

The functions $a_\ell(E)$ and $b_\ell(E)$ are holomorphic (i.e. single-valued and analytic) and therefore can be expanded in the Taylor series near any point E_0 within the domain $\mathcal{D}$ of the complex energy plane. The expansion around the point $E_0 = 0$ will give us the standard effective-range series. But we can also do such an expansion near an arbitrary point,

$$a_\ell(E) = \sum_{n=0}^{\infty} \alpha_n(\ell, E_0)(E - E_0)^n\,, \tag{16.102}$$

$$b_\ell(E) = \sum_{n=0}^{\infty} \beta_n(\ell, E_0)(E - E_0)^n\,. \tag{16.103}$$

It is not difficult to see that the expansion coefficients α_n and β_n can be found in the same way as it is described in Chap. 15 for the corresponding coefficients in various $3D$-problems. Indeed, the tilded functions $\tilde{A}_\ell(E,r)$ and $\tilde{B}_\ell(E,r)$ whose asymptotic values give $a_\ell(E)$ and $b_\ell(E)$ are also holomorphic in E and obey the differential equations (16.90), (16.91) that are the same as the corresponding equations (15.23), (15.24) of the $3D$-problem. The only difference is that the coefficients in Eqs. (16.90), (16.91) are the functions $\tilde{j}_\lambda$ and $\tilde{n}_\lambda$ with half-integer λ. However, despite this, they are still analytic and therefore can be expanded in the Taylor series

$$\tilde{j}_\lambda(E,r) = \sum_{n=0}^{\infty} s_n(\lambda, E_0, r)\,(E - E_0)^n\,, \tag{16.104}$$

$$\tilde{n}_\lambda(E,r) = \sum_{n=0}^{\infty} c_n(\lambda, E_0, r)\,(E - E_0)^n\,, \tag{16.105}$$

near an arbitrary point E_0 similarly to the expansions (15.29), (15.30). A recipe for finding the coefficients s_n and c_n is given in Appendix D.2.2. The same way as in Sect. 15.1.2, we look for the solutions of differential equations (15.23), (15.24) in the form of the power series,

$$\tilde{A}_\ell(E,r) = \sum_{n=0}^{\infty} \tilde{\alpha}_n(\ell, E_0, r)(E - E_0)^n \,, \tag{16.106}$$

$$\tilde{B}_\ell(E,r) = \sum_{n=0}^{\infty} \tilde{\beta}_n(\ell, E_0, r)(E - E_0)^n \,, \tag{16.107}$$

and repeating the same derivations, obtain the system of differential equations

$$\begin{aligned} \partial_r \tilde{\alpha}_n &= - \sum_{i+j+k=n} c_i V(s_j \tilde{\alpha}_k - c_j \tilde{\beta}_k) \,, \\ \partial_r \tilde{\beta}_n &= - \sum_{i+j+k=n} s_i V(s_j \tilde{\alpha}_k - c_j \tilde{\beta}_k) \end{aligned} \tag{16.108}$$

with the boundary conditions

$$\tilde{\alpha}_n(\ell, E_0, 0) = \delta_{n0} \,, \qquad \tilde{\beta}_n(\ell, E_0, 0) = 0 \,, \qquad \forall n \,. \tag{16.109}$$

Although the coefficients s_n and c_n in these equations differ from φ_n and γ_n in the corresponding Eq. (15.33), it is clear that everything what is said at the end of Sect. 15.1.2 remains valid here as well. In particular, it is necessary to integrate Eq. (16.108) along a complex path if the parameter E_0 is complex. This is because the exponential asymptotic behaviour of the Riccati–Hankel functions,

$$h_\lambda^{(\pm)}(kr) \underset{|kr|\to\infty}{\longrightarrow} e^{\pm i\left(kr - \frac{\lambda+1}{2}\pi\right)} \,, \tag{16.110}$$

remains valid for any λ (integer, or half-integer, or even complex), and thus the statements (15.35), (15.36) remain valid as well.

Therefore, if the $2D$-potential is known, the coefficients of the expansions (16.102), (16.103) can be found by a numerical solution of Eq. (16.108) from $r = 0$ up to a sufficiently large distance (along a complex path, if necessary) where the limits

$$\alpha_n(\ell, E_0) = \frac{1}{2} \lim_{r\to\infty} \tilde{\alpha}_n(\ell, E_0, r) \qquad \text{and} \qquad \beta_n(\ell, E_0) = \frac{1}{2} \lim_{r\to\infty} \tilde{\beta}_n(\ell, E_0, r) \tag{16.111}$$

are reached. Alternatively, if the potential is unknown, but there are some experimental data available, a finite number of terms of the expansions (16.102), (16.103) can be used to approximate the S-matrix (16.100), where the coefficients α_n and β_n serve as the fitting parameters.

The same as in the $3D$-problem, the standard effective-range expansion is obtained if $E_0 = 0$. Repeating the same derivations that are given in Sect. 15.1.1, we obtain

$$k^{2\lambda+1}\left[\cot\delta_\ell(E) - \mathfrak{h}(k)\right] = -\frac{1}{\mathsf{a}_\ell} + \frac{\mathsf{r}_\ell}{2}k^2 - \mathsf{P}_\ell\mathsf{r}_\ell^3k^4 + \mathsf{Q}_\ell\mathsf{r}_\ell^5k^6 - \mathsf{L}_\ell\mathsf{r}_\ell^7k^8 + \cdots \tag{16.112}$$

where the parameters on the right hand side have the same names and meaning as in Eq. (15.19). They can be found using the same formulae given in Sect. 15.1.1. The only difference from the $3D$-problem is the presence of the logarithmic function $\mathfrak{h}(k)$ on the left hand side of Eq. (16.112). It should be noted that the definition (16.82) of $\mathfrak{h}(k)$ involves an arbitrary parameter R. Apparently, the values of the effective-range parameters on the right hand side of Eq. (16.112) depend on the choice of R. To avoid such an uncertainty, we may agree that $R = 1$ (unit of the length) and consider the effective-range parameters with such a choice as the standard.

16.3.5 $2D$-Scattering

The scattering process on a plane looks similarly to the $3D$-scattering, but all the corresponding formulae have different forms. Instead of the spherical waves that diverge from the target, here we have circular waves. And the plane wave describing the incident projectile particles looks differently as well. The scattering cross section is measured in the units of length.

16.3.5.1 Plane and Circular Waves

Consider a two-dimensional plane wave normalized to the δ-function

$$\langle\vec{r}|\vec{k}\rangle = \frac{e^{i\vec{k}\vec{r}}}{2\pi}, \qquad \langle\vec{k}'|\vec{k}\rangle = \delta\left(\vec{k}' - \vec{k}\right) = \frac{1}{k}\delta(k'-k)\delta(\varphi'-\varphi), \tag{16.113}$$

where φ is the polar angle of the momentum $\hbar\vec{k}$. This plane wave can be expanded over the full set $\{\mathcal{Y}\}$ of the angular functions (16.64),

$$\frac{e^{i\vec{k}\vec{r}}}{2\pi} = \frac{e^{ikr\cos\varphi}}{2\pi} = \sum_{\ell m} a_m(kr)\mathcal{Y}_m(\varphi), \tag{16.114}$$

where the x-axis is directed along the coordinate vector $\vec{r}$. The expansion coefficients

$$a_m(kr) = \frac{1}{(2\pi)^{3/2}} \int_0^{2\pi} e^{i(kr\cos\varphi - m\varphi)} d\varphi \tag{16.115}$$

can be found using well-known integral representation of the Bessel function

$$J_m(z) = \frac{1}{\pi i^m} \int_0^{\pi} e^{iz\cos\varphi} \cos(m\varphi) d\varphi = \frac{1}{2\pi i^m} \int_0^{\pi} e^{iz\cos\varphi} \left(e^{im\varphi} + e^{-im\varphi}\right) d\varphi$$

$$= \frac{1}{2\pi i^m} \int_{-\pi}^{\pi} e^{i(z\cos\varphi - m\varphi)} d\varphi = \frac{i^m}{2\pi} \int_0^{2\pi} e^{i(-z\cos\varphi - m\varphi)} d\varphi \,. \tag{16.116}$$

Comparing Eq. (16.115) with (16.116) and using the symmetry property of the Bessel function $J_m(-z) = (-1)^m J_m(z)$, we see that

$$a_m(kr) = \frac{i^m}{\sqrt{2\pi}} J_m(kr) \tag{16.117}$$

and thus

$$\frac{e^{i\vec{k}\vec{r}}}{2\pi} = \frac{1}{2\pi} \sum_{\ell m} i^m e^{im\varphi} J_m(kr) = \frac{1}{2\pi} \sum_{-\infty}^{+\infty} i^m e^{im\varphi} J_m(kr)$$

$$= \frac{1}{\sqrt{2\pi}} \sum_{-\infty}^{+\infty} i^m J_m(kr) \mathcal{Y}_m(\varphi) \,. \tag{16.118}$$

Using another symmetry property, $J_{-m}(z) = (-1)^m J_m(z)$, we see that the product $i^m J_m(kr)$ does not depend on the sign of m, and thus this expansion can be rewritten as

$$\frac{e^{i\vec{k}\vec{r}}}{2\pi} = \frac{1}{2\pi} \left[J_0(kr) + \sum_{\ell=1}^{\infty} i^\ell \left(e^{i\ell\varphi} + e^{-i\ell\varphi}\right) J_\ell(kr) \right]$$

$$= \frac{1}{2\pi} \sum_{\ell=0}^{\infty} \epsilon_\ell i^\ell \cos(\ell\varphi) J_\ell(kr) \,, \tag{16.119}$$

where ϵ_ℓ is the "multiplicity" of an ℓ-state, i.e. is the analog of the factor $(2\ell + 1)$ of the 3D-case,

$$\epsilon_\ell = \begin{cases} 1 \,, \ \ell = 0 \,, \\ 2 \,, \ \ell > 0 \,. \end{cases} \tag{16.120}$$

Expressing the Bessel function via the Riccati–Hankel functions,

$$J_\ell(z) = \sqrt{\frac{1}{2\pi z}} \left[h^{(-)}_{\ell-1/2}(z) + h^{(+)}_{\ell-1/2}(z) \right] ,$$

we obtain the following decomposition of the $2D$ plane wave in the incoming $(-)$ and outgoing $(+)$ circular waves

$$\frac{e^{i\vec{k}\vec{r}}}{2\pi} = \frac{1}{2\pi\sqrt{kr}} \sum_{\ell=0}^{\infty} \sum_{m=\pm\ell} i^\ell \left[h^{(-)}_\lambda(kr) + h^{(+)}_\lambda(kr) \right] \mathcal{Y}_m(\varphi) , \qquad (16.121)$$

where λ is defined by Eq. (16.72).

In the above, we assumed that vector $\vec{r}$ was directed along the x-axis. If this is not the case, then the dot product

$$\vec{k}\vec{r} = kr(\cos\varphi_k \cos\varphi_r + \sin\varphi_k \sin\varphi_r) = kr \cos(\varphi_r - \varphi_k)$$

depends on the two polar angles. In this general case, the plane wave is expanded over two sets of functions $\{\mathcal{Y}(\varphi_k)\}$ and $\{\mathcal{Y}(\varphi_r)\}$ depending on the angles of the momentum and coordinate vectors. In a similar way, as we did it above, it is not difficult to show that

$$\frac{e^{i\vec{k}\vec{r}}}{2\pi} = \sum_{-\infty}^{+\infty} i^m J_m(kr) \mathcal{Y}^*_m(\varphi_r) \mathcal{Y}_m(\varphi_k) = \sum_{\ell=0}^{\infty} i^\ell J_\ell(kr) \sum_{m=\pm\ell} \mathcal{Y}^*_m(\varphi_r) \mathcal{Y}_m(\varphi_k)$$

$$= \frac{1}{2\pi} \sum_{\ell=0}^{\infty} \epsilon_\ell i^\ell \cos\left[\ell(\varphi_r - \varphi_k)\right] J_\ell(kr)$$

$$= \frac{1}{\sqrt{2\pi kr}} \sum_{\ell m} i^\ell \left[h^{(-)}_\lambda(kr) + h^{(+)}_\lambda(kr) \right] \mathcal{Y}^*_m(\varphi_r) \mathcal{Y}_m(\varphi_k)$$

$$= \frac{1}{\sqrt{2\pi kr}} \sum_{\ell m} i^\ell u^{(0)}_\ell(E, r) \mathcal{Y}^*_m(\varphi_r) \mathcal{Y}_m(\varphi_k) , \qquad (16.122)$$

where

$$u^{(0)}_\ell(E, r) = h^{(-)}_\lambda(kr) + h^{(+)}_\lambda(kr) = 2 j_\lambda(kr) \qquad (16.123)$$

is the physical solution of the radial Schrödinger equation (16.71) for the case $V(r) \equiv 0$. It is seen that

$$u_\ell^{(0)}(E,r) = \frac{\phi_\ell(E,r)}{1/2} = \frac{\phi_\ell(E,r)}{f_\ell^{(\mathrm{in})}(E)} \tag{16.124}$$

because for the free motion $\phi_\ell(E,r) \equiv j_\lambda(kr)$ and $f_\ell^{(\mathrm{in})}(E) = 1/2$.

The above partial-wave decompositions can be conveniently written using the following symbolic formulae:

$$|\vec{k}\rangle = \sum_{\ell m} |k\ell m\rangle \mathcal{Y}_m(\varphi_k)\,, \qquad |\vec{r}\rangle = \sum_{\ell m} |r\ell m\rangle \mathcal{Y}_m(\varphi_r)\,, \tag{16.125}$$

$$\langle r\ell m|k\ell' m'\rangle = \delta_{\ell\ell'}\delta_{mm'} i^\ell J_\ell(kr) = \delta_{\ell\ell'}\delta_{mm'} i^\ell \sqrt{\frac{2}{\pi kr}}\, j_\lambda(kr)\,, \tag{16.126}$$

$$\langle\vec{r}|k\ell m\rangle = i^\ell \sqrt{\frac{2}{\pi kr}}\, j_\lambda(kr)\mathcal{Y}_m^*(\varphi_r)\,, \qquad \langle\vec{k}|r\ell m\rangle = (-i)^\ell \sqrt{\frac{2}{\pi kr}}\, j_\lambda(kr)\mathcal{Y}_m^*(\varphi_k)\,, \tag{16.127}$$

$$\langle k\ell m|k'\ell' m'\rangle = \frac{1}{k}\delta(k-k')\delta_{\ell\ell'}\delta_{mm'}\,, \qquad \langle r\ell m|r'\ell' m'\rangle = \frac{1}{r}\delta(r-r')\delta_{\ell\ell'}\delta_{mm'}\,, \tag{16.128}$$

$$\int_0^\infty \sum_{\ell m} |k\ell m\rangle\langle k\ell m| k\,dk = 1\,, \qquad \int_0^\infty \sum_{\ell m} |r\ell m\rangle\langle r\ell m| r\,dr = 1\,. \tag{16.129}$$

It is instructive to compare them with the corresponding $3D$-formulae given in Appendix A.

16.3.5.2 Scattering Wave Function

The plane wave (16.122) is a scattering wave function $\psi_{\vec{k}}(\vec{r})$ for the particular case of $V(r) \equiv 0$. Apparently, the structure of its partial-wave decomposition should be the same for all potentials

$$\psi_{\vec{k}}(\vec{r}\,) = \frac{1}{\sqrt{2\pi kr}\, f_\ell^{(\mathrm{in})}(E)} \sum_{\ell m} i^\ell \phi_\ell(E,r)\mathcal{Y}_m^*(\varphi_r)\mathcal{Y}_m(\varphi_k)\,.$$

The first purpose of all the factors here is to always have exactly the same normalization, namely,

$$\langle\psi_{\vec{k}}|\psi_{\vec{k}'}\rangle = \delta\left(\vec{k}-\vec{k}'\right)\,. \tag{16.130}$$

The second purpose is that at large distances the wave function $\psi_{\vec{k}}(\vec{r}\,)$ should consist of the two parts: the initial (incident) wave $\psi_{\vec{k}}^{(0)}(\vec{r}\,)$ and a scattered circular wave that goes in all directions with certain amplitude F. If $\psi_{\vec{k}}(\vec{r}\,)$ is properly normalized, then $\psi^{(0)}_{\vec{k}}(\vec{r}\,) \equiv e^{i\vec{k}\vec{r}}/(2\pi)$. This means that the radial wave function at large distances should also be split into the corresponding two parts. In doing such a splitting of the function (16.75), we obtain

$$\phi_\ell(E, r) \underset{r\to\infty}{\longrightarrow} f_\ell^{(\text{in})} \left[h_\lambda^{(-)} + h_\lambda^{(+)} + \left(\frac{f_\ell^{(\text{out})}}{f_\ell^{(\text{in})}} - 1 \right) h_\lambda^{(+)} \right] ,$$

where the terms $h_\lambda^{(-)} + h_\lambda^{(+)}$ give the plane wave.

16.3.5.3 Cross Section

Defining the partial-wave S-matrix and the amplitude,

$$S_\ell(E) = \frac{f_\ell^{(\text{out})}(E)}{f_\ell^{(\text{in})}(E)} , \qquad f_\ell(E) = \frac{S_\ell(E) - 1}{\sqrt{2\pi i k}} ,$$

and using

$$h_\lambda^{(+)}(kr) \underset{r\to\infty}{\longrightarrow} -i \exp\left[i(kr - \lambda\pi/2)\right] ,$$

as well as the fact that

$$\sum_m \mathcal{Y}_m^*(\varphi_r)\mathcal{Y}_m(\varphi_k) = \begin{cases} \dfrac{1}{2\pi} & , \ \ell = 0 \\ \dfrac{1}{2\pi}\left(e^{-i\ell\varphi_r} e^{i\ell\varphi_k} + e^{i\ell\varphi_r} e^{-i\ell\varphi_k}\right) & , \ \ell > 0 \end{cases}$$

$$= \frac{\epsilon_\ell}{2\pi} \cos\left[\ell(\varphi_k - \varphi_r)\right] ,$$

we can write the asymptotic behaviour of the scattering wave function as

$$\boxed{\psi_{\vec{k}}(\vec{r}\,) \underset{r\to\infty}{\longrightarrow} \frac{1}{2\pi}\left[e^{i\vec{k}\vec{r}} + \mathcal{F}(E, \varphi)\frac{e^{ikr}}{\sqrt{r}} \right]} \qquad (16.131)$$

where $\varphi = \varphi_k - \varphi_r$ is the scattering angle, and the total scattering amplitude has the following partial-wave expansion:

$$\mathcal{F}(E, \varphi) = \sum_{\ell=0}^{\infty} \epsilon_\ell f_\ell(E) \cos(\ell\varphi) . \qquad (16.132)$$

The cross section for a two-dimensional scattering has the units of length. The number of particles scattered into the angle spanned by the arc $r\,d\varphi$ is the product of the flux in that radial direction and the length of the arc. The corresponding cross section $d\sigma$ is defined as such a length that after its multiplication by the total incoming flux, it gives the same number of particles, i.e.

$$\left|\vec{j}^{\,(\mathrm{in})}\right| d\sigma = j_r^{(\mathrm{out})}(\varphi) r\, d\varphi \,.$$

Using standard definition of the particle flux, $\vec{j} = \hbar/(2i\mu)(\psi^* \vec{\nabla}\psi - \psi \vec{\nabla}\psi^*)$, with the operator $\vec{\nabla}$ given by Eq. (16.66), it is not difficult to find that the incoming (corresponding to the first term of the wave function (16.131)) and outgoing (obtained from the second term of the same wave function) fluxes are

$$\vec{j}^{\,(\mathrm{in})} = \frac{\hbar\vec{k}}{(2\pi)^2\mu}\,, \qquad j_r^{(\mathrm{out})}(\varphi) = \frac{\hbar k\,|\mathcal{F}(E,\varphi)|^2}{(2\pi)^2\mu r}\,,$$

and thus the differential cross section is

$$\frac{d\sigma}{d\varphi} = |\mathcal{F}(E,\varphi)|^2 \,.$$

Using the integral

$$\int_0^{2\pi} \cos(\ell\varphi)\cos(\ell'\varphi)d\varphi = \left\{ \begin{array}{l} 0 \;\;\, , \; \ell \neq \ell' \\ 2\pi \;, \; \ell = \ell' = 0 \\ \pi \;\;\, , \; \ell = \ell' \neq 0 \end{array} \right\} = \frac{2\pi}{\epsilon_\ell}\delta_{\ell\ell'},$$

the total cross section can be written as follows:

$$\sigma = \int_0^{2\pi} |\mathcal{F}(E,\varphi)|^2\, d\varphi = \sum_\ell \sigma_\ell \,,$$

$$\sigma_\ell = 2\pi\epsilon_\ell\,|f_\ell(E)|^2 = \frac{\epsilon_\ell}{k}\,|S_\ell(E) - 1|^2 \,, \tag{16.133}$$

where σ_ℓ is the partial-wave cross section.

***

An example of using the $2D$-theory presented here can be found in Ref. [73].

17 Miscellaneous Extensions of the Jost Function Approach

17.1 Many-Body Problems

The Jost functions and matrices are rigorously defined and have clear physical meaning for the systems with binary channels, i.e. when we deal with two interacting bodies, even if they change their nature during the interaction process,

$$a + b \longrightarrow c + d \,.$$

In such a problem, the asymptotic behaviour of the wave function (even in the multi-channel case) is rather simple: it is a superposition of the incoming and outgoing waves (in each channel), and the coefficients in this superposition (i.e. the amplitudes of the waves) are the Jost functions (or the elements of the Jost matrices).

When more than two bodies can appear as a result of the interaction,

$$a + b \longrightarrow c + d + e \,,$$

the wave function has a much more complicated asymptotic behaviour because there are more than one radial variable (the inter-particle distances $|\vec{r}_1|$, $|\vec{r}_2|$,...) that can go to infinity independently. It is not clear how the amplitudes of the waves (the Jost functions) can be introduced in such a case.

A space configuration of an N-body system can be specified by a set of $(N-1)$ Jacobi vectors, $\vec{r} = \{\vec{r}_1, \vec{r}_2, \ldots, \vec{r}_{N-1}\}$. An example of them is shown in Fig. 17.1 for a three-body system. Formally, we can consider $\vec{r}$ as a vector in the $3(N-1)$-dimensional space. In the space of any dimension, it is possible to specify a point by its distance from the origin and by a set of angles that give the direction towards the point. The distance, $r = |\vec{r}\,|$, varies from zero to infinity, while all the angles have finite ranges. We therefore can try and define the Jost matrices by considering the asymptotic behaviour of the N-body wave function when the variable r goes to infinity.

S. A. Rakityansky, *Jost Functions in Quantum Mechanics*,
https://doi.org/10.1007/978-3-031-07761-6_17

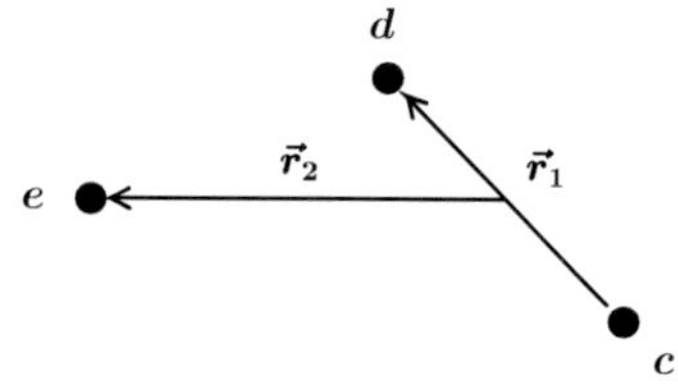

Fig. 17.1 Jacobi vectors specifying the space configuration of the three particles, c, d, and e

However, in general, such an asymptotic behaviour strongly depends on the direction (on the angles) in which r is increasing. In the three-body example depicted in Fig. 17.1, the growth of either r_1 or r_2 (while the other remains finite) leads to an increase of r, while the directions in the six-dimensional space for these two choices are different. If the particles c and d can form a bound state, then at large r_1 the three-body wave function becomes zero, but it becomes a product of their bound-state wave function and the incoming and outgoing waves along variable r_2, when r_1 is finite and $r_2 \to \infty$. This simple example shows the difficulties that preclude the development of a general theory of many-body Jost matrices.

There is, however, a particular class of many-body problems where the asymptotic behaviour of the wave function can be rather accurately approximated by a simple superposition of the waves depending on a single radial variable. These are the problems associated with the systems of particles that cannot form any bound clusters. An example of such a system is a multi-neutron system. Neutrons can collide and form a multi-neutron resonance, but eventually all of them go free,

$$n + n + n \longrightarrow n + n + n \,. \tag{17.1}$$

We can also imagine an N-particle system that can form a bound state of all N particles, but no less, i.e. none of its subsystems is bound. A hypothetical example of such a system could be a multi-neutron system with few Λ-particles added to it.

The asymptotic behaviour of the wave function of a many-body system, which does not form bound clusters, is rather simple. We can assume that it is the same in all directions in the multi-dimensional space. If a bound state of whole system is possible, then at the corresponding energy the wave function at large distance vanishes in all directions. At the (complex) energy of an N-particle resonance, the wave function exponentially grows in all directions. At the other energies, it is a superposition of the collective incoming and outgoing waves.

It should be emphasized that such a simplification of the asymptotic behaviour of the many-body wave function is an approximation. Strictly speaking, that asymptotic form should include the terms describing, for example, the process when the particles c and d (of Fig. 17.1) are still colliding, while particle e is already far away. However, intuitively we understand that in an unbound system all the particles go away more or less at the same time.

It should also be noted that the theory developed further down can, in principle, be applied for describing the multi-particle collision processes of the type (17.1), i.e. when several particles collide simultaneously. However, the probabilities of such collisions are very low even in very dense matter. We therefore restrict our consideration to finding possible bound and resonant states of an N-body system that cannot form any bound clusters.

17.1.1 Hyperspherical Expansion

Consider a system of N particles. Let $\vec{\rho} = \{\vec{\rho}_1, \vec{\rho}_2, \dots, \vec{\rho}_{N-1}\}$ be the complete set of Jacobi vectors specifying the space configuration of the system. The operator of the total kinetic energy of the internal motion in such a system is

$$H_0 = -\frac{\hbar^2}{2\mu_1}\Delta_{\vec{\rho}_1} - \frac{\hbar^2}{2\mu_2}\Delta_{\vec{\rho}_2} - \cdots - \frac{\hbar^2}{2\mu_{N-1}}\Delta_{\vec{\rho}_{N-1}} , \tag{17.2}$$

where μ_i is the reduced mass associated with the motion along the i-th Jacobi coordinate. For example, for the system depicted in Fig. 17.1,

$$\mu_1 = \frac{m_c m_d}{m_c + m_d} \qquad \text{and} \qquad \mu_2 = \frac{m_e(m_c + m_d)}{m_e + (m_c + m_d)} .$$

Alternatively, we can specify the same space configuration by the set $\vec{r} = \{\vec{r}_1, \vec{r}_2, \dots, \vec{r}_{N-1}\}$ of the scaled vectors defined as

$$\vec{r}_i = \sqrt{\frac{\mu_i}{\mu}}\,\vec{\rho}_i , \qquad i = 1, 2, \dots, N-1 , \tag{17.3}$$

where μ is the reduced mass of the whole system.[1] The purpose of the scaling (17.3) is to make the same all the coefficients in all terms of H_0,

$$H_0 = -\frac{\hbar^2}{2\mu}\left(\Delta_{\vec{r}_1} + \Delta_{\vec{r}_2} + \cdots + \Delta_{\vec{r}_{N-1}}\right) = -\frac{\hbar^2}{2\mu}\Delta , \tag{17.4}$$

where Δ is the Laplacian in the space of the dimension $D = 3(N-1)$. Each point in this space can be uniquely specified by its distance from the origin,

$$r = \sqrt{r_1^2 + r_2^2 + \cdots + r_{N-1}^2} , \tag{17.5}$$

[1] Actually, the value of μ can be chosen arbitrarily. The results of calculations of any observables do not depend on the choice of μ.

which is called hyperradius, and by the set of $(D-1)$ hyperangles, $\vec{r} = \{r, \Omega\}$. It can be shown that[2]

$$\Delta = \partial_r^2 + \frac{D-1}{r}\partial_r - \frac{\Lambda^2(\Omega)}{r^2}, \tag{17.6}$$

where Λ^2 is the operator involving the second-order derivatives over the hyperangles. The specific form of Λ^2 depends on the number of particles N. The eigenfunctions, $Y_{[\mathcal{L}]}(\Omega)$, of this operator are called hyperspherical harmonics (or simply hyperharmonics). They are solutions of the equation

$$\Lambda^2 Y_{[\mathcal{L}]} = \hbar^2 L(L+D-2) Y_{[\mathcal{L}]} \,. \tag{17.7}$$

These harmonics are explicitly known for any number of particles. There are $(D-1)$ hyperangles, which implies that the full set of the quantum numbers, denoted here by the multi-index $[\mathcal{L}]$, is a collection of $(D-1)$ elements. These numbers include the following:

- Orbital angular momentum ℓ_i and its third component m_i for each Jacobi coordinate, $i = 1, 2, \ldots, N-1$.
- Partial grand orbital numbers, L_i, for the subsystems of the particles, and the grand orbital number, L, for the whole system:

$$L_i = L_{i-1} + \ell_i + 2n_i, \qquad i = 2, 3, \ldots, N-1, \qquad n_i = 0, 1, 2, \ldots,$$

  where $L_1 = \ell_1$ and $L_{N-1} = L$ is the grand orbital number.

The hyperspherical harmonics constitute an ortho-normal basis in the space of the hyperspherical angles, which implies that

$$\int Y^*_{[\mathcal{L}']}(\Omega) Y_{[\mathcal{L}]}(\Omega) d\Omega = \delta_{[\mathcal{L}'][\mathcal{L}]} \,. \tag{17.8}$$

In the particular case of the two-body problem ($N = 2$ and $D = 3$), the hyperspherical harmonics are the well-known spherical harmonics $Y_{\ell m}(\theta, \varphi)$, for which $[\mathcal{L}] = \{\ell, m\}$ and $\Omega = \{\theta, \varphi\}$.

It is always more convenient to work with the basis whose elements correspond to the states with definite values of the (conserving) total angular momentum. Therefore, using the appropriate Clebsch–Gordan coefficients, we can construct linear combinations of the hyperspherical harmonics, such that they describe the states with total angular momentum J and its third component M (we denote

[2] The reader can easily find the corresponding derivations and proofs in the literature on the hyperspherical expansions; for example, in Ref. [74].

these combinations as $\mathcal{Y}_{[L]}^{JM}(\Omega)$). There are many ways of making such linear combinations. They are determined by the chosen sequence of adding of the angular momenta. If the particles have non-zero spins, then they should also be taken into account. In such a case, the basis functions $\mathcal{Y}_{[L]}^{JM}$ become ortho-normal vectors in the spin-angular space. Of course, the multi-index $[L]$ is different from $[\mathcal{L}]$. The total number of elements (quantum numbers) in $[L]$ plus two conserving quantum numbers is the same as in $[\mathcal{L}]$.

To make things more clear, let us consider a simple example, namely, the three-body system depicted in Fig. 17.1, and assume that all the particles are spinless. If ℓ_{cd} and m_{cd} are the orbital angular momentum and its third component of the relative cd-motion, while ℓ_e and m_e are the corresponding quantities for the motion of e relative to the centre of mass of the (cd)-pair, then the multi-index $[\mathcal{L}]$ consists of the following five quantum numbers:

$$[\mathcal{L}] = \{L,\ \ell_{cd},\ m_{cd},\ \ell_e,\ m_e\}\ .$$

After coupling the angular momenta $\vec{\ell}_{cd}$ and $\vec{\ell}_e$, we obtain new basis functions:

$$\mathcal{Y}_{[L]}^{JM}(\Omega) = \sum_{m_{cd}\ m_e} Y_{[\mathcal{L}]}(\Omega) C_{\ell_{cd}\, m_{cd}\, \ell_e\, m_e}^{JM}\ ,$$

where the subscript $[L]$ consists of a smaller set of (three) quantum numbers:

$$[L] = \{L,\ \ell_{cd},\ \ell_e\}\ .$$

The remaining two of the full set of the numbers are the conserving J and M.

Since the functions $\mathcal{Y}_{[L]}^{JM}$ constitute a basis in the space of hyperangles (as well as in the space of the spins, if they are non-zero), the N-body wave function, $\Psi^{JM}(E, \vec{r}\,)$, can be expanded over this basis with unknown coefficients, $u_{[L]}^{J}(E, r)$, depending on the hyperradius,

$$\Psi^{JM}(E, \vec{r}\,) = r^{2-3N/2} \sum_{[L]=[L_{\min}]}^{\infty} \mathcal{Y}_{[L]}^{JM}(\Omega) u_{[L]}^{J}(E, r)\ . \tag{17.9}$$

The summation in this equation begins, generally speaking, from a non-zero $L_{\min}$. The minimal value of the grand orbital number is determined by chosen value of J and the permutation symmetry of the physical system under consideration. For example, if in the three-body system depicted in Fig. 17.1, all the particles are neutrons, then according to the Pauli principle only one of the two angular momenta can be zero. With the minimal value of $J = 1/2$ (with spins taken into account), this implies that $L_{\min} = 1$.

Since L is not conserving, the summation in the expansion (17.9) formally goes to infinity. However, in a particular calculation, we can always find some physical reasons to truncate this sum.

Equation (17.9) involves the factor $r^{2-3N/2}$, which for $N = 2$ becomes familiar $1/r$ that is used to get rid of the first derivative in the two-body radial Schrödinger equation. For an arbitrary N, the factor $r^{2-3N/2}$ serves for the same purpose.

17.1.2 Hyperradial Equation and the Jost Matrices

Let

$$U(\vec{r}\,) = \sum_{i<j} U_{ij}(\vec{r}\,)$$

be the sum of all two-body potentials[3] that describe the interactions between the particles i and j. Then the N-body wave function obeys the following Schrödinger equation:

$$\left(-\frac{\hbar^2}{2\mu}\Delta + U - E\right)\Psi^{JM}(E,\vec{r}\,) = 0\,. \tag{17.10}$$

The same way as we did it in Sect. 9.1.1 for a non-central potential, here for the hyperradial parts, $u^J_{[L]}(E,r)$, of the wave function, we obtain the system of coupled hyperradial equations,

$$\left[\partial_r^2 + k^2 - \frac{\lambda(\lambda+1)}{r^2}\right] u^J_{[L]}(E,r) = \sum_{[L']} V^J_{[L][L']}(r)\, u^J_{[L']}(E,r)\,, \tag{17.11}$$

where the parameter $k = \sqrt{2\mu E/\hbar^2}$ can be called hypermomentum, and the parameter

$$\lambda = L + \frac{3}{2}(N-2)\,, \tag{17.12}$$

which is integer for an even number of particles and half-integer for an odd N, is the generalization of the orbital angular momentum. The elements of the potential matrix are obtained by integrating $U(\vec{r}\,)$ with the hyperharmonics over all the hyperangles:

$$V^J_{[L][L']}(r) \equiv \frac{2\mu}{\hbar^2}\langle \mathcal{Y}^{JM}_{[L]}|U(\vec{r}\,)|\mathcal{Y}^{JM}_{[L']}\rangle\,. \tag{17.13}$$

[3] It is clear that, if necessary, we can add here any many-body potentials as well.

It should be noted that there is the complete analogy of the system of equations (17.11) with the corresponding system (10.16) obtained for the two-body problem with a non-central potential. This analogy allows us to use here most of the relations derived in Chap. 9. We only need to replace in them the symbol $[\ell]$ with $[L]$. A significant difference of the hyperspherical expansion (17.9) and the system (17.11) from the corresponding equations of the two-body problem are that the summations in (17.9) and (17.11) formally run to infinity. However, in any practical application of the hyperspherical expansion, we always truncate it at some $[L_{\max}]$, which makes the number $\mathcal{N}$ of equations in the system (17.11) finite and each of its solutions becomes a column-matrix of a finite length $\mathcal{N}$. Then all the reasoning of Chap. 9 become applicable here.

In particular, we should only consider the regular column-solutions, which are determined by the same boundary conditions described in Sect. 9.3.4. There are $\mathcal{N}$ linearly independent regular columns that can be combined in a square $(\mathcal{N} \times \mathcal{N})$-matrix. This matrix ϕ asymptotically behaves as a superposition of the incoming and outgoing hypespherical waves,

$$W^{(\text{in/out})}(E,r) = \begin{pmatrix} h^{(\mp)}_{\lambda_1}(kr) & 0 & \cdots & 0 \\ 0 & h^{(\mp)}_{\lambda_2}(kr) & \cdots & 0 \\ \vdots & \vdots & \vdots & \vdots \\ 0 & 0 & \cdots & h^{(\mp)}_{\lambda_{\mathcal{N}}}(kr) \end{pmatrix} , \tag{17.14}$$

i.e.

$$\phi^J(E,r) \underset{r\to\infty}{\longrightarrow} W^{(\text{in})}(E,r) f^{(\text{in})J}(E) + W^{(\text{out})}(E,r) f^{(\text{out})J}(E) , \tag{17.15}$$

where $f^{(\text{in/out})J}(E)$ are the N-body Jost matrices for the states with the total angular momentum J. They can be found as the asymptotic values,

$$f^{(\text{in/out})J}(E) = \lim_{r\to\infty} F^{(\text{in/out})J}(E,r) , \tag{17.16}$$

of the functions $F^{(\text{in/out})J}(E,r)$ that are defined in the same way as in Eq. (10.24) of the two-body problem, i.e.

$$\phi^J(E,r) = W^{(\text{in})}(E,r) F^{(\text{in})J}(E,r) + W^{(\text{out})}(E,r) F^{(\text{out})J}(E,r) , \tag{17.17}$$

and obey the same system of differential equations,

$$\partial_r F^{(\text{in})J} = -\frac{1}{2ik} W^{(\text{out})} V^J \left[W^{(\text{in})} F^{(\text{in})J} + W^{(\text{out})} F^{(\text{out})J} \right] , \tag{17.18}$$

$$\partial_r F^{(\text{out})J} = \frac{1}{2ik} W^{(\text{in})} V^J \left[W^{(\text{in})} F^{(\text{in})J} + W^{(\text{out})} F^{(\text{out})J} \right] , \tag{17.19}$$

with the same boundary conditions

$$\lim_{r\to 0}\frac{j_\lambda(kr)F^{(\mathrm{in/out})J}_{[L][L']}(E,r)}{j_{\lambda'}(kr)}=\frac{1}{2}\delta_{[L][L']}\,. \tag{17.20}$$

And of course, for complex values of E, the differential equations (17.18, 17.19) need to be integrated along a path in the complex r-plane, as is explained in Sect. 9.5.1.

It is easy to see that the analysis given in Sect. 9.7.1 for the problem of finding bound and resonant states generated by a non-central potential is applicable here. In other words, the discrete states of an N-body system can be found as the roots of the following equation:

$$\det f^{(\mathrm{in})J}(E)=0\,. \tag{17.21}$$

In the same way as it was done in Sect. 9.3.3, we can derive the alternative form of differential equations that can be used for calculating the Jost matrices. To this end, we introduce the linear combinations of the matrices (17.14),

$$j=\mathrm{diag}\left\{j_{\lambda_1}(kr),\,j_{\lambda_2}(kr),\ldots,\,j_{\lambda_{\mathcal{N}}}(kr)\right\}\,, \tag{17.22}$$

$$n=\mathrm{diag}\left\{n_{\lambda_1}(kr),n_{\lambda_2}(kr),\ldots,n_{\lambda_{\mathcal{N}}}(kr)\right\}\,, \tag{17.23}$$

as well as the combinations of the unknown matrices,

$$A^J(E,r)=F^{(\mathrm{in})J}(E,r)+F^{(\mathrm{out})J}(E,r)\,, \tag{17.24}$$

$$B^J(E,r)=i\left[F^{(\mathrm{in})J}(E,r)-F^{(\mathrm{out})J}(E,r)\right]\,, \tag{17.25}$$

which obey the alternative system of differential equations,

$$\partial_r A^J=-\frac{1}{k}nV^J\left(jA^J-nB^J\right)\,, \tag{17.26}$$

$$\partial_r B^J=-\frac{1}{k}jV^J\left(jA^J-nB^J\right)\,, \tag{17.27}$$

with the boundary conditions

$$\lim_{r\to 0}\frac{j_\lambda(kr)A^J_{[L][L']}(E,r)}{j_{\lambda'}(kr)}=\delta_{[L][L']}\,,\qquad \lim_{r\to 0}\frac{j_\lambda(kr)B^J_{[L][L']}(E,r)}{j_{\lambda'}(kr)}=0\,. \tag{17.28}$$

17.1.3 Analytic Structure of the Jost Matrices

The analytic structure of the N-body Jost matrices can be established similarly to how it was done in Sect. 13.4 for the two-body multi-channel problem. Introducing the diagonal matrices

$$D = \text{diag}\left\{k^{\lambda_1+1}, k^{\lambda_2+1}, \ldots, k^{\lambda_{\mathcal{N}}+1}\right\}, \tag{17.29}$$

$$\tilde{j} = \text{diag}\left\{\tilde{j}_{\lambda_1}(E,r), \tilde{j}_{\lambda_2}(E,r), \ldots, \tilde{j}_{\lambda_{\mathcal{N}}}(E,r)\right\}, \tag{17.30}$$

$$\tilde{n} = \text{diag}\left\{\tilde{n}_{\lambda_1}(E,r), \tilde{n}_{\lambda_2}(E,r), \ldots, \tilde{n}_{\lambda_{\mathcal{N}}}(E,r)\right\}, \tag{17.31}$$

and using the representations (7.58), (7.59) of the Riccati–Bessel and Riccati–Neumann functions for an integer order, as well as their representations (16.80, 16.81) for a half-integer order, we can obtain the following representations of the matrices j and n, defined by Eqs. (17.22), (17.23), which are valid in both cases:

$$j = D\tilde{j}, \qquad n = kD^{-1}\tilde{n} + \mathfrak{h}D\tilde{j}, \tag{17.32}$$

where[4]

$$\mathfrak{h} = \begin{cases} 0 & , \ \lambda \text{ is integer}, \\ \dfrac{2}{\pi}\ln\left(\dfrac{kR}{2}\right) & , \ \lambda \text{ is half-integer}. \end{cases} \tag{17.33}$$

Consider the following tilded matrices:

$$\tilde{A}^J = DA^JD^{-1} - \mathfrak{h}DB^JD^{-1}, \tag{17.34}$$

$$\tilde{B}^J = kD^{-1}B^JD^{-1}. \tag{17.35}$$

Multiplying Eqs. (17.26), (17.27) by the matrices D and D^{-1} as is given by (17.34) and (17.35), we obtain the following differential equations for the tilded matrices:

$$\partial_r\tilde{A}^J = -\tilde{n}V^J\left(\tilde{j}\tilde{A}^J - \tilde{n}\tilde{B}^J\right), \tag{17.36}$$

[4] R can be taken as the unit of length.

$$\partial_r \tilde{B}^J = -\tilde{j} V^J \left(\tilde{j} \tilde{A}^J - \tilde{n} \tilde{B}^J \right) . \tag{17.37}$$

These equations do not involve any multi-valued functions (neither powers of k nor $\mathfrak{h}(k)$). The boundary conditions for the matrix functions $\tilde{A}^J(E, r)$ and $\tilde{B}^J(E, r)$ at $r = 0$ are the same as given by Eq. (17.28), i.e. energy independent. Based on the Poincaré theorem (see Appendix G), we conclude that at any finite r they are single-valued analytic functions of complex variable E. For their asymptotic values, denoted as

$$\tilde{A}^J(E, r) \underset{r\to\infty}{\longrightarrow} 2a^J(E) , \qquad \tilde{B}^J(E, r) \underset{r\to\infty}{\longrightarrow} 2b^J(E) , \tag{17.38}$$

we can use the same proof that was given in Sect. 7.7.2 to show that they are analytic and single-valued within certain domain $\mathcal{D}$ of the complex E-plane. It is also seen that for real E the matrices $a^J(E)$ and $b^J(E)$ are real.

Expressing A^J and B^J via $\tilde{A}^J$ and $\tilde{B}^J$, and using the relations $F^{(\text{in/out})J} = (A^J \mp i B^J)$, we obtain the following structure of the N-body Jost matrices:

$$f^{(\text{in/out})J}(E) = D^{-1} a^J(E) D \mp i k^{-1} D b^J(E) D + \mathfrak{h}(k) b^J(E) , \tag{17.39}$$

or in the explicit form for the matrix elements:

$$\boxed{f^{(\text{in/out})J}_{[L'][L]}(E) = k^{\lambda-\lambda'} a^J_{[L'][L]}(E) + \left[\mathfrak{h}(k) \mp i k^{\lambda'+\lambda+1} \right] b^J_{[L'][L]}(E)} \tag{17.40}$$

where the matrices a^J and b^J are single-valued analytic functions of variable E. It should be noted that the logarithmic function $\mathfrak{h}(k)$ is only present in these equations if N is odd, i.e. when the effective angular momentum λ is half-integer, while for an even number of particles, $\mathfrak{h}(k) \equiv 0$.

It is clear that the Riemann surface, which the Jost matrices are defined on, has different topologies for systems with different N values. When we deal with an even number of particles, this surface consists of just two sheets that are cut and interconnected in exactly the same way as is shown in Figs. 3.1 and 3.2, where the threshold energy, $E = 0$, is the square-root branch point. For a system of an odd number of particles, due to the presence of the logarithmic function $\mathfrak{h}(k)$, the number of Riemann sheets becomes infinite with the spiral topology, similarly to the two-body problems with charged particles and to the $2D$-problem.

Analysing the symmetry properties of the matrices (17.40) on the Riemann surface, we see that the change $k \to -k$ causes incoherent changes in their different matrix elements. The reason is that (in contrast to the two-body case, where due to the parity conservation the angular momenta ℓ and ℓ' always have the same parity) the powers $\lambda \pm \lambda'$ of k can be even or odd for different matrix elements. Therefore, in general, for the N-body problem, there is no "vertical" symmetry for the Jost matrices. At the same time, as it is easy to check, the "diagonal" symmetry,

$$f^{(\text{in/out})J}(E^*) = \left[f^{(\text{out/in})J}(E)\right]^* , \tag{17.41}$$

which follows from the Schwarz reflection principle, remains valid for any number of particles N, even or odd.

* * *

Examples of using the hyperspherical many-body Jost matrices presented here can be found in Refs. [75–77] where this theory is applied for the analysis of some nuclear and hypernuclear systems.

17.2 Non-local Potential

The non-local potentials are used in describing various types of quantum systems. For example, they appear in nuclear and atomic physics when the interactions between two nuclei or two atoms are modelled by some effective two-body potentials. Since these interacting bodies are composed of identical particles (nucleons or electrons), in addition to the direct-interaction potential, there is the so-called exchange potential that is non-local and appears due to the antisymmetrization of the wave function.

Actually, if we were able to solve the many-body Schrödinger equation directly, then in most cases we would do it without using the non-local potentials. However, when introducing some effective nucleus–nucleus or atom–atom two-body potentials, we have to use non-local interactions to make the approximation more accurate. In fact, the interaction between composed particles depends not only on their relative distance, but also on their relative velocity. It can be shown (see Sect. 17.2.2) that such a velocity dependence is equivalent to the non-locality of the interaction.

17.2.1 Schrödinger Equation

The action of a non-local operator U (the potential) on the wave function ψ_E in the coordinate representation is not a simple multiplication, but the integral over all possible space positions of the particle (hence the name non-local),

$$\langle\vec{r}\,|U|\psi_E\rangle = \langle\vec{r}\,|U\left(\int|\vec{r}\,'\rangle\langle\vec{r}\,'|\,d^3r'\right)|\psi_E\rangle = \int\langle\vec{r}\,|U|\vec{r}\,'\rangle\psi_E(\vec{r}\,')\,d^3r' , \tag{17.42}$$

with the kernel $\langle\vec{r}\,|U|\vec{r}\,'\rangle$ depending on the two space variables. The two-body Schrödinger equation with such a potential becomes integro-differential,

$$\left(\frac{\hbar^2}{2\mu}\Delta + E\right)\psi_E(\vec{r}\,) = \int \langle\vec{r}\,|U|\vec{r}\,'\rangle\psi_E(\vec{r}\,')\,d^3r'\,. \tag{17.43}$$

For the sake of clarity, we assume that the potential is spherically symmetric, i.e. that it does not mix states with different angular momenta.[5] Formally, this assumption means that (see Eq. (A.8) of Appendix A)

$$\langle\vec{r}\,|U|\vec{r}\,'\rangle = \sum_{\ell m\ell' m'} Y_{\ell m}(\hat{\vec{r}}\,)\langle r\ell m|U|r'\ell' m'\rangle Y^*_{\ell' m'}(\hat{\vec{r}}\,')\,, \tag{17.44}$$

where the matrix

$$\langle r\ell m|U|r'\ell' m'\rangle = \frac{U_\ell(r,r')}{rr'}\delta_{\ell\ell'}\delta_{mm'} \tag{17.45}$$

is diagonal in $[\ell] = \{\ell, m\}$ and does not depend on m. The factor $(rr')^{-1}$ is introduced here to make the potential function U_ℓ measured in its proper units (energy). The need for the factor $(rr')^{-1}$ can also be seen from another point of view. Indeed, for a local potential on the right hand side of this equation, we would have the δ-function $\delta(r - r')$, and then the radial integration $(r'^2 dr')$ would give the product $U_\ell\psi_E$, as it should be.

Substituting the expansion (17.44) together with the partial- wave decomposition for the wave function (9.1) into Eq. (17.43) multiplied from the left by $Y^*_{[\ell]}(\hat{\vec{r}}\,)$, and integrating over all the spherical angles, $\hat{\vec{r}}$ and $\hat{\vec{r}}\,'$, we obtain the radial integro-differential Schrödinger equation,

$$\left[\partial_r^2 + k^2 - \frac{\ell(\ell+1)}{r^2}\right]u_\ell(E,r) = \int_0^\infty V_\ell(r,r')u_\ell(E,r')\,dr'\,, \tag{17.46}$$

where the reduced potential (in the units of $[\text{length}]^{-2}$) is given by

$$V_\ell(r,r') = \frac{2\mu}{\hbar^2}U_\ell(r,r')\,. \tag{17.47}$$

Since the operator U is a part of the Hamiltonian, it is Hermitian, i.e. $U^\dagger = U$. Taking into account the fact that the potential is real, we conclude that it is symmetric with respect to variables r and r':

$$\langle\vec{r}\,|U|\vec{r}\,'\rangle = \langle\vec{r}\,'|U|\vec{r}\,\rangle^* \qquad \Rightarrow \qquad V_\ell(r,r') = V_\ell(r',r)\,.$$

[5] If necessary, the reader can generalize all the derivations for a non-central potential, using the same approach that is described in Chap. 9.

17.2.2 Non-locality and Velocity Dependence

Let U be a non-local potential. Changing variable in the integral (17.42),

$$\vec{r}\,' = \vec{r} + \vec{\rho}\,,$$

we obtain

$$\langle \vec{r}\,|U|\psi_E\rangle = \int U(\vec{r}, \vec{r} + \vec{\rho}\,)\psi_E(\vec{r} + \vec{\rho}\,)\,\mathrm{d}^3\rho\,. \tag{17.48}$$

Wave function $\psi_E(\vec{r} + \vec{\rho}\,)$ can be expanded in the Taylor series:

$$\psi_E(\vec{r} + \vec{\rho}\,) = \psi_E(\vec{r}\,) + \vec{\rho}\cdot\left[\vec{\nabla}\,\psi_E(\vec{r}\,)\right] + \dots\,.$$

Substituting this expansion in Eq. (17.48) and performing the integration, we obtain

$$\langle \vec{r}\,|U|\psi_E\rangle = \left[U_0(\vec{r}\,) + \vec{U}_1(\vec{r}\,)\cdot\vec{p} + \cdots\right]\psi_E(\vec{r}\,)\,, \tag{17.49}$$

where $\vec{p} = (\hbar/i)\,\vec{\nabla}$ and

$$U_0(\vec{r}\,) = \int U(\vec{r}, \vec{r} + \vec{\rho}\,)\,\mathrm{d}^3\rho\,,$$

$$\vec{U}_1(\vec{r}\,) = \frac{i}{\hbar}\int U(\vec{r}, \vec{r} + \vec{\rho}\,)\vec{\rho}\,\mathrm{d}^3\rho\,,$$

etc. ....

It is clear that in Eq. (17.49) the action of the non-local potential U is equivalent to the action of the local potential

$$\mathcal{U}(\vec{r}\,) = U_0(\vec{r}\,) + \vec{U}_1(\vec{r}\,)\cdot\vec{p} + \cdots$$

that explicitly depends on the momentum $\vec{p}$ (and therefore on the corresponding velocity and the energy).

17.2.3 Jost Functions

To simplify things, we assume that $V_\ell(r, r')$ is a short-range potential, i.e. at large values of any of the two radial variables, the kernel of the potential vanishes faster than the centrifugal potential. Therefore, when $r \to \infty$, the right hand side of Eq. (17.46) becomes negligible, and its regular solution can be written as a linear combination of the incoming and outgoing spherical waves (the same as for a local potential),

$$\phi_\ell(E,r) \underset{r\to\infty}{\longrightarrow} h_\ell^{(-)}(kr) f_\ell^{(\mathrm{in})}(E) + h_\ell^{(+)}(kr) f_\ell^{(\mathrm{out})}(E)\,, \tag{17.50}$$

where $f_\ell^{(\mathrm{in/out})}(E)$ are the Jost functions. The same way as for a local potential (see Sect. 2.2.1.1), we look for the regular solution at all distances in the form:

$$\phi_\ell(E,r) = h_\ell^{(-)}(kr) F_\ell^{(\mathrm{in})}(E,r) + h_\ell^{(+)}(kr) F_\ell^{(\mathrm{out})}(E,r) \tag{17.51}$$

with the Lagrange condition (2.13) imposed on the unknown functions $F_\ell^{(\mathrm{in/out})}(E,r)$. It is not difficult to repeat all the derivations of Sect. 2.2.1.1, where instead of the simple product Vu_ℓ, we have the integral $\int_0^\infty V_\ell(r,r')u_\ell(E,r')\,dr'$, and to obtain the following system of integro-differential equations for the unknown functions:

$$\begin{aligned}
\partial_r F_\ell^{(\mathrm{in})}(E,r) = -\frac{h_\ell^{(+)}(kr)}{2ik}\int_0^\infty V_\ell(r,r')\Big[h_\ell^{(-)}(kr')F_\ell^{(\mathrm{in})}(E,r') \\
+ h_\ell^{(+)}(kr')F_\ell^{(\mathrm{out})}(E,r')\Big]\,dr'\,, \\
\partial_r F_\ell^{(\mathrm{out})}(E,r) = \frac{h_\ell^{(-)}(kr)}{2ik}\int_0^\infty V_\ell(r,r')\Big[h_\ell^{(-)}(kr')F_\ell^{(\mathrm{in})}(E,r') \\
+ h_\ell^{(+)}(kr')F_\ell^{(\mathrm{out})}(E,r')\Big]\,dr'\,,
\end{aligned} \tag{17.52}$$

with the same boundary conditions

$$F_\ell^{(\mathrm{in/out})}(E,0) = \frac{1}{2}\,. \tag{17.53}$$

For practical use (i.e. for numerical calculations), it is more convenient to transform these equations into pure integral form. This is done by integrating both sides of Eqs. (17.52) from zero to r. Then, taking into account the values (17.53) of the unknown functions at zero, we obtain the following integral equations:

$$
\begin{aligned}
F_\ell^{(\mathrm{in})}(E,r) &= \frac{1}{2} - \frac{1}{2ik}\int_0^r dr''\int_0^\infty dr'\, h_\ell^{(+)}(kr'')V_\ell(r'',r')\times \\
&\quad \times\left[h_\ell^{(-)}(kr')F_\ell^{(\mathrm{in})}(E,r') + h_\ell^{(+)}(kr')F_\ell^{(\mathrm{out})}(E,r')\right], \\
F_\ell^{(\mathrm{out})}(E,r) &= \frac{1}{2} + \frac{1}{2ik}\int_0^r dr''\int_0^\infty dr'\, h_\ell^{(-)}(kr'')V_\ell(r'',r')\times \\
&\quad \times\left[h_\ell^{(-)}(kr')F_\ell^{(\mathrm{in})}(E,r') + h_\ell^{(+)}(kr')F_\ell^{(\mathrm{out})}(E,r')\right].
\end{aligned}
\tag{17.54}
$$

In many cases, the potential involves both local (V_ℓ^{L}) and non-local (V_ℓ^{N}) terms. In the above equations, the local part of the potential should appear with the δ-function,

$$
V_\ell(r,r') = V_\ell^{\mathrm{L}}(r)\delta(r-r') + V_\ell^{\mathrm{N}}(r,r').
$$

If we deal with charged particles, then in the total Hamiltonian it is always possible to separate a pure Coulomb local potential, as is illustrated in Fig. 8.1. In such a case, the spherical waves $h_\ell^{(\pm)}$ become Coulombic, i.e. the Riccati–Hankel functions should be replaced with the functions $H_\ell^{(\pm)}(\eta,kr)$ defined by Eq. (8.12).

The same as for pure local potentials, the Jost functions are obtained as the asymptotic values of the functions $F_\ell^{(\mathrm{in/out})}(E,r)$,

$$
f_\ell^{(\mathrm{in/out})}(E) = \lim_{r\to\infty} F_\ell^{(\mathrm{in/out})}(E,r). \tag{17.55}
$$

It is not difficult to check that all the reasoning given in Sect. 7 about convergence of the limits (17.55) are applicable in the case of the integral equations (17.54) as well. This means that for complex values of the energy the integrations in these equations should be performed along a complex path shown in Fig. 7.3.

17.2.3.1 Short-Distance Behaviour of Non-local Potentials

When studying local potentials, we established that near the origin ($r \to 0$) they should be less singular than the centrifugal potential,[6] i.e.

$$
V_\ell^{\mathrm{L}}(r) \underset{r\to 0}{\longrightarrow} \sim \frac{1}{r^{2-\epsilon}}, \qquad \epsilon > 0.
$$

[6] Local potentials can be singular if they are repulsive near the origin (see Sect. 16.1). For non-local potentials, this is not the case.

Is this requirement still sufficient in the case of non-local potentials? The answer is "No". The reason can be seen from Eqs. (17.54), which can be written in the form:

$$F_\ell^{\text{(in/out)}}(E,r) = \frac{1}{2} \mp \frac{1}{2ik}\int_0^r dr'' \int_0^\infty dr'\, h_\ell^{(\pm)}(kr'')V_\ell(r'',r')\phi_\ell(E,r') \,. \tag{17.56}$$

In contrast to the case of local potentials, here in the integrand the Riccati–Hankel functions, $h_\ell^{(\pm)}(kr'')$, and the regular solution, $\phi_\ell(E,r')$, depend on different radial variables, r'' and r'. This means that the divergence of the functions

$$h_\ell^{(\pm)}(kr) \underset{r\to 0}{\longrightarrow} \sim \frac{1}{r^\ell}$$

cannot be compensated by vanishing off the regular solution,

$$\phi_\ell(E,r) \underset{r\to 0}{\longrightarrow} \sim r^{\ell+1} \,.$$

The only way to avoid the divergence of the integrals is to require that the potential is vanishing near the origin as

$$V_\ell(r,r') \underset{r\to 0}{\longrightarrow} \sim r^{\ell-1+\epsilon} \,, \qquad \epsilon > 0 \,. \tag{17.57}$$

This condition looks strange, but the whole concept of non-local potential is counterintuitive. For example, contrary to our usual understanding, it is not possible to say (based on the sign of the potential) which non-local potential is attractive and which is repulsive because the product $V_\ell u_\ell$ actually sums up the contributions from different locations in the space.

In fact, the condition (17.57) is fulfilled for all non-local potentials actually used to model the interactions among composite particles. To demonstrate this, let us consider the typical example of constructing a phenomenological non-local potential.

First of all, we notice that the non-locality has its range. This means that the potential $\langle \vec{r}\,|U|\vec{r}\,'\rangle$ cannot relate the points $\vec{r}$ and $\vec{r}\,'$ that are too far away from each other. In other words, the potential tends to zero when the distance $|\vec{r} - \vec{r}\,'|$ grows. Therefore, when constructing a phenomenological potential, it is usually introduced a Gaussian factor with the non-locality range parameter, β,

$$\langle \vec{r}\,|U|\vec{r}\,'\rangle = \frac{1}{\beta^3\pi^{3/2}} \exp\left[-\frac{(\vec{r}-\vec{r}\,')^2}{\beta^2}\right] \mathcal{U}(\vec{r},\vec{r}\,') \,, \tag{17.58}$$

where $\mathcal{U}(\vec{r},\vec{r}\,')$ is the non-local potential itself. The potential $\mathcal{U}(\vec{r},\vec{r}\,')$ as well as the non-locality range β is chosen based on the physical properties of the system under consideration. When the non-locality range tends to zero, $\beta \to 0$, the potential (17.58) becomes local,

$$\lim_{\beta\to 0^+}\langle\vec{r}\,|U|\vec{r}\,'\rangle = \delta(\vec{r}-\vec{r}\,')\mathcal{U}(\vec{r},\vec{r}\,')\ .$$

This is because one of the definitions of the δ-function is via the limit

$$\delta(x) = \lim_{\beta\to 0^+}\frac{1}{\beta\sqrt{\pi}}\exp\left(-\frac{x^2}{\beta^2}\right)\ .$$

An analog of a central local potential (which conserves angular momentum) is the non-local potential $\mathcal{U}$ that does not depend on the directions of $\vec{r}$ and $\vec{r}\,'$,

$$\mathcal{U}(\vec{r},\vec{r}\,') = \mathcal{U}(r,r\,')\ .$$

For such a potential, the partial-wave expansion (17.44) gives

$$\langle r\ell m|U|r'\ell'm'\rangle = \int d\Omega_{\vec{r}}\,d\Omega_{\vec{r}\,'}\,Y^*_{\ell m}(\hat{\vec{r}}\,)\langle\vec{r}\,|U|\vec{r}\,'\rangle Y_{\ell'm'}(\hat{\vec{r}}\,')$$

$$= \frac{\mathcal{U}(r,r\,')}{\beta^3\pi^{3/2}}\exp\left(-\frac{r^2+r'^2}{\beta^2}\right)\int d\Omega_{\vec{r}}\,d\Omega_{\vec{r}\,'}\,Y^*_{\ell m}(\hat{\vec{r}}\,)\exp\left(\frac{2\vec{r}\cdot\vec{r}\,'}{\beta^2}\right)Y_{\ell'm'}(\hat{\vec{r}}\,')\ .$$

The exponential function under this integral can be expanded over the spherical harmonics (using the same formula as for the plane wave; see Eq (A.4) of Appendix A):

$$\exp\left(\frac{2\vec{r}\cdot\vec{r}\,'}{\beta^2}\right) = \exp\left(i\frac{2\vec{r}\cdot\vec{r}\,'}{i\beta^2}\right) = 2\pi i\beta^2\sum_{\ell m}\frac{i^\ell}{rr'}j_\ell\left(\frac{2rr'}{i\beta^2}\right)Y_{\ell m}(\hat{\vec{r}}\,)Y^*_{\ell'm'}(\hat{\vec{r}}\,')\ .$$

Substituting this expansion into the integral, we obtain

$$\langle r\ell m|U|r'\ell'm'\rangle = \delta_{\ell\ell'}\delta_{mm'}\frac{2i^{\ell+1}}{\beta\sqrt{\pi}}\exp\left(-\frac{r^2+r'^2}{\beta^2}\right)\frac{j_\ell\left(\frac{2rr'}{i\beta^2}\right)}{rr'}\mathcal{U}(r,r\,')$$

$$= \frac{U_\ell(r,r')}{rr'}\delta_{\ell\ell'}\delta_{mm'}\ , \qquad (17.59)$$

where

$$U_\ell(r,r') = \frac{2i^{\ell+1}}{\beta\sqrt{\pi}} \exp\left(-\frac{r^2+r'^2}{\beta^2}\right) j_\ell\left(\frac{2rr'}{i\beta^2}\right) \mathcal{U}(r,r')\,.$$

Since

$$j_\ell\left(\frac{2rr'}{i\beta^2}\right) \underset{r\to 0}{\longrightarrow} \sim r^{\ell+1}\,,$$

this factor in the potential compensates the divergence of the Riccati–Hankel functions in the integral (17.56), and therefore, the restriction on the short-range behaviour of $\mathcal{U}(r,r')$ is exactly the same as for local potentials, namely,

$$\mathcal{U}(r,r') \underset{r\to 0}{\longrightarrow} \sim \frac{1}{r^{2-\epsilon}}\,, \qquad \epsilon > 0\,. \tag{17.60}$$

17.2.3.2 Symmetry and Analytic Structure of the Jost Functions

The symmetry properties of the Jost functions on the Riemann surface, established in Sect. 3.1 and schematically depicted in Fig. 3.6, remain the same when the potential is non-local. This can be easily proved using the same reasoning as in Sect. 3.1, when the transformation $k \to -k$ and the complex conjugation are applied to the integral equations (17.54).

In order to check if the factorized representations (7.65) for the Jost functions remain valid for non-local potentials, we transform Eqs. (17.54) into the alternative form for the following linear combinations of the unknown functions:

$$A_\ell(E,r) = F_\ell^{(\mathrm{in})}(E,r) + F_\ell^{(\mathrm{out})}(E,r)\,, \tag{17.61}$$

$$B_\ell(E,r) = i\left[F_\ell^{(\mathrm{in})}(E,r) - F_\ell^{(\mathrm{out})}(E,r)\right]\,. \tag{17.62}$$

In the corresponding combinations of Eqs. (17.54), the functions $j_\ell(kr)$ and $n_\ell(kr)$ appear in accordance with Eqs. (2.25), (2.26):

$$\begin{aligned} A_\ell(E,r) &= 1 - \frac{1}{k}\int_0^r dr'' \int_0^\infty dr'\, n_\ell(kr'')V_\ell(r'',r')\times \\ &\quad \times\left[j_\ell(kr')A_\ell(E,r') - n_\ell(kr')B_\ell(E,r')\right]\,, \\ B_\ell(E,r) &= \quad -\frac{1}{k}\int_0^r dr'' \int_0^\infty dr'\, j_\ell(kr'')V_\ell(r'',r')\times \\ &\quad \times\left[j_\ell(kr')A_\ell(E,r') - n_\ell(kr')B_\ell(E,r')\right]\,. \end{aligned} \tag{17.63}$$

As it is done for a local potential in Sect. 7.7.1, we factorize the Riccati–Bessel and Riccati–Neumann functions as is given by Eqs. (7.58, 7.59), as well as introduce the same tilded unknown functions

$$A_\ell(E,r) = \tilde{A}_\ell(E,r) , \qquad B_\ell(E,r) = k^{2\ell+1}\tilde{B}_\ell(E,r) ,$$

which leads to the following equations for them:

$$\begin{aligned}
\tilde{A}_\ell(E,r) = 1 - &\int_0^r dr'' \int_0^\infty dr'\, \tilde{n}_\ell(E,r'')V_\ell(r'',r')\times \\
&\times\left[\tilde{j}_\ell(E,r')\tilde{A}_\ell(E,r') - \tilde{n}_\ell(E,r')\tilde{B}_\ell(E,r')\right] , \\
\tilde{B}_\ell(E,r) = \quad - &\int_0^r dr'' \int_0^\infty dr'\, \tilde{j}_\ell(E,r'')V_\ell(r'',r')\times \\
&\times\left[\tilde{j}_\ell(E,r')\tilde{A}_\ell(E,r') - \tilde{n}_\ell(E,r')\tilde{B}_\ell(E,r')\right] .
\end{aligned} \tag{17.64}$$

As is seen, in these equations, there is no dependence on any odd power of k, which means that the tilded functions $\tilde{A}_\ell(E,r)$ and $\tilde{B}_\ell(E,r)$ are single-valued with respect to variable E. Therefore, the same as for local potentials, the Jost functions can be written as

$$f_\ell^{(\text{in/out})}(E) = a_\ell(E) \mp ik^{2\ell+1}b_\ell(E) , \tag{17.65}$$

where the single-valued functions of the energy, $a_\ell(E)$ and $b_\ell(E)$, are equal to the limits:

$$a_\ell(E) = \lim_{r\to\infty}\frac{1}{2}\tilde{A}_\ell(E,r) , \qquad b_\ell(E) = \lim_{r\to\infty}\frac{1}{2}\tilde{B}_\ell(E,r) . \tag{17.66}$$

The existence of these limits and analyticity of the functions $a_\ell(E)$ and $b_\ell(E)$ depend on the behaviour of the potential at large distances. Intuitively, we can guess that the faster the potential vanishes the more likely that the limits (17.66) exist and the wider is the analyticity domain $E \in \mathcal{D}$.

It is easy to see that the set of equations (17.64) is a matrix Fredholm integral equation of the second kind and can be written as

$$\begin{bmatrix}\tilde{A}_\ell(E,r)\\ \tilde{B}_\ell(E,r)\end{bmatrix} = \begin{bmatrix}1\\0\end{bmatrix} + \int_0^\infty \begin{bmatrix}K_{11}(r,r') & K_{12}(r,r')\\ K_{21}(r,r') & K_{22}(r,r')\end{bmatrix}\begin{bmatrix}\tilde{A}_\ell(E,r)\\ \tilde{B}_\ell(E,r)\end{bmatrix} dr' , \tag{17.67}$$

where the kernel has the following elements:

$$K_{11}(r,r') = -\int_0^r \tilde{n}_\ell(E,r'')V_\ell(r'',r')\tilde{j}_\ell(E,r')\,dr'' ,$$

$$K_{12}(r,r') = \int_0^r \tilde{n}_\ell(E,r'')V_\ell(r'',r')\tilde{n}_\ell(E,r')\,dr'' ,$$

$$K_{21}(r,r') = -\int_0^r \tilde{j}_\ell(E,r'')V_\ell(r'',r')\tilde{j}_\ell(E,r')\,dr'' ,$$

$$K_{22}(r,r') = \int_0^r \tilde{j}_\ell(E,r'')V_\ell(r'',r')\tilde{n}_\ell(E,r')\,dr'' .$$

As is shown in Appendix F, the iterative series for Eq. (17.67) converges if

$$\max_{i,r\in[0,\infty]} \int_0^\infty \left\{|K_{i1}(r,r')| + |K_{i2}(r,r')|\right\} dr' < 1 . \tag{17.68}$$

When the iterations converge, the functions $\tilde{A}_\ell(E,r)$ and $\tilde{B}_\ell(E,r)$ analytically depend on E. The condition (17.68) has to be checked for each specific potential.

17.2.4 Separable Potential

It is known (see Appendix F.1) that an integral equation with a separable kernel can be solved in the closed form. This is why separable potentials are sometimes used in the traditional scattering theory based on the integral Lippmann–Schwinger equation. It is natural to expect that the Jost functions can also be found in the closed form if the potential is separable.

* * *

Consider a one-term separable potential in the partial wave ℓ,

$$V_\ell(r,r') = \lambda_\ell\, v_\ell(r)v_\ell(r') , \tag{17.69}$$

where λ_ℓ is the strength parameter of the potential. Substituting it into the integral equations (17.54), we have[7]

$$F_\ell^{(\text{in/out})}(E,r) = \frac{1}{2} \mp \frac{\lambda_\ell}{2ik}\int_0^r h_\ell^{(\pm)}(kr')v_\ell(r')C\,dr' , \tag{17.70}$$

[7] We assume here that the particles are not charged, and therefore, $W_\ell^{(\text{in/out})} = h_\ell^{(\mp)}$.

where the unknown constant is defined by

$$C=\int_0^\infty v_\ell(r)\left[h_\ell^{(-)}(kr)F_\ell^{(\mathrm{in})}(E,r)+h_\ell^{(+)}(kr)F_\ell^{(\mathrm{out})}(E,r)\right]dr\,.$$

Substituting $F_\ell^{(\mathrm{in/out})}$ given by Eq. (17.70), into this integral, we obtain a linear equation for finding C:

$$C=\int_0^\infty dr\,v_\ell(r)\left\{h_\ell^{(-)}(kr)\left[\frac{1}{2}-\frac{\lambda_\ell C}{2ik}\int_0^r dr'\,v(r')h_\ell^{(+)}(kr')\right]+\right. \tag{17.71}$$

$$\left.+\;h_\ell^{(+)}(kr)\left[\frac{1}{2}+\frac{\lambda_\ell C}{2ik}\int_0^r dr'\,v(r')h_\ell^{(-)}(kr')\right]\right\}.$$

Solving this equation, we find the following expression for the unknown constant C:

$$C=\frac{\displaystyle\int_0^\infty v_\ell(r)j_\ell(kr)\,dr}{\displaystyle 1+\frac{\lambda_\ell}{2ik}\int_0^\infty dr\,v_\ell(r)\int_0^r dr'\,v_\ell(r')\left[h_\ell^{(-)}(kr)h_\ell^{(+)}(kr')-h_\ell^{(+)}(kr)h_\ell^{(-)}(kr')\right]}\,. \tag{17.72}$$

The denominator on the right hand side of this equation involves the free Green's function (2.47) of the regular solution. Therefore,

$$C=\frac{\displaystyle\int_0^\infty v_\ell(r)j_\ell(kr)\,dr}{\displaystyle 1-\int_0^\infty\!\!\int_0^\infty dr'\,dr''\,g_\ell(r',r'')V_\ell(r'',r')}=\frac{\displaystyle\int_0^\infty v_\ell(r)j_\ell(kr)\,dr}{1\quad \mathrm{Sp}\,(g_\ell V_\ell)}\,. \tag{17.73}$$

After substitution of this expression into Eq. (17.70), we obtain the functions $F_\ell^{(\mathrm{in/out})}(E,r)$ for any finite r in the closed form:

$$F_\ell^{(\mathrm{in/out})}(E,r)\;=\;\frac{1}{2}\mp\frac{\displaystyle\frac{\lambda_\ell}{2ik}\int_0^r h_\ell^{(\pm)}(kr')v_\ell(r')\,dr'\int_0^\infty v_\ell(r'')j_\ell(kr'')\,dr''}{1-\mathrm{Sp}\,(g_\ell V_\ell)}\,. \tag{17.74}$$

The corresponding Jost functions are their limits when $r\to\infty$,

$$f_\ell^{(\mathrm{in/out})}(E)\;=\;\frac{1}{2}\mp\frac{1}{2}\cdot\frac{\displaystyle\frac{\lambda_\ell}{ik}\int_0^\infty\!\!\int_0^\infty dr\,dr'\,j_\ell(kr)h_\ell^{(\pm)}(kr')v_\ell(r')v_\ell(r)}{1-\mathrm{Sp}\,(g_\ell V_\ell)}\,. \tag{17.75}$$

If the energy E is complex, all the integrations involved in Eqs. (17.74), (17.75) should be performed along an appropriate complex path.

The products $j_\ell(kr)h_\ell^{(\pm)}(kr')$ in the integrand on the right hand side of Eq. (17.75) look like the free Green's functions $g_\ell^{(\pm)}(r,r')$:

$$g_\ell^{(\pm)}(r,r') = \pm\frac{1}{ik}j_\ell(kr_<)h_\ell^{(\pm)}(kr_>)\,. \tag{17.76}$$

However, in contrast to (17.76), in the integral (17.75), the function $j_\ell(kr)$ always depends on r, while $h_\ell^{(\pm)}(kr')$—on r', no matter which of the two variables is greater. We can transform the integral over r' in (17.75) in such a way that it involves difference of two Green's functions. To this end, we split this integral into two: over $[0,r]$ and over $[r,\infty)$,

$$\frac{1}{ik}\int_0^\infty dr'\, j_\ell(kr)h_\ell^{(\pm)}(kr')V_\ell(r',r)$$

$$= \frac{1}{ik}\int_0^r dr'\, j_\ell(kr)h_\ell^{(\pm)}(kr')V_\ell(r',r) + \frac{1}{ik}\int_r^\infty dr'\, j_\ell(kr)h_\ell^{(\pm)}(kr')V_\ell(r',r)\,.$$

Then we add and subtract the integral

$$\frac{1}{ik}\int_0^r dr'\, h_\ell^{(\pm)}(kr)j_\ell(kr')V_\ell(r',r)\,.$$

Combining the four terms in pairs, we obtain the numerator on the right hand side of Eq. (17.75) in the form:

$$\pm\int_0^\infty\int_0^\infty dr\,dr'\, g_\ell^{(\pm)}(r,r')V_\ell(r',r) \mp \int_0^\infty\int_0^\infty dr\,dr'\, g_\ell(r,r')V_\ell(r',r)\,,$$

and thus

$$f_\ell^{(\mathrm{in/out})}(E) = \frac{1}{2} + \frac{1}{2}\cdot\frac{\mathrm{Sp}\,(g_\ell V_\ell) - \mathrm{Sp}\left[g_\ell^{(\pm)}V_\ell\right]}{1-\mathrm{Sp}\,(g_\ell V_\ell)}$$

$$= \frac{1}{2}\cdot\frac{1-\mathrm{Sp}\left[g_\ell^{(\pm)}V_\ell\right]}{1-\mathrm{Sp}\,(g_\ell V_\ell)}\,. \tag{17.77}$$

The corresponding S-matrix is

$$S_\ell(E) = \frac{1 - \mathrm{Sp}\left[g_\ell^{(-)} V_\ell\right]}{1 - \mathrm{Sp}\left[g_\ell^{(+)} V_\ell\right]} \tag{17.78}$$

17.2.5 Generalized Jost–Pais Theorem

As an alternative approach to finding the Jost functions, one can use the Fredholm method (see Sect. 6.7), i.e. can calculate (or just mathematically analyse) the corresponding Fredholm determinants (defined in Appendix F.3.2). When potential is non-local, the Fredholm method, however, needs some modification. Instead of being equal to a single Fredholm determinant (as is given by Eq. (6.89)), the Jost function is the ratio of two determinants. This is what is shown in this section and what is known as the generalized Jost–Pais theorem.

* * *

Applying the Fredholm method to a non-local potential, we follow the same derivations as in Sect. 6.7 because they are given there in the general operator form that is equally applicable for both local and non-local interactions. However, starting from Eq. (6.81), we have to remember that the potential is an integral operator. Therefore, Eqs. (6.81) and (6.82) should be written as

$$\phi_\ell'(E,r) = \iint_0^\infty dr' dr''\, G_\ell(r,r') V(r',r'') \phi_\ell(E,r'') \,, \tag{17.79}$$

$$f_\ell^{(\mathrm{in})} = \frac{1}{2} - \frac{\lambda}{2ik} \iint_0^\infty dr\, dr'\, h_\ell^{(+)}(kr) V(r,r') \phi_\ell(E,r') \,. \tag{17.80}$$

Here, λ is (as before) the strength of the potential, which is finally should be put to unity. Following the same steps as in Sec. 6.7, we obtain (instead of Eq. (6.84))

$$\frac{d \ln f_\ell^{(\mathrm{in})}}{d\lambda} = -\frac{1}{2ik} \iint_0^\infty dr\, dr'\, \chi_\ell^{(+)}(E,r') V(r',r) u_\ell(E,r) \,. \tag{17.81}$$

From this point on, our derivations have to be different from the case of local potentials. The problem is that here (in contrast to Eq. (6.84)) in the product $u_\ell(E,r)\chi_\ell^{(+)}(E,r')$, the functions $u_\ell(E,r)$ and $\chi_\ell^{(+)}(E,r')$ depend on different radial variables. We could try and relate this product to the Green's function $G_\ell^{(+)}(r,r')$ (see Eq. (6.14)), but the product $u_\ell(E,r)\chi_\ell^{(+)}(E,r')$ under the integral

(17.81) is the same for the whole integration quadrant, $(r, r') \in [0, \infty) \otimes [0, \infty)$, independently of which of the two variables is greater.

As a first step to circumvent this difficulty, we split the integration interval $r' \in [0, \infty)$ into two subintervals, $r' \in [0, r]$ and $r' \in [r, \infty)$:

$$\frac{d \ln f_\ell^{(\mathrm{in})}}{d\lambda} = -\frac{1}{2ik}\int_0^\infty dr \left[\int_0^r dr'\, u_\ell(E,r)\chi_\ell^{(+)}(E,r')V(r',r) + \int_r^\infty dr'\, u_\ell(E,r)\chi_\ell^{(+)}(E,r')V(r',r)\right].$$

Then, in the square brackets, we add and subtract the integral

$$\int_0^r dr'\, u_\ell(E,r')\chi_\ell^{(+)}(E,r)V(r',r)\,.$$

This makes the expression in the square brackets equal to

$$\int_0^r dr'\, u_\ell(E,r')\chi_\ell^{(+)}(E,r)V(r',r) + \int_r^\infty dr'\, u_\ell(E,r)\chi_\ell^{(+)}(E,r')V(r',r)$$
$$-\int_0^r dr' \left[\chi_\ell^{(+)}(E,r)u_\ell(E,r') - u_\ell(E,r)\chi_\ell^{(+)}(E,r')\right]V(r',r)\,,$$

where the first two terms correctly swap the variables $r_<$ and $r_>$, in accordance with Eq. (6.14), while the second pair of the terms gives the total Green's function of the regular solution (see Eq. (6.37) or Fig. 6.4). Therefore,

$$\frac{d \ln f_\ell^{(\mathrm{in})}}{d\lambda} = -\int_0^\infty dr \int_0^\infty dr' \left[G_\ell^{(+)}(r,r') - G_\ell(r,r')\right]V(r',r) \tag{17.82}$$
$$= -\left[\mathrm{Sp}\left(G_\ell^{(+)}V\right) - \mathrm{Sp}\left(G_\ell V\right)\right].$$

Comparing this with Eq. (6.85), we see that the non-locality of the potential results in the appearance of an additional term,

$$\mathrm{Sp}\left(G_\ell V\right) = \int_0^\infty dr \int_0^\infty dr'\, G_\ell(r,r')V(r',r)\,. \tag{17.83}$$

It is clear that this term vanishes, if the potential is local, i.e. $V \sim \delta(r - r')$, because $G_\ell(r,r) \equiv 0$.

Both $G_\ell^{(+)}$ and G_ℓ depend on the strength parameter λ. These dependences can be obtained explicitly from Eqs. (6.18) and (6.40):

$$G_\ell^{(+)} = (1 - \lambda g_\ell^{(+)} V)^{-1} g_\ell^{(+)} , \tag{17.84}$$

$$G_\ell = (1 - \lambda g_\ell V)^{-1} g_\ell . \tag{17.85}$$

Integrating both sides of Eq. (17.82) over λ from 0 to 1, we have

$$\ln f_\ell^{(\mathrm{in})}\Big|_{\lambda=1} - \ln f_\ell^{(\mathrm{in})}\Big|_{\lambda=0} = -\iint_0^\infty dr\, dr' \int_0^1 d\lambda \left[G_\ell^{(+)}(r,r') - G_\ell(r,r')\right] V(r',r) .$$

According to our choice of the Jost function normalization,

$$f_\ell^{(\mathrm{in})}\Big|_{\lambda=0} = \frac{1}{2} .$$

Therefore,

$$f_\ell^{(\mathrm{in})}\Big|_{\lambda=1} = \frac{1}{2} \cdot \frac{\exp\left\{-\int_0^1 d\lambda\, \mathrm{Sp}\left[(1 - \lambda g_\ell^{(+)} V)^{-1} g_\ell^{(+)} V\right]\right\}}{\exp\left\{-\int_0^1 d\lambda\, \mathrm{Sp}\left[(1 - \lambda g_\ell V)^{-1} g_\ell V\right]\right\}} .$$

Finally, taking into account Eq. (F.16) of Appendix F.3.2, we find that the Jost function for a non-local potential is the ratio of two Fredholm determinants,

$$\boxed{f_\ell^{(\mathrm{in})} = \frac{1}{2} \cdot \frac{\Delta}{\Lambda}} \tag{17.86}$$

where Δ and Λ are the Fredholm determinants of the integral equations for the physical scattering wave function u_ℓ (see Eq. (6.5)) and for the regular solution ϕ_ℓ (see Eq. (2.52), respectively. As we have seen, for a local potential, the trace of the operator product $G_\ell V$ is zero. This implies that $\Lambda = 1$,

$$V \sim \delta(r - r') \Rightarrow \mathrm{Sp}\,(G_\ell V) = 0 \Rightarrow \Lambda = 1 ,$$

which brings us back to Eq. (6.89).

17.3 Off-Shell Jost Functions

In quantum mechanics in general and in scattering theory in particular, it is very useful and productive to allow various variables to assume non-physical values. For example, considering the S-matrix at complex energies and complex angular momenta reveals information about possible bound and resonant states of a quantum system. Yet another deviation from physical energies that is also very useful consists in abandoning the energy-conservation law. Of course, when calculating the scattering observable quantities, this law is always adhered to. However, in various integrals that involve functions depending on the initial momentum, k, final momentum, k', and the total energy E, it is sometimes needed to consider them as independent variables, not bound by the conservation law, i.e.

$$\frac{(\hbar k')^2}{2\mu} \neq E \neq \frac{(\hbar k)^2}{2\mu} . \tag{17.87}$$

In such a case, it is said that the functions of these variables are considered off the energy shell (or simply off-shell). The name comes from the fact that the tip of the physical vector $\vec{k}$ in the $3D$ momentum space should be on the spherical shell $\vec{k}^2 = 2\mu E/\hbar^2$. If in Eq. (17.87) only one of the two momenta is off the shell, then the values of the corresponding function are called half-off-shell. If both momenta are off the energy shell, then the function is fully off-shell.

In this section, it is introduced the off-shell Jost function depending on two energy variables (instead of one) that are independent of each other. Using such a function, it is possible to obtain the half-off-shell T-matrix (i.e., loosely speaking, the scattering amplitude). In its turn, the off-shell amplitude is used when considering collisions in many-body systems, where the energy in a two-body subsystem does not conserve because the other particles can either give or take away some energy. The off-shell Jost functions are the generalizations and become the ordinary Jost functions when their two energy variables tend to each other.

* * *

In scattering theory,[8] in addition to the S-matrix, it is introduced the T-matrix that differs from S by the identity term, I,

$$S(E) = I - 2\pi i T(E) , \tag{17.88}$$

and obeys the Lippmann–Schwinger (integral) equation

$$T(E) = U + U g^{(+)}(E) T(E) , \tag{17.89}$$

[8] See, for example, book [4].

where the energy variable, E, in the free Green's operator

$$g^{(+)}(E) = (E + i0 - H_0)^{-1} \tag{17.90}$$

is shifted with an infinitely small positive imaginary term $+i0$ (which in all final formulae becomes zero). This shift is needed, when E is real and positive, to bypass the pole at the corresponding eigenvalue of the free Hamiltonian H_0. A detailed discussion of this shift as well as the explicit expression for the coordinate representation of the operator $g^{(+)}$ is given in Sect. 6.5.

Consider the following operator

$$\psi(E) = I + g^{(+)}(E)T(E) . \tag{17.91}$$

If it is multiplied from the left by the potential U, then, taking into account Eq. (17.89), we obtain

$$U\psi = U + Ug^{(+)}T = T . \tag{17.92}$$

Therefore,

$$\psi = I + g^{(+)}U\psi , \tag{17.93}$$

from where the physical meaning of the operator ψ can be understood. Indeed, in the mixed representation, i.e. when "sandwich" between the bra-vector $\langle\vec{r}\,|$ and the ket-vector $|\vec{k}\,\rangle$, this equation becomes the Lippmann–Schwinger equation (6.66) for the scattering wave function

$$\psi_{\vec{k}}(\vec{r}\,) = \langle\vec{r}\,|\psi|\vec{k}\,\rangle . \tag{17.94}$$

The operator ψ can also be viewed as the Møller operator Ω_+ defined by Eq. (1.40). Indeed, it acts on the incoming plane wave and transforms it into the scattering state whose projection onto the bra-vector $\langle\vec{r}\,|$ gives the corresponding wave function. Here we tacitly assumed that the momentum $\vec{k}$ is on the energy shell, i.e. $(\hbar k)^2/(2\mu) = E$. In Eq. (17.94), the dependence on E is not explicitly shown because the magnitude of k determines the value of E as well.

In principle, we can use the same Møller operator and act with it on the plane wave with an arbitrary momentum $\vec{q}$, not related to the energy variable. Then the resulting wave function,

$$\varphi_{\vec{q}}(E, \vec{r}\,) = \langle\vec{r}\,|\psi|\vec{q}\,\rangle , \tag{17.95}$$

depends on both $\vec{q}$ and E, which are independent of each other. This is what is called off-shell wave function. It is intuitively clear that the off-shell Jost functions should be defined using the radial components of $\varphi_{\vec{q}}(E,\vec{r})$ in a way that is similar to how the ordinary (on-shell) Jost functions are defined via the radial components of $\psi_{\vec{k}}(\vec{r})$.

First of all, we need to find out which differential equation is satisfied by the function $\varphi_{\vec{q}}(E,\vec{r})$. To this end, we multiply Eq. (17.91) from the left by $(E-H_0-U)$. This gives

$$\begin{aligned}(E-H_0-U)\psi &= (E-H_0-U)(I+g^{(+)}T) \\ &= E-H_0+(E-H_0)(E-H_0)^{-1}T-U\psi \\ &= E-H_0\,.\end{aligned} \tag{17.96}$$

The mixed representation of both sides of Eq. (17.96), i.e. its "sandwich" between the bra-vector $\langle\vec{r}\,|$ and the ket-vector $|\vec{q}\,\rangle$, has the following form:

$$\langle\vec{r}\,|(E-H_0-U)\psi|\vec{q}\,\rangle = \left[E-\frac{(q\hbar)^2}{2\mu}\right]\langle\vec{r}\,|\vec{q}\,\rangle\,,$$

which, after multiplying by $2\mu/\hbar^2$, reads

$$\left(\Delta+k^2-V\right)\langle\vec{r}\,|\psi|\vec{q}\,\rangle = \left(k^2-q^2\right)\langle\vec{r}\,|\vec{q}\,\rangle\,. \tag{17.97}$$

It is seen that, when $q=k$, this becomes the ordinary three-dimensional Schrödinger equation. In general, when q is off the energy shell, the right hand side of Eq. (17.97) is non-zero. We can call (17.97) the off-shell Schrödinger equation. The same way as for the on-shell one, we do the partial-wave decomposition of this equation. To this end, the following representations of the functions involved in it are used:[9]

$$\langle\vec{r}\,|\psi|\vec{q}\,\rangle = \frac{1}{\sqrt{2\pi}qr}\sum_{\ell m} i^\ell\omega_\ell(E,Q,r)Y^*_{\ell m}(\hat{\vec{q}}\,)Y_{\ell m}(\hat{\vec{r}}\,)\,, \tag{17.98}$$

$$\langle\vec{r}\,|\vec{q}\,\rangle = \sqrt{\frac{2}{\pi}}\frac{1}{qr}\sum_{\ell m} i^\ell j_\ell(qr)Y^*_{\ell m}(\hat{\vec{q}}\,)Y_{\ell m}(\hat{\vec{r}}\,)\,, \tag{17.99}$$

where ω_ℓ is the radial part of the off-shell scattering wave function, depending on the two energy variables,

$$E=\frac{(\hbar k)^2}{2\mu}\qquad\text{and}\qquad Q=\frac{(\hbar q)^2}{2\mu}\,.$$

[9] The structure of the function $\langle\vec{r}\,|\psi|\vec{q}\,\rangle$ is chosen in exactly the same way as for the scattering wave function (4.5).

Substituting the expansions (17.98) and (17.99) in Eq. (17.97), then multiplying whole equation by $Y^*_{\ell m}(\hat{\vec{r}})$, and integrating over the spherical angles $\Omega_{\vec{r}}$, we obtain the corresponding radial equation:

$$\left[\partial_r^2 + k^2 - \frac{\ell(\ell+1)}{r^2} - V_\ell(r)\right]\omega_\ell(E, Q, r) = 2\left(k^2 - q^2\right) j_\ell(qr) , \tag{17.100}$$

where we assumed that the potential is spherically symmetric,

$$\frac{2\mu}{\hbar^2}\int d\Omega_{\vec{r}}\, Y^*_{\ell m}(\hat{\vec{r}})U(r)Y_{\ell' m'}(\hat{\vec{r}}) = \delta_{\ell\ell'}\delta_{mm'}\, V_\ell(r) .$$

Consider the following two differential equations:

$$\left[\partial_r^2 + k^2 - \frac{\ell(\ell+1)}{r^2} - V_\ell(r)\right]\chi_\ell^{(\pm)}(E, Q, r) = \left(k^2 - q^2\right) h_\ell^{(\pm)}(qr) , \tag{17.101}$$

which differ from Eq. (17.100) by their right hand sides. If we impose the boundary conditions

$$\chi_\ell^{(\pm)}(E, Q, r) \underset{r\to\infty}{\longrightarrow} h_\ell^{(\pm)}(qr) ,$$

then $\chi_\ell^{(\pm)}(E, Q, r)$ can be called the off-shell Jost solutions. Apparently, for $q = k$, they become the on-shell Jost solutions,

$$(q = k) \;\Rightarrow \qquad \chi_\ell^{(\pm)}(E, E, r) = \chi_\ell^{(\pm)}(E, r) . \tag{17.102}$$

Since $h_\ell^{(+)}(qr) + h_\ell^{(-)}(qr) = 2j_\ell(qr)$, the sum of any two solutions of equations (17.101) with plus and minus signs obeys Eq. (17.100),

$$\left[\partial_r^2 + k^2 - \frac{\ell(\ell+1)}{r^2} - V_\ell(r)\right]\left[\chi_\ell^{(+)}(E, Q, r) + \chi_\ell^{(-)}(E, Q, r)\right] = 2\left(k^2 - q^2\right) j_\ell(qr) .$$

Therefore, the sum of the off-shell Jost solutions,

$$\omega_\ell^{\text{part}}(E, Q, r) = \chi_\ell^{(+)}(E, Q, r) + \chi_\ell^{(-)}(E, Q, r) ,$$

is a particular solution of Eq. (17.100). As is known in the theory of differential equations, the general solution of a linear equation is equal to the general solution of the corresponding homogeneous equation plus any particular solution of the total (inhomogeneous) equation. Therefore, the general solution of Eq. (17.100) can be written as

$$\omega_\ell^{\text{gen}}(E, Q, r) = u_\ell^{\text{gen}}(E, r) + \chi_\ell^{(+)}(E, Q, r) + \chi_\ell^{(-)}(E, Q, r) , \tag{17.103}$$

where $u_\ell^{\text{gen}}(E, r)$ is the general solution of the homogeneous equation:

$$\left[\partial_r^2 + k^2 - \frac{\ell(\ell+1)}{r^2} - V_\ell(r)\right] u_\ell^{\text{gen}}(E, r) = 0 . \tag{17.104}$$

What we actually need is not the general solution of (17.100) but its particular solution that somehow is related to the scattering process (see footnote 9). It should obey certain boundary conditions that we can impose on the "flexible" part of the solution (17.103), i.e. on the function $u_\ell^{\text{gen}}(E, r)$. What are these conditions? To answer this question, we need to consider asymptotic behaviour of the off-shell wave function (17.95).

Using Eq. (17.92), we replace $U\psi$ in Eq. (17.93) with T and then write (17.93) in the mixed representation:

$$\langle \vec{r}\,|\psi|\vec{q}\,\rangle = \langle \vec{r}\,|\vec{q}\,\rangle + \int d^3r'\, \langle \vec{r}\,|g^{(+)}|\vec{r}\,'\rangle\langle \vec{r}\,'|T|\vec{q}\,\rangle .$$

In this equation, we substitute the partial-wave expansions (17.98) and (17.99) as well as (6.73, 6.74) and

$$\langle \vec{r}\,|T|\vec{q}\,\rangle = \sum_{\ell m} \langle r\ell m|T|q\ell m\rangle Y_{\ell m}(\hat{\vec{r}}\,) Y^*_{\ell m}(\hat{\vec{q}}\,) .$$

After performing the angular integration, this gives

$$\omega_\ell(E, Q, r) = 2j_\ell(qr) - \frac{2\mu}{\hbar^2} i^{1-\ell} q\sqrt{2\pi} \int_0^\infty r'^2 dr'\, \frac{j_\ell(kr_<)h_\ell^{(+)}(kr_>)}{kr'} \langle r'\ell m|T|q\ell m\rangle . \tag{17.105}$$

If we let $r \to \infty$, then $r > r'$ and

$$(r \to \infty) \;\Rightarrow\; \qquad j_\ell(kr_<)h_\ell^{(+)}(kr_>) = j_\ell(kr')h_\ell^{(+)}(kr) .$$

Using Eq. (A.25) of Appendix A, we obtain

$$\int_0^\infty r'^2 dr'\, \frac{j_\ell(kr')}{kr'} \langle r'\ell m|T|q\ell m\rangle = \sqrt{\frac{\pi}{2}}\, i^\ell \langle k\ell m|T|q\ell m\rangle .$$

Therefore,

$$\omega_\ell(E, Q, r) \underset{r\to\infty}{\longrightarrow} 2j_\ell(qr) - \frac{2\mu}{\hbar^2} i\pi q \langle k\ell m|T|q\ell m\rangle h_\ell^{(+)}(kr) . \tag{17.106}$$

This is the condition at infinity that we are looking for. It allows us to specify the term $u_\ell^{\text{gen}}(E, r)$ in Eq. (17.103). Since

$$\left[\chi_\ell^{(+)}(E, Q, r) + \chi_\ell^{(-)}(E, Q, r)\right] \underset{r\to\infty}{\longrightarrow} 2j_\ell(qr)\,,$$

the second term in Eq. (17.106) can be obtained if the general solution u_ℓ^{gen} is replaced with the following particular solution:

$$u_\ell^{\text{gen}}(E, r): \qquad u_\ell^{\text{part}}(E, r) = -\frac{2\mu}{\hbar^2} i\pi q \langle k\ell m|T|q\ell m\rangle \chi_\ell^{(+)}(kr)\,,$$

where $\chi_\ell^{(+)}(kr)$ is the on-shell Jost solution. In this way, from the general solution (17.103), we obtain the particular solution

$$\omega_\ell(E, Q, r) = \chi_\ell^{(+)}(E, Q, r) + \chi_\ell^{(-)}(E, Q, r) - \frac{2\mu}{\hbar^2} i\pi q \langle k\ell m|T(E)|q\ell m\rangle \chi_\ell^{(+)}(kr)\,, \tag{17.107}$$

which is the off-shell scattering wave function. As we see, it is related to the half-off-shell T-matrix, $\langle k\ell m|T(E)|q\ell m\rangle$.

At this point, we need to introduce the half-off-shell Jost functions. It is not clear how they could be defined via the asymptotic behaviour of $\omega_\ell(E, Q, r)$. However, we can use the alternative definition, the same as for the on-shell Jost functions, via the behaviour of the Jost solutions near $r = 0$ (see Eq. (2.78)). Therefore, we define the half-off-shell Jost functions as the limits

$$f_\ell^{(\text{in/out})}(E, Q) = \lim_{r\to 0} \frac{\pm i(kr)^\ell}{2(2\ell - 1)!!} \chi_\ell^{(\pm)}(E, Q, r) \tag{17.108}$$

It is clear that thanks to the relations (17.102), the off-shell Jost functions become the on-shell ones, when $q = k$,

$$(q = k) \Rightarrow \qquad f_\ell^{(\text{in/out})}(E, E) = f_\ell^{(\text{in/out})}(E)\,. \tag{17.109}$$

To find the relation among the off-shell Jost functions and the off-shell T-matrix, we notice that $\omega_\ell(E, Q, r)$ is a physical solution and therefore is zero at $r = 0$. This can be formally proven using Eq. (17.105), where

$$(r \to 0) \to \qquad j_\ell(kr_<)h_\ell^{(+)}(kr_>) = j_\ell(kr)h_\ell^{(+)}(kr')\,.$$

Therefore, multiplying all terms of Eq. (17.107) by the factor $\pm i(kr)^{\ell}/[2(2\ell - 1)!!]$ and letting $r \to 0$, we obtain

$$0 = f_{\ell}^{(\mathrm{in})}(E, Q) + f_{\ell}^{(\mathrm{out})}(E, Q) - \frac{2\mu}{\hbar^2} i\pi q \langle k\ell m|T(E)|q\ell m\rangle f_{\ell}^{(\mathrm{in})}(E)\,,$$

from where we finally obtain

$$\boxed{\langle k\ell m|T(E)|q\ell m\rangle = \frac{\hbar^2}{2\mu} \cdot \frac{f_{\ell}^{(\mathrm{in})}(E, Q) + f_{\ell}^{(\mathrm{out})}(E, Q)}{i\pi q f_{\ell}^{(\mathrm{in})}(E)}} \tag{17.110}$$

18 Some Exactly Solvable Potential Models

For some simple potentials, the radial Schrödinger equation can be solved analytically. Unfortunately, in most cases, this can only be done for $\ell = 0$. This restriction, however, does not diminish the importance of such solutions since they demonstrate all the features that are discussed in this book. Moreover, the exact solutions can be used for testing the accuracy and efficiency of the numerical procedures for calculating the Jost functions.

18.1 Exponential Potential

An important example of a potential that allows an exact solution of the S-wave Schrödinger equation,

$$\left[\frac{d^2}{dr^2} + k^2 - V(r)\right] u(E,r) = 0\,, \tag{18.1}$$

is the pure exponential potential[1]

$$V(r) = -V_0 e^{-\beta r}\,, \qquad V_0 > 0\,. \tag{18.2}$$

With the help of the variable change

$$\zeta = \frac{2}{\beta}\sqrt{V_0}e^{-\beta r/2}, \tag{18.3}$$

[1] Note that this is the so-called reduced potential defined by Eq. (2.5).

S. A. Rakityansky, *Jost Functions in Quantum Mechanics*,
https://doi.org/10.1007/978-3-031-07761-6_18

the Schrödinger equation (18.1) is reduced to the Bessel equation

$$u'' + \frac{1}{\zeta}u' + \left[1 - \frac{(ip)^2}{\zeta^2}\right]u = 0\,, \tag{18.4}$$

where "prime" means derivative with respect to ζ, and the parameter p is defined as

$$p = \frac{2k}{\beta}\,.$$

The two linearly independent solutions of Eq. (18.4) are the Bessel functions $J_{ip}(\zeta)$ and $J_{-ip}(\zeta)$, defined by the series

$$J_\nu(\zeta) = \sum_{m=0}^{\infty}(-1)^m\left(\frac{\zeta}{2}\right)^{\nu+2m}\frac{1}{m!\Gamma(m+\nu+1)}\,. \tag{18.5}$$

The transformation (18.3) maps the infinite interval $r \in [0, \infty)$ onto the finite segment $\zeta \in [2\sqrt{V_0}/\beta, 0)$, i.e.

$$r = 0 \quad\longleftrightarrow \zeta = \frac{2\sqrt{V_0}}{\beta}\,,$$

$$r = \infty \longleftrightarrow \zeta = 0\,.$$

Therefore, the wave function $u(E, r)$, written as a linear combination of the Bessel functions,

$$u(E, r) = C_1 J_{ip}(\zeta) + C_2 J_{-ip}(\zeta)\,, \tag{18.6}$$

is regular at $r = 0$ with the following choice of the combination constants

$$u(E, r) = C\left[J_{-ip}(\zeta_0)J_{ip}(\zeta) - J_{ip}(\zeta_0)J_{-ip}(\zeta)\right]\,, \tag{18.7}$$

where

$$\zeta_0 = \frac{2\sqrt{V_0}}{\beta},$$

and C is determined by the normalization. As we emphasized many times, the normalization is not important. However, in order to be consistent, let us normalize the wave function (18.7) in such a way that its short-range behaviour is given by Eq. (2.36) with $\ell = 0$, i.e.

$$u(E, r) \equiv \phi_0(E, r) \underset{r\to 0}{\longrightarrow} kr\,. \tag{18.8}$$

With such a normalization, the wave function (18.7) coincides with the function (2.12), where $F_0^{(\text{in/out})}(E, r)$ obeys Eq. (2.21) and the boundary conditions (2.24).

In order to find the normalization constant C, we expand the wave function in the Taylor series near the point ζ_0 (i.e. near $r = 0$),

$$u = u\Big|_{\zeta=\zeta_0} + u'\Big|_{\zeta=\zeta_0} \cdot (\zeta - \zeta_0) + O[(\zeta - \zeta_0)^2] \,.$$

Since $u\Big|_{\zeta=\zeta_0} = 0$ and $\zeta - \zeta_0 = -\sqrt{V_0}\, r + O(r^2)$, we obtain

$$u \underset{\zeta\to\zeta_0}{\longrightarrow} -\sqrt{V_0}\, r\, u'\Big|_{\zeta=\zeta_0} + O(r^2) \,.$$

Knowing the Wronskian of the Bessel functions

$$W\left[J_{-\nu}(\zeta), J_\nu(\zeta)\right] = \frac{2\sin(\nu\pi)}{\pi\zeta} \,, \tag{18.9}$$

we can calculate the derivative

$$u'\Big|_{\zeta=\zeta_0} = C\left[J_{-ip}(\zeta_0)J'_{ip}(\zeta_0) - J_{ip}(\zeta_0)J'_{-ip}(\zeta_0)\right]$$

$$= \frac{2C\sin(i\pi p)}{\pi\zeta_0} \,.$$

Therefore, near the point $r = 0$, we have

$$-C\sqrt{V_0}\,\frac{2\sin(i\pi p)}{\pi\zeta_0}\, r = kr \,,$$

which gives the following normalization constant:

$$C = -\frac{\pi k}{\beta\sin\left(i\dfrac{2\pi k}{\beta}\right)} \,, \tag{18.10}$$

and thus

$$u(E, r) = -\frac{\pi k}{\beta\sin\left(i\dfrac{2\pi k}{\beta}\right)}\left[J_{-ip}(\zeta_0)J_{ip}(\zeta) - J_{ip}(\zeta_0)J_{-ip}(\zeta)\right] \,. \tag{18.11}$$

Because of the chosen normalization (18.8), this is, actually, the S-wave regular solution, $u(E, r) = \phi_0(E, r)$. In order to find the Jost functions $f_0^{(\text{in/out})}(E)$

that correspond to the wave function (18.11), we make use of the equation (see Sect. 2.2.3)

$$f_0^{\text{(in/out)}}(E) = \pm\frac{1}{2ik}W_r\left[u,\chi_0^{(\pm)}\right], \tag{18.12}$$

where $\chi_0^{(\pm)}(E,r)$ are the Jost solutions of Eq. (18.1) with the boundary conditions at infinity,

$$\chi_0^{(\pm)}(E,r) \underset{r\to\infty}{\longrightarrow} \mp i e^{\pm ikr}. \tag{18.13}$$

In Eq. (18.12), we put the subscript r in the notation of the Wronskian. This is done to indicate that the derivatives in this Wronskian are taken with respect to r. Since

$$\frac{d}{dr} = \frac{d\zeta}{dr}\cdot\frac{d}{d\zeta} = -\frac{\beta\zeta}{2}\frac{d}{d\zeta},$$

it is related to the corresponding Wronskian with respect to ζ as follows:

$$W_r\left[u,\chi_0^{(\pm)}\right] = -\frac{\beta\zeta}{2}W_\zeta\left[u,\chi_0^{(\pm)}\right]. \tag{18.14}$$

The functions $\chi_0^{(\pm)}(E,r)$ and $J_{\pm ip}(\zeta)$ are solutions of the same equation and thus can be related to each other. This is not difficult to do since $r\to\infty$ corresponds to $\zeta\to 0$, and therefore at large distances, the functions $J_{\pm ip}(\zeta)$ can be represented by the first term of the series (18.5),

$$J_{\pm ip}(\zeta) \underset{r\to\infty}{\longrightarrow} \frac{1}{\Gamma(1\pm ip)}\left(\frac{\zeta}{2}\right)^{\pm ip} = \left(\frac{\sqrt{V_0}}{\beta}\right)^{\pm ip}\frac{e^{\mp ikr}}{\Gamma(1\pm ip)}. \tag{18.15}$$

Comparing this with Eq. (18.13), we see that

$$\chi_0^{(\pm)}(E,r) = \mp i\Gamma(1\mp ip)\left(\frac{\sqrt{V_0}}{\beta}\right)^{\pm ip}J_{\mp ip}(\zeta). \tag{18.16}$$

Substituting the wave function (18.11) and the Jost solutions (18.16) into Eq. (18.12) and taking into account the relation (18.14), we have

$$f_0^{\text{(in/out)}}(E) = -\frac{\pi\zeta\Gamma(1\mp ip)}{4\sin(\pi ip)}\left(\frac{\sqrt{V_0}}{\beta}\right)^{\pm ip}\left\{J_{-ip}(\zeta_0)W_\zeta\left[J_{ip}(\zeta),J_{\mp ip}(\zeta)\right]-\right.$$

$$\left.-J_{ip}(\zeta_0)W_\zeta\left[J_{-ip}(\zeta),J_{\mp ip}(\zeta)\right]\right\}.$$

And with the help of Eq. (18.9), we finally obtain the following explicit formulae for the S-wave Jost functions corresponding to the exponential potential (18.2)

$$f_0^{(\text{in/out})}(E) = \frac{1}{2}\Gamma(1 \mp ip)\left(\frac{\sqrt{V_0}}{\beta}\right)^{\pm ip} J_{\mp ip}(\zeta_0) \tag{18.17}$$

It is easy to see that these Jost functions possess the symmetry properties shown in Fig. 3.5. The S-matrix is their ratio

$$S_0(E) = \frac{\Gamma\left(1 + \frac{2ik}{\beta}\right) J_{+2ik/\beta}\left(\frac{2\sqrt{V_0}}{\beta}\right)}{\Gamma\left(1 - \frac{2ik}{\beta}\right) J_{-2ik/\beta}\left(\frac{2\sqrt{V_0}}{\beta}\right)} \left(\frac{V_0}{\beta^2}\right)^{-2ik/\beta} . \tag{18.18}$$

The spectral points k_n are those at which the Jost function $f_0^{(\text{in})}(E)$ is zero. They can therefore be found as the roots of the equation

$$J_{-2ik_n/\beta}\left(\frac{2\sqrt{V_0}}{\beta}\right) = 0 . \tag{18.19}$$

As can be seen from Eq. (18.17), the Jost functions $f_0^{(\text{in})}(E)$ and $f_0^{(\text{out})}(E)$ have singularities at the points where the argument of the gamma-function is a negative integer number. For $f_0^{(\text{in})}(E)$, these points are on the negative imaginary axis of the k-plane:

$$k = -i\frac{\beta}{2}(n+1) , \qquad n = 1, 2, 3, \dots . \tag{18.20}$$

The singularities of $f_0^{(\text{out})}(E)$ are located symmetrically, on the positive half of the imaginary k-axis. The other factors on the right hand side of Eq. (18.17) are analytic functions of k on the whole plane. Therefore, the Jost function $f_0^{(\text{in})}$ for the exponential well (18.2) is a meromorphic function of k with an infinite series of poles (18.20) on the negative imaginary axis. This is in accordance with the generic properties of the exponentially decaying potentials (see Sect. 7.4.1). Indeed, by varying the rotation angle θ in Eq. (7.52) within the interval $-\pi/2 < \theta < +\pi/2$, the domain of analyticity of the Jost function $f_0^{(\text{in})}$ can be extended to almost whole k-plane except for the line starting at $k = -i\beta/2$ and going down to infinity along the negative imaginary axis (see Fig. 7.9).

18.2 Single-Channel Square-Well Potential

Consider a particle of mass μ moving with $\ell = 0$ in a spherically symmetric potential

$$U(r) = \begin{cases} U_0 \,, & r \leqslant R \,, \\ 0 \,, & r > R \,. \end{cases} \tag{18.21}$$

On the interval $0 \leqslant r \leqslant R$, the regular solution of the radial Schrödinger equation,

$$\left[\partial_r^2 + k^2 - \frac{2\mu}{\hbar^2} U(r)\right] \phi_0(E, r) = 0 \,, \tag{18.22}$$

is

$$\phi_0(E, r) = \sin(\kappa r) \,, \qquad 0 \leqslant r \leqslant R \,, \tag{18.23}$$

where

$$k^2 = \frac{2\mu}{\hbar^2} E \,, \tag{18.24}$$

and

$$\kappa^2 = \frac{2\mu}{\hbar^2}(E - U_0) \,. \tag{18.25}$$

At $r > R$, Eq. (18.22) has two linearly independent solutions, $\sin(kr)$ and $\cos(kr)$. Its general solution is their linear combination,

$$\phi_0(E, r) = \sin(kr)x(E) + \cos(kr)y(E) \,, \qquad r > R \,. \tag{18.26}$$

The combination coefficients, $x(E)$ and $y(E)$, can be found from the smooth matching of the functions (18.23) and (18.26) at the point $r = R$,

$$\sin(\kappa R) = \sin(kR)x + \cos(kR)y \,, \tag{18.27}$$

$$\kappa \cos(\kappa R) = k \cos(kR)x - k \sin(kR)y \,. \tag{18.28}$$

From these matching equations, it is easy to find that

$$x(E) = \sin(\kappa R) \sin(kR) + \frac{\kappa}{k} \cos(\kappa R) \cos(kR) \,, \tag{18.29}$$

$$y(E) = \sin(\kappa R) \cos(kR) - \frac{\kappa}{k} \cos(\kappa R) \sin(kR) \,. \tag{18.30}$$

The Jost functions are defined as the asymptotic amplitudes of the spherical waves,

$$\phi_\ell(E,r) \underset{r\to\infty}{\longrightarrow} h_\ell^{(-)}(kr) f_\ell^{(\mathrm{in})}(E) + h_\ell^{(+)}(kr) f_\ell^{(\mathrm{out})}(E)\,, \tag{18.31}$$

where $h_\ell^{(\pm)}(z)$ are the Riccati–Hankel functions. Using the explicit representations, $h_0^{(\pm)}(z) = \mp i\exp(\pm iz)$, it is not difficult to see that

$$f_0^{(\mathrm{in/out})}(E) = \frac{x(E) \mp iy(E)}{2}\,. \tag{18.32}$$

This gives the following S-wave Jost functions

$$\boxed{f_0^{(\mathrm{in/out})}(E) = \frac{e^{\pm ikR}}{2}\left[\frac{\kappa}{k}\cos(\kappa R) \mp i\sin(\kappa R)\right]} \tag{18.33}$$

and the corresponding S-matrix

$$S_0(E) = \frac{f_0^{(\mathrm{out})}(E)}{f_0^{(\mathrm{in})}(E)} = \frac{\dfrac{\kappa}{k}\cos(\kappa R) + i\sin(\kappa R)}{\dfrac{\kappa}{k}\cos(\kappa R) - i\sin(\kappa R)}\exp(-2ikR)\,. \tag{18.34}$$

18.2.1 Analytic Structure

As we know, the S-wave single-channel Jost functions have the following analytic structure (see Eq. (7.65)):

$$f_0^{(\mathrm{in/out})}(E) = a_0(E) \mp ikb_0(E)\,, \tag{18.35}$$

where $a_0(E)$ and $b_0(E)$ are single-valued analytic functions of the energy variable E. Let us check if we can reduce the expression (18.33) to the form (18.35). Actually, this is already done in Eq. (18.32), from which we immediately find that

$$a_0(E) = \frac{\kappa}{2k}\left[\cos(kR)\cos(\kappa R) + \frac{k}{\kappa}\sin(kR)\sin(\kappa R)\right]\,, \tag{18.36}$$

$$b_0(E) = \frac{\kappa}{2k}\left[\frac{1}{\kappa}\cos(kR)\sin(\kappa R) - \frac{1}{k}\sin(kR)\cos(\kappa R)\right]\,. \tag{18.37}$$

Are the functions $a_0(E)$ and $b_0(E)$ single-valued? To check this, we take notice of the fact that

$$\kappa = k\sqrt{1 - \frac{2\mu U_0}{\hbar^2 k^2}}\,, \tag{18.38}$$

which implies that, if k changes sign to the opposite ($k \to -k$), then κ changes sign as well, i.e. the ratio κ/k remains the same for $+k$ and for $-k$. In Eqs. (18.36, 18.37), the expressions in the square brackets do not depend on the choices of the signs of k and κ. Therefore, $a_0(E)$ and $b_0(E)$ are single-valued functions of variable E (as it should be). Their analyticity is obvious.

18.3 Single-Channel Square-Well Potential with the Coulomb Tail

Let the potential be the same as in Sect. 18.2 for $r \leqslant R$ but has the Coulomb tail for $r > R$, i.e.

$$U(r) = \begin{cases} U_0 \,, \; r \leqslant R \,, \\ \dfrac{\gamma}{r} \,, \; r > R \,. \end{cases} \tag{18.39}$$

This potential is depicted in Fig. 18.1. If we still consider the state with $\ell = 0$, then inside the sphere of the radius R, the regular solution is the same as is given by Eq. (18.23). In the outside region, however, we should replace Eq. (18.31) with its Coulomb analog (8.67). Therefore,

$$\phi_0(E,r) = \sin\kappa r \,, \qquad r \leqslant R \,, \tag{18.40}$$

$$\phi_0(E,r) = H_0^{(-)}(\eta, kr)e^{i\sigma_0} f_0^{(\mathrm{in})}(E) + H_0^{(+)}(\eta, kr)e^{-i\sigma_0} f_0^{(\mathrm{out})}(E) \,, \quad r > R \,, \tag{18.41}$$

where $H_0^{(\pm)}(\eta, kr) = F_0(\eta, kr) \mp iG_0(\eta, kr)$, $\eta = \mu\gamma/(k\hbar^2)$, and σ_0 is the S-wave Coulomb phase shift. At the border, $r = R$, between the regions, the functions (18.40) and (18.41) must smoothly match. This gives

$$\sin\kappa R = \mathcal{H}^{(-)} f_0^{(\mathrm{in})} + \mathcal{H}^{(+)} f_0^{(\mathrm{out})} \,, \tag{18.42}$$

$$\kappa\cos\kappa R = \mathcal{H}^{(-)\prime} f_0^{(\mathrm{in})} + \mathcal{H}^{(+)\prime} f_0^{(\mathrm{out})} \,, \tag{18.43}$$

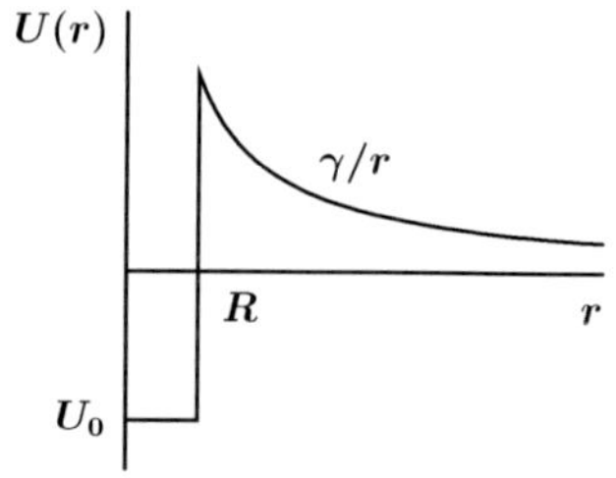

Fig. 18.1 The potential (18.39) consisting of the square well (for $r \leqslant R$) and the Coulomb tail (for $r > R$)

where

$$\mathcal{H}^{(\pm)} = H_0^{(\pm)}(\eta, kr)e^{\mp i\sigma_0}\,, \qquad \mathcal{H}^{(\pm)\prime} = \Big[\partial_r H_0^{(\pm)}(\eta, kr)\Big]_{r=R} e^{\mp i\sigma_0}\,.$$

Using the Wronskian (see Eq. (E.14) of Appendix E)

$$H_\ell^{(-)}(\eta, kr)\partial_r H_\ell^{(+)}(\eta, kr) - H_\ell^{(+)}(\eta, kr)\partial_r H_\ell^{(-)}(\eta, kr) = 2ik\,, \tag{18.44}$$

we find the following solutions of the linear equations (18.42, 18.43):

$$f_0^{(\text{in/out})}(E) = \pm\frac{e^{\mp i\sigma_0(\eta)}}{2ik}\Big[H_0^{(\pm)\prime}(\eta, kR)\sin\kappa R - H_0^{(\pm)}(\eta, kR)\kappa\cos\kappa R\Big] \tag{18.45}$$

18.3.1 Analytic Structure

The decomposition (8.100) of the Jost functions for a system of charged particles in the case $\ell = 0$ takes the form:

$$f_0^{(\text{in/out})}(E) = \frac{e^{\mp i\sigma_0}}{C_0(\eta)}\Big\{a_0(E) - \Big[2\eta h(\eta) \pm iC_0^2(\eta)\Big]\,k\,b_0(E)\Big\}\,. \tag{18.46}$$

Let us see how the explicitly found Jost functions (18.45) can be reduced to (18.46). To this end, we recall the structural representations of the Coulomb functions (E.24) and (E.30) (see Appendix E), which in the S-wave case read

$$F_0(\eta, kr) = C_0(\eta)k\tilde{F}_0(\eta, kr), \tag{18.47}$$

$$G_0(\eta, kr) = \frac{2\eta h(\eta)k}{C_0(\eta)}\tilde{F}_0(\eta, kr) + \frac{1}{C_0(\eta)}\tilde{G}_0(\eta, kr)\,, \tag{18.48}$$

where $\tilde{F}_0$ and $\tilde{G}_0$ are single-valued analytic functions of E. It is clear that the derivatives $\partial_r\tilde{F}_0$ and $\partial_r\tilde{G}_0$ (which we denote by prime) are single-valued and analytic as well. Therefore, the functions $H_0^{(\pm)}$ and their derivatives $H_0^{(\pm)\prime}$ can be represented as

$$H_0^{(\pm)} = C_0k\tilde{F}_0 \mp i\left(\frac{2\eta hk}{C_0}\tilde{F}_0 + \frac{1}{C_0}\tilde{G}_0\right)\,, \tag{18.49}$$

$$H_0^{(\pm)\prime} = C_0 k \tilde{F}_0' \mp i \left(\frac{2\eta hk}{C_0} \tilde{F}_0' + \frac{1}{C_0} \tilde{G}_0' \right) , \tag{18.50}$$

where all the factors responsible for branching of the Riemann surface are given explicitly. Substituting Eqs. (18.49, 18.50) in Eq. (18.45) and doing some simple algebra, we obtain the representation (18.46) with

$$a_0(E) = \frac{1}{2k} \left[\tilde{G}_0(\eta, kR) \kappa \cos \kappa R - \tilde{G}_0'(\eta, kR) \sin \kappa R \right] , \tag{18.51}$$

$$b_0(E) = \frac{1}{2k} \left[\tilde{F}_0'(\eta, kR) \sin \kappa R - \tilde{F}_0(\eta, kR) \kappa \cos \kappa R \right] . \tag{18.52}$$

It is not difficult to see that $a_0(E)$ and $b_0(E)$ are single-valued functions of E (as it should be). Indeed, as was already said, the functions $\tilde{F}_0$ and $\tilde{G}_0$ as well as their derivatives $\tilde{F}_0'$ and $\tilde{G}_0'$ are single-valued. And according to Eq. (18.38), κ changes sign together with k, which means that the ratios κ/k and $\sin \kappa R/k$ are even functions of k, i.e. the functions of k^2, and thus are also single-valued functions of E.

18.4 Two-Channel Square-Well Potential

To avoid the unnecessary complications, we assume that our two-channel square-well potential has the same radius R in both channels, and in its non-diagonal elements,

$$U(r) = \begin{cases} \begin{pmatrix} U_{11} & U_{12} \\ U_{21} & U_{22} \end{pmatrix} , & r \leqslant R , \\ 0 & , r > R , \end{cases} \tag{18.53}$$

where $U_{12} = U_{21}$. If μ_i and E_i are the channel reduced masses and the thresholds, then

$$k_i = \sqrt{\frac{2\mu_i}{\hbar^2}(E - E_i)} , \qquad i = 1, 2, \tag{18.54}$$

are the channel momenta and

$$\begin{pmatrix} V_{11} & V_{12} \\ V_{21} & V_{22} \end{pmatrix} = \begin{cases} \begin{pmatrix} \frac{2\mu_1}{\hbar^2} U_{11} & \frac{2\mu_1}{\hbar^2} U_{12} \\ \frac{2\mu_2}{\hbar^2} U_{21} & \frac{2\mu_2}{\hbar^2} U_{22} \end{pmatrix} , & r \leqslant R , \\ 0 & , r > R , \end{cases} \tag{18.55}$$

is the so-called reduced potential (where, generally speaking, $V_{12} \neq V_{21}$). Let $u_1(E, r)$ and $u_2(E, r)$ be the radial wave functions in the first and second channels; then the radial Schrödinger equation,

$$\begin{pmatrix} \partial_r^2 + k_1^2 & 0 \\ 0 & \partial_r^2 + k_2^2 \end{pmatrix} \begin{pmatrix} u_1 \\ u_2 \end{pmatrix} = \begin{pmatrix} V_{11} & V_{12} \\ V_{21} & V_{22} \end{pmatrix} \begin{pmatrix} u_1 \\ u_2 \end{pmatrix}, \tag{18.56}$$

can be re-written as

$$\begin{cases} u_1'' + \kappa_1^2 u_1 - V_{12} u_2 = 0\,, \\ u_2'' + \kappa_2^2 u_2 - V_{21} u_1 = 0\,, \end{cases} \tag{18.57}$$

where

$$\kappa_i^2 = \frac{2\mu_i}{\hbar^2}(E - E_i - U_{ii})\,, \qquad i = 1, 2\,. \tag{18.58}$$

We are only interested in the regular solution of Eq. (18.56). Let us look for it in the following form:

$$\begin{pmatrix} u_1 \\ u_2 \end{pmatrix} = \begin{bmatrix} \alpha_1 \sin(pr) + \beta_1 \sin(qr) \\ \alpha_2 \sin(pr) + \beta_2 \sin(qr) \end{bmatrix}, \tag{18.59}$$

where α_i and β_i are some constants and p and q are unknown momenta, such that $p \neq q$. Substituting the ansatz (18.59) in Eq. (18.57), we obtain

$$\begin{cases} [(\kappa_1^2 - p^2)\alpha_1 - V_{12}\alpha_2]\sin(pr) + [(\kappa_1^2 - q^2)\beta_1 - V_{12}\beta_2]\sin(qr) = 0\,, \\ [(\kappa_2^2 - p^2)\alpha_2 - V_{21}\alpha_1]\sin(pr) + [(\kappa_2^2 - q^2)\beta_2 - V_{21}\beta_1]\sin(qr) = 0\,. \end{cases} \tag{18.60}$$

These equations are valid identically for any r. Since $p \neq q$, this is only possible if the constant multipliers of the functions $\sin(pr)$ and $\sin(qr)$ are zero,

$$\begin{aligned} (\kappa_1^2 - p^2)\alpha_1 - V_{12}\alpha_2 &= 0\,, \\ (\kappa_1^2 - q^2)\beta_1 - V_{12}\beta_2 &= 0\,, \\ (\kappa_2^2 - p^2)\alpha_2 - V_{21}\alpha_1 &= 0\,, \\ (\kappa_2^2 - q^2)\beta_2 - V_{21}\beta_1 &= 0\,. \end{aligned}$$

This is a homogeneous system of linear equations for the unknown coefficients α_1, α_2, β_1, and β_2. It is seen that this system can be split into two independent pairs of equations, namely,

$$\begin{pmatrix} \kappa_1^2 - p^2 & -V_{12} \\ -V_{21} & \kappa_2^2 - p^2 \end{pmatrix} \begin{pmatrix} \alpha_1 \\ \alpha_2 \end{pmatrix} = \begin{pmatrix} 0 \\ 0 \end{pmatrix} , \tag{18.61}$$

$$\begin{pmatrix} \kappa_1^2 - q^2 & -V_{12} \\ -V_{21} & \kappa_2^2 - q^2 \end{pmatrix} \begin{pmatrix} \beta_1 \\ \beta_2 \end{pmatrix} = \begin{pmatrix} 0 \\ 0 \end{pmatrix} . \tag{18.62}$$

These homogeneous systems may have non-trivial solutions, if and only if the determinants of the corresponding matrices are zero. This condition gives the same quadratic equations for p^2 and for q^2,

$$(p^2)^2 - (\kappa_1^2 + \kappa_2^2)p^2 + \kappa_1^2\kappa_2^2 - V_{12}V_{21} = 0 ,$$

$$(q^2)^2 - (\kappa_1^2 + \kappa_2^2)q^2 + \kappa_1^2\kappa_2^2 - V_{12}V_{21} = 0 .$$

When solving these equations, we find two roots. In order to have the values of p and q different, we have to choose for them different signs in front of the discriminant (the choice is arbitrary),

$$p^2 = \frac{1}{2}\left[\kappa_1^2 + \kappa_2^2 + \sqrt{(\kappa_1^2 - \kappa_2^2)^2 + 4V_{12}V_{21}}\right] , \tag{18.63}$$

$$q^2 = \frac{1}{2}\left[\kappa_1^2 + \kappa_2^2 - \sqrt{(\kappa_1^2 - \kappa_2^2)^2 + 4V_{12}V_{21}}\right] . \tag{18.64}$$

Now, we know the momenta p and q in the solution (18.59), but the combination coefficients α_1, α_2, β_1, and β_2 remain unknown. The homogeneous systems (18.61) and (18.62) have an infinite number of solutions and can only give us the ratios (x and y):

$$\frac{\alpha_1}{\alpha_2} = \frac{V_{12}}{\kappa_1^2 - p^2} = \frac{\kappa_2^2 - p^2}{V_{21}} = x , \tag{18.65}$$

$$\frac{\beta_1}{\beta_2} = \frac{V_{12}}{\kappa_1^2 - q^2} = \frac{\kappa_2^2 - q^2}{V_{21}} = y . \tag{18.66}$$

Matrix differential equation (18.56) has four linearly independent solutions, two of which are regular. These two regular solutions have the same form, namely, given by Eq. (18.59), but with different choices of the combination coefficients α_1, α_2,

β_1, and β_2. The two regular columns can be combined in a square matrix $\phi(E, r)$, obeying the following boundary conditions (see Eqs. (12.3) and (9.31)):

$$\phi(E, r) \underset{r \to 0}{\longrightarrow} \begin{bmatrix} k_1 r & o(k_2 r) \\ o(k_1 r) & k_2 r \end{bmatrix} . \tag{18.67}$$

These boundary conditions together with the ratios (18.65, 18.66) uniquely determine the combination coefficients α_{ij} and β_{ij} (where the second subscript indicated the first or second column) in the matrix of the regular solutions:

$$\phi(E, r) = \begin{bmatrix} \alpha_{11} \sin(pr) + \beta_{11} \sin(qr) \ , \ \alpha_{12} \sin(pr) + \beta_{12} \sin(qr) \\ \alpha_{21} \sin(pr) + \beta_{21} \sin(qr) \ , \ \alpha_{22} \sin(pr) + \beta_{22} \sin(qr) \end{bmatrix} . \tag{18.68}$$

The conditions (18.67) are fulfilled, if the leading terms of the Taylor series for the sin-functions are either combined into $k_1 r$ and $k_2 r$ (for the diagonal elements) or cancelled (for the off-diagonal elements):

$$\begin{aligned} \alpha_{11} p + \beta_{11} q &= k_1 \ , \\ \alpha_{22} p + \beta_{22} q &= k_2 \ , \\ \alpha_{12} p + \beta_{12} q &= 0 \ , \\ \alpha_{21} p + \beta_{21} q &= 0 \ , \\ \alpha_{11}/\alpha_{21} = \alpha_{12}/\alpha_{22} &= x \ , \\ \beta_{11}/\beta_{21} = \beta_{12}/\beta_{22} &= y \ . \end{aligned}$$

Here we have eight equations for the eight combination coefficients. These equations can be written in the following matrix form:

$$\begin{pmatrix} p & 0 & 0 & 0 & q & 0 & 0 & 0 \\ 0 & 0 & 0 & p & 0 & 0 & 0 & q \\ 0 & p & 0 & 0 & 0 & q & 0 & 0 \\ 0 & 0 & p & 0 & 0 & 0 & q & 0 \\ 1 & 0 & -x & 0 & 0 & 0 & 0 & 0 \\ 0 & 1 & 0 & -x & 0 & 0 & 0 & 0 \\ 0 & 0 & 0 & 0 & 1 & 0 & -y & 0 \\ 0 & 0 & 0 & 0 & 0 & 1 & 0 & -y \end{pmatrix} \begin{pmatrix} \alpha_{11} \\ \alpha_{12} \\ \alpha_{21} \\ \alpha_{22} \\ \beta_{11} \\ \beta_{12} \\ \beta_{21} \\ \beta_{22} \end{pmatrix} = \begin{pmatrix} k_1 \\ k_2 \\ 0 \\ 0 \\ 0 \\ 0 \\ 0 \\ 0 \end{pmatrix} . \tag{18.69}$$

Since the matrix is very sparse, the system can be solved actually via simple substitutions. In this way, we find

$$\alpha_{11} = \frac{x k_1}{p(x - y)} \ , \qquad \alpha_{12} = \frac{x y k_2}{p(y - x)} \ , \tag{18.70}$$

$$\alpha_{21} = \frac{k_1}{p(x-y)}\,, \qquad \alpha_{22} = \frac{yk_2}{p(y-x)}\,, \tag{18.71}$$

$$\beta_{11} = \frac{yk_1}{q(y-x)}\,, \qquad \beta_{12} = \frac{xyk_2}{q(x-y)}\,, \tag{18.72}$$

$$\beta_{21} = \frac{k_1}{q(y-x)}\,, \qquad \beta_{22} = \frac{xk_2}{q(x-y)}\,, \tag{18.73}$$

where x and y are given by Eqs. (18.65, 18.66). Therefore, the matrix of the regular solutions (18.68) is found. Using it, we can now find the Jost matrices. To this end, we consider the regular solution at the point $r = R$, beyond which this matrix reaches its asymptotic form:

$$\begin{pmatrix} \phi_{11} & \phi_{12} \\ \phi_{21} & \phi_{22} \end{pmatrix} \underset{r\to\infty}{\longrightarrow} \begin{bmatrix} h_0^{(-)}(k_1 r) & 0 \\ 0 & h_0^{(-)}(k_2 r) \end{bmatrix} \begin{bmatrix} f_{11}^{(\text{in})} & f_{12}^{(\text{in})} \\ f_{21}^{(\text{in})} & f_{22}^{(\text{in})} \end{bmatrix} + \begin{bmatrix} h_0^{(+)}(k_1 r) & 0 \\ 0 & h_0^{(+)}(k_2 r) \end{bmatrix} \begin{bmatrix} f_{11}^{(\text{out})} & f_{12}^{(\text{out})} \\ f_{21}^{(\text{out})} & f_{22}^{(\text{out})} \end{bmatrix}, \tag{18.74}$$

where $h_0^{(\pm)}(z) = \mp i \exp(\pm iz)$. All eight elements of the Jost matrices $f^{(\text{in/out})}(E)$ can be found by the smooth matching of $\phi(R,r)$ with its asymptotic form at the point $r = R$. Equalizing the functions and their derivatives, we have the necessary eight equations:

$$\begin{pmatrix} \phi_{11} & \phi_{12} \\ \phi_{21} & \phi_{22} \end{pmatrix}_{r=R} = \begin{bmatrix} ie^{-ik_1R} & 0 \\ 0 & ie^{-ik_2R} \end{bmatrix} \begin{bmatrix} f_{11}^{(\text{in})} & f_{12}^{(\text{in})} \\ f_{21}^{(\text{in})} & f_{22}^{(\text{in})} \end{bmatrix} + \begin{bmatrix} -ie^{ik_1R} & 0 \\ 0 & -ie^{ik_2R} \end{bmatrix} \begin{bmatrix} f_{11}^{(\text{out})} & f_{12}^{(\text{out})} \\ f_{21}^{(\text{out})} & f_{22}^{(\text{out})} \end{bmatrix}, \tag{18.75}$$

$$\partial_r \begin{pmatrix} \phi_{11} & \phi_{12} \\ \phi_{21} & \phi_{22} \end{pmatrix}_{r=R} = \begin{bmatrix} k_1e^{-ik_1R} & 0 \\ 0 & k_2e^{-ik_2R} \end{bmatrix} \begin{bmatrix} f_{11}^{(\text{in})} & f_{12}^{(\text{in})} \\ f_{21}^{(\text{in})} & f_{22}^{(\text{in})} \end{bmatrix} + \begin{bmatrix} k_1e^{ik_1R} & 0 \\ 0 & k_2e^{ik_2R} \end{bmatrix} \begin{bmatrix} f_{11}^{(\text{out})} & f_{12}^{(\text{out})} \\ f_{21}^{(\text{out})} & f_{22}^{(\text{out})} \end{bmatrix}. \tag{18.76}$$

In explicit form, this gives

$$\alpha_{11}\sin(pR)+\beta_{11}\sin(qR)=ie^{-ik_1R}f_{11}^{(\mathrm{in})}-ie^{ik_1R}f_{11}^{(\mathrm{out})}\,, \tag{18.77}$$

$$\alpha_{11}p\cos(pR)+\beta_{11}q\cos(qR)=k_1e^{-ik_1R}f_{11}^{(\mathrm{in})}+k_1e^{ik_1R}f_{11}^{(\mathrm{out})}\,, \tag{18.78}$$

$$\alpha_{12}\sin(pR)+\beta_{12}\sin(qR)=ie^{-ik_1R}f_{12}^{(\mathrm{in})}-ie^{ik_1R}f_{12}^{(\mathrm{out})}\,, \tag{18.79}$$

$$\alpha_{12}p\cos(pR)+\beta_{12}q\cos(qR)=k_1e^{-ik_1R}f_{12}^{(\mathrm{in})}+k_1e^{ik_1R}f_{12}^{(\mathrm{out})}\,, \tag{18.80}$$

$$\alpha_{21}\sin(pR)+\beta_{21}\sin(qR)=ie^{-ik_2R}f_{21}^{(\mathrm{in})}-ie^{ik_2R}f_{21}^{(\mathrm{out})}\,, \tag{18.81}$$

$$\alpha_{21}p\cos(pR)+\beta_{21}q\cos(qR)=k_2e^{-ik_2R}f_{21}^{(\mathrm{in})}+k_2e^{ik_1R}f_{21}^{(\mathrm{out})}\,, \tag{18.82}$$

$$\alpha_{22}\sin(pR)+\beta_{22}\sin(qR)=ie^{-ik_2R}f_{22}^{(\mathrm{in})}-ie^{ik_2R}f_{22}^{(\mathrm{out})}\,, \tag{18.83}$$

$$\alpha_{22}p\cos(pR)+\beta_{22}q\cos(qR)=k_2e^{-ik_2R}f_{22}^{(\mathrm{in})}+k_2e^{ik_1R}f_{22}^{(\mathrm{out})}\,. \tag{18.84}$$

It is seen that the above eight equations are split into four independent pairs. Each pair involves $f_{mn}^{(\mathrm{in})}$ and $f_{mn}^{(\mathrm{out})}$ with the same subscripts. It is sufficient to solve one such pair of equations. For the other pairs, the solutions are of the same form. In this way, we obtain

$$\boxed{\begin{aligned}f_{mn}^{(\mathrm{in/out})}(E)=\frac{e^{\pm ik_mR}}{2}\Bigg[&\alpha_{mn}\left(\frac{p}{k_m}\cos pR\mp i\sin pR\right)\\&+\beta_{mn}\left(\frac{q}{k_m}\cos qR\mp i\sin qR\right)\Bigg]\end{aligned}} \tag{18.85}$$

This explicit formula generalizes Eq. (18.33) that was derived for the single-channel case. The corresponding S-matrix can be found as

$$S(E)=f^{(\mathrm{out})}(E)\left[f^{(\mathrm{in})}(E)\right]^{-1}\,, \tag{18.86}$$

and its spectral points are the roots of the equation

$$\det f^{(\text{in})}(E) = 0\,. \tag{18.87}$$

18.4.1 Analytic Structure

Since in both channels the angular momentum is zero, the general representation of the multi-channel Jost matrices, given by Eq. (13.24), is reduced to

$$f_{mn}^{(\text{in/out})}(E) = \frac{k_n}{k_m}a_{mn}(E) \mp ik_n b_{mn}(E)\,. \tag{18.88}$$

To convince ourselves that Eq. (18.85) is consistent with (18.88), we take notice of the fact that both α_{mn} and β_{mn} are proportional to k_n, i.e.

$$\alpha_{mn} = \tilde{\alpha}_{mn}k_n \qquad \text{and} \qquad \beta_{mn} = \tilde{\beta}_{mn}k_n\,,$$

where the constant matrices $\tilde{\alpha}$ and $\tilde{\beta}$ can be found from Eqs. (18.70–18.73). Therefore, Eq. (18.85) can be rewritten in the form

$$f_{mn}^{(\text{in/out})}(E) = \frac{e^{\pm ik_m R}}{2}\left[\frac{k_n}{k_m}\left(\tilde{\alpha}_{mn}p\cos pR + \tilde{\beta}_{mn}q\cos qR\right)\right.$$
$$\left.\mp ik_n\left(\tilde{\alpha}_{mn}\sin pR + \tilde{\beta}_{mn}\sin qR\right)\right]\,.$$

Substituting here $e^{\pm ik_m R} = \cos k_m R \pm i \sin k_m R$, and separating the terms multiplied by i, we obtain the Jost matrices in the form (18.88), where

$$a_{mn}(E) = \frac{\cos k_m R}{2}\left(\tilde{\alpha}_{mn}p\cos pR + \tilde{\beta}_{mn}q\cos qR\right)$$
$$+\frac{k_m \sin k_m R}{2}\left(\tilde{\alpha}_{mn}\sin pR + \tilde{\beta}_{mn}\sin qR\right)\,, \tag{18.89}$$

$$b_{mn}(E) = \frac{\cos k_m R}{2}\left(\tilde{\alpha}_{mn}\sin pR + \tilde{\beta}_{mn}\sin qR\right)$$
$$-\frac{\sin k_m R}{2k_m}\left(\tilde{\alpha}_{mn}p\cos pR + \tilde{\beta}_{mn}q\cos qR\right)\,. \tag{18.90}$$

It is not difficult to see that the right hand sides of Eqs. (18.89, 18.90) do not depend on the choice of the signs of the channel momenta k_m. They also do not depend on the choice of the signs of the momenta p and q because these momenta are in the denominators of the matrices α and β, respectively. Therefore, the matrices a and b are single-valued functions of the energy (as it should be).

18.5 Two-Channel Square-Well Potential with the Coulomb Tail

Let us add the Coulomb tails to the potential (18.53):

$$U(r) = \begin{cases} \begin{pmatrix} U_{11} & U_{12} \\ U_{21} & U_{22} \end{pmatrix}, & r \leqslant R, \\ \begin{pmatrix} \dfrac{\gamma_1}{r} & 0 \\ 0 & \dfrac{\gamma_2}{r} \end{pmatrix}, & r > R. \end{cases} \tag{18.91}$$

Apparently, for the distances less than R, the matrix $\phi(E, r)$ of the regular solutions is exactly the same as in the previous section. Outside the sphere of the radius R, instead of (18.74), we have

$$\phi(E, r) \underset{r\to\infty}{\longrightarrow} H^{(-)}(E, r)\sigma^{(+)} f^{(\text{in})}(E) + H^{(+)}(E, r)\sigma^{(-)} f^{(\text{out})}(E) , \tag{18.92}$$

with

$$H^{(\pm)}(E, r) = \begin{bmatrix} F_0(\eta_1, k_1 r) \mp i G_0(\eta_1, k_1 r) & 0 \\ 0 & F_0(\eta_2, k_2 r) \mp i G_0(\eta_2, k_2 r) \end{bmatrix} , \tag{18.93}$$

$$\sigma^{(\pm)} = \begin{bmatrix} e^{\pm i\sigma_0(\eta_1)} & 0 \\ 0 & e^{\pm i\sigma_0(\eta_2)} \end{bmatrix} , \tag{18.94}$$

where F_0 and G_0 are the standard Coulomb functions with $\ell = 0$ (see Appendix E),

$$\eta_n = \frac{\mu_n \gamma_n}{k_n \hbar^2} , \qquad n = 1, 2 , \tag{18.95}$$

and the S-wave Coulomb phase shift is given by

$$\sigma_0(\eta) = \frac{1}{2i} \left[\ln \Gamma(1 + i\eta) - \ln \Gamma(1 - i\eta)\right] . \tag{18.96}$$

In order to find the Jost matrices, we only need to perform the smooth matching of the regular solution $\phi(E, r)$ with its asymptotic form (18.92) at the point $r = R$,

$$\phi(E, R) = H^{(-)}(E, R)\sigma^{(+)} f^{(\text{in})}(E) + H^{(+)}(E, R)\sigma^{(-)} f^{(\text{out})}(E) , \tag{18.97}$$

$$[\partial_r \phi(E,r)]_{r=R} = \Big[\partial_r H^{(-)}(E,r)\sigma^{(+)} f^{(\text{in})}(E) + \partial_r H^{(+)}(E,r)\sigma^{(-)} f^{(\text{out})}(E)\Big]_{r=R} . \tag{18.98}$$

If we denote

$$\mathcal{H}^{(\pm)} = \sigma^{(\mp)} H^{(\pm)}(E,R) , \qquad \text{and} \qquad \mathcal{H}'^{(\pm)} = \sigma^{(\mp)} \Big[\partial_r H^{(\pm)}(E,r)\Big]_{r=R} , \tag{18.99}$$

then the matching equations in explicit form read

$$\begin{cases} \alpha_{11} \sin(pR) + \beta_{11} \sin(qR) = \mathcal{H}_{11}^{(-)} f_{11}^{(\text{in})} + \mathcal{H}_{11}^{(+)} f_{11}^{(\text{out})} , \\ \alpha_{11} p \cos(pR) + \beta_{11} q \cos(qR) = \mathcal{H}_{11}'^{(-)} f_{11}^{(\text{in})} + \mathcal{H}_{11}'^{(+)} f_{11}^{(\text{out})} , \end{cases} \tag{18.100}$$

$$\begin{cases} \alpha_{12} \sin(pR) + \beta_{12} \sin(qR) = \mathcal{H}_{11}^{(-)} f_{12}^{(\text{in})} + \mathcal{H}_{11}^{(+)} f_{12}^{(\text{out})} , \\ \alpha_{12} p \cos(pR) + \beta_{12} q \cos(qR) = \mathcal{H}_{11}'^{(-)} f_{12}^{(\text{in})} + \mathcal{H}_{11}'^{(+)} f_{12}^{(\text{out})} , \end{cases} \tag{18.101}$$

$$\begin{cases} \alpha_{21} \sin(pR) + \beta_{21} \sin(qR) = \mathcal{H}_{22}^{(-)} f_{21}^{(\text{in})} + \mathcal{H}_{22}^{(+)} f_{21}^{(\text{out})} , \\ \alpha_{21} p \cos(pR) + \beta_{21} q \cos(qR) = \mathcal{H}_{22}'^{(-)} f_{21}^{(\text{in})} + \mathcal{H}_{22}'^{(+)} f_{21}^{(\text{out})} , \end{cases} \tag{18.102}$$

$$\begin{cases} \alpha_{22} \sin(pR) + \beta_{22} \sin(qR) = \mathcal{H}_{22}^{(-)} f_{22}^{(\text{in})} + \mathcal{H}_{22}^{(+)} f_{22}^{(\text{out})} , \\ \alpha_{22} p \cos(pR) + \beta_{22} q \cos(qR) = \mathcal{H}_{22}'^{(-)} f_{22}^{(\text{in})} + \mathcal{H}_{22}'^{(+)} f_{22}^{(\text{out})} . \end{cases} \tag{18.103}$$

It is easy to see that these eight equations are split into four independent pairs, and in general, such a pair can be written as

$$\begin{bmatrix} \mathcal{H}_{mm}^{(-)} & \mathcal{H}_{mm}^{(+)} \\ \mathcal{H}_{mm}'^{(-)} & \mathcal{H}_{mm}'^{(+)} \end{bmatrix} \begin{bmatrix} f_{mn}^{(\text{in})} \\ f_{mn}^{(\text{out})} \end{bmatrix} = \begin{pmatrix} \alpha_{mn} \sin pR + \beta_{mn} \sin qR \\ \alpha_{mn} p \cos pR + \beta_{mn} q \cos qR \end{pmatrix} . \tag{18.104}$$

This equation can be solved by the direct inversion of the matrix on its left hand side:

$$\begin{bmatrix} \mathcal{H}_{mm}^{(-)} & \mathcal{H}_{mm}^{(+)} \\ \mathcal{H}_{mm}'^{(-)} & \mathcal{H}_{mm}'^{(+)} \end{bmatrix}^{-1} = \frac{1}{\Delta} \begin{bmatrix} \mathcal{H}_{mm}'^{(+)} & -\mathcal{H}_{mm}^{(+)} \\ -\mathcal{H}_{mm}'^{(-)} & \mathcal{H}_{mm}^{(-)} \end{bmatrix} ,$$

where the determinant, Δ, is equal to the Wronskian

$$\Delta = W\left[\mathcal{H}_{mm}^{(-)}, \mathcal{H}_{mm}^{(+)}\right] = 2ik_m .$$

Therefore,

$$\begin{bmatrix} f_{mn}^{(\text{in})} \\ f_{mn}^{(\text{out})} \end{bmatrix} = \frac{1}{2ik_m} \begin{bmatrix} \mathcal{H}_{mm}^{\prime(+)} & -\mathcal{H}_{mm}^{(+)} \\ -\mathcal{H}_{mm}^{\prime(-)} & \mathcal{H}_{mm}^{(-)} \end{bmatrix} \begin{pmatrix} \alpha_{mn} \sin pR + \beta_{mn} \sin qR \\ \alpha_{mn} p \cos pR + \beta_{mn} q \cos qR \end{pmatrix} \tag{18.105}$$

or in another form

$$f_{mn}^{(\text{in/out})} = \frac{\mp i e^{\mp i\sigma_0(\eta_m)}}{2k_m} \Big[H_{mm}^{\prime(\pm)}(\eta_m, k_m R)\,(\alpha_{mn} \sin pR + \beta_{mn} \sin qR) - H_{mm}^{(\pm)}(\eta_m, k_m R)\,(\alpha_{mn} p \cos pR + \beta_{mn} q \cos qR) \Big] \tag{18.106}$$

Here, the matrices α and β as well as the constant momenta p and q are the same as defined in Sect. 18.4. The S-matrix and its spectral points can be found using the same equations (18.86) and (18.87).

18.5.1 Analytic Structure

Since $\ell_1 = \ell_2 = 0$, the diagonal elements $H_{mm}^{(\pm)}$ and $H_{mm}^{\prime(\pm)}$ of the two-channel Coulomb spherical waves and their derivatives in Eq. (18.106) can be replaced by the representations (18.49, 18.50), where all the multi-valued factors are given explicitly. Substituting them in (18.106) and doing some simple but lengthy algebra (which we skip here), we obtain the Jost functions in the form:

$$f_{mn}^{(\text{in/out})}(E) = \frac{\exp[\mp i\sigma_0(\eta_m)]}{C_0(\eta_m)} \left\{ \frac{k_n}{k_m} a_{mn}(E) - \left[\frac{2\eta_m h(\eta_m)}{C_0^2(\eta_m)} \pm i \right] C_0^2(\eta_m) k_n\, b_{mn}(E) \right\} , \tag{18.107}$$

where

$$a_{mn}(E) = \frac{\tilde{G}_0(\eta_m, k_m R)}{2k_n} (\alpha_{mn} p \cos pR + \beta_{mn} q \cos qR) - \frac{\tilde{G}_0'(\eta_m, k_m R)}{2k_n} (\alpha_{mn} \sin pR + \beta_{mn} \sin qR) , \tag{18.108}$$

$$b_{mn}(E) = \frac{\tilde{F}_0'(\eta_m, k_m R)}{2k_n}(\alpha_{mn}\sin pR + \beta_{mn}\sin qR) -$$
$$-\frac{\tilde{F}_0(\eta_m, k_m R)}{2k_n}(\alpha_{mn}p\cos pR + \beta_{mn}q\cos qR) \,. \qquad (18.109)$$

It is not difficult to see that the matrices $a_{mn}(E)$ and $b_{mn}(E)$ are single-valued functions of the complex variable E. Indeed, the tilded functions $\tilde{F}_0$ and $\tilde{G}_0$ as well as their derivatives, $\tilde{F}_0'$ and $\tilde{G}_0'$, are single-valued and analytic with respect to E. The remaining factors involve α_{mn} and β_{mn}, which are proportional to the ratios k_n/p and k_n/q, respectively (see Eqs. (18.70–18.73)),

$$\alpha_{mn} \sim \frac{k_n}{p}\,, \qquad \beta_{mn} \sim \frac{k_n}{q}\,.$$

This makes the functions (18.108, 18.109) independent on the choices of the signs of the momenta k_n, p, and q and thus single-valued with respect to E.

The representation (18.107) is the particular (S-wave) case of the general expression (14.70).

* * *

A detailed analysis of the distribution of the spectral points generated by the potentials used in Sects. 18.2, 18.3, 18.4, and 18.5 is given in Ref. [15], where the movement of these points on the Riemann sheets is shown for a continuous change of the depths of the potential wells.

18.6 Bargmann Potentials

This section is devoted to a wide class of the potentials for which the exact solution of the corresponding S-wave Schrödinger equation can be found if the problem is solved in the inverse direction. This means that we start from postulating the Jost function and then try to reconstruct the potential that leads to such a Jost function. This is an example of the so-called inverse scattering problem, which is a separate branch of the scattering theory.

The main motivation for the development of the inverse scattering theory was the need for extracting the information on the forces (i.e. the potential) acting between the colliding particles, from the observed cross sections. Nowadays, it became a sophisticated mathematical theory where not only the methods for reconstructing the potential are based on the exact integral equations, but the uniqueness of such a reconstruction is rigorously analysed.

Despite the fact that the Jost functions are widely used in solving the inverse scattering problem, this theory is not described in the present book. The reason for the omission is mainly the huge volume of the material that would need to be included. In order to keep the focus on the Jost functions, this material would need to be significantly compressed here. Such a condensed presentation (without mathematical proofs and lengthy derivations) would be of little use, while there are many comprehensive books covering the inverse scattering theory in full. See, for example, Refs. [78–85].

It is however worthwhile to mention here the most general facts (theorems) established in the inverse scattering theory, concerning the uniqueness of the reconstruction of the potential. To avoid unnecessary complications, we assume that the potential is of a short range.

First of all, one should understand that the potential cannot be rigorously reconstructed directly from an experimentally measured scattering cross section. As a preliminary step, the so-called partial-wave analysis of the data is done. In other words, the cross section is decomposed in a sum of the contributions from the states with different values of the angular momentum. Then each partial cross section is parametrized via the corresponding scattering phase shift. As a result, the original data are represented by an equivalent set of the phase shifts. The partial-wave analysis is a rather complicated procedure and constitutes a separate field of research (a comprehensive description of it can be found in Ref. [18]). In the inverse scattering theory, it is simply assumed that the phase shifts are given.

There are two possible ways of deducing the potential from a given set of data:[2]

- At a fixed ℓ, using the phase shift $\delta_\ell(E)$ given for all energies, $E \in [0, \infty)$. In this case, the potential can be uniquely reconstructed if it does not support any bound states or else if the energies of all bound states (for a given ℓ) and the corresponding asymptotic normalization constants are known. Otherwise, there is a possibility to reproduce a given phase shift function $\delta_\ell(E)$ using many completely different potentials. Such potentials are called phase-equivalent.
- At a fixed E, using the phase shifts $\delta_\ell(E)$ given for all angular momenta, $0 \leqslant \ell < \infty$. Similarly to the fixed-ℓ case, this way of reconstructing the potential may also give a non-unique result.

It should also be noted that even if a unique solution of the inverse scattering problem is found, there is no guarantee that one obtains the same potential when the input data (the phase shifts) are taken at a different value of ℓ (for a fixed ℓ approach) or at a different energy (for a fixed E approach).[3] This fact diminishes the significance of the inverse scattering theory for solving the problem from which it originated, i.e. for deducing the interaction potential from experimental data. Practically, all the potentials that are actually used in describing real physical systems were constructed in the simple way of trial and error. Sometimes the inverse scattering theory can help in choosing the initial guess for such a trial-and-error procedure, but it almost never can produce a reliable potential by itself.

---

[2] It is clear that in practical calculations the contributions from very high energies and angular momenta can be ignored because they quickly diminish when either E or ℓ grows.

[3] This becomes obvious if the reader recollects the discussion in Sect. 17.2 concerning non-locality of the interaction between composed bodies, as well as the fact that a non-local potential is equivalent to an energy-dependent potential (see Sect. 17.2.2).

This however does not mean that the inverse scattering theory is useless. Quite contrary! First of all, as any other rigorous mathematical theory, it has value by itself. Besides that, it sheds new light on whole quantum mechanics by giving us a deeper understanding of the relationships among the observable quantities and the underlying interactions.

There is another very wide field of research where the methods developed in the inverse scattering theory turned out to be not only useful but very important and in some cases even indispensable. This is the theory of nonlinear differential equations in general and the theory of solitary waves (solitons) in particular. This theory is far outside the scope of the present book. There are many other publications devoted to this subject. See, for example, Refs. [86, 87].

The potentials that are described in the present section were suggested by Valentine Bargmann [89]. He was the first who demonstrated the existence of phase-equivalent potentials and found a systematic way of constructing them. Nowadays, the potentials of this type are constructed either using the integral equations of the inverse scattering theory with separable kernels (see Ref. [79]) or employing the modern approach based on the so-called super symmetric transformations where the Hamiltonian is factorized via the raising and lowering operators.[4] Despite the development of these modern approaches, the Bargmann potentials are constructed in the present book using the original method suggested in Ref. [89]. The reason is that such a simple (old style) presentation does not require additional lengthy explanations and derivations of either the integral equations or the super symmetric methods of the inverse scattering theory.

Similarly to the single-channel problems, the multi-channel potentials can also be phase-equivalent. A famous example of a family of such potentials was suggested by Joseph Cox [90]. The potentials named after him generalize the Bargmann potentials. In order to save space, the lengthy and tedious derivations of the Cox potentials are not given in the present book. An interested reader can find them elsewhere, for example, in Ref. [91].

18.6.1 "Linear" Type Bargmann Potentials

Let a particle of mass μ moves with the angular momentum $\ell = 0$ in the field of a single-channel spherically symmetric potential $U(r) = \dfrac{\hbar^2}{2\mu}V(r)$. Following Bargmann [89], we try to find such a potential that generates the S-wave Jost solution, $\chi_0^{(+)}$, which is factorized as follows:

$$\chi_0^{(+)}(E, r) = h_0^{(+)}(kr)\frac{\theta(E, r)}{\theta(E, \infty)}, \tag{18.110}$$

[4] An introduction to super symmetry in quantum mechanics (sometimes denoted as SUSY) can be found in Ref. [88].

where $h_0^{(+)}(kr) = -i\exp(ikr)$ and the unknown function $\theta(E,r)$ has a finite and non-zero limit at infinity,

$$\theta(E,\infty) = \lim_{r\to\infty} \theta(E,r)\,. \tag{18.111}$$

It is clear that (independently of the value of this limit) such a factorization gives the correct asymptotic behaviour of the Jost solution, defined by Eq. (2.67). Let us look for the function $\theta(E,r)$ in the following special form:

$$\theta(E,r) = k + i\zeta(r)\,, \tag{18.112}$$

i.e. as being linearly dependent on the wave number k, where $\zeta(r)$ is some unknown function of the coordinate.

The value of the S-wave Jost solution at $r = 0$ gives the corresponding Jost function (see Eq. (2.79)),

$$f_0^{(\text{in})}(E) = \frac{i}{2}\chi_0^{(+)}(E,0) = \frac{1}{2}\cdot\frac{\theta(E,0)}{\theta(E,\infty)} = \frac{1}{2}\cdot\frac{k+i\zeta(0)}{k+i\zeta(\infty)}\,. \tag{18.113}$$

As we see, the linear ansatz (18.112) is equivalent to the assumption that the Jost function is rational and has one zero and one pole on the k-plane. If in Eq. (18.112), instead of linear function, we used a polynomial of the order N, then the corresponding Jost function would also be rational, but with N zeros and N poles.[5] By choosing the degree N, we can in principle fix the number of the spectral points, i.e. the number of zeros of the function $f_0^{(\text{in})}(E)$. The zeros of $\theta(E,\infty)$, i.e. the singularities of $f_0^{(\text{in})}(E)$, are the points where the Jost function does not exist.

By definition, the Jost solution (18.110) obeys the radial Schrödinger equation,

$$\left[\partial_r^2 + k^2 - V(r)\right]\chi_0^{(+)}(E,r) = 0\,. \tag{18.114}$$

Substituting the factorized representation (18.110) into this equation, we find that

$$\theta'' + 2ik\theta' = V\theta\,, \tag{18.115}$$

where the "prime" denotes the derivative with respect to r. Replacing here the function $\theta(E,r)$ with the linear ansatz (18.112), we have

$$i\zeta''(r) - 2k\zeta'(r) = V(r)k + iV(r)\zeta(r)\,. \tag{18.116}$$

[5] An interested reader can find the description of the quadratic case ($N = 2$) in Ref. [89].

In this equation, there are terms depending only on r and those depending on both r and k. Since these two variables can vary independently, the equality can be maintained for all r and k only if

$$\zeta'' = V\zeta \qquad \text{and} \qquad V = -2\zeta' . \tag{18.117}$$

Substituting the second of these equations into the first one, we obtain

$$\zeta'' = -2\zeta'\zeta . \tag{18.118}$$

Since the potential V is assumed to be real, from the second equation of (18.117), it follows that ζ' is also real. The derivative ζ'' of a real function is real as well. Then Eq. (18.118) implies that ζ is real.

Integrating both sides of Eq. (18.118), we have

$$\zeta' = -\zeta^2 + c , \tag{18.119}$$

where c is the constant of integration. This is a nonlinear differential equation. In order to solve it, we introduce new unknown function,

$$\xi(r) = \exp\left[\int_0^r \zeta(t)\, dt\right] , \tag{18.120}$$

which is real and positive, $\xi(r) > 0$, for all r. Its derivative is

$$\xi' = \zeta\xi , \tag{18.121}$$

which gives

$$\zeta = \frac{\xi'}{\xi} \tag{18.122}$$

and thus

$$V = -2\zeta' = -2\,\frac{\xi''\xi - \xi'^2}{\xi^2} . \tag{18.123}$$

If we find the function $\xi(r)$ and its first two derivatives, then the potential that we are looking for can be easily obtained. Therefore, we need a differential equation that determines this function. Such an equation follows from Eq. (18.119):

$$\frac{\xi''\xi - \xi'^2}{\xi^2} = -\frac{\xi'^2}{\xi^2} + c \qquad \Rightarrow$$

$$\Rightarrow \qquad \xi'' - c\xi = 0 . \tag{18.124}$$

In contrast to Eq. (18.118), the corresponding equation (18.124) for the function $\xi(r)$ is linear. This is the reason why this new unknown function was introduced via the relation (18.120).

The reader should note that we do not fix any boundary conditions for this second-order differential equation. This means that by finding the general solution of Eq. (18.124), which depends on two arbitrary parameters, we actually find a whole family of the potentials. Each member of this family corresponds to a particular choice of these two parameters. All such potentials correspond to a rational Jost function with a single zero and a single pole. Of course, the positions of these poles depend on the parameters, and not all the potentials of the family are phase-equivalent.

Looking at Eq. (18.124), we see that its two linearly independent solutions can be chosen in the form of the exponential functions $\exp(\pm i\sqrt{-c}r)$. Do we have any restrictions on possible values of the constant c? Yes, we do. As it follows from Eqs. (18.123) and (18.119), we have

$$V = 2\zeta^2 - 2c\,.$$

This implies that $c = \zeta^2(\infty)$ because $V(\infty) = 0$. Since ζ is real, this means that $c \geqslant 0$. The two different cases, namely, $c = 0$ and $c > 0$, give two different types of the potential V.

18.6.1.1 Potentials with $\sim r^{-2}$ Tail

If $c = 0$, then Eq. (18.124) takes the form

$$\xi'' = 0, \tag{18.125}$$

and its general solution is the linear function:

$$\xi(r) = \alpha r + \beta\,, \tag{18.126}$$

where α and β are some real constants (parameters of the potential). Actually they cannot be negative because the function (18.120) is positive. Substituting (18.126) into Eq. (18.123), we find the potential,

$$V(r) = \frac{2\alpha^2}{(\alpha r + \beta)^2}\,. \tag{18.127}$$

If we exclude the trivial identically zero potential, then $\alpha \neq 0$. Since we are considering the S-wave motion, the case $\beta = 0$ is also excluded because it gives the centrifugal potential, $2/r^2$. Dividing the numerator and denominator on the right hand side of Eq. (18.127) by α^2, we see that in fact this is a family of the potentials depending on a single parameter $\gamma = \beta/\alpha > 0$,

$$\boxed{V(r) = \frac{2}{(r+\gamma)^2}} \qquad \gamma > 0\,. \tag{18.128}$$

This potential is completely repulsive and therefore cannot support bound states. However, according to our initial ansatz, the Jost function should have one zero and one pole. Let us find out where these points are.

The Jost function is given by Eq. (18.113) that involves the values $\zeta(0)$ and $\zeta(\infty)$. These values can be found using Eqs. (18.122) and (18.126), which gives

$$\zeta(r) = \frac{\alpha}{\alpha r + \beta} = \frac{1}{r+\gamma} \qquad \Rightarrow \qquad \zeta(0) = \frac{1}{\gamma}\,, \quad \zeta(\infty) = 0\,.$$

Therefore, the potential (18.128) corresponds to the following S-wave Jost function:

$$\boxed{f_0^{(\mathrm{in})}(E) = \frac{1}{2}\left(1 + \frac{i}{\gamma k}\right)} \tag{18.129}$$

It has zero at $k = -i/\gamma$. This means that the potential (18.128) generates a virtual state. The Jost function (18.129) is not defined at a single point, $k = 0$, and, as it should be, tends to the constant value, $1/2$, in all directions on the complex k-plane (see Fig. 7.14).

In order to find the phase shift, we use the values of $\mathrm{Re}\, f_0^{(\mathrm{in})} = 1/2$ and $\mathrm{Im}\, f_0^{(\mathrm{in})} = 1/(2\gamma k)$ for real positive k. Since $f_0^{(\mathrm{in})} = \left|f_0^{(\mathrm{in})}\right| \exp(-i\delta_0)$, we obtain the following S-wave phase shift generated by the potential (18.128):

$$\delta_0(E) = -\arctan\frac{\mathrm{Im}\, f_0^{(\mathrm{in})}}{\mathrm{Re}\, f_0^{(\mathrm{in})}} = -\arctan\frac{1}{\gamma k}\,. \tag{18.130}$$

As is known in the inverse scattering theory, a potential that does not support bound states with a given ℓ is uniquely determined by a given phase shift function $\delta_\ell(E)$. In other words, there are no phase-equivalent potentials corresponding to the phase shift (18.130).

* * *

Analytic Structure

Once again we recollect that the S-wave single-channel Jost function (18.129) should have the following analytic structure (see Eq. (7.65)):

$$f_0^{(\mathrm{in})}(E) = a_0(E) - ikb_0(E) \tag{18.131}$$

with $a_0(E)$ and $b_0(E)$ being single-valued analytic functions of E for all values of the energy variable where the Jost function exists. It is easy to see that this is so indeed. To make it obvious, we divide and multiply the second term in Eq. (18.129) by $\hbar^2 k/(2\mu)$, which gives

$$f_0^{(\mathrm{in})}(E) = \frac{1}{2} + ik\frac{\hbar^2}{4\mu\gamma E} \tag{18.132}$$

with $a_0(E) = 1/2$ and $b_0(E) = -\hbar^2/(4\mu\gamma E)$.

18.6.1.2 Eckart Potential

If in the differential equation (18.124) the arbitrary constant c is non-zero (it can only be positive), then the general solution of this equation is a linear combination of the functions $\exp(\pm|\sqrt{c}\,|r)$. Denoting

$$|\sqrt{c}\,| = \frac{\lambda}{2}\,, \qquad \lambda > 0\,,$$

we can write such a combination as

$$\xi(r) = \alpha\left(e^{\lambda r/2} + \beta e^{-\lambda r/2}\right)\,. \tag{18.133}$$

The definition (18.120) of the function $\xi(r)$ implies that it is positive for all r and $\xi(0) = 1$. Because of the condition $\xi(0) = 1$, the two parameters, α and β, are not independent. Indeed,

$$\xi(0) = 1 \qquad \Rightarrow \qquad \alpha = \frac{1}{1+\beta}\,.$$

Since at large values of r the first term in the combination (18.133) dominates, the function $\xi(r)$ can only remain positive when $r \to \infty$, if $\alpha > 0$, i.e. if $\beta > -1$.

Substituting the function $\xi(r)$ and its first two derivatives,

$$\xi' = \frac{\alpha\lambda}{2}\left(e^{\lambda r/2} - \beta e^{-\lambda r/2}\right)\,,$$

$$\xi'' = \frac{\alpha\lambda^2}{4}\left(e^{\lambda r/2} + \beta e^{-\lambda r/2}\right) = \frac{\lambda^2}{4}\xi\,,$$

in Eq. (18.123), we obtain the potential:

$$\begin{aligned}
V(r) &= -\frac{2}{\xi^2}\left[\frac{\lambda^2}{4}\xi^2 - \frac{\alpha^2\lambda^2}{4}\left(e^{\lambda r/2} - \beta e^{-\lambda r/2}\right)^2\right] \\
&= -\frac{\lambda^2}{2} + \frac{\lambda^2}{2}\cdot\frac{\left(e^{\lambda r/2} - \beta e^{-\lambda r/2}\right)^2}{\left(e^{\lambda r/2} + \beta e^{-\lambda r/2}\right)^2} \\
&= -\frac{\lambda^2}{2}\left[1 - \frac{\left(1 - \beta e^{-\lambda r}\right)^2}{\left(1 + \beta e^{-\lambda r}\right)^2}\right] = \boxed{-2\lambda^2\beta\frac{e^{-\lambda r}}{\left(1 + \beta e^{-\lambda r}\right)^2}}
\end{aligned} \tag{18.134}$$

$$\lambda > 0\,, \qquad \beta > -1\,.$$

The Schrödinger equation for a potential of this kind was solved (using the hypergeometric functions) in 1930 by Carl Eckart [92]. Since then, it bears his name.

The potential (18.134) can be either repulsive (if $\beta < 0$) or attractive (if $\beta > 0$). By construction, it can support only one bound state. In order to find the corresponding energy, we need to obtain the explicit expression for the Jost function. This is easily done using Eqs. (18.113), (18.122), and (18.133). With

$$\zeta(0) = \frac{\lambda}{2}\cdot\frac{1-\beta}{1+\beta} \qquad \text{and} \qquad \zeta(\infty) = \frac{\lambda}{2},$$

we obtain the following expression:

$$\boxed{f_0^{(\mathrm{in})}(E) = \frac{1}{2}\left(k + i\frac{\lambda}{2}\cdot\frac{1-\beta}{1+\beta}\right)\left(k + i\frac{\lambda}{2}\right)^{-1}} \tag{18.135}$$

It is seen that if $\beta > 1$, then this Jost function has zero at $k = i\kappa$ ($\kappa > 0$) on the positive half of the imaginary axis of the k-plane, where

$$\kappa = \frac{\lambda}{2}\cdot\frac{\beta-1}{\beta+1}, \qquad (\beta > 1 \Rightarrow \kappa > 0)\ . \tag{18.136}$$

The corresponding energy of the bound state,

$$E_b = -\frac{\hbar^2\kappa^2}{2\mu} = -\frac{\hbar^2\lambda^2}{8\mu}\left(\frac{\beta-1}{\beta+1}\right)^2, \tag{18.137}$$

is uniquely determined by the two parameters, λ and β. Another quantity that is also determined by these parameters is the ANC of the corresponding wave function. As it is given by Eq. (4.55), the ANC, $\mathcal{A}_0$, can be expressed via the residue, Q_0, of the S-matrix at the spectral point $k = i\kappa$. The S-matrix involves the Jost function $f_0^{(\mathrm{out})}$, which in accordance with the diagonal symmetry shown in Fig. 3.5, can be obtained from $f_0^{(\mathrm{in})}$ by changing sign of the momentum, $k \to -k$. Therefore,

$$\begin{aligned} S_0 &= \frac{f_0^{(\mathrm{out})}}{f_0^{(\mathrm{in})}} = \frac{(-k - i\kappa)(k + i\lambda/2)}{(k - i\kappa)(-k + i\lambda/2)} \\ &\underset{k\to i\kappa}{\longrightarrow} \frac{2i\kappa(\kappa + \lambda/2)}{(k - i\kappa)(\kappa - \lambda/2)} \qquad \Rightarrow \qquad Q_0 = \frac{2i\kappa(\kappa + \lambda/2)}{\kappa - \lambda/2}\ . \end{aligned}$$

This gives the ANC[6] (see Eq. (4.55)):

$$\mathcal{A}_0^2 = \frac{2\kappa(\kappa + \lambda/2)}{\kappa - \lambda/2} = 2\lambda\beta\frac{\beta - 1}{\beta + 1} . \tag{18.138}$$

Therefore, there is one to one correspondence between the pairs of the parameters:

$$(E_b, \mathcal{A}_0) \quad \Leftrightarrow \quad (\lambda, \beta) .$$

The phase shift function, $\delta_0(E)$, is given by Eq. (18.130). For the real positive momenta, $k > 0$, the required real and imaginary parts of the Jost function (18.135) are

$$\mathrm{Re}\, f_0^{(\mathrm{in})} = \frac{1}{2(4k^2 + \lambda^2)}\left(4k^2 + \lambda^2\frac{1-\beta}{1+\beta}\right) ,$$

$$\mathrm{Im}\, f_0^{(\mathrm{in})} = -\frac{1}{4k^2 + \lambda^2} \cdot \frac{2\lambda\beta k}{1+\beta} .$$

Therefore,

$$\delta_0(E) = \arctan\frac{4\lambda\beta k}{4k^2(1+\beta) + \lambda^2(1-\beta)} . \tag{18.139}$$

It is seen that $\delta_0(0) = 0$ and $\delta_0(\infty) = 0$. This is as it should be if no bound states exist. However, if the potential supports one bound state, then, according to Levinson's theorem, we should have $\delta_0(0) = \pi$.

In our case, a bound state appears if $\beta > 1$. With such values of the parameter β, the denominator on the right hand side of Eq. (18.139) has zero somewhere at $k > 0$. As a result, the argument of the arctan-function at this point jumps from $-\infty$ to $+\infty$, and therefore, δ_0 jumps from $-\pi/2$ to $+\pi/2$. The phase shift, however, must be continuous. To overcome this contradiction, we use the fact that arctan is a multi-valued function with an infinite number of branches that differ from each other by an integer number of π's:

$$\tan(a + \pi n) = x \quad \Rightarrow \quad \arctan x = a + \pi n\,, \quad n = 0, \pm 1, \pm 2, \ldots ,$$

where $-\pi/2 \leqslant a \leqslant +\pi/2$ is the principal value of the arctan-function. In order to keep the phase shift continuous, we have to assume that, when passing the discontinuity point, we go from one branch of the arctan-function to the other. In

[6] The sign of $\mathcal{A}_\ell$ is a matter of convention because the sign of the wave function can be chosen arbitrarily.

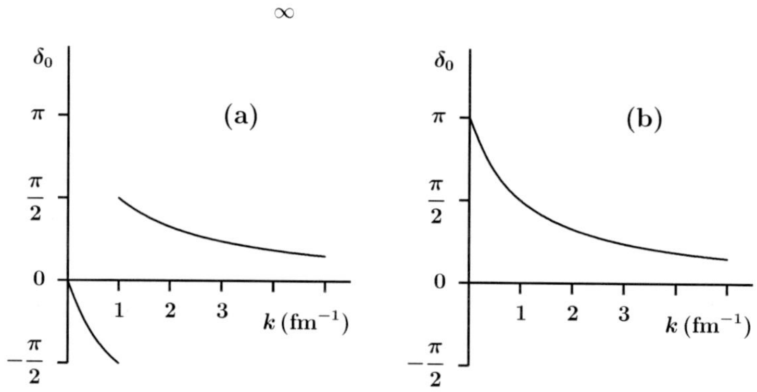

Fig. 18.2 S-wave phase shift (18.139) for the Eckart potential (18.134) with the parameters $\lambda = 4\,\text{fm}^{-1}$ and $\beta = 5/3$. Removing the discontinuity of δ_0 at $k = 1\,\text{fm}^{-1}$ (left part of the figure) results in the threshold value $\delta_0(0) = \pi$ (right part of the figure) in accordance with Levinson's theorem

other words, on the left or right hand side of such a point, we add or subtract π to or from the phase shift function.

Such a procedure of restoring the continuity as well as reconciling the threshold behaviour of the phase shift with the requirement of Levinson's theorem is illustrated in Fig. 18.2 for the Eckart potential (18.134) with the parameters $\lambda = 4\,\text{fm}^{-1}$ and $\beta = 5/3$. This choice of the parameters results in the discontinuity of the phase shift at $k = 1\,\text{fm}^{-1}$, as is shown in Fig. 18.2a. After adding π to the left hand side of the curve, we remove the discontinuity and obtain the correct value of δ_0 at zero energy at the same time, which is shown in Fig. 18.2b.

At this point, an attentive reader should have a question of the kind: is it possible to reproduce the phase shift shown in Fig. 18.2b with different potentials? In other words, can we construct the phase-equivalent potentials in this case? Generally speaking, the answer should be "yes" because here we deal with a physical system that has a bound state, and therefore (as it was mentioned before), such an answer follows from the rigorous inverse scattering theory.

However, looking at Eq. (18.139), it is not difficult to see that the curve shown in Fig. 18.2b corresponds to a unique potential (18.134). Indeed, such a curve is uniquely determined by the two constants, c_1 and c_2, defined via the following equation:

$$\delta_0(E) = \arctan \frac{k}{c_1 k^2 + c_2}\,, \tag{18.140}$$

where $c_1 > 0$ and $c_2 < 0$. Comparing (18.140) with Eq. (18.139), we see that

$$c_1 = \frac{1 + \beta}{\lambda\beta}\,, \tag{18.141}$$

$$c_2 = \frac{\lambda}{4\beta}(1 - \beta) . \tag{18.142}$$

Multiplying these two equations, we find how β can be obtained when c_1 and c_2 are given,

$$\beta = \pm \frac{1}{\sqrt{1 + 4c_1 c_2}} . \tag{18.143}$$

As we have seen, a bound state can only be supported if $\beta > 1$. Therefore, we can only choose the positive sign in front of the square root in Eq. (18.143). This means that β is uniquely determined by c_1 and c_2. As it follows from Eq. (18.141), the other parameter of the potential, λ, is also uniquely determined by them.

It seems that this conclusion contradicts the following general rule known in the inverse scattering theory: if there are N bound states in a particular partial wave, then there is always an N-parametric family of phase-equivalent potentials that generate exactly the same phase shift curve in that partial wave. There is, however, an explanation of such a puzzle. The reason of the contradiction is that when obtaining β and λ from a given phase shift curve (i.e. from c_1 and c_2), we are trying to reproduce the phase shift using a specific functional form of the potential, namely, the potential parametrized as is given by Eq. (18.134). If our search of the potential were more "flexible", i.e. not restricted to the simple form (18.134), we could find different potentials corresponding to the same phase shift.

Such a family of one-parametric phase-equivalent potentials that generate the phase shift (18.139) was found in Ref. [91]. This family, depending on the parameter γ, has rather complicated functional form:

$$V(r) = 2(\kappa^2 - \kappa_0^2)\frac{\dfrac{\kappa_0^2}{\sinh^2(\kappa_0 r)} + \dfrac{4\gamma\kappa^2}{\left[\exp(\kappa r) + \gamma\exp(-\kappa r)\right]^2}}{\left[\kappa_0\coth(\kappa_0 r) - \kappa\dfrac{\exp(\kappa r) - \gamma\exp(-\kappa r)}{\exp(\kappa r) + \gamma\exp(-\kappa r)}\right]^2} , \tag{18.144}$$

where κ is given by Eq. (18.136) and $\kappa_0 = \lambda/2$. The parameter γ can be chosen arbitrarily under the condition[7] $\gamma > -1$. Each choice of this parameter gives a different potential, but all of them correspond to the same γ-independent phase shift function (18.139). It is not difficult to check that our "simple" potential (18.134) is a particular case of (18.144) for $\gamma = 0$.

The potential (18.144) is obtained in Ref. [91] using a sequence of two super symmetric (SUSY) transformations applied to the identical zero potential. Since SUSY is outside the scope of the present book, we skip the derivations. It is not

[7] Actually, in Ref. [91], the range of possible values of γ is more wide. It also includes $\gamma < -1$, but for $\kappa_0 < \kappa$, which does not correspond to our choice of the ranges of β and λ.

clear how such complicated potential could be obtained in the traditional way, i.e. similarly to what we did for the Eckart potential (18.134). Most probably, it would be impossible (or, at least, extremely difficult) to find the potential (18.144) without using SUSY.

* * *

Analytic Structure

It is easy to check that the S-wave Jost function (18.135) has the analytic structure given by Eq. (7.65). Indeed, after simple algebraic transformations, this Jost function can be written as

$$f_0^{(\mathrm{in})}(E) = \frac{E - \dfrac{\hbar^2}{2\mu}\kappa\kappa_0}{2E + \dfrac{\hbar^2}{\mu}\kappa_0^2} - ik \cdot \frac{\dfrac{\hbar^2}{2\mu}(\kappa + \kappa_0)}{2E + \dfrac{\hbar^2}{\mu}\kappa_0^2}\,, \tag{18.145}$$

where both fractions are single-valued functions of the energy. This Jost function is analytic everywhere except for a single negative value, $E = -\hbar^2\kappa_0^2/(2\mu)$. As it follows from (18.135), the corresponding momentum has a negative imaginary part, and therefore, this point is on the non-physical sheet of the Riemann surface.

Partial-Wave Expansion

A

When doing various derivations involving decomposition of quantum states in partial waves, it is very helpful to use formal relations that are listed below.[1] Here $|\vec{k}\,\rangle$ denotes the state with definite momentum $\vec{k}$. When it is projected onto the state $\langle\vec{r}\,|$ with definite position in space, we obtain the plane wave. The states $|\vec{k}\,\rangle$ and $|\vec{r}\,\rangle$ can be expanded over the spherical harmonics. These expansions involve the states $|k\ell m\rangle$ and $|r\ell m\rangle$ with definite values of $k = |\vec{k}\,|$ and $r = |\vec{r}\,|$ as well as of the angular momentum ℓ and its third component m. There are many useful relations among these states, spherical harmonics, δ-functions, etc.:

$$\langle\vec{k}'\,|\vec{k}\,\rangle = \delta(\vec{k}' - \vec{k}\,)\,, \tag{A.1}$$

$$\langle\vec{r}'\,|\vec{r}\,\rangle = \delta(\vec{r}' - \vec{r}\,)\,, \tag{A.2}$$

$$\langle\vec{r}\,|\vec{k}\,\rangle = (2\pi)^{-3/2}\exp\left(i\vec{k}\vec{r}\right)\,, \tag{A.3}$$

$$\exp\left(i\vec{k}\vec{r}\right) = \frac{4\pi}{kr}\sum_{\ell m} i^{\ell} j_\ell(kr) Y^*_{\ell m}(\hat{\vec{k}}\,) Y_{\ell m}(\hat{\vec{r}}\,)\,, \tag{A.4}$$

$$j_\ell(z) = \sqrt{\frac{\pi z}{2}} J_{\ell+1/2}(z)\,, \qquad j_0(z) = \sin z\,, \tag{A.5}$$

$$\int_0^\infty j_\ell(kr) j_\ell(k'r)dr = \frac{\pi}{2}\delta(k' - k)\,, \tag{A.6}$$

[1] All the corresponding derivations can be found in Ref. [5].

S. A. Rakityansky, *Jost Functions in Quantum Mechanics*,
https://doi.org/10.1007/978-3-031-07761-6

$$|\vec{k}\,\rangle = \sum_{\ell m} |k\ell m\rangle Y^*_{\ell m}(\hat{\vec{k}}\,)\,, \tag{A.7}$$

$$|\vec{r}\,\rangle = \sum_{\ell m} |r\ell m\rangle Y^*_{\ell m}(\hat{\vec{r}}\,)\,, \tag{A.8}$$

$$\langle \vec{r}\,|\vec{k}\,\rangle = \sum_{\ell m \ell' m'} Y_{\ell m}(\hat{\vec{r}}\,)\langle r\ell m|k\ell' m'\rangle Y^*_{\ell' m'}(\hat{\vec{k}}\,)\,, \tag{A.9}$$

$$\langle r\ell m|k\ell' m'\rangle = \delta_{\ell\ell'}\delta_{mm'}\frac{1}{kr}\sqrt{\frac{2}{\pi}}\,i^\ell j_\ell(kr)\,, \tag{A.10}$$

$$\langle \vec{r}\,|k\ell m\rangle = \frac{1}{kr}\sqrt{\frac{2}{\pi}}\,i^\ell j_\ell(kr) Y_{\ell m}(\hat{\vec{r}}\,)\,, \tag{A.11}$$

$$\langle \vec{k}\,|r\ell m\rangle = \frac{1}{kr}\sqrt{\frac{2}{\pi}}(-i)^\ell j_\ell(kr) Y_{\ell m}(\hat{\vec{k}}\,)\,, \tag{A.12}$$

$$|\vec{r}\,\rangle = |r, \Omega_{\vec{r}}\rangle\,, \qquad |\vec{k}\,\rangle = |k, \Omega_{\vec{k}}\rangle\,, \tag{A.13}$$

$$\langle \Omega_{\vec{r}}\,|\ell m\rangle = Y_{\ell m}(\hat{\vec{r}}\,)\,, \qquad \langle \Omega_{\vec{k}}|\ell m\rangle = Y_{\ell m}(\hat{\vec{k}}\,)\,, \tag{A.14}$$

$$\langle r|k\rangle = \frac{1}{kr}\sqrt{\frac{2}{\pi}}\,i^\ell j_\ell(kr)\,, \tag{A.15}$$

$$\langle \vec{k}'\,|k\ell m\rangle = \int \langle \vec{k}'\,|\vec{r}\,\rangle\langle \vec{r}\,|k\ell m\rangle \mathrm{d}^3 r = \frac{\delta(k'-k)}{k^2} Y_{\ell m}(\hat{\vec{k}}'\,)\,, \tag{A.16}$$

$$\langle \vec{r}'\,|r\ell m\rangle = \int \langle \vec{r}'\,|\vec{k}\,\rangle\langle \vec{k}\,|r\ell m\rangle \mathrm{d}^3 k = \frac{\delta(r'-r)}{r^2} Y_{\ell m}(\hat{\vec{r}}'\,)\,, \tag{A.17}$$

$$\langle k'|k\rangle = \frac{\delta(k'-k)}{k^2}\,, \qquad \langle r'|r\rangle = \frac{\delta(r'-r)}{r^2}\,, \tag{A.18}$$

$$\langle k'\ell' m'|k\ell m\rangle = \delta_{\ell'\ell}\delta_{m'm}\frac{\delta(k'-k)}{k^2}\,, \tag{A.19}$$

$$\langle r'\ell'm'|r\ell m\rangle = \delta_{\ell'\ell}\delta_{m'm}\frac{\delta(r'-r)}{r^2}\ , \tag{A.20}$$

$$\int |\vec{k}\,\rangle\langle\vec{k}\,| \mathrm{d}^3k = 1\ , \tag{A.21}$$

$$\int |\vec{r}\,\rangle\langle\vec{r}\,| \mathrm{d}^3r = 1\ , \tag{A.22}$$

$$\sum_{\ell m}\int_0^\infty |k\ell m\rangle\langle k\ell m|\, k^2\, dk = 1\ , \tag{A.23}$$

$$\sum_{\ell m}\int_0^\infty |r\ell m\rangle\langle r\ell m|\, r^2\, dr = 1\ , \tag{A.24}$$

$$\langle k\ell m| = \int_0^\infty \langle k\ell m|r\ell m\rangle\langle r\ell m|\, r^2\, dr = \sqrt{\frac{2}{\pi}}i^{-\ell}\int_0^\infty \frac{j_\ell(kr)}{kr}\langle r\ell m|\, r^2\, dr\ , \tag{A.25}$$

$$\delta(\vec{x}-\vec{y}\,) = \frac{1}{x^2}\delta(x-y)\sum_{\ell m} Y^*_{\ell m}(\hat{\vec{x}}\,)Y_{\ell m}(\hat{\vec{y}}\,)\ , \tag{A.26}$$

$$\sum_{\ell m} Y^*_{\ell m}(\hat{\vec{x}}\,)Y_{\ell m}(\hat{\vec{y}}\,) = \delta(\Omega_{\vec{x}}-\Omega_{\vec{y}})\ , \tag{A.27}$$

$$\int Y_{\ell m}(\hat{\vec{r}}\,)Y^*_{\ell' m'}(\hat{\vec{r}}\,)d\Omega_{\vec{r}} = \delta_{\ell\ell'}\delta_{mm'}\ , \tag{A.28}$$

$$Y^*_{\ell m}(\hat{\vec{r}}\,) = (-1)^m Y_{\ell,-m}(\hat{\vec{r}}\,)\ , \tag{A.29}$$

$$\int Y_{\ell m}(\hat{\vec{r}}\,)Y_{\ell' m'}(\hat{\vec{r}}\,)d\Omega_{\vec{r}} = (-1)^m\delta_{\ell\ell'}\delta_{-mm'}\ , \tag{A.30}$$

$$\int Y_{\ell_1 m_1}(\hat{\vec{r}})Y_{\ell_2 m_2}(\hat{\vec{r}}\,)Y^*_{\ell_3 m_3}(\hat{\vec{r}}\,)d\Omega_{\vec{r}} = \sqrt{\frac{(2\ell_1+1)(2\ell_2+1)}{4\pi(2\ell_3+1)}}C^{\ell_3 0}_{\ell_1 0\ell_2 0}C^{\ell_3 m_3}_{\ell_1 m_1\ell_2 m_2}\ , \tag{A.31}$$

$$\int \exp(i\vec{k}\vec{r}\,)\frac{j_\ell(qr)}{qr}Y_{\ell m}(\hat{\vec{r}}\,)\mathrm{d}^3r = \frac{2\pi^2}{k^2}i^\ell\delta(k-q)Y_{\ell m}(\hat{\vec{k}}\,)\ , \tag{A.32}$$

$$\frac{\exp(ik|\vec{r}-\vec{r}\,'|)}{|\vec{r}-\vec{r}\,'|} = \frac{4\pi i}{krr'} \sum_{\ell m} j_\ell(kr_<) h_\ell^{(+)}(kr_>) Y_{\ell m}^*(\Omega_{\vec{r}_<}) Y_{\ell m}(\Omega_{\vec{r}_>}) \,, \qquad \text{(A.33)}$$

where $r_< = \min(r, r')$ and $r_> = \max(r, r')$.

Since the Legendre polynomials, $P_\ell(\cos\theta)$, are real, the complex conjugation in the so-called scalar product of two spherical harmonics can be interchanged:

$$\frac{2\ell+1}{4\pi} P_\ell\left[\cos(\widehat{\vec{r},\vec{r}\,'})\right] = \sum_m Y_{\ell m}^*(\Omega_{\vec{r}\,'}) Y_{\ell m}(\Omega_{\vec{r}}) = \sum_m Y_{\ell m}(\Omega_{\vec{r}\,'}) Y_{\ell m}^*(\Omega_{\vec{r}}) \,. \qquad \text{(A.34)}$$

Basics of Complex Analysis

B

Complex numbers, z, are represented by the points of the complex plane with the Cartesian coordinates $x = \mathrm{Re}\, z$ and $y = \mathrm{Im}\, z$. Alternatively, the same points can be specified using the polar coordinates,

$$z = |z|e^{i\,\arg(z)}\ , \tag{B.1}$$

where the polar angle,

$$\arg(z) = \varphi + 2\pi n\ , \qquad n = 0, \pm 1, \pm 2, \dots\ , \tag{B.2}$$

is not uniquely defined. The same point is reached with any number of additional full circles around the origin (see Fig. B.1).

A complex function $f(z)$ of a complex variable[1] $z = x + iy$ can be presented as the sum of two real functions, u and v, depending on two real variables, x and y,

$$f(z) = u(x, y) + iv(x, y).$$

It is said that $f(z)$ has the derivative, $f'(z)$, at the point z, if the limit

$$f'(z) = \lim_{\Delta z \to 0} \frac{f(z + \Delta z) - f(z)}{\Delta z}$$

exists and does not depend on the direction on the complex z-plane from which the point $z + \Delta z$ approaches z. The independence of $f'(z)$ on Δz can be expressed in

[1] There are many textbooks on complex analysis where all the theorems and statements given here are proved. See, for example, Refs. [93, 94].

S. A. Rakityansky, *Jost Functions in Quantum Mechanics*,
https://doi.org/10.1007/978-3-031-07761-6

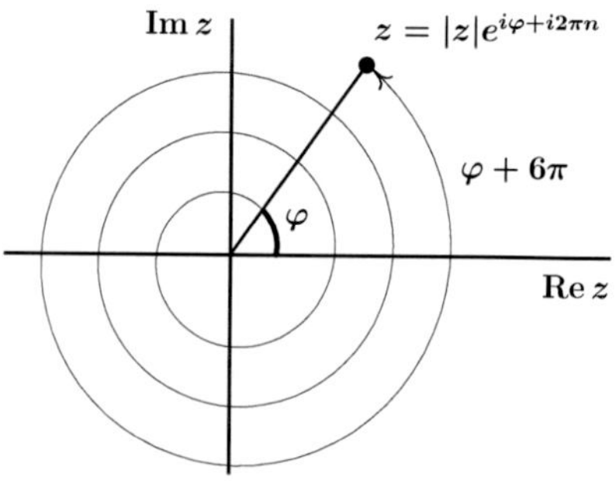

Fig. B.1 Polar representation of a complex number z with three additional full circles around the origin

the form of the Cauchy–Riemann conditions on the components of $f(z)$:

$$\frac{\partial u}{\partial x} = \frac{\partial v}{\partial y} \quad \text{and} \quad \frac{\partial v}{\partial x} = -\frac{\partial u}{\partial y}\,. \tag{B.3}$$

If $f(z)$ has the derivative for all z in a domain D of the complex plane, it is called analytic (or holomorphic) in that domain. It can be shown that an analytic function has the derivatives of any order within the domain of its analyticity, i.e. the existence of just the first derivative implies that all the derivatives do exist as well,

$$\boxed{\text{if} f'(z) \quad \text{exists}} \Rightarrow \boxed{f''(z),\ f'''(z),\ \ldots \quad \text{also exist}} \tag{B.4}$$

This means that an analytic function can be expanded in the Taylor series around an arbitrary point $z_0 \in D$,

$$f(z) = f(z_0) + \frac{f'(z_0)}{1!}(z - z_0) + \frac{f''(z_0)}{2!}(z - z_0)^2 + \cdots\,. \tag{B.5}$$

In other words, the analyticity is equivalent to the possibility of expanding the function in Taylor series. This series is convergent within a circle whose radius is equal to the distance from z_0 to the nearest singularity of the function $f(z)$, as is shown in Fig. B.2.

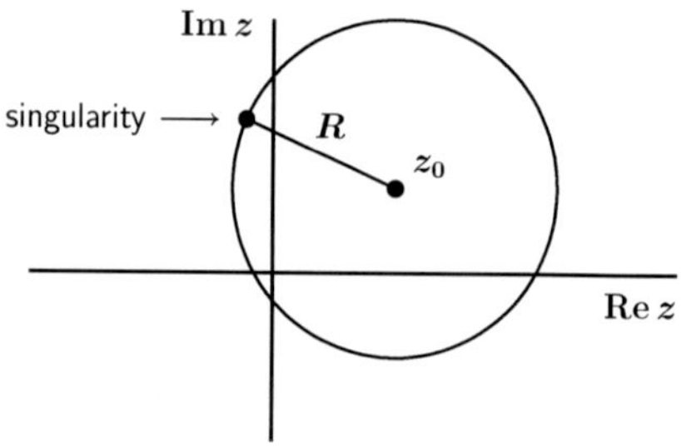

Fig. B.2 Circle of convergence of the Taylor series for the function $f(z)$ whose nearest singularity is at a distance R from the center of expansion z_0

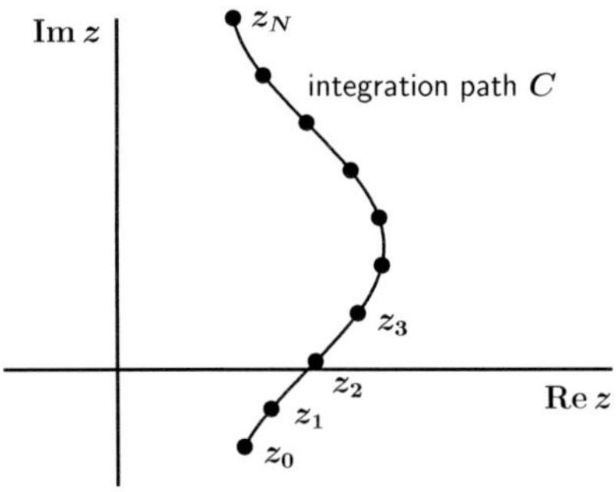

Fig. B.3 The path C and some discrete points z_n that are used to define the integral of a complex function

The integral of a complex function along a curve in the z-plane is a generalization of the real integral. It is defined as the limit

$$\int_C f(z)\, dz = \lim_{N\to\infty} \sum_{n=1}^{N} f(\tilde{z}_n)(z_n - z_{n-1})$$

where $\tilde{z}_n$ is any point on the straight line connecting z_{n-1} and z_n, and C denotes the integration path shown in Fig. B.3. For an analytic function, the integral has the same value for any path between the same points as long as the path is within the analyticity domain.

The integral along a closed contour is denoted by the symbol $\oint$. By convention, the positive direction of a closed contour integral is the anti-clockwise one. If $f(z)$ is analytic in the domain D including the contour C that encloses that domain, then the integral of $f(z)$ over whole contour is zero (the Cauchy theorem),

$$\oint_C f(z)\, dz = 0\,. \tag{B.6}$$

There is an important corollary of this theorem, which is known as Cauchy's integral formula. Let $f(z)$ be analytic in the domain D and on its border C, then

$$f(z_0) = \frac{1}{2\pi i} \oint_c \frac{f(z)}{z - z_0} dz\,. \tag{B.7}$$

This formula allows us to calculate the value of $f(z)$ at any point z_0 inside the domain, if we know its values on the border. Cauchy's integral formula can be generalized for the derivatives of an analytic function,

$$f^{(n)}(z_0) = \frac{n!}{2\pi i} \oint_c \frac{f(z)}{(z - z_0)^{n+1}} dz \tag{B.8}$$

This can be viewed not only as a useful formula but also as a proof that an analytic function has derivatives of any order because it gives a simple recipe how to find any derivative in the domain of analyticity.

Using Cauchy's integral formula, it can be proved Liouville's theorem:

If $f(z)$ is analytic on the whole complex plane and its absolute value is uniformly limited, i.e. $|f(z)| < M$ for all z, then $f(z)$ is constant.

(B.9)

For the analytic continuation of complex functions as well as for extending the experimental information from real to complex energies, it is important the so-called identity theorem:

If two functions $f(z)$ and $g(z)$ are analytic in a domain D and they coincide on a segment of a curve C belonging to D, then they are identical everywhere in D.

(B.10)

The function $f(z)$ is said to have a pole of the order n at the point z_0, if in the immediate vicinity of that point it can be written as

$$f(z) \underset{z \to z_0}{\longrightarrow} \frac{g(z)}{(z - z_0)^n}, \tag{B.11}$$

where $g(z)$ is analytic at z_0. Alternatively, we can define a pole of the order n as such a point where the limit

$$\lim_{z \to z_0} (z - z_0)^n f(z) = a \tag{B.12}$$

exists and is a finite non-zero complex number. The function that is analytic everywhere in D except for some isolated points, where it has poles, is called meromorphic.

A function that has a pole at z_0, is obviously not analytic there and thus cannot be expanded in Taylor series around that point. However, we can expand the function

$$g(z) = (z - z_0)^n f(z),$$

which is analytic. After obtaining such an expansion, we can divide it by $(z - z_0)^n$. Therefore

$$g(z) = b_0 + b_1(z - z_0) + b_2(z - z_0)^2 + \cdots,$$

$$f(z) = \frac{b_0}{(z-z_0)^n} + \frac{b_1}{(z - z_0)^{n-1}} + \frac{b_2}{(z-z_0)^{n-2}} + \cdots + b_n + b_{n+1}(z-z_0) + \cdots.$$

If we denote $a_k = b_{k+n}$, then

$$f(z) = \sum_{k=-n}^{\infty} a_k(z - z_0)^k. \tag{B.13}$$

This is a generalization of Taylor series for the case when there is a pole at z_0. This generalized series is called Laurent series. It includes both the negative and positive powers of $(z - z_0)$. The highest negative power corresponds to the order of the singularity of $f(z)$. The terms with the negative powers of $(z - z_0)$ in the Laurent series are collectively called the principal part, and all the other terms constitute the analytic part of the function,

$$f(z) = \underbrace{\sum_{k=-n}^{-1} a_k(z - z_0)^k}_{\text{principal part}} + \underbrace{\sum_{k=0}^{\infty} a_k(z - z_0)^k}_{\text{analytic part}}. \tag{B.14}$$

If we calculate the integral of this function along the closed contour around the point z_0 and choose the contour small enough such that no other singularities are included inside it, then it can be shown that among all the terms of the Laurent series, only one term gives a non-zero contribution, namely the term with the number $k = -1$,

$$\oint f(z)\,dz = 2\pi i a_{-1}.$$

The Laurent coefficient a_{-1} is called the residue of the function $f(z)$ at its pole z_0. The name originates from the fact that this coefficient is the only one which remains after the integration. There is a special notation for the residues,

$$\text{Res}[f, z_0] = a_{-1}.$$

If at the point z_0 the function $f(z)$ has a pole of the order m, then the residue can be found as the following limit:

$$\boxed{\text{Res}[f, z_0] = \lim_{z \to z_0} \left\{ \frac{1}{(m-1)!} \frac{d^{m-1}}{dz^{m-1}} \left[(z - z_0)^m f(z) \right] \right\}} \tag{B.15}$$

The most often used cases are

- $m = 1$ (simple pole)

$$\boxed{\text{Res}[f, z_0] = \lim_{z \to z_0} (z - z_0) f(z)} \tag{B.16}$$

- $m = 2$ (second order pole)

$$\boxed{\text{Res}[f, z_0] = \lim_{z \to z_0} \left\{ \frac{d}{dz} \left[(z - z_0)^2 f(z) \right] \right\}} \tag{B.17}$$

The most important application of the notion or residue stems from the residue theorem which states that a closed contour integral of a meromorphic function is equal to the sum of its residues at the enclosed poles, multiplied by $2\pi i$, i.e.

$$\boxed{\oint f(z)\, dz = 2\pi i \sum_{n=1}^{N} \text{Res}[f, z_n]} \tag{B.18}$$

where the sum is taken over all poles enclosed by the integration contour.

When examining the symmetry properties of the Jost functions, we use the so-called Schwartz reflection principle, which reads

> If the function $f(z)$ is analytic in a domain D that includes a line segment L and $f(z)$ is real on L, then this function can be analytically continued across L into the domain $\overline{D}$, which is a mirror reflection of D relative to the line L, and the values of $f(z)$ at any two points symmetric with respect to L are complex conjugates of each other.

(B.19)

Perhaps it is easier to look at Fig. B.4, from where the meaning of this principle becomes clear.

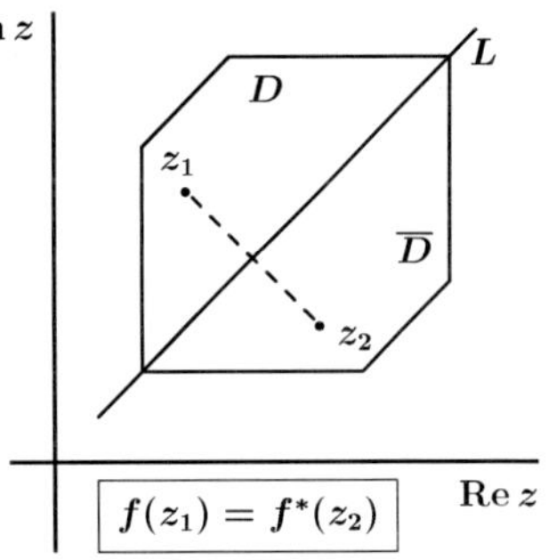

Fig. B.4 Illustration of the Schwartz reflection principle

Yet another theorem that we need in analyzing the properties of the Jost functions states

> If $\varphi(z, t)$ is an analytic function of z for any t, both φ and its derivative $\partial\varphi/\partial z$ are continuous functions of both z and t, and the integral $\int_a^\infty \varphi(z, t)dt$ converges uniformly, then this integral is an analytic function of z.

(B.20)

Of course, there is a rigorous proof of this statement, but it is not difficult to understand the theorem intuitively. Indeed, an integral is a sum. What is stated here can be formulated (in the oversimplified way) as: a convergent infinite sum of analytic functions is also analytic.

Complex functions are generalizations of the real functions or rather analytic extensions of them from the "narrow railway" of the real axis to the "vast fields" of the nearby plain. Their properties are therefore more rich and diverse. In particular, many of them are multi-valued. This means that at a single z, the function $f(z)$ may have several (or even infinitely many) different values. For example, the function

$$f(z) = z^{1/m} = \left[|z|e^{i\arg(z)}\right]^{1/m} = |z|^{1/m} \exp\left(i\frac{\varphi}{m} + i\frac{2\pi n}{m}\right) \tag{B.21}$$

has m different values. When n runs over the integer numbers, $n = 0, \pm1, \pm2, \ldots$, the function $f(z)$ repeatedly "jumps" from one value to the other over exactly m distinct complex numbers. The cubic root, $f(z) = z^{1/3}$, has three different values. For instance, at $z = 1$, they are (see Fig. B.5)

$$\sqrt[3]{1} = \begin{cases} 1 \\ e^{+i2\pi/3} \\ e^{-i2\pi/3}. \end{cases}$$

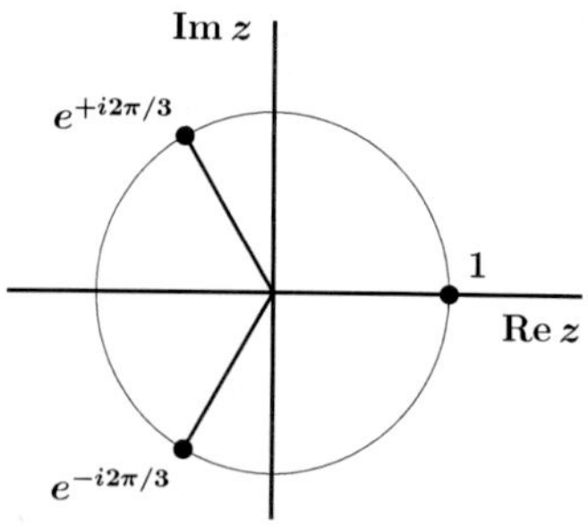

Fig. B.5 Complex values of the function $f(z) = z^{1/3}$ for $z = 1$

In order to avoid multi-valuedness, we can take m copies of the same complex z-plane (for the cubic root $m = 3$) and put them in stack one over the other. A vertical line passing through the stack of the planes crosses them at the points $z^{(0)}, z^{(1)}, \ldots, z^{(m)}$ that correspond to the same complex number, i.e.

$$z^{(0)} = z^{(1)} = \cdots = z^{(m)} = z \,.$$

However, we assign to them different values of the function,

$$f(z^{(0)}) = f_0 \,, \; f(z^{(1)}) = f_1 \,, \ldots \,, f(z^{(m)}) = f_m \,. \tag{B.22}$$

If we consider the function $f(z)$ as being defined not on the single z-plane, but on the stack of m parallel planes, then on such a more complicated domain this function becomes single-valued.

The planes of the stack cannot be completely isolated. In all cases, there are such points on them, where some of the values (B.22) are the same. For example, all m values of the function (B.21) are zero at $z = 0$. At these points, the planes, on which the values are the same, coincide, i.e. touch each other. To make this, we deform the planes by "pushing" (like a rubber film) towards each other at those points. As a result, the planes are not flat anymore and we call them sheets or layers.

In addition to the touching, the sheets must have another type of interconnection. Indeed, if we make a full circle around the origin and come back to the same z, the value of the function $f(z)$ changes. For example,

$$e^{i2\pi} z^{(0)} = z^{(1)} \quad \Rightarrow \quad f_0 \to f_1 \,.$$

This means that going around some points (point zero in this example), we find ourselves on a different sheet. This is only possible if the sheets are cut along certain lines and the edges of the cuts are interconnected ("glued"). The points, going around which brings us to another sheet, are called branch points, and the multi-layered surface thus constructed is called the Riemann surface.

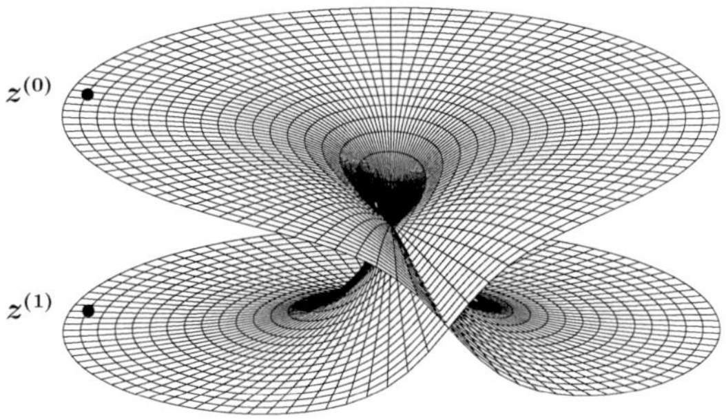

Fig. B.6 Two-sheet Riemann surface for the function $f(z) = \sqrt{z}$. The points $z^{(0)}$ and $z^{(1)}$ differ by any number of the factors $e^{i2\pi} e^{i2\pi} \dots$, i.e. by any number of the full circles around the branch point, $z = 0$

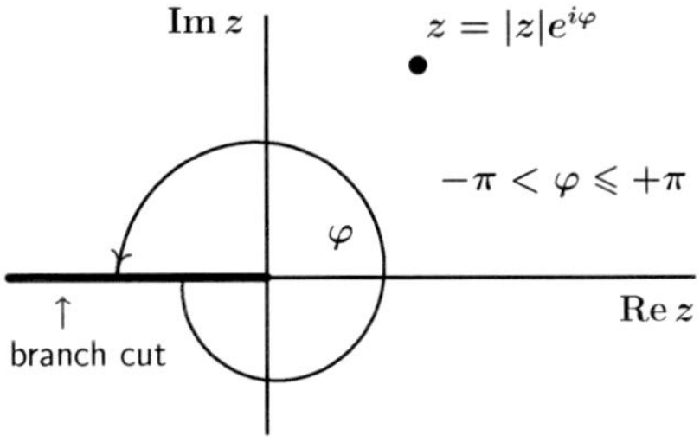

Fig. B.7 Principal sheet of the Riemann surface for the function $f(z) = \ln z$. It is cut along the negative part of the real axis from the branch point at $z = 0$ to $-\infty$

The simplest Riemann surface (which plays an important role in scattering theory) is the one for the square-root function, $f(z) = \sqrt{z}$. It consists of just two sheets with $z = 0$ being the branch point, as is shown in Fig. B.6. It should be clearly understood that on the line where the two sheets seemingly cross each other, they actually do not have common points. In other words, when moving through this "crossing" up or down the slope, we "do not notice" the other sheet. The positive half of the real axis is usually chosen along this "crossing". There are two copies of this axis belonging to each of the sheets. These axes have only one common point, at $z = 0$.

Yet another complex function that is important in scattering theory is the logarithm, $f(z) = \ln(z)$. It has infinitely many values for each single z. Indeed, if we take the independent variable z in the polar representation (B.1), then

$$\ln z = \ln|z| + i\varphi + i2\pi n \,, \qquad n = 0, \pm1, \pm2, \dots \,. \tag{B.23}$$

As is seen, if φ is monotonically increasing or decreasing, then $\ln z$ never comes back to the same value (in contrast to $\sqrt{z}$). It is a convention to take as the main (principal) branch of the logarithm, its values corresponding to $n = 0$ and $-\pi < \varphi \leqslant +\pi$. This range of the polar angle covers one (principal) sheet of the Riemann surface, as is shown in Fig. B.7. Each sheet is cut along the negative part of the real axis (see Fig. B.8) and the edges of the cuts of the neighbouring sheets (which differ by one full circle around the branch point) are connected to form the spiral Riemann surface with an infinite number of sheets as is depicted in Fig. B.9.

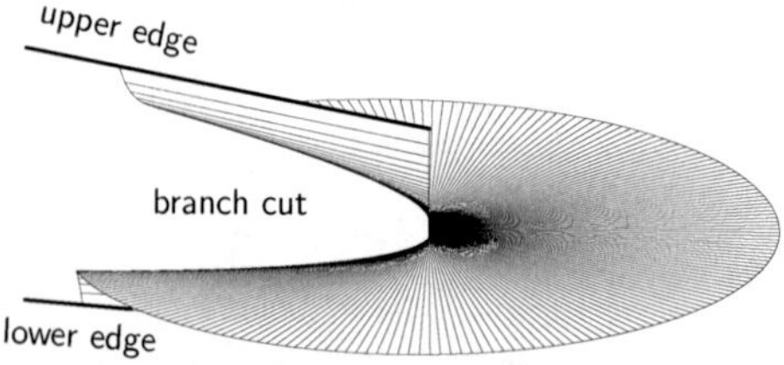

Fig. B.8 Every sheet of the Riemann surface for the function $f(z) = \ln z$ is cut in the same way along the negative part of the real axis

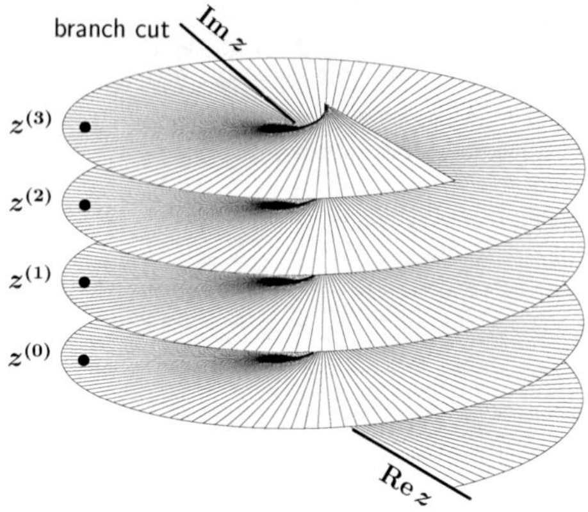

Fig. B.9 Infinite spiral Riemann surface for the function $f(z) = \ln z$. The points $z^{(0)}$, $z^{(1)}$, $z^{(2)}$, ... differ by the factor $e^{i2\pi}$, i.e. by the full circles around the branch point, $z = 0$

In this figure there are shown one real and one imaginary axis (on different sheets). Actually, each sheet has its own real and imaginary axes. On different sheets, they are parallel and are located above (or under) each other.

Wronskian of Scalar and Matrix Functions

C

The notion of Wronskian has its roots in the task of establishing linear independence of a set of N functions (see, for example, Ref. [9]). In general, in such a task, it is considered an arbitrary number N of the functions. However, in the present book for our purposes, it is sufficient to deal with just two functions. If necessary, the generalization of all the formulae for an arbitrary number of functions is straightforward.

Let we are given two analytic functions, $f(z)$ and $g(z)$, defined in a complex domain $\mathcal{D}$. These functions are linearly independent, if their linear combination,

$$c_1 f(z) + c_2 g(z) \equiv 0 \,, \qquad z \in \mathcal{D} \,, \tag{C.1}$$

is identical zero in the whole domain $\mathcal{D}$ only with $c_1 = c_2 = 0$. How to check if no other coefficients make a zero combination? To answer such a question, let us differentiate Eq. (C.1). This gives us the system of equations (in general, we need to differentiate $(N-1)$-times):

$$c_1 f(z) + c_2 g(z) = 0 \,, \tag{C.2}$$

$$c_1 f'(z) + c_2 g'(z) = 0 \,. \tag{C.3}$$

This is a degenerate linear system of equations for the unknown coefficients c_1 and c_2, which has non-trivial (i.e. non-zero) solution if and only if the determinant

$$W(f,g) = \det \begin{pmatrix} f & g \\ f' & g' \end{pmatrix} = f(z)g'(z) - f'(z)g(z) \tag{C.4}$$

is non-zero at least at one point in $\mathcal{D}$. This determinant is called Wronskian of the functions $f(z)$ and $g(z)$.

S. A. Rakityansky, *Jost Functions in Quantum Mechanics*,
https://doi.org/10.1007/978-3-031-07761-6

It is easy to show that for any two solutions $f(z)$ and $g(z)$ of the equation

$$\left[\frac{d^2}{dz^2} + P(z)\right] f(z) = 0,$$

the Wronskian is independent of z. Indeed,

$$\frac{d}{dz} W(f, g) = f'g' + fg'' - f''g - f'g' = fg'' + Pfg - (f''g + Pfg)$$

$$= (g'' + Pg)f - (f'' + Pf)g = 0 ,$$

which means that $W(f, g) = \text{const}$.

The notion of Wronskian of two functions can be generalized for the case when both f and g are matrices. In such a case, the first matrix in the products involved in the Wronskian should be transposed:

$$W(f, g) = f^T(z)g'(z) - \left[f'(z)\right]^T g(z) . \tag{C.5}$$

In the most simple way, the necessity of the transposition is seen when f and g are the column-solutions of a system of N coupled differential equations. For example, if $N = 2$, we have

$$f = \begin{pmatrix} f_1 \\ f_2 \end{pmatrix} , \qquad g = \begin{pmatrix} g_1 \\ g_2 \end{pmatrix} .$$

It is clear that there is no way to construct the Wronskian without transposing the column f. For these columns, the definition (C.5) gives

$$W(f, g) = (f_1, f_2) \begin{pmatrix} g_1' \\ g_2' \end{pmatrix} - (f_1', f_2') \begin{pmatrix} g_1 \\ g_2 \end{pmatrix}$$

$$= f_1 g_1' - f_1' g_1 + f_2 g_2' - f_2' g_2 \;=\; W(f_1, g_1) + W(f_2, g_2) .$$

As we see, the Wronskian of the two columns is equal to the sum of Wronskians of their corresponding elements. Therefore, for the independence of the two columns, it is sufficient that at least one pair of the corresponding elements is independent (which is expected intuitively).

If f and g are (2×2)-matrices,

$$f = \begin{pmatrix} f_{11} & f_{12} \\ f_{21} & f_{22} \end{pmatrix} , \qquad g = \begin{pmatrix} g_{11} & g_{12} \\ g_{21} & g_{22} \end{pmatrix} ,$$

then it is not difficult to see that

$$W(f,g) = \begin{bmatrix} W(f_{11},g_{11}) + W(f_{21},g_{21}) & W(f_{11},g_{12}) + W(f_{21},g_{22}) \\ W(f_{12},g_{11}) + W(f_{22},g_{21}) & W(f_{12},g_{12}) + W(f_{22},g_{22}) \end{bmatrix} = \begin{pmatrix} W_{11} & W_{12} \\ W_{21} & W_{22} \end{pmatrix},$$

where W_{mn} is the Wronskian of the column m of the matrix f and the column n of the matrix g. Thanks to the transposition, we always test the independence of only the corresponding elements of different columns and never mix the elements occupying different positions in the columns. When we deal with matrix differential equations, each column is a separate solution. Therefore, only those elements of the Wronskian are non-zero, which involve linearly independent column-solutions.

It is easy to obtain the following useful property of the matrix Wronskian:

$$[W(f,g)]^T = -W(g,f)\,. \tag{C.6}$$

For a scalar radial Schrödinger equation, the Wronskian of any two solutions does not depend on r (see the proof given above). What about the matrix equation? Is the Wronskian still r-independent? To answer this question, consider the following differential equation:

$$\left[\frac{d^2}{dz^2} + P(z)\right] f(z) = 0\,, \tag{C.7}$$

where $P(z)$ is a matrix function of the dimension $(N \times N)$. Let f and g be the column-solutions of Eq. (C.7). The same as we did it for the scalar equation, consider the derivative of their Wronskian,

$$\begin{aligned} \frac{d}{dz}W(f,g) &= \left[f'\right]^T g' + f^T g'' - \left[f''\right]^T g - \left[f'\right]^T g' = f^T g'' - \left[f''\right]^T g \\ &= f^T g'' + f^T Pg - f^T Pg - \left[f''\right]^T g \\ &= f^T(g'' + Pg) - (f^T P + \left[f''\right]^T)g\,. \end{aligned} \tag{C.8}$$

Since g is a solution of Eq. (C.7), we have $g'' + Pg = 0$. The second term in (C.8) is zero under the condition that $P^T = P$, i.e. if the matrix P is symmetric. Indeed, the transposition of Eq. (C.7) gives $\left[f''\right]^T + f^T P^T = 0$. Therefore

$$P^T = P \quad \Rightarrow \quad \frac{d}{dz}W(f,g) = 0\,. \tag{C.9}$$

As far as the radial Schrödinger equation is concerned, the requirement $P^T = P$ means that the potential must be represented by a symmetric matrix (which is the case in the most problems).

Bessel Functions

D

D.1 Definition and Some Properties

The Bessel and Neumann functions,[1] $J_\nu(z)$ and $N_\nu(z)$, are two linearly independent solutions of the differential equation

$$\left[z^2\frac{d^2}{dz^2} + z\frac{d}{dz} + (z^2 - \nu^2)\right]\mathcal{J}_\nu(z) = 0 \tag{D.1}$$

with the following asymptotic behaviour (which can be considered as the boundary conditions that uniquely determine them):

$$J_\nu(z) \underset{|z|\to\infty}{\longrightarrow} \sqrt{\frac{2}{\pi z}}\cos\left(z - \frac{\nu\pi}{2} - \frac{\pi}{4}\right), \tag{D.2}$$

$$N_\nu(z) \underset{|z|\to\infty}{\longrightarrow} \sqrt{\frac{2}{\pi z}}\sin\left(z - \frac{\nu\pi}{2} - \frac{\pi}{4}\right). \tag{D.3}$$

For quantum mechanical problems in three-dimensional space, the other two functions, $j_\ell(kr)$ and $n_\ell(kr)$ (called the Riccati–Bessel and Riccati–Neumann functions), are more important. They can be expressed via the functions (D.2, D.3),

$$j_\ell(kr) = \sqrt{\frac{\pi kr}{2}}J_{\ell+1/2}(kr)\,, \qquad n_\ell(kr) = \sqrt{\frac{\pi kr}{2}}N_{\ell+1/2}(kr)\,, \tag{D.4}$$

[1] A comprehensive description of all the properties of Bessel functions can be found in Ref. [95].

S. A. Rakityansky, *Jost Functions in Quantum Mechanics*,
https://doi.org/10.1007/978-3-031-07761-6

and are two independent solutions of the "free" radial Schrödinger equation,

$$\left[\frac{d^2}{dr^2} + k^2 - \frac{\ell(\ell+1)}{r^2}\right] u_\ell(k,r) = 0 \,, \tag{D.5}$$

with the asymptotic behaviour

$$j_\ell(kr) \underset{|kr|\to\infty}{\longrightarrow} \sin\left(kr - \frac{\ell\pi}{2}\right) , \tag{D.6}$$

$$n_\ell(kr) \underset{|kr|\to\infty}{\longrightarrow} -\cos\left(kr - \frac{\ell\pi}{2}\right) . \tag{D.7}$$

The linear combinations of these functions (called the Riccati–Hankel functions),

$$h_\ell^{(\pm)}(kr) = j_\ell(kr) \pm i n_\ell(kr) \,, \tag{D.8}$$

asymptotically behave like the outgoing (+) and incoming (−) spherical waves:

$$h_\ell^{(+)}(kr) \underset{|kr|\to\infty}{\longrightarrow} -i \exp\left(+ikr - i\frac{\ell\pi}{2}\right) = e^{+i\left(kr - \frac{\ell+1}{2}\pi\right)} , \tag{D.9}$$

$$h_\ell^{(-)}(kr) \underset{|kr|\to\infty}{\longrightarrow} +i \exp\left(-ikr + i\frac{\ell\pi}{2}\right) = e^{-i\left(kr - \frac{\ell+1}{2}\pi\right)} . \tag{D.10}$$

They transform into each other when the momentum changes sign,

$$h_\ell^{(\pm)}(-kr) = -e^{\mp i\ell\pi} h_\ell^{(\mp)}(kr) = (-1)^{\ell+1} h_\ell^{(\mp)}(kr) \,, \tag{D.11}$$

and under the operation of complex conjugation,

$$\left[h_\ell^{(\pm)}(z)\right]^* = h_\ell^{(\mp)}(z^*) \,. \tag{D.12}$$

It is easy to see that for any two functions, say, f and g, obeying the differential equation (D.5), the so-called Wronskian, which is defined as (where the prime denotes the derivative with respect to the independent variable, r)

$$W(f, g) = fg' - f'g \,, \tag{D.13}$$

does not depend on r and therefore can be found, using their asymptotic forms. This allows us to find that

$$W\big[j_\ell(kr), n_\ell(kr)\big] = k \,, \tag{D.14}$$

and

$$W\left[h_\ell^{(-)}(kr), h_\ell^{(+)}(kr)\right] = 2ik\,. \tag{D.15}$$

These are explicit formulae for the first few Riccati–Hankel functions:

$$h_0^{(\pm)}(z) = \mp i e^{\pm iz}\,, \tag{D.16}$$

$$h_1^{(\pm)}(z) = -e^{\pm iz}\left(1 \mp \frac{2}{2iz}\right), \tag{D.17}$$

$$h_2^{(\pm)}(z) = \pm i e^{\pm iz}\left(1 \mp \frac{6}{2iz} + \frac{12}{(2iz)^2}\right), \tag{D.18}$$

$$h_3^{(\pm)}(z) = \quad e^{\pm iz}\left(1 \mp \frac{16}{2iz} + \frac{60}{(2iz)^2} \mp \frac{120}{(2iz)^3}\right). \tag{D.19}$$

For an integer order, ℓ, both Riccati–Bessel and Riccati–Neumann functions have rather simple representations in the form of absolutely convergent Taylor series:

$$j_\ell(z) = \frac{z^{\ell+1}}{1\cdot 3\cdot 5\cdots(2\ell+1)}\left[1 - \frac{z^2/2}{1!(2\ell+3)} + \frac{(z^2/2)^2}{2!(2\ell+3)(2\ell+5)} - \cdots\right], \tag{D.20}$$

$$n_\ell(z) = -\frac{1\cdot 3\cdot 5\cdots(2\ell-1)}{z^\ell}\left[1 - \frac{z^2/2}{1!(1-2\ell)} + \frac{(z^2/2)^2}{2!(1-2\ell)(3-2\ell)} - \cdots\right]. \tag{D.21}$$

The Riccati–Bessel and Riccati–Neumann functions transform to each other when the order is inverted, as is given by the relation

$$n_\ell(z) = (-1)^{\ell+1} j_{-\ell-1}(z), \qquad \ell = 0, \pm 1, \pm 2, \ldots \tag{D.22}$$

These functions obey the relations useful for calculating their derivatives:

$$\frac{d}{dz} y_\ell(z) = y_{\ell-1} - \frac{\ell}{z} y_\ell(z)\,, \tag{D.23}$$

$$\frac{d}{dz} y_\ell(z) = \frac{\ell+1}{z} y_\ell(z) - y_{\ell+1}\,, \tag{D.24}$$

where $y_\ell(z)$ stands for either $j_\ell(z)$, $n_\ell(z)$, or $h_\ell^{(\pm)}(z)$.

The following bounds are true in the entire complex plane (for complex variables z and k):

$$|j_\ell(z)| \leqslant C e^{|\mathrm{Im}\, z|} \left(\frac{|z|}{1+|z|}\right)^{\ell+1}, \tag{D.25}$$

$$|n_\ell(z)| \leqslant C e^{|\mathrm{Im}\, z|} \left(\frac{|z|}{1+|z|}\right)^{-\ell}, \tag{D.26}$$

$$|h_\ell^{(\pm)}(z)| \leqslant C e^{\mp \mathrm{Im}\, z} \left(\frac{|z|}{1+|z|}\right)^{-\ell}, \tag{D.27}$$

$$|j_\ell(kr) n_\ell(kr') - n_\ell(kr) j_\ell(kr')| \leqslant \tag{D.28}$$

$$\leqslant C e^{|\mathrm{Im}\, k|(r-r')} \left(\frac{|kr|}{1+|kr|}\right)^{\ell+1} \left(\frac{|kr'|}{1+|kr'|}\right)^{-\ell}, \qquad r > r',$$

where C is a constant (which is different for each of the above equations).

D.2 Analytic Structure of $j_\ell(kr)$ and $n_\ell(kr)$

When considered as the functions of the energy variable, E, which gives two values for the momentum, $k = \pm\sqrt{2\mu E/\hbar^2}$, the Riccati–Bessel and Riccati–Neumann functions, $j_\ell(kr)$ and $n_\ell(kr)$, (generally speaking) are not single-valued. For our purposes, it is needed to split and factorize them in such a way that all the factors responsible for their multi-valuedness are given explicitly. Such representations of these functions look differently for an integer and half-integer order.

D.2.1 Integer Order

The representations (D.20, D.21) can be written as

$$j_\ell(kr) = \left(\frac{kr}{2}\right)^{\ell+1} \sum_{n=0}^{\infty} \frac{(-1)^n \sqrt{\pi}}{\Gamma(\ell + 3/2 + n) n!} \left(\frac{kr}{2}\right)^{2n}, \tag{D.29}$$

$$n_\ell(kr) = \left(\frac{2}{kr}\right)^{\ell} \sum_{n=0}^{\infty} \frac{(-1)^{n+\ell+1} \sqrt{\pi}}{\Gamma(-\ell + 1/2 + n) n!} \left(\frac{kr}{2}\right)^{2n}. \tag{D.30}$$

In both of these representations, the infinite sums are absolutely convergent for any complex value of k. However, the factors in front of the sums are the sources of "trouble" for both of them. If ℓ is even, then the factor $k^{\ell+1}$ is an odd power of the wave number, k. As a result, the function $j_\ell(kr)$ becomes double-valued, if considered as a function of the energy, $E = (\hbar k)^2/(2\mu)$, because for a single value of the energy, the wave number may have two different values,

$$k = \pm\sqrt{2\mu E/\hbar^2}\,, \tag{D.31}$$

depending on the choice of the sign in front of the square root. The function $n_\ell(kr)$ has the same "trouble" for the odd values of the angular momentum, ℓ. Moreover, for $\ell \neq 0$, the function $n_\ell(kr)$ is singular at $k = 0$ because of that factor.

In order to isolate the "troublesome" factors, we can introduce the following auxiliary functions:

$$\tilde{j}_\ell(E,r) = \frac{j_\ell(kr)}{k^{\ell+1}} = \left(\frac{r}{2}\right)^{\ell+1} \sum_{n=0}^{\infty} \frac{(-1)^n\sqrt{\pi}}{\Gamma(\ell+3/2+n)n!}\left(\frac{\mu r^2}{2\hbar^2}\right)^n E^n\,, \tag{D.32}$$

$$\tilde{n}_\ell(E,r) = k^\ell n_\ell(kr) = \left(\frac{2}{r}\right)^{\ell} \sum_{n=0}^{\infty} \frac{(-1)^{n+\ell+1}\sqrt{\pi}}{\Gamma(-\ell+1/2+n)n!}\left(\frac{\mu r^2}{2\hbar^2}\right)^n E^n\,, \tag{D.33}$$

which are entire (analytic single-valued) functions of the complex variable E. The series (D.32) and (D.33) are the Maclaurin expansions of these functions around the point $E = 0$.

When doing the power-series expansions of the Jost functions, we need the general Taylor series representations of the tilded functions $\tilde{j}_\ell(E,r)$ and $\tilde{n}_\ell(E,r)$ expanded around an arbitrary complex point E_0,

$$\tilde{j}_\ell(E,r) = \sum_{n=0}^{\infty} \varphi_n(\ell, E_0, r)(E-E_0)^n\,, \tag{D.34}$$

$$\tilde{n}_\ell(E,r) = \sum_{n=0}^{\infty} \gamma_n(\ell, E_0, r)(E-E_0)^n\,. \tag{D.35}$$

The coefficients, φ_n and γ_n, in these expansions can be found via the corresponding derivatives,

$$\varphi_n(\ell, E_0, r) = \frac{1}{n!}\left[\frac{d^n}{dE^n}\,\tilde{j}_\ell(E,r)\right]_{E=E_0}\,, \tag{D.36}$$

$$\gamma_n(\ell, E_0, r) = \frac{1}{n!}\left[\frac{d^n}{dE^n}\,\tilde{n}_\ell(E,r)\right]_{E=E_0}\,. \tag{D.37}$$

Since

$$\frac{d}{dE} = \frac{dk}{dE} \cdot \frac{d}{dk} = \frac{\mu}{\hbar^2 k} \cdot \frac{d}{dk} ,$$

we have

$$\frac{d}{dE} \tilde{j}_\ell(E,r) = \frac{\mu}{\hbar^2 k} \cdot \frac{d}{dk} \cdot \frac{j_\ell(kr)}{k^{\ell+1}} = \frac{\mu r^{\ell+2}}{\hbar^2 k} \cdot \frac{d}{d(kr)} \cdot \frac{j_\ell(kr)}{(kr)^{\ell+1}} .$$

At this point, we can use the following property of the Riccati–Bessel function:

$$\frac{d}{dz}\left[\frac{j_\ell(z)}{z^{\ell+1}}\right] = -\frac{j_{\ell+1}(z)}{z^{\ell+1}} , \tag{D.38}$$

which is valid for any complex z. This gives

$$\frac{d}{dE} \tilde{j}_\ell(E,r) = -\frac{\mu r}{\hbar^2} \tilde{j}_{\ell+1}(E,r) . \tag{D.39}$$

Applying this recurrence relation n times, we finally obtain

$$\varphi_n(\ell, E_0, r) = \frac{1}{n!}\left[\left(-\frac{\mu r}{\hbar^2}\right)^n \tilde{j}_{\ell+n}(E,r)\right]_{E=E_0} . \tag{D.40}$$

Similarly, for the derivative that is used in Eq. (D.37), we have

$$\frac{d}{dE} \tilde{n}_\ell(E,r) = \frac{\mu}{\hbar^2 k r^{\ell-1}} \cdot \frac{d}{d(kr)}\left[(kr)^\ell n_\ell(kr)\right] .$$

Together with the known property of the Riccati–Neumann function,

$$\frac{d}{dz}\left[z^\ell n_\ell(z)\right] = z^\ell n_{\ell-1}(z) , \tag{D.41}$$

we obtain the following recurrence relation:

$$\frac{d}{dE} \tilde{n}_\ell(E,r) = \frac{\mu r}{\hbar^2} \tilde{n}_{\ell-1}(E,r) . \tag{D.42}$$

Using this relation in Eq. (D.37) n times, we obtain

$$\gamma_n(\ell, E_0, r) = \frac{1}{n!}\left[\left(\frac{\mu r}{\hbar^2}\right)^n \tilde{n}_{\ell-n}(E,r)\right]_{E=E_0} . \tag{D.43}$$

D.2.2 Half-Integer Order

For a half-integer order, $\lambda = \ell - 1/2$, the Riccati–Bessel function is represented in the same way as is given by Eq. (D.29), namely,

$$j_\lambda(kr) = k^{\lambda+1} \sum_{n=0}^{\infty} k^{2n} f_n^{(\lambda)}(r) , \tag{D.44}$$

where

$$f_n^{(\lambda)}(r) = \frac{\sqrt{\pi}(-1)^n}{n!\Gamma(n+\lambda+3/2)} \left(\frac{r}{2}\right)^{2n+\lambda+1} . \tag{D.45}$$

However, the Riccati–Neumann function has a more complicated representation than that given by Eq. (D.30):

$$n_\lambda(kr) = k^{-\lambda} \sum_{n=0}^{\infty} k^{2n} g_n^{(\lambda)}(r) + \mathfrak{h}(k) j_\lambda(kr) , \tag{D.46}$$

where

$$g_n^{(\lambda)}(r) = \begin{cases} -\dfrac{(\lambda-n-1/2)!}{\sqrt{\pi}n!} \left(\dfrac{r}{2}\right)^{2n-\lambda} & , 0 \leqslant n \leqslant \lambda - \frac{1}{2} , \\ \dfrac{2}{\pi} \ln\left(\dfrac{r}{R}\right) f_{n-\lambda-\frac{1}{2}}^{(\lambda)}(r) - & \\ -\dfrac{(-1)^{n-\lambda-\frac{1}{2}} \left[\psi(n+1) + \psi(n-\lambda+\frac{1}{2})\right]}{\sqrt{\pi}n!(n-\lambda-\frac{1}{2})!} \left(\dfrac{r}{2}\right)^{2n-\lambda} & , \lambda + \frac{1}{2} \leqslant n < \infty , \end{cases} \tag{D.47}$$

$$\mathfrak{h}(k) = \frac{2}{\pi} \ln\left(\frac{kR}{2}\right) . \tag{D.48}$$

Here R is an arbitrary number (in the units of length). It can be chosen arbitrarily because any increase or decrease of $\mathfrak{h}(k)$, cased by the change of R, is compensated by the corresponding change in the first term of Eq. (D.47). The parameter R is introduced to separate the r and k dependences in the original term containing $\ln(kr/2) = \ln(kR/2) + \ln(r/R)$ and to have dimensionless products under the logarithm. In practical calculations, the parameter R can always be taken as the unit of the length, i.e. $R = 1$. The ψ-function in Eq. (D.47) is defined as follows:

$$\psi(n) = \frac{\Gamma'(n)}{\Gamma(n)} = \begin{cases} -\gamma & , n = 1 , \\ -\gamma + \sum_{m=1}^{n-1} m^{-1} & , n \geqslant 2 , \end{cases} \tag{D.49}$$

where $\gamma = 0.577\ldots$ is the Euler constant.

Equations (D.44, D.46) represent the Riccati–Bessel and Riccati–Neuman functions via the infinite series. Each term of these series is a product of a function depending on k and another function depending on r, i.e. the k and r dependences are given in a separable form. These series involve only even powers of k, i.e. the powers of the energy $k^2 = 2\mu E/\hbar^2$. Since for any finite r these series are absolutely and uniformly convergent on the whole complex plane of E, they define some entire functions of E, i.e.

$$j_\lambda(kr) = k^{\lambda+1}\tilde{j}_\lambda(E, r) , \tag{D.50}$$

$$n_\lambda(kr) = k^{-\lambda}\tilde{n}_\lambda(E, r) + k^{\lambda+1}\mathfrak{h}(k)\tilde{j}_\lambda(E, r) , \tag{D.51}$$

where these entire functions are denoted as

$$\tilde{j}_\lambda(E, r) = \sum_{n=0}^{\infty} \left(\frac{2\mu E}{\hbar^2}\right)^n f_n^{(\lambda)}(r) , \tag{D.52}$$

$$\tilde{n}_\lambda(E, r) = \sum_{n=0}^{\infty} \left(\frac{2\mu E}{\hbar^2}\right)^n g_n^{(\lambda)}(r) . \tag{D.53}$$

Since these functions are entire (holomorphic) with respect to the energy variable E, we can expand them in the Taylor series,

$$\tilde{j}_\lambda(E, r) = \sum_{n=0}^{\infty} s_n(\lambda, E_0, r)\,(E - E_0)^n , \tag{D.54}$$

$$\tilde{n}_\lambda(E, r) = \sum_{n=0}^{\infty} c_n(\lambda, E_0, r)\,(E - E_0)^n , \tag{D.55}$$

near an arbitrary point E_0. The expansion coefficients,

$$s_n(\lambda, E_0, r) = \frac{1}{n!}\frac{\partial^n}{\partial E^n}\tilde{j}_\lambda(E, r)\bigg|_{E=E_0} , \tag{D.56}$$

$$c_n(\lambda, E_0, r) = \frac{1}{n!}\frac{\partial^n}{\partial E^n}\tilde{n}_\lambda(E, r)\bigg|_{E=E_0}, \tag{D.57}$$

are expressed via the corresponding derivatives. In order to find them, we note that

$$E = \frac{\hbar^2 k^2}{2\mu} \qquad \Longrightarrow \qquad \frac{\partial}{\partial E} = \frac{\mu}{\hbar^2 k}\frac{\partial}{\partial k} \tag{D.58}$$

and also make use of known recurrence relations

$$\frac{d}{dz}\left[\frac{\mathcal{J}_\lambda(z)}{z^{\lambda+1}}\right] = -\frac{\mathcal{J}_{\lambda+1}(z)}{z^{\lambda+1}}, \tag{D.59}$$

$$\frac{d}{dz}\left[z^\lambda \mathcal{J}_\lambda(z)\right] = z^\lambda \mathcal{J}_{\lambda-1}(z), \tag{D.60}$$

where $\mathcal{J}_\lambda(z)$ stands for either $j_\lambda(z)$ or $n_\lambda(z)$. Therefore

$$\frac{\partial}{\partial E}\tilde{j}_\lambda(E, r) = \frac{\mu}{\hbar^2 k}\cdot\frac{\partial}{\partial k}\left[\frac{j_\lambda(kr)}{k^{\lambda+1}}\right] = \frac{\mu r^{\lambda+2}}{\hbar^2 k}\cdot\frac{\partial}{\partial(kr)}\left[\frac{j_\lambda(kr)}{(kr)^{\lambda+1}}\right] =$$

$$= -\frac{\mu r^{\lambda+2}}{\hbar^2 k}\frac{j_{\lambda+1}(kr)}{(kr)^{\lambda+1}} = -\frac{\mu r}{\hbar^2}\tilde{j}_{\lambda+1}(E, r)$$

and thus

$$\frac{\partial^n}{\partial E^n}\tilde{j}_\lambda(E, r) = \left(-\frac{\mu r}{\hbar^2}\right)^n \tilde{j}_{\lambda+n}(E, r), \tag{D.61}$$

$$\begin{aligned} s_n(\lambda, E_0, r) &= \frac{1}{n!}\left(-\frac{\mu r}{\hbar^2}\right)^n \left[\frac{j_{\lambda+n}(kr)}{k^{\lambda+n+1}}\right]_{E=E_0} \\ &= \frac{1}{n!}\left(-\frac{\mu r}{\hbar^2}\right)^n \sqrt{\frac{\pi r}{2}}\left[\frac{J_{\ell+n}(kr)}{k^{\ell+n}}\right]_{E=E_0}. \end{aligned} \tag{D.62}$$

As it should be (since $\tilde{j}_\lambda(E, r)$ is single-valued), the expansion coefficients $s_n(\lambda, E_0, r)$ do not depend on the choice of the sign of the momentum $k_0 = \pm\sqrt{2\mu E_0/\hbar^2}$. Indeed,

$$\frac{j_{\lambda+n}(kr)}{k^{\lambda+n+1}} = \sqrt{\frac{\pi kr}{2}}\frac{J_{\lambda+n+1/2}(kr)}{k^{\lambda+n+1}} = \sqrt{\frac{\pi r}{2}}\frac{J_{\lambda+n+1/2}(kr)}{k^{\lambda+n+1/2}}, \tag{D.63}$$

and since

$$J_\nu(ze^{i\pi}) = \left(e^{i\pi}\right)^\nu J_\nu(z)\,, \tag{D.64}$$

the numerator and denominator in Eq. (D.63) acquire the same phase factor when k changes its sign.

Finding the derivative $\partial_E^n \tilde{n}_\lambda(E,r)$ is a little bit more complicated. The first derivative can be written as

$$\frac{\partial}{\partial E}\tilde{n}_\lambda(E,r) = \frac{\mu r^{1-\lambda}}{\hbar^2 k}\left\{\frac{\partial}{\partial(kr)}\left[(kr)^\lambda n_\lambda(kr)\right] - \frac{\partial}{\partial(kr)}\left[\mathfrak{h}(k)(kr)^\lambda j_\lambda(kr)\right]\right\}.$$

Using Eq. (D.60) and explicit form of the function $\mathfrak{h}(k)$ given by Eq. (D.48), we obtain

$$\begin{aligned}
\frac{\partial}{\partial E}\tilde{n}_\lambda(E,r) &= \frac{\mu r^{1-\lambda}}{\hbar^2 k}\left[(kr)^\lambda n_{\lambda-1}(kr) - \mathfrak{h}(k)(kr)^\lambda j_{\lambda-1}(kr) - (kr)^\lambda j_\lambda(kr)\frac{2}{\pi kr}\right] \\
&= \frac{\mu r}{\hbar^2}\left[k^{\lambda-1} n_{\lambda-1}(kr) - k^{\lambda-1}\mathfrak{h}(k) j_{\lambda-1}(kr) - \frac{2}{\pi r}k^{\lambda-2} j_\lambda(kr)\right] \\
&= \frac{\mu r}{\hbar^2}\tilde{n}_{\lambda-1}(E,r) - \frac{2\mu}{\pi\hbar^2}k^{\lambda-2} j_\lambda(kr) \\
&= \frac{\mu r}{\hbar^2}\tilde{n}_{\lambda-1}(E,r) - \frac{2\mu}{\pi\hbar^2}q_{\lambda 1}(kr)\,,
\end{aligned} \tag{D.65}$$

where we introduced an auxiliary function

$$q_{\lambda n}(k,r) = k^{\lambda-2n} j_\lambda(kr) = k^{\lambda+\frac{1}{2}-2n}\sqrt{\frac{\pi r}{2}}J_{\lambda+1/2}(kr)\,, \tag{D.66}$$

whose derivatives can be found using the following recurrence relation:

$$\begin{aligned}
\frac{\partial}{\partial E}\left[k^{\lambda-2n} j_\lambda(kr)\right] &= \frac{\mu}{\hbar^2 k}r^{2n-\lambda+1}\frac{\partial}{\partial(kr)}\left[\frac{1}{(kr)^{2n}}(kr)^\lambda j_\lambda(kr)\right] \\
&= \frac{\mu}{\hbar^2 k}r^{2n-\lambda+1}\left[-\frac{2n}{(kr)^{2n+1}}(kr)^\lambda j_\lambda(kr)\right.
\end{aligned}$$

$$+\frac{1}{(kr)^{2n}}(kr)^{\lambda} j_{\lambda-1}(kr)\Bigg]$$

$$= -\frac{2n\mu}{\hbar^2}k^{\lambda-2(n+1)} j_\lambda(kr) + \frac{\mu r}{\hbar^2}k^{\lambda-1-2n} j_{\lambda-1}(kr)\,,$$

i.e.

$$\frac{\partial}{\partial E}q_{\lambda n} = -\frac{2n\mu}{\hbar^2}q_{\lambda,n+1} + \frac{\mu r}{\hbar^2}q_{\lambda-1,n}\,. \tag{D.67}$$

Repeatedly using the relations (D.65) and (D.67), we can calculate any number of the derivatives $\partial_E^n \tilde{n}_\lambda(E,r)$ needed for finding the expansion coefficients (D.57).

Coulomb Wave Functions

E

E.1 Definitions and the Main Properties

The standard Coulomb wave functions,[1] $F_\ell(\eta, kr)$ and $G_\ell(\eta, kr)$, are defined as two linearly independent solutions of the radial Schrödinger equation,

$$\left[\partial_r^2 + k^2 - \frac{\ell(\ell+1)}{r^2} - \frac{2k\eta}{r}\right] u_\ell(E, r) = 0\,, \tag{E.1}$$

with the pure Coulomb potential, $2k\eta/r$, where the energy parameter E is proportional to the square of the wave number k,

$$E = \frac{\hbar^2 k^2}{2\mu}\,, \tag{E.2}$$

and the (dimensionless) Sommerfeld parameter η is inversely proportional to k,

$$\eta = \frac{\mu e^2 Z_1 Z_2}{\hbar^2 k} = \frac{a}{k} \tag{E.3}$$

with the charge-dependent constant

$$a = \frac{\mu e^2 Z_1 Z_2}{\hbar^2}\,. \tag{E.4}$$

The differential equation (E.1) is singular at the point $r = 0$. This singularity is, however, not severe because the multiplication by r^2 makes all the coefficients

[1] All the proofs and derivations of the relations listed here can be found in Ref. [96].

S. A. Rakityansky, *Jost Functions in Quantum Mechanics*,
https://doi.org/10.1007/978-3-031-07761-6

regular. In the theory of differential equations, such points are called regular singular points. The fact that there is a singularity (although a mild one) has an implication. As a consequence of this singularity, only one of any two linearly independent solutions of Eq. (E.1) can be finite at $r = 0$, while the other is always singular at this point. The functions F_ℓ and G_ℓ are such a pair of the regular and irregular solutions,

$$F_\ell(\eta, kr) \underset{r\to 0}{\longrightarrow} 0\,, \qquad G_\ell(\eta, kr) \underset{r\to 0}{\longrightarrow} \infty\,. \tag{E.5}$$

Specific forms and normalizations of these functions are uniquely determined by their asymptotic behaviour at large values of r,

$$F_\ell(\eta, kr) \underset{r\to\infty}{\longrightarrow} \sin\left[kr - \eta \ln(2kr) - \frac{\ell\pi}{2} + \sigma_\ell\right], \tag{E.6}$$

$$G_\ell(\eta, kr) \underset{r\to\infty}{\longrightarrow} \cos\left[kr - \eta \ln(2kr) - \frac{\ell\pi}{2} + \sigma_\ell\right], \tag{E.7}$$

where σ_ℓ is the Coulomb phase shift,

$$\sigma_\ell = \frac{1}{2i}\left[\ln\Gamma(\ell + 1 + i\eta) - \ln\Gamma(\ell + 1 - i\eta)\right]. \tag{E.8}$$

Equations (E.6) and (E.7) can be considered as the boundary conditions that uniquely define the two solutions of the differential equation (E.1).

Near the origin ($r \to 0$), the Coulomb functions behave as

$$F_\ell(\eta, kr) \underset{r\to 0}{\longrightarrow} \frac{C_\ell(\eta)}{(2\ell + 1)!!}(kr)^{\ell+1}[1 + O(kr)] \tag{E.9}$$

and

$$G_\ell(\eta, kr) \underset{r\to 0}{\longrightarrow} \frac{(2\ell - 1)!!}{C_\ell(\eta)}(kr)^{-\ell}\left\{1 + \begin{array}{r} O[kr\ln(kr)]\,,\ \ell = 0 \\ O(kr)\,,\ \ell \neq 0 \end{array}\right\}, \tag{E.10}$$

where the function

$$C_\ell(\eta) = \frac{e^{-\pi\eta/2}}{\Gamma(\ell + 1)} \exp\left\{\frac{1}{2}\left[\ln\Gamma(\ell + 1 + i\eta) + \ln\Gamma(\ell + 1 - i\eta)\right]\right\} \tag{E.11}$$

is called the Coulomb barrier factor.

Since the Wronskian of any two solutions of a differential equation of the type (E.1) does not depend on r, it can be found either at infinity, i.e. using Eqs. (E.6,

E.7), or near the origin, i.e. using the leading terms of Eqs. (E.9, E.10). It is easy to check that in both cases the same result is obtained, namely,

$$G_\ell(\eta, kr)\partial_r F_\ell(\eta, kr) - F_\ell(\eta, kr)\partial_r G_\ell(\eta, kr) = k \; . \tag{E.12}$$

The choice of a pair of linearly independent solutions of Eq. (E.1) is not unique and, in principle, can be done in an infinite number of ways. In addition to the pair of the functions F_ℓ and G_ℓ, there is an alternative choice that is also very convenient and is often used. This choice consists in the following linear combinations of F_ℓ and G_ℓ:

$$H_\ell^{(\pm)}(\eta, kr) = F_\ell(\eta, kr) \mp i G_\ell(\eta, kr) \tag{E.13}$$

with the Wronskians:

$$W\left[H_\ell^{(-)}, H_\ell^{(+)}\right] = H_\ell^{(-)}(\eta, kr)\partial_r H_\ell^{(+)}(\eta, kr) - H_\ell^{(+)}(\eta, kr)\partial_r H_\ell^{(-)}(\eta, kr) = 2ik \, , \tag{E.14}$$

$$W\left[H_\ell^{(-)}, H_\ell^{(-)}\right] = W\left[H_\ell^{(+)}, H_\ell^{(+)}\right] = 0 \; . \tag{E.15}$$

They are useful because at large distances the functions $H_\ell^{(-)}$ and $H_\ell^{(+)}$ behave as the incoming and outgoing spherical waves modified by the Coulomb forces,

$$H_\ell^{(\pm)}(\eta, kr) \underset{r\to\infty}{\longrightarrow} \mp i \exp\left\{\pm i \left[kr - \frac{\ell\pi}{2} - \eta \ln(2kr) + \sigma_\ell\right]\right\} \; . \tag{E.16}$$

The functional forms and the normalizations of the Coulomb functions are chosen in such a way that in the zero-charge limit (i.e. when $\eta \to 0$), they become the corresponding Riccati–Bessel, Riccati–Neumann,[2] and Riccati–Hankel functions,

$$F_\ell(\eta, kr) \underset{\eta\to 0}{\longrightarrow} j_\ell(kr) \, , \tag{E.17}$$

$$G_\ell(\eta, kr) \underset{\eta\to 0}{\longrightarrow} -n_\ell(kr) \, , \tag{E.18}$$

$$H_\ell^{(\pm)}(\eta, kr) \underset{\eta\to 0}{\longrightarrow} h_\ell^{(\pm)}(kr) \; . \tag{E.19}$$

The Coulomb phase shift in such a limit vanishes, and the barrier factor becomes unity,

$$\sigma_\ell(\eta) \underset{\eta\to 0}{\longrightarrow} 0 \, , \qquad C_\ell(\eta) \underset{\eta\to 0}{\longrightarrow} 1 \; . \tag{E.20}$$

[2] Please take note of the minus sign in Eq. (E.18).

E.2 Analytic Structure

In quantum mechanics, where the Coulomb functions are used, the energy parameter E is treated as a complex variable. The corresponding wave number (momentum) k and thus the Sommerfeld parameter η are therefore complex as well. This makes $F_\ell(\eta, kr)$ and $G_\ell(\eta, kr)$ complex functions of the complex variable E.

It is easy to see that these functions are multi-valued, i.e. for every single value of E, there are many different values of F_ℓ and G_ℓ. This is seen from the boundary conditions (E.6) and (E.7) that make the Coulomb functions depending on the odd powers of $k \sim \sqrt{E}$ as well as on $\ln k$. For a single value of the energy, any odd power of k has two different values, depending on the choice of the sign in front of the square root

$$k = \pm\sqrt{(2\mu/\hbar^2)E} \,. \tag{E.21}$$

As a result, the functions F_ℓ and G_ℓ become double-valued. Further increase of the number of their values stems from their dependence on the complex function $\ln k$, which has infinitely many different values,

$$\ln k = \ln\left(|k|e^{i[\arg(k)+2\pi m]}\right) = \ln|k| + i\arg(k) + i2\pi m\,, \quad m = 0, \pm 1, \pm 2, \dots\,, \tag{E.22}$$

which correspond to different choices of the arbitrary integer number m.

In other words, the functions $F_\ell(\eta, kr)$ and $G_\ell(\eta, kr)$, when considered as complex functions of E, are defined on a "spiral" Riemann surface with the point $E = 0$ being the square-root and logarithmic branch point at the same time. It is possible, however, to write these functions in such a factorized form, where the "troublesome" factors responsible for the branching are given explicitly, while the other factors are single-valued analytic functions of E. The proofs of some statements given below can be found in the work published in 1984 by J. Humblet[97].[3]

For the regular function, such a factorization is given as

$$F_\ell(\eta, kr) = (2\ell+1)!C_\ell(\eta)(kr)^{\ell+1}\Phi_\ell(E, r)\,, \tag{E.23}$$

where the function $\Phi_\ell(E, r)$ is an entire (holomorphic) function of E, i.e. is analytic and single-valued on the whole plane of complex E. As is seen, the odd powers of k (and hence the double-valuedness) are present in $C_\ell(\eta)$ and (for even ℓ) in $(kr)^{\ell+1}$. The factor $(2\ell+1)!$, in principle, can be included in the holomorphic function $\Phi_\ell(E, r)$. This is not done for the sake of convenience. If we are going

[3] See also Refs. [35,98].

to consider ℓ also as a complex variable, then this factor should be replaced with the appropriate Γ-function which is singular at the negative integer values.

Therefore, the regular Coulomb function can be written as

$$\boxed{F_\ell(\eta, kr) = C_\ell(\eta) k^{\ell+1} \tilde{F}_\ell(\eta, kr)} \tag{E.24}$$

where the tilded function

$$\tilde{F}_\ell(\eta, kr) = (2\ell+1)!\, r^{\ell+1} \Phi_\ell(E, r) \tag{E.25}$$

is an entire function of the energy variable E.

The irregular function has a more complicated structure,

$$G_\ell(\eta, kr) = \frac{(2\eta)^{2\ell+1}(kr)^{\ell+1}}{C_\ell(\eta)(2\ell+1)!} \left[h(\eta) u_\ell(E) \Phi_\ell(E, r) + u_\ell(E) \Psi_\ell(E, r)\right] , \tag{E.26}$$

with

$$h(\eta) = \frac{1}{2}\psi(1 - i\eta) + \frac{1}{2}\psi(1 + i\eta) - \ln(\eta) , \tag{E.27}$$

$$\psi(z) = \Gamma'(z)/\Gamma(z) , \tag{E.28}$$

and

$$u_\ell(E) = \left[\frac{(2\ell+1)! C_\ell(\eta)}{(2\eta)^\ell C_0(\eta)}\right]^2 . \tag{E.29}$$

It can be proved that the functions $u_\ell(E)$, $\Phi_\ell(E, r)$, and $\Psi_\ell(E, r)$ are entire functions of the variable E. This means that in the square brackets on the right hand side of Eq. (E.26), the only non-holomorphic function is $h(\eta)$.

Substituting Eq. (E.29) in (E.26), we obtain

$$\begin{aligned} G_\ell &= \frac{(2\eta)^{2\ell+1}(kr)^{\ell+1}}{(2\ell+1)! C_\ell} \left[\frac{(2\ell+1)! C_\ell}{(2\eta)^\ell C_0}\right]^2 h\Phi_\ell + \frac{(2\eta)^{2\ell+1} k^{2\ell+1} r^{\ell+1}}{(2\ell+1)! C_\ell k^\ell} u_\ell \Psi_\ell \\ &= \frac{2\eta (kr)^{\ell+1}(2\ell+1)! C_\ell}{C_0^2} h\Phi_\ell + \frac{(2a)^{2\ell+1} r^{\ell+1}}{(2\ell+1)! C_\ell k^\ell} u_\ell \Psi_\ell . \end{aligned}$$

This can be written as

$$G_\ell(\eta, kr) = \frac{2\eta h(\eta) C_\ell(\eta) k^{\ell+1}}{C_0^2(\eta)} \tilde{F}_\ell(\eta, kr) + \frac{1}{C_\ell(\eta) k^\ell} \tilde{G}_\ell(\eta, kr) \tag{E.30}$$

where the holomorphic function of the energy, $\tilde{F}_\ell(\eta, kr)$, is defined by Eq. (E.25) and

$$\tilde{G}_\ell(\eta, kr) = \frac{(2a)^{2\ell+1} r^{\ell+1}}{(2\ell+1)!} u_\ell(E) \Psi_\ell(E, r) \,. \tag{E.31}$$

Since both u_ℓ and Ψ_ℓ are holomorphic functions of E, the tilded function $\tilde{G}_\ell$ is holomorphic as well.

The holomorphic parts, $\tilde{F}_\ell$ and $\tilde{G}_\ell$, of the Coulomb functions can be expanded in the Taylor series around an arbitrary energy, E_0,

$$\tilde{F}_\ell(E, r) = \sum_{n=0}^{\infty} \varphi_n(\ell, E_0, r)(E - E_0)^n \,, \tag{E.32}$$

$$\tilde{G}_\ell(E, r) = \sum_{n=0}^{\infty} \gamma_n(\ell, E_0, r)(E - E_0)^n \,. \tag{E.33}$$

The corresponding expansion coefficients for $E_0 = 0$ can be derived from the formulae of Ref. [97], where it can also be found that for real E_0 the coefficients φ_n and γ_n are real as well. If $E_0 \neq 0$, the simplest way of obtaining these coefficients could be just the numerical differentiation.

E.3 Coulomb-Related Functions Near the Spectral Points

An attractive Coulomb potential ($Z_1 Z_2 < 0$) always generates the infinite set of (hydrogen-like) bound states,

$$E_n^c = -\frac{|Z_1 Z_2|^2 \mu e^4}{2\hbar^2 n^2} \,, \tag{E.34}$$

where $n = 1, 2, 3, \ldots$ is the principal quantum number. It is composed of the radial quantum number $n_r = 0, 1, 2, \ldots$ and the orbital number $\ell = 0, 1, 2, \ldots$,

$$n = n_r + \ell + 1 \,. \tag{E.35}$$

At the bound states, the Sommerfeld parameter is pure imaginary,

$$\eta_n = in \,, \tag{E.36}$$

and the following Γ-function (involved in various formulae) is singular:

$$\Gamma(\ell + 1 + i\eta_n) = \Gamma(-n_r) = \infty \,. \tag{E.37}$$

In some cases, it is necessary to know how fast this function increases when the energy approaches a spectral point (E.34),

$$E = E_n^c + \delta \,, \qquad \delta \to 0 \,, \tag{E.38}$$

where δ is an arbitrary complex number. This energy corresponds to

$$k = \sqrt{\frac{2\mu(E_n^c + \delta)}{\hbar^2}} \approx k_n + \frac{k_n \delta}{2E_n^c}$$

and

$$\eta = \frac{\mu e^2 Z_1 Z_2}{\hbar^2 k_n \left(1 + \dfrac{\delta}{2E_n^c}\right)} \approx \eta_n \left(1 - \frac{\delta}{2E_n^c}\right) = i(n + \epsilon) \,, \qquad \epsilon \to 0 \,,$$

where

$$\epsilon = -\frac{n}{2E_n^c}\delta \,. \tag{E.39}$$

For the Γ-function, this gives

$$\Gamma(\ell + 1 + i\eta) = \Gamma(\ell + 1 - n_r - \ell - 1 - \epsilon) = \Gamma(-n_r - \epsilon) \,.$$

In order to get rid of the negative integer, $-n_r$, in this Γ-function, the recurrency relation

$$\Gamma(z) = \frac{1}{z}\Gamma(z + 1)$$

is applied n_r times. This gives

$$\begin{aligned} \Gamma(-n_r - \epsilon) &= \frac{1}{-n_r - \epsilon}\Gamma(1 - n_r - \epsilon) \\ &= \frac{1}{(-n_r - \epsilon)(1 - n_r - \epsilon)}\Gamma(2 - n_r - \epsilon) = \cdots \end{aligned}$$

$$= \frac{1}{(-n_r-\epsilon)(1-n_r-\epsilon)\cdots(n_r-n_r-\epsilon)}\Gamma(n_r+1-n_r-\epsilon)$$

$$= \frac{(-1)^{n_r+1}}{n_r!}\cdot\frac{1}{\epsilon}\,. \tag{E.40}$$

Therefore the Γ-function (E.37) diverges as $\sim 1/\delta$,

$$\Gamma(\ell+1+i\eta)|_{E=E_n^c+\delta} \underset{\delta\to 0}{\sim} \frac{1}{\delta}\,, \tag{E.41}$$

and thus the pure Coulomb Jost function (8.32) linearly vanishes with δ at each point E_n^c,

$$f_\ell^{\mathrm{c\,(in)}}(E_n^c+\delta) = \left.\frac{e^{\pi\eta/2}(2\ell)!!}{2^{\ell+1}\Gamma(\ell+1\pm i\eta)}\right|_{E=E_n^c+\delta}$$

$$\underset{\delta\to 0}{\longrightarrow} \frac{e^{\pi\eta_n/2}(2\ell)!!}{2^{\ell+1}}(-1)^{n_r+1}n_r!\epsilon \underset{\delta\to 0}{\sim} \delta\,. \tag{E.42}$$

Similarly, the rate of the divergence of the Coulomb barrier factor (8.31) can be obtained,

$$C_\ell(\eta)|_{E=E_n^c+\delta} = \left.\frac{2^\ell e^{-\pi\eta/2}}{(2\ell)!!}\sqrt{\Gamma(\ell+1+i\eta)\Gamma(\ell+1-i\eta)}\right|_{E=E_n^c+\delta}$$

$$\underset{\delta\to 0}{\longrightarrow} \frac{2^\ell e^{-\pi\eta_n/2}}{(2\ell)!!}\sqrt{\frac{(-1)^{n_r+1}(2\ell+1+n_r)!}{n_r!\epsilon}} \underset{\delta\to 0}{\sim} \frac{1}{\sqrt{\delta}}\,. \tag{E.43}$$

Another Coulomb-related quantity that is singular at the spectral points is the function (8.90)

$$h(\eta) = \frac{1}{2}\left[\psi(i\eta)+\psi(-i\eta)\right] - \ln\eta\,, \tag{E.44}$$

where the Digamma function $\psi(z)$ is singular when z is a non-positive integer. The only quantity in (E.44) that becomes singular at a Coulomb spectral point is

$$\psi(i\eta) = \psi(-n-\epsilon)\,. \tag{E.45}$$

To find out how fast it goes to infinity with $\epsilon \to 0$, we apply $(n+1)$ times the recurrence relation

$$\psi(z) = \psi(z+1) - \frac{1}{z}\,.$$

As a result, we increment the argument of the function (E.45) until it becomes positive,

$$\begin{aligned}\psi(-n-\epsilon) &= \psi(1-n-\epsilon) - \frac{1}{-n-\epsilon}\\ &= \psi(2-n-\epsilon) - \frac{1}{1-n-\epsilon} - \frac{1}{-n-\epsilon} = \ldots\\ &= \psi(1-\epsilon) - \frac{1}{-\epsilon} - \frac{1}{-1} - \frac{1}{-2} - \cdots - \frac{1}{-n}\,.\end{aligned}$$

Since $\psi(1) = -\gamma = -0.5772156649\ldots$ (Euler's constant), we obtain

$$\psi(-n-\epsilon) = \frac{1}{\epsilon} - \gamma + \left(1 + \frac{1}{2} + \frac{1}{3} + \cdots + \frac{1}{n}\right),$$

and thus

$$h(\eta) \underset{\epsilon\to 0}{\longrightarrow} \frac{1}{2\epsilon} \underset{\delta\to 0}{\sim} \frac{1}{\delta}\,. \tag{E.46}$$

Integral Equations

F

There are many different kinds of integral equations. In this book, we only consider linear equations. This class of integral equations is still very wide and can be divided into four subclasses (see Ref. [9]):

1. Fredholm equation of the first kind

$$0 = g(x) + \int_a^b K(x,t)f(t)dt \tag{F.1}$$

2. Fredholm equation of the second kind

$$f(x) = g(x) + \int_a^b K(x,t)f(t)dt \tag{F.2}$$

3. Volterra equation of the first kind

$$0 = g(x) + \int_a^x K(x,t)f(t)dt \tag{F.3}$$

4. Volterra equation of the second kind

$$f(x) = g(x) + \int_a^x K(x,t)f(t)dt \tag{F.4}$$

where $f(x)$ is the unknown, while the kernel $K(x,t)$ and the inhomogeneous term $g(x)$ are given functions on the interval $[a,b]$. Generally speaking, all the functions and variables can be complex. However, if the integration runs along a complex contour, we can always introduce real parameters x and t that determine the position

S. A. Rakityansky, *Jost Functions in Quantum Mechanics*,
https://doi.org/10.1007/978-3-031-07761-6

of a point z on this contour. Therefore, without compromising generality, we may assume that the variables x and t in the above equations are real.

F.1 Separable Kernel

It is clear that there are no general recipes for finding solutions of the integral equations in the closed (analytical) form. There is, however, a particular class of problems for which such a recipe can be given. These are the problems with separable kernels, i.e. the kernels of the type:

$$K(x,t) = \sum_{i=1}^{N} p_i(x)q_i(t) \,, \tag{F.5}$$

where the two variables, (x,t), are "separated". It is not difficult to show that with such a kernel any of the equations (F.1, F.2, F.3, F.4) can be solved analytically. To give the idea, we demonstrate it here for Eq. (F.2) and for the simplest case of just one term ($N = 1$) in the kernel (F.5),

$$K(x,t) = p(x)q(t) \,. \tag{F.6}$$

Substituting this kernel in Eq. (F.2), we obtain

$$f(x) = g(x) + p(x)C \,, \tag{F.7}$$

where

$$C = \int_a^b q(x)f(x)dx \,. \tag{F.8}$$

Now we put $f(x)$ given by Eq. (F.7), into the integral (F.8),

$$C = \int_a^b q(x)g(x)dx + C\int_a^b q(x)p(x)dx$$

and find C,

$$C = \frac{\int_a^b q(x)g(x)dx}{1 - \int_a^b q(x)p(x)dx} \,,$$

which finally gives the solution of the integral equation (F.2):

$$f(x) = g(x) + \frac{p(x)\displaystyle\int_a^b q(x)g(x)dx}{1 - \displaystyle\int_a^b q(x)p(x)dx} . \tag{F.9}$$

F.2 Numerical Solution

In numerical calculations, the integrals are replaced with finite sums. As a result, an integral equation is reduced to a system of linear algebraic equations determining the values of $f(x)$ at certain discrete points in the interval $[a, b]$. Consider, for example, Eq. (F.2). Using an appropriate quadrature method, we choose N mesh points x_n covering the interval $[a, b]$ and to each of them assign a weight factor W_n such that

$$\int_a^b \phi(x)dx \approx \sum_{n=1}^{N} W_n \phi(x_n) .$$

Then, writing Eq. (F.2) at each of the mesh points in its approximate form

$$f_n = g_n + \sum_{m=1}^{N} K_{nm} W_m f_m ,$$

where $f_n = f(x_n)$, $g_n = g(x_n)$, and $K_{nm} = K(x_n, x_m)$, we obtain the linear system of equations,

$$\begin{pmatrix} 1 - K_{11}W_1 & -K_{12}W_2 & \dots & -K_{1N}W_N \\ -K_{21}W_1 & 1 - K_{22}W_2 & \dots & -K_{2N}W_N \\ \vdots & \vdots & \vdots & \vdots \\ -K_{N1}W_1 & -K_{N2}W_2 & \dots & 1 - K_{NN}W_N \end{pmatrix} \begin{pmatrix} f_1 \\ f_2 \\ \vdots \\ f_N \end{pmatrix} = \begin{pmatrix} g_1 \\ g_2 \\ \vdots \\ g_N \end{pmatrix} , \tag{F.10}$$

determining $f(x_n)$. This system can be solved using any of the methods of linear algebra.

F.3 Fredholm Theory

Erik Fredholm developed his theory from the idea of discretization outlined in the preceding section (Sect. F.2). The linear system (F.10) approximates the integral equation (F.2). The more discrete points are taken, the more accurate this approximation becomes. Fredholm derived the theorems bearing his name, by considering the limit when the number of these discrete points tends to infinity. The only requirements that he used were the finiteness of the interval $[a, b]$ and the continuity (or even a finite number of discontinuities) of the kernel. Actually, the same result can be obtained for an infinite interval as well, but for a bounded kernel. Giving here all these theorems and their lengthy proofs would be outside the scope of this book. There are many textbooks on integral equations (for example, Refs. [99–101]) where the reader can find them. We only mention two Fredholm's findings that are the most important for the theory of the Jost functions.

F.3.1 Fredholm Alternative

The inhomogeneous system (F.10) can only have solution if the determinant of the square matrix on its left hand side is nonzero. At the same time, the corresponding homogeneous system (with $g_1 = g_2 = \cdots = g_N = 0$) can only have (non-trivial) solutions if this determinant is zero. Therefore, either the homogeneous or inhomogeneous system can have solution (not both). In the case of homogeneous system, there is an infinite number of solutions (if they exist), while for the inhomogeneous system the solution is unique.

Fredholm derived the same result (the alternative) for the integral equation (F.2).

F.3.2 Fredholm Determinant and Resolvent

Equation (F.2) can be written in operator form as

$$f = g + \lambda K f \,, \tag{F.11}$$

where K is the integral operator acting in the space of functions where f and g belong to, and λ is a parameter ("strength" of the kernel) which we finally put to unity, $\lambda = 1$. Formally, the solution of this equation can be written as

$$f = A^{-1} g \,,$$

where the operator A is

$$A = 1 - \lambda K \,.$$

If the integral equation is discretized, then A is represented by a square matrix of a finite dimension. Its inverse matrix can be obtained as

$$A^{-1} = \frac{M}{\Delta} \,, \tag{F.12}$$

where $\Delta = \det A$ is the determinant of A and the square matrix M is composed of the transposed minors (i.e. determinants of a smaller dimension) of the corresponding elements of A. The determinant Δ depends on the parameter λ, and we can find its derivative with respect to this parameter. To this end, we take partial derivative of Δ over each matrix element and multiply it by the full derivative of this element over λ,

$$\frac{d\Delta}{d\lambda} = \sum_{ij} \frac{\partial \Delta}{\partial A_{ij}} \frac{dA_{ij}}{d\lambda} .$$

To find the partial derivatives, we expand the determinant over the corresponding line,

$$\Delta = \sum_{j} A_{ij} M_{ji} ,$$

which gives

$$\frac{\partial \Delta}{\partial A_{ij}} = M_{ji}$$

and brings us to well-known rule of differentiating determinants:

$$\frac{d\Delta}{d\lambda} = \mathrm{Sp}\left(M \cdot \frac{dA}{d\lambda} \right) , \tag{F.13}$$

where Sp means the trace of a matrix,

$$\mathrm{Sp}\, A = \sum_{i} A_{ii} .$$

For a kernel of an integral operator, the trace is defined as the integral:

$$\mathrm{Sp}\, K = \int_a^b K(x, x)\, dx .$$

Dividing Eq. (F.13) by Δ, we obtain

$$\frac{1}{\Delta} \frac{d\Delta}{d\lambda} = \mathrm{Sp}\left(\frac{1}{\Lambda} M \cdot \frac{dA}{d\lambda} \right) ,$$

which can be written as[1]

$$\frac{d\ \ln \Delta}{d\lambda} = \mathrm{Sp}\left(A^{-1} \cdot \frac{dA}{d\lambda}\right) = \mathrm{Sp}\left(\frac{d\ \ln A}{d\lambda}\right) = -\mathrm{Sp}\left[(1-\lambda K)^{-1}K\right]. \quad \text{(F.14)}$$

Now, we integrate both sides of this equation from 0 to λ. In doing this, we take into account that $A\big|_{\lambda=0} = I$ and thus $\Delta\big|_{\lambda=0} = 1$. We also use the fact that the trace of a sum of matrices is the sum of their traces and an integral is a sum. This means that the operations of taking trace and integrating can be interchanged, for example,

$$\int_a^b \mathrm{Sp}\, A(\lambda) d\lambda = \mathrm{Sp}\left[\int_a^b A(\lambda) d\lambda\right].$$

Therefore

$$\Delta(\lambda) = \exp\left\{\mathrm{Sp}\left[\ln A(\lambda)\right]\right\}. \quad \text{(F.15)}$$

In terms of the kernel of the integral equation (F.11), this can also be written as

$$\Delta(\lambda) = \exp\left\{-\int_0^\lambda \mathrm{Sp}\left[(1-\lambda' K)^{-1}K\right] d\lambda'\right\} = \exp\left\{\mathrm{Sp}\left[\ln(1-\lambda K)\right]\right\}. \quad \text{(F.16)}$$

All these formal derivations are not sufficient to solve the integral equation, because, in addition to the determinant Δ, we also need the operator M in Eq. (F.12). Starting from the discretized form, Fredholm managed to rigorously transform the above matrix formulae to the corresponding relations for the continuous integral operators. He also developed practical algorithm (see further down in this section) for calculating both the determinant and the operator M.

Actually, instead of the inverse operator A^{-1}, it is more convenient to construct the resolvent, R, which differs from A^{-1} by the unit operator, $R = A^{-1} - I$. With the help of the resolvent, the solution of Eq. (F.11) can be written as

$$f = g + Rg. \quad \text{(F.17)}$$

Comparing Eqs. (F.11) and (F.17), we see that

$$Rg = Kf. \quad \text{(F.18)}$$

Multiplying Eq. (F.17) from the left by K and using the relation (F.18), we obtain

$$Rg = Kg + KRg,$$

[1] A function of a matrix, such as $\ln A$, is understood as the corresponding Taylor series where the powers of the matrix are well defined.

which should be valid for any g, and thus

$$R = K + KR\,. \tag{F.19}$$

Similarly, multiplying Eq. (F.11) from the left by R, we find that

$$R = K + RK\,. \tag{F.20}$$

In the explicit form, the operator relations (F.19) and (F.20) become the following (equivalent) integral equations for the resolvent:

$$R(x,t) = K(x,t) + \int_a^b K(x,s)R(s,t)ds\,, \tag{F.21}$$

$$R(x,t) = K(x,t) + \int_a^b R(x,s)K(s,t)ds\,. \tag{F.22}$$

Using the discretization of the integrals in these equations and considering the limit when the number of the discrete points tends to infinity, Fredholm showed that the resolvent can be found as the ratio

$$R(x,t) = \frac{\mathcal{M}(x,t)}{\Delta}\,, \tag{F.23}$$

where both the numerator and denominator are represented by absolutely convergent infinite series:

$$\mathcal{M}(x,t) = \sum_{n=0}^{\infty} \frac{(-1)^n}{n!} M_n(x,t)\,, \tag{F.24}$$

$$\Delta = \sum_{n=0}^{\infty} \frac{(-1)^n}{n!} \Delta_n\,. \tag{F.25}$$

The coefficients in terms of these series are defined as

$$M_0(x,t) = K(x,t)\,, \tag{F.26}$$

$$M_n(x,t) = \int_a^b \cdots \int_a^b \det \begin{bmatrix} K(x,t) & K(x,t_1) & \cdots & K(x,t_n) \\ K(t_1,t) & K(t_1,t_1) & \cdots & K(t_1,t_n) \\ \vdots & \vdots & \vdots & \vdots \\ K(t_n,t) & K(t_n,t_1) & \cdots & K(t_n,t_n) \end{bmatrix} dt_1 dt_2 \cdots dt_n, \tag{F.27}$$

$$\Delta_0 = 1 , \tag{F.28}$$

$$\Delta_n = \int_a^b \cdots \int_a^b \det \begin{bmatrix} K(t_1,t_1) & K(t_1,t_2) & \cdots & K(t_1,t_n) \\ K(t_2,t_1) & K(t_2,t_1) & \cdots & K(t_2,t_n) \\ \vdots & \vdots & \vdots & \vdots \\ K(t_n,t_1) & K(t_n,t_2) & \cdots & K(t_n,t_n) \end{bmatrix} dt_1 dt_2 \cdots dt_n . \tag{F.29}$$

The ratio (F.23) originates from the initial discretization procedure where the elements of the inverse matrix are found as the ratios of the corresponding minors and the determinant. This is why the function $\mathcal{M}(x,t)$ and the quantity Δ are called Fredholm minor and Fredholm determinant.

No doubts that formulae (F.27) and (F.29) look cumbersome and impractical for numerical calculations. Fortunately, the coefficients of the series (F.24, F.25) obey the following recurrency relations:

$$\Delta_n = \int_a^b M_{n-1}(s,s)ds , \tag{F.30}$$

$$M_n(x,t) = \Delta_n K(x,t) - n \int_a^b K(x,s) M_{n-1}(s,t)ds , \tag{F.31}$$

which allow one to find any number of them without much effort.

F.4 Contraction Mapping Principle

The integral equations can be solved iteratively if the kernel is subject to certain restrictions. In order to obtain the conditions under which the corresponding iterative series are convergent, we apply the contraction mapping principle (see, for example, Refs. [102, 103]). It is a very powerful and general tool which is used for analyzing various iterative techniques.

Let $\mathcal{S}$ be a compact mathematical set and $d(s_1, s_2)$ an appropriately defined distance between any two of its elements $s_1 \in \mathcal{S}$ and $s_2 \in \mathcal{S}$, i.e. $\mathcal{S}$ is a metric space. Consider a mapping F of the set $\mathcal{S}$ on itself

$$F: \quad \mathcal{S} \longrightarrow \mathcal{S} .$$

This means that F transforms an element of $\mathcal{S}$ into another element of the same set. The mapping F is said to be a contraction mapping if there exists a number $0 \leqslant \alpha < 1$ such that

$$d(F(s_1), F(s_2)) \leqslant \alpha d(s_1, s_2) . \tag{F.32}$$

In other words, this mapping reduces (contracts) the distance between the elements. The importance of the contraction mapping becomes obvious when one deals with an iterative procedure. Any such procedure can be considered as the search for solution of the equation

$$s = F(s) , \tag{F.33}$$

which is also known as the fixed point equation. If the iterations,

$$s_{n+1} = F(s_n) , \tag{F.34}$$

are convergent, the self-consistency is achieved when $n \to \infty$. The limit of the sequence $s_1, s_2, \ldots$ is called the fixed point (because the mapping F does not move it).

Stefan Banach proved that a contraction mapping on a complete metric space has a unique fixed point and that convergence of successive approximations to the fixed point occurs independently of the choice of the initial approximation. The proof of this extremely powerful theorem is amazingly simple. Indeed, applying the condition (F.32) n times, we obtain

$$d(s_{n+1}, s_n) \leqslant \alpha^n d(s_1, s_0) , \tag{F.35}$$

where s_0 is an arbitrary point in $\mathcal{S}$. Now, if $m > n$, then the triangular inequality for the metric gives

$$\begin{aligned} d(s_m, s_n) &\leqslant d(s_m, s_{m-1}) + d(s_{m-1}, s_{m-2}) + \cdots + d(s_{n+2}, s_{n+1}) + d(s_{n+1}, s_n) \\ &\leqslant (\alpha^{m-1} + \alpha^{m-2} + \cdots + \alpha^{n+1} + \alpha^n) d(s_1, s_0) \\ &= \alpha^n (1 + \alpha + \cdots + \alpha^{m-n-1}) d(s_1, s_0) \\ &\leqslant \alpha^n \left(\sum_{k=0}^{\infty} \alpha^k \right) d(s_1, s_0) = \frac{\alpha^n}{1-\alpha} d(s_1, s_0) \underset{n\to\infty}{\longrightarrow} 0 . \end{aligned}$$

The last means that $\{s_n\}$ is a Cauchy sequence and thus convergent. The uniqueness of its limit $\lim_{n\to\infty} s_n = \sigma$ is proved by contradiction. Suppose there is another limit σ', then both σ and σ' are the fixed points (i.e. immovable by the operation F). Therefore

$$d(\sigma, \sigma') = d(F(\sigma), F(\sigma')) \leqslant \alpha d(\sigma, \sigma') ,$$

which is only possible if $d(\sigma, \sigma') = 0$ because $\alpha < 1$. This completes the proof of the Banach theorem.

F.5 Contraction Mapping for Fredholm Equation

The integral equation (F.2) can be considered as a fixed point equation

$$f(x) = F[f(x)]$$

for the mapping

$$F[f(x)] = g(x) + \int_a^b K(x,t) f(t) dt \tag{F.36}$$

on the space of continuous functions with the metric (i.e. with the distance between any two elements of the space)

$$d(f_1, f_2) = \max_{t \in [a,b]} |f_1(t) - f_2(t)| \,. \tag{F.37}$$

Therefore in order to check if the iterative procedure for an integral equation is convergent, we only need to check if the corresponding mapping is a contraction mapping. Let the integral of the absolute value of the kernel be bounded, i.e.

$$\max_{x \in [a,b]} \int_a^b |K(x,t)| dt = B < \infty \,. \tag{F.38}$$

For two arbitrarily chosen continuous functions (continuity is needed to be bounded on $[a, b]$), we have

$$\begin{aligned} d(F(f_1), F(f_2)) &= \max_{x \in [a,b]} \left| \int_a^b K(x,t)[f_1(t) - f_2(t)] dt \right| \\ &\leqslant \max_{x \in [a,b]} \int_a^b |K(x,t)| \, |f_1(t) - f_2(t)| \, dt \\ &\leqslant B \max_{t \in [a,b]} |f_1(t) - f_2(t)| \,. \end{aligned}$$

Therefore

$$d(F(f_1), F(f_2)) \leqslant B d(f_1, f_2) \,, \tag{F.39}$$

which means that the iterations for the Fredholm equation are convergent if

$$\max_{x \in [a,b]} \int_a^b |K(x,t)| dt < 1 \,. \tag{F.40}$$

We can generalize all the derivations of this section for a matrix Fredholm equation. Let all the functions in Eq. (F.36) be $(N \times N)$-matrices. Then the distance (F.37) between any two matrix functions can be defined as the maximal difference between the corresponding matrix elements on the whole interval of the independent variable:

$$d(f_1, f_2) = \max_{i,j,t\in[a,b]} \left|[f_1]_{ij}(t) - [f_2]_{ij}(t)\right| . \tag{F.41}$$

If we also generalize the definition (F.38) of the kernel-norm as

$$B = \max_{i,x\in[a,b]} \int_a^b \sum_{j=1}^{N} |K_{ij}(x,t)| dt , \tag{F.42}$$

then it is not difficult to see that we obtain the same relation (F.39), and the iteration series is convergent under the same condition, $B < 1$.

F.6 Contraction Mapping for Volterra Equation

With the same assumption (F.38) as for the Fredholm equation, the mapping that corresponds to the Volterra equation,

$$f(x) = g(x) + \int_a^x K(x,t) f(t) dt , \tag{F.43}$$

has different properties, namely, it is always a contraction mapping, i.e. with any finite B. This is because the upper limit of the integral is variable. This can be shown as follows. When the upper limit is less than b, the bound in Eq. (F.38) cannot be greater than B. Moreover, when $x \to a$, the integral vanishes. This means that there is a number $C \leqslant B$ such that

$$\int_a^x |K(x,t)| dt \leqslant C \frac{x-a}{b-a} , \tag{F.44}$$

i.e. for all $x \in [a, b]$, the mean value of $|K(x, t)|$ in the interval $t \in [a, b]$ is less than $C/(b - a)$. Therefore

$$\begin{aligned}
\left|F[f_1(x)] - F[f_2(x)]\right| &\leqslant \int_a^x |K(x,t)|\,|f_1(t) - f_2(t)|\, dt \\
&\leqslant \max_{t\in[a,b]} |f_1(t) - f_2(t)| \int_a^x |K(x,t)| dt \\
&\leqslant C \frac{x-a}{b-a} d(f_1, f_2) ,
\end{aligned}$$

where the difference between $F[f_1(x)]$ and $F[f_2(x)]$ linearly depends on x. This dependence has very important implications. It can be shown by induction that

$$\left|F^n[f_1(x)] - F^n[f_2(x)]\right| \leqslant \left(C\frac{x-a}{b-a}\right)^n \frac{1}{n!} d(f_1, f_2)\,, \tag{F.45}$$

which is obviously true for $n = 1$. If we assume that (F.45) is valid for $n = k$, then for $n = k + 1$

$$\left|F^{k+1}[f_1(x)] - F^{k+1}[f_2(x)]\right| \leqslant \int_a^x |K(x,t)| \left|F^k[f_1(t)] - F^k[f_2(t)]\right| dt \leqslant$$

$$\leqslant \frac{C}{b-a}\int_a^x \left(C\frac{t-a}{b-a}\right)^k \frac{1}{k!} d(f_1, f_2)dt = \left(C\frac{x-a}{b-a}\right)^{k+1} \frac{1}{(k+1)!} d(f_1, f_2)\,.$$

Therefore

$$d\left(F^n[f_1], F^n[f_2]\right) \leqslant \frac{C^n}{n!} d(f_1, f_2)\,, \tag{F.46}$$

where, with any finite C, the contraction factor can always be made less than 1,

$$\frac{C^n}{n!} < 1\,, \tag{F.47}$$

if n is large enough. This means that under the above assumption about the kernel, the iterations for the Volterra equation are always convergent, independently of the choice of the initial approximation, which is usually taken as $f(x) \approx g(x)$.

Using the same definitions (F.41) and (F.42) of the distance in the space of the matrix-functions and of the kernel norm, we can easily generalize the above derivations for the matrix Volterra equation (F.43).

There is a corollary from the established convergence of the integral equations, which is very important for the scattering theory. If $g(x)$ or $K(x,t)$ (or both) analytically depends on a parameter, then $f(x)$ is also an analytic function of that parameter. Indeed, the iteration series for the integral equation converges uniformly with respect to the parameter because the condition (F.47) does not involve it. The integration of an analytic function produces another analytic function.

Poincaré Theorem

G

There is an important theorem in the theory of ordinary differential equations that is ascribed to Henri Poincaré:

> *A solution of a linear differential equation is an analytic function of a parameter E, if the coefficients of this equation analytically depend on E and the boundary conditions do not depend on this parameter.*

Usually, when this theorem is used in physics literature, it is given reference to the paper [104] (see, for example, theorem number 7 used in the book [105] or section 12.1 of the book [13]). However, in the cited paper by Poincaré, it is impossible to find either the above statement or its proof. Moreover, it is difficult to find an understandable proof of such a statement in any of well-known textbooks on differential equations. To fill this gap, it is given here a simple proof of a similar statement, which perhaps is not sufficiently rigorous, and is only applicable for a particular form of the differential equation that is relevant to the theory presented in this book.

Consider the following first-order differential equation for the unknown matrix-function $f(E, r)$:

$$\partial_r f(E, r) - K(E, r) f(E, r) = 0 , \tag{G.1}$$

with the boundary condition at $r = \varepsilon$,

$$f(E, \varepsilon) = g(E) , \tag{G.2}$$

where E is a complex parameter and g is a constant matrix depending on the parameter E. Integrating both terms of Eq. (G.1), we transform it in the integral

S. A. Rakityansky, *Jost Functions in Quantum Mechanics*,
https://doi.org/10.1007/978-3-031-07761-6

equation of Volterra type:

$$f(E,r) = g(E) + \int_{\varepsilon}^{r} K(E,r')f(E,r')\,dr' \,. \tag{G.3}$$

Let us assume that the kernel of this integral equation obeys the following condition:

$$\int_{\varepsilon}^{\infty} \sum_{ij} |K_{ij}(E,t)|dt < \infty \qquad \forall\, E \in \mathcal{D} \tag{G.4}$$

for all values of the parameter E within a complex domain $\mathcal{D}$. Let also the matrices $K(E,t)$ (for all $t \in [\varepsilon,\infty)$) and $g(E)$ be analytic within $\mathcal{D}$.

Equation (G.3) is a particular case of the Volterra equation (F.43) with x-independent kernel and x-independent inhomogeneous term, g. The condition (G.4) guaranties that the constant C in (F.46) is finite. Therefore the iteration series

$$\begin{aligned} f(E,r) &= g(E) + \int_{\varepsilon}^{r} dr' K(E,r')g(E) + \\ &+ \int_{\varepsilon}^{r} dr' K(E,r') \int_{\varepsilon}^{r'} dr'' K(E,r'')g(E) + \cdots \\ &= g(E) + \mathcal{F}_1(E,r) + \mathcal{F}_2(E,r) + \cdots \,, \end{aligned} \tag{G.5}$$

where

$$\mathcal{F}_n(E,r) = \int_{\varepsilon}^{r} K(E,r')\mathcal{F}_{n-1}(E,r')\,dr' \,, \tag{G.6}$$

is convergent. It converges uniformly with respect to the parameter $E \in \mathcal{D}$ because neither the condition (F.42) nor the relation (F.46) depends on E. Each term in the series (G.6) is analytic in $\mathcal{D}$ because they are the integrals (G.6) of analytic functions. Since the series is uniformly convergent, its sum, i.e. the solution $f(E,r)$ of the original differential equation (G.1), is analytic function of E within the domain $\mathcal{D}$.

This, actually, proves the Poincaré theorem in our particular case. As is seen, in contrast to Poincaré, we do not require E-independence of the boundary conditions, $g(E)$. This is because we assume that the differential equation (G.1) does not have singular points. To be precise, it does not have singularities for $r \geqslant \varepsilon$. The original Poincaré theorem remains valid even for the equation with a singular point.

Newton Method for Locating Zeros of a Complex Function

H

In search for bound states and resonances, we look for zeros of the Jost function or of the Jost matrix determinant (in the multi-channel problems). The most efficient numerical method for such a task is based on the algorithm which bears the name of Newton. Sometimes it is also called the Newton–Raphson method.[1] The idea of the method is based on the approximation of the function by the first two terms of the Taylor series:

$$f(z + \Delta z) \approx f(z) + f'(z)\Delta z \,. \tag{H.1}$$

Let z be our starting point on the complex plain and we need to move from it to another point, $z + \Delta z$, such that $f(z + \Delta z) = 0$. Then, according to Eq. (H.1), the step Δz should be taken as

$$\Delta z \approx -\frac{f(z)}{f'(z)} \,. \tag{H.2}$$

Since Eq. (H.1) is not exact, the move in the step Δz does not bring us exactly to the zero of the function, but the new point, $z + \Delta z$, becomes closer to the nearest zero of the function $f(z)$. If we repeat the procedure, starting from the new point, then we find ourselves even closer to the zero. In this way, each new point z_{i+1} is obtained from the previous one, z_i, by adding the corresponding step Δz:

$$z_{i+1} = z_i - \frac{f(z_i)}{f'(z_i)} \,. \tag{H.3}$$

As is seen, to find the necessary step, we need to calculate the values of the function and of its derivative at the previous point, z_i. It is assumed that the function itself

[1] See, for example, Ref. [106].

S. A. Rakityansky, *Jost Functions in Quantum Mechanics*,
https://doi.org/10.1007/978-3-031-07761-6

can be calculated for any given z. What about its derivative? Of course it can always be found numerically via the finite difference formula. This may be good for the first steps of the iteration procedure (H.3). However, when z_i approaches the exact zero of the function, the accuracy of the simple finite difference formula becomes insufficient. There is an alternative and robust method for finding $f'(z_i)$. This method is based on local approximation of the function $f(z)$ around the point z_i by a quadratic polynomial. Then the derivative is obtained analytically by differentiating this polynomial.

Suppose we found the values f_1, f_2, and f_3 of the function $f(z)$ at the three points z_1, z_2, and z_3. The quadratic polynomial coinciding with $f(z)$ at these three points can be constructed with the help of the classical Lagrange formula (see Ref. [106]):

$$Q(z) = \frac{(z-z_2)(z-z_3)}{(z_1-z_2)(z_1-z_3)} f_1 + \frac{(z-z_1)(z-z_3)}{(z_2-z_1)(z_2-z_3)} f_2 + \frac{(z-z_1)(z-z_2)}{(z_3-z_1)(z_3-z_2)} f_3 \,.$$

By the direct substitution of z_1, z_2, or z_3 in this formula, it is easy to see that $Q(z)$ indeed gives f_1, f_2, and f_3, respectively. The derivative of the polynomial, which approximates $f'(z)$, is found as

$$Q'(z) = \frac{2z-z_2-z_3}{(z_1-z_2)(z_1-z_3)} f_1 + \frac{2z-z_1-z_3}{(z_2-z_1)(z_2-z_3)} f_2 + \frac{2z-z_1-z_2}{(z_3-z_1)(z_3-z_2)} f_3 \,.$$

Therefore, to find a root of a complex function, we guess the starting point, z_1, and choose two additional (nearby) points, z_2 and z_3. Calculating the function at these three starting points, we construct the polynomial $Q(z)$ and obtain its derivative $Q'(z_3)$. Then, using the iteration formula (H.3), we find the new point z_4 (which is closer to the root). After that, we repeat the iteration with the new starting points, z_2, z_3, and z_4. Schematically, the algorithm is depicted in Fig. H.1.

In most cases this iterative procedure converges very rapidly to the root which is nearest to the initial starting point, z_1. Typically, ~10 iterations are sufficient to find the root with an accuracy of five decimal places. The more iterations are done, the

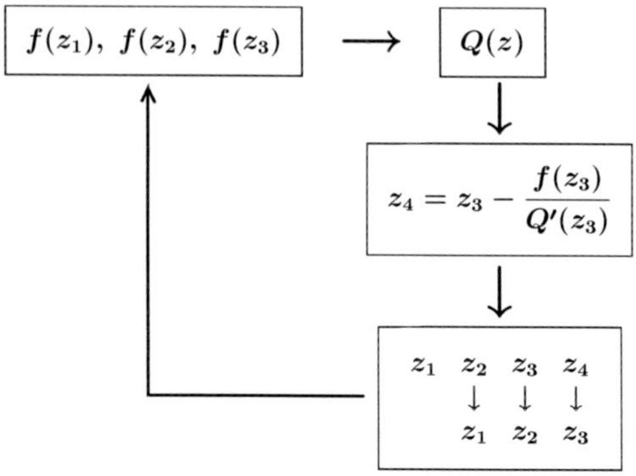

Fig. H.1 Algorithm of the Newton–Raphson method for finding roots of a complex function

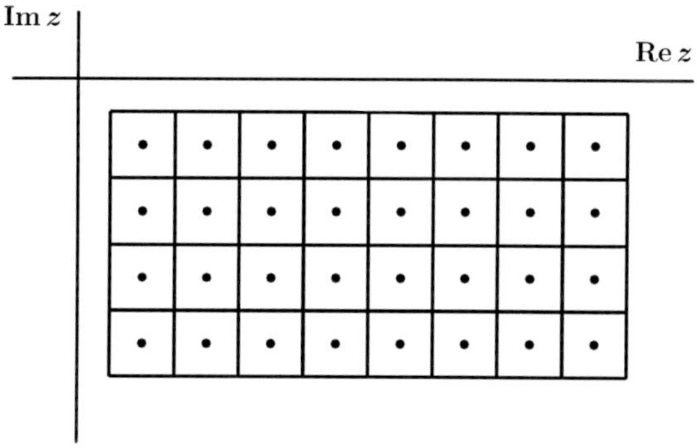

Fig. H.2 Dots are the starting points for the Newton–Raphson algorithm. All of them should be used for finding all possible roots of a complex function within the rectangular domain

more accurately the root is located. The iterations can be stopped when the absolute value of the function at the new point becomes less than a specified tolerance.

In some rare cases, when the initial guess is very wrong or the function is very rough, the iterations may go astray. Therefore, provisions should be made in the code to stop the iterations if z goes outside a chosen big circle.

Very often we do not know how many roots a given function has, if any. And our task is to find all of them within certain domain of the complex plain, for example, within the rectangular domain shown in Fig. H.2. To do it automatically, we can subdivide this rectangular domain into small squares. Then we go from one square to the other and repeat the search starting from the points in their centres. The Newton–Raphson algorithm converges to the nearest root. If we use sufficiently small squares, we cannot miss any of possible roots. Of course, some of them will be found several times. Therefore, after finishing the search, the code should compare all the roots and remove the repetitions.

Choice of the Units

Very often in research papers (especially in theoretical ones), the artificial units are used, such that, for example, $\mu = \hbar = 1$. This may seem as simplification of notation and formulae. However, when it comes to actual calculations, such a simplification may cause some troubles in putting in place all necessary coefficients. Therefore, throughout this book, in all the formulae that involve the mass and Planck's constant, they are written explicitly.

When doing numerical calculations, we always try to avoid extremely small as well as extremely large numbers. This can be achieved if the units appropriate for a particular problem are used. The choice of the units is determined by the scale of the physical system under consideration. First of all, we should look at the typical size of the system and the typical values of the energies that we are dealing with.

In all cases, when numerically solving the radial differential equations, it is needed to transform all the quantities into the units of the powers of length. This can be done with the help of the constant

$$\hbar c = 197.3269718\,\text{MeV} \cdot \text{fm}\,, \tag{I.1}$$

where the product of Planck's constant and the speed of light is given in the units that are most suitable for nuclear systems. Of course, if necessary, it can be transformed into any other units. Experimentalists never stop to do new measurements and to improve the accuracy of this product. So, it is advisable to search for its latest value before doing the calculations.

In the differential equations, the physical potential $U(r)$ is always present in the reduced form, i.e.

$$V(r) = \frac{2\mu}{\hbar^2} U(r)\,.$$

S. A. Rakityansky, *Jost Functions in Quantum Mechanics*,
https://doi.org/10.1007/978-3-031-07761-6

Therefore, in the computer code, it should be used as

$$V(r) = \frac{2\mu c^2}{(\hbar c)^2} U(r) ,$$

where the mass is replaced with its energy equivalent, μc^2. Since the physical potential is measured in the units of energy, the reduced one is in the units of $[\text{length}]^{-2}$. Similarly, the wave number (momentum) k

$$k = \sqrt{\frac{2\mu}{\hbar^2} E} = \sqrt{\frac{2\mu c^2}{(\hbar c)^2} E}$$

has the units of $[\text{length}]^{-1}$.

I.1 Nuclear Units

Scale: $r \sim 1\,\text{fm} = 10^{-13}\,\text{cm} , \qquad E \sim 1\,\text{MeV}$

The most simple units are used in nuclear physics, where the masses, potentials, and energies are usually given in MeV. So, in the code, it is sufficient to divide them by $\hbar c$ (and to forget about $\hbar$ in all the formulae), and then all the other quantities automatically become powers of fm. The final results, if they are, for example, the resonance energies and widths, should be transformed back to MeV by multiplication by $\hbar c$. The half-life of a resonance,

$$T_{1/2} = \frac{\hbar \ln 2}{\Gamma} ,$$

corresponding to the width Γ, can be found with the help of

$$\hbar = 6.582119569 \times 10^{-22}\,\text{MeV} \cdot \text{s} .$$

The calculated cross sections that are obtained in the units of fm^2 are usually presented in barns or millibarns, using the relation

$$1\,\text{fm}^2 = 10^{-2}\,\text{barn} .$$

I.2 Atomic Units

Scale: $r \sim 1\,\text{Å} = 10^{-8}\,\text{cm}\,, \qquad E \sim 1\,\text{eV}$

When dealing with atoms, it is convenient to take the mass of electron, m_e, and its charge, e, as the units of mass and charge, the Bohr radius

$$a_0 = \frac{\hbar^2}{m_e e^2} = 0.529177210903\,\text{Å}\,,$$

as the unit of length, and the Hartree energy,

$$E_h = \frac{m_e e^4}{\hbar^2} = 27.211386245988\,\text{eV}\,,$$

as the unit of energy. Half of this energy (which is equal to the binding energy of electron in the hydrogen atom) is called one Rydberg,

$$1\,\text{Ry} = E_h/2\,.$$

When doing numerical calculations in these units, the constants e and $\hbar$, appearing in the equations, should be put to unity, $e = \hbar = 1$. The mass μ should be divided by the electron mass, μ/m_e, and all the energies should be divided by E_h. Of course, the final results for the energies and cross sections should be converted back.

Bibliography

1. P.A.M. Dirac, *The Principles of Quantum Mechanics*, Revised Fourth edition (Oxford University Press, 1967)
2. E. Fermi, *Notes on Quantum Mechanics*, 2nd edn. (The University of Chicago Press, 1994)
3. R.P. Feynman, R.B. Leighton, M. Sands, *Feynman Lectures on Physics*, vol. 3 (Addison-Wesley Publishing Company, 1964)
4. J.R. Taylor, *Scattering Theory: The Quantum Theory on Nonrelativistic Collisions* (Wiley, New York, 1972)
5. D.A. Varshalovich, A.N. Moskalev, V.K. Khersonskii, *Quantum Theory of Angular Momentum* (World Scientific, Singapore, 1988)
6. H. Jeffreys, On certain approximate solutions of linear differential equations of the second order. Proc. Lond. Math. Soc. **s2–23**(1), 428–436 (1925)
7. N. Fröman, P.O. Fröman, *JWKB Approximation* (North Holland Publishing Company, Amsterdam, 1965)
8. E.T. Whittaker, G.N. Watson, *A Course of Modern Analysis*, 5th edn. (Cambridge University Press, Cambridge, 2021)
9. K.F. Riley, M.P. Hobson, S.J. Bence, *Mathematical Methods for Physics and Engineering*, 4th edn. (Cambridge University Press, Cambridge, 2006)
10. R. Jost, Über die falschen Nullstellen der Eigenwerte der S-Matrix. Helv. Phys. Acta **20**, 256–266 (1947)
11. B.A. Fuchs, B.V. Shabat, *Functions of a Complex Variable and Some of Their Applications*, vol. I (Pergamon Press, Oxford, 1964)
12. E.C. Titchmarsh, *The Theory of Functions*, 2nd edn. (Oxford University Press, London, 1939)
13. R.G. Newton, *Scattering Theory of Waves and Particles*, 2nd edn. (Springer Science+Business Media, New York, 1982)
14. E. Brändas, N. Elander (eds.), *Resonances*, Lecture Notes in Physics, vol. 325 (Springer, Berlin, Heidelberg, 1989)
15. S.N. Ershov, S.A. Rakityansky, Jost matrices for some analytically solvable potential models. Phys. Rev. **C103**, 024612 (2021)
16. R.P. Boas, *Invitation to Complex Analysis* (Random House, New York, 1992)
17. V. Bargmann, On the number of bound states in a central field of force. Proc. Natl. Acad. Sci. U. S. A **38**, 961–966 (1952)
18. F. Nichitiu, *Phase-Shift Analysis in Physics of Nuclear Interactions* (Mir, Moscow, 1983)
19. H. Jeffreys, B.S. Jeffreys, *Methods of Mathematical Physics*, 3rd edn. (Cambridge University Press, Cambridge, 1988)
20. N.G. van Kampen, S Matrix and causality condition. II. Nonrelativistic particles. Phys. Rev. **91**, 1267 (1953)
21. N. Michel, Precise Coulomb wave functions for a wide range of complex ℓ, η, and z. Comput. Phys. Commun. **176**, 232–249 (2007)
22. T. Regge, Introduction to complex orbital momenta. Nuovo Cimento **XIV**(5), 951–976 (1959)

S. A. Rakityansky, *Jost Functions in Quantum Mechanics*,
https://doi.org/10.1007/978-3-031-07761-6

23. S.C. Frautschi, *Regge Poles and S-Matrix Theory* (W.A. Benjamin, New York, 1963)
24. S. Franke-Arnold, S.M. Barnett, E. Yao, J. Leach, J. Courtial, M. Padgett, Uncertainty principle for angular position and angular momentum. New J. Phys. **6**, 103 (2004)
25. E. Merzbacher, *Quantum Mechanics*, 3rd edn. (Wiley, New York, 1998)
26. F.D. Gakhov, *Boundary Value Problems* (Pergamon Press, Oxford, 1966)
27. J. Liouville, Sur le développement des fonctions et séries. J. Math. Pures Appl. **1**, 16–35 (1837)
28. G. Green, On the motion of waves in a variable canal of small depth and width. Trans. Camb. Philos. Soc. **6**, 457–462 (1837)
29. M.V. Fedoryuk, *Asymptotic Analysis* (Springer, Berlin, Heidelberg, 1993)
30. S.A. Rakityansky, S.A. Sofianos, K. Amos, A method of calculating the Jost function for analytic potentials. Nuovo Cimento **B111**, 363–378 (1996)
31. S.A. Sofianos, S.A. Rakityansky, Exact method for locating potential resonances and Regge trajectories. J. Phys. **A30**, 3725–3737 (1997)
32. M. Rittby, N. Elander, E. Brändas, *On the Analytical Continuation of the Partial Wave S-matrix Using Complex Scaling Techniques*. Lecture Notes in Physics, vol. 325 (Springer, Berlin, Heidelberg, 1989), pp. 129–152
33. P.G. Burke, *Potential Scattering in Atomic Physics* (Plenum Press: New York, 1977)
34. A. Messiah, *Quantum Mechanics* (Dover Publications, Mineola, 2014)
35. D. Gaspard, J.M. Sparenberg, Effective-range function methods for charged particle collisions. Phys. Rev. C **97**, 044003 (2018)
36. S.A. Rakityansky, S.A. Sofianos, Jost function for coupled partial waves. J. Phys. A Math. Gen. **31**, 5149–5175 (1998)
37. J.M. Blatt, L.C. Biedenharn, Neutron-proton scattering with spin-orbit coupling. I. General expressions. Phys. Rev. **86**, 399–404 (1952)
38. H.P. Stapp, T.J. Ypsilantis, N. Metropolis, Phase-shift analysis of 310-MeV proton-proton scattering experiments. Phys. Rev. **105**, 302–310 (1957)
39. S.A. Rakityansky, N. Elander, Analyzing the contribution of individual resonance poles of the *S*-matrix to two-channel scattering. Int. J. Quantum Chem. **106**, 1105–1129 (2006)
40. S.A. Rakityansky, Partial widths of a multi-channel resonance. J. Phys. Conf. Seri. **915**, 012008 (2017)
41. R.M. Id Betan, A.T. Kruppa, T. Vertse, Shadow poles in coupled-channel problems calculated with the Berggren basis. Phys. Rev. **C97**, 024307 (2018)
42. J.B. Conway, *Functions of One Complex Variable*, 2nd edn. (Springer, New York, 1978)
43. P. Fröbrich, R. Lipperheide, *Theory of Nuclear Reactions* (Clarendon Press, Oxford, 1996)
44. M.D. Semon, J.R. Taylor, Scattering by potentials with Coulomb tails. Nuovo Cimento **26A**(1), 48–58 (1975)
45. H.A. Bethe, Theory of the effective range in nuclear scattering. Phys. Rev. **76**(1), 38–50 (1949)
46. M.L. Goldberger, K.M. Watson, *Collision Theory* (Wiley, New York, 1964)
47. Y.H. Uang, W.C. Stwalley, The possibility of a $^4He_2$ bound state, effective range theory, and very low energy He-He scattering. J. Chem. Phys. **76**(10), 5069–5072 (1982)
48. S.J. Buckman, J. Mitroy, Analysis of low-energy electron scattering cross sections via effective-range theory. J. Phys. B At. Mol. Opt. Phys. **22**, 1365–1384 (1989)
49. M. Marinescu, Computation of the scattering length and effective range in molecular physics. Phys. Rev. **A50**, 3177–3180 (1994)
50. F. Arnecke, H. Friedrich, J. Madronero, Effective-range theory for quantum reflection amplitudes. Phys. Rev. **A74**, 062702 (2006)
51. Z. Idziaszek, G. Karwasz, Applicability of modified effective-range theory to positron-atom and positron-molecule scattering. Phys. Rev. **A73**, 064701 (2006)
52. K. Willner, F.A. Gianturco, Low-energy expansion of the Jost function for long-range potentials. Phys. Rev. **A74**, 052715 (2006)
53. D. Minossi, E. Epelbaum, M. Pavón Valderrama, A. Nogga, Modified effective range expansion for nucleon-nucleon scattering. Eur. Phys. J. Web Conf. **3**, 05018 (2010)

54. K. Fedus, Z. Idziaszek, G.P. Karwasz, Positron and electron scattering on atoms and molecules - modified effective range theory revisited. Eur. Phys. J. Spec. Top. **222**, 2335–2344 (2013)
55. O.L. Ramírez Suárez, J.M. Sparenberg, Phase-shift parametrization and extraction of asymptotic normalization constants from elastic-scattering data. Phys. Rev. C **96**, 034601 (2017)
56. Z. Wang, S. Tan, Scattering hypervolume of spin-polarized fermions. Phys. Rev. A **104**, 043319 (2021)
57. W.H. Press, S.A. Teukolsky, W.T. Vetterling, B.P. Flannery, *Numerical Recipes in Fortran 77: The Art of Scientific Computing*, 2nd edn. (Press Syndicate of the University of Cambridge, New York, 1997)
58. S.A. Rakityansky, N. Elander, Generalized effective-range expansion. J. Phys. **A42**, 225302 (2009)
59. S.A. Rakityansky, N. Elander, Multi-channel analog of the effective-range expansion. J. Phys. **A44**, 115303 (2011)
60. S.A. Rakityansky, N. Elander, A method for extracting the resonance parameters from experimental cross-sections. Int. J. Mod. Phys. **E22**, 1350032 (2013)
61. S.A. Rakityansky, S.N. Ershov, Jost-matrix analysis of the resonance $^5\mathrm{He}^*(\frac{3}{2}^+)$ near the dt-threshold. Int. J. Mod. Phys. **E28**, 1950064 (2019)
62. P. Vaandrager, S.A. Rakityansky, Residues of the S-matrix for several $\alpha^{12}\mathrm{C}$ resonances from the Jost function analysis. Nucl. Phys. **A992**, 121627 (2019)
63. S.A. Rakityansky, S.N. Ershov, T.J. Tshipi, Resonance states 0^+ of the Boron isotope $^8\mathrm{B}$ from the Jost-matrix analysis of experimental data. Int. J. Mod. Phys. **E28**, 1950083 (2019)
64. R.M. Adam, S.A. Sofianos, Quantum bound states in a C–C60 system. Few-Body Syst. **56**, 133–138 (2015)
65. K. Waiblinger, J.R. Williams, J.P. D'Incao, Quenched magneto-association of ultracold Feshbach molecules. Phys. Rev. A **104**, 033310 (2021)
66. E.A. Koval, O.A. Koval, Aspects of arbitrarily oriented dipoles scattering in a plane: Short-range interaction influence. Phys. Rev. A **102**, 042815 (2020)
67. W.M. Frank, D.J. Land, Singular potentials. Rev. Mod. Phys. **43**, 36–98 (1971)
68. S.A. Sofianos, S.A. Rakityansky, S.E. Massen, Jost function for singular potentials. Phys. Rev. A **60**, 337–343 (1999)
69. E. Johnson, Low-dimensional semiconductor structures: fundamentals and device applications, ed. by K. Barnhamand, D. Vvedensky (Cambridge University Press, Cambridge, 2001)
70. H. Ünlü, N.J. Horing (ed.), *Low Dimensional Semiconductor Structures: Characterization, Modeling and Applications* (Springer, Heidelberg, 2013)
71. M. Mohammed, A.S. Verhulst, D. Verreck, M. Van de Put, E. Simoen, B. Sorée, B. Kaczer, R. Degraeve, A. Mocuta, N. Collaert, A. Thean, G. Groeseneken, Electric-field induced quantum broadening of the characteristic energy level of traps in semiconductors and oxides. J. Appl. Phys. **120**, 245704 (2016)
72. S.A. Rakityansky, Modified transfer matrix for nanostructures with arbitrary potential profile. Phys. Rev. B **70**, 205323 (2004)
73. S.A. Rakityansky, N. Elander, Analytic structure and power series expansion of the Jost function for the two-dimensional problem. J. Phys. A Math. Theory **45**, 135209 (2012)
74. L.S. Ferreira, A.C. Fonseca, L. Streit, *Models and Methods in Few-Body Physics*. Lecture Notes in Physics, vol. 273 (Springer, Berlin, 1989), pp. 283–323
75. S.A. Sofianos, S.A. Rakityanskyy, G.P. Vermaak, Subthreshold resonances in few-neutron systems. J. Phys. G Nucl. Part. Phys. **23**, 1619–1629 (1997)
76. V.B. Belyaev, S.A. Rakityansky, W. Sandhas, Three-body resonances Λnn and $\Lambda\Lambda n$. Nucl. Phys. A **803**, 210–226 (2008)
77. S.A. Rakityansky, I.M. Gopane, Effective-range parameters and vertex constants for Λ-nuclear systems. Int. J. Mod. Phys. E **26**(4), 1750014 (2017)
78. Z.S. Agranovich, V.A. Marchenko, *The Inverse Problem of Scattering Theory* (Gordon and Breach Science Publishers, New York, 1963)

79. K. Chadan, P.C. Sabatier, *Inverse Problems in Quantum Scattering Theory* (Springer, Berlin, 1977)
80. R.G. Newton, *Inverse Schrödinger Scattering in Three Dimensions* (Springer, Berlin, 1989)
81. B.N. Zakhariev, A.A. Suzko, *Direct and Inverse Problems. Potentials in Quantum Scattering* (Springer, Berlin, Heidelberg, 1990)
82. A.G. Ramm, *Multidimensional Inverse Scattering Problems* (Wiley, New York, 1992)
83. G.M.L. Gladwell, *Inverse Problems in Scattering: An Introduction* (Springer Science + Business Media, Dordrecht, 1993)
84. K. Chadan, D. Colton, L. Päivärinta, W. Rundell, *An Introduction to Inverse Scattering and Inverse Spectral Problems* (SIAM Monographs on Mathematical Modeling and Computation, Philadelphia, 1997)
85. H. Isozaki, *Inverse Spectral and Scattering Theory: An Introduction*. Springer Briefs in Mathematical Physics, vol. 38 (Springer, Singapore, 2020)
86. M.J. Ablowitz, P.A. Clarkson, *Solitons, Nonlinear Evolution Equations and Inverse Scattering* (Cambridge University Press, Cambridge, 1991)
87. B. Guo, X.F. Pang, Y.F. Wang, N. Liu, *Solitons* (Walter de Gruyter GmbH, Berlin, 2018)
88. A. Gangopadhyaya, J.V. Mallow, C. Rasinariu, *Supersymmetric Quantum Mechanics: An Introduction* (World Scientific, Singapore, 2011)
89. V. Bargmann, On the connection between phase shifts and scattering potential. Rev. Mod. Phys. **21**(3), 488–493 (1949)
90. J.R. Cox, Many-channel Bargmann potentials. J. Math. Phys. **5**(8), 1065–1069 (1964)
91. D. Baye, J.M. Sparenberg, A.M. Pupasov-Maksimov, B.F. Samsonov, Single- and coupled-channel radial inverse scattering with supersymmetric transformations. J. Phys. A Math. Theor. **47**, 243001 (75pp) (2014)
92. C. Eckart, The penetration of a potential barrier by electrons. Phys. Rev. **35**(11), 1303–1309 (1930)
93. G.F. Carrier, M. Krook, C.E. Pearson, *Functions of a Complex Variable: Theory and Technique* (SIAM, Philadelphia, 2005)
94. E. Freitag, R. Busam, *Complex Analysis* (Springer, Berlin, Heidelberg, 2005)
95. G.N. Watson, *A Treatise on the Theory of Bessel Functions*, 2nd edn. (Cambridge University Press, Cambridge, 1952)
96. M.H. Hull, G. Breit, Coulomb wave functions, in *Encyclopedia of Physics*, ed. by S. Flügge, vol. XLI/1 (Springer, Berlin, 1959), pp. 408–465
97. J. Humblet, Analytic structure and properties of coulomb wave functions for real and complex energies. Ann. Phys. **155**, 461–493 (1984)
98. D. Gaspard, Connection formulas between Coulomb wave functions. J. Math. Phys. **59**(11), 112104 (2018)
99. W.V. Lovitt, *Linear Integral Equations* (Dover Publications, New York, 1950)
100. R.P. Kanwal, *Linear Integral Equations: Theory and Technique* (Academic Press, New York, 1971)
101. H. Hochstadt, *Integral Equations* (Wiley, New York, 1973)
102. A. Granas, J. Dugundji, *Fixed Point Theory* (Springer, New York, 2003)
103. P. Agarwal, M. Jleli, B. Samet, *Fixed Point Theory in Metric Spaces* (Springer Nature, Singapore, 2018)
104. H. Poincaré, Sur les Groupes des Équations Linéaires. Acta Math. **4**, 201–312 (1884)
105. V. de Alfaro, T. Regge, *Potential Scattering* (North-Holland Publishing Company, Amsterdam, 1965)
106. W.H. Press, S.A. Teukolsky, W.T. Vetterling, B.P. Flannery, *Numerical Recipes in Fortran 77: The Art of Scientific Computing*, 2nd edn. (Press Syndicate of the University of Cambridge, New York, 1997)

Index

S. A. Rakityansky, *Jost Functions in Quantum Mechanics*,
https://doi.org/10.1007/978-3-031-07761-6